INTRACELLULAR pH:
ITS MEASUREMENT, REGULATION, AND UTILIZATION IN CELLULAR FUNCTIONS

THE KROC FOUNDATION SERIES

Volumes in the Series Published by Alan R. Liss, Inc.

Volume 10: Propranolol and Schizophrenia
Eugene Roberts and Peter Amacher, *Editors*

Volume 11: Sleep Apnea Syndromes
Christian Guilleminault and William C. Dement, *Editors*

Volume 12: Mechanisms of Intestinal Secretion
Henry J. Binder, *Editor*

Volume 13: Erythrocyte Mechanics and Blood Flow
Giles R. Cokelet, Herbert J. Meiselman, and Donald E. Brooks, *Editors*

Volume 14: Biochemistry of the Acute Allergic Reactions: Fourth International Symposium
Elmer L. Becker, Arlene Stolper Simon, and K. Frank Austen, *Editors*

Volume 15: Intracellular pH: Its Measurement, Regulation, and Utilization in Cellular Functions
Richard Nuccitelli and David W. Deamer, *Editors*

Kroc Foundation Series
Volume 15

INTRACELLULAR pH:
ITS MEASUREMENT, REGULATION, AND UTILIZATION IN CELLULAR FUNCTIONS

Proceedings of a conference held at the Kroc Foundation,
Santa Ynez Valley, California, July 20–24, 1981

Editors

Richard Nuccitelli
and
David W. Deamer
Zoology Department
University of California
Davis, California

Alan R. Liss, Inc. • New York

Address all Inquiries to the Publisher
Alan R. Liss, Inc., 150 Fifth Avenue, New York, NY 10011

Copyright © 1982 Alan R. Liss, Inc.

Printed in the United States of America.

Under the conditions stated below the owner of copyright for this book hereby grants permission to users to make photocopy reproductions of any part or all of its contents for personal or internal organizational use, or for personal or internal use of specific clients. This consent is given on the condition that the copier pay the stated per-copy fee through the Copyright Clearance Center, Incorporated, 21 Congress Street, Salem, MA 01970, as listed in the most current issue of "Permissions to Photocopy" (Publisher's Fee List, distributed by CCC, Inc.), for copying beyond that permitted by sections 107 or 108 of the US Copyright Law. This consent does not extend to other kinds of copying, such as copying for general distribution, for advertising or promotional purposes, for creating new collective works, or for resale.

Library of Congress Cataloging in Publication Data

Main entry under title:

Intracellular pH.

   (Kroc Foundation series; v. 15)
   Includes index.
   1. Hydrogen-ion concentration — Measurement — Congresses. 2. Cell metabolism — Congresses.
I. Nuccitelli, Richard. II. Deamer, D.W.
III. Kroc Foundation. IV. Series.
QP535.H1I57   574.87'6   81-18616
ISBN 0-8451-0305-9   AACR2

Contents

Contributors xix
Photograph of Meeting Participants xxiv
Preface
Richard Nuccitelli and David W. Deamer xxv
Plates I and II xxvii

I. TECHNIQUES FOR MEASURING INTRACELLULAR pH
A. Single Cell pH_i Measurement
1. pH MICROELECTRODES

pH Microelectrodes: Tips on Making the Recessed-Tip Type for Intracellular Use
Roger C. Thomas

I. Introduction 1
II. Materials and Methods 2
   A. Glass 2
   B. Apparatus 3
III. How to Make a Batch of Electrodes 4
   A. Inner pH-Sensitive Micropipettes 4
   B. Insulating Micropipettes 4
   C. Sealing the Inner to the Outer 4
   D. Filling and Conditioning the Completed Microelectrodes 5
IV. References 6

Double-Barreled Intracellular pH Electrode: Construction and Illustration of Some Results
A. de Hemptinne, R. Marrannes, and B. Vanheel

I. Introduction 7
II. Materials and Methods 8
   A. Construction of the Double-Barrel Micropipette 8
   B. Construction of the pH-Sensitive Part 9
III. Results 12
   A. Measurement of pH_i and pH_s in Rat Soleus Muscle ... 12
   B. Influence of Simulated Ischemia on pH_i and pH_s in Rat Soleus Muscle 12

C. Effect of an Organic Acid on pH_i and pH_s in the Rat Soleus Muscle 14
D. Influence of Simulated Ischemia on pH_i and pH_s in Sheep Purkinje Fiber 15
E. Effect of Organic Acids on pH_i and pH_s in Sheep Purkinje Fiber 16
IV. Discussion 18
V. References 19

2. IMPERMEANT FLUORESCENT PROBES

An Optical Technique for Measurement of Intracellular pH in Single Living Cells
Jeanne M. Heiple and D. Lansing Taylor

I. Introduction 22
   A. pH_i and Contractility 22
   B. pH_i and Endocytic Events 24
   C. Techniques of pH_i Measurements 24
II. Materials and Methods 29
   A. Materials 29
   B. Preparation of FTC-Ovalbumin 30
   C. Microinjections and Microfluorometric Measurements ... 30
   D. Induction of Endocytosis 33
III. Results 34
   A. Spectral Characterization of FTC-Ovalbumin pH-Sensitivity 34
   B. Cytoplasmic pH Measurements 39
   C. Phagosomal pH Measurements 44
IV. Discussion 46
V. References 49

B. Cell Population pH_i Measurement

1. WEAK ACID/BASE DISTRIBUTION

Estimation of Intracellular pH From Distribution of Weak Electrolytes
Albert Roos and David W. Keifer

I. Introduction 55
II. Advantages of Weak Acid/Base Distribution Methods 56
III. Limitations and Sources of Error 56
IV. Summary 59
V. References 59

2. ^{31}P NUCLEAR MAGNETIC RESONANCE

pH_i Measurements of Cardiac and Skeletal Muscle Using ^{31}P-NMR
David G. Gadian, George K. Radda, M. Joan Dawson, and Douglas R. Wilkie

I. Introduction 61
II. What Can ^{31}P-NMR Tell Us About Metabolism? 62
III. The Measurement of pH_i 65
IV. Applications of pH_i Measurements 68
   A. Muscular Fatigue 68

B. Creatine Kinase Activity in Muscle	71
C. Studies of Acidosis and Metabolic Control	72
V. Conclusions	75
VI. References	76

Intracellular pH Measured by NMR: Methods and Results
R.J. Gillies, J.R. Alger, J.A. den Hollander, and R.G. Shulman

I. Background	79
A. Origin of Chemical Shifts and Fast Exchange	79
B. Indicators of pH_i	82
II. Methods	85
A. General Considerations	85
B. Signal-to-Noise	86
C. High Density Cultures	86
D. Mammalian Cells	89
III. Results and Discussion	94
A. Bacteria	94
B. Compartmentation	95
C. Time Resolution	97
D. pH_i and the Control of Glycolysis in Yeast	100
IV. References	103

3. LIGHT ABSORBANCE AND FLUORESCENCE TECHNIQUES

Spectrophotometric Determination of Cytoplasmic and Mitochondrial pH Transitions Using Trapped pH Indicators
John A. Thomas, Peter C. Kolbeck, and Thomas A. Langworthy

I. Introduction	105
II. Methods and Materials	106
A. Cells	106
B. Fluorescein Dyes	106
C. Spectrophotometric Measurements	107
D. Loading the Cells With the Indicator	107
E. Assessment of Dye Leakage	108
F. Calibration of Internal Dye Spectral Responses	108
III. Results	109
A. Spectral Sensitivity of Carboxyfluorescein and Fluorescein to pH	109
B. Internal pH Measurements in a Leaky Cell	110
C. Measurements With Monolayer Cell Cultures	110
D. Cytoplasmic and Mitochondrial pH Transitions	114
IV. Discussion	118
V. Summary	122
VI. References	122

Determination of Intracellular pH Changes in Lymphocytes With 4-Methylumbelliferone by Flow Microfluorometry
Donald F. Gerson

I. Introduction	125
II. Methods	126
A. Intracellular pH	126
B. Other Methods	128

III. Results and Discussion 128
IV. References 132

The Use of Fluorescent Amines for the Measurement of pH_i: Applications in Liposomes, Gastric Microsomes, and Sea Urchin Gametes
Hon Cheung Lee, John G. Forte, and David Epel

I. Introduction 136
II. Methods....................................... 136
   A. Liposome Preparation 136
   B. Preparation of Gastric Microsomes 137
   C. Isolation of Sea Urchin Gametes 137
   D. Fluorescence Measurements 137
   E. Fluorescence Microscopy 137
   F. Flow Dialysis 137
   G. Assessment of Sperm Motility 138
   H. Scoring of Acrosome Reaction 138
   I. Media 138
III. Results and Discussion 138
   A. Probes That Exhibit Decrease in Quantum Yield in the Presence of a pH Gradient 138
   B. Probes That Exhibit pH-Dependent Shift in Fluorescence Spectra......................... 149
   C. Probes That Exhibit Concentration-Dependent Shift in Fluorescence Spectra......................... 154
IV. Conclusions 157
V. References 158

C. Advantages and Limitations of the Various pH_i Measurement Techniques

Intracellular pH Measurement Techniques: Their Advantages and Limitations
Richard Nuccitelli

I. Introduction 161
II. pH-Measurement in Cell Populations 164
   A. Spatial Resolution........................... 164
   B. Temporal Resolution......................... 164
   C. Amount of Material Needed 165
   D. Advantages and Limitations of These Techniques .. 165
III. pH_i Measurement in Single Cells................... 165
   A. pH-Sensitive Microelectrodes................... 166
   B. Fluorescein-Ovalbumin Injection 166
IV. Summary of Discussion Following Technique Paper Presentations 166
   A. Weak Acid/Base Distribution 166
   B. ^{31}P–NMR for pH_i Measurement 167
   C. pH-Sensitive Glass Microelectrodes.............. 168
   D. Optical Techniques for pH_i Measurement.......... 168
V. References 169

II. REGULATION OF INTRACELLULAR pH

A. Proton Permeability of Cells

Proton Permeability in Biological and Model Membranes
David W. Deamer

I. Introduction	173
II. Methods and Potential Artifacts	176
III. Proton Permeability of Model Membrane Systems	177
IV. Discussion	182
A. Proton Permeability of Biological Membranes	182
B. Relative Cation Permeabilities of Model Membranes	183
C. Mechanisms of Proton Flux Across Lipid Bilayers	184
V. Summary	185
VI. References	186

B. Studies of pH_i Regulation by Cells

Snail Neuron Intracellular pH Regulation
Roger C. Thomas

I. Introduction	189
A. Formal Introduction	189
B. Informal Introduction	190
II. Methods	190
A. Experimental Setup	190
B. Preparation	191
C. Lighting	192
D. Physiological Solutions	193
E. Microelectrodes	193
F. Calibration Solutions	194
G. Electrical Arrangements	194
H. Setting-up Procedure	194
I. Stumbling Points	194
J. Possible Mechanisms for pH_i Regulation	195
III. Results	196
A. Effect of pH_i Recovery on the Membrane Potential (E_m)	196
B. Removal of External K^+	197
C. Effect of Decreasing Internal Cl^-	197
D. Effect of Removing Bicarbonate	198
E. Effect of Removing External Na^+	199
F. Effect of Metabolic Inhibitors on pH_i Recovery	200
IV. Discussion	202
V. References	204

Regulation of Intracellular pH in Barnacle Muscle
Albert Roos and Walter F. Boron

I. Introduction	205
II. Results and Discussion	206
III. References	218

Intracellular pH Regulation in Squid Giant Axons
John M. Russell and Walter F. Boron

I. Introduction	221
II. Methods	222
A. General	222
B. Solutions	222
C. Measurement of pH_i and Net, Equivalent HCO_3^- Flux	223
D. Measurement of Na^+ and Cl^- Fluxes	226
III. Summary of Previous Findings	227
IV. Results	233
V. Discussion	236
VI. References	237

Chloride-Bicarbonate Exchange in the Sheep Cardiac Purkinje Fibre
R.D. Vaughan-Jones

I. Introduction	239
II. Methods	240
A. Preparation and Solutions	240
B. Electrodes and Calibrations	241
III. Results	242
A. The Existence of Cl^-–HCO_3^- Exchange in the Purkinje Fibre	242
B. The Coupling Ratio	245
C. Dependence on HCO_3^- and External Cl^-	246
D. The Effects of Ammonia on Cl^-–HCO_3^- Exchange	248
IV. Discussion	250
V. References	251

Hydrogen and Bicarbonate Transport by Salamander Proximal Tubule Cells
Walter F. Boron and Emile L. Boulpaep

I. Introduction	253
II. Methods	254
A. General	254
B. Solutions	255
C. Microelectrodes	256
III. Results	260
A. Control Studies	260
B. Normal Values	260
C. Mechanism of pH_i Regulation	260
D. Basolateral HCO_3^- Transport	262
IV. Discussion and Conclusions	266
V. References	267

The Effect of External Ions on pH_i in Sea Urchin Eggs
Sheldon S. Shen

I. Introduction	269
II. Materials and Methods	271

| A. Experimental System 271
| B. Solutions 271
| C. Electrophysiological Measurements 272
| D. H^+-Selective Microelectrodes 272
| E. Problems in Measuring Intracellular pH 272
| III. Results... 273
| A. E_m and pH_i of Unfertilized Eggs 273
| B. Effect of $(H^+)_o$ and $(Cl^-)_o$ on pH_i in Unfertilized Eggs 274
| C. Effect of $(Na^+)_o$ on Cytoplasmic Alkalinization 276
| D. Effect of NH_4Cl on pH_i of Unfertilized Eggs............... 278
| IV. Discussion 279
| V. References 281

Intracellular pH Regulation: A Summary of the Proposed Mechanisms and Meeting Discussion
William J. Moody, Jr.

| I. A Survey of pH_i-Regulating Mechanisms 283
| II. Summary of General Discussion of pH_i Regulation............. 285
| III. Cells in Which pH_i-Regulating Mechanisms Serve Additional
| Functions 286
| IV. Summary of Discussion on Renal Tubule Cells and Oocytes 287

III. UTILIZATION OF pH_i IN THE CONTROL OF CELLULAR FUNCTIONS

A. Activation of Egg Development

Intracellular pH Changes Accompanying the Activation of Development in Frog Eggs: Comparison of pH Microelectrodes and ^{31}P-NMR Measurements
Dennis J. Webb and Richard Nuccitelli

| I. Introduction 294
| II. Materials and Methods 295
| A. Animals and Eggs for Microelectrode Experiments 295
| B. pH and Voltage Electrodes............................. 296
| C. Animals and Eggs for ^{31}P-NMR Experiments 299
| D. Egg Extract Preparation 302
| E. Recording and Treatment of Spectra.................... 302
| III. Results... 304
| A. pH_i Changes Following Fertilization or Activation........ 304
| B. Evidence for Lack of Na^+-H^+ and Cl^--HCO_3^- Exchange . 306
| C. pH_i Changes Independent of pH_o....................... 309
| D. pH_i Oscillations Associated With the Cell Cycle 309
| E. Localized pH_i Changes Following Injury 312
| F. ^{31}P-NMR Spectra of Unfertilized Eggs 313
| G. ^{31}P-NMR Spectra of Fertilized Eggs 313
| IV. Discussion 316
| A. pH_i of the Unfertilized Egg 316
| B. Initial Transient Acidification 318
| C. Permanent pH_i Increase 319

  D. Nature of the Permanent pH_i Increase 320
  E. Manipulation of pH_i 320
  F. Cyclic pH_i Oscillations 321
  G. The Importance of the Permanent pH_i Increase for the
    Activation of Development 322
 V. References 322

Regulation of Protein Synthesis in Sea Urchin Eggs by Intracellular pH
Matthew M. Winkler

 I. Introduction 325
 II. Materials and Methods 326
  A. Preparation of the Cell-Free System 326
  B. Protein Synthesis in the Cell-Free System 327
  C. Solutions 327
 III. Results and Discussion 327
  A. Intracellular pH and Protein Synthesis 328
  B. Translational Efficiency 332
  C. Summary 339
 IV. References 339

B. pH_i Changes During the Cell Cycle

Intracellular pH and Proliferation in Yeast, Tetrahymena and Sea Urchin Eggs
Robert J. Gillies

 I. Introduction 341
  A. Regulation of Proliferation 341
  B. Cell Cycle 342
  C. Intracellular pH 342
 II. Intracellular pH and Growth 343
  A. Tetrahymena 343
  B. Yeast 344
 III. Intracellular pH and the Cell Cycle 344
  A. Tetrahymena 344
  B. Yeast 347
 IV. Intracellular pH Manipulation 347
 V. Mediation of pH Changes in Sea Urchin Eggs Following
  Fertilization 353
 VI. Conclusions 358
 VII. References 358

Changes in Intracellular pH of Physarum plasmodium During the Cell Cycle and in Response to Starvation
R.A. Steinhardt and M. Morisawa

 I. Introduction 361
 II. Materials and Methods 362
  A. Plasmodia 362
  B. Microelectrode Recordings 363
  C. Solutions 363

III. Results	364
A. Choice of Recording Method	364
B. Intracellular pH and the Cell Cycle	366
C. Delay of Mitosis by Lowering Intracellular pH	366
D. Decrease of Intracellular pH and the Delay of Mitosis During Starvation	369
E. Increase of Intracellular pH with Refeeding	370
F. Morphological Changes During Starvation and Sodium Acetate Treatment	371
IV. Discussion	372
V. References	373

The Relation Between Intracellular pH and DNA Synthesis Rate in Proliferating Lymphocytes
Donald F. Gerson

I. Introduction	375
II. Methods	376
A. Spleen Lymphocytes	376
B. Measurement Techniques	376
III. Results and Discussion	378
A. Intracellular pH	378
B. Correlation of pH_i and DNA Synthesis	380
IV. Conclusions	382
V. References	382

C. Control of Metabolism

The Role of Intracellular pH in Insulin Action
Richard D. Moore, Mark L. Fidelman, Jeffrey C. Hansen, and John N. Otis

I. Introduction	386
A. Model for Insulin Transduction	387
B. Effect of Insulin on $Na^+:H^+$ Exchange	388
C. Effect of Insulin on Glycolysis	389
II. Review of Methodologies	390
A. Ringer Solutions	390
B. Procedures	391
C. Materials	394
III. Review of Experimental Results	394
A. Effect of Insulin on $Na^+:H^+$ Exchange	394
B. Mechanism of Insulin Action on Glycolysis	398
IV. Discussion	406
A. The Insulin Transduction System Mediates the Acute Action of Insulin Upon Glycolysis	406
B. Blocking Cellular Effects Due to the Insulin Transduction System	407
C. Possible Role of the Insulin Transduction System in Regulation of Other Cellular Functions	408
D. Possible Nonionic Mechanism for Mediation of Insulin Action	409

E. General Implications for Cell Function	409
F. Implications for Studies of Intracellular Enzymes	411
G. The Proton as an Intracellular Signal	411
V. References	412

Cellular Dormancy and the Scope of pH_i-Mediated Metabolic Regulation
William B. Busa

I. Introduction	417
II. Three Cryptobiotic Systems Demonstrating Large pH_i Changes	418
A. Bacteria	418
B. Yeast	419
C. Crustaceans	419
III. Determination of pH_i in Artemia Cysts	419
A. The Organism	419
B. Materials and Methods	420
IV. Results and Discussion	422
V. Conclusions	424
VI. References	425

D. Control of Plasma Membrane Properties

The Effect of Decreased Intracellular pH on the Electrical Properties of Invertebrate Muscle Fibers and Oocytes
William J. Moody, Jr.

I. Introduction	427
II. Methods	429
A. pH Microelectrode Construction	429
B. The Experimental Setup	430
III. Results	430
A. Effect of Low pH_i on Ca^{2+} Spike Generation in Crayfish Slow Muscle Fibers	430
B. Effect of pH_i on Inwardly Rectifying K^+ Currents in Starfish Oocytes	437
IV. Discussion	441
V. References	442

Comparison of pH and Calcium Dependence of Gap Junctional Conductance
D.C. Spray, A.L. Harris, and M.V.L. Bennett

I. Introduction	445
II. Techniques	448
A. Measurements of Gap Junctional Conductance	448
B. Intracellular pH Measurement	450
III. Results	450
A. pH_i Electrode Measurements	452
B. Does the pH_i Change Affect Intracellular Ca^{2+}?	454
C. Increases in H_i^+ and Ca_i^{2+} Uncouple Independently	455
IV. Discussion	457

A. Mechanism of Gating by Hydrogen Ions	457
B. Conclusions	458
V. References	458

E. pH_i Changes During Stimulus-Response Coupling

Stimulus Response Coupling in Human Platelets: Thrombin-Induced Changes in pH_i
Elizabeth R. Simons, David B. Schwartz, and Nancy E. Norman

I. Introduction	463
II. Materials	464
A. Buffers	465
B. Platelets	465
III. Fluorometric Methods	465
A. 9-Aminoacridine	466
B. 6-Carboxyfluorescein Diacetate	471
C. Dimethyl 6-Carboxyfluorescein Diacetate	474
D. General Comment	476
IV. Results	477
V. Discussion	478
VI. References	480

The Role of Protons in Glucose-Induced Stimulus-Secretion Coupling in Pancreatic Islet B Cells
Caroline S. Pace, John T. Tarvin, and Joel S. Smith

I. Introduction	483
A. Influence of Protons on the B Cell Plasma Membrane	483
B. Influence of Glucose on the Proton Gradient in Secretory Granules of B Cells	485
II. Materials and Methods	485
A. Animals, Isolation, and Culture of Islets	485
B. Electrophysiological Studies	486
C. $^{86}Rb^+$ Efflux	488
D. Islet Perifusion System	489
E. Acridine Orange Experiments	490
F. Electron Microscopy	490
III. Results	491
A. Electrical Activity	491
B. $^{86}Rb^+$ Efflux	495
C. Insulin Release	497
D. Acridine Orange	498
IV. Discussion	502
A. Influence of Protons on Electrical Activity and $^{86}Rb^+$ Efflux	503
B. Influence of pH on Glucose-Induced Insulin Release	507
C. Glucose-Induced Proton Gradient in Secretory Granules	508
V. References	510

Chemotactic Stimuli-Induced Changes in the pH_i of Rabbit Neutrophils
R.I. Sha'afi, P.H. Naccache, T.F.P. Molski, and M. Volpi

I. Introduction	513

xvi / Contents

II. Materials and Methods	514
III. Results	515
A. Effect of f-Met-Leu-Phe on the pH_i of Rabbit Neutrophils	515
B. The Role of Calcium Ions in the Chemotactic Factor-Dependent Changes in pH_i	516
C. The Role of Sodium Ions in the Chemotactic Factor-Dependent Changes in pH_i	518
D. Effect of the Anion Transport Inhibitor DIDS	520
IV. Discussion	521
V. References	524

F. Tissue pH_i Measurements: Diagnostic Applications

Noninvasive pH_i Measurements of Human Tissue Using ^{31}P-NMR
Peter J. Bore, Lawrence Chan, David G. Gadian, George K. Radda, Brian D. Ross, Peter Styles, and Doris J. Taylor

I. Introduction	527
II. Recent Technical Developments	527
A. Whole Animal Studies	527
B. Studies of Human Metabolism	528
III. Studies of Kidney Preservation	529
IV. Studies of Human Forearm Muscle	529
V. Conclusions	534
VI. References	534

The Role of Intracellular pH in the Control of Normal and Ischemic Myocardial Contractility: A ^{31}P Nuclear Magnetic Resonance and Mass Spectrometry Study
William E. Jacobus, Ira H. Pores, Scott K. Lucas, Clayton H. Kallman, Myron L. Weisfeldt, and John T. Flaherty

I. Introduction	537
II. Materials and Methods	539
A. Nuclear Magnetic Resonance Methods	539
B. Calibration of NMR pH Measurements	541
C. Other Aspects of pH_i Assessment	547
D. Mass Spectrometry Methods	548
III. Results	549
A. Correlation of pH_i to Ventricular Performance in Normal Hearts	549
B. Contractility and pH_i During Ischemia	551
C. Mass Spectrometry Studies	554
IV. Discussion	557
V. References	562

G. Summary of the Evidence and Discussion Concerning the Involvement of pH_i in the Control of Cellular Functions

Summary of the Evidence and Discussion Concerning the Involvement of pH_i in the Control of Cellular Functions
Richard Nuccitelli and Jeanne M. Heiple

I. Introduction 567
II. Summary of Presentations and Discussions 568
    A. Activation of Egg Development 568
    B. pH_i Changes During the Cell Cycle 569
    C. Control of Metabolism 578
    D. Control of Plasma Membrane Properties 579
    E. Intracellular pH Changes During Stimulus-Response
       Coupling... 581
    F. Tissue pH_i Measurements: Diagnostic Applications 582
III. Conclusions 583
IV. References 585

Index.. 587

Contributors

J.R. Alger [79]
 Department of Molecular Biophysics and Biochemistry, Yale University, New Haven, CT 06511

M.V.L. Bennett [445]
 Division of Cellular Neurobiology, Department of Neuroscience, Albert Einstein College of Medicine, Bronx, NY 10461

Peter J. Bore [527]
 Department of Biochemistry, University of Oxford, South Parks Road, Oxford OX1 3QU, England

Walter F. Boron [205, 221, 253]
 Department of Physiology, Yale University School of Medicine, New Haven, CT 06510

Emile L. Boulpaep [253]
 Department of Physiology, Yale University School of Medicine, New Haven, CT 06510

William B. Busa [417]
 Zoology Department, University of California, Davis CA 95616

Lawrence Chan [527]
 Radcliffe Infirmary, Oxford, England

M. Joan Dawson [61]
 Department of Physiology, University College London, Gower Street, London WC1E 6BT, England

David W. Deamer [xxv, 173]
 Zoology Department, University of California, Davis CA 95616

A. de Hemptinne [7]
 Laboratory of Normal and Pathological Physiology, University of Gent, B-9000, Gent, Belgium

J.A. den Hollander [79]
 Department of Molecular Biophysics and Biochemistry, Yale University, New Haven, CT 06511

The boldface number in brackets following each contributor's name indicates the first page of that author's paper.

David Epel [135]
  Hopkins Marine Station, Stanford University, Pacific Grove, CA 93950
Mark L. Fidelman [385]
  Department of Physiology, Box 551, Medical College of Virginia, Virginia Commonwealth University, MCV Station, Richmond, VA 23298
John T. Flaherty [537]
  Division of Cardiology, The Johns Hopkins University School of Medicine, Baltimore, MD 21205
John G. Forte [135]
  Department of Physiology-Anatomy, University of California, Berkeley, CA 94720
David G. Gadian [61, 527]
  Department of Biochemistry, University of Oxford, South Parks Road, Oxford OX1 3QU, England
Donald F. Gerson [125, 375]
  Basel Institute for Immunology, 487 Grenzacherstrasse, CH-4005 Basel, Switzerland
Robert J. Gillies [79, 341]
  Department of Molecular Biophysics and Biochemistry, Yale University, New Haven, CT 06511
Jeffrey C. Hansen [385]
  Biophysics Laboratory, State University of New York, Plattsburgh, NY 12901
A.L. Harris [445]
  Division of Cellular Neurobiology, Department of Neuroscience, Albert Einstein College of Medicine, Bronx, NY 10461
Jeanne M. Heiple [21, 567]
  Boston University School of Medicine, Department of Biochemistry, 80 E. Concord Street, Boston, MA 02118 *formerly of* Cell and Developmental Biology, The Biological Laboratories, Harvard University, Cambridge, MA 02138
William E. Jacobus [537]
  Division of Cardiology, Johns Hopkins University School of Medicine, Baltimore, MD 21205
Clayton H. Kallman [537]
  Division of Cardiology, The Johns Hopkins University School of Medicine, Baltimore, MD 21205
David W. Keifer [55]
  Department of Botany, University of California, Davis, CA 95616
Peter C. Kolbeck [105]
  School of Medicine, Emory University, Atlanta, GA 30322 *formerly of* Department of Biochemistry, University of South Dakota School of Medicine, Vermillion, SD 57069
Thomas A. Langworthy [105]
  Department of Microbiology, University of South Dakota School of Medicine, Vermillion, SD 57069

Hon Cheung Lee [135]
Department of Physiology, Medical School, University of Minnesota, Minneapolis, MN 55455

Scott K. Lucas [537]
University of Oklahoma Health Center, Department of Surgery, P.O. Box 26901, Oklahoma City, OK 73190

R. Marrannes [7]
Laboratory of Normal and Pathological Physiology, University of Gent, B-9000, Gent, Belgium

T.F.P. Molski [513]
Department of Physiology, University of Connecticut Health Center, Farmington, CT 06032

William J. Moody, Jr. [283, 427]
Jerry Lewis Neuromuscular Research Center, University of California, Los Angeles, CA 90024

Richard D. Moore [385]
Biophysics Laboratory, State University of New York, Plattsburgh, NY 12901

M. Morisawa [361]
Ocean Research Institute, University of Tokyo, 15-1-1-Chome, Minamidai, Nakano-Ku, Tokyo 164, Japan

P.H. Naccache [513]
Department of Pathology, University of Connecticut Health Center, Farmington, CT 06032

Nancy E. Norman [463]
Department of Biochemistry, Boston University School of Medicine, Boston, MA 02118

Richard Nuccitelli [xxv, 161, 293, 567]
Zoology Department, University of California, Davis, CA 95616

John N. Otis [385]
Biophysics Laboratory, State University of New York, Plattsburgh, NY 12901

Caroline S. Pace [483]
Department of Physiology and Biophysics, University of Alabama in Birmingham, Diabetes Hospital, 1808 Seventh Avenue, South, Birmingham, AL 35294

Ira H. Pores [537]
Division of Cardiology, Johns Hopkins University School of Medicine, Baltimore, MD 21205, *presently at* Newark Beth Israel Medical Center, 201 Lyons Avenue, Newark, NJ 07112

George K. Radda [61, 527]
Department of Biochemistry, University of Oxford, South Parks Road, Oxford OX1 3QU, England

Albert Roos [55, 205]
Departments of Physiology and Biophysics and of Anesthesiology, 660 So. Euclid Avenue, Washington University School of Medicine, St. Louis, MO 63110

Brian D. Ross [527]
Radcliffe Infirmary, Oxford, England

John M. Russell [221]
Department of Physiology and Biophysics, University of Texas Medical Branch, Galveston, TX 77550

David B. Schwartz [463]
Department of Biochemistry, Boston University School of Medicine, Boston, MA 02118

R.I. Sha'afi [513]
Department of Physiology, University of Connecticut Health Center, Farmington, CT 06032

Sheldon S. Shen [269]
Department of Zoology, Iowa State University, Ames, IA 50011

R.G. Shulman [79]
Department of Molecular Biophysics and Biochemistry, Yale University, New Haven, CT 06511

Elizabeth R. Simons [463]
Department of Biochemistry, Boston University School of Medicine, Boston, MA 02118

Joel S. Smith [483]
Department of Physiology and Biophysics, University of Alabama in Birmingham, Diabetes Hospital, 1808 Seventh Avenue, South, Birmingham, AL 35294

D. C. Spray [445]
Division of Cellular Neurobiology, Department of Neuroscience, Albert Einstein College of Medicine, Bronx, NY 10461

R. A. Steinhardt [361]
Department of Zoology, University of California, Berkeley, CA 94720

Peter Styles [527]
Department of Biochemistry, University of Oxford, South Parks Road, Oxford OX1 3QU, England

John T. Tarvin [483]
Department of Physiology and Biophysics, University of Alabama in Birmingham, Diabetes Hospital, 1808 Seventh Avenue, South, Birmingham, AL 35294

Doris J. Taylor [527]
Department of Biochemistry, University of Oxford, South Parks Road, Oxford OX1 3QU, England

D. Lansing Taylor [21]
Cell and Developmental Biology, The Biological Laboratories, Harvard University, 16 Divinity Avenue, Cambridge, MA 02138

John A. Thomas [105]
Department of Biochemistry, University of South Dakota School of Medicine, Vermillion, SD 57069

Roger C. Thomas [1, 189]
Department of Physiology, University of Bristol, Bristol BS8 1TD, England

B. Vanheel [7]
Laboratory of Normal and Pathological Physiology, University of Gent, B-9000, Gent, Belgium

R. D. Vaughan-Jones [239]
University Laboratory of Pharmacology, Oxford University, South Parks Road, Oxford OX1 3QT, England

M. Volpi [513]
Department of Physiology, University of Connecticut Health Center, Farmington, CT 06032

Dennis J. Webb [293]
Zoology Department, University of California, Davis, CA 95616

Myron L. Weisfeldt [537]
Division of Cardiology, The Johns Hopkins University School of Medicine, Baltimore, MD 21205

Douglas R. Wilkie [61]
Department of Physiology, University College London, Gower Street, London WC1E 6BT, England

Matthew M. Winkler [325]
Department of Biological Chemistry, School of Medicine, University of California, Davis, CA 95616

MEETING PARTICIPANTS

Front Row: Walter F. Boron, Elizabeth R. Simons, Robert Kroc, Jeanne M. Heiple, Caroline S. Pace, Richard Nuccitelli, David W. Deamer, R. A. Steinhardt, Albert Roos

Second Row: Roger C. Thomas, A. de Hemptinne, Donald F. Gerson, R.D. Vaughan-Jones, D.C. Spray, William E. Jacobus, John A. Thomas, Matthew Winkler, Hon Cheung Lee

Third Row: Peter Amacher, David G. Gadian, William J. Moody, Jr., Dennis J. Webb, Robert J. Gillies, William Regelson, Sheldon S. Shen, Richard D. Moore, Thomas Parker

Preface

Ion concentration gradients across the plasma membrane are of fundamental importance for cellular function. In past years, most research effort has focused on sodium, potassium, and calcium ions, primarily because of their central role in nerve and muscle function. Protons are equally ubiquitous in the cell environment, but have received much less attention. This has been due in part to the difficulty of measuring intracellular proton concentrations (pH_i), and to a prevailing assumption that pH_i is under strict homeostatic control and therefore not very interesting.

This attitude changed dramatically as it became apparent during the past two decades that proton gradients were integral components of energy transduction mechanisms of the cell. Furthermore, during this same period technological advances permitted the development of new techniques which can provide accurate measurements of intracellular pH. The most significant of these include the introduction of the weak acid, DMO, in 1959 for use with the weak acid/base distribution technique; the development of the pH-sensitive glass microelectrode by Hinke in 1967 and the improved recessed-tip version by Thomas in 1974; the utilization of ^{31}P-NMR for noninvasive pH_i measurement in populations of cells beginning in 1973; and the recent techniques for cytoplasmic restriction of optical probes with pH_i-dependent absorbance or fluorescence described in 1979. As a result, investigators are now able to choose from several reliable techniques for pH_i measurement, and this has led to important advances in our understanding of intracellular pH. However, despite this burgeoning interest, there have been no recent conferences or books devoted to this topic.

On July 20–24, 1981, 24 scientists working in this area met at the Kroc Foundation in Santa Ynez, California, to evaluate and compare pH_i measurement techniques, discuss mechanisms of pH_i regulation, and consider what role pH_i might play in the control of cellular functions. A separate goal was to publish the conference proceedings in the form of an edited text which would present a timely sampling of the most significant advances in this field. The book is divided into three sections. The first contains papers on the most reliable methods for pH_i determination. Authors were encouraged to include detailed diagrams and photographs of their apparatus so that their tools would be readily accessible to other investigators wishing to make such measurements. The second section deals with mechanisms by which cells regulate pH_i, and the third section includes chapters related to the role of pH_i in the control of cell functions. In order to communicate some of the interchange of ideas that occurred during the conference, edited versions of the discussions are included in summary chapters at the end of each section. These same papers describe the main findings presented in that section in table format so that the reader might quickly identify points of interest.

No such undertaking could be successful without the help of many individuals. The organizers wish to express their sincere thanks to Drs. Roger Thomas, Bill Moody, Bob Gillies, Peter Amacher and Robert Kroc for their help in organizing the meeting and choosing participants, and Drs. Amacher and Kroc for their help with travel, logistics and local arrangements. We are particularly grateful to the Kroc Foundation for providing a superb meeting site, and for their generous financial aid in bringing together this remarkable group of scientists.

Richard Nuccitelli
David W. Deamer

Plates I and II

Plates I and II

Plate I. (opposite page) [from Lee, Forte, and Epel, this volume, page 156] Fluorescence micrographs of sea urchin eggs stained with acridine orange. Lytechinus pictus eggs (0.5%; A–C) were stained with 2 μM acridine orange and fertilized with a limited number of sperm. The mixture was incubated at 16°C with constant stirring. A) Thirty minutes after the addition of sperm. The fertilized eggs were identified by the presence of fertilization membranes visible in bright field (not shown). B) Staining pattern of fertilized eggs at the clear-streak stage (1.5 hours after the addition of sperm). C) A mixture of cells at different stages of development. The arrow points to the unfertilized egg. D) Strongylocentrotus purpuratus eggs (0.9%) were fertilized and washed once to remove excess sperm. An equal volume of unfertilized eggs was added and the mixture stained with 4 μM acridine orange. The mixture was incubated at 16°C with constant stirring for 10 min after the addition of sperm. Unfertilized eggs are indicated by arrows.

Plate II. (following page) [from Pace, Tarvin and Smith, this volume, page 498] Influence of glucose on monolayer culture of islet cells in the presence of 25 μM acridine orange photographed with fluorescence optics. A) The monolayer of islet cells in the resting state or absence of glucose. The cells appear yellowish-green and the cytoplasmic granules appear as distinct structures ($\times$ 1,200). B) Upon exposure to 27.8 mM glucose the yellowish-green fluorescence was replaced by the appearance of a red fluorescence. In some areas the fluorescence is distinctly observed to be present in the secretory granules. The nucleus of each cell remains green ($\times$ 1,200). C) The application of 10 μg/ml monensin to the medium containing 27.8 mM glucose rapidly dissipated the red fluorescence of the cell. This field was photographed after 10 minutes of exposure of the cells to monensin. Some of the granules are still red ($\times$ 1,300). D) After 30 minutes of exposure to monensin the cells appeared lime green. Note the apparent disappearance of distinct granules. It is not known whether the granules have actually burst or do not maintain a distinct color as compared to the cytoplasm ($\times$ 1,300). E) After exposure to 27.8 mM glucose, 1.0 mM benzylamine was added to the medium. After 10 minutes the granules appeared to be swollen and the red fluorescence remained ($\times$ 3,300).

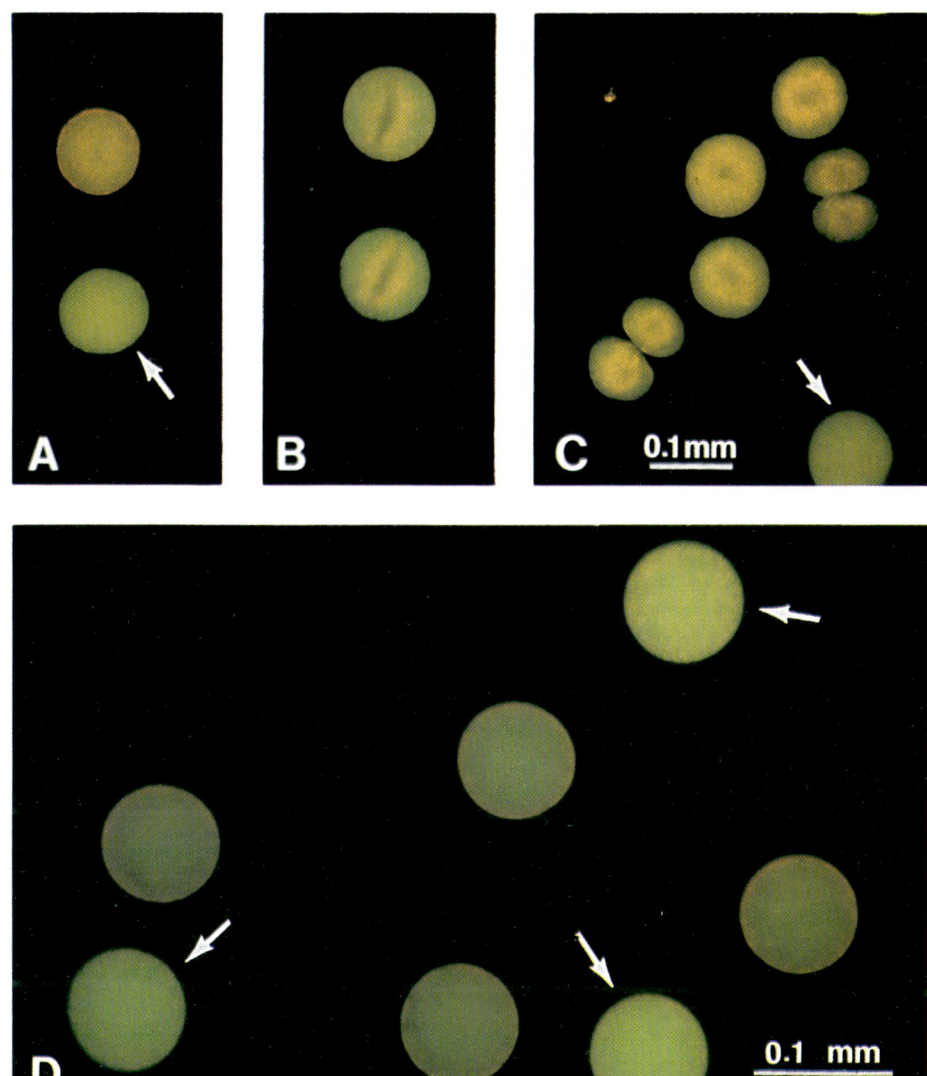

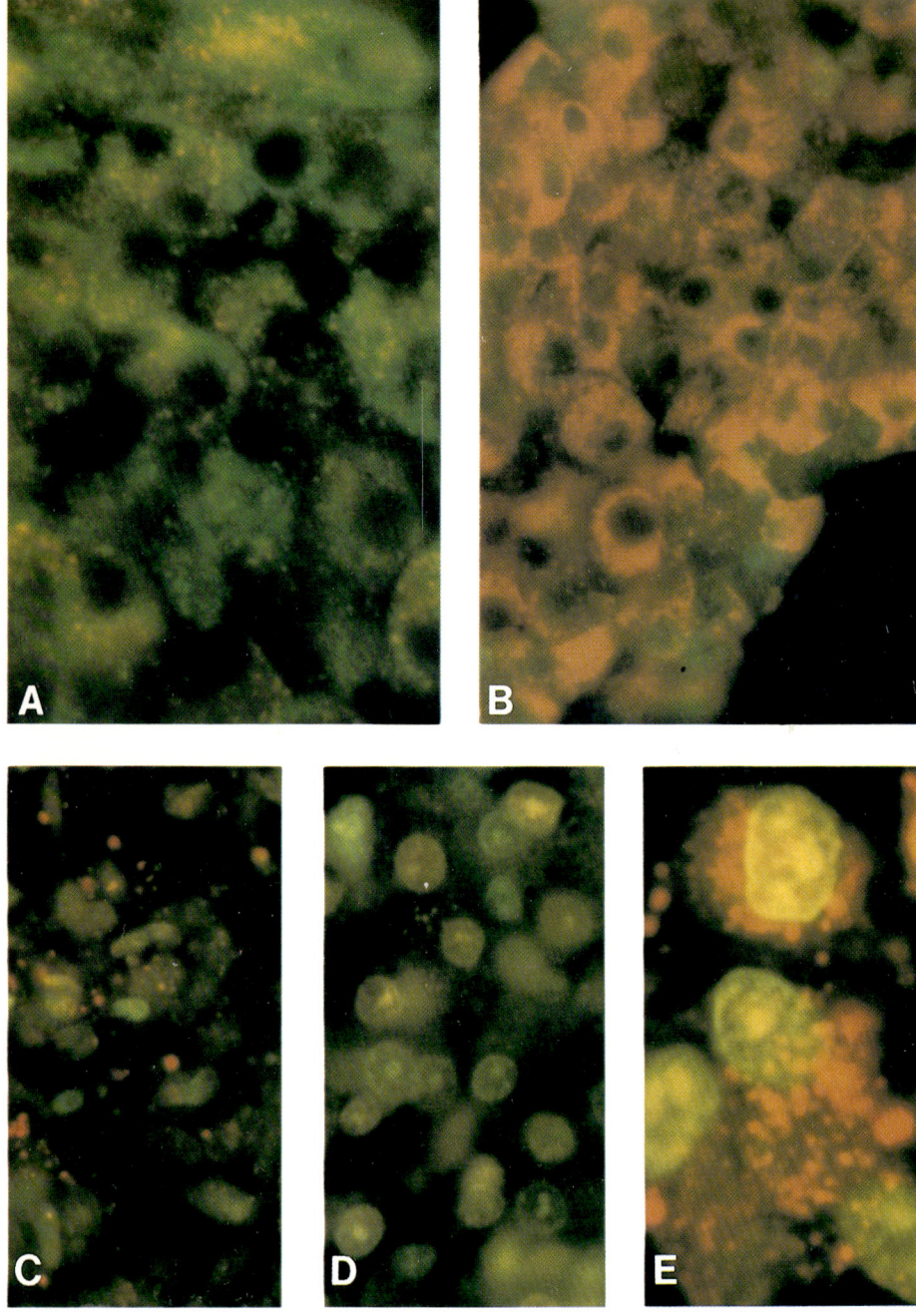

TECHNIQUES FOR MEASURING INTRACELLULAR pH

pH Microelectrodes: Tips on Making the Recessed-Tip Type for Intracellular Use

Roger C. Thomas
Department of Physiology, University of Bristol, Bristol BS8 ITD, England

I.	Introduction	1
II.	Materials and Methods	2
	A. Glass	2
	B. Apparatus	3
III.	How to Make a Batch of Electrodes	4
	A. Inner pH-Sensitive Micropipettes	4
	B. Insulating Micropipettes	4
	C. Sealing the Inner to the Outer	4
	D. Filling and Conditioning the Completed Microelectrodes	5
IV.	References	6

I. INTRODUCTION

Few would argue that the best way of measuring the pH of a liter of aqueous solution is to use a pH-sensitive glass electrode, a reference electrode, and a pH meter. Such equipment has been in routine use for over 30 years, and is probably used by all measurers of intracellular pH (pH_i) in the calibration of their techniques. It follows that of all the methods of measuring pH_i the easiest to understand is the pH-sensitive microelectrode; one simply puts a pH and reference microelectrode into the cell interior and reads off the pH.

A variety of different designs of intracellular pH electrode have been used since Caldwell [1954] first made glass pH electrodes just small enough to go into invertebrate giant axons and muscle fibers. The most successful have been the protruding-tip type (Fig. 1A) first developed by Hinke [1959] for sodium- and postassium-sensitive glass, and the recessed-tip type (Fig. 1B) originally made with sodium-sensitive glass [Thomas, 1970]. The protruding-tip design has been widely used in the last few years by Boron and co-workers to investigate pH_i in squid giant axons and crustacean muscle fibers, but is too large for other cells. The recessed-tip type of pH microelectrode, first described in 1974, can

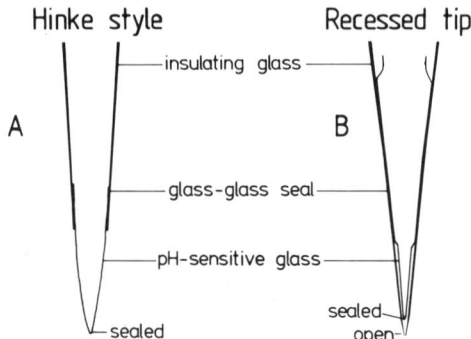

Fig. 1. Diagram of the sharp ends of the protruding-tip or Hinke style (A) and recessed-tip (B) pH-sensitive microelectrodes. About 0.3 mm is shown of the Hinke electrode, but only about 0.1 mm of the recessed-tip type.

be made with tip diameters of less than half a micron, and thus can be used on a wide variety of cells [Thomas, 1974]. It is even possible to make a double-barreled version [de Hemptinne, 1980, and this volume].

It is much easier to make liquid ion-sensor microelectrodes, but unfortunately there is as yet no reliable pH-sensitive liquid. One report describing liquid sensor pH microelectrodes has appeared [Harmon and Poole-Wilson, 1981], but as far as I know no one else has yet been able to make them work.

In this chapter I will describe how to make single-barreled recessed-tip microelectrodes, with special attention to recent improvements. For a detailed description of the technique of ion-sensitive microelectrodes in general, I strongly recommend my recent book [Thomas, 1978].

II. MATERIALS AND METHODS

A. Glass

Two sorts of glass are required to make pH-sensitive microelectrodes: insulating and pH-sensitive. The best insulating glass is aluminosilicate, Corning 1720 or Schott Supremax. It is much more durable than borosilicate or lead glass, and has a higher specific resistivity [Thomas, 1978]. To date only Supremax is readily available in small quantities of 2-mm tubing; it can be obtained from Clark Electromedical Instruments, P.O. Box 8, Pangbourne, Reading, RG8 7HU, England.

The only pH-sensitive glass available as 1-mm tubing is NaCaS 22-6 (Corning code 0150). Until recently it was sold by Corning, but at present is available only from Clark Electromedical or Micro-electrodes Inc, Oak Hill Park, Londonderry, NH 03053, U.S.A.

B. Apparatus

The main items of equipment required are a micropipette puller suitable for 1- and 2-mm glass tubing, and a microforge of some sort. (A micropipette is a glass tube pulled to a fine point; when filled with a conducting fluid it becomes a microelectrode.)

A micropipette puller is a machine that holds a length of glass tubing at each end with its center in a heating element. When the central region has softened the two ends are pulled apart, forming two micropipettes. A micropipette puller suitable for making pH microelectrodes can be either bought, borrowed, or made locally. For most purposes the easiest to use and most adaptable commercial machine I am familiar with is the Kopf 700C, but for very sharp micropipettes the Ensor design (available from Clark Electromedical) may be better.

Commercial microforges are usually too coarse for ion-sensitive microelectrodes. I prefer an arrangement like that illustrated in Figure 2, consisting of a

Fig. 2. Photograph of microforge used to make pH-sensitive recessed-tip microelectrodes. The left-hand Prior micromanipulator holds the stainless steel tube supplying gas to the pH micropipette, and the right-hand one holds the outer micropipette. The microforge element is held on a third, smaller manipulator hidden behind the left-hand Prior.

compound microscope, three micromanipulators, and an electrically heated loop of 50-μm diameter platinum-rhodium wire. The microscope should have at least $10\times$ and $40\times$ objectives, and be mounted with its visual axis horizontal. This allows the micropipettes to be manipulated vertically in the focal plan. A source of compressed gas at up to 700 lb/in^2 is required, and it is also helpful to have nearby a source of heat such as a small flame or old soldering iron (an old flame is more fun) suitable for melting wax.

III. HOW TO MAKE A BATCH OF ELECTRODES

The most time-consuming part of this process is finding the best way to make insulating and pH-sensitive micropipettes that fit tightly together near their tips. This can be done only by trial and error. Once this has been done, it takes only a few hours to make a dozen microelectrodes. The general procedure is as follows; details will depend on the characteristics of the pullers available.

A. Inner pH-Sensitive Micropipettes

These are best made with a micropipette puller using a single-stage process. At first you should make a batch as long and as sharp as is possible while still retaining a smooth taper down to the tip. Having made a batch, and noted their diameters at (say) 50 and 100 μm from their tips, seal the tips in the microforge. Do this by first positioning them (view wth the $40\times$ objective) with the tip 1–20 μm from and perpendicular to the heating element, and then turning up the heating current until the tip just begins to melt and appears to collapse very slightly. The length of the solid, sealed tip should be only 3–5 μm.

B. Insulating Micropipettes

These need to be sharp but not too long or acutely tapered; otherwise it will be impossible to fit the pH glass micropipettes inside them. I suggest a two-stage process, using a vertical puller such as the Kopf 700 with the solenoid switched off.

A long narrow coil gives better results than a short one. First heat the glass, allow it to be pulled out by 8–12 mm, and then cool and recenter it in the puller. Second, using minimal heat, allow the glass to soften and separate into two micropipettes. Make a number of micropipettes in this way, adjusting the second heats until the micropipettes inner dimensions match those of the pH glass micropipettes.

C. Sealing the Inner to the Outer

First, mount an outer micropipette in the microforge. Second, fix a matching, sealed inner micropipette to the end of a length of 0.5-mm (outer diameter) steel tubing with wax. The chosen inner should fit so tightly inside the terminal 40 μm of the outer that it almost fills the recess, as shown in Fig. 3A. The other

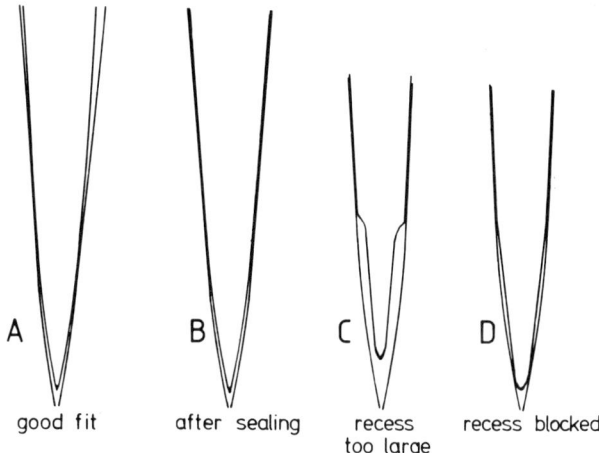

Fig. 3. Diagrams of the tips of pH-sensitive microelectrodes.

end of the steel tubing should be connected to a source of compressed N_2 or CO_2 at a pressure of about 50 bar (700 lb/sq. in). (This pressure is much higher than I originally used, as I have found that it allows the glass–glass seal to be made much closer to the tip of the pH glass). Third, turn on the gas, check for leaks, and then lower the pH glass carefully into the top of the outer until the two tips are correctly positioned, as in Figure 3A. Fourth, apply heat from the carefully positioned Pt loop until the two glasses are sealed together leaving an exposed length of pH-sensitive glass of about 20 μm, as shown in Figure 3B. Heat need only be applied from one side.

It is important to leave room inside the recess to allow rapid exchange of its contents with the outside, but the recess volume should be made as small as possible. Recess configurations giving slow responses are shown in Figure 3C and D. Once the glasses have been sealed together, the steel tube and pH glass above the seal should be withdrawn and the remaining pH glass discarded.

D. Filling and Conditioning the Completed Microelectrodes

Use a syringe to fill the microelectrodes with a 100 mM NaCl solution buffered to pH 6 with 100 mM Na citrate (start with tri-sodium citrate and add HCl until the pH reaches 6). With a cat's whisker coax the air bubbles from the tip. Then loosely cap the electrodes and incubate them (tips in air) overnight in a 70°C oven. Next day seal a Ag:AgCl wire into the top of the electrodes and store them temporarily with their tips dipped in chromic acid cleaning solution. (To make about 1 liter, take about 50 ml of saturated K_2CrO_4 and slowly add concentrated H_2SO_4 until a clear solution is obtained.)

If the electrodes have a low resistance and no response to a pH change when first tested, throw them away and use more heat to seal them next time. If they have a high resistance (over 10^{11} ohms) or are noisy and unstable, further aging often greatly improves them. A response that is too slow may be accelerated by carefully enlarging the tip by beveling, breakage, or etching in 4 M NaOH with 100 mM EDTA.

With care and an easy preparation, a single pH microelectrode can give many months of service. Between experiments its tip may be kept in chromic acid (if it needs cleaning) or simply in air. Occasionally material not dissolving in chromic acid can be removed by a brief treatment with the etching solution. Animal or plant life appearing inside the electrodes should be ignored.

IV. REFERENCES

Caldwell PC: An investigation of the intracellular pH of crab fibres by means of micro-glass and micro-tungsten electrodes. J Physiol 126:169, 1954.

de Hemptinne A: Intracellular pH and surface pH in skeletal and cardiac muscle measured with a double-barrelled pH microelectrode. Pflugers Arch 386:121, 1980.

Harmon MC, Poole-Wilson PA: A liquid ion-exchange intracellular pH microelectrode. J Physiol 315:1P, 1981.

Hinke JAM: Glass microelectrodes for measuring intracellular activities of sodium and potassium. Nature (Lond) 184:1257, 1959.

Thomas RC: New design for sodium-sensitive glass microelectrode. J Physiol 210:82P, 1970.

Thomas RC: Intracellular pH of snail neurones measured with a new pH-sensitive glass microelectrode. J Physiol 238:159, 1974.

Thomas RC: "Ion-Sensitive Intracellular Microelectrodes." London: Academic Press, 1978.

Double-Barreled Intracellular pH Electrode: Construction and Illustration of Some Results

A. de Hemptinne, R. Marrannes, and B. Vanheel
Laboratory of Normal and Pathological Physiology University of Gent, B-9000, Gent, Belgium

I.	Introduction	7
II.	Materials and Methods	8
	A. Construction of the Double-Barrel Micropipette	8
	B. Construction of the pH-Sensitive Part	9
III.	Results	12
	A. Measurement of pH_i and pH_s in Rat Soleus Muscle	12
	B. Influence of Simulated Ischemia on pH_i and pH_s in Rat Soleus Muscle	12
	C. Effect of an Organic Acid on pH_i and pH_s in the Rat Soleus Muscle	14
	D. Influence of Simulated Ischemia on pH_i and pH_s in Sheep Purkinje Fiber	15
	E. Effect of Organic Acids on pH_i and pH_s in Sheep Purkinje Fiber	16
IV.	Discussion	18
V.	References	19

I. INTRODUCTION

The recessed tip design for the construction of pH-sensitive microelectrodes as described by Thomas [1974] has been proved to be suitable for measuring the intracellular pH in nerve, skeletal muscle, and cardiac muscle [Thomas, 1974; Ellis and Thomas, 1976; Aickin and Thomas, 1977]. As the pH-sensitive electrode also measures the membrane potential, one has to subtract this potential from the composite pH signal. This can be done by measuring the membrane potential separately with a conventional KCl-filled microelectrode. The obligation to use two separate microelectrodes that have to penetrate the same cell can be overcome by making double-barreled micropipettes and making one of them pH-sensitive.

The construction, the calibration, and some results obtained using these electrodes are described in this contribution. Some of this work appeared in previous publications [de Hemptinne, 1979, 1980; de Hemptinne and Marrannes, 1979; Marrannes et al, 1979, 1981].

II. MATERIALS AND METHODS

A. Construction of the Double-Barrel Micropipette

Two glass capillaries (Hildenberg glastype AR Borosilicate glass) are held together at both ends with two short pieces of silastic tubing. One glass tube has a cross section o.d. of 2 mm and a length of 7 cm; the other, in which a glass filament is introduced, has a cross section o.d. of 1.5 mm and length of 4.5 cm.

Using a David Kopf (model 700 C) vertical micropipette puller with a fairly long heater coil (9 ½ turns), the thinner capillary is twisted 360° relative to the other by hand. A long metal needle (o.d. 1.2 mm) is introduced into the wider capillary to prevent it from bending during the heating and rotation procedure. This is illustrated in Figure 1. We learned by experience that a rotation of 360° and not less is important to maintain the circular cross section of the wider capillary in the middle of the twisted part, this to facilitate the penetration of the pH glass capillary in a later stage of construction.

The twisted capillaries are removed from the puller, and the extremities are glued to each other with rapid setting Araldite. The pipettes are then pulled using

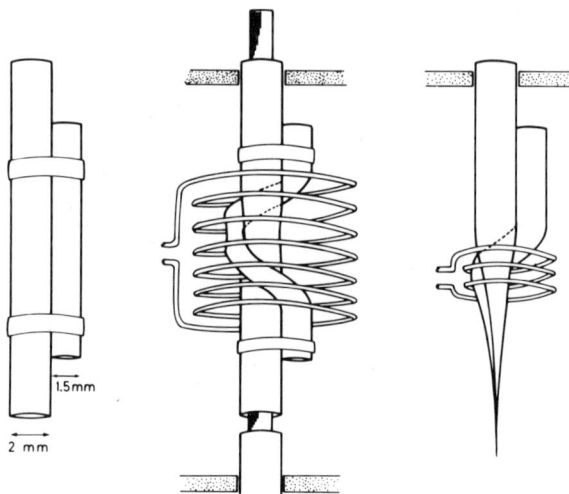

Fig. 1. Diagram showing the construction of a double-barreled micropipette. The explanation is given in the text.

the same puller with a short heater coil (2 ½ turns). A scanning micrograph of the tip of double-barreled micropipettes prepared in this way is shown in Figure 2.

The micropipettes are viewed under a microscope at a magnification of 150×, and the inner diameter of the wide capillary is measured at a distance of 520 μm from the tip. The prepull distance, the heat, and final pull force of the puller were adapted to obtain micropipettes with an inner diameter of about 15–20 μm at the given distance.

B. Construction of the pH-Sensitive Part

The technique described in detail by Thomas [1978] with small personal adaptations is applied to construct a recessed type pH-sensitive electrode. Corning 0150 pH glass tubing (o.d. varying between 0.5 and 1 mm) is used. Long and sharp capillaries are pulled using the Kopf puller with a long narrow heater coil (8 ½ turns). Prepull distance, heat, and pull force are adapted to obtain micro-

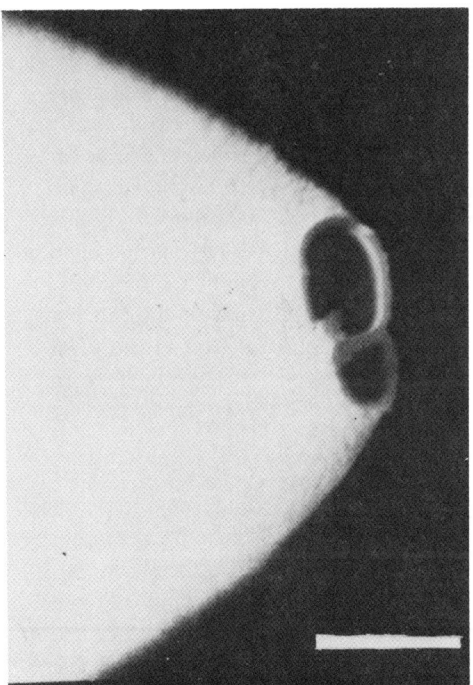
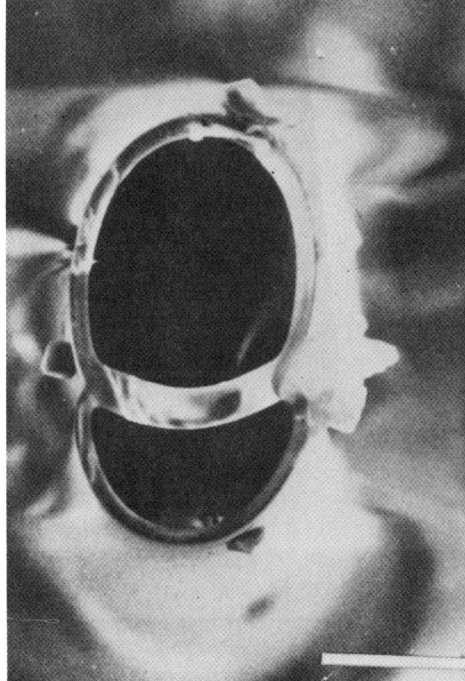

Fig. 2. Electron scanning micrograph of the tip of two different double-barreled micropipettes. The calibration marks are 1,8 μm (left) and 10,3 μm (right). The tip of the right pipette is broken to show the shape of the capillaries at the level where the pH glass capillary has to fit in. The cross section of the wider capillary should ideally be circular.

pipettes with an outer diameter of about 8–10.5 μm at a distance of 520 μm from the tip. The pH capillaries are mounted vertically in a microforge, and the sealing procedure of the tip is performed while pressure (15 bar) is applied from the back end of the capillary via a steel tube fixed with wax into that end. The purpose of applying pressure during the sealing procedure is to reduce the tendency of the glass to collapse over too long a distance when the glass of the pH capillary is heated at its tip.

A one turn minicoil platinum iridium wire (60 μm diameter) is used for heating. The two extremities of the heater wire are held between two pieces of glass fiberepoxy-printed circuit material, which assures electrical contact with a variable current device.

When the tip of the pH capillary, which is viewed at high magnification (× 675), is thought to be sealed, it is lowered centrally through the miniheater coil, and heating is again applied at a distance of about 12 μm from the tip. In case of correct sealing of the tip, as soon as the softening point of the pH glass is reached, the capillary tends to expand under the influence of the applied pressure. If the tip of the pH capillary is not properly sealed, it tends to collapse when heated at that place. As soon as the tendency to expand is noticed the heat is turned down, and one can be assured that the tip is correctly sealed.

No special innovation is introduced for the assembly of the outer and inner capillaries. The double-barrel micropipette is mounted with the tip centrally located through the miniheater coil. The sealed pH capillary is descended as far as possible within the outer capillary and the glass–glass seal is obtained on applying heat at the given pressure. Above the glass–glass seal the pH glass is fractured by withdrawing the remaining part. The pH-sensitive capillary is filled with a solution containing 0.1 M NaCl buffered to pH 6 with 0.1 M citrate buffer. Removal of air bubbles in the recessed tip is facilitated by introducing in the capillary a cat's whisker and carefully applying local heat at the level where the pH capillary broke off. The microelectrodes are kept overnight in an oven at 65°C with the open end of the pH-sensitive barrel covered to prevent evaporation of the fluid.

On the next day, the electrodes are found suitable for use. The reference electrode barrel, which contains a glass filament, is filled with a solution containing, for experiments on cardiac preparations, 3 M KCl, 10 mM K citrate, and 10 mM EDTA buffered to pH 6; and, for experiments on skeletal muscle, saturated K_2SO_4 instead of the KCl. We found that the addition of calcium chelators to the filling solution of the reference barrel is useful to reduce the problems of instability that can arise with high resistance microelectrodes. The resistance of the reference barrel is of the order of 20–100 MΩ.

Before use, the tip of the filled double-barrel microelectrodes is dipped in a solution containing 50% ethanol and water and subsequently in a concentrated chromic acid solution. When the microelectrode is then dipped in the perfusion chamber, its pH sensitivity and the speed of its response can immediately be judged from the large and rapid drift of the electrical signal that reflects the shift from the initial, extremely low pH in the recessed chamber of the pH electrode (the pH of chromic acid) to the pH of the superfusion solution.

An Analog Devices 311 J Varactor bridge operational amplifier, (input impedance 10^{14} Ω working at unity gain) is used as preamplifier. It is fixed to a micromanipulator and is connected to the pH-sensitive barrel via a short Ag/AgCl wire. A driven shield system is used connecting the output potential of the preamplifier to aluminum foil that is wrapped around the double-barrel pipette.

Single-barrel, pH-sensitive electrodes were also constructed to measure the pH at the surface of cells. The tip was intentionally broken and fire-polished before the introduction of the pH-sensitive glass.

Experiments were performed on isolated rat soleus muscle and sheep Purkinje fibers. The preparations were disposed horizontally in a perfusion chamber, which permitted a front view with a binocular microscope through a glass window while the microelectrodes penetrated the preparation vertically.

A photograph of the perfusion chamber with a double- and single-barrel electrode connected to the preamplifiers is shown in Figure 3. The potential of

Fig. 3. Photograph of the experimental chamber and three preamplifiers. The one on the left is connected to a single pH-sensitive micropipette. The two preamplifiers on the right are connected to a double-barrel, pH-sensitive micropipette. The superfusion chamber is illuminated from both sides with a cold light source (glass fiber optics). A black background is used to improve the visualization of the micropipettes through the horizontally disposed binocular microscope.

the experimental chamber was clamped to ground using an operational amplifier in a closed loop configuration. This amplifier delivered the current required to equalize the potential of the chamber to that of the ground.

The solution used to superfuse the soleus muscle contained in mmol/l: NaCl 121; KCl 4; $CaCl_2$ 10; $MgCl_2$ 1; $NaHCO_3$ 21; Na_2HPO_4 0.1; glucose 10; and for the Purkinje fiber NaCl 120.5; KCl 4; $CaCl_2$ 2.5; $MgCl_2$ 1.2; $NaHCO_3$ 25; Na_2HPO_4 1.2; glucose 5. The solutions were equilibrated with a 95% O_2 and 5% CO_2 gas mixture and warmed to 37°C.

A roller pump assured a constant flow rate of saline through the perfusion chamber. The fluid dripped out of the bath through a piece of polyethylene tubing whose terminal outflow level, acting like a siphon, could be adjusted to regulate the level of the fluid in the perfusion chamber. A few drops of paraffin oil were layed on the superfusion fluid in the experimental chamber.

III. RESULTS

A. Measurement of pH_i and pH_e in Rat Soleus Muscle

Several examples of repeated measurements of surface pH and intracellular pH in rat soleus muscle and sheep Purkinje fibers have been shown elsewhere [de Hemptinne, 1979, 1980].

One of these examples is reproduced in Figure 4. Two consecutive cell penetrations were performed with a double-barrel pH electrode. The bulk pH (pH_o) is measured as soon as the electrode tip comes into contact with the superfusion solution. A somewhat lower pH (usually about 0.05–0.1 pH units) can be recorded when the tip of the electrode touches the surface of the cell, and a still lower pH value is recorded on penetrating the cell. On withdrawing the electrode from the cell, bulk pH is again measured. The response speed of the electrode to a change in pH can be judged from the time constant of the pH shift seen after withdrawal of the electrode. It is usually of the order of a few seconds. On superfusing the preparation during a short period with a solution containing 50 mEq/l K^+ in replacement of Na^+, a partial and transient depolarization is measured by both the reference and the pH barrel. This shows that the difference signal, which is proportional to pH, is not sensitive to a change in membrane potential.

B. Influence of Simulated Ischemia on pH_i and pH_e in Rat Soleus Muscle

In order to simulate ischemia, the superfusion was interrupted, and the fluid level in the superfusion bath was lowered to have the preparation completely

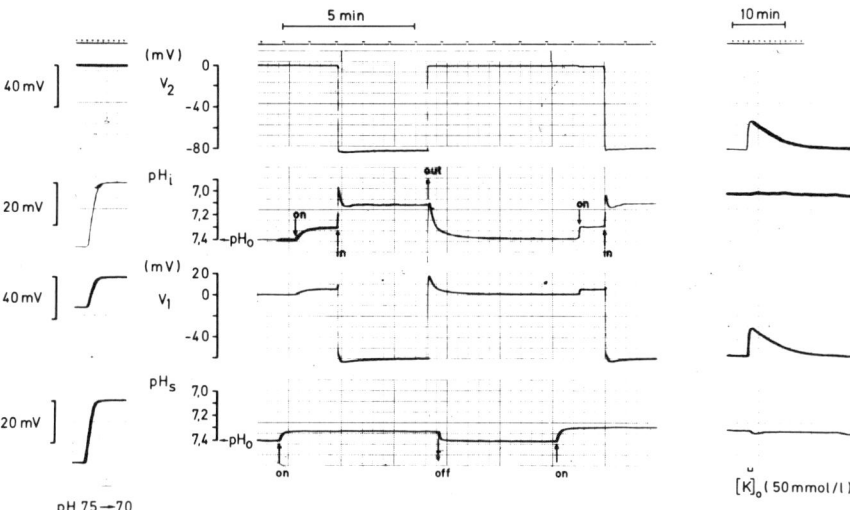

Fig. 4. Intracellular (pH_i) surface (pH_s) pH as measured with a double-barreled, pH-sensitive microelectrode in rat soleus muscle during two successive cell penetrations. The reference potential (V_2) is measured with the reference electrode barrel. V_1 is the electrical potential measured by the pH-sensitive electrode barrel. pH_i is obtained by substracting V_2 from V_1. The pH_i trace gives the pH of the bulk solution (pH_o), the surface (on) and the cell (in). pH_s is also measured with a single-barrel, pH-sensitive electrode giving the surface pH when its tip touches the surface of the cel (on). The pH sensitivity of the electrodes is shown on the left (about 30 mV for 0.5 pH unit change). The effect of a short perfusion with high potassium (50 mmol/l) is shown on the right.

immersed in paraffin oil. In these conditions, a progressive acid shift of the intracellular pH can be seen (Fig. 5). On withdrawing the electrode from the cell and measuring the pH in the thin fluid film surrounding the preparation, a much more acidic pH is measured. The speed of acidification of that compartment can be seen from the records obtained with the surface pH electrode. On reperfusing the preparation, a very rapid normalization of the intracellular pH with a slight overshoot is observed. The most likely explanation is that during simulated ischemia there is still some CO_2 and presumably also lactic acid produced by the muscle cell, which cause acidification. The acidification is, however, limited by the buffer power of the cells. As CO_2 and lactic acid diffuse in the fluid film that surrounds the preparation, greater acidification is induced in this compartment since the buffer power of the extracellular fluid is much smaller than that of the cell.

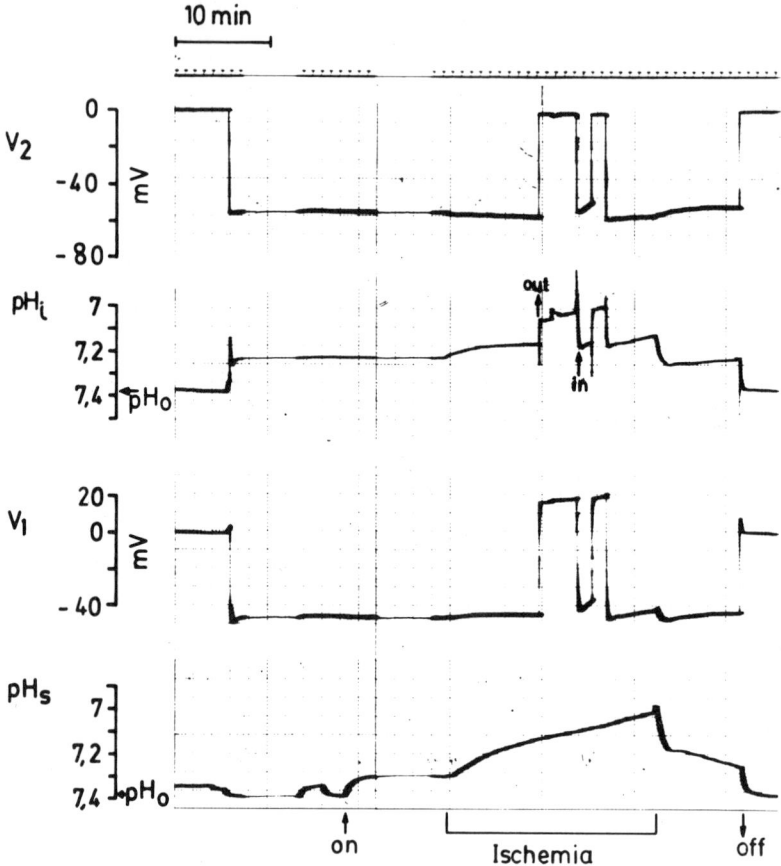

Fig. 5. Membrane potential (V_2), electrical potential measured by the pH-sensitive electrode barrel ($V_1 - V_2$) and surface pH (pH_s) measured with a double-barreled, pH-sensitive microelectrode and single-barrel pH electrode in rat soleus muscle. Three cell penetrations were performed. Ischemia was simulated as described in the text. During ischemia, the pH_i trace gives the values of the surface pH when the electrode comes out of the cell. This happened twice. Notice the larger acidification of pH_s as compared to that of pH_i during ischemia, whereas the membrane potential shows a small hyperpolarization.

C. Effect of an Organic Acid on pH_i and pH_s in the Rat Soleus Muscle

Figure 6 shows the effect of 20 mEq/l propionate on the membrane potential, pH_i and pH_s. In this experiment, the fiber was superfused with a relatively acid solution (pH_o = 6.8) obtained by reducing the HCO_3^- concentration to 5 mEq/l at constant P_{CO_2}. The propionate was added in replacement of 20 mEq/l Cl^-.

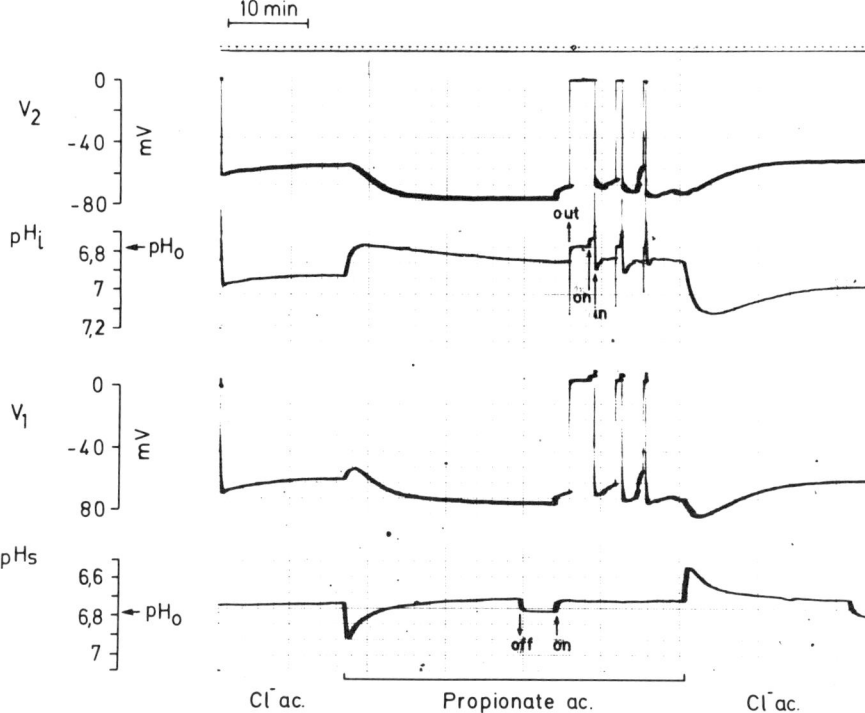

Fig. 6. Influence of propionate (20 mmol/l) at pH 6.8 in rat soleus muscle. V_2, pH_i, V_1 and pH_s are measured as explained in Figure 3. Notice the hyperpolarization, the intracellular acidification, and transient surface alcalinization under influence of the organic acid.

As propionic acid (which is much more permeable than propionate as anion) penetrates into the muscle cell, a rapid intracellular acid shift can be seen, whereas at the surface of the cells the pH becomes transiently more alkaline. An intracellular pH regulation system has to be postulated to account for the progressive shift of pH_i to a higher steady-state value, which is reached at the end of the perfusion period with propionate, since it occurs against both unfavorable proton and electrical gradients. On removing the propionate the intracellular pH becomes rapidly more alkaline, since propionic acid also leaves the cell in combined form and removes intracellular protons. At the extracellular surface of the cells, however, dissociation of propionic acid causes locally a measurable transient acidification.

D. Influence of Simulated Ischemia on pH_i and pH_s in Sheep Purkinje Fiber

As in rat soleus muscle, ischemia was simulated by interrupting the superfusion and having the preparation immersed in paraffin oil. In these conditions,

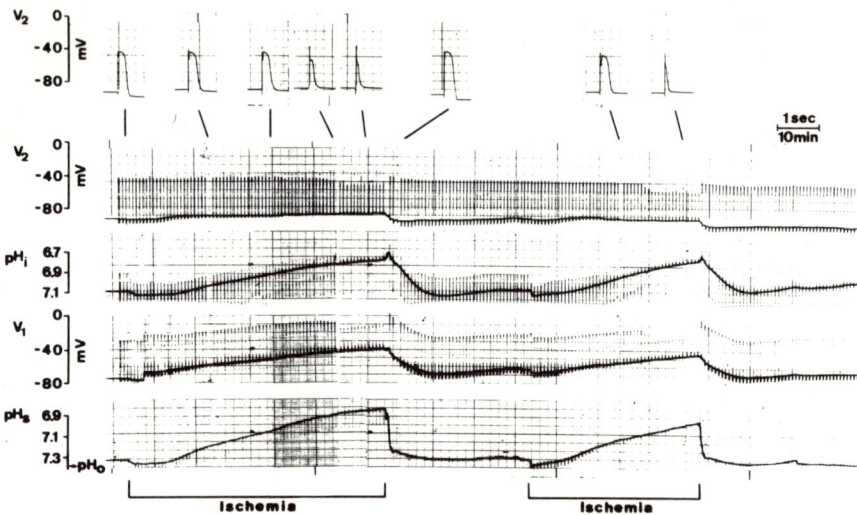

Fig. 7. Influence of two periods of simulated ischemia on the shape of the action potential (top trace), membrane potential, intracellular (pH$_i$), and surface pH (pH$_s$) in a sheep Purkinje fiber. Notice the depolarization, reduction of pacemaker potential, and loss of "plateau" formation caused by ischemia. Both pH$_i$ and pH$_s$ change to more acid values. The second period of ischemia causes a more rapid change in pH and a faster alteration of the electrophysiological parameters.

a progressive acid shift of both the intracellular and surface pH can be seen (Fig. 7). Contrary to soleus muscle where the surface pH became rapidly more acidic than the intracellular pH, we never observed in sheep Purkinje fibers a reversal of the initial pH gradient between the surface and the intracellular medium.

Associated with the acidification, we observed some reduction of the resting membrane potential, a marked decrease in the duration of the action potentials with loss of "plateau" formation. These functional electrophysiological alterations are completely reversible on reperfusing the trabeculum. Complete normalization of both the surface and the intracellular pH can be seen when the trabeculum is reperfused.

When several periods of ischemia are imposed in succession, separated by periods of reperfusion, we observed that the speed of intracellular acidification increased, and that the functional electrophysiological alterations appeared sooner. This observation is also illustrated in Figure 6.

E. Effect of Organic Acids on pH$_i$ and pH$_s$ in Sheep Purkinje Fiber

Figure 8 shows the effect on replacing 20 mEq/l Cl$^-$ by several organic anions on pH$_i$ and pH$_s$. All the organic acids that can penetrate the cells (pK not too

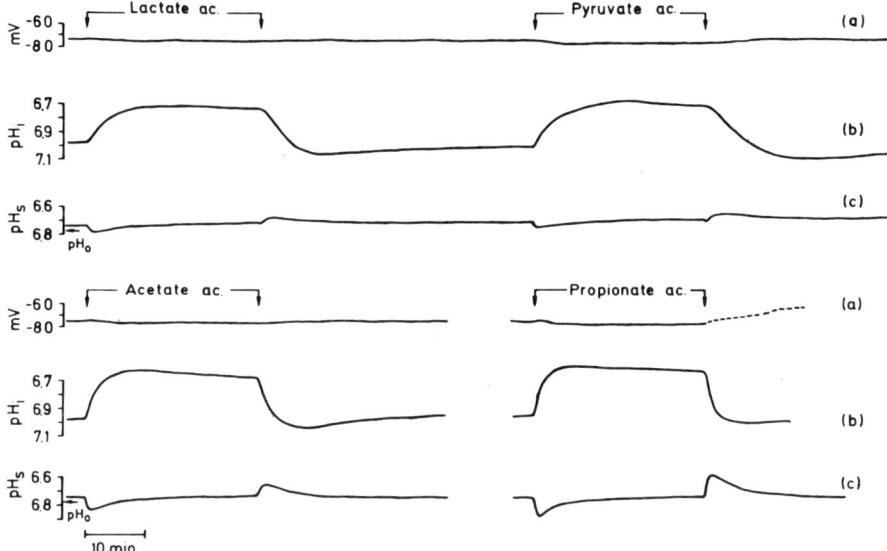

Fig. 8. Influence of lactate, pyruvate, acetate, and propionate (20 mmol/l) at pH 6.78 on membrane potential (a), intracellular pH (b), and surface pH (c) in a sheep Purkinje fiber. Notice the larger surface pH transients and the faster pH_i changes with propionate > acetate > lactate ≃ pyruvate.

low so that some acid is present in undissociated form) cause intracellular acidification and transient extracellular changes in the other direction. We found that the faster the acid can penetrate into the cell, in accordance to its pK and lipid:water partition ratio, the faster the intracellular acidification occurs and the greater the surface pH transients. All these acids cause a slight hyperpolarization. Salycilic acid also apparently penetrates readily into the cells and causes a relatively rapid intracellular acidification, but this is associated with depolarization (Fig. 9). The larger and the more prolonged surface pH transient, seen under influence of salycilate rather than under influence of propionate, although the latter acid causes a faster intracellular acidification, can perhaps be explained by the higher liposolubility of salycilic acid, which can eventually dissolve in all the intracellular lipid constituents, whereas penetration of propionic acid would be more restricted to the aqueous intracellular phase. If the intracelluar volume in which salycilic acid can penetrate is significantly larger than that of propionic acid, a more prolonged influx of salycilic acid can be expected.

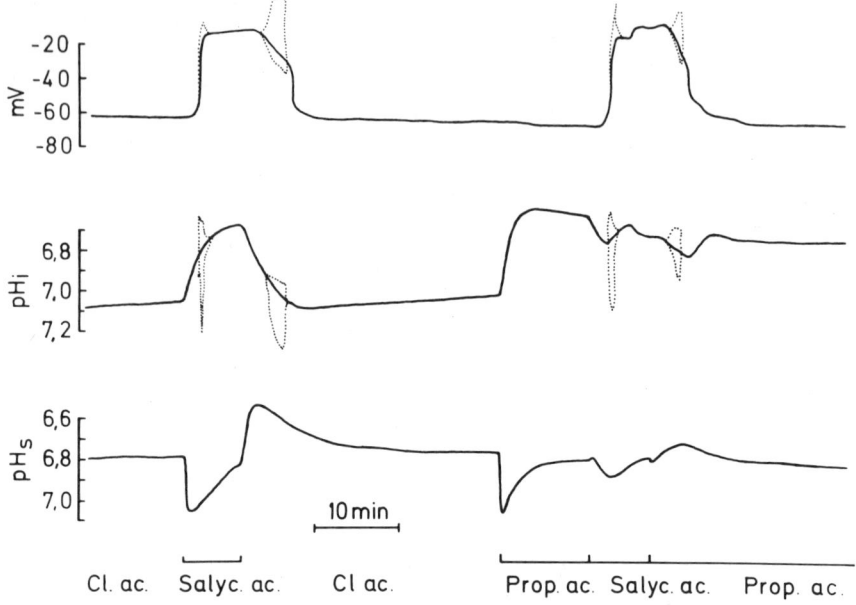

Fig. 9. Influence of salycilate and propionate (20 mmol/l) at pH 6.8 on membrane potential, intracellular pH (pH_i) and surface pH (pH_s). Salycilate causes depolarization and induces repetitive activity. The envelope of the peak of the action potentials is given by the dotted line. Notice the faster intracellular acidification caused by propionate as compared to salycilate. Notice also the large and more prolonged surface pH transient caused by salycilate as compared to propionate.

IV. DISCUSSION

The results show that a double-barreled, pH-sensitive electrode can be used to obtain successive and reproducible measurements of pH_i in rat soleus muscle and sheep heart Purkinje fiber. The obvious advantage of using a double-barreled microelectrode is that the reference potential is measured at the same location as the pH-dependent signal. This can be important for cardiac work where electrical uncoupling of neighboring cells can occur in a variety of conditions such as ischemia and intracellular acidosis [Weingart and Reber, 1979; Wojtczak, 1979].

Our results on the effects of organic acids on the intracellular pH can be interpreted in the same way as the results obtained by Boron and De Weer [1976] on the squid giant axon, and recently by Keifer and Roos [1981] on the barnacle muscle; intracellular acidification results from penetration of the organic acid in combined form and causes an intracellular acid load as it dissociates inside the cell. In the unstirred layer surrounding the cell, the anionic form of the acid, remaining in excess, induces a transient alkaline shift.

Knowing the buffer power of the intracellular medium, and the surface-to-volume ratio of the cells, the influx rate of the acid can be calculated from the measured intracellular acidification rate.

Our results with simulated ischemia are informative because they give some information on what can perhaps occur in a real physiopathological ischemic condition. It is known that ischemia induces intracellular and interstitial acidification associated with an increase of P_{CO_2} [Poole-Wilson, 1978]. Although the acids are produced by the cellular metabolism, the fall in pH can be larger in the interstitial compartment than in the cytosol presumably because of the difference in buffer power.

These electrodes can also be used to penetrate smaller cells. We performed recently with success long-lasting (about 1 hour) pH measurements in atrial trabeculae isolated from the guinea pig heart.

V. REFERENCES

Aickin CC, Thomas RC: Microelectrode measurement of the intracellular pH and buffering power of mouse soleus muscle fibers. J Physiol 267:791, 1977.

Boron WF, De Weer P: Intracellular pH transients in squid giant axons caused by CO_2, NH_3 and metabolic inhibitors. J Gen Physiol 67:91, 1976.

de Hemptinne A: A double-barrel pH micro-electrode for intracellular use. J Physiol 295:5P, 1979.

de Hemptinne A: Intracellular pH and surface pH in skeletal and cardiac muscle measured with a double-barrelled pH microelectrode. Pflüg Arch 386:121, 1980.

de Hemptinne A, Marrannes R: Effect of propionate on intracellular and surface pH in the rat soleus muscle. J Physiol (Lond) 295:22P, 1979.

Ellis D, Thomas RC: Direct measurement of the intracellular pH of mammalian cardiac muscle. J Physiol (Lond) 262:755, 1976.

Keifer DW, Roos A: Membrane permeability to the molecular and ionic forms of DMO in barnacle muscle. Am J Physiol 240:C73, 1981.

Marrannes R, de Hemptinne A, Leusen I: Correlation between conduction velocity transients in isolated heart fibers and pH changes. Arch Int Physiol 87:770, 1979.

Marrannes R, de Hemptinne A, Leusen I: pH Aspects of transient changes in conduction velocity in isolated heart fibers after partial replacement of chloride with organic anions. Pflüg Arch 389:199, 1981.

Poole-Wilson PA: Measurement of myocardial intracellular pH in pathological states (editorial). J Mol Cell Cardiol 10:511, 1978.

Thomas RC: Intracellular pH of snail neurones measured with a new pH-sensitive glass microelectrode. J Physiol (Lond) 238:159, 1974.

Thomas RC: "Ion-sensitive Intracellular Microelectrode." London: Academic Press, 1978.

Weingart R, Reber W: Influence of internal pH on r_i of Purkinje fibers from mammalian heart. Experientia 35:929, 1979.

Wojtczak J: Contractures and increase in internal longitudinal resistance of cow ventricular muscle induced by hypoxia. Circ Res 44:88, 1979.

Intracellular pH: Its Measurement, Regulation, and
Utilization in Cellular Functions, pages 21–54
© 1982 Alan R. Liss, Inc., 150 Fifth Avenue, New York, NY 10011

An Optical Technique for Measurement of Intracellular pH in Single Living Cells

Jeanne M. Heiple* and D. Lansing Taylor
Cell and Developmental Biology, The Biological Laboratories, Harvard University, 16 Divinity Avenue, Cambridge, Massachusetts 02138

I.	Introduction	22
	A. pH_i and Contractility	22
	B. pH_i and Endocytic Events	24
	C. Techniques of pH_i Measurement	24
II.	Materials and Methods	29
	A. Materials	29
	B. Preparation of FTC-Ovalbumin	30
	C. Microinjections and Microfluorometric Measurements	30
	D. Induction of Endocytosis	33
III.	Results	34
	A. Spectral Characterization of FTC-Ovalbumin pH-Sensitivity	34
	B. Cytoplasmic pH Measurements	39
	1. Intracellular distribution of the probe	39
	2. Standard curve	40
	3. Average cytoplasmic pH and variations	42
	4. Wound-healing response	42
	C. Phagosomal pH Measurements	44
	1. Intracellular distribution of the probe	44
	2. Standard curve	44
	3. Phagosomal pH changes	45
IV.	Discussion	46
V.	References	49

*Present address: Boston University School of Medicine, Department of Biochemistry, 80 E. Concord Street, Boston, MA 02118.

I. INTRODUCTION

Our interest in the potential role of intracellular pH changes as a stimulus-response coupling mechanism stems from work in our laboratory and in others implicating Ca^{++} and/or pH in the regulation of amoeboid movement and of endocytosis (see below). The technique presented here represents our efforts to meet a specific challenge. That is, the precise, continuous quantitation of pH and pH changes within specific subcellular compartments of a single, moving eucaryotic cell. The emphasis here will be on the development of the method, with sufficient attention to background and to applications to provide a sense of the potential usefulness of the technique and of the motivation behind its development. A number of excellent recent reviews have compared and evaluated methods of intracellular pH (pH_i) measurements [Waddell and Bates, 1969; Gillies and Deamer, 1979b; Gerson, 1978; Rottenberg, 1979; Roos and Boron, 1981]; have summarized evidence for the regulation of pH_i by plant cells [Smith and Raven, 1979; Davies, 1973] and by animal cells [Boron, 1980; Roos and Boron, 1981; Gillies and Deamer, 1979b; Gillies, 1981]; and have surveyed evidence for the regulation of cell processes by pH_i in plant cells [Smith and Raven, 1979] and in animal cells [Gillies and Deamer, 1979b; Gerson, 1978; Gillies, 1981; Roos and Boron, 1981]. A brief outline of some of the evidence for cell regulation by pH_i is presented below. The reader is referred to later chapters in this volume, and to the reviews cited above, for more detailed information. Our intention here is to place into context our own interest in the role of pH_i in regulation of amoeboid movement and of endocytosis.

A. pH_i and Contractility

Stimulus-response coupling in a variety of cells and cell processes involves interaction of external stimuli with the cell surface, transduction of a signal across the plasma membrane, and mobilization of intracellular responses. The coordination of these responses may involve changes in cytoplasmic concentrations of ions and/or of small molecules, including monovalent cations, divalent cations, and cyclic nucleotides [see Kaplan, 1978; Rasmussen and Goodman, 1978]. Notable examples include the following: (1) excitation-contraction coupling in muscle [Ebashi, 1976]; (2) stimulus-response coupling in postsynaptic junctions [Puszkin and Kochwa, 1974] in exocrine cells [Rubin, 1974], in platelets [Horne et al, 1981; Bennet et al, 1979], in macrophages [Gallin et al, 1975; Gallin and Gallin, 1977; Lew and Stossel, 1979], and in leukocytes [Whitin et al, 1980; Naccache et al, 1979; Weissmann, 1978]; and (3) stimulus-response coupling in egg and sperm activation during fertilization [Epel, 1978; and this volume] and in amoeboid movement [Hitchcock, 1977; Taylor and Condeelis, 1979; Nuccitelli et al, 1977; Prusch and Hannafin, 1979a, b; Gawlitta et al, 1980; Taylor et al, 1980a] (see below). The possibility that pH may also serve as a secondary messenger, or stimulus-response coupler, is suggested by accumulating evidence that intracellular pH changes accompany many of these processes and are closely coupled to fluxes in other pleiotropic factors [Gillies, 1981] (see later chapters, this volume).

Ion fluxes (including ΔpH_i) during stimulus-response coupling may exert their effect, in part, by altering the structure, activity, or interactions of the contractile proteins implicated in each event. For example, it has been demonstrated that gelation and contraction of single-cell models [Taylor et al, 1973] and of bulk cytoplasmic extracts from free-living amoebae are regulated by changes in free Ca^{++} concentration in the micromolar range [Pollard, 1976; Condeelis and Taylor, 1977; Hellewell and Taylor, 1979; Yin and Stossel, 1979] and by pH [Condeelis and Taylor, 1977; Hellewell and Taylor, 1979]. Regulatory pH changes in this system are on the order of $\Delta 0.2$–0.4 U, in the physiological range, with high pH (> 7.0) favoring solation and contraction. During extension of the acrosomal process in echinoderm sperm, alkalinization may trigger actin polymerization, whereas Ca^{++} appears to be necessary for membrane fusion events [Tilney et al, 1978]. Similarly, during egg activation, early cortical events, subsequent to alkalinization and increased free Ca^{++} concentration, include fusion of cortical granules and elongation of actin-filled microvilli [Epel, 1978]. In unfertilized egg cortices, a rise in pH, similar to that measured during fertilization, can induce the transformation of amorphous cortical actin to a filamentous form, and this transformation is not sensitive to changes in free Ca^{++} concentration [Begg and Rhebun, 1979]. Proton efflux during fertilization has also been correlated with actin polymerization and formation of the fertilization cone [Tilney and Jaffe, 1980]. Studies with Triton-extracted muscle cells have shown that raising the pH from 7.0 to 7.7 at subthreshold Ca^{++} concentrations can induce contractions as effectively as raising the free Ca^{++} concentration to the threshold level [Taylor, 1976]. Oddly, pH-induced muscle contractions are oscillatory. Natural oscillatory contractions occur in the acellular slime mold, P. polycephalum [Wohlfarth-Bottermann, 1979], and are correlated with ion fluxes [Ridgeway and Durham, 1976]. However, the mechanism of regulation of contractility in this system remains unknown. The importance of the pH_i regulation to contractility is also suggested by in vitro studies demonstrating the pH-sensitivity of a variety of protein interactions that have been implicated in contractile events. These include myosin filament formation [Hinssen and D'Haese, 1974; Craig and Megerman, 1977]; microtubule polymerization [Regula et al, 1981]; and interactions between actin and filamin [Nunnally and Craig, 1980]. In vitro assembly of clathrin, the major protein comprising the coat around certain endocytic vesicles [Goldstein et al, 1979], has also been shown to be pH-sensitive [Woodward and Roth, 1978]. The studies above show that, at pH 6.5, "side-polar" myosin filaments are favored, whereas at pH 7.5 "bipolar" myosin filaments are favored; that microtubule depolymerization is progressively favored as pH is raised from 6.35 to 7.75; that actin–filamin interactions are favored between 7.5 and 7.7; and that clathrin assembles into coats at pH 6.5 and disassembles at pH 7.5. All of these in vitro regulatory pH changes are within the physiological range.

The precise mechanisms, whether direct or indirect, by which pH_i exerts regulatory control over the systems described above remain entirely unknown. Speculations range from specific effects on protein assembly to modulatory effects on coordinated enzyme systems. The development of a method by which to monitor pH within specific subcellular microenvironments in a single cell may shed some light on these possibilities.

B. pH_i and Endocytic Events

Endocytosis in mammalian cells [Allison and Davies, 1974; Stossel, 1974; Silverstein, 1977; Babior, 1978] and in protozoa [Holter, 1959; Chapman-Andresen, 1977; Stockem, 1977] has long provided a model system for stimulus-response coupling in eucaryotic cells. The emphasis in studies of pinocytosis or phagocytosis has most often been on events mediated by the cell surface (eg, ligand–receptor interactions, changes in membrane permeability) and on the mechanism of internalization itself. Little is known of the ionic regulation of later events, including movements of endosomal vesicles, fusion with other organelles, and the eventual fate of the internalized substances. Several studies suggest that changes in intravesicular pH may be partially responsible for regulation of events subsequent to internalization. Some of this evidence is indirect, such as the inhibitory effect of amines, which raise lysosomal pH [Ohkuma and Poole, 1978], on phagosome–lysosome fusion in mouse peritoneal macrophages [Gordon et al, 1980] and on mannose-specific receptor recycling in macrophages [Stahl and Schlesinger, 1980]. Alterations in the normal course of endosomal pH changes in leukocytes [Mandell, 1970; Jensen and Bainton, 1973; Jacques and Bainton, 1978] have been correlated with impaired bactericidal function in diseased cells [Segal et al, 1981]. Finally, certain toxic or pathologic substances, such as diphtheria toxin [Sandvig and Olsnes, 1980] and Semliki Forest Virus [Helenius et al, 1980] appear to rely on acidification within some endosome-derived subcellular compartment to exert their effects. One important, unresolved question is whether lysosome–endosome fusion precedes (and is responsible for) or follows acidification. The answer will clearly be facilitated by pH_i-measuring techniques with a high degree of spatial and temporal resolution.

C. Techniques of pH_i Measurement

Currently available techniques for measuring pH_i include direct pH measurements on cell homogenates, assays of pH-sensitive reaction rates, distribution of weak acids or of weak bases, insertion of pH-sensitive microelectrodes, ^{31}P-nuclear magnetic resonance (NMR), and visual observations or spectrophotometric quantitation of pH-sensitive dyes. The relative strengths and weaknesses of each of these techniques have been thoroughly reviewed [Waddell and Bates, 1969; Boron and Roos, 1976; Gerson, 1978; Rottenberg, 1979; Gillies and Deamer, 1979b; Roos and Boron, 1981; this volume], and will not be considered in detail here.

Ideally, any intracellular pH measuring technique will meet certain criteria. The most important of these are specificity for and sensitivity to pH, nondestructiveness, localization to a specific subcellular compartment (spatial resolution), and the ability to make continuous, rapid measurements throughout a cellular process (temporal resolution). Many of the available techniques satisfy more than one of these requirements. To date, however, none has satisfied all of them. In fact, the state of the art led Gerson [1978, p 107] to remark that "at this stage in the technology of intracellular pH measurements, there is no suitable method of accurately measuring the pH of the intracellular compartments of most cells . . . ."* The choice of technique depends on the requirements and limitations of the process and system under study.

Detection and quantitation of pH and pH changes within specific subcellular compartments of a single, motile, eucaryotic cell, such as an amoeba, requires more sensitive and nonperturbing techniques than those previously applied to the question of pH_i. Measurement of the distribution of weak acids or of weak bases across the cell membrane, while a powerful tool for studying large populations of cells or for studying pH gradients in isoated vesicle populations, is impossible to apply to continuous measurements in single living cells. Measurements via insertion of microelectrodes are unsatisfactory, as moving cells crawl off the electrodes, open wounds are introduced during measurements, and it is impossible to make rapid measurements at more than one place in the cell without further injury or disturbance.

Numerous investigators have estimated the pH_i of a variety of cells by qualitative observation of the color of cell-incorporated indicator dyes [reviewed in Waddell and Bates, 1969; Chambers and Chambers, 1961]. Sulfonated indicators of the Clark and Lubs' series [Chambers and Chambers, 1961], of which phenol red is one, are usually employed. These do not penetrate the plasma membrane after microinjection, exhibit color transitions in the physiological range, and exhibit negligible protein-binding errors [Needham and Needham, 1925; Lisman and Strong, 1979]. In various amoeboid cells, cytoplasmic and organellar pH has been estimated after microinjection of dye [Needham and Needham, 1925;

---

*To avoid confusion, an important distinction must be made between "intracellular" and "cytoplasmic." Throughout this chapter, "intracellular" will be used to denote that region of the cell bounded by the plasma membrane, including cytoplasm and organelles, with no distinction made among the various compartments. "Cytoplasmic," on the other hand, will be used to denote only the intracellular ground substance, excluding organelles.

Chambers et al, 1927; Marshak, 1944; Wiercinski, 1944], or after uptake of dye during pinocytosis or phagocytosis [eg, Mast, 1942; Mandell, 1970; Jensen and Bainton, 1973]. However, evaluations of dye color by eye are subject to serious errors, such as those arising from changes in path length and in dye concentration, and from interactions of dyes with cellular constituents [eg, Mast, 1942]. Many of the limitations of the weak acid or weak base distribution technique apply equally to the distribution of pH-sensitive indicator dyes within the cell, most of which freely partition among and accumulate within subcellular compartments of heterogeneous pH. In addition, the pH_i values thus obtained (6.5–7.5, ± 0.2 U) are imprecise and can obscure pH_i changes of a regulatory nature, so more precise measurements are clearly needed.

pH-sensitive absorbance changes have been quantitated spectrophotometrically after incorporation of indicator dyes into tissues or into suspensions of cells or of organelles, such as frog skeletal muscle [MacDonald and Jobsis, 1976; MacDonald et al, 1977], spinach chloroplasts [Auslander and Junge, 1975; Pick and Avron, 1976], and tumor cell suspensions [Thomas et al, 1979]. Recently a laser pulse technique has been described for quantitation of rapid pH jumps in suspensions of indicator dye-containing liposomes [Gutman and Huppert, 1979]. Extension of these quantitative techniques to the unicellular level has depended upon the development of sophisticated microphotometry [see Wied, 1966]. The first quantitative microphotometer for measurement of cytochemical reactions was described by Pollister and Ris [1948]. Since that time, much of the impetus for the development and improvement of quantitative microphotometry has come from cytochemical studies (eg, chromatin quantitation from the Feulgen reaction [see Swift, 1966; Garcia and Iorio, 1966]), from studies on cytochrome c [eg, Chance, 1962], and from research on visual pigments [eg, Brown, 1961]. The consequent potential for precise microphotometric quantitation and localization of ionic parameters in single living cells has only begun to be explored. For example, in single Limulus photoreceptors, the absorption of microinjected Arsenazo III [Brown et al, 1977] or of phenol red [Lisman and Strong, 1979] has been used to measure light-induced intracellular calcium and pH_i fluxes, respectively. Similar experiments have quantitated Ca^{++} fluxes after microinjection of Arsenazo III into single Aplysia pacemaker neurones [Thomas and Gorman, 1977; Gorman and Thomas, 1978]. Since absorbance of Arsenazo III has been shown to be pH-sensitive as well as Ca^{++}-sensitive [Ogan and Simons, 1979], the importance of careful quantitation of both parameters with other probes is clear. Recently, quantitation of aequorin luminescence after microinjection into single motile amoebae has been used to map free Ca^{++} in these cells [Taylor et al, 1980a].

Incorporation of pH-sensitive fluorescent probes has been successfully applied to cell suspensions [Gerson and Burton, 1977; Ohkuma and Poole, 1978; Segal et al, 1981; Geisow et al, 1981; Horne et al, 1981] and to suspensions of subcellular particles [Deamer et al, 1972; Eisenbach et al, 1978; Kano and

Fendler, 1978; Thomas and Johnson, 1975; Lee and Forte, 1980, and this volume]. Visser et al [1979] have estimated intracellular pH of individual cells in a flow cytometer by analyzing excitation spectra from rat bone marrow cells treated with fluorescein diacetate (also see Gerson, this volume). However, flow cytometry does not permit either prolonged continuous pH recording from a single cell or the measurement of localized pH changes within a single cell. Recently, organellar pH in mast cell granules [Johnson et al, 1980] and in the hamster sperm acrosome [Meizel and Deamer, 1978] has been estimated from the distribution of fluorescent amines within these cell types, using a combination of quantitative macrofluorometry of cell suspensions and qualitative microfluorometry of single cells.

The application of quantitative microfluorometry to single cells has been largely confined to measurements of fluorescent reaction products in fixed cells or tissues [eg, nucleic acid staining, Ploem, 1977; Wied, 1966; quantitative immunofluorescence, Haaijman and van Dalen, 1974] and to the study of fluorescent reaction products in living cells [ie, fluorochromasia, Sernetz, 1973; Rotman and Papermaster, 1966]. Rapid technical developments in microfluorometry over the past several years [Ploem, 1977; Ploem et al, 1974; Sernetz, 1973; Sernetz and Thaer, 1973; Haaijman and van Dalen, 1974; Reynolds and Taylor, 1980] have vastly expanded the potential uses of this technique. Increased sensitivity of photodetection systems and advances in electronics have opened the door to precise quantitation of fluorescence and of fluorescence changes from fluorescent probes incorporated into single living cells. Such probes include those sensitive to the ionic environment of the cell, as described below [Heiple, 1981; Heiple and Taylor, 1980a, b], as well as the reintroduction of fluorescently labeled functional cellular constituents [Taylor and Wang, 1980; Taylor et al, 1980b].

A potential problem in interpreting any optical signal, such as fluorescence or absorbance, from moving eucaryotic cells, is that of constantly changing optical path length and probe concentration. These variations occur from cell to cell and also within a single cell. This problem is surmounted by looking at the ratio of fluorescence or absorbance at two different wavelengths ($\lambda 1$, $\lambda 2$), a value that is conveniently independent of both path length and of concentration [MacDonald et al, 1977; Ohkuma and Poole, 1978; Thomas et al, 1979]. A derivation of the path length and concentration independence for a pH-sensitive indicator dye follows.

Assume, as is true for each of the probes used here, that the dye is in simple equilibrium between a more protonated (acid, a) form and a less protonated (basic, b) form, each of which has a different absorbance maximum. Let A_1 = absorbance at λ_1; A_2 = absorbance at λ_2; ξ_1^a and ξ_2^a = molar extinction coefficient of acid form in solution at λ_1 and λ_2, respectively; ξ_1^b and ξ_2^b = molar extinction coefficient of basic form in solution at λ_1 and λ_2, respectivelyy; C^a = concentration of acid form in solution; C^b = concentration of basic form in solution; L = path length through sample. To obtain the total absorbance of the dye at each wavelength, the absorbances of each species are added, using the Lambert-Beer law [Piller, 1977]:

$$A_1 = \xi_1^a C^a L + \xi_1^b C^b L,$$

and

$$A_2 = \xi_2^a C^a L + \xi_2^b C^b L$$

The ratio of these two absorbances is then:

$$\frac{A_1}{A_2} = \frac{\xi_1^a C^a L + \xi_1^b C^b L}{\xi_2^a C^a L + \xi_1^b C^b L}$$

Dividing by $C^b L$ gives:

$$\frac{A_1}{A_2} = \frac{(\xi_1^a C^a/C^b) + \xi_1^b}{(\xi_2^a C^a/C^b) + \xi_2^b}$$

From the law of mass action, for a simple acid-base equilibrium $HI \rightleftharpoons H^+ + I^-$, where a = HI, b = I$^-$, and K is the equilibrium constant of the reaction,

$$\frac{[H^+]}{K} = \frac{C^a}{C^b}$$

Substituting this value for C^a/C^b into the last ratio equation gives

$$\frac{A_1}{A_2} = \frac{\xi_1^a \dfrac{[H^+]}{K} + \xi_1^b}{\xi_2^a \dfrac{[H^+]}{K} + \xi_2^b}$$

The absorbance ratio, at any two wavelengths in the absorbance spectrum of the probe, is therefore dependent only upon a set of constants and pH. It is not necessary to know any of the extinction coefficients or the equilibrium constant in order to make practical use of this ratio, as long as the signals are calibrated under conditions equivalent to those of the sample to be measured. These equations hold equally for fluorescence, with the same precautions.

This chapter describes a novel application of microfluorometry to the determination of cytoplasmic and endosomal pH in the motile amoeba, Chaos carolinensis. Portions of this work have been published recently [Heiple and Taylor,

1980a, b]. Our approach has been to use optical techniques to quantitate signals obtained from pH-sensitive probes, after incorporation by microinjection or by endocytosis, into single living amoebae. We have taken advantage of the path length and concentration independence of absorbance ratio in the experimental design. Advantages of these techniques include increased sensitivity to and specificity for pH, minimal perturbation of the cell, and improved spatial and temporal resolution. Precise subcellular localization of the signal is possible, and continuous measurements of freely moving or of experimentally manipulated cells are possible without further injury to or disturbance of the cells.

The giant, free-living, freshwater amoeba, Chaos carolinensis, provides an ideal model system for the study of a variety of fundamental processes that also occur in other cell types. Both the physiology and morphology of free-living amoebae has been well studied [Jeon, 1973], and extensive analysis of contractility and of endocytic mechanisms in these cells has been carried out over the last 50 years [reviewed by Chapman-Andersen, 1977; Taylor and Condeelis, 1979]. A major advantage of free-living amoebae is that their extracellular environment (which, in nature, is pond water) can be controlled and varied at will, and that they tolerate a wide range of culture conditions [Jeon, 1973]. The exceptionally large size of C. carolinensis (up to 600 μm in diameter) makes it ideal for macromanipulation of, and microquantitation from, single cells. It can be conveniently studied by a variety of techniques, with the potential for detailed subcellular localization of cell activity even at rather crude levels of resolution. Contractile proteins from C. carolinensis have been partially characterized biochemically [Condeelis, 1977], and the distribution of fluorescently labeled actin has been followed after microinjection into living cells [Taylor et al, 1980b]. Moreover, the concentration of free Ca^{++} has been mapped in these cells via luminescence of microinjected aequorin [Taylor et al, 1980a]. As mentioned above, the experiments described herein help to lay the groundwork for investigations of the possible relationships among pH, pCa, contractility, and endocytosis in C. carolinensis.

II. MATERIALS AND METHODS

A. Materials

Chaos carolinensis, obtained from Carolina Biological Supply, Burlington, North Carolina, are cultured in Marshall's medium (5×10^{-5} M $MgSO_4$, 5×10^{-4} M $CaCl_2$, 1.47×10^{-4} M K_2HPO_4, 1.1×10^{-4} M KH_2PO_4, pH 7.0) with mixed ciliates, as previously described [Taylor et al, 1973]. The pH of all solutions is measured with a VanLab Ag/AgCl Combination Microprobe Electrode (VWR Scientific, San Francisco) on a Corning model 10 pH meter (Corning Glass Works, Science Products Division, Corning, New York). Glass-distilled water and reagent grade chemicals are used throughout this work.

B. Preparation of FTC-Ovalbumin

Fluorescein-thiocarbamyl (FTC)-ovalbumin is prepared by mixing chick ovalbumin (Worthington Biochemicals, Freehold, New Jersey, 2× crystallized) with fluorescein isothyocyanate (FITC), isomer I (Sigma, St Louis) in 50 mM carbonate-bicarbonate buffer, pH 9.5. Initial concentrations of ovalbumin and FITC are 10 mg/ml and 0.4 mg/ml, respectively. After 1 hour at room temperature, the reaction is stopped by addition of excess (10 mg/ml) glycine in carbonate-bicarbonate buffer, pH 9.5. The solution is clarified by centrifugation at 100,000g for 30 minutes (40,000 rpm, Sorvall OTD-65 Ultracentrifuge, 865.1 rotor, 4°C). To separate labeled protein from unlabeled protein and from free dye, the solution is then passed over a Sephadex G-25-150 medium (50–150 μ, Sigma) column. Protein is eluted slowly from the 1 × 60 cm column (bed volume 37 ml, pressure head 70 cm, 0.67 ml/min), to effect this separation. One-milliliter fractions are collected and assayed spectrophotometrically (Cary 219, Varian Instruments, Waltham, Massachusetts) for the presence of protein and dye. A typical column profile (OD_{495} and protein concentration vs fraction number) is illustrated in Figure 1. Profiles from different preparations are virtually superimposable. The unusual separation of labeled from unlabeled protein has been reported previously [George and Walton, 1961]. Peak fractions (Fig. 1, 18–25, inclusive) of labeled ovalbumin are pooled and dialyzed extensively against 2.5 mM PIPES, pH 7.0, and are stored at −20°C in 100 λ aliquots. Solutions are protected from light throughout these procedures. The dye-to-protein molar ratio of each preparation is estimated using a molar extinction coefficient of 68,000 M^{-1} cm^{-1} at 495 nm, pH 8.0, for bound fluorescein. Different preparations have dye–protein ratios ranging from 0.92 to 1.47. Final protein concentrations vary from 2.5 to 4.0 mg/ml, as determined using an $A_{280}^{1\%}$ of 7.4 for ovalbumin and correcting for absorbance of bound dye at this wavelength ($A_{280} - (A_{495} \times 0.31)$). Spectral characteristics of the probe in solution are measured with a SPEX Fluorolog (Spex Industries, Metuchen, New Jersey) and microfluorometrically (see below).

C. Microinjections and Microfluorometric Measurements

Amoebae are pressure-microinjected, as described elsewhere [Taylor et al, 1980b], with approximately one-tenth cell volume of 3.0 mg/ml FTC-ovalbumin in a carrier solution of 2.5 mM PIPES, pH 6.95. The apparatus for measuring fluorescence intensity from the injected cells was designed in collaboration with Bob Zeh of Custom Instrumentation, New York, and is diagrammed schematically in Figure 2. On a Zeiss Photomicroscope equipped with epiillumination, light from a 12 V/60 W quartz-halogen lamp (QH) passes through one of two interchangeable narrow-band excitation filters (EF1, EF2) and is reflected by the dichroic mirror (DM) to the sample (SA). The excitation filters (496 or 452 nm

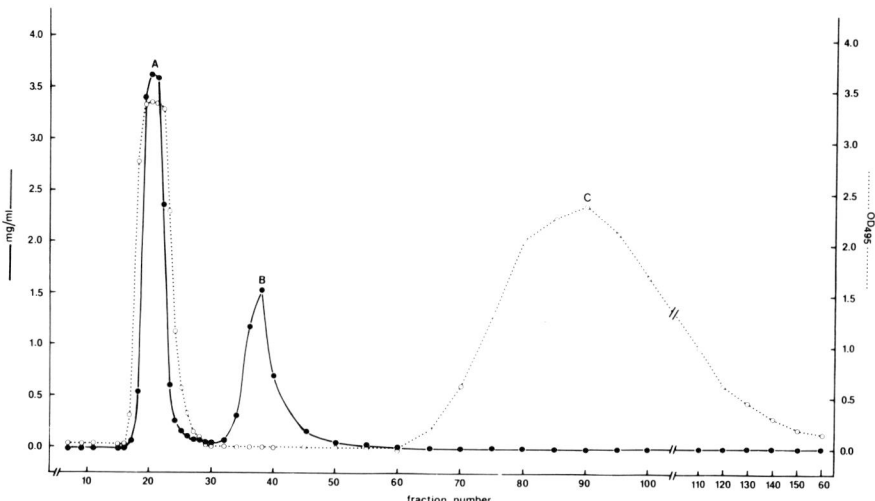

Fig. 1. Gel chromatography of FTC-ovalbumin preparation on Sephadex G-25 med column (see text for further details). OD_{495} ○·····○ and mg/ml ●——● vs fraction number. A) labeled protein; B) unlabeled protein; C) free dye.

interference filters, band width at 50% T_{max} 11 and 14 nm, respectively, Feuer Optics, Upper Montclair, New Jersey) are rapidly changed (0.5 second at each position, 0.3 second change-time) by a Superior Slo-Syn stepping motor (Superior Electric, Bristol, Connecticut) controlled by a KIM I Microprocessor. Fluorescent light collected by the objective (OBJ) from the sample then passes through the dichroic mirror, a variable aperture (AP), and through a broad-band barrier filter (BF, 520–560 nm, Zeiss LP520 and KP560), and impinges on a photomultiplier tube (RCA Solid State, Sommerville, New Jersey) where the signal is converted into a deflection on a chart recorder (Hewlett-Packard Co., Palo Alto, California, model 7132A). Note that the barrier filter has been placed above the eyepiece (OC) in front of the photomultiplier. This allows simultaneous fluorescence measurement and observation of the cell in transmitted bright-field light. A 621-nm or 645-nm interference filter (IF, PTR Optics, Waltham, Massachusetts) is placed over the substage light source (TU) so that the barrier filter can prevent this light from reaching the photomultiplier. Measurements reported here are made with a Zeiss 25 × Pol Neofluar objective (NA 0.6) and an aperture, or spot size, of ~50 or 75 μm in diameter (as specified in the text). The arrangement of the changeable excitation filters, on the microscope, is illustrated in Figure 3.

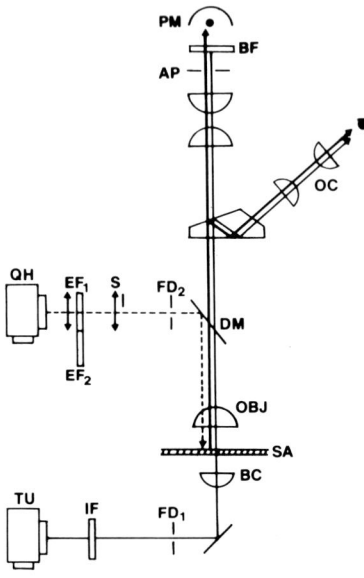

Fig. 2. Schematic representation of the microfluorometer used to quantitate fluorescent signals. The dotted line represents the epiexcitation light from a quartz-halogen lamp (QH), the heavy solid line represents the path of fluorescent light from the sample (SA), and the lighter solid line represents light from a substage tungsten lamp (TU). A substage interference filter (IF) selects a wavelength of light, for bright-field observations, that will not pass through the barrier filter (BF) to the photomultiplier (PM). Only fluorescent light from the sample passes through the barrier filter. Two interchangeable excitation filters (EF_1 and EF_2) select the wavelength of the excitation light. These are powered by a microprocessor-controlled stepping motor (not shown). A moveable shutter (S) allows the excitation light to be blocked when desired. Other symbols are as follows: FD_1 and FD_2, field diaphragms; DM, dichroic mirror; OBJ, objective; BC, bright-field condenser; OC, ocular; AP, variable aperture. The photomultiplier is connected to a chart recorder (not shown). See text for further details. (Reproduced with permission from Heiple and Taylor [1980a]. Copyright © 1980, Rockefeller University Press.)

In each case, the amoeba and observation chamber are first measured for background fluorescence ($\leq$ 10% final signal). The amoeba is then injected and allowed to recover for a few minutes. The fluorescence intensity at the two excitation wavelengths is measured while the cell is unperturbed and moving freely or after experimental manipulation. After correction for background fluorescence at each wavelength, the ratio of fluorescence intensities, Ex_{496}/Ex_{452}, is calculated for each measurement. Calibration of these ratios is described in Results. Image intensification and recording is carried out as described elsewhere [Taylor et al, 1980b].

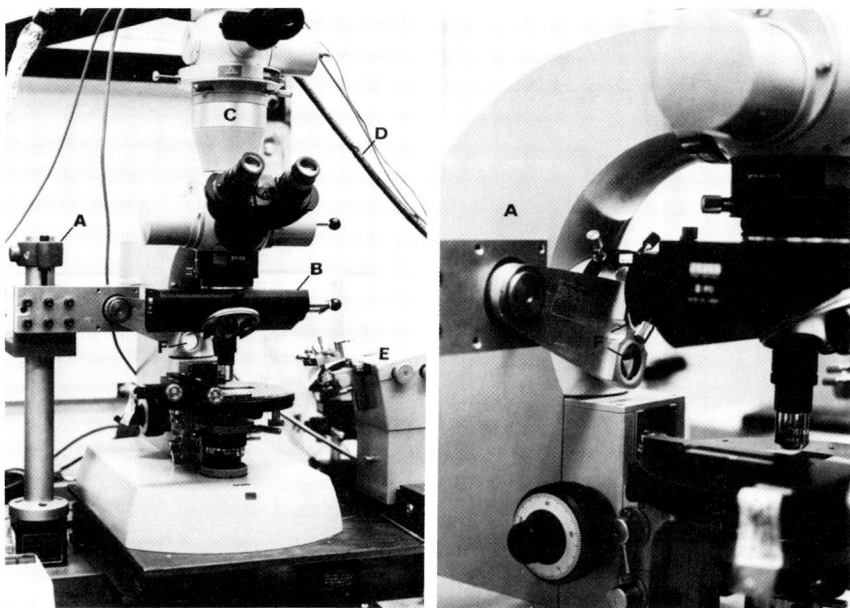

Fig. 3. Arrangement of changeable excitation filters on the Zeiss Universal microscope. Left: face view of entire setup; right: closeup of filter holder in path of epiilluminator. A) Excitation filter assembly-motor and filter-holder mounted on magnetic chuck; B) Zeiss epiillumination filter system; C) Zeiss photometer head; D) light pipe to photometer; E) Leitz micromanipulator; F) excitation filters. Rotation back and forth through 16.2° brings each filter into the correct position (they have been moved slightly out of alignment for this photo).

D. Induction of Endocytosis

Starved amoebae are induced to pinocytose FTC-ovalbumin by incubation in acid medium containing the probe (0.3 mg/ml) for varying lengths of time. Amoebae, immersed in neutral medium containing probe and ciliates, form fluorescent phagosomes. In either case, extracellular probe is washed away, and measurements are corrected for background fluorescence [Heiple and Taylor, 1980b; submitted to J Cell Biol]. Whereas single phagosomes can be detected and measured in freely moving amoebae, the size, heterogeneity, and rapid movements of pinosomes require that cells be chilled and centrifuged before measurements can be made.

III. RESULTS

A. Spectral Characterization of FTC-Ovalbumin pH-Sensitivity

In aqueous solution, fluorescein may exist in equilibrium among six different species, depending on pH [Martin and Lindqvist, 1975]. Each species, diagrammed in Figure 4, has a different absorption (and therefore, excitation) spectrum. In the pH range 5.0–12.0, and therefore in the range of physiologic pHs (6.0–8.0) [see Waddell and Bates, 1969], the monoanion and dianion of the dye predominate (pK_a = 6.7). These species are shown again in Figure 5, which presents the excitation spectra of 5 μg/ml FTC-ovalbumin in 10 mM Tris-maleate at pH 5.0 and at pH 9.0. At the lower pH, a shoulder at Ex_{450} appears, owing primarily to the monoanion of the dye, and the fluorescence of the probe at Ex_{495} is reduced. At pH 9.0, the fluorescence intensity at Ex_{450} decreases and that at Ex_{495}, owing primarily to the dianion of the dye, increases. Thus, the ratio of fluorescence intensity, Ex_{495}/Ex_{450} (Em_{520}) is a sensitive pH indicator, as shown in Figure 6, where the ratio is plotted against pH (3.0–9.0). As discussed above, this ratio is independent of path length and of concentration of the probe (also see below).

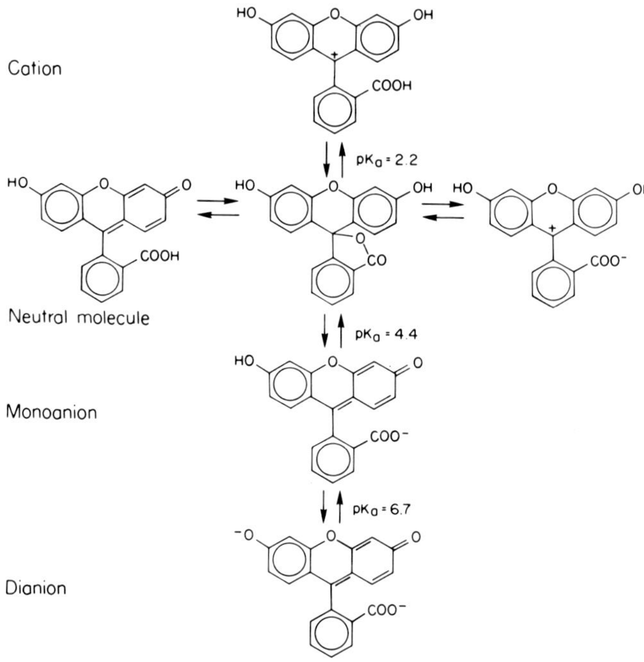

Fig. 4. Acid–base equilibria of fluorescein in aqueous solution.

Control experiments indicate the specificity for pH of the fluorescence intensity ratio of FTC-ovalbumin [Heiple, 1981]. The ratio is not significantly affected by type of buffer (100 mM PIPES, HEPES, Tris-maleate, or Na-phosphate, data not shown), or by the presence of 100 mM K^+-acetate (Fig. 6, solid triangles) or of 50 mM $(NH_4)_2SO_4$ (Fig. 6, open circles). In the system described here, the ratio is also unaffected by changing aperture diameter, down to the current practical limit of ~ 50 μm (data not shown). Neither is the ratio significantly affected by 0–1.0 mM $MgCl_2$ (Fig. 7a), 0–200 mM KCl (Fig. 7b), 10^{-8} M – 10^{-5} M $CaCl_2$ (Ca^{++}/EGTA varied between 0.05 and 0.9) (Fig. 7c), or by photobleaching in the microfluorometer over a period of 20 minutes of continuous measurement (Fig. 7d). Further controls indicate that standard curves in the presence of 10^{-5} M or 10^{-8} M $CaCl_2$ are identical (pH 6.4–7.4, data not shown).

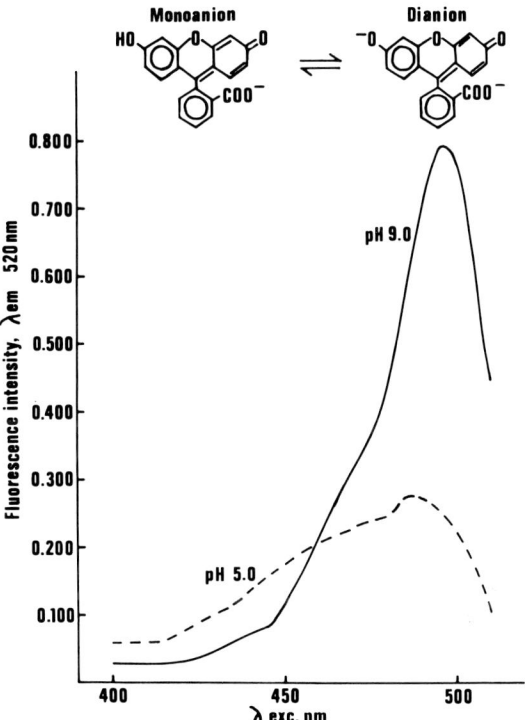

Fig. 5. Excitation spectra of FTC-ovalbumin (5 μg/ml) in 10 mM Tris-malete, pH 5.0 ——————— (monoanion) and pH 9.0 ——————— (dianion). SPEX Fluorolog, E/R mode, 5 nm steps, 5 nm bandpass, $E \times_{400-510}, Em_{520}$. pH 5.0 curve at 10× sensitivity pH 9.0 curve.

Though theoretically independent of probe concentration (see Introduction), the fluorescence intensity ratio may be affected by interactions of adjacent fluorophores at high concentrations (self-quenching), or by the presence of high concentrations of protein, which may alter the hydrophobicity of the probe microenvironment and, therefore, its fluorescent characteristics (Bridges, 1968; Barltrop, 1975). It has been shown [Schauenstein et al., 1978] that the fluorescence emission of fluorescein diacetate, free FITC, and FTC-γ-globulin decreases at dye concentrations > 5 μg/ml (10 μM). However, the fluorescence intensity *ratio* of FTC-ovalbumin is independent of probe concentration from 0.1–1.5 mg/ml (see Table I). Assuming a dye–protein ratio of about 1.0 (as is true for preparations used here), this represents a range of dye concentrations from 2.5 to 40 μM. Cells injected with 1/10 cell volume of 3.0 mg/ml FTC-ovalbumin will contain a cytoplasmic probe concentration of 0.3 mg/ml (8 μM fluorescein), assuming that all of the intracellular space is accessible to the probe. It is possible, in free-living amoebae, for up to half of the cell volume to consist of organelles [Jeon, 1973], which will exclude FTC-ovalbumin. This would double its cytoplasmic concentration (0.6 mg/ml ovalbumin, 16 μM fluorescein), but the latter still falls within a range shown to have no significant effect on the fluorescence intensity ratio. Moreover, conjugation of each fluorescein moiety to an ovalbumin molecule, a globular protein with molecular weight 43,000 and diameter ~ 50Å [Taborsky, 1974], will prevent self-quenching, which requires excimer formation at intermolecular distances of ~ 3.5 Å [Barltrop, 1975]. At the cytoplasmic concentrations estimated above, intermolecular distances of FTC-ovalbumin

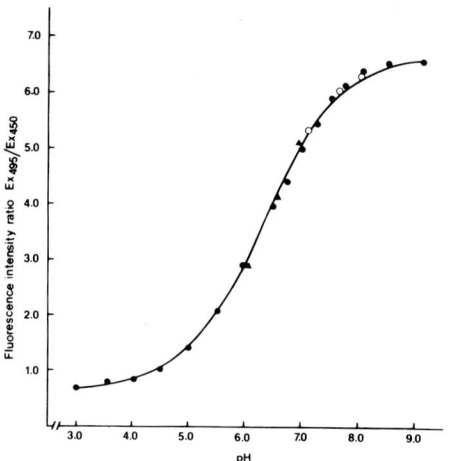

Fig. 6. Fluorescence intensity ratio (Ex_{495}/Ex_{450}, Em_{520}) vs pH for FTC-ovalbumin (5 μg/ml) in 10 mM Tris-maleate (●), 10 mM Tris-maleate + 50 mM $(NH_4)_2SO_4$ (○), 10 mM Tris-maleate + 100 mM KAc (▲). SPEX Fluorolog, as in Figure 5.

molecules will average from 730 to 780 Å (half cell volume accessible—entire cell volume accessible, respectively). At these distances self-quenching or energy transfer is highly unlikely.

The effect of unlabeled ovalbumin on fluorescence intensity ratio of FTC-ovalbumin has also been examined (Fig. 8). In the first experiment (Fig. 8, solid circles), a concentrated solution of unlabeled ovalbummin is dialyzed overnight against a large excess of 1.0 mM PIPES, pH 7.0. Serial dilutions (0–100 mg/ml) are prepared, using the dialysis buffer, and 10 µg/ml FTC-ovalbumin is

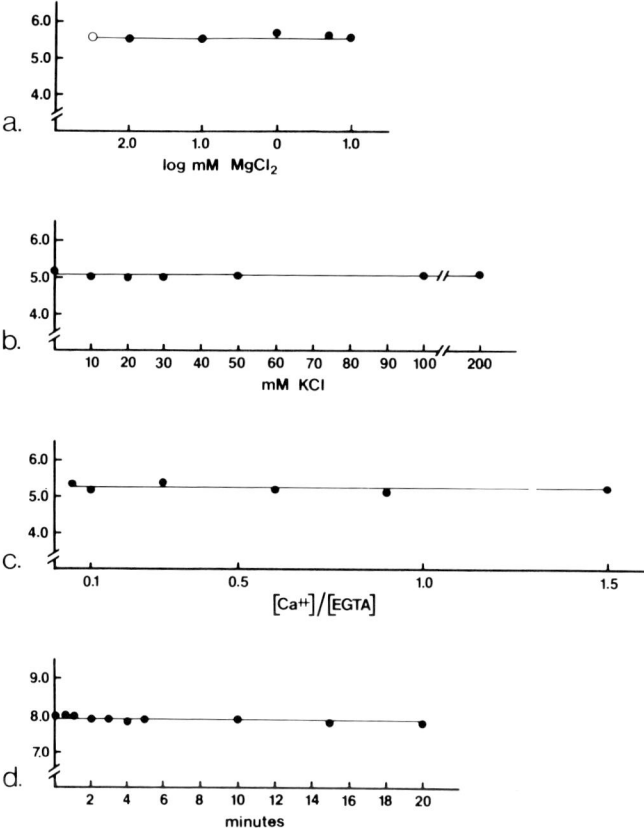

Fig. 7. The effect of ionic environment and of photobleaching on fluorescence intensity ratio of FTC-ovalbumin in solution. a–c) SPEX Fluorolog, E/R mode, Ex_{495}/Ex_{450} (band pass 1 nm), Em_{520} (band pass 6 nm), 10 µg/ml FTC-ovalbumin in 10 mM PIPES, pH 7.00. a) Effect of Mg Cl_2 (○ = 0 mM); b) effect of KCl; c) effect of $CaCl_2$. d) Effect of photobleaching, microfluorometer, Ex_{496}/Ex_{452}, $Em_{520-560}$, 0.3 mg/ml FTC-ovalbumin in 10 mM Tris-maleate, pH 8.0, 0.3 mm i.d. path length silica glass microcuvettes. Aperture 75 µm diameter. Solid lines represent mean value in each experiment.

added to each. As shown in Figure 8 and as previously reported [Sernetz and Thaer, 1972], performing the experiment in this fashion results in an apparent concentration (or solvent) effect. However, the Δ ratio/Δ fluorescence intensity Ex_{495} is precisely the same here as it is for a pH titration of the probe. This would be highly unlikely if fluorescence quenching were occurring, and is an artifact of the buffering capacity of the ovalbumin itself (pI = 4.53) at the high concentrations used. Note that as the protein concentration decreases, the buffering capacity of the buffer raises the pH and raises the fluorescence intensity ratio (Fig. 8, solid circles). This apparent concentration effect disappears when the experiment is repeated with 10 mM PIPES, pH 7.0, and sample volumes large enough to permit careful pH adjustments before fluorescence quantitation

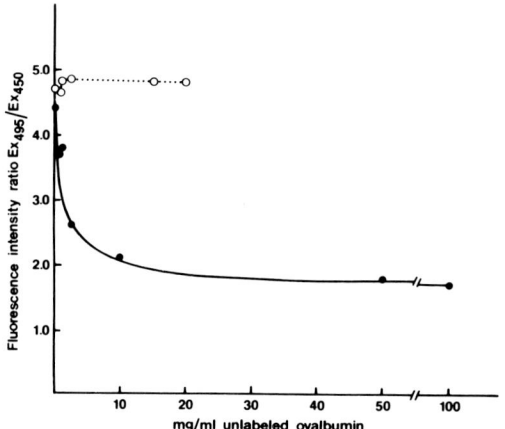

Fig. 8. Fluorescence intensity ratio (Ex_{495}/Ex_{450}, Em_{520}) of 10 µg/ml FTC-ovalbumin vs concentration of added unlabeled ovalbumin (0–100 mg/ml). SPEX Fluorolog, as in Figure 5. (●) 1 mM PIPES, serial dilutions; (○) 10 mM PIPES, pH-adjusted to 7.00, after dilution.

TABLE I. Effect on Fluorescence Intensity Ratio of the Concentration of FTC-Ovalbumin in 250 mM Tris-maleate, pH 7.01, 0.3 mm i.d. Path Length Cuvettes, 75µm Diameter Spot. (Mean ± SE duplicate measurements)

Concentration (mg/ml)	Fluorescence intensity ratio (Ex_{496}/Ex_{452})
1.5	9.33 (± 0.08)
0.75	9.13 (± 0.05)
0.37	9.00 (± 0.09)
0.18	9.17 (± 0.09)
0.09	9.08 (± 0.09)

(Fig. 8, open circles). At the low probe concentrations used in the following experiments, the buffering capacity of the ovalbumin itself is trivial.

Although fluorescence intensity is somewhat sensitive to temperature [Bridges, 1968, p 103], the fluorescence intensity ratio of FTC-ovalbumin incorporated into amoebae is not significantly affected by temperature changes from 4°C to 22°C (data not shown).

B. Cytoplasmic pH Measurements

1. Intracellular distribution of the probe.
Knowledge of the intracellular distribution of any absorbant or fluorescent probe is critical to interpretation of signals obtained from it. The distribution of FTC-ovalbumin, in a typical, microinjected amoeba, is shown in Figure 9. The fluorescent image of this freely

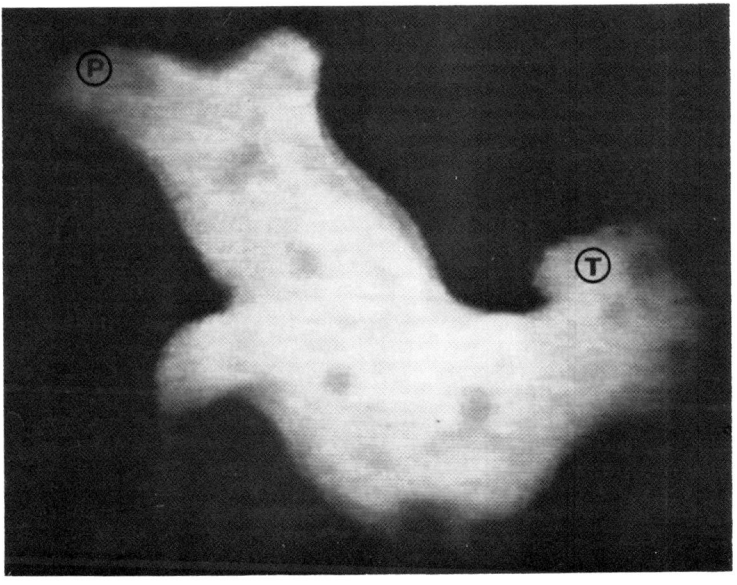

Fig. 9. A specimen of Chaos carolinensis injected with FTC-ovalbumin, photograph of intensified video image. The fluorescent probe is evenly distributed throughout the cytoplasm and is excluded from the organelles, which appear dark. The amoeba is moving toward the upper left-hand corner. Drawings of the measuring spot are superimposed upon the photograph in typical positions in an advancing pseudopod tip (P) and in the tail (T). Bar, 100 μm. (Reproduced with permission from Heiple and Taylor [1980a]. Copyright © 1980, Rockefeller University Press.)

moving cell is from the video screen of an image intensifier. An advancing pseudopod (P) and the tail region (T) are indicated by scale drawings of the 50 μm diameter measuring spot. Note the even cytoplasmic distribution of FTC-ovalbumin fluorescence and its exclusion from organelles, which appear dark against the bright cytoplasm. Rapid cytoplasmic streaming establishes this distribution immediately upon injection of the cell. Occassionally, one or more bright vesicles form at the injection site. However, amoebae maintained in Marshall's medium, for up to a week after injection, retain this pattern of cytoplasmic fluorescence, confirming the slow rate of autophagy of fluorescently labeled ovalbumin observed after its injection into other cell types [Stacey and Allfrey, 1977]. Further evidence for the unbound, random distribution of FTC-ovalbumin in these cells comes from the observation that it diffuses freely out of ruptured amoebae [Taylor et al, 1980b].

2. Standard Curve. Standard curves of the ratio of fluorescence intensity vs physiological pH [Waddell and Bates, 1969] for FTC-ovalbumin in situ (a) and in vitro (b) are shown in Figure 10. As the pH increases from 6.0 to 8.0, the ratio (Ex_{496}/Ex_{452}) also increases (on the in situ curve, from 7.00 to 10.7 and on the in vitro curve, from 3.90 to 9.60). Signals from cell-incorporated dyes or probes are usually calibrated by standard curves generated in solution, despite the well-known sensitivity of many dyes and probes to unpredictable and often undetermined parameters of the cell environment. Indeed, we find that the standard curve of FTC-ovalbumin in solution is a poor approximation of the one generated in situ by collapsing ΔpH across the membrane of an injected amoeba with weak acid or base, as suggested by the work of Pollard et al [1979]. Although similar in shape to the in situ curve, significantly different ratios are obtained for the range of pH values tested. Near neutrality, errors of up to 0.7 pH U can be incurred by calibrating ratios from injected cells with the standard curve generated in solution. Therefore, fluorescence intensity ratios from injected cells are calibrated with the in situ standard curve, shown in Figure 10a.

The in situ curve is generated in two distinct ways. First (solid symbols in Fig. 10a), the internal pH of an injected amoeba is equilibrated with an external pH determined by 10 mM PIPES, 10 mM HEPES, or 10 mM Tris-maleate buffers using 100 mM K^+-acetate or 50 mM $(NH_4)_2 SO_4$ as described previously [Pollard et al, 1979]. At these high concentrations, weak acids or bases can be used to equilibrate internal and external pH [Pollard et al, 1979]. The mechanism of pH equilibration, across a membrane-bound compartment, by weak acid or base has been discussed in detail elsewhere [Pollard et al, 1979; Jacobs, 1940; McLaughlin and Dilger, 1980] (see Roos and Keiffer, this volume).

An important assumption in the pH equilibration technique is that the contribution of cellular constituents to the acid–base equilibria is negligible compared with the large excess of acid or base used to collapse the gradient. To corroborate

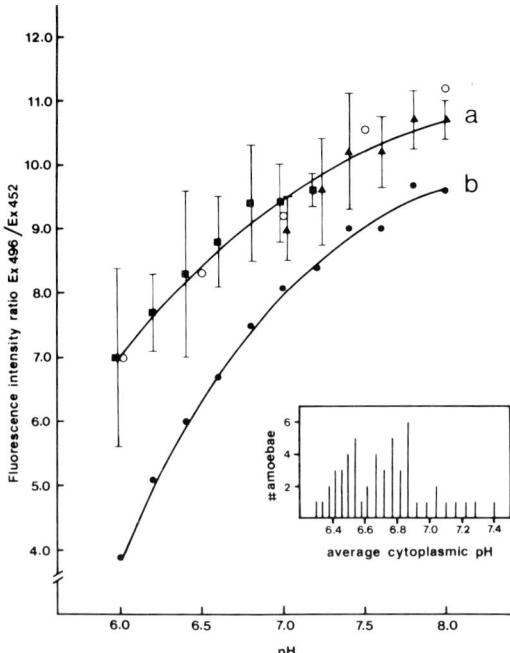

Fig. 10. Standard curves of fluorescence intensity ratio (Ex_{496}/Ex_{452}, $Em_{520-560}$) vs pH for FTC-ovalbumin in situ (a) and in solution (b), as measured with the microfluorometer diagrammed in Figure 2. ■, FTC-ovalbumin-injected amoebae equilibrated in 10 mM Tris-maleate (pH 6.0–7.2), 100 mM K-acetate or in 10 mM PIPES (pH 6.4–7.2), 100 mM K-acetate. ▲, FTC-ovalbumin-injected amoebae equilibrated in 10 mM Tris-maleate (pH 7.0–8.0), 50 mM $(NH_4)_2SO_4$ or 10 mM HEPES (pH 7.0–7.8), 50 mM $(NH_4)_2SO_4$. Time courses show that the pH stabilizes at the new value within 2 minutes with ammonia and within 7 minutes with acetate (data not shown); therefore, the latter value is used as a minimum incubation time. The postalkalinization acidification observed in muscle cells after treatment with ammonia [Aickin and Thomas, 1977] is not seen in these experiments. Each solid symbol in (a) represents an average of 6–13 amoebae, with the single exception of three amoebae at pH 6.2, for a total of 83 amoebae. Vertical bars represent SD for these points. ○, FTC-ovalbumin-injected amoebae injected with ≥ 1/10 cell volume of 100 mM Tris-maleate (pH 6.0–8.0), 100 mM HEPES (pH 8.0), or 100 mM PIPES (pH 7.0). Each point represents an average of 2–4 amoebae; SD have been omitted for clarity, but in all cases they are smaller than those obtained by equilibration (average SD ± 0.48 ratio units). ●, Solutions of 0.3 mg/ml FTC-ovalbumin in 10 mM Tris-maleate (pH 6.0–8.0) in silica-glass microcuvettes (Vitro-Dynamics, Rockaway, New Jersey), averages of three samples each (average SD ± 0.55 ratio unit). Inset, a bar graph of the distribution of average cytoplasmic pH among 52 FTC-ovalbumin-injected amoebae measured by this technique (using the in situ curve (a)). The values appear to form two groups, one clustered about pH 6.5, the other clustered about pH 6.8. (Reproduced with permission from Heiple and Taylor [1980a]. Copyright © 1980, Rockefeller University Press.)

these values, some amoebae containing FTC-ovalbumin are injected with a minimum of 1/10 cell volume of 100 mM Tris-maleate, 100 mM PIPES, or 100 mM HEPES buffers at a known pH, then measured (Fig. 10a, open circles). More frequently, one-third cell volume or more is injected, as estimated from visual inspection of increased cell size, and decreased fluorescence intensity, when buffers at pH 7.0 are injected. These values concur with the acid–base equilibration method. Preliminary attempts to generate this standard curve using the proton ionophores nigericin/K^+ or carbonyl cyanide m-chlorophenyl hydrazone (CCCP) were unsuccessful. Ionophores have been applied to in situ probe calibration in other cell types with great success [eg, Thomas et al, 1979] and are highly recommended, where applicable. The buffering capacity of amoeboid cytoplasm is estimated to be ~11 meq H^+ ions/U pH change/liter, from titration of a homogenate of Dictyostelium discoideum amoebae. As the buffering capacity of 100 mM Tris-maleate is only 50 meq H^+ ions/U pH change/liter (pH 6.0–8.0), buffer injections are repeated with 500 mM Tris-maleate at pH 6.0, 7.0, and 8.0 (at least three amoebae at each point, see Table II). The average ratios obtained agree very well with those obtained by the injection of 100 mM buffer, indicating that the buffering capacity of the cell has been overcome in these experiments. Concentrations of buffer > 500 mM cannot be microinjected into living cells because of the violently disruptive effect of such high ionic strength.

Amoebae react in a characteristic fashion to injection of (or to equilibration in) buffers of various pHs. At pH 6.0–6.8, amoeba cytoplasm forms a solid, gelled mass of great tensile strength. As buffer pH is increased from 6.8 to 8.0, cytoplasmic contractions of increasing intensity and speed are induced and amoebae become increasingly fragile. This reaction is reminiscent of that produced by calcium injections [Taylor et al, 1980a].

3. Average cytoplasmic pH and variations. The inset in Figure 10 shows the distribution of average cytoplasmic pH among 52 individual amoebae measured using the fluorescence technique. Each cell is sampled thoroughly ($\geq$ 9 measurements, 45–75 μm diameter spot). The cumulative average obtained is pH 6.75 (SD among amoebae ± 0.30) with a suggestion of clustering about pH 6.5 and pH 6.8. The pH of a given spot, relative to the morphology of a moving amoeba, tends to remain fairly constant (± 0.08 pH U over approximately half a minute of constant measurement). However, two different spots in one cell may differ in pH by as much as 0.4 U. Varying extracellular pH (Marshall's medium) from 6.0 to 8.0 has no detectable effect on cytoplasmic pH in these cells, even after overnight incubation at these values.

4. Wound-healing response. The microfluorometric technique is ideally suited to the detection and quantitation of rapid, continuous, local pH changes within a single cell. This is dramatically illustrated in Figure 11, which shows

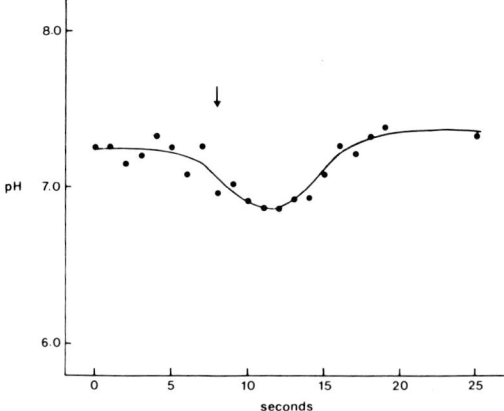

Fig. 11. Cytoplasmic pH vs time during the wound-healing response in a typical experiment. The spot observed is a 50 μm diameter area on the side of an advancing pseudopod of an FTC-ovalbumin-injected amoeba. Upon damage by puncture (arrow), the pH drops from 7.25 to 6.90 within 5 seconds, then recovers to its original value within another 5 seconds. External medium: Marshall's, pH ~7.0. (Reproduced with permission from Heiple and Taylor [1980a]. Copyright © 1980, Rockefeller University Press.)

TABLE II. Injection of 500 mM Tris-maleate into FTC-Ovalbumin-Containing Amoebae, 75 μm diameter spot

pH	Fluorescence intensity ratio* (Ex_{496}/Ex_{452})
6.00	6.28 (± 0.35)
7.00	8.51 (± 0.91)
8.00	9.76 (± 0.68)

*Mean ± SD (≥ 3 amoebae each).

pH vs time during the wound-healing response in C. carolinensis. Measurements are made in a 45–50-μm diameter spot on the side of an advancing pseudopod of a typical, preinjected, amoebae as it is punctured by a clean glass microneedle, which is quickly removed. There is a slight, transient drop in pH at the wound site upon puncture (Fig. 11, arrow) that increases in extent with the degree of damage. In the experiment shown here, $\Delta pH = -0.35$ U, and the pH of the measured spot recovers to its original value within 5 seconds. Decreases in pH vary in magnitude and duration, with respect to the degree of damage. If the damaged area is small, confined to the cortex, and lies outside the measuring spot, the pH change is not detected. Correct placement of the measuring spot is critical to the successful detection of small, localized pH changes.

C. Phagosomal pH Measurements

1. Intracellular distribution of the probe. After endocytosis of FTC-ovalbumin, either by pinocytosis or by phagocytosis, the probe remains confined to endosomes or to vesicles derived from them. Fluorescence is never seen in other organelles (nuclei, mitochondria), and free cytoplasmic fluorescence is never observed. A typical fluorescent phagosome is illustrated in Figure 12. The tail region of a single amoeba is shown in Nomarski (left) and fluorescence (right) images from the video screen of the image intensifier. As is often the case, the phagosome is virtually undetectable, at this depth of focus, in the Nomarski image. Subdivision of the phagosome into several vesicles, after internalization of a ciliate, is common.

2. Standard curve. For calibration of signals obtained from endocytosed probe [Heiple and Taylor, 1980b; submitted to J Cell Biol], crude preparations of FTC-ovalbumin-containing pinosomes are exposed to acid–based equilibration

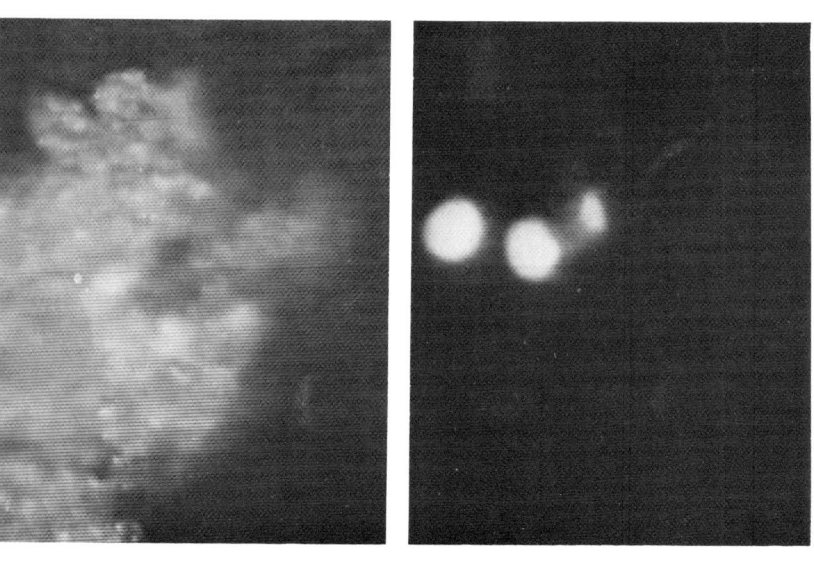

Fig. 12. Phase contrast (left) and fluorescent (right) images of the tail region of a moving amoeba about 1 hour after phagocytic vacuole formation around ciliates immersed in FTC-ovalbumin. Photographs from image intensifier. Bar = 100 μm.

solutions similar to these described above. The fluorescence intensity ratios obtained are similar to those from cytoplasmically localized probe (data not shown). Phagosomes in intact amoebae are also exposed to equilibration solutions (see Table III).

3. Phagosomal pH changes. Using the fluorescent technique, pH changes within a single phagosome, in situ, in a freely moving amoeba, can be followed. A consistent sequence of ΔpH, involving a rapid acidification (complete within 15–20 minutes) and a slower realkalinization (over several hours) is observed. A representative series of measurements from one such phagosome is shown in Table III. As previously reported [Mast, 1942], motion of the ciliate within the phagosome stops when acidification is complete (at ~ 20 minutes in this experiment). Measurements of pinosomes (not shown), which may undergo pH changes by a similar mechanism, suggest that the realkalinized endosome can attain intravesicular pH values (7.5–8.0) higher than that of the cytoplasm. Note that, at the end of this experiment, the amoeba was treated with 50 mM $(NH_4)_2 SO_4$, 10 mM Tris-maleate, pH 8.01. The new, stable fluorescence intensity ratio obtained agrees favorably with previously measured values at a presumptive pH of 8.0, in situ (see Fig. 10a).

TABLE III. pH Changes in a Single Phagosome Containing FTC-Ovalbumin, 45 μm Diameter Spot

Minutes	Ratio	pH
2.0	8.83	6.70
2.5	7.81	6.26
3.0	7.04	6.00
5.0	7.38	6.12
6.0	7.13	6.05
7.0	7.21	6.07
10.0	6.36	5.78
12.5	5.26	5.48
15.0	3.73	5.14
17.5	3.30	5.04
20.0	3.05	< 5.00
22.5	2.96	< 5.00
25.5	3.13	< 5.00
26.0	2.71	< 5.00
30.0	2.71	< 5.00
55.0	2.62	< 5.00
90.0	4.83	5.38
120	4.58	5.31
150	6.62	5.87
Incubate 20 min in 50 mM $(NH_4)SO_4$, 10 mM Tris-maleate, pH 8.01	11.2	

IV. DISCUSSION

Quantitation of fluorescence from a cell-incorporated pH-sensitive probe makes possible the precise, continuous determination of pH and of pH changes within a single living cell. Moreover, utilization of a fluorescent protein conjugate allows localization of the probe to a specific subcellular compartment (ie, cytoplasm, pinosomes, phagosomes). The technique is sensitive, specific, and nondestructive, and it affords a high degree of spatial and temporal resolution. Use of a fluorescence or absorbance ratio renders the measured values independent of path length and of probe concentration. The sensitivity and precision of the technique are indicated, in part, by the standard curve generated in solution. Microfluorometric measurements of standardized solutions (Fig. 10b) have an average SD of ± 0.55 fluorescence intensity ratio unit. This corresponds roughly to ± 0.15 pH unit (depending on the region of the curve). The variation is not due to instrument noise, as any given measurement is stable to ± 0.02 pH U, or better. Some of the variation may be due to light scattering from the thick glass microcuvettes, as substitution of coverslip chambers improves the reproducability of the measurements. A number of improvements on the prototype device described here are now underway in this laboratory. With L. Tanasugarn, we are computerizing the system and have replaced the side-window photomultiplier tube with a much more sensitive end-window phototube, operating in the photon-counting mode. In the experiments described here, a number of practical factors limit our spatial resolution to ~45 μm (diameter of measuring spot) and our temporal resolution to 1.3 seconds (time to make a pair of measurements—one at each excitation wavelength). These values should not be taken as inherent limitations of this approach; they are by no means absolute. With the improved system described above, measurements of 1 μm diameter spots in 300 msec are readily attainable. Clearly, there will always be a trade-off between spatial and temporal resolution, and one may have to be sacrificed in order to improve the other. The technique is extremely sensitive and is easily capable of detecting pH changes of a regulatory nature, which are often as great as ± 0.5 U. Detection of ΔpH during wound healing (Fig. 11) and within single phagosomes (Table III) supports this contention, as does the successful application of the microfluorometric technique to other cell types. For example, we have measured cytoplasmic alkalinizations (ΔpH + 0.4–0.5 U) upon fertilization of sea urchin eggs after microinjection of FTC-ovalbumin (Heiple and Taylor, manuscript in preparation). Furthermore, the fluorescence intensity ratio of FTC-ovalbumin is shown to be reasonably specific for pH (Figs. 6–8). The one-to-one conjugation of fluorescein to a soluble protein and subsequent careful characterization of the probe minimizes the possibility of artifactual fluorescence changes.

Utilizing this technique, we have determined average cytoplasmic pH, pH changes during wound healing, and pH changes in pinosomes and phagosomes in single living specimens of the freshwater amoeba, Chaos carolinensis. The average cytoplasmic pH of 6.75 (SD among amoebae ± 0.30 U) determined from fluorescence quantitation corresponds to that shown in vitro to regulate calcium-sensitivity of gelation and contraction in amoeba extracts [Hellewell and Taylor, 1979; Taylor et al, 1973]. The differential response of amoebae exposed to high (>6.8) or low (<6.8) pHs during generation of the standard curve and the range of average cytoplasmic pH values (Fig. 10, inset) support the view that cytoplasmic pH variations may play a regulatory role in fine-tuning the motile state of the cell, perhaps by shifting the "set-point" for calcium sensitive contractile proteins, or by directly altering their supramolecular structures. Dual regulation of amoeboid movement by Ca^{++} and pH is suggested by the in vitro experiments cited above, by the persistently low level of free calcium in these cells [Taylor et al, 1980a], and by extracellular current measurements [Nuccitelli et al, 1977], as we have discussed in detail elsewhere [Heiple and Taylor, 1980a].

Confidence in the absolute value of a given pH measurement depends on the accuracy of the standard curve used to calibrate the fluorescence intensity ratios. At the current state of the art, a great deal of uncertainty remains with respect to the absolute pH values as calibrated from the in situ standard curve (Fig. 10a). However, several factors increase our confidence in these values. First, similar results are obtained when the in situ standard curve is generated in two different ways (equilibration vs buffer injection), each of which differs from the standard curve in solution, suggesting that the former is more appropriate for calibration purposes. Second, similar values for cytoplasmic pH in amoebae are obtained from microphotometric quantitation of microinjected phenol red (Heiple and Taylor, manuscript in preparation). Phenol red standard curves, generated in situ by injection of strong buffers, correspond precisely to those generated in solution. The agreement of pH values, in spite of the different in situ behavior of each probe, increases our confidence in them, and emphasizes the importance of in situ calibration of optical signals. This point has been made by others as well [Pollard et al, 1979; Thomas et al, 1979]. Previous qualitative estimations of pH_i in free-living amoeba [Chambers and Chambers, 1961; Chambers et al, 1927; Marshak, 1944; Wiercinski, 1944; Jeon, 1973] using absorbant dyes as indicators, agree with our measurements only when care was taken to measure hyaline cytoplasm alone [Chambers and Chambers, 1961; Wiercinski, 1944], without contamination of the visual field by dye-containing organelles of various pHs. The final, satisfactory resolution of this calibration problem will obviously depend upon the corroboration of the standard curve values using an entirely independent technique (eg, microelectrode insertion). This problem does not, however, detract from the sensitivity of the technique to localized pH changes.

The relative constancy of cytoplasmic pH within a given cell ($\pm$ 0.08 U from fluorescence quantitation) suggests that cytoplasmic pH is strictly regulated in these cells. Resistance of cytoplasmic pH to changes in extracellular pH and its rapid recovery during wound healing (Fig. 11) indicates the powerful buffering capacity of these cells. Occasional variations of up to 0.4 U within a single cell, as measured microfluorometrically, are not yet understood. Synchronized cultures may help to elucidate the meaning of average cytoplasmic pH variations among different cells and their possible relation to growth control [eg, Gerson and Burton, 1977; Gillies and Deamer, 1979a, b]. Correlation of ΔpH with specific points in the cell cycle, as observed in Tetrahymena [Gillies and Deamer, 1979a], would be extremely interesting. At the current level of microfluorometric spatial (45 μm) and temporal (1.3 sec) resolution, significant differences between cytoplasmic pH in tails and advancing pseudopod tips of locomoting amoebae were not observed. However, improvements in the microfluorometric technique, including the more sensitive photodetection system and computer interfacing mentioned above, will enable much more rapid (300 msec vs 1.3 sec) and more finely localized (down to 1 μm vs 45 μm) measurements to be made. Reexamination of the "tail vs pseudopod" question with this new system will be of interest. Another approach to this question involves computer analysis of videotapes from the image intensifier during excitation of microinjected FTC-ovalbumin (experiments performed in collaboration with Dr. Elisha Haas, Bar Ilan University, Israel). Point-by-point analysis of fluorescence intensity ratio will yield a "topographical map" of cytoplasmic pH and, especially with the improved detection system, may turn up subtle pH gradients that have been missed by the current system. Many unanswered questions remain regarding regulation of and by cytoplasmic pH in these amoebae. Ion substitution experiments, treatments with metabolic inhibitors, and treatments with agents thought to inhibit specific ion fluxes (eg, lanthanum, amiloride) will help to elucidate the mechanism of cytoplasmic pH regulation. Many other important phenomena remain to be examined. For example, does cytoplasmic pH change during pinocytosis or during chemotaxis? Does enucleation of A. proteus, which halts locomotion [Jeon, 1973], have any effect on the pH of the remaining cytoplasm?

Microfluorometric measurements of FTC-ovalbumin in the endocytic vacuoles of C. carolinensis reveal consistent features of continuous pH change in these organelles. Rapid acidification is followed by gradual alkalinization to final values higher than those observed for cytoplasmic pH. The ability of phagosomally localized probe to to respond as expected to ammonia treatment at high pH (Table III) suggests that artifactual concentration quenching is not responsible for the fluorescence changes observed. Moreover, similar results are obtained using the absorbant dye, phenol red (Heiple and Taylor, in preparation). Next to nothing is known of the role of these pH changes in the processing of the internalized substance. The possibility that endosomal pH drops to very low values *before* lysosomal fusion occurs [Chapman-Andresen and Lagunoff, 1966] suggests a role for endosomal pH (or ΔpH) in directing fusion events. As substances internalized via receptor-mediated endocytosis in other cell types [Goldstein et al, 1979] are not immediately degraded, elucidation of pH changes in these endosomes will be extremely interesting, and the microfluorometric technique described here is clearly applicable to this problem. Although pH measurements from internalized fluorescent probes have been carried out on a population of cells in suspension [Ohkuma and Poole, 1978; Segal et al, 1981; Geisow et al, 1981], this represents the first quantitation of pH changes within a single phagosome in situ in a motile cell. In fact, there is no other conceivable way to make such measurements. More detailed information about individual vesicles can be obtained through higher resolution measurements, fractionation of fluorescently loaded vesicles, and "cell sorting" to facilitate pH measurements of single vesicles. Recovery and characterization of internalized probe from cells that can be grown in quantity (eg, D. discoideum) will also be important.

Elucidation of cytoplasmic and endosomal pH in C. carolinensis contributes necessary information to the picture of regulation of contractile and endocytic cellular events. Freshwater amoebae remain one of the best model systems for studying these processes within single cells. Moreover, their large size and ease of handling has much simplified the task of characterizing the optical technique described above. This technique has a great deal of built in flexibility. It is applicable to the quantitation of a variety of intracellular parameters, given the appropriate choice of probe, filters, and cell system. Microinjection can be bypassed in smaller cells by employing diffusible, hydrolyzable probes such as 6-carboxyfluorescein diacetate [Thomas et al, 1979; Horne et al, 1981]. However, careful characterization and calibration of each probe in each new system cannot be overemphasized.

V. REFERENCES

Aickin CC, Thomas RC: Microelectrode measurement of the intracellular pH and buffering power of mouse soleus muscle fibres. J Physiol 267:791, 1977.
Allison AC, Davies P: Mechanisms of endocytosis and exocytosis. Symp Soc Exptl Biol 28:419, 1974.
Ausländer W, Junge W: Neutral red, a rapid indicator for pH changes in the inner phase of thylkoids. FEBS Lett 59:310, 1975.
Babior BM: Oxygen-dependent microbial killing by phagocytes. N Engl J Med 298:659, 1978.
Barltrop J: Excited states: Time-dependent phenomena, and quenching of excited states. In "Excited States in Organic Chemistry. In Barltrop JA, Coyle JD (eds): New York: Wiley, 1975, p 64.
Begg DA, Rebhun LI: pH regulates the polymerization of actin in the sea urchin egg cortex. J Cell Biol 83:241, 1979.
Bennett WF, Belville JS, Lynch G: A study of protein phosphorylation in shape change and Ca^{++}-dependent serotonin release by blood platelets. Cell 18:1015, 1979.
Boron WF: Intracellular pH regulation. In Bronner F, Kleinzeller A (eds): "Current Topics in Membranes and Transport." New York: Academic Press, Vol 13:3, 1980.

Boron WF, Roos A: Comparison of microelectrode, DMO, and methylamine methods for measuring intracellular pH. Am J Physiol 231:799, 1976.

Bridges JW: Fluorescence of organic compounds. In Bowen EJ (ed): "Luminescence in Chemistry." E.J. Bowen, editor. Princeton, New Jersey: Van Nostrand, 6:77, 1968.

Brown JE, Brown PK, Pinto LH: Detection of light-induced changes of intracellular ionized calcium concentration in Limulus ventral photoreceptors using arsenazo III. J Physiol 267:299, 1977.

Brown PK: A system for microspectrophotometry employing a commercial recording spectrophotometer. J Opt Soc Am 51:1000, 1961.

Chambers R, Chambers EL: Hydrogen-ion concentration of cell components. In: "Explorations into the Nature of the Living Cell." Cambridge, Massachusetts: Harvard University Press, 141, 1961.

Chambers R, Pollack H, Hiller S: The protoplasmic pH of living cells. Proc Soc Exp Biol Med 24:760, 1927.

Chance B: Kinetics of enzyme reactions within single cells. Ann NY Acad Sci 97:431, 1962.

Chapman-Andresen C: Endocytosis in freshwater amebas. Physiol Rev 57:371, 1977.

Chapman-Andresen C, Lagunoff D: The distribution of acid phosphatase in the amoeba Chaos chaos L. Comp Rend Trav Lab Carls 35:419, 1966.

Condeelis JS: The isolation of microquantities of myosin from amoeba proteus and Chaos carolinensis. Anal Biochem 78:374, 1977.

Condeelis JS, Taylor DL: The contractile basis of amoeboid movement. V. The control of gelation, solation, and contraction in extracts from Dictyostelium discoideum. J Cell Biol 74:901, 1977.

Craig R, Megerman J: Assembly of smooth muscle myosin into side-polar filaments. J Cell Biol 75:990, 1977.

Davies DD: Control of and by pH. Symp Soc Exp Biol 27:513, 1973.

Deamer DW, Prince RC, Crofts AR: The response of fluorescent amines to pH gradients across liposome membranes. Biochim Biophys Acta 274:323, 1972.

Ebashi S: Excitation–contraction coupling. Ann Rev Physiol 38:293, 1976.

Eisenbach M, Garly H, Bakker EP, Klemperer G, Rottenberg H, Caplan SR: Kinetic analysis of light-induced pH changes in bacteriorhodopsin-containing particles from Halobacterium holobium. Biochemistry 17:4691, 1978.

Epel D: Mechanisms of activation of sperm and egg during fertilization of sea urchin gametes. Dev Biol 12:185, 1978.

Gallin EK, Gallin JI: Interaction of chemotactic factors with human macrophages. J Cell Biol 75:277, 1977.

Gallin EK, Wiederhold ML, Lipsky PE, Rosenthal AS: Spontaneous and induced membrane hyperpolarizations in macrophages. J Cell Physiol 86:653, 1975.

Garcia A, Iorio R: Potential sources of error in two-wavelength cytophotometry. In Weid (ed): "Introduction to Quantitative Cytochemistry." New York: Academic Press, 1966, p 216.

Gawlitta W, Stockem W, Wehland J, Weber K: Pinocytosis and locomotion of amoebae. XV. Visualization of Ca^{++}-dynamics by cholortetracycline (CTC) fluorescence during induced pinocytosis in living amoeba proteus. Cell Tissue Res 213:9, 1980.

Geisow MJ, D'Arcy Hart P, Young MR: Temporal changes of lysosome and phagosome pH during phagolysosome formation in macrophages: Studies by fluorescence spectroscopy. J Cell Biol 89:645, 1981.

George W, Walton KW: Purification and concentration of dye–protein conjugates by gel filtration. Nature (Lond) 192:1188, 1961.

Gerson DF: Intracellular pH and the mitotic cycle in Physarum and mammalian cells. In: Cell Cycle Regulation. New York: Academic Press, 1978, p 105.

Gerson DF, Burton AC: The relation of cycling of intracellular pH to mitosis in the acellular slime mould, Physarum polycephalum. J Cell Physiol 91:297, 1977.

Gillies RJ: Intracellular pH and growth control in eukaryotic cells. In Cameron I, Pool T (eds): "The Transformed Cell." New York: Academic Press, 1981, p 347.

Gillies RJ, Deamer DW: Intracellular pH changes during the cell cycle in Tetrahymena. J Cell Physiol 100:23, 1979a.

Gillies RJ, Deamer DW: Intracellular pH: Methods and applications. Curr Topics Bioenerg 9:63, 1979b.

Goldstein JL, Anderson RGW, Brown M: Coated pits, coated vesicles, and receptor-mediated endocytosis. Nature (Lond) 279:679, 1979.

Gordon AH, D'Arcy Hart P, Young MR: Ammonia inhibits phagosome-lysosome fusion in macrophages. Nature (Lond) 286:79, 1980.

Gorman ALF, Thomas MV: Changes in the intracellular concentration of free calcium ions in a pacemaker neurone, measured with the metallochromic indicator dye arsenazo III. J Physiol (Lond) 275:357, 1978.

Gutman M, Huppert D: Rapid pH and $\Delta\mu H^+$ jump by short laser pulse. J Bioch Biophys Meth 1:9, 1979.

Haaijman JJ, van Dalen JPR: Quantitation in immunofluorescence microscopy. A new standard for fluorescein and rhodamine emission measurements. J Immunol Meth 5:359, 1974.

Heiple JM: Intracellular pH in single living cells. Doctoral Dissertation, Harvard University, 1981.

Heiple JM, Taylor DL: Intracellular pH in single motile cells. J Cell Biol 86:885, 1980a.

Heiple JM, Taylor DL: Intracellular pH changes during phagocytosis and pinocytosis in single amoebae. J Cell Biol 87:222a, 1980b.

Heiple JM, Taylor DL: pH changes in pinosomes and phagosomes of single motile cells. Submitted to J Cell Biol.

Helenius A, Karlenbeek J, Simons K, Fries E: On the entry of Semliki forest virus into BHK-21 cells. J Cell Biol 84:404, 1980.

Hellewell, SB, Taylor DL: The contractile basis of amoeboid movement. VI. The solation-contraction coupling hypothesis. J Cell Biol 83:633, 1979.

Hinssen H, Haese JD: Filament formation by slime mould myosin isolated at low ionic strength. J Cell Sci 15:113, 1974.

Hitchcock SE: Regulation of motility in nonmuscle cells. J Cell Biol 74:1, 1977.

Holter H: Pinocytosis. Int Rev Cytol 8:481, 1959.

Horne WC, Tanoue LE, Laband AM, Norman NE, Schwartz DB, Simons ER: Correlation of the thrombin-induced changes in the transmembrane potential, transmembrane pH gradient and serotinin secretion of human platelets. Eur J Biochem (in press) 1981.

Jacobs, MH: Some aspects of cell permeability to weak electrolytes. Cold Spring Harbor Symp Quant Biol 8:30, 1940.

Jacques VV, Bainton DF: Changes in pH within the phagocytic vacuoles of human neutrophils and monocytes. Lab Invest 39:179, 1978.

Jensen MS, Bainton DF: Temporal changes in pH within the phagocytic vacuole of the polymorphonuclear neutrophilic leukocyte. J Cell Biol 56:379, 1973.

Jeon KW (ed): "The Biology of Amoeba." New York: Academic Press, 1973.

Johnson RG, Carty SE, Fingerhood BJ, Scarpa A: The internal pH of mast cell granules. FEBS Lett 120:75, 1980.

Kano K, Fendler JH: Pyranine as a sensitive pH probe for liposome interiors and surfaces. pH gradients across phospholipid vesicles. Biochim Biophys Acta 509:289, 1978.

Kaplan JG: Membrane cation transport and the control of proliferation of mammalian cells. Ann Rev Physiol 40:19, 1978.

Lee AC, Forte JG: A novel method for measurement of intravesicular pH using fluorescent probes. Biochim Biophys Acta 601:152, 1980.

Lew DP, Stossel TP: Calcium pump activity of the macrophage plasma membrane. J Cell Biol 83:288a, 1979.

Lisman JE, Strong JA: The initiation of excitation and light adaptation in Limulus ventral photoreceptors. J Gen Physiol 73:219, 1979.

MacDonald VW, Jöbsis FF: Spectrophotometric studies on the pH of frog skeletal muscle. pH change during and after contractile activity. J Gen Physiol 68:179, 1976.

MacDonald VW, Keizer JH, Jöbsis FF: Spectrophotometric measurements of metabolically induced pH changes in frog skeletal muscle. Arch Biochem Biophys 184:423, 1977.

Mandell GL: Intraphagosomal pH of human polymorphonuclear neutrophils. Proc Soc Exp Biol Med 134:447, 1970.

Marshak A: Changes in the apparent cytoplasmic hydrogen ion concentration of Amoeba dubia on injection of egg albumin. J Gen Physiol 28:95, 1944.

Martin MM, Lindqvist L: The pH dependence of fluorescein fluorescence. J Luminescence 10:381, 1975.

Mast SO: The hydrogen ion concentration of the content of the food vacuoles and the cytoplasm in amoeba and other phenomena concerning the food vacuoles. Biol Bull 83:173, 1942.

McLaughlin SGA, Dilger JP: Transport of protons across membranes by weak acids. Phys Rev 60:825, 1980.

Meizel S, Deamer DW: The pH of the hamster sperm acrosome. J Histochem Cytochem 26:98, 1978.

Naccache PH, Volpi M, Showell HJ, Becker EL, Sha'afi RI: Chemotactic factor-induced release of membrane calcium in rabbit neutrophils. Science 203:461, 1979.

Needham J, Needham DM: The hydrogen ion concentration and the oxidation-reduction potential of the cell interior. Proc R Soc Lond Ser B 98:259, 1925.

Nuccitelli R, Poo M, Jaffe LF: Relations between ameboid movement and membrane-controlled electrical currents. J Gen Physiol 69:743, 1977.

Nunnally MH, Craig SW: Small changes in pH within the physiological range cause large changes in the consistency of actin–filamin mixtures. J Cell Biol 87:218a, 1980.

Ogan K, Simons ER: The influence of pH on Arsenazo III. Anal Biochem 96:70, 1979.

Ohkuma S, Poole B: Fluorescence probe measurement of the intralysosomal pH in living cells and the perturbation of pH by various agents. Proc Natl Acad Sci USA 75:3327, 1978.

Pick U, Avron CM: A method for measuring the internal pH in illuminated chloroplasts based on the stimulation of proton uptake by amines. Eur J Biochem 70:569, 1976.

Piller H: "Microscope Photometry." New York: Springer-Verlag, 1977.

Ploem JS: Quantitative fluorescence microscopy. In Meek GA, Elder HY (eds): "Analytical and Quantitative Methods in Microscopy." Cambridge: Cambridge University Press, 1977.

Ploem JS, DeSterke JA, Bonnet J, Wasmund H: A microspectrofluorometer with epi-illumination operated under computer control. J Histochem Cytochem 22:668, 1974.

Pollard HB, Shindo H, Creutz CE, Paxoles CJ, Cohen JS: Internal pH and state of ATP in adrenergic chromaffin granules determined by ^{31}P nuclear magnetic resonance spectroscopy. J Biol Chem 254:1170, 1979.

Pollard TD: The role of actin in the temperature-dependent gelation and contraction of extracts of Acanthamoeba. J Cell Biol 68:579, 1976.

Pollister AW, Ris H: Nucleoprotein determinations in cytological preparations. Cold Spring Harb Symp Quant Biol 12:147, 1948.

Prusch RD, Hannafin J: Calcium distribution in Amoeba proteus. J Gen Physiol 74:511, 1979a.

Prusch RD, Hannafin J: Sucrose uptake by pinocytosis in Amoeba proteus and the influence of external calcium. J Gen Physiol 74:523, 1979b.

Puszkin S, Kochwa S: Regulation of neurotransmitter release by a complex of actin with relaxing protein isolated from rat brain synaptosomes. J Biol Chem 249:7711, 1974.

Rasmussen H, Goodman DBP: Relationships between calcium and cyclic nucleotides in cell activation. Physiol Rev 57:421, 1978.
Regula CS, Pfeiffer JR, Berlin RD: Microtubule assembly and disassembly at alkaline pH. J Cell Biol 89:45, 1981.
Reynolds GT, Taylor DL: Image intensification applied to light microscopy. BioScience 30:586, 1980.
Ridgeway EB, Durham ACH: Oscillations of Ca^{++} ion concentrations in P. polycephalum. J Cell Biol 69:223, 1976.
Roos A, Boron WF: Intracellular pH. Physiol Rev 61:296, 1981.
Rotman B, Papermaster BW: Membrane properties of living mammalian cells as studied by enzymatic hydrolysis of fluorogenic esters. Proc Natl Acad Sci USA 55:134, 1966.
Rottenberg H: The measurement of membrane potential and pH gradient in cells, organelles, and vesicles. Meth Enzymol 55:547, 1979.
Rubin RP (ed): "Calcium and the Secretory Process." New York: Plenum Press, 1974.
Sandvig K, Olsnes S: Diptheria toxin entry into cells is facilitated by low pH. J Cell Biol 87:828, 1980.
Schauenstein K, Schauenstein E, Wick G: Fluorescence properties of free and protein bound fluorescein dyes. I. Macrospectrofluorometric measurements. J Histochem Cytochem 26:277, 1978.
Segal AW, Geisow M, Garcia R, Harper A, Miller R: The respiratory burst of phagocytic cells is associated with a rise in vacuolar pH. Nature 290:406, 1981.
Sernetz M: Microfluorometric investigations of the intracellular turnover of fluorogenic substrates. In Thaer A, Sernetz (eds): "Fluorescence Techniques in Cell Biology." New York: Springer-Verlag, 1973, p 243.
Sernetz M, Thaer A: Microcapillary fluorometry and standardization for microscope photometry. In Thaer A, Sernetz M (eds): "Fluorescence Techniques in Cell Biology." New York: Springer-Verlag, 1973, p 41.
Sernetz M, Thaer A: Microfluorometric binding studies of fluorescein-albumin conjugates and determination of fluorescein-protein conjugates in single fibroblasts. Anal Biochem 50:98, 1972.
Silverstein SC, Steinman RM, Cohn ZA: Endocytosis. Ann Rev Biochem 46:668, 1977.
Smith FA, Raven JA: Intracellular pH and its regulation. Ann Rev Plant Physiol 30:289, 1979.
Stacey DW, Allfrey CG: Evidence for the autophagy of microinjected proteins in HeLa cells. J Cell Biol 75:807, 1977.
Stahl PD, Schlesinger PH: Receptor-mediated pinocytosis of mannose/N-acetylglucosamine-terminated glycoproteins and lysosomal enzymes by macrophages. Trends Biochem Sci 5:194, 1980.
Stockem W: Endocytosis. In Jamieson GA, Robison DM (eds): "Mammalian Cell Membranes 2." London: Butterworths, 1977, p 151.
Stossel TP: Phagocytosis. New Eng J Med 290:717, 1974.
Swift H: Analytical microscopy of biological materials. In Wied GL (ed): "Introduction to Quantitative Cytochemistry." New York: Academic Press, 1966, p 1.
Taborsky G: Phosphoproteins. Adv Prot Chem 28:1, 1974.
Taylor DL: Quantitative studies on the polarization optical properties of striated muscle. J Cell Biol 68:497, 1976.
Taylor DL, Blinks JR, Reynolds G: Contractile basis of ameboid movement. VIII. Aequorin luminescence during ameboid movement, endycytosis, and capping. J Cell Biol 86:599, 1980a.
Taylor DL, Condeelis JS: Cytoplasmic structure and contractility in amoeboid cells. Int Rev Cytol 56:57, 1979.
Taylor DL, Condeelis JS, Moore PL, and Allen RD: The contractile basis of amoeboid movement. I. Chemical control of motility in isolated cytoplasm. J Cell Biol 59:378, 1973.
Taylor DL, Wang Y-L: Fluorescently labeled molecules as probes of the structure and function of living cells. Nature (Lond) 284:405, 1980.

Taylor DL, Wang Y-L, Heiple JM: Contractile basis of ameboid movement. VII. The distribution of fluorescently labeled actin in living amebas. J Cell Biol 86:590, 1980b.

Thomas JA, Buchsbaum RN, Zimniak A, Racker E: Intracellular pH measurements in Ehrlich ascites tumor cells utilizing spectroscopic probes generated in situ. Biochemistry 18:2210, 1979.

Thomas JA, Johnson DL: Fluorescein conjugates of cytochrome c as internal pH probes in submitochondrial particles. Biochem Biophys Res Commun 65:931, 1975.

Thomas MV, Gorman ALF: Internal calcium changes in a bursting pacemaker neuron measured with Arsenazo III. Science 196:531, 1977.

Tilney LG, Jaffe LA: Actin, microvilli, and the fertilization cone of sea urchin eggs. J Cell Biol 87:771, 1980.

Tilney LG, Kiehart DP, Sardet C, Tilney M: Polymerization of actin. IV. Role of Ca^{++} and H^+ in the assembly of actin and in membrane fusion in the acrosomal reaction of echinoderm sperm. J Cell Biol 77:536, 1978.

Visser JWM, Jongeling AAM, Tanke HT: Intracellular pH—Determination by fluorescence measurements. J Histochem Cytochem 27:32, 1979.

Waddell WJ, Bates RG: Intracellular pH. Physiol Rev 49:285, 1969.

Weissmann G: Lysosomes. New Engl J Med 273:1084, 1965.

Whitin JC, Chapman CE, Simons ER, Chovaniec ME, Cohen HJ: Correlation between membrane potential changes and superoxide production in human granulocytes stimulated by phorbol myristate acetate. J Biol Chem 255:1874, 1980.

Wied GL: "Introduction to Quantitative Cytochemistry." New York: Academic Press, 1966.

Wiercinski FJ: An experimental study of protoplasmic pH determination. I. Amoebae and Arbacia punctulata. Biol Bull 86:98, 1944.

Wolfarth-Bottermann KE: Oscillatory contraction activity in Physarum. J Exp Biol 81:15, 1979.

Woodward MP, Roth TF: Coated vesicles: Characterization, selective dissociation, and reassembly. Proc Natl Acad Sci USA 75(9):4394, 1978.

Yin HL, Stossel TP: Control of cytoplasmic actin gel-sol transformation by gelsolin, a calcium-dependent regulatory protein. Nature (Lond) 281:583, 1979.

Estimation of Intracellular pH From Distribution of Weak Electrolytes

Albert Roos and David W. Keifer*

Departments of Physiology and Biophysics and of Anesthesiology, Washington University School of Medicine, 660 South Euclid Avenue, St. Louis, Missouri 63110

I.	Introduction	55
II.	Advantages of Weak Acid/Base Distribution Methods	56
III.	Limitations and Sources of Error	56
IV.	Summary	59
V.	References	59

I. INTRODUCTION

The weak acid and base methods rest on the assumption that only the uncharged form of the pH indicators is permeant. Their steady-state distribution can then be characterized as an equilibrium of the neutral form across the membrane, in which its concentrations in the intracellular and extracellular water are the same. The concentrations of the charged forms in the two media are determined by intracellular and extracellular pH. The extracellular pH and total indicator concentration in the intracellular water can be measured. With the intracellular concentration of the uncharged partner set equal to its known extracellular concentration, the intracellular pH (pH_i) can then be calculated by applying the mass action law to the intracellular equilibrium between the two partners. Such a calculation for a weak acid was first carried out by Jacobs [1940]. That for a weak base is given in the paper by Boron and Roos [1976] and in the recent review by Roos and Boron [1981] to which the reader is also referred for further details and references of various aspects of these methods.

We shall limit ourselves here to a brief survey of the advantages and, especially, the limitations and sources of error of the methods.

---

*Present address: Department of Botany, University of California, Davis, CA 95616.

II. ADVANTAGES OF WEAK ACID/BASE DISTRIBUTION METHODS

The main advantages of weak acid and base methods are the simplicity of their execution (especially when isotopically labeled compounds are used), and the ease with which they can be applied to the various tissues of the intact animal, to isolated tissues, or to cells and organelles too small to be impaled by microelectrodes. Thus measurements have been made in mitochondria (DMO), chromaffin granules (methylamine), chloroplasts (methylamine), sarcoplasmic reticulum vesicles (DMO), and lysosomes (methylamine) [Roos and Boron, 1981].

III. LIMITATIONS AND SOURCES OF ERROR

1) Only one single measurement can be made.

2) This, and the relative slowness of the indicator's distribution make it impossible to follow pH_i transients accurately. When the surface:volume ratio of the cell is relatively great, indicator equilibration will be relatively rapid, but so will be the pH_i transients produced by transmembrane events. Roos and Boron [1978] have been able to estimate by means of DMO (5,5-dimethyl-2,4-oxazolidine dione) transient changes in pH_i of rat diaphragm at 30-minute intervals. These changes were produced by first acid-loading the muscles with CO_2 or DMO, pH_o being maintained at 7.4, and then returning them to the original solution. The general time course of the pH_i agreed with that measured more precisely with microelectrodes in single cells of other types.

3) Errors are introduced because of binding or metabolic transformation of the indicators. There is no evidence that this objection holds for the weak acids CO_2 or DMO in animal cells, although in plant cells DMO is metabolized [Kurkdjian and Guern, 1978]. Weak bases such as morphine, atropine, and trimethylamine are also not metabolized, but the case for nicotine in mammalian and amphibian muscle is doubtful [for a critical discussion on this subject see Roos and Boron, 1981, p 304].

4) Uncertainty is introduced because of the need to estimate the volume of the extracellular fluid included in the sample. Extracellular markers (inulin, sorbitol, etc) give somewhat different values. At very low distribution ratios (inside/outside) of the indicators (ie, with $pH_i \ll pH_o$ when weak acids are used, or $pH_i \gg pH_o$ for weak bases), the effect of small changes in extracellular space on derived pH_i can be considerable.

5) The dissociation constant of the indicator in the intracellular fluid is unknown. It is customarily assumed to be the same as that in the external fluid. Any error in intracellular pK produces an identical error in the derived pH_i.

6) The methods yield some average pH_i. Whenever there is inhomogeneity of intracellular pH (and this is probably the case in all cells other than nonnu-

cleated red cells), all weak acids give the same pH_i, but this value is higher than that obtained with weak bases. The expression for the weak acid-derived pH_i is

$$\overline{pH}_{acid} = \log \sum_{j=1}^{n} f_j 10^{pH_j}$$

where f_j is the fractional volume of the jth compartment. The expression for the weak base-derived average pH_i is

$$\overline{pH}_{base} = -\log \sum_{j=1}^{n} f_j 10^{-pH_j}$$

These two values for pH_i bracket the volume-weighted average pH_i, defined as

$$\overline{pH}_i = \sum_{j=1}^{n} f_j pH_j$$

Therefore, the following relationship holds:

$$\overline{pH}_{acid} > \overline{pH}_i > \overline{pH}_{base}$$

[Roos and Boron, 1981]. It should be stressed that $\overline{pH}_i$ is not identical with the average pH_i obtained from an external fluid-free cell homogenate; the latter is affected by the relative buffering powers of the various compartments as well as by their pHs.

7) As stated at the beginning, the charged partners of the indicators are assumed to be impermeant. If this assumption is incorrect, both acids and bases will yield an erroneously low pH_i. The reason for this error is that ionic permeability leads to a reduction of total intracellular acid, and to an increase in total intracellular base (Fig. 1). It is important to emphasize that, at a particular ratio of ionic to molecular permeability, the error is greater the greater the divergence between the indicator's pK_a and the prevailing pH_o and pH_i. The precise steady-state transmembrane distributions resulting from membrane permeability to both partners have been derived in previous papers from our laboratory [Roos, 1965, 1971; Boron and Roos, 1976; Roos and Boron, 1981].

Theoretically, the combination of ionic and molecular permeability, which amounts to the shuttling of H ions across the membrane (see Fig. 1) leads to progressive erosion of the normally inward-directed electrochemical H^+ gradient, until eventually H^+ would be at equilibrium across the membrane. The acid extrusion mechanism that is present in nearly all cells will prevent such progressive acidification. Thus, a new steady state will be reached in which the rate of acid extrusion exactly balances that of shuttling-engendered proton accumulation.

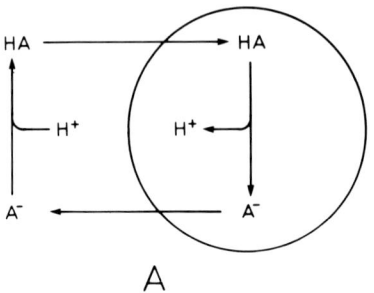 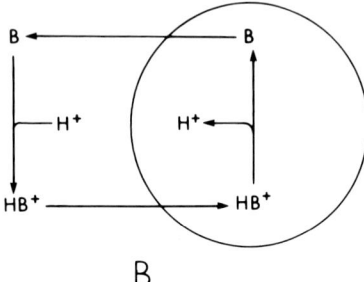

Fig. 1. Passive movements of charged and uncharged forms of weak acids and bases. A) Weak acids. Under normal conditions of V_m, pH_i, and pH_o, the electrochemical gradient of the anionic partner (A^-) of the conjugate weak acid (HA) favors net efflux of A^-. As $(A^-)_i$ decreases, the intracellular equilibrium HA $\rightleftharpoons$ H^+ + A^- shifts to the right, thereby lowering $(HA)_i$ and leading to influx of HA. Net result is inward shuttling of protons. A steady state can be maintained if H^+ is extruded as rapidly as it is shuttled in; then the influx of HA exactly balances the efflux of A^-. If H^+ is in equilibrium across the membrane, A^- is also in equilibrium, and there is no net flux of eiher A^- or HA. B) Weak bases. Under normal circumstances, the electrochemical gradient of the cationic partner (HB^+) of the conjugate weak base (B) favors net influx of HB^+. As $(HB^+)_i$ increases, the intracellular equilibrium HB^+ $\rightleftharpoons$ H^+ + B shifts to the right, leading to efflux of B. Net effect is inward shuttling of protons. Steady-state and equilibrium considerations are analogous to the case for weak acid. (From Roos and Boron [1981].)

Fortunately, the permeability of the plasma membrane to the ionic form of the most commonly used weak acid indicator, DMO, seems to be very low. In barnacle muscle fibers, Keifer and Roos [1980] measured a value, about 1/1,000 that of the neutral form, which should lead to an underestimation of pH_i of only a few hundredths of a unit. In the case of the weak acid CO_2, even though a significant permeability to HCO_3^- has been observed in a number of cells [Roos and Boron, 1981, p 351], it is doubtful that this would vitiate the pH_i measurement based on the distribution of the acid in view of the probably very high permeability to molecular CO_2. (Actually, no precise quantitative information is available on this point!) The use of the weak bases methylamine and, especially, ammonia as pH_i indicators is questionable, because the ionic partners of both bases are permeant in a number of cells. This is particularly serious because their pK_as are very high (about 12 and 9, respectively). Trimethylamine, morphine, and possibly some other bases might be acceptable, but each candidate should be carefully evaluated.

8) It should not be forgotten that weak acids and bases can directly affect pH_i by two mechanisms. In the first place, the entry of the neutral form of a weak acid decreases pH_i, whereas that of a weak base increases it. The magnitude of

these effects depends on concentration, pH_o, and pH_i. At least in the case of weak acid entry, the cell will respond by extruding acid so that the net intracellular acidification may actually be very small, even at very high indicator concentrations [Roos, 1975]. In the second place, as mentioned above, the efflux of the anionic form of weak acids or the influx of the cationic form of weak bases produces a fall in pH_i. These effects are dependent on concentration, membrane permeability, and membrane voltage, and are, of course, also opposed by acid extrusion.

IV. SUMMARY

In summary, even now, when direct electrical and optical methods are available for the measurement of intracellular pH, the weak electrolyte approach can play a useful role when single measurements and average values are acceptable, and when time is allowed for reaching a steady state. These older approaches are especially valuable in the case of organelles, and as standards against which to evaluate more modern techniques. In applying these methods, one should keep their limitations firmly in mind.

ACKNOWLEDGMENTS

This work was supported by Public Health Service grant HL-00082. A. Roos is the recipient of Research Career Award HR-19608.

V. REFERENCES

Boron WF, Roos A: Comparison of microelectrode, DMO, and methylamine methods for measuring intracellular pH. Am J Physiol 231:799–809, 1976.

Jacobs MH: Some aspects of cell permeability to weak electrolytes. Cold Spring Harbor Symp Quant Biol 8:30–39, 1940.

Keifer DW, Roos A: Membrane permeability to the molecular and ionic forms of DMO in barnacle muscle. Am J Physiol 240 (Cell Physiol 9):C73–C79, 1980.

Kurkdjian A, Guern J: Intracellular pH in higher plant cells. I. Improvements in the use of the 5,5-dimethyloxazolidine-2(^{14}C),4-dione distribution technique. Plant Sci Lett 11:337–344, 1978.

Roos A: Intracellular pH and intracellular buffering power of the cat brain. Am J Physiol 209:1233–1246, 1965.

Roos A: Intracellular pH and buffering power of rat muscle. Am J Physiol 221:182–188, 1971.

Roos A: Intracellular pH and distribution of weak acids across cell membranes. A study of D- and L-lactate and of DMO in rat diaphragm. J Physiol (Lond) 249:1–25, 1975.

Roos A, Boron WF: Intracellular pH transients in rat diaphragm muscle measured with DMO. Am J Physiol 235 (Cell Physiol 4):C49–C54, 1978.

Roos A, Boron WF: Intracellular pH. Physiol Rev 61:296–434, 1981.

pH$_i$ Measurements of Cardiac and Skeletal Muscle Using ^{31}P-NMR

David G. Gadian, George K. Radda, M. Joan Dawson, and Douglas R. Wilkie

Department of Biochemistry, University of Oxford, South Parks Road, Oxford OX1 3QU, England (D.G.G., G.K.R.), and Department of Physiology, University College London, Gower Street, London WC1E 6BT, England (M.J.D., D.R.W.)

I.	Introduction	61
II.	What Can ^{31}P-NMR Tell Us About Metabolism?	62
III.	The Measurement of pH$_i$	65
IV.	Applications of pH$_i$ Measurements	68
	A. Muscular Fatigue	68
	B. Creatine Kinase Activity in Muscle	71
	C. Studies of Acidosis and Metabolic Control	72
V.	Conclusions	75
VI.	References	76

I. INTRODUCTION

Phosphorus nuclear magnetic resonance (^{31}P-NMR) was first used to study cellular and tissue metabolism almost 10 years ago [Moon and Richards, 1973; Hoult et al, 1974], and since then there has been increasing interest in this non-invasive method of studying metabolism in vivo [for reviews see Burt et al, 1979; Gadian et al, 1979; Griffiths and Iles, 1980; Gadian and Radda, 1981; Gadian, 1981, 1982]. One useful feature of the method is that it provides a measure of intracellular pH. In this article, we shall first describe in general terms the type of information that is available from ^{31}P-NMR, concentrating in particular on the measurement of pH, and we shall then discuss some specific applications to skeletal and cardiac muscle.

A brief introduction to NMR should be useful to those who are unfamiliar with the method. NMR is a spectroscopic technique that is probably most widely used to study the structures of molecules in solution. For these structural studies, a solution containing the molecules of interest is placed within a strong magnetic field. The field interacts with those atomic nuclei within the sample (such as ^{1}H, ^{13}C, and ^{31}P) that possess intrinsic magnetic properties, and the precise nature of the interaction can be examined by observing the characteristic frequencies at which the nuclei absorb, and subsequently re-emit, radio frequency radiation. The strength of the magnetic interaction is affected by the local chemical environment of the various nuclei, and therefore nuclei in different chemical environments absorb radiation (and hence give rise to signals) at slightly different frequencies. Because the frequencies, or chemical shifts, of these signals are sensitive to chemical environment, information can be obtained about molecular structure.

NMR studies of metabolism are based on the same principles. However, relatively little structural information is available from the signals; instead, the spectra provide a means of identifying the presence of certain molecules within the sample, and of evaluating their concentrations and intracellular environments. The studies described in this article make use of signals from ^{31}P nuclei (^{31}P is the naturally occurring isotope of phosphorus). Different phosphorus-containing molecules produce signals at different frequencies, and, simply by monitoring how the signal intensities vary with time, it is possible to monitor the interconversions of the various molecules; ie, to follow their metabolism. Moreover, the intracellular pH can be measured from the frequency of the inorganic phosphate signal, which is sensitive to pH variations in the physiological range.

Perhaps the main difference between structural and metabolic studies is in the nature of the sample and of the probe (see Fig. 2) in which it is mounted. The sample is not in the form of a simple solution, but instead may be a cellular suspension, a perfused tissue or organ, a whole animal, or a human. Certain modifications to instrumentation are therefore required [see Gadian, 1982], some of which are mentioned briefly in this article.

II. WHAT CAN ^{31}P-NMR TELL US ABOUT METABOLISM?

The type of information that ^{31}P-NMR can provide is illustrated by Figure 1, which shows a ^{31}P-NMR spectrum obtained from a perfused rat heart [Grove et al, 1980]. The heart is held within a specially designed probe, as shown in Figure 2, which is positioned within the 10 cm diameter bore of a cylindrical superconducting magnet. The spectrum of Figure 1 was accumulated over a period of 8 minutes, and contains signals that can readily be assigned to the β, α, and γ phosphates of ATP, phosphocreatine, and inorganic phosphate. The spectrum is remarkably simple, the reason being that signals are observed only from mobile phosphorus-containing compounds that are present at concentrations of above 0.2–0.5 mM. ADP, if present in mobile form at sufficiently high concentration, would generate two signals overlapping with the signals from the γ and α phosphates of ATP. In fact, it is of considerable interest to note that

ADP generally makes no detectable contribution to the spectra of well-oxygenated intact tissues, suggesting that the concentration of free ADP is very much lower than the total amounts that are estimated by the technique of freeze-clamping. NAD produces signals that can often be detected as a shoulder just to the right (ie, to low frequency) of the signal from the α phosphate of ATP.

A particularly useful feature of the spectra is that the frequency of the inorganic phosphate signal is sensitive to pH variations in the normal physiological range. This signal therefore provides a monitor of intracellular pH. The sensitivity to pH arises because inorganic phosphate has a pK_a of about 6.75, and it accounts for the observation that the signal in the spectrum of Figure 1 is split into two components; inorganic phosphate is present within both the intracellular space and the perfusion medium, and these two environments have different pH values. On the basis of calibration curves performed in vitro (see Fig. 3), it can be concluded that the intracellular pH is 7.11. The signal from the external inorganic phosphate has a frequency that reflects the pH of 7.37 within the perfusion medium.

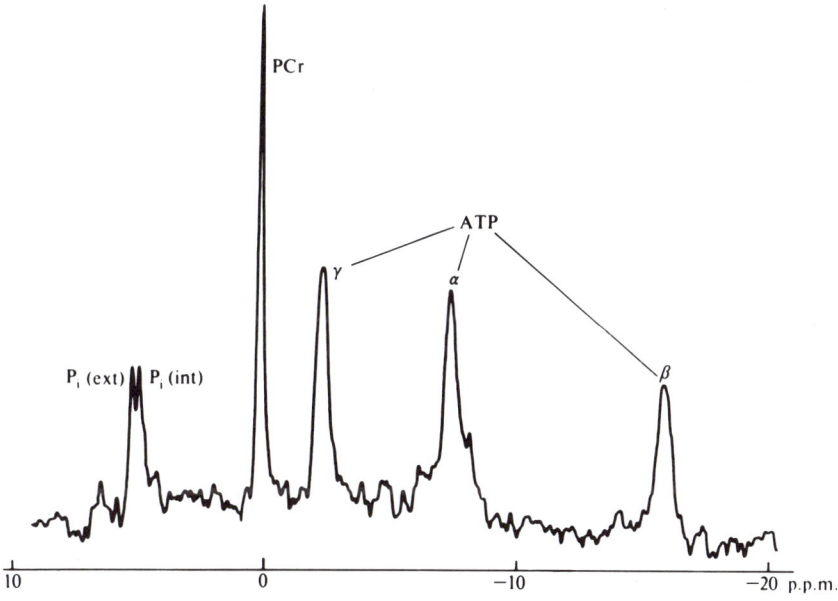

Fig. 1. ^{31}P-NMR spectrum obtained in 8 minutes at 73.8 MHz from a Langendorff-perfused rat heart. The signals are assigned as shown. PCr refers to phosphocreatine, and P_i (int) and P_i (ext) to inorganic phosphate in the intracellular space and the perfusion medium respectively. [From Grove et al, 1980.]

The frequencies of the ATP signals are sensitive to the binding of divalent metal ions [Cohn and Hughes 1962] (see also Gadian et al [1979] for pH titrations of free ATP and MgATP performed at physiological concentrations). It can be concluded from the spectrum shown in Figure 1 that the ATP in the heart is predominantly complexed to divalent metal ions, presumably Mg^{2+} ions, and similar conclusions regarding the state of ATP have been reached for other tissues. The binding of Mg^{2+} to ATP lowers the pK_a of this compound from about 6.6 to 5.1, and for this reason the frequencies of the ATP signals are not at all sensitive to pH changes in the region of neutrality.

The concentrations of the various metabolites are, under certain conditions, proportional to their signal areas, and therefore metabolic processes, together with any accompanying pH changes, can be followed simply by monitoring how the spectra vary with time. It should be noted, however, that careful controls are necessary in order to quantify the concentrations of the various metabolites [see, for example, Dawson et al, 1977]. The heart spectrum of Figure 1 was obtained in 8 minutes, but adequate signal-to-noise ratios can usually be obtained in times much shorter than this, and the typical time resolution for kinetic studies

 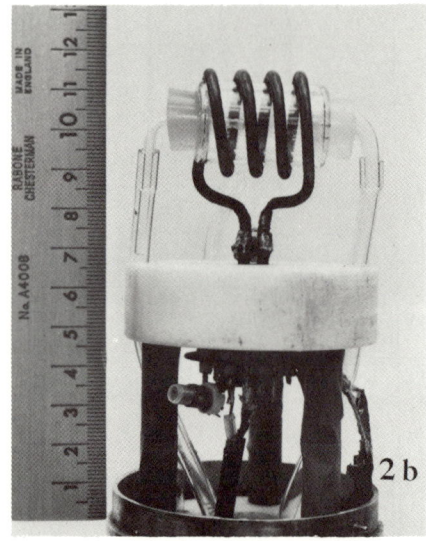

Fig. 2. An NMR probe built in the Oxford laboratory for ^{31}P-NMR studies of perfused rat hearts [see Garlick et al, 1979]. a) The complete probe, except that the top half of the surrounding aluminium alloy tube has been removed to reveal the sample chamber and radiofrequency coil; b) a close-up showing in more detail the sample chamber, the solenoidal radiofrequency coil, and the perfusion fluid inflow and outflow. It should be pointed out that this is not a conventional type of probe; it was designed specifically for studying rat hearts.

is about 1 minute. For some experiments, the time resolution can be enhanced considerably by synchronising the collection of data with physiological activity [Dawson et al, 1977; Fossel et al, 1980].

NMR can readily be used in this way to study the changes in pH$_i$ and metabolite levels that are associated with muscular contraction, ischaemia, etc [for reviews see Ackerman et al, 1980a; Dawson et al, 1980a; Griffiths and Iles, 1980; Gadian and Radda, 1981]. However, before discussing some examples of these studies, we shall first describe in more detail the principles and problems associated with pH measurements using NMR.

III. THE MEASUREMENT OF pH$_i$

NMR was first used to measure intracellular pH by Moon and Richards [1973] in their studies of red blood cells. Their measurements were based on the observation that 2,3-diphosphoglycerate and inorganic phosphate generate ^{31}P-NMR signals whose frequencies are sensitive to pH variations in the normal physiological range. In principle, any signal whose frequency is sensitive to pH

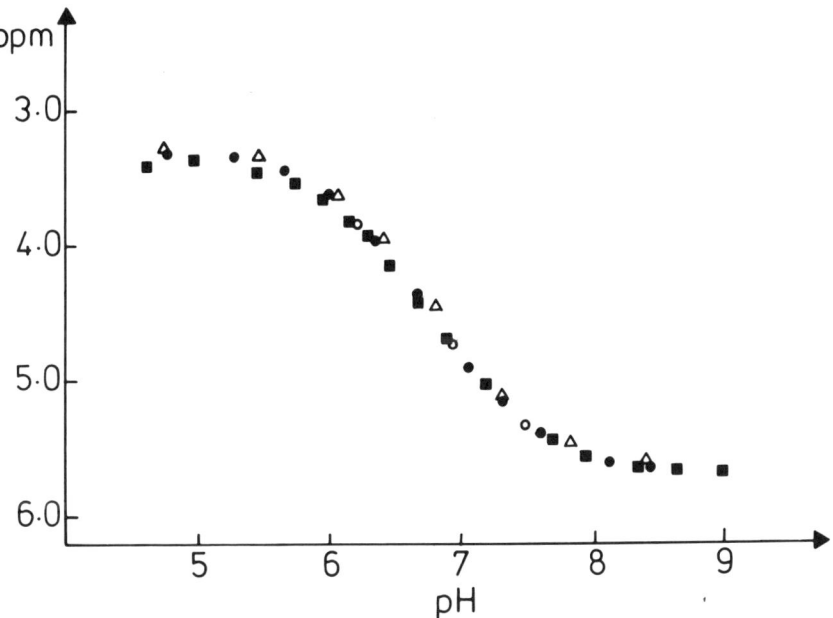

Fig. 3. Variation of the chemical shift of inorganic phosphate with solution pH at 37°C. The chemical shift is expressed relative to phosphocreatine at 7.0. Solutions containing 10 mM inorganic phosphate and 10 mM phosphocreatine were adjusted to various ionic strengths by addition of KCl or NaCl, and titrated at 37°C by addition of HCl and KOH or NaOH. Symbols: △ 120 mM KCl; ■ 160 mM KCl; ● 200 mM KCl; ○ 160 mM NaCl. [Adapted from Garlick et al, 1979].

can provide an indication of pH, but in practice the inorganic phosphate signal is most commonly used, because it is readily observable in the majority of ^{31}P spectra, and because its frequency is particularly sensitive to pH changes in the region of neutrality.

Inorganic phosphate exists mainly as HPO_4^{2-} and $H_2PO_4^-$ around neutral pH. If these two species did not exchange with each other, they would give rise to two signals separated from each other by about 2.4 ppm. In solution, however, the two species exchange with each other very rapidly, and therefore the observed spectrum consists of a single signal, the frequency of which is determined by the relative amounts of the two species; in fact, the frequency measured as a function of pH gives the standard type of pH curve, as shown in Figure 3. It should thus be possible to determine pH_i simply by measuring the chemical shift of the inorganic phosphate signal in vivo, and determining from a standard titration curve performed on a solution the pH to which this chemical shift corresponds. However, there are several potential difficulties that have to be overcome.

First, there is bound to be some lack of knowledge of the precise intracellular conditions, and therefore it is essential to ascertain whether factors other than pH might affect the chemical shift of the inorganic phosphate signal. Fortunately, control experiments such as those shown in Figure 3 have indicated that the effects of *likely* variations in ionic strength or metal ion binding within living systems are small, and can in general be ignored. Nevertheless, since *large* changes in these parameters can produce significant shifts [Gadian et al, 1979; Jacobson and Cohen, 1981], it is important to ensure that the ionic composition of the standard solution resembles as closely as possible the conditions expected in vivo. Of course the same uncertainties attend the use of intracellular glass electrodes, which should also be calibrated under simulated intracellular conditions. It is far more difficult to take into account the binding of inorganic phosphate to macromolecules, which could also affect chemical shifts. However, it has been found that inorganic phosphate titration curves are similar in simple aqueous solution and in homogenised dog heart preparations [Hollis, 1979], which suggests that phosphate-protein binding interactions have little effect on the observed chemical shifts of inorganic phosphate. Additional evidence that such binding has little effect is largely circumstantial, and is based on the similarity between pH values obtained by NMR and by other means (see below).

A second problem is that the chemical shift of any signal is measured by comparing the frequency of that signal with that of a reference compound. It is therefore necessary for a suitable reference compound to be present. In the case of ^{31}P-NMR, phosphocreatine, when present, provides a very convenient reference, for the pK_a of this compound is about 4.6 and its signal is therefore insensitive to pH changes in the region of neutrality. Thus for many whole tissue studies, the intracellular pH is determined from the chemical shift difference between the inorganic phosphate and phosphocreatine signals. When phosphocreatine is not present, the ^{1}H-NMR signal from the water within the tissue provides an acceptable frequency standard [Ackerman et al, 1981], and this will probably become increasingly used in the future.

A final problem relates to the location of inorganic phosphate within the various intracellular compartments. For example, mitochondria occupy about 40% of the intracellular volume of the heart, and therefore it is conceivable that one might be measuring mitochondrial, rather than cytoplasmic pH. However, experiments utilising 2-deoxyglucose (see below) indicate that in the rat heart ^{31}P-NMR does indeed monitor cytoplasmic pH.

In view of these various problems, it is important to establish the reliability of pH measurements using NMR. In fact, there is no longer any doubt that ^{31}P-NMR provides a reliable method of measuring intracellular pH. First, pH measurements made by ^{31}P-NMR in a variety of samples under a variety of conditions agree well with anticipated values. Second, notwithstanding the uncertainties attending both methods, some specific experiments have been performed that verify that the NMR method and intracellular pH electrodes give similar results. Thus the pH of the giant barnacle muscle cell was found to be 7.35 using ^{31}P-NMR; the pH of similar cells from the same barnacle was measured to be 7.25 using a microelectrode (Bagshaw, Gadian, Radda, and Vaughan-Jones, unpublished observations). Nuccitelli et al [1981, and this volume] have used ^{31}P-NMR to measure the average pH_i of unfertilized, fertilized, and activated eggs of Xenopus laevis. They obtained pH values of 7.42, 7.66, and 7.64, respectively, which are almost identical to the values that were obtained using pH-sensitive glass microelectrodes. Third, the pH of heart and skeletal muscle is the same regardless of whether phosphocreatine or water is the reference compound [Ackerman et al, 1981]. Moreover, in cases where two different signals provide independent measures of pH (for example, in experiments on hearts perfused with 2-deoxyglucose (see below), for which both the inorganic phosphate signal and the signal from 2-deoxyglucose-6-phosphate provide estimates of intracellular pH), the values obtained using the two signals are very similar.

For all of these reasons, it is clear that ^{31}P-NMR provides a convenient and reliable means of estimating pH_i. Nevertheless, there are some uncertainties. In particular, the standard titration curve is generally obtained by plotting the chemical shift of inorganic phosphate in solution against the solution pH measured with a combined glass electrode. Illingworth [1981] has recently pointed out that such electrodes can produce pH measurements that are in error by as much as 0.2 pH units or even more. This could account for the fact that different pH titration curves that have been reported in the literature differ slightly from each other, and, in view of this, it is unwise to hope for *absolute* accuracy of better than 0.1 pH unit when measuring pH_i by NMR. However, *changes* in pH can often be measured to better than 0.05 pH unit.

It is interesting to note that NMR can also provide information about heterogeneity of pH_i. For example, the inorganic phosphate signal is sometimes split into two components, implying that the sample contains two environments of differing pH (see Fig. 1; see also Busby et al [1978]). Alternatively, if there is a range of environments within a sample of differing pH values, the sample will generate a range of inorganic phosphate signals that are slightly shifted from each other, the net effect being that the observed signal is broadened. In many ^{31}P spectra of skeletal muscle, the inorganic phosphate signal is indeed broader than the signal from phosphocreatine, and this has been interpreted in terms of a distribution of pH_i within the muscle [Seeley et al, 1976].

In certain circumstances, intracellular pH can also be measured by ^1H-NMR, for the ^1H chemical shifts of histidine residues are sensitive to pH. The pH-dependence of the histidine signals from haemoglobin has been carefully characterised by Brown and Campbell [1976]. From observations of the haemoglobin signals in red blood cells, it is therefore possible to determine the pH_i of these cells, as described by Brown et al [1977].

IV. APPLICATIONS OF pH_i MEASUREMENTS

A. Muscular Fatigue

The mechanical output of skeletal muscle declines after a sufficiently prolonged and intense period of exercise, and this decline, or fatigue, has long been a subject for experiment. As early as 1807 Berzelius believed that "the amount of free lactic acid in a muscle is proportional to the extent to which it has been previously exercised" [quoted in Lehman, 1850]. We now know a lot more about the importance of anaerobic processes, such as lactic acid formation, in brief intense exercise, and of oxidative metabolism in more prolonged exercise. However, there is no universal agreement about the fundamental mechanism of fatigue [see CIBA Symposium, 1981].

^{31}P-NMR is ideally suited to investigating the metabolic events associated with fatigue, for it is the only technique that enables mechanical function, pH_i and metabolic state to be monitored continuously within a single preparation. The biochemical basis of fatigue has been studied in this way in frog gastrocnemius muscles maintained at 4°C under anaerobic conditions [Dawson et al, 1978, 1980a]. Figure 4 shows the results of a typical experiment, in which frog gastrocnemii were stimulated repetitively for 5 seconds every 5 minutes. The force measurements shown at the right of each spectrum reveal the typical signs of fatigue; the force gradually declines in magnitude, and the rate of mechanical relaxation becomes progressively slower. The spectra indicate the expected decline in phosphocreatine, increase in inorganic phosphate, and decline in pH_i and the time course of the various metabolic changes is plotted in Figure 5. The concentration of free ADP is deduced from the creatine kinase equilibrium, as discussed in more detail below.

The decline in pH_i is a result of the accumulation of lactic acid within the muscle, and indeed the quantity of lactic acid that is produced can be estimated from the pH_i change and the known buffering capacity of the muscle. The measurement of lactic acid is of considerable interest, not only because of its suggested role in the onset of fatigue, but also because under anaerobic conditions it is the end product of glycogenolysis, and so provides an estimate of the rate of breakdown of glycogen. Interesting conclusions can therefore be reached about the control of glycogen breakdown in vivo [Dawson et al, 1980] (see also the studies of heart described below).

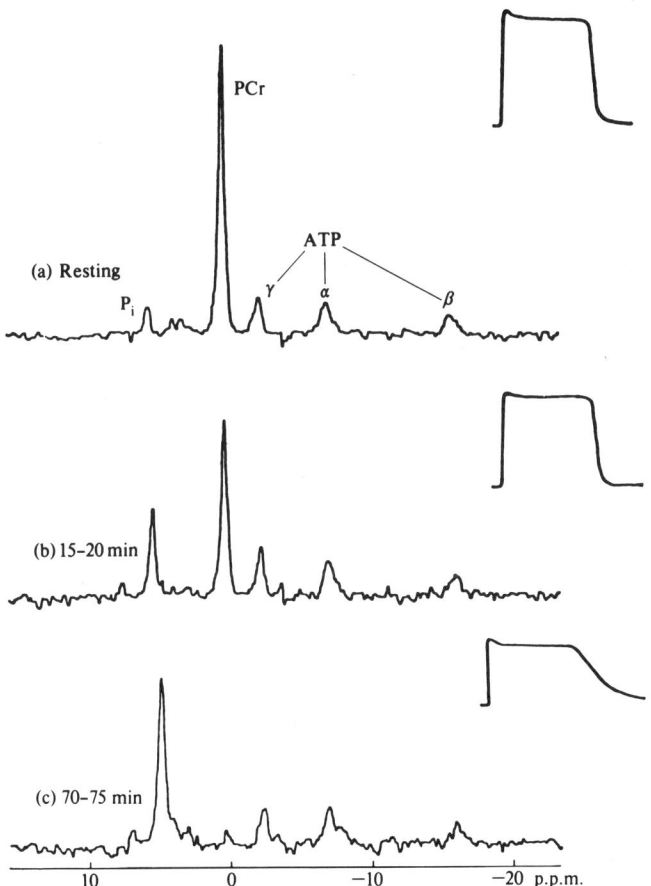

Fig. 4. ^{31}P-NMR spectra obtained at 129 MHz from anaerobic frog gastrocnemii at 4°C during a fatiguing series of 5-second contractions repeated every 5 minutes. Adjacent to the spectra are mechanical records showing the time course of isometric force development; the force progressively declines and mechanical relaxation becomes slower. [Adapted from Dawson et al, 1980a.]

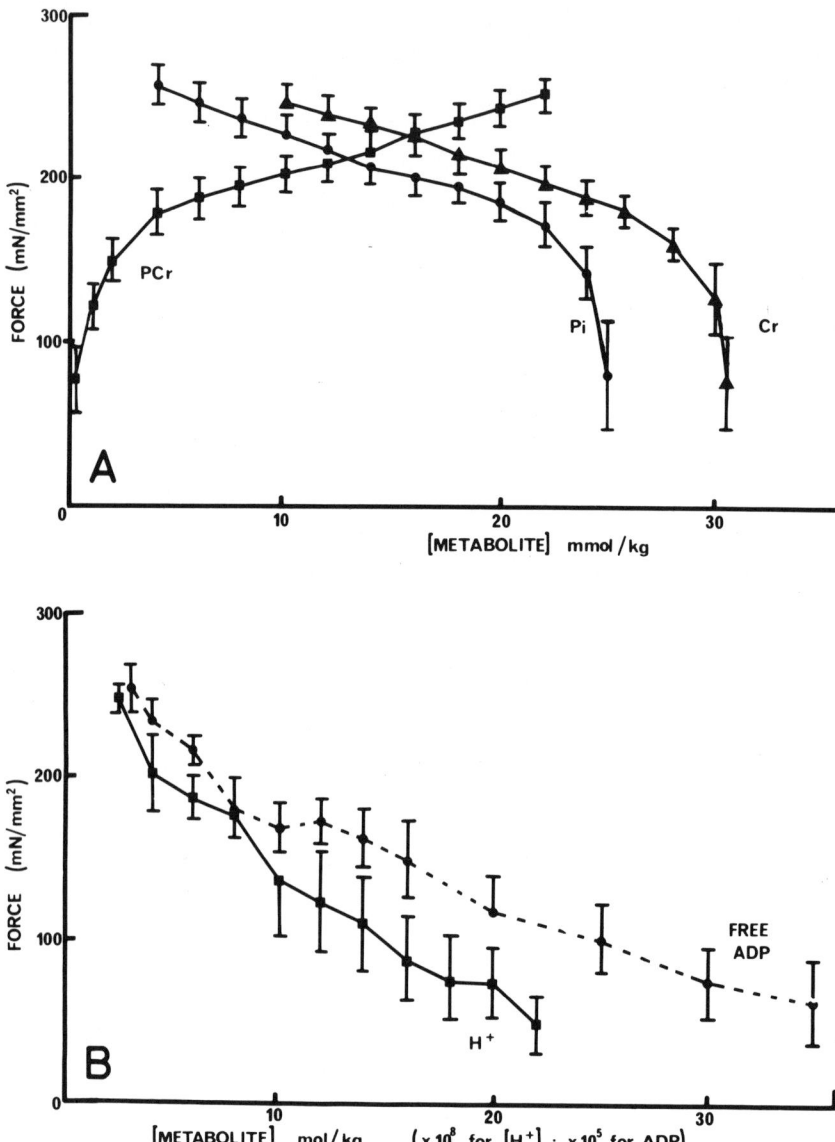

Fig. 5. Force developed as a function of metabolite levels; averaged results of six experiments. [From Dawson et al, 1978, where further details are given.]

The main problem in intepreting the data shown in Figure 4 is that a large number of biochemical changes accompany fatigue, and it is not immediately apparent which of these, if any, is actually responsible for the development of fatigue. In particular, despite lactic acid formation, there is nevertheless a net conversion of high-energy phosphates to low-energy phosphates, and it is difficult on the basis of these experiments to distinguish between the effects these processes have on fatigue. However, it is interesting to note that patients with McArdle syndrome (a defect in glycogen metabolism) generate very little lactic acid, but fatigue rapidly during exercise [Ross et al, 1981]. This suggests that the changing phosphorylation state of the muscles [CIBA Foundation Symposium, 1981, p 82] may be the important factor in the onset of fatigue. In fact, one of the possible mechanisms that would be consistent with the NMR results is that the build up of ADP, inorganic phosphate, and H^+ (ie, the products of ATP hydrolysis) could, by product inhibition, slow down the rate of ATP hydrolysis at the cross bridges and hence cause a reduction in the force development.

Another manifestation of fatigue is the decline in the rate of mechanical relaxation following electrical stimulation, which can clearly be seen in the force records of Figure 4. The free energy change for ATP hydrolysis can be calculated from the NMR measurements (a knowledge of pH_i is required for this calculation), and it was suggested that the decline in free energy of ATP hydrolysis that is observed to accompany fatigue causes a reduction in the rate of Ca^{2+} uptake into the sarcoplasmic reticulum, which in turn causes a slowing down in the rate of mechanical relaxation [Dawson et al, 1980b].

B. Creatine Kinase Activity in Muscle

It is now possible to use NMR to measure the activities of certain enzymes in vivo under steady-state conditions. One of the methods that is employed is a type of "magnetic labeling" experiment known as saturation transfer NMR. This technique has been used to measure the activity of creatine kinase in skeletal muscle and in heart, and the principles of the method are explained in the paper of Gadian et al [1981] describing their studies on frog gastrocnemius muscles.

Creatine kinase catalyses the reaction

$$\text{Phosphocreatine} + \text{ADP} + H^+ \rightleftharpoons \text{ATP} + \text{creatine}$$

and its role in skeletal muscle is to maintain the ATP level constant during contraction. In resting frog gastrocnemius muscles at 4°C, saturation transfer NMR studies have shown that, as anticipated, the creatine kinase reaction is close to equilibrium [Gadian et al 1981]. It follows that this reaction can be used to deduce the concentration of ADP, provided that the other reactants can be measured. Of the other reactants, ATP, phosphocreatine, and H^+ can be estimated directly from the ^{31}P-NMR spectra, and creatine can be deduced by making use of the data that are available from freeze-clamping studies. An estimate of the ADP concentration can therefore be made, and it is of great interest that the value that is obtained is as low as 20 μM in resting muscle. This is the concentration of ADP that is available to interact with creatine kinase, and is therefore the concentration of ADP that is free in solution. The low level of free ADP (it is only a few percent of the total ADP measured by other techniques) could have some important kinetic and thermodynamic implications. For example, several glycolytic reactions involve ADP, either as a substrate or as an allosteric regulator, and this low concentration of free ADP could profoundly affect our understanding of these reactions.

It is also interesting to note that the concentrations of AMP, ADP, and ATP are linked by the adenylate kinase reaction:

$$AMP + ATP \rightleftharpoons 2\ ADP$$

If this reaction is at equilibrium, the known concentrations of free ADP and ATP enable us to calculate the level of free AMP. The calculated value is about 0.1 μm, which is about two orders of magnitude lower than the values normally quoted for total AMP. Again, this is an important observation, for AMP is believed to be a powerful regulator of enzymes such as phosphorylase and phosphofructokinase [see, for example, Newsholme and Start, 1973].

These deductions regarding the ADP and AMP levels rely on a number of measurements available from NMR experiments, among them being the measurement of intracellular pH, for H^+ is directly involved in the creatine kinase reaction. Additional interesting information that is available from these saturation transfer studies is discussed by Gadian et al [1981], Ackerman et al [1980a], Nunnally and Hollis [1979], and Gadian and Radda [1981].

C. Studies of Acidosis and Metabolic Control

There is much interest in the effects of acidosis on heart contractility, the recovery of cardiac function from such acidosis, and the control of excitation by pH effects. ^{31}P-NMR is a particularly suitable technique for examining the role of H^+ ions in ischaemic cardiac tissue.

Figure 6 shows the time course of changes in intracellular pH in total global ischaemia of the rat heart at 37°C. The circles represent control hearts, whereas the squares represent hearts that have been depleted of their glycogen content by 70% [Garlick et al, 1979]. In the control hearts the pH_i declines from 7.05 to about 6.2 after 13 minutes, whereas the pH_i change in the hearts depleted of glycogen is much smaller. It therefore seems that the pH_i change in ischaemia is related to the glycogen content of the heart prior to the ischaemic period. (A similar effect in skeletal muscle was shown by Claude Bernard in 1855; see Needham 1971, p.87,367.)

In further experiments, 2-deoxyglucose was added to the perfusion medium prior to ischaemia [Bailey et al, 1981]. This compound is known to enter muscle tissue, where it is phosphorylated and accumulates within the cytoplasm. During ischaemia, it was found that the pH_i of hearts treated with 2-deoxyglucose decreased more slowly than in the control hearts, reaching an average value of 6.40 after 15 minutes in comparison with 6.13 in the absence of 2-deoxyglucose. This difference can be interpreted in terms of a reduction in the rate of glycogen breakdown during ischaemia. It is interesting that experiments performed in vitro showed that 2-deoxyglucose-6-phosphate specifically inhibits the b form of phosphorylase. Therefore, it was suggested that the reduced glycogen breakdown is caused by inhibition of phosphorylase b by the 2-deoxygluxose 6-phosphate that accumulates within the heart. This implies that the b form of phosphorylase makes a significant contribution to glycogen breakdown in ischaemia; indeed, after about 5 minutes of ischaemia, the contributions of the a and b forms of phosphorylase are similar. This conclusion, which is based primarily on the noninvasive measurement of pH_i probably provides the best estimate currently

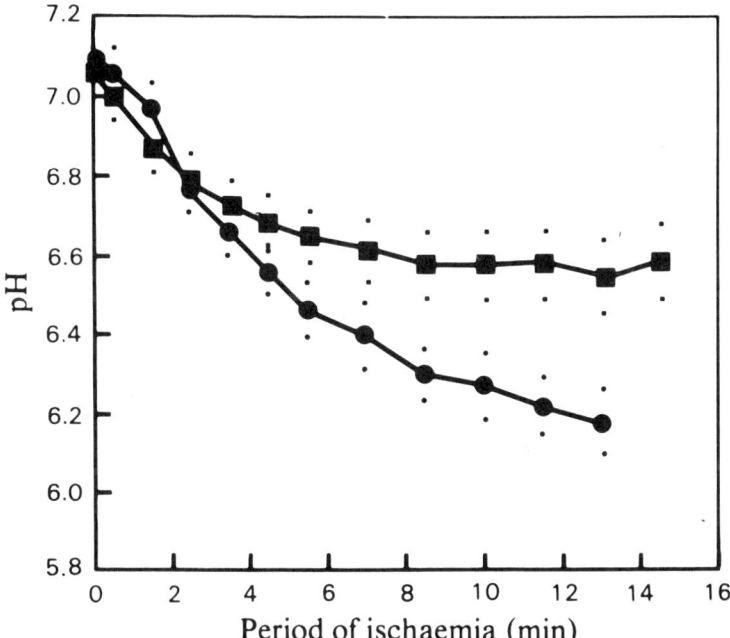

Fig. 6. Time course of changes in intracellular pH in total global ischaemia. Symbols: ● control hearts (seven hearts); ■ glycogen-depleted hearts (six hearts). The small symbols represent SEM values. [From Garlick et al, 1979.]

available of the activity of phosphorylase b during ischaemia, for there are great problems in making these measurements using more conventional extraction procedures.

In interesting contrast to these results on heart muscle, NMR experiments on a human forearm have shown that resting human *skeletal* muscle goes slightly *alkaline* by about 0.1 pH unit when it is made anoxic by cutting off the circulation at least for the first 50 minutes of occlusion [Dawson et al, submitted]. The reason for this is that phosphocreatine is demonstrably hydrolysed and this process *absorbs* protons at around physiological pH; the inorganic phosphate that is formed on hydrolysis of phosphocreatine has a pK_a of 6.75, whereas the corresponding pK_a for phosphocreatine is 4.6. This experimental result shows that in humans, lactic acid production is turned on only weakly by anoxia, so that its effect on pH_i is more than neutralized by the hydrolysis of phosphocreatine. Following active contraction it is quite another story [see Bore et al, this volume, for studies of human exercise]; the pH_i falls rapidly, sometimes to levels below 6.1, showing that whatever switches on glycolysis is intimately associated with contraction itself; it is not sufficient that there be elevated levels of substances such as ADP and AMP. We had come to the same conclusion in experiments on isolated frog muscle [Dawson et al, 1980b]. These results on resting skeletal muscle in no way conflict with those on heart muscle, which is never at rest.

Another important aspect of the 2-deoxyglucose experiments is that 2-deoxyglucose-6-phosphate provides a marker for the myocardial cytoplasm. This compound has a pK_a of about 6.15, and it can therefore be used to estimate the cytoplasmic pH. It was found that the cytoplasmic pH measured from the 2-deoxyglucose-6-phosphate signal was identical to that derived from the inorganic phosphate signal under both normoxic and ischaemic conditions. It can be concluded that the inorganic phosphate signal essentially reflects the cytoplasmic pH. Moreover, the similarity of the pH values deduced from the two signals suggests that effects such as binding to membranes and proteins have little effect on the chemical shift of the inorganic phosphate signal.

Additional studies have been performed to examine the protective effects of perfusing hearts with a medium of high buffering capacity prior to ischaemia [Garlick et al, 1979]. Hearts were perfused for 12 minutes before ischaemia with a Krebs-Henseleit buffer supplemented with 100 mM HEPES. As shown in Figure 7, both the rate and extent of cellular acidification were significantly reduced in comparison with the controls, and the final pH_i was 6.8. It was also shown that HEPES had a significant protective effect on the ischaemic heart. After 14 minutes of ischaemia, spectra of hearts perfused with Krebs-Henseleit buffer contained signals only from inorganic phosphate and sugar phosphate, whereas both phosphocreatine and ATP could be observed in hearts perfused with the buffer supplemented with HEPES. Reperfusion of the hearts for 5 minutes with the original buffers after 30 minutes of total ischaemia resulted in complete recovery of phosphocreatine and ATP levels in hearts perfused with Krebs-plus-HEPES solution, but resulted in no recovery in the controls.

V. CONCLUSIONS

The intracellular pH of tissues and organs can be measured using ^{31}P-NMR. The measurements are totally noninvasive, and therefore the pH, together with other metabolic parameters and physiological function, can be monitored as a function of time within a single experiment. This makes it possible to make thorough investigations of the role of pH_i in the control and function of cellular processes in vivo.

Furthermore, recent technical developments [Ackerman et al, 1980b; Gordon et al, 1980] now make it possible to study the metabolic state of selected tissues and organs within whole animals, and with the development of wide bore superconducting magnets, these studies can be further extended in order to investigate human metabolism [Cresshull et al, 1981a, b; Ross et al, 1981; Dawson et al, submitted; see also Bore et al, this volume].

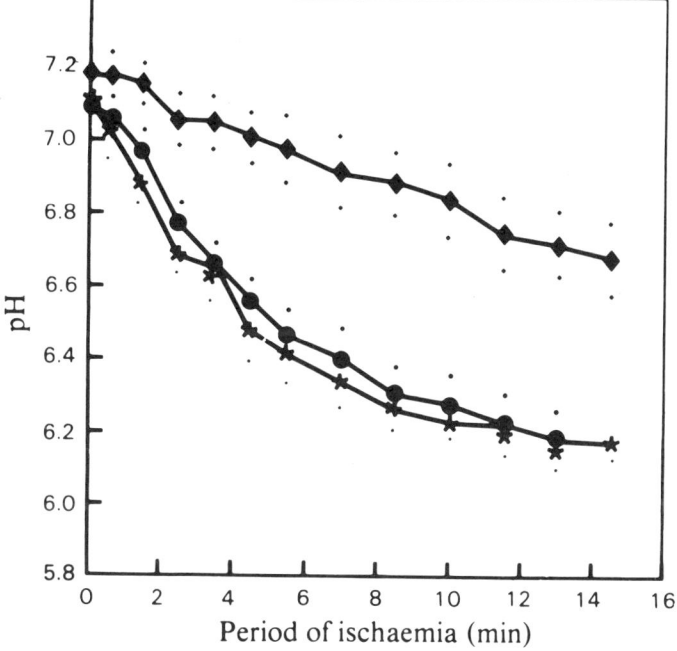

Fig. 7. Time course of changes in intracellular pH in total global ischaemia. Symbols: ● control hearts (seven hearts); ★ hearts preperfused with Krebs-Henseleit buffer + 50 mM mannitol (three hearts); ◆ hearts preperfused with Krebs-Henseleit buffer + 100 mM HEPES (seven hearts). The small symbols represent SEM values. [From Garlick et al, 1979.]

ACKNOWLEDGMENTS

This work is supported by the Science Research Council, the British Heart Foundation, and the U.S. National Institutes of Health (grant 18708-02S1).

VI. REFERENCES

Ackerman JJH, Bore PJ, Gadian DG, Grove TH, Radda GK: Studies of metabolism in perfused organs by NMR. Phil Trans R Soc Lond B 289:425, 1980a.

Ackerman JJH, Grove TH, Wong GG, Gadian DG, Radda GK: Mapping of metabolites in whole animals by ^{31}P NMR using surface coils. Nature 283:167, 1980b.

Ackerman JJH, Gadian DG, Radda GK, Wong GG: Observation of ^{1}H NMR signals with receiver coils tuned for other nuclides. The optimization of B_o homogeneity and a multinuclear chemical shift reference. J Mag Reson 42:498, 1981.

Bailey IA, Williams SR, Radda GK, Gadian DG: The activity of phosphorylase in total global ischaemia in the rat heart: A ^{31}P NMR study. Biochem J 196:171, 1981.

Bore PJ, Chan L, Gadian DG, Radda GK, Ross BD, Styles P, Taylor DJ: Non-invasive pH measurements of human tissue using ^{31}P NMR. This volume.

Brown FF, Campbell ID: Cross correlation of titrating histidines in oxy- and deoxyhaemoglobin; an NMR study. FEBS Lett 65:322, 1976.

Brown FF, Campbell ID, Kuchel PW, Rabenstein DC: Human erythrocyte metabolism studies by ^{1}H spin echo NMR. FEBS Lett 82:12, 1977.

Burt CT, Cohen SM, Barany M: Analysis of intact tissue with ^{31}P NMR. Ann Rev Biophys Bioeng 8:1, 1979.

Busby SJW, Gadian DG, Radda GK, Richards RE, Seeley PJ: Phosphorus nuclear magnetic resonance studies of compartmentation in muscle. Biochem J 170:103, 1978.

CIBA Foundation symposium 82. "Human Muscle Fatigue: Physiological Mechanisms." London: Pitman Medical, 1981.

Cohn M, Hughes TR: Nuclear magnetic resonance spectra of adenosine di- and tri-phosphate. II. Effect of complexing with divalent metal ions. J Biol Chem 237:176, 1962.

Cresshull ID, Gordon RE, Hanley P, Shaw D, Gadian DG, Radda GK, Styles P: ^{31}P NMR studies of human forearm muscle. Bull Magn Reson 2:426, 1981a.

Cresshull I, Dawson MJ, Edwards RHT, Gadian DG, Gordon RE, Radda GK, Shaw D, Wilkie DR: Human muscle analysed by ^{31}P nuclear magnetic resonance in intact subjects. J Physiol 317:18P, 1981b.

Dawson MJ, Edwards RHT, Gordon RE, Shaw D, Wilkie DR: ^{31}P NMR studies of muscle in intact human subjects. Submitted to Science.

Dawson MJ, Gadian DG, Wilkie DR: Contraction and recovery of living muscles studied by ^{31}P nuclear magnetic resonance. J Physiol 267:703, 1977.

Dawson MJ, Gadian DG, Wilkie DR: Muscular fatigue investigated by phosphorus nuclear magnetic resonance. Nature 274:861, 1978.

Dawson MJ, Gadian DG, Wilkie DR: Mechanical relaxation rate and metabolism studied in fatiguing muscle by phosphorus nuclear magnetic resonance (^{31}P NMR). J Physiol 299:465, 1980a.

Dawson MJ, Gadian DG, Wilkie DR: Studies of the biochemistry of contracting and relaxing muscle by the use of ^{31}P NMR in conjunctions with other techniques. Phil Trans R Soc Lond B 289:445, 1980b.

Fossel ET, Morgan HE, Ingwall JS: Measurement of changes in high-energy phosphates in the cardiac cycles by using gated ^{31}P nuclear magnetic resonance. Proc Natl Acad Sci USA 77:3654, 1980.

Gadian DG: Metabolic studies of whole animals and humans using phosphorus nuclear magnetic resonance. Biosci Rep 1:449, 1981.

Gadian DG: Nuclear magnetic resonance and its applications to living systems. Oxford: Oxford University Press, 1982.

Gadian DG, Radda GK: NMR studies of tissue metabolism. Ann Rev Biochem 50:69, 1981.

Gadian DG, Radda GK, Brown TR, Chance EM, Dawson MJ, Wilkie DR: The activity of creatine kinase in frog skeletal muscle studied by saturation-transfer nuclear magnetic resonance. Biochem J 194:215, 1981.

Gadian DG, Radda GK, Richards RE, Seeley PJ: ^{31}P NMR in living tissue. In Shulman RG (ed): "Biological Applications of Magnetic Resonance." New York: Academic Press, 1979, p 463.

Garlick PB, Radda GK, Seeley PJ: Studies of acidosis in the ischaemic heart by phosphorus nuclear magnetic resonance. Biochem J 184:547, 1979.

Gordon RE, Hanley P, Shaw D, Gadian DG, Radda GK, Styles P, Bore PJ, Chan L: Localization of metabolites in animals using ^{31}P topical magnetic resonance. Nature 287:736, 1980.

Griffiths JR, Iles RA: Nuclear magnetic resonance—A 'magnetic eye' on metabolism. Clin Sci 59:225, 1980.

Grove TH, Ackerman JJH, Radda GK, Bore PJ: Analysis of rat heart in vivo by phosphorus nuclear magnetic resonance. Proc Natl Acad Sci USA 77:299, 1980.

Hollis DP: Nuclear magnetic resonance studies of cancer and heart disease. Bull Magn Reson 1:27, 1979.

Hoult DI, Busby SJW, Gadian DG, Radda GK, Richards RE, Seeley PJ: Observations of tissue metabolites using ^{31}P nuclear magnetic resonance. Nature 252:285, 1974.

Illingworth JA: A common source of error in pH measurements. Biochem J 195:259, 1981.

Jacobson L, Cohen JS: Improved technique for investigation of cell metabolism by ^{31}P NMR spectroscopy. Biosci Rep 1:141, 1981.

Lehman CF: Lehrbuch der physiologischen Chemie 1850; and page 98, Ed 2, translated by Day GE. London: Cavendish Society, 1851.

Moon RB, Richards JH: Determination of intracellular pH by ^{31}P magnetic resonance. J Biol Chem 248:7276, 1973.

Needham DM: "Machina Carnis." Cambridge: Cambridge University Press, 1971.

Newsholme EA, Start C: "Regulation in Metabolism." London: John Wiley and Sons, 1973.

Nuccitelli R, Webb DJ, Lagier ST, Matson GB: ^{31}P NMR reveals increased intracellular pH after fertilization in Xenopus eggs. Proc Natl Acad Sci USA 78:4421, 1981.

Nunnally RL, Hollis DP: Adenosine triphosphate compartmentation in living hearts: A phosphorus nuclear magnetic resonance saturation transfer study. Biochemistry 18:3642, 1979.

Ross BD, Radda GK, Gadian DG, Rocker G, Esiri M, Falconer-Smith J: Examination of a case of suspected McArdle's syndrome by ^{31}P nuclear magnetic resonance. New Engl J Med 304:1338, 1981.

Seeley PJ, Busby SJW, Gadian DG, Radda GK, Richards RE: A new approach to compartmentation in muscle. Biochem Soc Trans 4:62, 1976.

Intracellular pH Measured by NMR: Methods and Results

R.J. Gillies, J.R. Alger, J.A. den Hollander, and R.G. Shulman

Department of Molecular Biophysics and Biochemistry, Box 6666, Yale University, New Haven, Connecticut 06511

I.	Background		79
	A.	Origin of Chemical Shifts and Fast Exchange	79
	B.	Indicators of pH_i	82
		1. Inorganic phosphate; 2. Methyl phosphonate; 3. Other indicators	82
II.	Methods		85
	A.	General Considerations	85
	B.	Signal-to-Noise	86
	C.	High Density Cultures	86
	D.	Mammalian Cells	89
III.	Results and Discussion		94
	A.	Bacteria	94
	B.	Compartmentation	95
	C.	Time Resolution	97
	D.	pH_i and the Control of Glycolysis in Yeast	100
IV.	References		103

I. BACKGROUND

A. Origin of Chemical Shifts and Fast Exchange

Most of this chapter discusses the application of ^{31}P-NMR spectroscopy to the measurement of intracellular pH in living systems. ^{31}P resonance frequencies are sensitive to the nature of the molecule of which the phosphorous nucleus is a part. The entire spectrum of ^{31}P chemical shifts is about 700 parts per million (ppm) in width (Fig. 1). However, physiological phosphorous-containing metabolites are usually phosphates, which resonate in the region from 10 to -30 ppm relative to phosphoric acid. An example of a ^{31}P spectrum of glycolyzing yeast is shown in Figure 2. For comparison, the spectrum of a perchloric acid

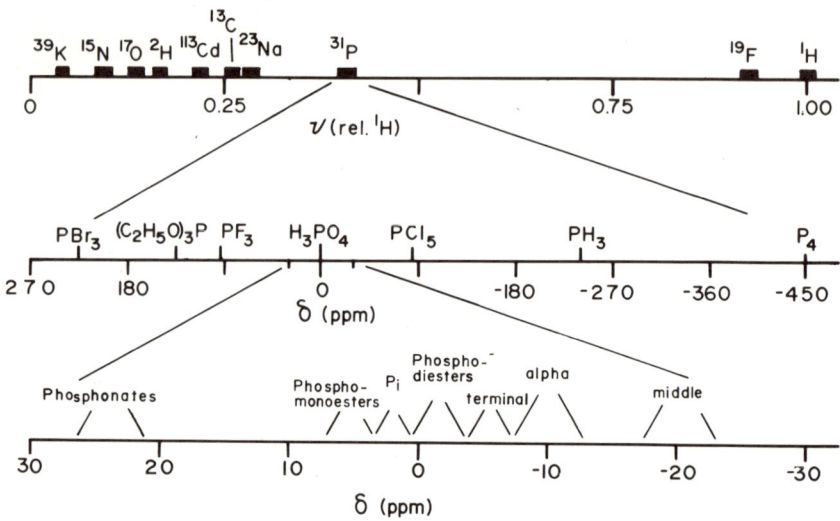

Fig. 1. Frequency spectra of NMR sensitive radio frequency region, usually from 10 to 400 MHz. Upper spectrum indicates resonance frequencies of selected nuclei expressed relative to proton resonance frequency. Middle spectrum indicates resonance frequencies of selected phosphorous-containing compounds, expressed as ppm relative to 80% phosphoric acid. Lower spectrum illustrates resonance frequency ranges of phosphorous-containing groups of biological interest. P_i = inorganic orthophosphate; terminal = end phosphate of phosphate chain (n > 2); alpha = phosphodiester where R_2 = phosphate group; middle phosphate of phosphate chains (n ≥ 3).

extract of the cells is also shown in the figure. This extract spectrum can be used to assign peaks to metabolites by standard chemical methods such as adding the suspected compound, by titrations, or by studying the effects of specific enzymatic action. Specific ^{31}P resonances have been assigned to highly concentrated intermediates such as nucleoside tri- and diphosphates, inorganic phosphate, phosphomonoesters of sugars, NAD^+, and polyphosphates. Notice that the cellular spectrum shows three distinct orthophosphate (P_i) resonances, whereas the extract spectrum shows only one. P_i resonance frequencies are pH-sensitive, and each of the cellular P_i resonances corresponds to a compartment having a different pH.

pH measurements are possible because H_3PO_4, $H_2PO_4^-$, HPO_4^{2-}, and PO_4^{3-} all have unique electronic structures and therefore unique chemical shifts. In the range of physiological pH, $H_2PO_4^-$ (δ = 0.58 ppm) and HPO_4^{2-} (δ = 3.14 ppm) are the important chemical species. Any particular ^{31}P nucleus exchanges between $H_2PO_4^-$ and HPO_4^{2-} with a rate much faster than the frequency difference between the two resonances. In this situation the exchange is said to be rapid on an NMR time scale, and a single resonance is observed at a position between 0.58 and

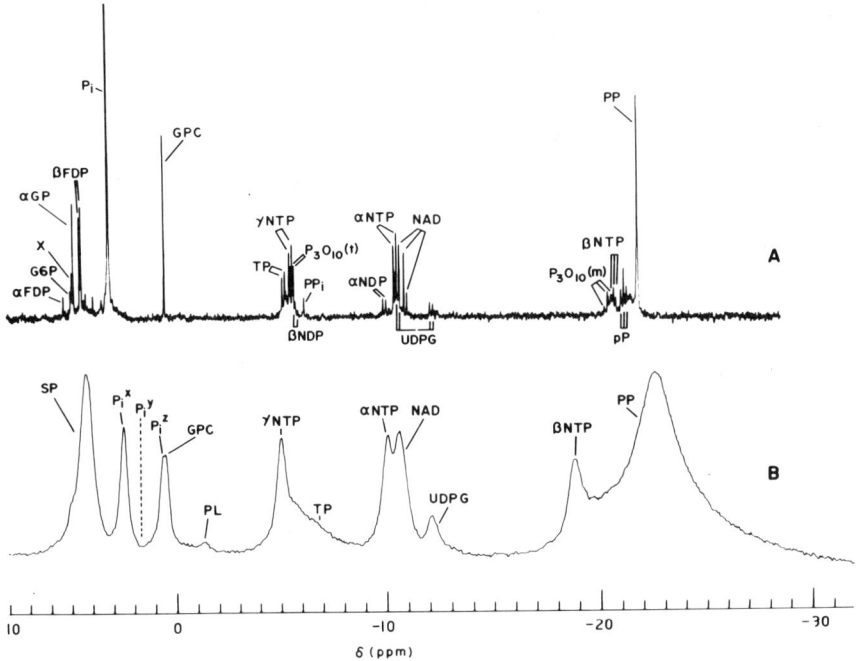

Fig. 2 Perchloric acid extract and in vivo 145.78 MHz ^{31}P spectra of S. cerevisiae (NCYC 239). A = 100 min accumulation of extract representing sum of 3000 FIDs arising from 50° tipping pulses with a rep rate of 2.0 secs. B = 20 min accumulation of 50% anaerobic cell suspension in the absence of external phosphate. Accumulation is sum of 2,000 FIDs arising from 68° tipping pulses with a rep rate of 0.6 secs. *Abbreviations*: αGP = glycerol phosphate; G6P = glucose-6 = phosphate; α,β FDP = anomers of fructose-1,6-bisphosphate; X = tentatively identified as trehalose phosphate; GPC = glycerol phosphoryl choline (endogenous) used as chemical shift reference at 0.49 ppm; TP = terminal phosphate of oligo- and polyphosphate; P_3O_{10} (t,m) = terminal and middle phosphate of triphosphate; α,β,γ NTP/NDP = phosphates of nucleoside tri- and diphosphate; PP$_i$ = pyrophosphate; UDPG = uridine diphosphoryl glucose; PP = middle phosphates of oligo- and polyphosphate; SP = sugar phosphates; PL = phospholipids; $P_i^{x,y,z}$ = compartmentalized orthophosphate at different pHs. Note that dotted line of P_i^y represents position of external phosphate peak which was depleted prior to accumulation of spectrum.

3.14 ppm, which is determined by the weighted average of the two species. The relative amounts of the species depend upon the pH, so the observed chemical shift can be used as a measure of pH after suitable standardization.

In addition to P_i, several other ^{31}P resonances have been used to measure intracellular pH. These other pH indicators are now discussed.

B. Indicators of pH_i

1. Inorganic phosphate. The ^{31}P resonance most often used to measure intracellular pH is from inorganic orthophosphate (P_i^{in}). Because of its pK_a of 6.9 (I = 200 mM), the chemical shift of P_i^{in} is very sensitive to pH changes in the range of pH = 6 to pH = 8. Because P_i^{in} is an endogenous compound, this technique can be called truly noninvasive. However, since it is an endogenous compound, its concentration can vary [den Hollander et al, 1981] and care must be taken to insure that it is not below the limit of detection. In addition, because the line widths of cellular peaks are on the order of 30–50 Hz at 145 MHz, differences in pH between internal and external compartments of less than 0.2 pH units are difficult to resolve.

As mentioned earlier, the chemical shift of P_i^{in} must be compared with suitable standard curves run under conditions which approximate the intracellular environment, since the pK_a of P_i is affected by ionic strength (I). This is shown in Figure 3, where the chemical shift of P_i is shown as a function of pH at different ionic strengths. Notice in the figure that the end points of the titrations are not changed by ionic strength, being 0.58 for $H_2PO_4^-$ and 3.14 for HPO_4^{2-}. Given the end points and pK_as, then, a suitable titration curve can be generated at any ionic strength between 50 mM and 500 mM.

2. Methyl phosphonate. The signal from P_i does not provide optimum resolution at the high pHs sometimes observed under physiological conditions. For experiments in which it is wished to resolve somewhat alkaline values, the ^{31}P signal from methylphosphonate (MP) has been used [Slonczewski et al, 1981]. At physiological ionic strength MP has a pK_a of 7.65, which is rather insensitive to changes in ionic strength (Fig. 4). Its chemcial shift is quite different from those of naturally occuring phosphate metabolites. MP titrates from around 25 ppm in the monobasic form to about 21 ppm in the dibasic form. Note that the direction of titration is opposte from that of P_i and that the titration range is appreciably larger. In bacterial studies it was found that MP is slowly accumulated by aerobic E. coli cells grown on glycerol as a carbon source, but not by glucose or succinate-grown cells. Transport of MP into other types of cells remains to be investigated. Even when the cell does not transport MP it could still serve as a good indicator of external pH and allow the use of a low phosphate buffer. Furthermore, MP is innocuous to cellular metabolism as shown by the fact that MP at concentrations observable by NMR (5 mM) does not affect cell-doubling times. An example of the use of MP in bacterial cytoplasmic pH determination is discussed in section III A.

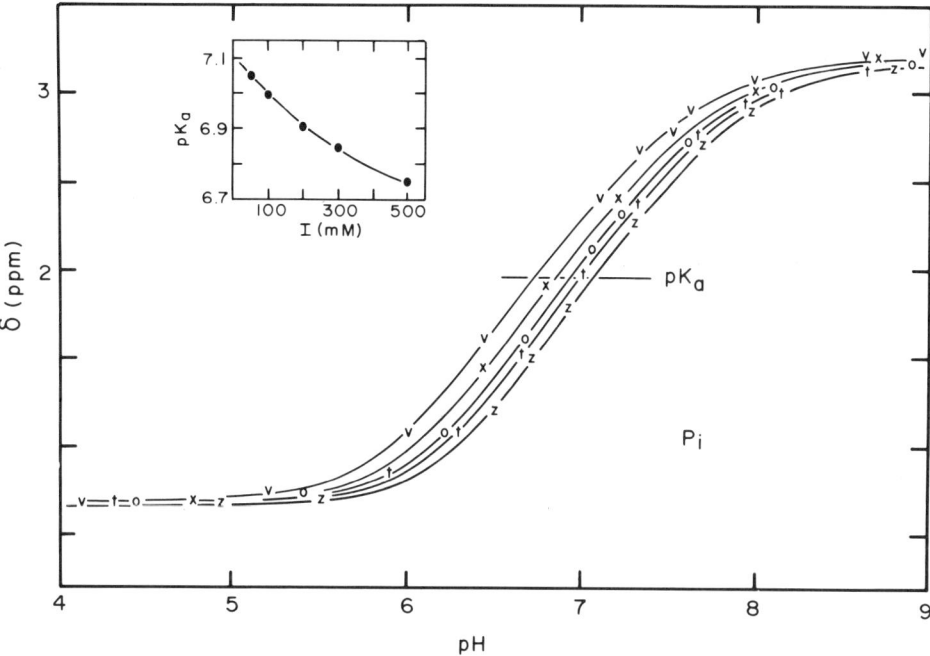

Fig. 3. Chemical shift of orthophosphate as a function of pH at different ionic strengths. One milliliter of buffer containing 5 mg GPC, 100 mM Na_2HPO_4, 100 mM $NaCH_3PO_3$, 100 mM 2-deoxyglucose-6-phosphate was combined with various amounts of 50 mM HEPES in 1 M KCl to give final ionic strengths of 50 (z), 100 (+), 200 (o), 300 (x) and 500 (v) mM in a volume of 10 ml. Solutions were titrated with (0.1, 0.2, 0.4, 0.6, and 1.0 N) HCl, and chemical shifts were determined by ^{31}P-NMR at 145.78 MHz; shifts were referenced to GPC at 0.49 ppm relative to phosphoric acid. pK_a was determined as pH at 1.84 ppm, which is halfway between end points of titrations. Inset shows effect of ionic strength (I) on pK_a. For comparison see Roberts et al [1980].

3. Other indicators. As mentioned earlier, most phosphate shifts are pH-dependent. However, only the sugar phosphates (phosphomonoesters) have pKs near physiological pH. The titration curves of many phosphate compounds are presented in an earlier paper by Navon et al [1979]. Using endogenous resonances of sugarphosphates has the advantages of being noninvasive and being more concentrated in the cytoplasm, but has the disadvantages of variable signal and relative low sensitivity. Navon et al [1977] showed that 2-deoxyglucose (2dGlu) was taken up by Ehrlich ascites tumor cells, phosphorylated to 2-deoxyglucose-6-phosphate (2dG6P), and gave a pH-dependent NMR peak that they suggested could be used to monitor pH. This has actively been done by Bailey et al [1981],

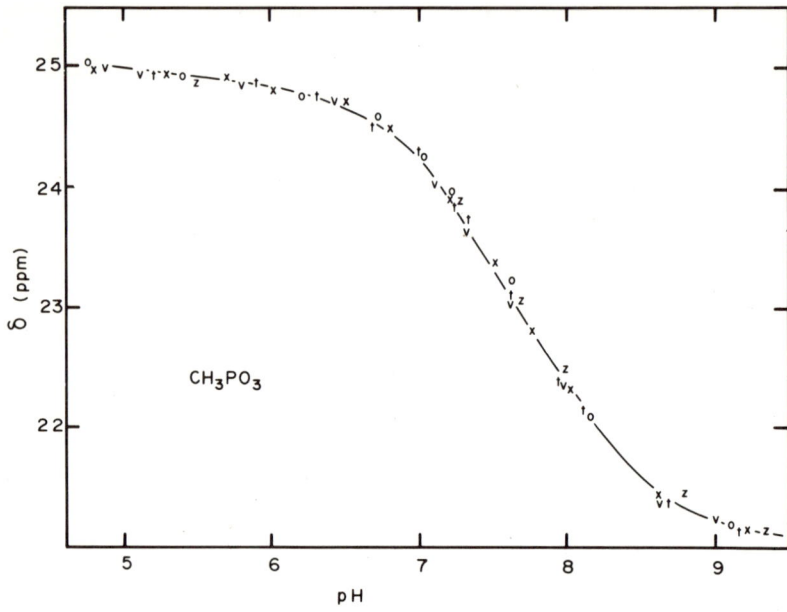

Fig. 4. Chemical shift of methyl phosphonate as a function of pH at various ionic strengths. Conditions were identical to those described for Figure 3. Note that there is little, if any, effect of ionic strength on chemical shifts.

using rat heart. In our studies using Ehrlich ascites cells, we have also estimated intracellular pH from the chemical shift of 2-deoxyglucose-6-phosphate (R. Gillies, T. Ogino, and R. Shulman, unpublished observations). The titration curves for 2dG6P are illustrated in Figure 5, and show that although the pK_a is lower, the chemical shift of 2dG6P is more sensitive to changes in pH than is that of P_i. The disadvantage of this technique is that deoxyglucose is a poison and possibly affects the pH and ATP levels of the cells.

Navon et al [1979] have shown that G6P and FBP give well-resolved peaks in the ^{31}P NMR spectrum of glycolyzing yeast cells. The positions of these peaks were used to determine the cytoplasmic pH, which was used to assign one of the three P_i peaks to the cytoplasm, by showing that the pHs determined from all three peak positions agreed to within experimental error. In a similar determination of cytosolic pH, fructose added to suspensions of liver cells was phosphorylated to yield an intense fructose-1-phosphate peak. The pH determined from this peak was used to determine which of the two P_i^{in} peaks was from the cytosolic compartment, the remaining being assigned to the mitochondria [Iles et al 1980].

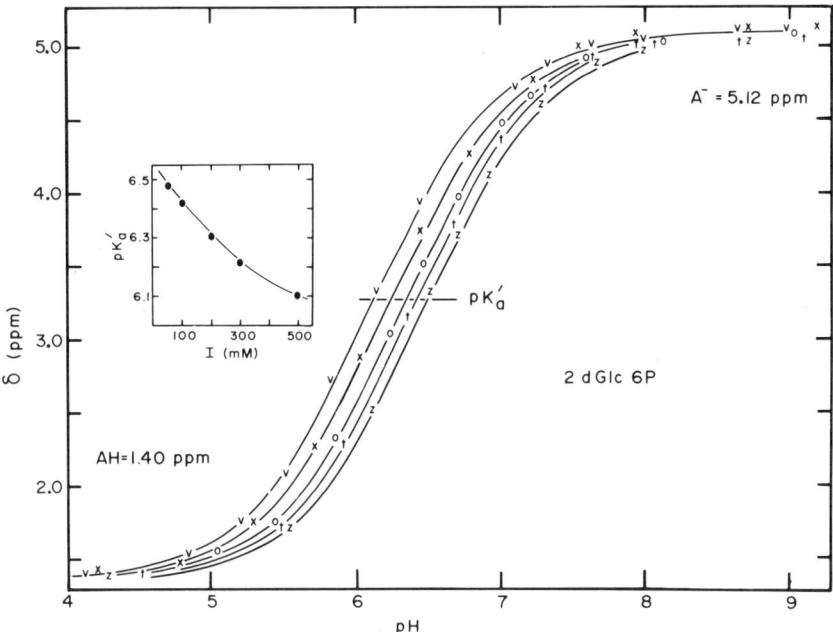

Fig. 5. Chemical shift of 2-deoxyglucose-6-phosphate as a function of pH at different ionic strength. Conditions were identical to those described in Figure 3. pK_a was determined as pH at chemical shift of 3.26 ppm.

II. METHODS

A. General Considerations

There are several important considerations that arise with NMR studies of intact cells. The high density cultures required by the technique present difficulties in aeration, feeding, and removal of metabolic wastes. In this section we discuss our approaches to these problems.

Preparing suspensions of microorganisms for ^{31}P-NMR measurement is now a well-established procedure. Cells are grown to mid-log or stationary phase in liquid culture. They are then cooled to below 10°C and pelleted by centrifugation for several minutes. The dense suspensions are then washed twice with ice-cold buffer, usually composed of zwitterionic (and therefore impermeant) synthetic buffer ions chosen to provide suitable buffering capacities and pHs for the particular experiment. The density of the final suspension used for NMR analysis depends on the signal-to-noise ratio (S/N) and time resolution requirements of the particular experiment. The sample temperature is maintained at the desired value during NMR data acquisition by the variable temperature apparatuses supplied with the NMR spectrometer.

B. Signal-to-Noise

One problem in using NMR is the low S/N generally observed. Operationally, a S/N of ~ 5 is necessary to measure the presence of a peak accurately, and it should be higher in order to assign chemical shifts reproducibly.

In recent years several approaches have been taken to increase S/N. Large static magnetic fields from superconducting solenoids are used to obtain more favorable Bolzmann distributions. Large sample volumes (tubes up to 20 mm in diameter) and highly concentrated solutions are also used to increase the number of nuclei in the NMR sensitive region. Solenoid radiofrequency coils which give a 2–3-fold improvement in S/N are being used for cellular studies. Signal averaging over time is then used to obtain the S/N to suit the needs of the experiment. S/N increases as the square root of accumulation time; hence a factor of two reduction in sample concentration requires a 4-fold increase in the accumulation time to achieve the same S/N. Below a certain metabolite concentration the accumulation times become prohibitively long.

C. High Density Cultures

The above signal-to-noise considerations must be balanced against the maximum allowable concentration for each given system. The following paragraphs will discuss our investigations into the application of NMR to yeast, and illustrate how we arrived at conditions required for the NMR sensitivity that are also physiologically acceptable for the cells.

As shown in Figure 2, yeast is very rich in phosphate metabolites. P_i is present intracellularly at 1 mM to 30 mM depending on conditions. The division time of a yeast cell is 2.5 hours. To examine dynamic aspects of metabolism it is desirable to have a time resolution 2 orders of magnitude smaller than this. To achieve this time resolution we must be able to identify several mM intracellular P_i in 1 or 2 minutes of acquisition time. Since yeast can easily be grown in large quantities, we are able to obtain enough sample to use the largest tube available (20 mm diameter). Using optimal pulsing conditions [Becker et al, 1979], the Bruker WH-360 wide bore NMR spectrometer can get a good S/N ratio from 5 mM P_i standard solution in one minute. However, since the area (and not the height) of the peak is proportional to the concentration, any increase in line width such as is observed in cells results in a decrease in peak height, which adversely affects sensitivity. Line widths of cellular spectra are typically an order of magnitude larger than buffer spectra. Because of this, we have empirically determined that the minimum concentration of cells necessary in order to observe 5 mM intracellular P_i in 1 minute is approximately 20%. This results in a S/N in the neighborhood of 20:1.

Can this concentration of cells be adequately oxygenated? Proper oxygenation of samples has been determined empirically. Through much trial and error, we have settled on a "double bubble" aeration manifold [Ugurbil et al, in press; den Hollander et al, in press], shown in Figure 6. As indicated, the apparatus consists of lower bubblers of quartz glass drawn out to a diameter of approximately 100 μm and upper bubblers drawn out to approximately 0.5 mm. At a flow rate of 10–20 ml min^{-1}, the lower bubblers keep the cells in suspension but are not the main source of oxygen. The presence of bubbles within the sample does not appreciably increase the line widths of the NMR signals of the cell suspensions. The upper bubblers are situated above the coil, and a flow rate greater than 200 ml min^{-1} will oxygenate the suspension, provided there is an adequate (1–2 cm) column of medium above the bubblers themselves.

Figure 7 illustrates the oxygen tension of yeast suspensions as determined in an O_2-electrode chamber designed to mimic NMR conditions. As shown in the figure, yeast suspensions (20% v/v) can be adequately oxygenated at 25°C. Under these conditions, the O_2 consumption rate (V_{max}) is 0.88 μmoles min^{-1} (g wet cells)$^{-1}$ in the absence of glucose. In the presence of glucose the V_{max} for O_2 consumption is greater than 4.4 μmoles min^{-1} (g wet cells)$^{-1}$. At this point it should be noted that these measurements depend upon temperature. Higher temperatures decrease O_2 solubility and increase cellular metabolism (hence, O_2 demand). These effects are cumulative and make oxygenation of cells more difficult at higher temperatures.

Glucose consumption rates should also be determined empirically for each system to be used. It can be determined directly by feeding the suspension ^{13}C-labeled glucose and following consumption in vivo by ^{13}C-NMR. This is relatively straightforward, and the glucose consumption rate has been determined anaerobically at 20°C to be 16 μmoles min^{-1} (g wet cells)$^{-1}$ (glucose-grown S. cerevisiae strain BG9A [den Hollander et al, 1979].

Alternatively, the glucose consumption can also be determined using ^{31}P-NMR. Immediately after glucose addition the level of sugar phosphates increases, whereas the level of P_i decreases. At the moment of glucose exhaustion the level of sugar phosphates starts to decrease, whereas P_i increases. These changes can be monitored by ^{31}P-NMR. From these observations it is possible to determine the time in which a certain amount of added glucose is used up, hence the glucose consumption rate. Using this assay, glucose consumption rate has been determined to be 20 μmoles min^{-1} (g wet cells)$^{-1}$ at 25°C (S. cerevisiae, strain NCYC 239). From this number, it can be determined that 10 ml of a 10% suspension of yeast cells will consume 1.2 mmoles of glucose in 60 minutes. This amount is easily deliverable through the infusion tube in the aeration manifold.

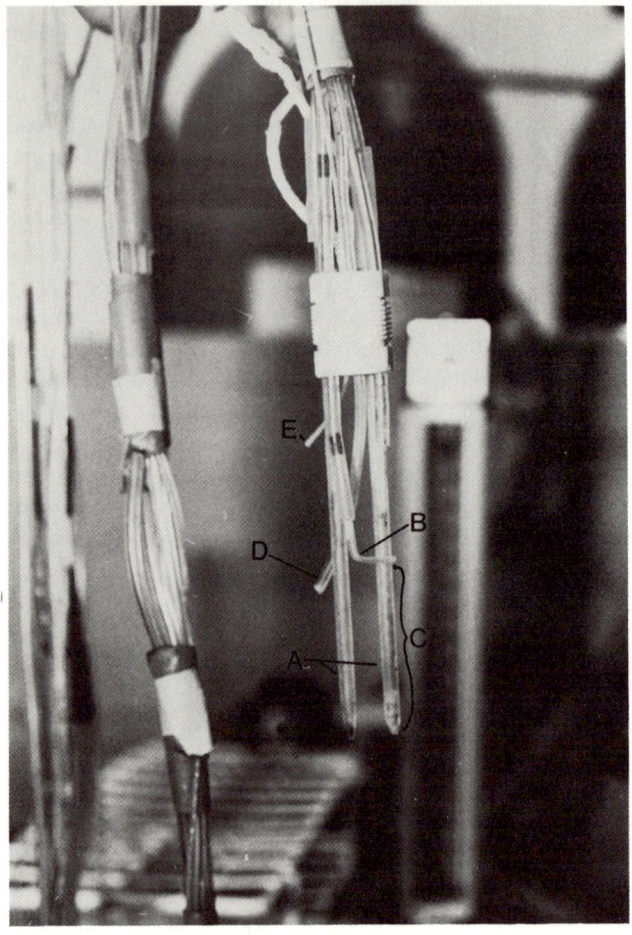

Fig. 6. "Double bubble" aeration manifold designed for use with cellular suspensions in a 20 mm NMR sample tube. Manifold is constructed out of a 20 mm Teflon plug fitted with six holes of 1–3 mm diameter. Through two of these holes pass the lower bubblers (A) of 3 mm quartz glass drawn out to approximately 100 μm. The gas mixture is passed through these at a rate of 10–20 ml min^{-1} and serves to maintain the sample in suspension. The upper bubblers (B) are constructed out of 0.5 mm borosilicate glass fitted in to 3 mm tygon tubing. The gas mixture is passed through these at a flow rate greater than 200 ml min^{-1} and serves to aerate the suspension. Care must be taken to ensure that the coil volume (C) rests between upper and lower bubbler termini. The remaining two holes in the manifold are used for large (D) and small (E) diameter infusion tubes through which compounds may be added or samples taken.

The buildup of toxic metabolic byproducts is a more serious and less identifiable problem. It is known that buildup of toxic wastes by microorganisms in culture will inhibit cell growth (Gillies, this volume). The end products of glucose metabolism are ethanol, CO_2, glycerol, and other by-products. Because of its volatility, CO_2 is readily removed from the suspension by rapid bubbling. Ethanol is not toxic to growth in yeast until it reaches concentrations above 5% [Rose and Harrison, 1971], and it therefore rarely affects the physiology of the cells under our experimental condition. For most other systems, however, the glycolytic end product is lactate, which is much more toxic to the cells than ethanol. Control of lactate buildup can be achieved through perfusion. Glycerol is a minor and relatively innocuous byproduct.

Taking all of the above into account, it would appear that NMR can be used to study yeast in conditions in which it is well energized and maintained over a prolonged period of time. A major test of this was to determine if yeast could actually grow under the conditions described above. This is illustrated in Figure 8 for a 10% (v/v) synchronous suspension of NCYC 239. As shown, the cells not only bud and replicate their DNA, but also do so *synchronously*. This confirms that yeast is able to be maintained at these high densities under conditions adequate to promote and sustain cell growth.

D. Mammalian Cells

The application of NMR to cultured mammalian cells is of great interest, and many attempts have been made to place them physiologically at NMR densities. Mammalian cells generally fall into two distinct categories: those grown in suspension and those grown in monolayer, each of which requires different procedures. Suspension cells that have been studied include lymphocytes, platelets, erythrocytes, and certain tumor cells, notably erythroleukemia cells, Ehrlich ascites tumor cells, and HeLa cells. Generally speaking, these cells have been treated in a manner similar to bacteria or yeast—ie, collection by centrifugation, resuspension, and oxygenation via bubbling. This last point bears some mention, as mammalian cells are very sensitive and can be lysed by vigorous bubbling. However, because of their slower metabolism, mammalian cells also have lower oxygen consumption rates. We have determined that a bubbling rate greater than 20 ml min^{-1} will induce cell lysis in Ehrlich ascites cells (R.J. Gillies, T. Ogino, and R.G. Shulman, unpublished observations). Ehrlich ascites tumor cells bubbled with 95% O_2: 5% CO_2 at 15ml min^{-1} give well-resolved ^{31}P spectra, as indicated in Figure 9a. Note that the S/N for ATP is high and that there is no split in the P_i peaks, indicating that either there is no pH gradient across the plasma membrane or that internal P_i is very low. Hence, to determine definitively the intracellular pH of Ehrlich ascites cells, an alternative compound to inorganic phosphate must be used.

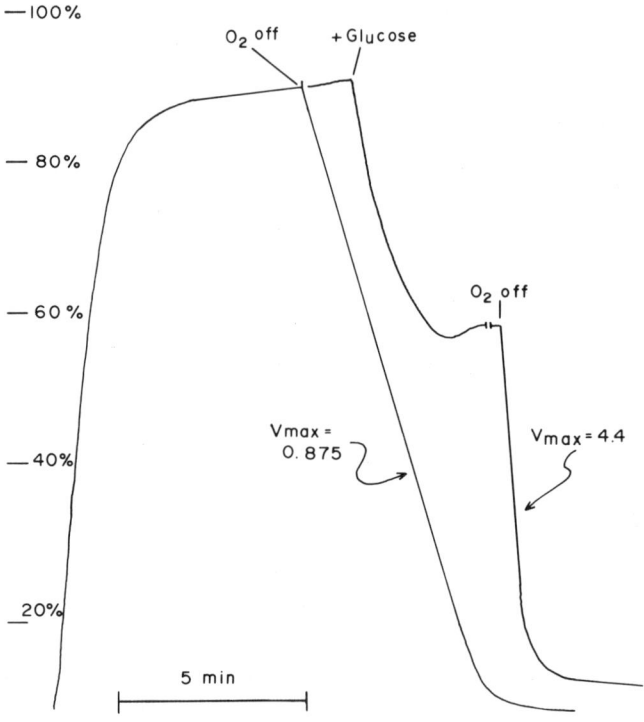

Fig. 7. O_2 tension in 20% suspension of glucose-repressed S cerevisiae (A364a) oxygenated with 95% O_2:5% CO_2 through double bubble manifold (see text). Scale represents percentage O_2 saturation at 23°C and is approximately 1 mM O_2 at 100%. First curve represents cells respiring on endogenous carbon source. O_2 consumption is visualized when O_2 supply is cut off and V_{max} = 0.875 μmoles min^{-1} g $cells^{-1}$. Addition of glucose leads to lower steady-state levels of O_2 when O_2 supply is turned off. V_{max} is determined to be 4.4 μmoles min^{-1} g $cells^{-1}$. This figure is taken to be a lower limit for V_{max}, owing to the relatively slow response time of the O_2 electrode.

Monolayer cells include fibroblasts and epithelial cells, which are essential to the study of transformation. Most of these cells exhibit a strong substrate attachment requirement, and are therefore under decidedly nonphysiological conditions while in suspension. The presence of an appropriate substratum in the NMR tube has presented serious technical difficulties. The main problem is

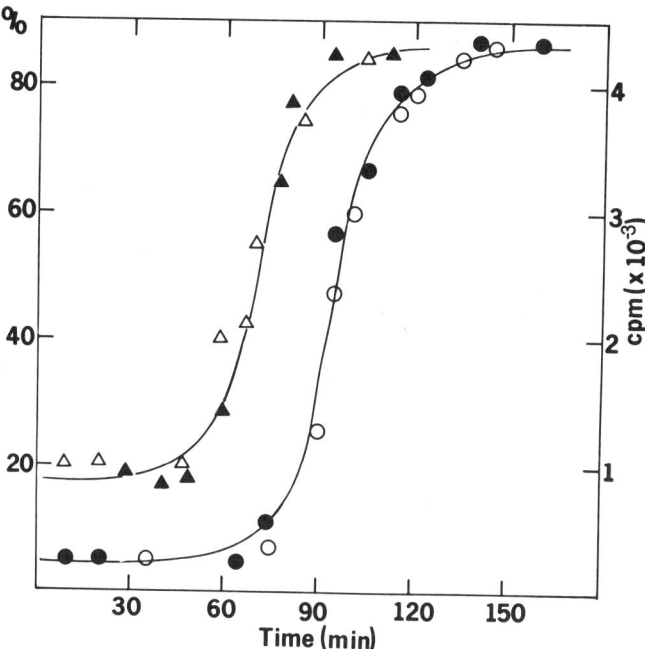

Fig. 8. DNA synthesis (△, △) and budding index (○, ○) as a function of time after addition of food to synchronous S. cerevisiae NCYC 239. Cultures were synchronized by sequential glucose deprivation and resuspended in starvation buffer to a density of 2×10^8 cells per ml. Then, 17 ml of this stock suspension was placed in a 20-mm NMR tube at 25°C, and either N_2 (●, ▲) or O_2 (○, △) was passed through the suspension for 30 minutes prior to feeding. At time 0, O_2 was passed through the suspension at 300 ml/min, and 4 ml of infusion medium was added to the cultures; thereafter, medium was injected at 3.3 ml/h. Also at time 0, 75 µCi of ^{3}H-uracil was added to the suspension. At times indicated, 100-µl aliquots were removed for determination of either budding index or ^{3}H incorporation into DNA. Budding index is expressed as percentage cells with buds; DNA synthesis is expressed as cpm as determined by liquid scintillation counting of base- and acid-extracted cellular precipitate.

associated with the introduction of foreign material into the coil volume. This decreases the effective volume of the sample, can lead to magnetic field heterogeneity, and decreases the availability of oxygen and nutrients to the cells. Two techniques are currently employed which should, in the future, circumvent these problems and allow high resolution spectra to be obtained from mammalian cells attached to a substratum.

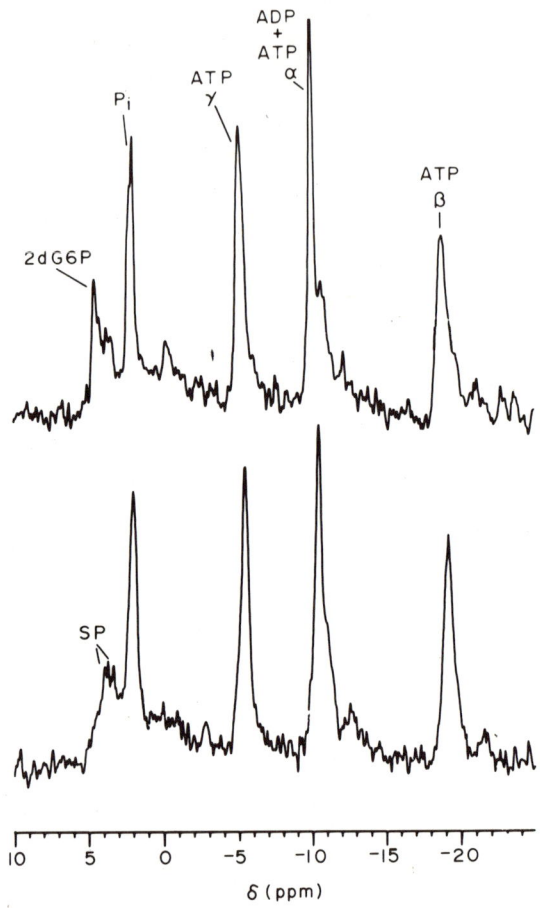

Fig. 9. 145.78 MHz ^{31}P-NMR spectra of 10% suspension of Ehrlich ascites tumor cells at 37°C. Data represent sum of 8,790 free-induction decays arising from 57° pulse angle with a rep rate of 0.1024 seconds (15 minutes total accumulation). Lower spectrum was taken from cells fed glucose, whereas upper spectrum represents cells that were fed with 2-deoxyglucose. The resultant 2-deoxyglucose-6-phosphate can be used to estimate intracellular pH.

The first technique involves growing the cells on derivatized plastic beads, which are available commercially. The use of hydrated beads is discouraged since the intrabead volume has a different magnetic susceptibility than that of the bulk medium, which leads to unacceptably broad lines. Once cells near confluency on the plastic beads they are easily collected by settling and are placed in a sample perfusion chamber (Fig. 10). Fresh, aerated medium is

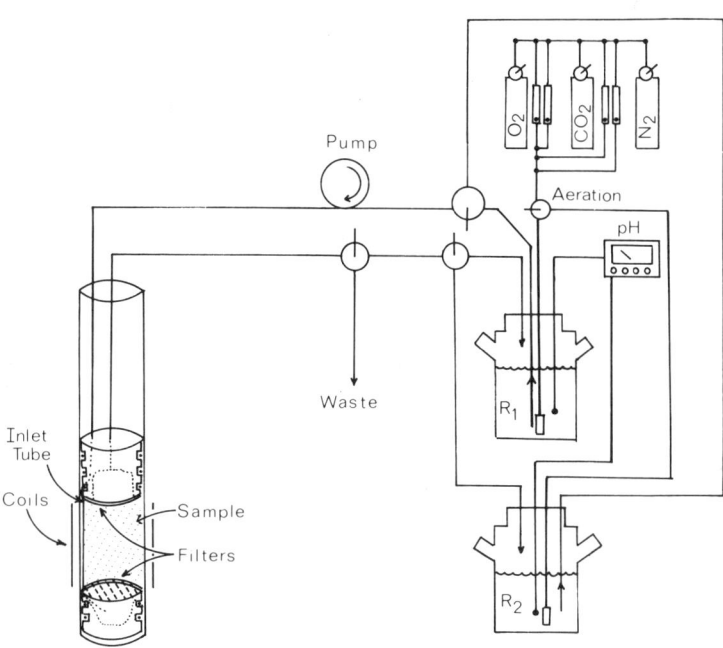

Fig. 10. Diagrammatic representation of perfusion system for mammalian cells. Medium is stored in sterile flasks (R_1 and R_2) into which are placed pH electrodes, aeration tubes, arteries, and media return tubes. Aeration is accomplished by bubbling variable $N_2:O_2:CO_2$ mixture through bubbling stone in media. Excessive foaming is prevented by addition of 0.01% Antifoam B to media, which is forced, via a peristaltic pump through T-50 intramedic tubing. Tubing passes through upper lucite plug, around 30 μm nylon cloth filter, through sample, around lower filter, and into chamber of lower lucite plug. Medium is forced across lower filter, washes over settled beads (containing cells), and is forced across upper filter, into chamber, and out through waste tube. Waste is either discarded (^{31}P experiments) of is recycled (^{13}C experiments). Coils represent RF coils of spectrometer.

perfused over the cell suspension, and cells can be kept viable for up to 18 hours. This preparation contains approximately 4% cell volume and, consequently, an ATP S/N of 4 is achieved after 30 minutes (R.J. Gillies, unpublished observations). If this sample is placed in a solenoid coil instead of the standard Helmholtz pair, a factor of 2.7 in signal is gained (K. Ugurbil, personal communication). Because of the slow doubling time of mammalian cells (12–24 hours), this time resolution is acceptable, yet it poses stringent limits on the type of experiments that can be performed.

A second, more promising, technique utilizes cell culture units designed by Gullino and Knazek [1979] in conjunction with solenoid coils. Under these conditions, cells grow to tissue density, and because of this the ATP S/N is approximately 50 in 10 minutes acquisition time (R. Balaban, personal communication).

III. RESULTS AND DISCUSSION

A. Bacteria

Of the many cellular systems currently under NMR investigation, bacteria are the least complex from the standpoint of defining their cytoplasmic pH. These microorganisms have only a single intracellular compartment, and the plasma membrane transport systems are, in many cases, well defined. The regulation of cytoplasmic pH is an important aspect of the cellular physiology of these free-living organisms because they are likely to encounter some variety in their environmental pH. The extent to which bacteria can maintain their cytoplasmic pH against changes in the medium pH has, for the past few years, been an area of great interest. Studies in which the distribution of weak acids were measured, and, where the chemical shift of the ^{31}P-NMR signal from P_i was monitored, have shown that the cytoplasmic pH stayed constant whereas the external pH varied from neutrality to pH 5.0. There remains considerable controversy, however, concerning the behavior of cytoplasmic pH when the external pH is above neutrality, although it is clear that many bacteria grow under such conditions. The maintenance of a cytoplasmic pH more acid than the external medium represents an energetically unfavorable situation from the standpoint of the cell and the variable results reported may partially stem from the difficulty of keeping cells adequately energized at alkaline external pH. In addition, there are data that suggest that methylamine, which is commonly used for weak base distribution experiments, may be actively accumulated. ^{31}P shift measurements of P_i are insensitive to pHs above 7.5 for reasons discussed above. For this reason methylphosphonate, which is an ideal pH indicator for this range, was used.

In order to provide a definitive result concerning the capability of bacteria to maintain cytoplasmic pH against changes in the external pH, ^{31}P-NMR was used continuously to monitor cytoplasmic pH in E. coli as external pH was varied from 5.6 to 8.7 [Slonczewski et al, 1981]. This experiment utilized the ^{31}P chemical shift of P_i for measurements in the acid range. For the alkaline range the chemical shift of methylphosphonate (Fig. 4) was used. The results (Fig. 11) show that E. coli are capable of maintaining a high degree of pH homeostasis, even when the external pH is above neutrality and the cell is required to maintain an energetically unfavorable pH gradient. Although the pH homeostasis is good, it is not perfect, with the pH_{in} systematically ranging from 7.4 at pH_{ex} 5.6 to 7.8 at pH_{ex} 8.7 with $\Delta pH = 0$ at pH 7.6. These small deviations are clear when

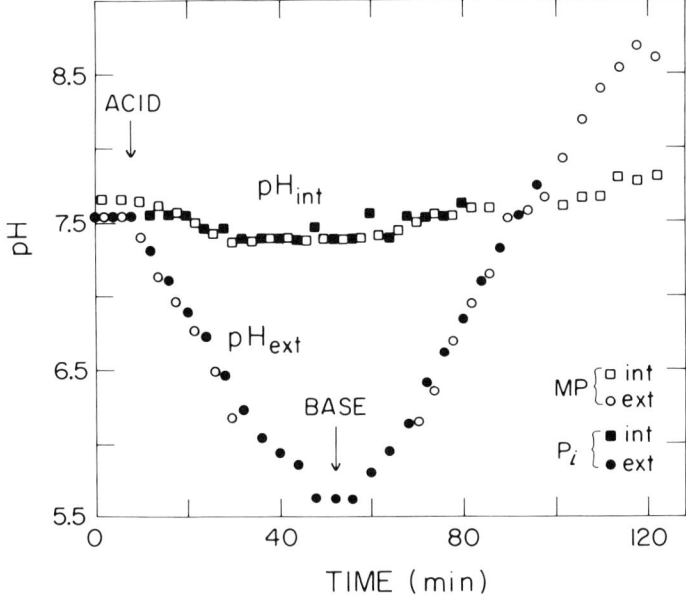

Fig. 11. Dependence of pH$_{in}$ of E. coli upon pH$_{ex}$. pH values were measured by the chemical shift of both MP and P$_i$. At the time indicated by the first arrow, pumping of 1 M HCl was begun; at the second arrow, the pump was switched to 1 M NaOH. Data were accumulated continuously and stored in 2-minute blocks; in the figure, data points represent two-block (4-minute) averages, staggered for the two indicators for clarity of presentation. Data points based on MP below pH 6.1 and P$_i$ above pH 7.8 have been omitted since they lie outside the useful titration range of the respective indicators. ○, internal P$_i$; ●, internal MP; □, external P$_i$; ■, external MP.

pH values observed for all the data points (irrespective of time) shown in Figure 11 are plotted against the external pH, as shown in Figure 12. In cells maintained for prolonged periods at extreme acid or alkaline pH$_{ex}$, homeostasis tended to break down, and deenergization through oxygen deprivation also caused the ΔpH to collapse.

B. Compartmentation

As a eukaryotic organism, yeast possesses several intracellular compartments: cytoplasm, mitochondria, vacuoles, etc. There are several proton-translocating ATPases identified in yeast: one associated with the plasmamembrane [Willsky, 1979], another with the inner mitochondrial membrane [Serrano, 1978], and one associated with the vacuolar membrane [Ohsumi and Araku, 1981]. Therefore,

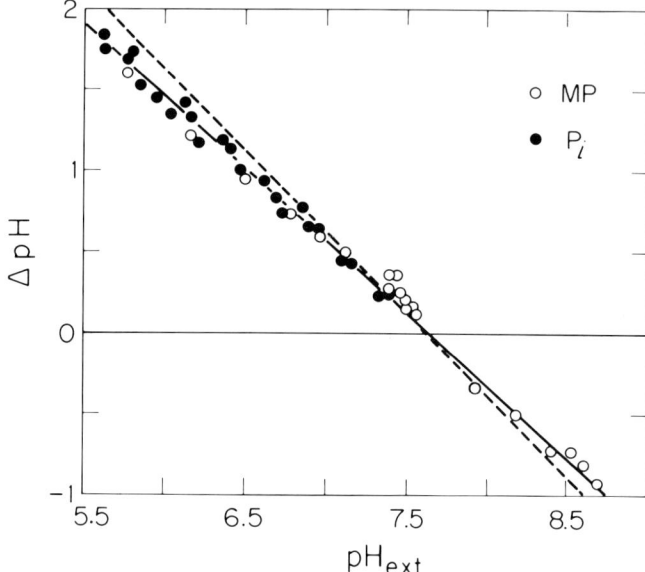

Fig. 12. Transmembrane pH difference (ΔpH) in E. coli add a function of pH_{ex}. Data are the same as used for Figure 11, and combine measurements made as pH_{ex} was being raised and lowered. The linear regression (continuous line) was made on the data from both MP and P_i; it has a slope of −0.89, a zero crossing of 7.63, and a correlation coefficient of 0.997. The dashed line, with a slope of −1, represents perfect homeostasis with pH_{in} constant at 7.63 at all pH_{ex}. ○, P_i; ●, MP.

yeast possesses in principle the means to maintain distinct pH in each of these compartments. In particular there is evidence that the vacuolar space, which in resting yeast cells is the largest intracellular compartment, is considerably more acidic than the cytoplasm. This compartmentation poses severe problems for the conventional techniques of measuring intracellular pH.

^{31}P-NMR techniques avoid these complications. If the pH of each compartment is sufficiently different, it is possible in principle to obtain separate NMR signals which measure the pH for each compartment. This also requires that the metabolite used as an indicator is sufficiently abundant to be observed in the ^{31}P-NMR spectrum. When compartmentation occurs, a pH-dependent peak such as P_i can be split into several components, as illustrated in Figure 2, and assignment of resonances to specific compartments is a problem. The easiest compartment to assign is the external volume, which is identified by simply adding P_i to the suspension and looking for a peak height increase. As shown in the figure, this corresponds to peak Y at 1.67 ppm, indicating an external pH of 6.45. The pH of the medium can be measured to test this assignment. Assignment of other P_i resonances to specific compartments is more difficult, and involves correlation of peaks with expected behavior and values of compartmental pH. Peak X at 2.25 ppm, indicating a pH of 7.10, has been assigned to cytosolic P_i. Evidence for this assignment is as follows. In glycolyzing yeast cells, high concentrations of cytosolic FBP and G6P are built up. These compounds also titrate in the physiological range [Navon et al, 1977], and using them as pH indicators the intracellular pH was shown to be 7.1, which corresponded to the pH found from peak X. In addition, the absolute position of peak X is relatively independent of external pH and generally behaves in a manner consistent with our knowledge of intracellular pH and P_i levels.

The third P_i peak (Z) in spectrum 2 has been tentatively assigned to vacuola. P_i. Evidence for this assignment is scanty and relies mainly on the observation that the chemical shift of peak Z at 0.9 is relatively invariant and corresponds to a pH of 6.0 consistent with low vacuolar pH. In addition, this peak is reduced in actively growing cells, corresponding to decreased vacuolar volume [Stevens, 1977].

In addition to the three P_i peaks described above in yeast cells, mitochondrial and cytosolic P_i have been identified in isolated rat liver cells [Cohen et al, 1979]. Assignment of mitochondrial P_i was made according to chemical shift, to its behavior toward agents such as valinomycin and FCCP and in comparison with behavior of P_i in isolated mitochondria. The cytosolic P_i peak was assigned on the basis of its giving the same pH as was deduced from the cytosolic fructose-1-phosphate peak.

In ascospores of yeast, two cellular P_i peaks were observed [Barton et al, 1980]. One of the two compartments corresponded to a pH of 5.5–6.3, whereas the other was between 5.0 and 5.8 when the external pH was varied between 3 and 10. On the basis of comparisons with isolated single spores and germination experiments, the more acidic pH was assigned to the epiplasmic space, and the higher pH was assigned to the ascospores inside the asci.

C. Time Resolution

Another advantage of the NMR technique is that it is possible to measure the pH with a time resolution of 15 seconds or less. This is particularly valuable if one is wishes to study changes in pH_i brought about by particular metabolic perturbations imposed on the cells. We have followed the pH_i with high time resolution after the addition of glucose to deenergized yeast cells, and observed rapid changes in the pH_i following that perturbation.

In the first ^{31}P-NMR study of suspensions of intact yeast cells, pH_i in deenergized yeast cells was determined, using the naturally occurring orthophosphate as a pH indicator [Salhany et al, 1975]. It was found that at extracellular

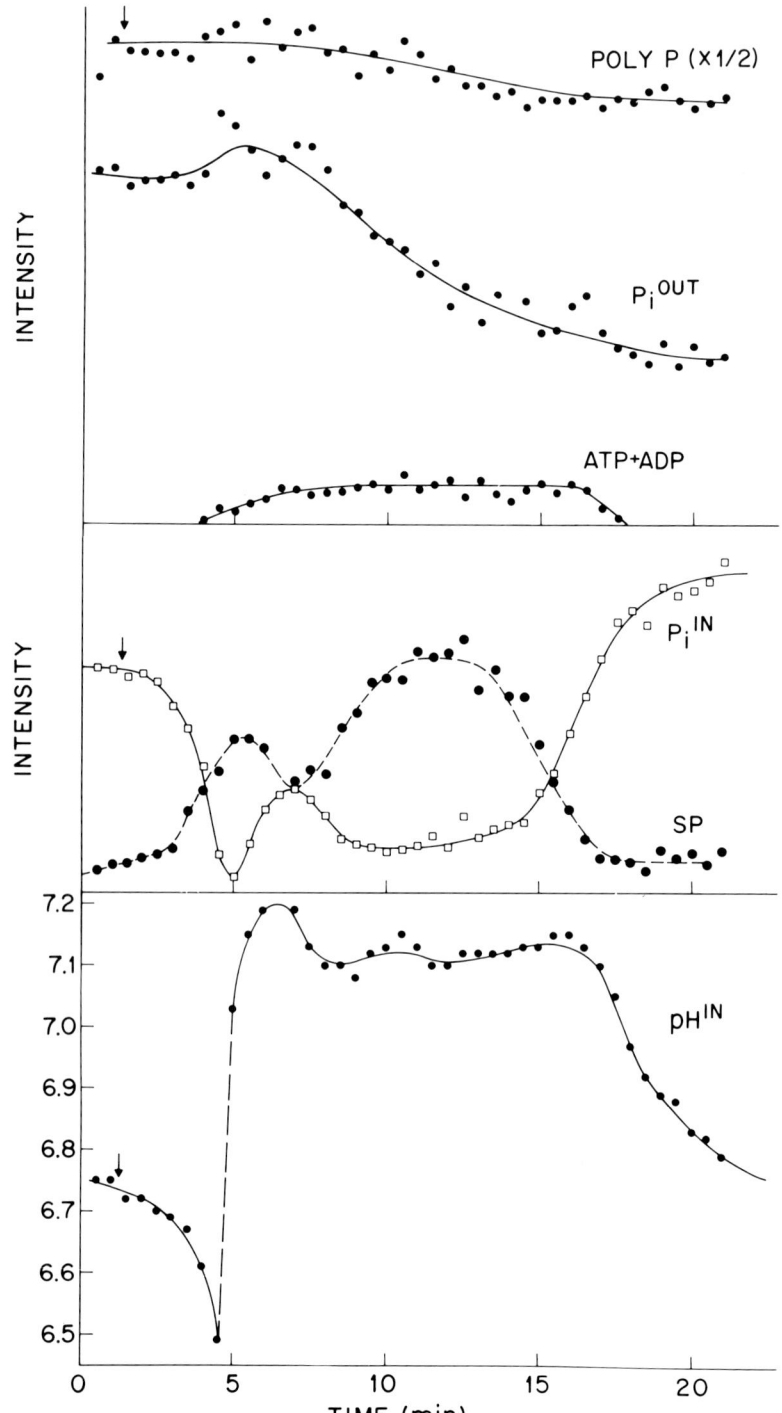

pH values below 6, pH_i in the compartment of the observable intracellular orthophosphate signal was about 5.8–6.0. The pH_i varied by only one pH unit when the extracellular pH was varied from 3.5 to 9.1. These experiments showed that in deenergized yeast cells pH_i is slightly acidic.

However, when yeast cells are energized, pH_i increases to above 7 and is quite independent of external pH [den Hollander et al, 1981]. The transient from the relatively low pH_i of deenergized yeast cells to the higher pH_i observed upon energization was followed in our recent report with a time resolution of 30 seconds. Figure 13 shows the time courses of the various ^{31}P-NMR signals in this experiment as a function of time, together with pH_i. Glucose was added (75 mM) to the suspension 1 minute after the onset of the experiment. The first points in the time course, obtained before the addition of glucose, show the slightly acidic pH_i of deenergized cells (6.7–6.8), a high P_i^{in} signal (25 mM) and a low sugarphosphate signal. ATP is too low to be observed. After the addition of the glucose the P_i^{in} starts to decrease, reaches a minimum at 5 minutes after glucose addition, goes through a maximum after 7 minutes, and then settles down at a stable, low concentration of about 5 mM. After the exhaustion of glucose the P_i^{in} returns to the high value for deenergized yeast cells. The sugarphosphate signal follows a time course that is complementary to the behavior of the P_i^{in} signal. The time course of the pH_i change correlates with the behavior of the intracellular orthophosphate signal. After the addition of the glucose, pH_i starts to decrease to about 6.5. Then, at the same moment that the intracellular orthophosphate signal reaches its minimum, pH_i suddenly increases to 7.2, and remains stable at about 7.1 until the glucose is exhausted, after which the pH_i gradually decreases to the slightly acidic value of deenergized cells. The ATP level as determined in the intact cell spectra is high (about 5 mM during the period of high pH_i) and after glucose exhaustion it diminishes. During the course of the experiment extracellular phosphate is being taken up by the yeast cells, whereas the polyphosphate signal remains approximately constant. This experiment shows that there is a close connection between the energization of yeast cells and their pH_i. The sudden increase in pH_i about 4 minutes after glucose addition is most likely intimately connected to the control of glycolysis. The activity of phosphofructokinase is known to be high at low pH and high orthophosphate levels [Bañuelos et al, 1977], and this high activity could be responsible for the low ATP level observed during the first few minutes after glucose

Fig. 13. Glucose-repressed cells were resuspended in the minimal medium, at a density of 50% wet weight. Glucose was added to the yeast suspension, and ^{31}P-NMR spectra were accumulated in consecutive 0.5-minute blocks. The first two points in these time courses were obtained before adding glucose to the suspension. Time courses measured from these spectra given for pH^{in} (determined from the peak position of P_i^{in}) and the levels of P_i^{in}, P_i^{out}, polyphosphate and sugarphosphate.

addition. During this period intracellular phosphate is incorporated into the sugar phosphates. After a few minutes the orthophosphate is decreased to a few mM, and the PFK activity diminishes, enabling the ATP concentration to increase. This ATP can then be used to drive the proton-translocating ATP-ase of the plasma membrane, leading to a higher pH_i. At the higher pH_i thus attained, the glycolytic flux and the level of glycolytic intermediates reach a stationary state until glucose is exhausted.

This experiment shows that about 10 minutes after glucose addition pH_i reaches a plateau value. The limiting factor for the duration of this constant pH_i value was the glucose supply; by the addition of more glucose this period can be extended. This steady state period has been studied extensively in subsequent experiments.

D. pH_i and the Control of Glycolysis in Yeast

We have studied pH_i by ^{31}P-NMR under a variety of conditions. For our studies of aerobic and anaerobic glycolysis it was necessary to grow the yeast cells under a variety of conditions [Perlman and Mahler, 1974]. The rationale for this is that if yeast cells are grown in a glucose-containing medium the enzymes needed for respiration are repressed, and the yeast cells show a diminished respiration. These cells derive their metabolic energy from fermentation, and are expected to be relatively insensitive to the presence of oxygen. Therefore, these cells will show only minor effects of oxygen upon the rate of fermentation—ie, they have a small Pasteur effect. To study the Pasteur effect, and the role of pH_i in the Pasteur effect, it has been necessary to obtain derepressed cells. These cells are obtained by growing the yeast cells in the absence of glucose, or of other highly repressive carbohydrates such as fructose and mannose. Yeast cells grown with ethanol or acetate as carbon sources have an active respiration, and are expected to show, upon oxygenation, a large effect on their rate of fermentation. They possess less of the glycolytic enzymes and have active gluconeogenic enzymes, fructose-1,6-bisphosphatase and PEP carboxykinase. Raffinose causes an effect that is somewhat in between these two extremes: Cells grown on this carbohydrate have an active respiration, but raffinose is being utilized via the glycolytic pathway. Figure 14 shows the experiments performed on glucose-repressed cells. The lower spectrum in the figure was obtained between 8 and 24 minutes after glucose addition, while bubbling N_2/CO_2 gas mixture through the suspension. The pH_i as derived from the P_i peak position was 7.30. The upper spectrum was obtained in a parallel experiment, but here O_2/CO_2 gas mixture was bubbled through the suspension. In this experiment pH_i was also 7.30. It therefore appears that in repressed yeast cells the presence of oxygen has little effect on pH_i during glycolysis. Also the level of intracellular P_i, ATP, and sugarphosphates differs by only a small amount. A different picture is obtained when investigating derepressed cells (Fig. 15). The pH_i during aerobic

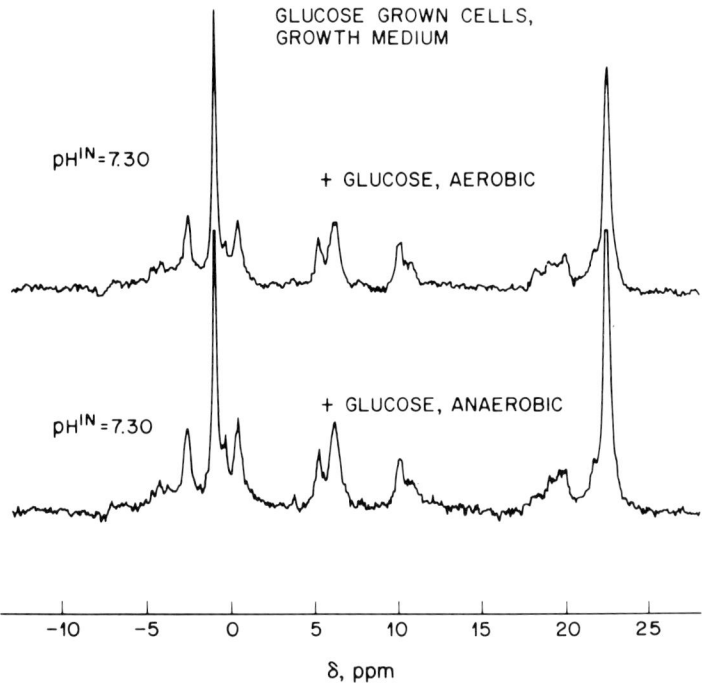

Fig. 14. ^{31}P-NMR spectra of aerobic and anaerobic suspension of yeast cells during glycolysis. Glucose-repressed yeast cells were resuspended in growth medium at a density of 10% wet weight. Glucose was added, and the NMR spectra were recorded subsequently. A) Sixteen-minute accumulation obtained between 10 and 16 minutes after glucose addition to the yeast suspension, while O_2 was bubbled. B) Same as (A) except here N_2 was bubbled instead of O_2.

glycolysis was 7.47, while during anaerobic glycolysis it was 7.25. In addition to this different pH$_i$ we observe other differences in the ^{31}P-NMR spectra as well. The level of intracellular P$_i$ is lower by a factor 3 aerobically, and the level and distribution of sugarphosphate also differ. There is a somewhat lower level of nucleotide triphosphates in the anaerobic case. Also note that the level of polyphosphate has decreased, in particular in the anaerobic case.

The difference in pH$_i$ between anaerobic and aerobic glycolysis has been observed for derepressed cells obtained under a variety of conditions. The pH$_i$ measured for derepressed cells are collected in Table I. The table shows that for derepressed cells, pH$_i$ is lower by 0.2–0.4 pH units during anaerobic glycolysis than it is during aerobic glycolysis. This difference in pH$_i$ is, however, not observed for repressed cells. The higher pH$_i$ we observed during aerobic gly-

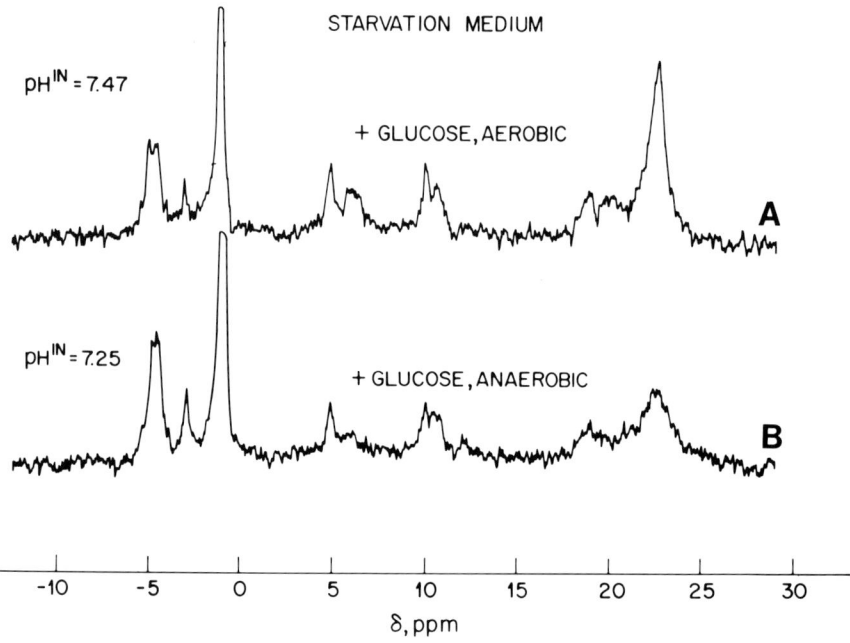

Fig. 15. Yeast cells were grown into saturation in 2% bactopeptone, 1% yeast extract, and 2% glucose. The cells were resuspended in the minimal medium. (A) gives the spectra obtained between 10 and 26 minutes after glucose addition to an aerobic suspension. (B) was obtained for an anaerobic suspension under otherwise identical conditions.

colysis has been shown to be related to the control of glycolysis. In vitro experiments have shown that glycolytic enzymes, and most notably phosphofructokinase (PFK), exhibit a pH dependence for their rates. PFK has a very high activity at low pH values (pH 6.50), and a low activity at high pH (7.5) [Banuelos et al, 1977]. Therefore, the higher pH observed during aerobic glycolysis will tend to decrease the rate through PFK, and thereby contribute to the control of that enzyme in the Pasteur effect. Other factors can also contribute to the control of PFK; in particular it is known that the level of P_i also affects the rate of PFK, and we have shown here that in derepressed cells the level of intracellular P_i is considerably lower during aerobic glycolysis than it is during anaerobic glycolysis. It is interesting to note that the pH profile of the gluconeogenic enzyme fructose-1,6-diphosphatase exhibits a pH profile that is opposite to that of PFK: It has a high activity at high pH and a low activity at low pH [Tortora et al, 1981]. It therefore appears that the high pH observed during aerobic conditions

TABLE I. The Effect of Oxygen on the Intracellular pH of Saccharomyces cerevisiae Cells Grown on Different Carbon Sources and Fed Glucose*

Carbon source used for growth	Intracellular pH	
	Aerobic	Anaerobic
Glucose (repressed)[a,c]	7.28	7.25
Glucose (repressed)[b,c]	7.30	7.30
Glucose (derepressed)[a,d]	7.52	7.23
Raffinose[a]	7.34	7.21
Raffinose[b]	7.35	7.19
Glycerol[a]	7.51	7.38
Ethanol[a]	7.35	6.99
Acetate[b]	7.45	7.03
Acetate[e]	7.58	7.13
Acetate[f]	7.50	7.15

*Cells were grown, harvested, and resuspended in either the minimal or the enriched medium. Subsequently, they were fed glucose either in the presence of O_2 (aerobic) or N_2 (anaerobic).
[a]Suspension in minimal medium at pH 6.0.
[b]Suspension in enriched medium at pH 6.0.
[c]Cells were harvested at a density which was 25% of the full saturation value. At the time of harvest these cultures still contain unused exogenous glucose and are therefore *catabolite-repressed*.
[d]Cells were harvested after 24 hours of growth and were fully saturated; these cells are *catabolite-derepressed*.
[e]Suspension in minimal medium buffered with 50 mM mesotartrate at pH 5.3.
[f]Suspension in minimal medium buffered with 50 mM tartrate at pH 3.5.

tends to increase the rate of gluconeogenesis, whereas the lower pH observed during anaerobic conditions will tend to increase the rate of glycolysis. The shift in pH_i can be a means by which the yeast cells suppress futile cycling through these two enzymes when switching to anaerobic glycolysis.

IV. REFERENCES

Bañuelos JM, Gancedo C, Gancedo JM: Activation by phosphate of yeast phosphofructokinase. J Biol Chem 252:6394–6398, 1977.

Bailey I, Radda GK, Williams SK, Gadian DG: The activity of phosphorylase in total global ischemia in the rat heart. Biochemistry (in press), 1981.

Barton JK, den Hollander JA, Lee TM, MacLaughlin A, Shulman RG: Measurement of the internal pH of yeast spores by ^{31}P nuclear magnetic resonance. Proc Natl Acad Sci USA 77:2470–2473, 1980.

Becker ED, Ferretti JA, Gambhir PN: Selection of optimum parameters for pulse Fourier transform NMR. Anal Chem 51:1413–1420, 1979.

Cohen SM, Ogawa S, Rottenberg H, Glynn P, Yamane T, Brown TR, Shulman RG, Williamson JR: ^{31}P nuclear magnetic resonance studies of isolated rat liver cells. Nature 273:554–555, 1978.

den Hollander JA, Brown TR, Ugurbil K, Shulman RG: ^{13}C nuclear magnetic resonance studies of anaerobic glycolysis in yeast cells. Proc Natl Acad Sci USA 76:6096–6100, 1979.

den Hollander JA, Ugurbil K, Brown TR, Shulman RG: ^{31}P NMR studies of the effect of oxygen upon glycolysis in yeast. Biochemistry (in press), 1981.

Gullino PM, Knazek RA: Tissue culture on artificial capillaries. Meth Enzymol 58:178–184, 1979.

Iles RA, Griffiths JR, Stevens AN, Gadian DG, Porteous R: Effects of fructose on the energy metabolism and acid–base status of the perfused–starved rat liver. Biochem J 192:191–202, 1980.

Navon G, Ogawa S, Shulman RG, Yamane T: ^{31}P nuclear magnetic resonance studies of Ehrlich ascites tumor cells. Proc Natl Acad Sci USA 74:87–91, 1977.

Navon G, Shulman RG: Yamane T, Eccleshall TR, Lam KB, Baronofsky JJ, Marmur J: ^{31}P NMR studies of wild-type and glycolytic pathway mutants of Saccharomyces cerevisiae. Biochemistry 18:4487–4489, 1979.

Ohsumi Y, Anraku Y: Active transport of basic amino acids driven by a proton motive force in vacuolar membrane vesicles of S. cerevisiae. J Biol Chem 256:2079–2082, 1981.

Perlman PS, Mahler HR: Derepression of mitochondria and their enzymes in yeast. Arch Biochem Biophys 162:248–271, 1974.

Roberts JKM, Ray PK, Wade-Jardetzky N, Jardetzky O: Estimation of cytoplasmic and vacuolar pH in higher plant cells by ^{31}P NMR. Nature 283:870–872, 1980.

Rose AH, Harrison JS: "The Yeasts." London and New York: Academic Press, 1971.

Salhany JM, Yamane T, Shulman RG, Ogawa S: High resolution ^{31}P nuclear magnetic resonance studies of intact yeast cells. Proc Natl Acad Sci USA 72:4966–4970, 1975.

Serrano R: Characterization of the plasma membrane ATPase of S. cerevisiae. Mol Cell Biochem 22:51–63, 1978.

Slonczewski JL, Rosen BP, Alger JR, Macnab RM: pH homeostasis in E. coli. Proc Natl Acad Sci USA 78: (in press), 1981.

Stevens BJ: Variation in number and volume of the mitochondria in yeast according to growth conditions. Biol Cell 28:37–55, 1977.

Tortora P, Birtel M, Lenz AG, Holzer H: Glucose-dependent metabolic interconversion of fructose-1,6-bisphosphatase in yeast. Biochem Biophys Res Commun 10:688–695, 1981.

Willsky GR: Characterization of the plasma membrane Mg-ATPase from the yeast, S. cerevisiae. J Biol Chem 254:3326–3332, 1979.

Spectrophotometric Determination of Cytoplasmic and Mitochondrial pH Transitions Using Trapped pH Indicators

John A. Thomas, Peter C. Kolbeck,* and Thomas A. Langworthy

Departments of Biochemistry (J.A.T., P.C.K.) and Microbiology (T.A.L.), University of South Dakota School of Medicine, Vermillion, South Dakota 57069

I.	Introduction	105
II.	Methods and Materials	106
	A. Cells	106
	B. Fluorescein Dyes	106
	C. Spectrophotometric Measurements	107
	D. Loading the Cells With the Indicator	107
	1. Fluorescein diacetate; 2. 6-Carboxyfluorescein diacetate; 3. Control cells	107
	E. Assessment of Dye Leakage	108
	F. Calibration of Internal Dye Spectral Responses	108
III.	Results	109
	A. Spectral Sensitivity of Carboxyfluorescein and Fluorescein to pH	109
	B. Internal pH Measurements in a Leaky Cell	110
	C. Measurements With Monolayer Cell Cultures	110
	D. Cytoplasmic and Mitochondrial pH Transitions	114
IV.	Discussion	118
V.	Summary	122
VI.	References	122

I. INTRODUCTION

The introduction of impermeant pH indicators into cells or organelles for the purpose of measuring internal pH has a long history [for a review, see Roos and Boron, 1981]. In spite of many attempts, none of these techniques has received widespread acceptance. The main difficulty in this approach is in finding a suitable means for introducting the impermeant dye behind the membrane without

---

*Current address: School of Medicine, Emory University, Atlanta, GA 30322.

damaging it or otherwise affecting the system to be measured. Theoretically, such a technique would provide a method of great sensitivity and rapid response characteristics for monitoring internal pH.

In the variation of this technique presented here, the impermeant dye is synthesized in situ from a permeant precursor by intracellular enzymes. Fluorescein diacetate (FA_2), or its analog 6-carboxyfluorescein diacetate (CFA_2), diffuses through the cell membrane into the cytoplasm, where intracellular enzymes remove the acetate groups, releasing fluorescein (F) or 6-carboxyfluorescein (CF). Whereas the permeant precursors are colorless and uncharged, the products are highly charged, and are strong chromophores, exhibiting spectral changes over the pH range from 7.5 to 3. The high negative charge of the released chromophores inhibits their release from the cell. The intracellular retention of fluorescein by this procedure was initially established as a tool for monitoring the intactness of the cell membrane [Rotman and Papermaster, 1966]; any damage of the cell membrane results in the rapid release of the trapped chromophore. Similarly, this method has been used to separate live from dead cells in cell sorting devices [Bonner et al, 1972].

Since both the absorbance and fluorescence excitation spectrum of fluorescein are highly dependent on pH, it seemed a natural extension of these studies to utilize the trapped indicator to monitor intracellular pH. The technique presented is applicable to both prokaryotic and eukaryotic cells. In addition, by the use of both CFA_2 and FA_2, one may distinguish between pH changes occurring in mitochondrial and cytosolic compartments.

II. METHODS AND MATERIALS

A. Cells

Ehrlich ascites tumor cells were maintained in mice, as described by Scholnick et al [1973]. Cells were harvested after 7–10 days of growth and washed twice with a pH 7.4 buffer containing 110 mM NaCl, 5 mM KCl, 1 mM $MgCl_2$, 4 mM sodium phosphate, and 50 mM HEPES (N-2-hydroxyethylpiperazine-N' ethanesulfonic acid).

A culture of Bacillus acidocaldarius (isolate 104-A) was kindly provided by T.D. Brock, University of Wisconsin, Madison. Cells were grown at pH 3, 60°C, in a basal medium consisting of inorganic salts and a rich source of glucose and yeast extract, as described elsewhere [Langworthy et al, 1976]. Cells were harvested in midlog growth, at a level of approximately 1 g wet weight per liter medium.

B. Fluorescein Dyes

The diacetate derivative of 6-carboxyfluorescein was synthesized by refluxing 6-carboxyfluorescein (Eastman Organic Chemicals, Rochester, New York) with acetic anhydryde, as described previously [Thomas et al, 1979].* Fluorescein diacetate was purchased from Sigma. Stock solutions of both diacetate compounds were prepared in dimethyl sulfoxide at 10 mM concentration. These solutions may be stored in the refrigerator for a month without significant decomposition, but must be kept tightly covered, since dimethylsulfoxide is hygroscopic, and these derivatives slowly hydrolzye in aqueous media.

C. Spectrophotometric Measurements

Absorbance measurements were made in an Aminco DW-2 spectrophotometer equipped with a magnetic stirrer and thermostated cell compartment, using a 3-nm bandpass. This instrument is designed to minimize the turbidity problems associated with cellular suspensions. To measure spectra, the instrument was operated in the split beam mode, with untreated cells used in the reference cuvette. For monitoring kinetic changes in intracellular pH, the instrument was used in the dual wavelength mode. In this mode, the instrument measures the difference in absorbance between two different wavelengths of light passed through the same cuvette. One monochromator is set at a reference wavelength where no spectral changes are expected (465 nm), and the second monochromator is at a wavelength sensitive to pH changes (490 nm). Perkin Elmer also sells an instrument capable of these measurements. Fluorescence measurements were determined in an Aminco Bowman SPF equipped with a thermostatted cuvette compartment. Excitation spectra were measured with a #15 Wratten filter in front of the photomultiplier, using 1 mm slits on both excitation and emission monochromators.

D. Loading the Cells With the Indicator

The procedure that follows has been found suitable for Ehrlich ascites tumor cells, and will provide a reasonable starting point for experimentation with other cell types. The objective is to trap enough of the dye within the cells to provide sufficient spectral sensitivity for pH measurements without altering the internal pH and other cellular functions. For ascites cells, a final intracellular indicator concentration of 0.2 mM (based on internal volume measurements) provides good sensitivity for absorbance measurements without altering the internal pH [Thomas et al, 1979] or glycolytic rates [Belt et al, 1979].

1. Fluorescein diacetate. After centrifugation at 2,000g for 5 minutes, ascites cells are diluted with 12 volumes of isotonic buffer, pH 7.4. This corresponds to a cell count of approximately 50×10^6/ml, or 6 mg protein/ml (Lowry). FA_2 is added directly from the stock solution to a final concentration of 6 μM. The cells are incubated at 10–15°C for about 10 minutes, during which

*CFA_2 may be purchased from Molecular Probes, Plano, Texas.

an obvious yellow color develops. The cells are then diluted with three volumes of ice-cold buffer, and centrifuged at 2,000g for 5 minutes at 4°C. The supernatant is decanted and the cells are gently resuspended in ice-cold buffer and recentrifuged. At this point, the supernatant should be colorless and the pellet yellow. After decanting the supernatant, the final pellet is suspended in a minimal amount of ice-cold buffer, and kept on ice until use to minimize leakage. For spectrophotometric measurements, cells are diluted to a final concentration of 20×10^6 cells/ml in buffer at the desired temperature.

If it is undesirable to store the cells on ice, loaded cells may be maintained at higher temperatures and recentrifuged immediately prior to use to remove extracellular fluorescein.

2. 6-Carboxyfluorescein diacetate. Loading cells with CFA_2 requires a couple of modifications of the above procedure. Unlike FA_2, CFA_2, possesses a free carboxyl group that is ionized at neutral pH. To facilitate entry through the cell membrane, the initial incubation medium is lowered to pH 6.2 to increase the concentration of the uncharged, permeant species. In addition, the concentration of CFA_2 is 35 µM to provide similar incubation times as with 6 µM FA_2. Once the initial incubation is complete, the succeeding washing steps are performed with pH 7.4 buffer.

3. Control cells. For absorbance spectral measurements, it is important to prepare control cells (no indicator) by the same procedure, substituting dimethylsulfoxide for the indicator stock solution. The control cells are used in the reference cuvette.

E. Assessment of Dye Leakage

The amount of dye leakage can be assessed from the quantity of indicator present in the supernatant (or filtrate) after centrifugation (filtration) of the cells. The fluorescence is measured at 490 nm excitation/540 nm emission after adjusting the pH to 8 or above. One hundred percent leakage is equated to the amount of fluorescence obtained in the supernatant from cells lysed with detergent. Typical results obtained with Ehrlich ascites tumor cells are seen in Table I.

F. Calibration of Internal Dye Spectral Responses

The intracellular pH may be determined from the spectrum of the trapped indicator. In our experience, the internal dye has similar but not equivalent spectral sensitivity to pH as dye in buffers. Hence, calibration of the spectral responses should be determined by an independent method for determining intracellular pH, such as the distribution ratio of a weak acid. With the Ehrlich ascites tumor cells, the addition of 10 µg/ml nigericin in 130 mM KCl buffers resulted in approximate equilibration of internal and external pH within 5 minutes.

TABLE I. Leakage of CF and F From Ehrlich Ascites Tumor Cells at 20°C

	Percent leakage for the intracellular indicators			
	$pH_i = 7.4$		$pH_i = 6.2$	
Time (min)	CF	F	CF	F
0	10%	18%	12%	20%
5	12	29	17	68
10	15	43	24	89
30	22	77	32	96
60	26	92	36	98

Cells were suspended at 2.5 mg protein/ml in either 50 mM HEPES (pH 7.4) or Mes (pH 6.2) buffers containing 110 mM NaCl, 5 mM KCl, 1 mM $MgCl_2$, and 10 μg nigericin/ml. Two-milliliter aliquots were discharged at time intervals into test tubes on ice and immediately centrifuged at 2,000g for 2 minutes at 4°C. The amount of indicator in the supernatant was measured by fluorescence (490 nm excitation/520 nm emission) after adjustment to pH 10. The calculation of the percent leakage was based on the total amount of indicator initially present in the cells.

III. RESULTS

A. Spectral Sensitivity of Carboxyfluorescein and Fluorescein to pH

Figure 1 illustrates the molecular alterations CFA_2 undergoes when intracellular esterases remove the acetate groups from this compound. FA_2 differs only in lacking the asterisked carboxyl group. Removal of the two acetate groups produces a chromophore with three ionizable groups (two for F), with the largest pK for both F and CF at 6.5, and the others below 5. Once formed, the slow rate of diffusion of these dyes out of the cell is probably the result of the high negative charge they possess at neutral pH.

Both the fluorescence excitation spectra and the absorbance spectra of these dyes are pH-sensitive, as is seen in Figures 2A and 3A respectively. The fluorescence spectra in Figure 2A have been normalized to the same relative intensity to emphasize the pH-dependent spectral shifts that are observed. Over the pH range depicted (pH 7 to 3), the fluorescence intensity actually decreases about 20-fold as the pH is lowered. The absorbance spectra (Fig. 3A) are not normalized, and the general decrease in absorbance and the blue wavelength shift with decreasing pH are evident.

Although the wavelength shifts could be used as an indicator of pH for either fluorescence or absorbance measurements, the relative shape of the spectra serves as a more sensitive pH monitor [Thomas et al, 1976, 1979; Ohkuma and Poole, 1978]. With our spectrophotofluorometer, the ratio of fluorescence intensities at 485 nm versus 465 ($F_{485} \div F_{465}$) serves as a useful monitor between pH 7.5

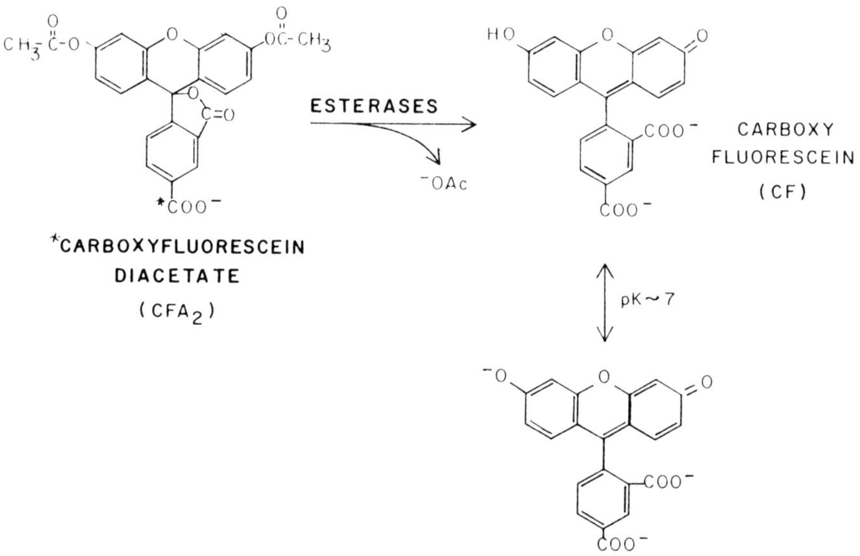

Fig. 1. Structures of 6-carboxyfluorescein diacetate and intracellular esterase reaction products.

and 5.5, whereas below pH 5 the ratio of intensities at 435 nm versus 465 nm ($F_{435} \div F_{465}$) is more effective (Fig. 2B). For absorbance spectra the ratio of the absorbance value at 490 nm to that at 465 nm (isosbestic point above pH 6) is convenient (Fig. 3B).

B. Internal pH Measurements in a Leaky Cell

Figure 4 illustrates results obtained with a relatively leaky cell, Bacillus acidocaldarius. This organism is a thermophilic, acidophilic, spore-forming rod isolated from acidic sulfur hot springs. It grows optimally at a pH of 2 to 3, and a temperature of 60–70°C. Figure 4 shows the fluorescence excitation spectra at time intervals after the direct addition of FA_2 to an undiluted growing suspension of this bacillus. The shape of the F spectra at early time intervals indicates a relatively neutral internal pH in spite of being in pH 3 growth medium. At longer times, the fluorescence at 435 nm rises rapidly because of the rapid rate of leakage of the indicator out of the cell. Thus, as time proceeds, the observed spectra are a combination of the internal (pH 7) and external (pH 3) spectral forms.

C. Measurements With Monolayer Cell Cultures

We have previously shown that the trapped indicator method may be used to monitor internal pH changes accompanying lactate transport in suspensions

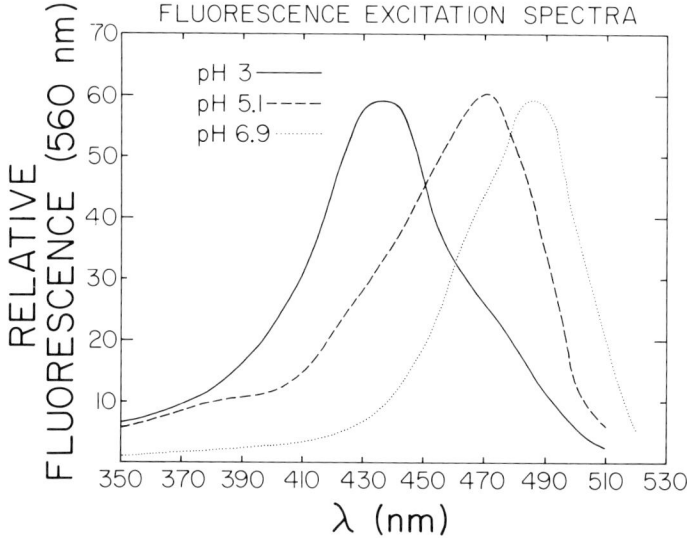

Fig. 2A. Normalized fluorescence excitation spectra of fluorescein in various buffers. The fluorescein concentration was 2 μM in 0.1 M phosphate-citrate buffer at the indicated pH.

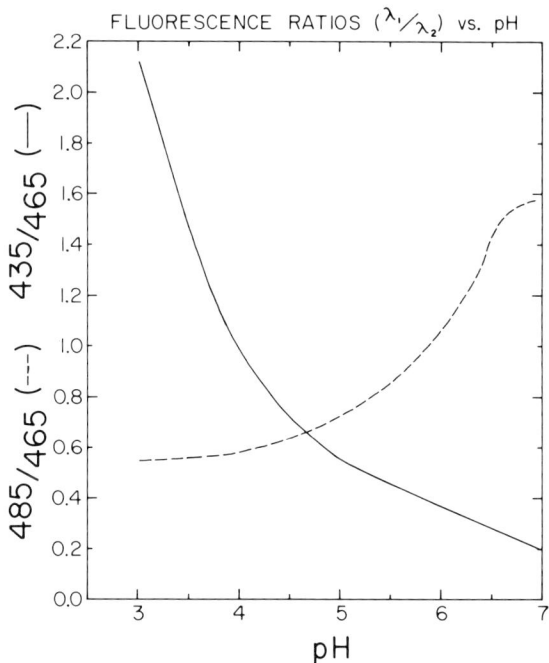

Fig. 2B. Calibration curve relating spectral shape to pH. Data from Figure 2A.

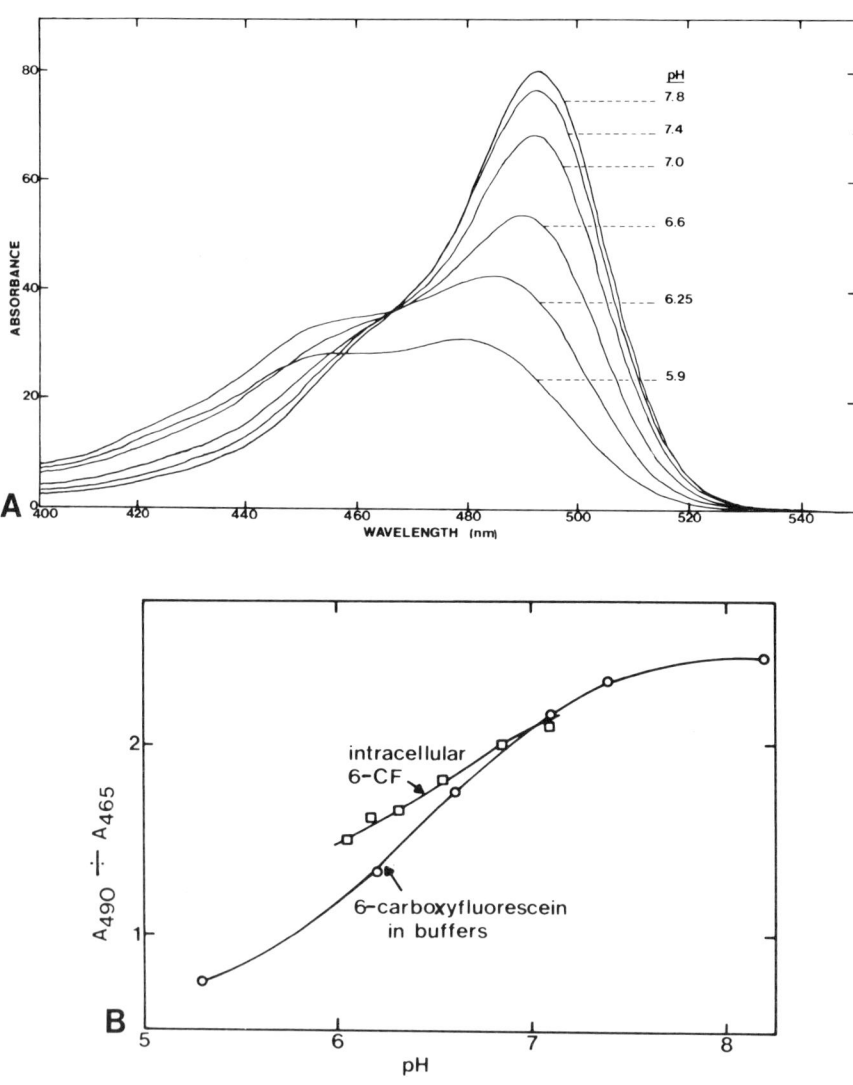

Fig. 3. A)—pH Dependence of 6-carboxyfluorescein spectra. The spectra of 9 μM 6-carboxyfluorescein were measured in 100 mM Mops buffers at the indicated pH. B) Calibration curves relating shape of absorbance spectra to pH. (○) 6-Carboxyfluorescein (6 μM) in the indicated buffers (50 mM) containing 11 mM NaCl, 5 mM KCl, and 1 mM MgCl: Tricine (pH 8.2); HEPES (pH 7.1); Mops (pH 7.1); Pipes (pH 6.6); and Mes (pH 6.2 and 5.3). (□) Ascites cells loaded with 6-carboxyfluorescein and suspended in 30 mM buffers containing 130 mM KCl and 1 mM Mg Cl_2 after a 5-minute equilibration with 10 μg/ml of nigericin: Mes (pH 6.54 and below) and Mops (above 6.54; 20°C).

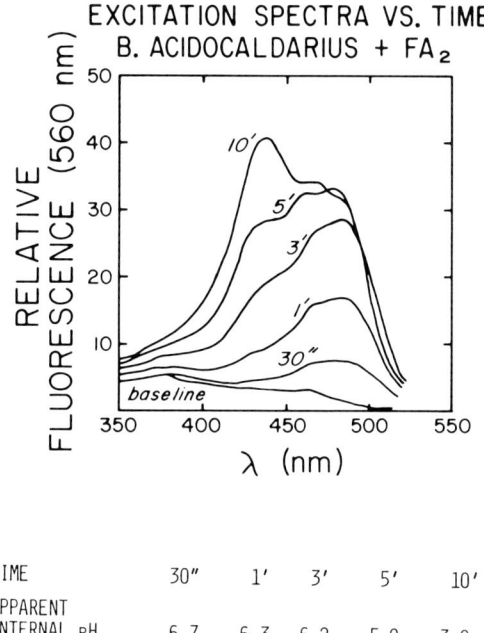

TIME	30"	1'	3'	5'	10'
APPARENT INTERNAL pH	6.7	6.3	6.2	5.9	3.8

Fig. 4. Fluorescence excitation spectra at various times after the addition of 2 μM fluorescein diacetate to a log phase culture of Bacillus acidocaldarius. The table indicates the approximate internal pH derived from the spectral shape using the calibration curve in Figure 2B.

of Ehrlich ascites tumor cell [Thomas et al, 1979]. Figure 5 demonstrates that this technique may be used to follow lactate transport in cell cultures grown in monolayer, using absorbance measurements [Bunow and Chien, 1981; Bunow and Kaplan, in preparation]. NMU-3 cells, a line derived from hepatocytes and transformed by nitrosomethylurea, and an untransformed cell line of the same origin (TRL 12–13) were grown to confluence on plastic slides in Leighton tubes (Co-Star). The slides were incubated for 10 minutes at 12°C in saline medium at pH 6.3 (130 mM NaCl, 5 mM KCl, 1 mM $MgCl_2$, 1 mM $CaCl_2$, 0.5 mM K_2PO_4, and 20 mM MES) containing 100 μM CFA_2, followed by rinsing in the same medium lacking CFA_2. Spectral measurements were obtained by inserting the slides into standard cuvettes, with the back of the slide in contact with the side of the cuvette (see Fig. 6). The data indicate that lactate is cotransported with protons in both cell lines, and is inhibited by the bioflavanoid quercetin. These results are consistent with those obtained with Ehrlich ascites tumor cells. Calibration spectra for the intracellular indicator were obtained by equilibration

Kinetics of pH Changes of Cytosol During Lactate Uptake and Release at 27° C and Effect on Quercetin on Kinetics

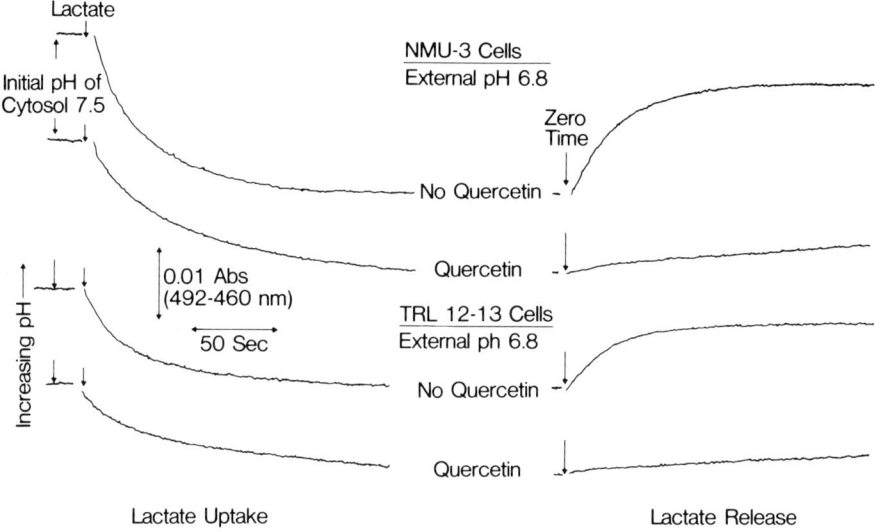

M.R. Bunow and A.E. Kaplan, 1981
National Cancer Institute
National Institutes of Health
Bethesda, MD 20205

Fig. 5. Kinetics of pH changes during lactate uptake and release from NMU-3 hepatocytes, transformed with nitrosomethylurea, and TRL 12-13 hepatocytes, a control line in cell culture. The pH changes were followed using the absorbance difference 492–460 nm to monitor CF-labeled cells. Release of lactate from lactate-loaded cells was initiated by inserting the slide into a cuvette containing fresh, lactate-free solution. The rate of release, but not of uptake of lactate, is markedly reduced in the presence of 5 μg/ml of the lactate transport inhibitor, quercetin; the inhibition is reflected in the slower rate of change of intracellular pH.

of the internal and external pH with nigericin in buffers of various pH (see Fig. 3B), and containing 135 mM KCl. The intracellular concentration of CF was estimated to be 250 μM for these experiments.

The use of plastic Leighton tube slides as supports for cells that are not easily put into suspension should extend the trapped indicator method to a wider variety of cell types. In addition, the slides allow rapid transfer of cells from medium to medium without exposure to mechanical damage, and provide a simple means of quickly removing external indicator.

D. Cytoplasmic and Mitochondrial pH Transitions

Intracellular CF has two distinct advantages over intracellular F for monitoring

Hepatocytes are Grown on Plastic Slides in Leighton Tubes (a). The Monolayers of Cells are Dyed and Transfered Intact on Slides to Cuvettes (b) for Spectrophotometric Studies of pH.

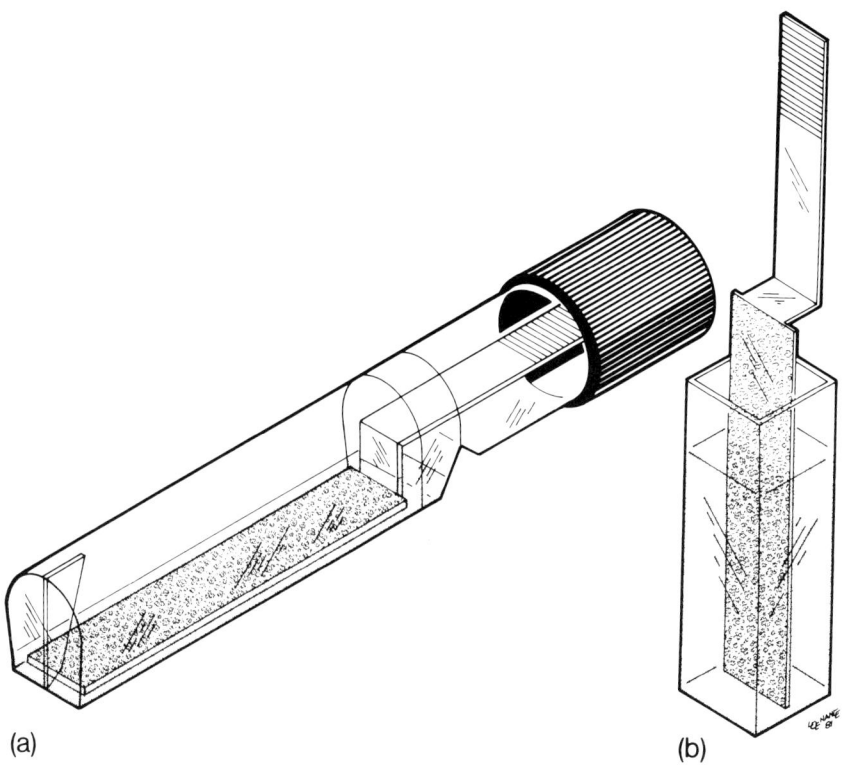

(a)   (b)

M.R. Bunow and A.E. Kaplan, 1981
National Cancer Institute
National Institutes of Health
Bethesda, MD 20205

Fig. 6. Schematic view of hepatocytes growing in monolayer on plastic slides in (a) Leighton tubes, transferred to a CFA_2-containing solution for uptake of pH indicator (not shown) and then to (b) cuvette for spectrophotometric study.

cytoplasmic pH. The rate of leakage of CF out of the cell is much slower than F [Thomas et al, 1979]. Secondly, CF is not responsive to mitochondrial pH transitions, whereas F is. This can be seen in Figure 7. CF- and F-loaded Ehrlich ascites tumor cells were placed in pH 6.4 medium. The subsequent addition of glucose results in an internal acidification owing to the high rate of lactate

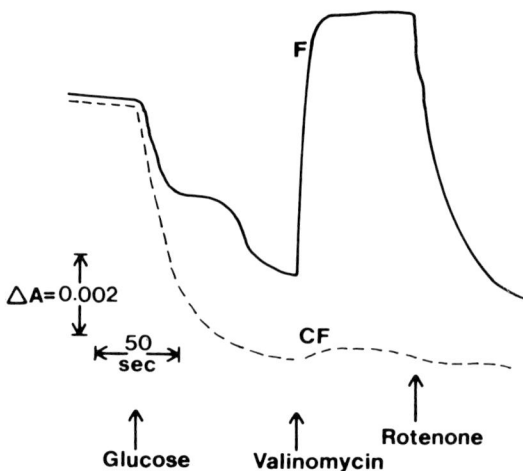

Fig. 7. Spectral responses of carboxyfluorescein (CF)- or fluorescein (F)-loaded Ehrlich ascites tumor cells to cytoplasmic and mitochondrial pH transitions. CF or F cells were suspended at 2.5 mg protein/ml in 50 mM Mes buffer (pH 6.4) containing 110 mM NaCl, 5 mM KCl, and 1mM $MgCl_2$. Absorbance was monitored at 490 nm (464 nm reference). Final concentrations: glucose, 10 mM; valinomycin, 1 µM; rotenone, 12 µM.

production in these cells. The subsequent addition of valinomycin induces a large pH gradient across the mitochondrial membrane, which is detected by intracellular F, but not by CF, Similar results were observed with isolated mitochondria [Thomas et al, 1979]. F added to mitochondrial suspensions reported pH transitions across the mitochondrial membrane, whereas CF was unresponsive.

Thus CF reports cytoplasmic pH transitions, whereas F reports both cytoplasmic and mitochondrial pH. A better understanding of this phenomenon is gained by observing these responses by fluorescence microscopy (Fig. 8A–D). Figure 8A shows the appearance of F-loaded cells incubated under the same conditions as in Figure 7, after the addition of glucose. A slight asymmetry in the distribution of the dye is seen in the cells. After the addition of valinomycin, inducing a large basic pH transition in the mitochondrial matrix, this asymmetrical distribution becomes quite dramatic (Fig. 8B). The dye distribution corresponds well with the location of mitochondria in these cells, as shown by electron microscopy (Fig. 9A, B). These cells have a large nucleus, with the mitochondria generally occupying the outer rim of the cell (Fig. 9A), or a crescent shaped area (Fig. 9B) at one side of the cell. Addition of the uncoupler 1799, which collapses the mitochondrial pH gradient, results in an even distribution of dye throughout the cell (Fig. 8C).

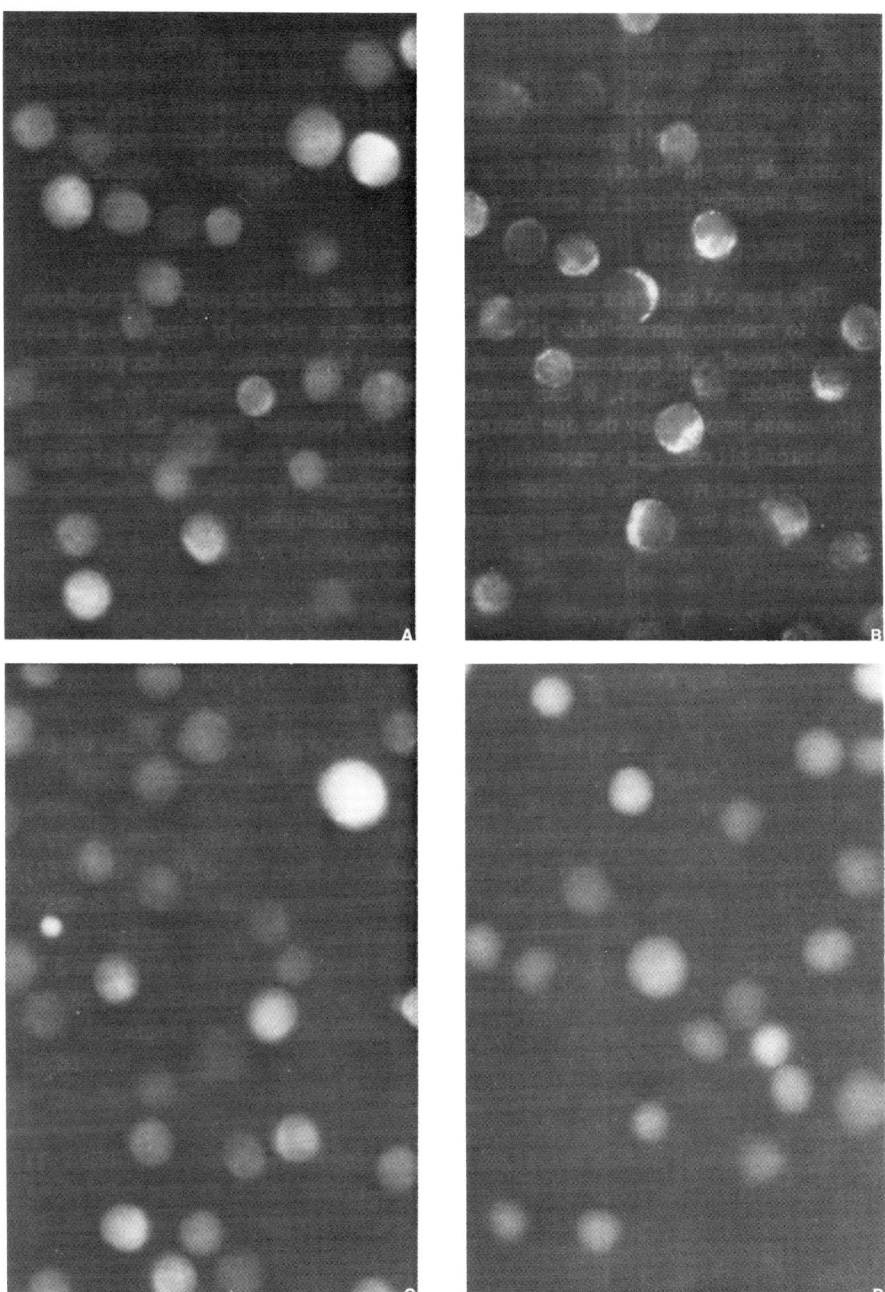

Fig. 8A–D. Fluorescence micrographs of cells incubated under conditions in Figure 7. A) F cells after glucose addition; B) F cells after glucose plus valinomycin; C) F cells plus glucose, valinomycin and the uncoupler 1799 (0.5 µM); D) CF cells plus glucose and valinomycin.

The behavior of CF-loaded cells contrasts markedly with that of F-loaded cells. Regardless of incubation conditions, the CF cells show an even distribution of dye. Figure 9D shows the appearance of CF cells under the identical conditions as in Figure 9B, where a large mitochondrial pH gradient is present. The presence of a mitochondrial pH gradient causes F to concentrate in the basic mitochondrial matrix, as might be expected for a weak acid. CF, apparently because of its higher negative charge, is unable to do so.

IV. DISCUSSION

The trapped indicator method offers several advantages over other methods used to monitor intracellular pH. The technique is relatively simple, and may be performed with equipment available to most laboratories. Besides providing intracellular pH values, it can monitor pH continuously, subject to the time limitations imposed by the dye leakage rate. The response time of the indicators to internal pH changes is essentially instantaneous, and the sensitivity for single measurements rivals the distribution ratio technique. The method is nondestructive and may be applied to large populations or individual cells.

One of the unique advantages of this technique is the possibility of distinguishing pH changes in the cytoplasmic and mitochondrial compartments. CF acts as a reporter of cytoplasmic pH, whereas F reports both the cytoplasmic and mitochondrial compartments. Although the enzyme releasing both indicators is predominantly cytoplasmic [Thomas et al, 1979], F is permeant across the mitochondrial membrane, whereas CF is not. Like other weak acids, the relative distribution of F in the cytoplasmic and mitochondrial spaces depends on the magnitude of the pH gradient across the mitochondrial membrane. Since F has two acidic groups ($pK_1 \simeq 4$, $pK_2 = 6.5$), a ΔpH of 1 across the mitochondrial membrane ($7 \rightarrow 8$) could theoretically result in an 80-fold higher concentration of F in the mitochondria than in the cytoplasm. We have found [Kolbeck, 1981] that isolated mitochondria incubated under the conditions of Figure 7 develop a pH gradient slightly greater than 1, with approximately the predicted fluorescein distribution ratio. This explains the localized fluorescence pattern observed in in Figure 8B; the dye is highly concentrated in the mitochondria. Since the mitochondrial volume appears to be approximately 20% of the cytoplasmic volume in these cells, a ΔpH of 1 would result in 95% of the F to be located in mitochondria. Thus, under conditions producing large mitochondrial pH gradients, F essentially reports only the mitochondrial compartment.

The usefulness of this technique relies on four criteria being satisfied: 1) permeability of the membrane to the colorless precursor, 2) the presence of an appropriate intracellular esterase to release the chromophore, 3) impermeability

---

Fig. 9A, B.   Electron micrographs of Ehrlich ascites tumor cells ($\times$ 6,000). Cells were fixed in 2.5% glutaraldehyde and stained with osmium tetroxide.

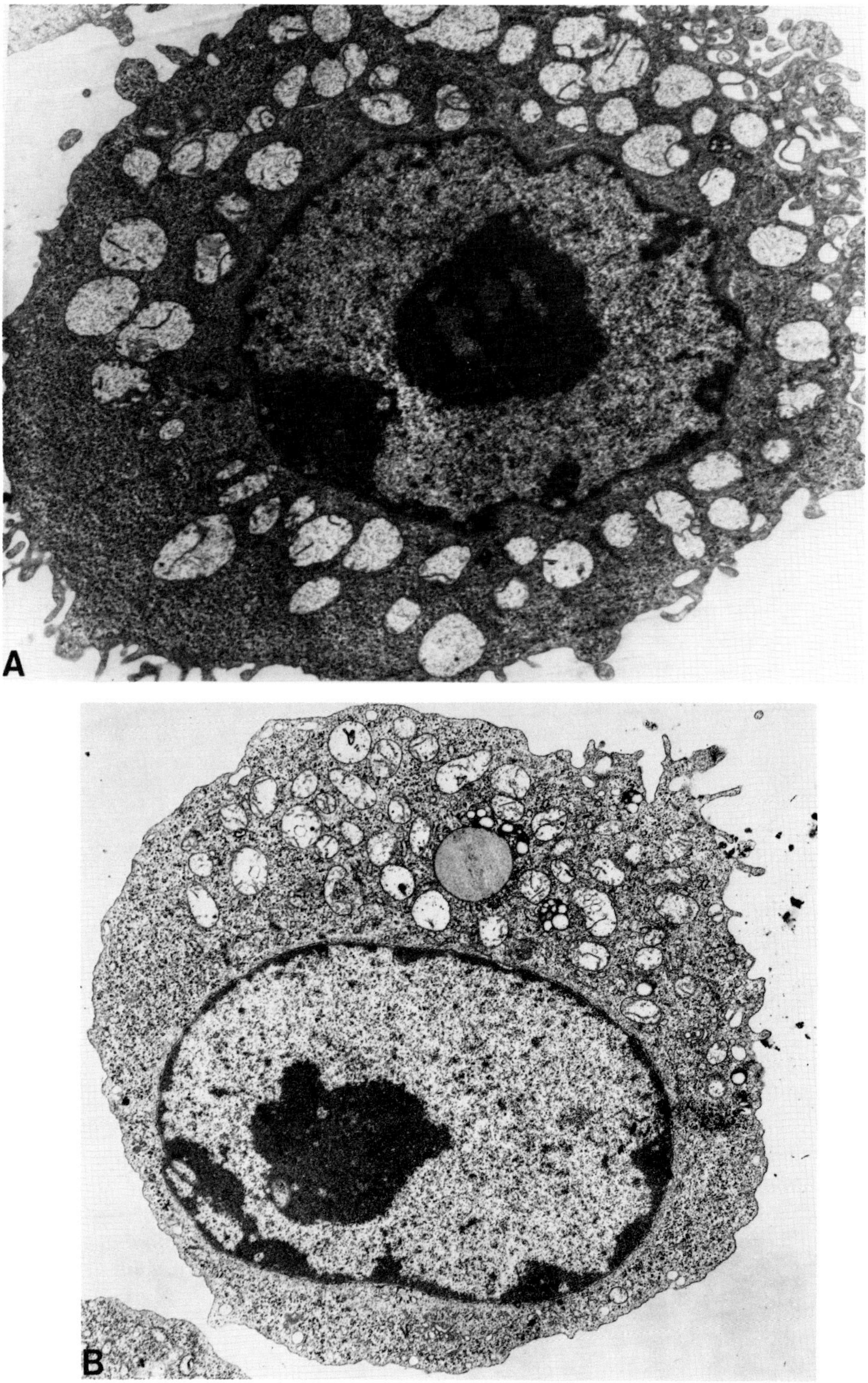

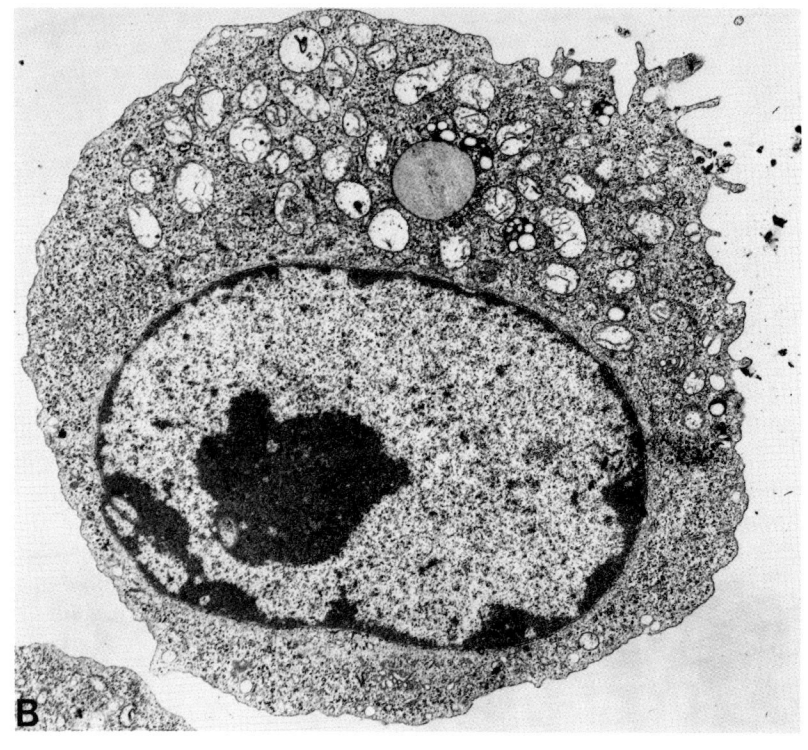

of the membrane to the negatively charged indicator, and 4) a suitable means of calibrating the responses.

Permeability to the colorless precursor molecule does not seem to be a problem with most cell types, with the possible exception of brown fat cells from hamster. An appropriate esterase seems to be present in most mammalian cell lines [Rotman and Papermaster, 1966], although the situation is more variable in protists [Medzon and Brady, 1969]. Thus, the first two criteria will be satisifed for most cells.

The permeability of the negatively charged indicator to membranes is quite variable, and must be evaluated for each cell type studied. The half-time for leakage varies from several weeks for CF in liposome vesicles [Weinstein et al, 1977] to less than a minute, as reported here for F in Bacillus acidocaldarius. When leakage is a problem, CF is definitely the indicator of choice over F because of its slower leakage rates. Both CF and F show increased leakage rates as the intracellular pH is lowered [Thomas et al, 1979]; therefore conditions favoring elevated intracellular pH will favor retention of the indicators. The external dye concentration must be minimized by centrifugation, rapid filtration, continous perfusion, dialysis, etc, in order for the spectral data to provide accurate estimates of the internal pH.

Calibration of the indicator responses should be performed in a medium that simulates the internal environment of the cell. Ideally, calibration is accomplished with the indicator inside the cell under conditions where the internal pH is known. This may be achieved by equilibrating the internal and (known) external pH with ionophores or, ideally, by calibrating the spectral responses with an independent method, such as the distribution ratio method, under conditions providing a suitable range of intracellular pH values. For the ascites cell, the addition of nigericin in 130 mM K^+ buffers was found approximately to equilibrate the internal and external pH, as determined by the distribution ratio technique [Thomas et al, 1979]. Addition of protonophores, such as 1799 or FCCP (carbonylcyanide p-trifluoromethoxyphenylhydrazone), or tributyltin (chloride–hydroxide exchange) gave no additional pH equilibration or indicator response over that observed with nigericin.

A simpler procedure for determining intracellular pH with carboxyfluorescein has been proposed recently (Babcock, 1981). Cells loaded with carboxyfluorescein were suspended in media buffered at various pH values. An increase or decrease in absorbance at 490 nm (465 reference) was observed with intracellular dye released by the addition of digitonin, depending on whether the extracellular medium was more acidic or basic, respectively, than the intracellular environment. The intracellular pH is thus determined by extrapolating the spectral responses to the pH at which no response is observed upon addition of digitonin. This calibraion technique gave reasonable intracellular pH values, although the values were not corroborated by an independent method. This approach has the advantage that extracellular dye does not contribute to the observed spectral response. It assumes, however, that no spectral change occurs upon cell lysis when both internal pH and external pH are identical.

The internal indicator concentration is a key to the success of this technique. An internal concentration of 0.2 mM (internal volume is 5 μl/mg protein, or 5 μl/8 × 10^6 cells) has given good results with Ehrlich ascites tumor cells, bovine spermatozoa [Babcock, 1981], and normal and transformed cell lines cultured from rat hepatocytes [Bunow and Chien, 1981; Bunow and Kaplan, in preparation]. Much higher concentrations are to be avoided because of possible effects on internal pH, as the formation of the dye releases acetic acid. Considerably lower concentrations are also not desirable; as the internal concentration is lowered, the proportion of dye bound to cellular elements increases. Bound indicator is relatively unresponsive to pH, and is also quenched with respect to its fluorescence. Bound dye is characterized by long wavelength shifts in both the absorbance and fluorescence emission spectra; when F is 50% bound to BSA, the absorbance maximum at pH 7.2 shifts from 492 nm to 512 nm [Udkoff and Norman, 1979]. Loaded cells exhibiting absorbance maxima beyond 495 nm are to be avoided; the remedy is to increase the internal dye concentration. This is easily accomplished by increasing the concenration of the diacetate derivative or decreasing the cell content in the initial incubation mixture.

Either the fluorescence or absorbance properties of these indicators may be used to monitor internal pH. In practice, the choice will likely depend on the quality of the available instruments. In general, the safer choice is absorbance, using the Aminco DW-2 spectrophotometer, which is designed to handle turbid solutions. Absorbance measurements are subject to fewer possible artifacts than fluorescence, especially where dye movement (F) and binding changes may be involved.

Nonetheless, fluorescence has some definite advantages for determining internal pH values from the spectral shape. Besides offering greater sensitivity for this type of measurement, the fluorescence spectra do not require a reference sample with exactly the same turbidity, as is necessary for absorbance spectra.

For continuous monitoring of internal pH (pH vs time), the dual wavelength absorbance technique is the method of choice. This is especially true when observing fluorescein responses in cells with mitochondria. By fluorescence, fluorescein movement into the basic mitochondrial matrix would be expected to result in increased fluorescence, but may in fact produce a fluorescence decrease if the concentration quenching effect is sufficiently large. The absorbance technique is not subject to concentration quenching. In addition, with fluorescence, small changes in fluorescence are measured against a large background signal, thereby reducing the sensitivity. The dual wavelength technique measures deviations from a null balance point and as shown by Bunow and Kaplan (Fig. 5), offers sufficient sensitivity to detect small pH changes in cells grown in monolayer cultures.

V. SUMMARY

The strength of the trapped indicator method lies in monitoring small cytoplasmic (CF) pH changes on a real time basis; changes as small as 0.01 pH may be easily followed. Calibration of the spectral responses of the internal dye should be performed by comparison to a second method. such as the distribution ratio method utilizing DMO, since the spectral characteristics of internal and extracellular dye differ. Interpretation of results with CF are simpler than with F, since the latter also reports mitochondrial as well as cytoplasmic pH transitions and has a considerably higher rate of leakage out of the cell.

ACKNOWLEDGMENTS

We are indebted to Mr. Richard Duman (Microbiology Department, Univeristy of South Dakota) for his assistance with the fluorescence microscopy work, and Dr. Daniel Neufeld (Anatomy Department, University of South Dakota) for the electron microscopy pictures. In addition, we would like to thank Drs. Ann E. Kaplan and Margaret R. Bunow (Laboratory of Carcinogen Metabolism, Division of Cancer Cause and Prevention, National Cancer Institute) for making their results (Figures 5 and 6) available to us for this manuscript.

VI. REFERENCES

Babcock DF: Internal ion concentrations coupled to intracellular fluorescein chromophore with ionophores. Fed Proc 40:1785 (Abstract), 1981.

Belt JA, Thomas JA, Buchsbaum RN, Racker E: Inhibition of lactate transport and glycolysis in Ehrlich ascites tumor cells by bioflavanoids. Biochemistry 18:3506, 1979.

Bonner WA, Hulett HR, Sweet RG, Herzenberg LA: Fluorescence activated cell sorting. Rev Sci Inst 43:404, 1972.

Bunow MR, Chien NC: Acidification of the cytosol accompanying lactate production and transport in nitrosomethylurea-transformed (NMU-3) and control (TRL 12-13) rat hepatocytes in culture. Fed Proc 40:1688 (Abstract), 1981.

Bunow MR, Kaplan AE: Shifts in cytosolic pH accompanying lactate production and release in control and chemically transformed epithelial cell lines. Manuscript in preparation.

Kolbeck PC: Use of fluorescent probes to monitor mitochondrial pH gradients in living cells. MA thesis, University of South Dakota, 1981.

Langworthy TA, Mayberry WR, Smith PF: A sulfonolipid and novel glucosamidyl glycolipids from the extreme thermoacidophile Bacillus acidocaldarius. Biochim Biophys Acta 431:550, 1976.

Medzon EL, Brady ML: Direct measurement of acetylesterase in living protist cells. J Bacteriol 97:402, 1969.

Ohkuma S, Poole B: Fluorescence probe measurement of the intralysosomal pH in living cells and the perturbation of pH by various agents. Proc Natl Acad Sci USA 75:3327, 1978.

Roos A, Boron WF: Intracellular pH. Physiol Rev 61:296, 1981.

Rotman B, Papermaster BW: Membrane properties of living mammalian cells as studied by enzymatic hydrolysis of fluorogenic esters. Proc Natl Acad Sci USA 55:134, 1966.

Scholnick P, Lang D, Racker E: Regulatory mechanisms in carbohydrate metabolism IX. Stimulation of aerobic glycolysis by energy-linked ion transport and inhibition by dextran sulfate. J Biol Chem 248:5175, 1973.

Thomas JA, Buchsbaum RN, Zimniak A, Racker E: Intracellular pH measurements in Ehrlich ascites tumor cells utilizing spectroscopic probes generated in situ. Biochemistry 18:2210, 1979.

Thomas JA, Cole RE, Langworthy TA: Intracellular pH measurements with a spectroscopic probe generated in situ. Fed Proc 35:1455 (Abstract), 1976.

Udkoff R, Norman A: Polarization of fluorescein fluorescence in single cells. J Histochem Cytochem 27:49, 1979.

Weinstein JN, Yoshikami S, Henkart P, Blumenthal R, Hagins WA: Lysosome–cell interaction: Transfer and intracellular release of a trapped fluorescent marker. Science 195:489, 1977.

Determination of Intracellular pH Changes in Lymphocytes With 4-Methylumbelliferone by Flow Microfluorometry

Donald F. Gerson
Basel Institute for Immunology, 487 Grenzacherstrasse, CH-4005 Basel, Switzerland

I.	Introduction	125
II.	Methods	126
	A. Intracellular pH	126
	B. Other Methods	128
III.	Results and Discussion	128
IV.	References	132

I. INTRODUCTION

Colorometric and fluorometric indicators of pH, in conjunction with sophisticated optical and electronic devices offer the possibility of highly sensitive, continuous monitoring of intracellular pH. The characteristics of a useful indication dye are that it be nontoxic, have a pK close to 7.0, and be relatively impermeant. Colorometric pH indicators (litmus, neutral red, or bromthymol blue) require accurate measurement of small changes in absorbance, whereas fluorometric pH indicators (fluorescein, 7-hydroxycoumarins, or pyranine) require accurate measurement of changes in low levels of fluorescence. The latter has often been easier to accomplish.

Fluorescent, pH-sensitive dyes have been used to study the internal pH of many membrane delimited systems. Grünhagen and Witt [1970] used umbelliferone (7-hydroxycoumarin) fluorescence as an indicator for pH changes in spinach chloroplasts. Deamer et al [1972] used the quenching of 9-aminoacridine fluorescence to determine pH gradients across liposome membranes. Gerson and Burton [1976] used methylesculetin (6,7-dihydroxy, 4-methylcoumarin) fluorescence to determine relative changes in the intracellular pH of Physarum. Ohkuma and Poole [1978] used fluorescein conjugated to dextran to measure intralysosomal pH in mouse peritoneal macrophages. Kano and Fendler [1978] have used pyranine to determine pH of the interior of liposomes. Recently, Heiple and Taylor [1980, and this volume] have used fluorescein conjugated to ovalbumin to study intracellular pH in single cells of the amoeba Chaos carolinensis. Other examples of the use of fluorescent dyes for intracellular pH measurement are given by Gerson [1977] and Roos and Boron [1981].

A major advance in the study of mixed populations of cells has been the advent of flow microfluorometry. Flow microfluorometry allows the collection of data on the fluorescence and light scattering characteristics of many thousands of cells in a few minutes. Most commonly, the technique is used to study subpopulations that may be distinguished by staining with fluorescent antibodies specific for cell surface marker proteins. Light scatter is an ill-defined function of cell size and shape.

Physiological properties may also be determined by flow microfluorometry. The quantities of cellular constituents may be determined if appropriately specific fluorescent stains are available. For instance, the quantities of DNA, RNA, and protein [Horan and Wheeless, 1977] have been determined by flow microfluorometry.

There have been two attempts to determine intracellular pH (pH_i) by flow microfluorometry. Visser et al [1979] used fluorescein as the pH-sensitive fluorochrome. This dye has the disadvantage that two excitation wavelengths are required for pH determination. The authors thus were constrained to the comparison of sequential runs of cells through their flow microfluorometer. Flow microfluorometers with two lasers are available, but have not yet been used successfully for this purpose. Valet et al [1981] used 2,3-dicyano-hydroquinone for the fluorometric determination of intracellular pH. This dye absorbs at 350 nm, and the ratio of the fluorescence emission intensities at 430 nm and 540 nm is a function of pH. This appears to be a successful approach to the problem, but the required ethyl ester of the dye is, as yet, difficult to obtain. In the study reported here, a method has been developed to measure the relative intracellular pH of lymphocytes by flow microfluorometry using 4-methylumbelliferone (4MU, 7-hydroxy-4-methyl coumarin) as the pH-sensitive fluorometric indicator.

II. METHODS

A. Intracellular pH

The fluorescence of 7-hydroxy coumarin derivatives depends largely on the concentration of the anionic form of the molecule, and thus on the pH (Fig. 1), as has been demonstrated by Grünhagen and Witt [1970], Gomes [1971], and Nakashima et al [1972]. In principle, it is thus possible to determine pH from the fluorescence emission intensity resulting from excitation at only one wavelength, if total concentration is constant or known. This is in contrast to fluorescein, which requires excitation at two wavelengths [Visier et al 1979; Heiple and Taylor, 1980]. A number of 7-hydroxy coumarins (umbelliferones) were examined for suitability as intracellular pH indicators, and 4-methylumbelliferone (4MU) was chosen because of its large change in fluorescence with pH, its relatively low toxicity, and its uniform intracellular distribution. In addition, it may be excited by an available laser line at 350 nm, and is moderately impermeable to the cell membrane (see below).

Variations in cell size and in dye uptake make it desirable to correct the fluorescence intensity measurement for the amount of dye in the cell. This can be done by taking the ratio of the emission intensities at two different wavelengths. The ratio of fluorescence emission intensities of 4MU at 450 nm and 550 nm is approximately linear over the pH range pH 7 to pH 8 (Fig. 1), and is independent of concentration up to the limit of solubility at the most acid pH (~ 0.2 mM). The data of Figure 1 were obtained with a conventional spectrofluorometer (Kontron FM-23); but the result obtained depends on both the bandwidths and spectral sensitivities of the instrument at each wavelength. Since these bandwidths and sensitivities cannot be replicated in a flow microfluoro-

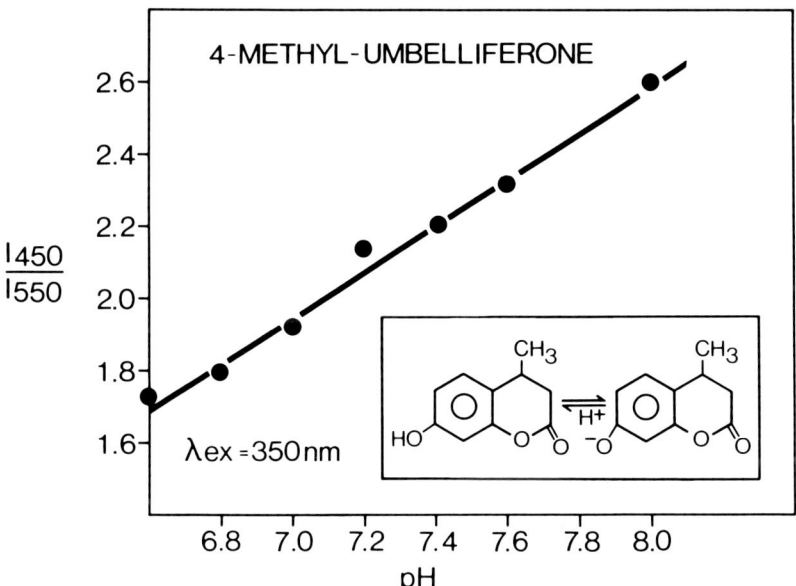

Fig. 1. The relation between the fluorescence emission intensities at 450 nm and 550 nm (I450/I550, × 0.1), at an excitation wavelength of 350 nm, for the dye 4-methylumbelliferone (inset) vs pH.

meter, it is difficult to obtain a suitable calibration allowing quantitative pH determination independent of a second method, such as DMO. We have thus used the emission intensity ratio only as a qualitative measure of pH_i.

Cells to be measured by flow microfluorometry were suspended in Iscove's modified Dulbecco's medium (IMDM, Gibco No. 785220), pH 7.3, plus 10% (v/v) fetal calf serum, 10 mM HEPES, 10 mM TRICINE and 10 mM PIPES (Sigma). An aliquot of cell suspension was diluted 1:1 with the same medium containing 1 mM 4MU, and the suspension was incubated at 37°C in 5% CO_2/95% air for 30–45 minutes prior to analysis. Washed cells were then studied with a FACS II (Becton-Dickinson) flow microfluorometer, with Dulbecco's phosphate buffered saline as sheath fluid. Excitation was provided by a Spectra-Physics model 164 laser operating at 350 nm. Emission was filtered with Balzers interference filters at 451 ± 5 nm and 550–580 nm with EMI 9798B photomultipliers. The log of the emission signal was taken on line by an amplifier designed and constructed by Dr. H. Koller, F. Hoffmann-La Roche, Basel. The difference between the log output at each wavelength gives the log of the fluorescence emission ratio, I450/I550.

Intracellular pH was also determined with ^{14}C-5,5-dimethyl-2,4-oxazolidizedione (DMO, New England Nuclear). Cells were equilibrated with DMO and ^3H_2O for 30 minutes; a parallel sample was equilibrated with ^{14}C-inulin (Amersham) and ^3H_2O to determine cell water volume. Cells were separated from the incubation mixture by the silicone-oil-microfuge technique described by Kiefer et al [1980]. The method of calculating pH_i from DMO distribution data is given by Gillies and Deamer [1979], but this was modified to allow correction for dead cells. It was independently shown that trypan blue-admitting (dead) cells had the same pH as the bulk medium. Trypan blue (0.02%) was also used to counterstain for dead cells on the FACS.

B. Other Methods

Cells used in this study were mouse (Balb/c) spleen lymphocytes obtained in the usual manner [Schreier, 1979]. These were cultured at an initial cell density of 1 × 10⁶/ml in IMDM as described above. Lymphocytes were stimulated to blastogenesis by the addition of concanavalin A (Con A, Pharmacia, 5 μg/ml). ^3H-thymidine incorporation was determined by incubating the cells with ^3H-thymidine (2 μg/ml, Amersham) for 1 hour, collecting the cells on glass fiber filters, and washing in distilled water. Results are expressed as uptake rate per 10⁶ live (trypan blue-excluding) cells.

III. RESULTS AND DISCUSSION

Con A-stimulated lymphocytes reach their maximum rate of DNA synthesis and intracellular pH approximately 48 hours following stimulation (see Gerson and Kiefer, 1981 and Gerson, this volume, section III). Table I gives values for

TABLE I. Intracellular pH, Determined with DMO, and the Rate of ^{3}H-Thymidine Uptake for Unstimulated and Con A-Stimulated Murine Spleen Lymphocytes

Lymphocytes (48 h)	Thymidine uptake rate (10^3 cpm/10^6 cells/h)	Intracellular pH (DMO method)
Unstimulated	2	7.22
Con A-stimulated	63	7.48

the rate of DNA synthesis and the intracellular pH (DMO method) for lymphocytes stimulated 48 hours previously with Con A, and for a control, unstimulated population. This system was used for the development of the flow microfluorometric determination of pH_i, since it provided a mixed population having two approximately equal subpopulations widely differing in intracellular pH and size.

Lymphocytes were stained by equilibration with 4MU, then analyzed with the FACS. Figure 2 illustrates the type of raw data obtained. Figure 2a is the pattern obtained with 48-hour, Con A-stimulated lymphocytes: The vertical axis is the log of the fluorescence intensity ratio; the horizontal axis is the light scatter signal and is roughly proportional to cell size. Each cell is represented by a single point, and data from 10,000 cells were accumulated in each display. The large cells with high fluorescence (pH_i) are the Con A blasts, the small cells with low fluorescence (pH_i) are the unstimulated B cells. Cell size increases appreciably during Con A-stimulated blastogenesis, and the lack of appreciable size dependence in the fluorescence signal indicates that the fluorescence intensity ratio technique has successfully corrected for the quantity of dye in each cell. Figure 2b gives the pattern obtained with unstimulated spleen lymphocytes: Cell size has remained small and the fluorescence (pH_i) is low, much as the B cell subpopulation analyzed in Figure 2a. The data of Figure 2 obtained with 4MU and flow microfluorometry confirm the data of Table I, obtained with the DMO method. Both demonstrate that the average intracellular pH of the rapidly growing Con A-stimulated lymphocyte population is significantly greater than the pH_i of unstimulated lymphocytes. The 4MU technique further demonstrates that, as expected, only half of the total population is actually stimulated to blast formation, and that this half of the population has a high intracellular pH, whereas the unstimulated half of the population has a low pH_i.

It is also possible to introduce 4MU into the cells by using 4MU-acetate. The acetate diffuses into cells more rapidly than does free 4MU, and is subsequently hydrolyzed by intracellular esterases in a manner analogous to the uptake and hydrolysis of fluorescein diacetate [Rotman, 1973]. The insolubility of 4MU-acetate limits the amount of staining that can be achieved to somewhat less than that obtained by equilibration with free 4MU. But the results obtained by both methods are quite similar, as can be seen by comparison of Figures 2 and 3.

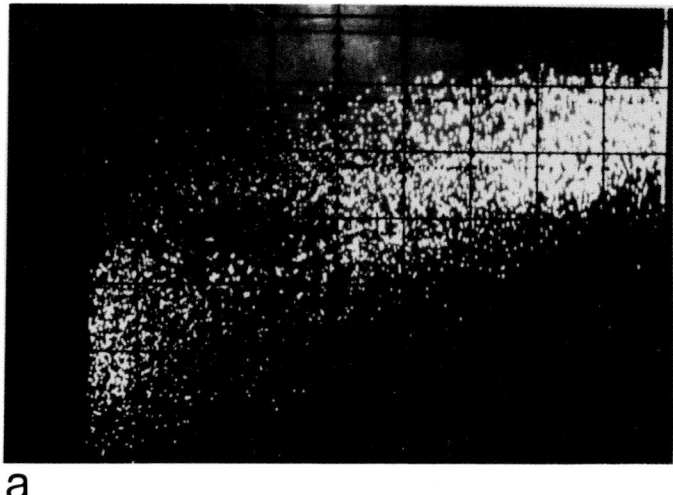

Fig. 2. Display from the FACS of data for 10,000 individual cells stained with 4-methylumbelliferone. Each dot is the datum from one cell. Vertical axis: log I450/I550, or intracellular pH; horizontal axis: light scatter, or cell size. a) Con A-stimulated murine spleen lymphocytes (48 h). b) Unstimulated lymphocytes.

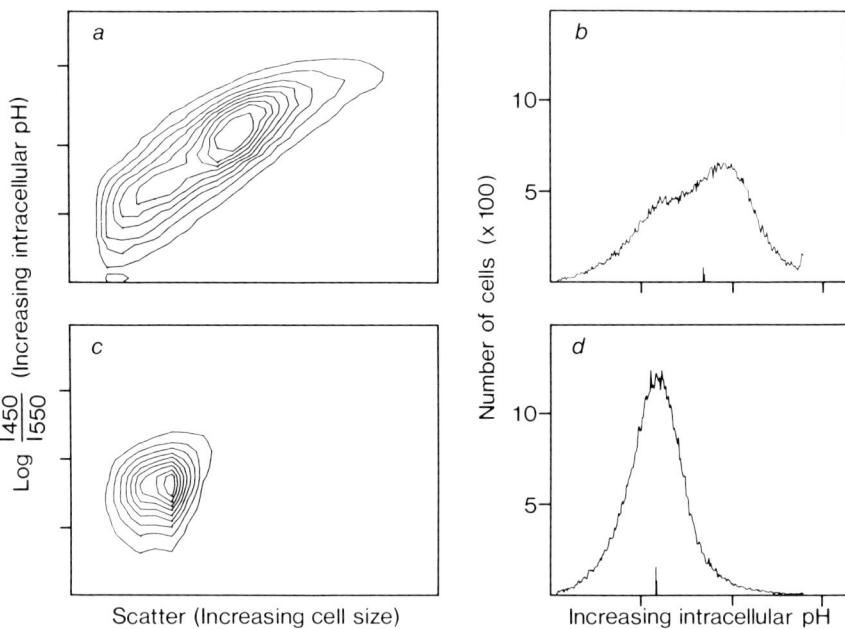

Fig. 3. Computer-generated contour plots (a, c) and histograms (b, d) of data for 100,000 cells stained with 4-methylumbelliferone acetate. a, b) Con A-stimulated murine spleen lymphocytes (48 h). c, d) Unstimulated lymphocytes. The tic mark below the histograms indicates the median of the population.

Figure 3 gives the results obtained with 4MU-acetate staining of both unstimulated and Con A-stimulated mouse spleen lymphocytes. Figures 3a and 3c are comparable to Figures 2a and 2b, and give the relation between the log of the fluorescence intensity ratio (vertical axis) and cell size (horizontal axis) for each lymphocyte population. Each distribution consists of the data obtained from 10^5 cells. The stimulated population has two components, one large and of high pH_i and one small and of low pH_i, whereas the unstimulated population has remained small and has a low pH_i. Figures 3b and 3d are histograms of the numbers of cells (vertical axis) having a given log fluorescence intensity ratio (pH_i, horizontal axis). The median pH_i has increased following Con A stimulation, in accordance with the DMO results of Table I. Figure 3b also demonstrates that pH_i is bimodally distributed in the mitogen-stimulated population, as expected.

Flow microfluorometry is most ideally suited to the measurement of mixed populations of cell types since it allows rapid analysis of 10^4–10^6 cells. This is in sharp distinction from the other favored methods for measuring pH_i. The DMO technique gives a population-average pH_i, also for a large number of cells (10^6–10^7), but with no possibility of identifying or measuring subpopulations. Micro-pH electrodes allow measurement of individual cells, but their use is cumbersome, limiting the number of cells that can be measured. Both DMO and pH microelectrodes can be used on tissues, nerves, and muscle fibers, which cannot be studied by flow microfluorometry. By comparison to an external method, such as DMO, and the courageous application of some assumptions concerning the behavior of the dye, it is possible to assign pH values to the fluorescence intensity ratios obtained on the FACS. This method for measuring intracellular pH with a FACS thus fills a gap in the previously existing techniques.

The most useful aspect of the method described here is that populations of suspended cells can be rapidly analyzed for changes in intracellular pH, even if these changes occur in only a small subpopulation. This subpopulation can be subsequently sorted on the basis of pH_i (or cell size or a secondary fluorescence marker) and separately recultured. This allows rapid screening of experimental situations which result in physiological alterations of intracellular pH.

ACKNOWLEDGMENTS

Expert technical assistance was provided throughout this study by Tim Knaak and Wolfgang Eufe. The advice and encouragement of Dr. C. Steinberg and Dr. H. Kiefer is greatly appreciated.

The Basel Institute for Immunology was founded and is supported by Hoffmann-La Roche and Co., Limited, CH-4002 Basel, Switzerland.

IV. REFERENCES

Deamer DW, Prince RC, Crofts AR: The response of fluorescent amines to pH gradients across liposome membranes. Biochim Biophys Acta 274:323, 1972.

Gerson DF, Burton AC: The relation of cycling of intracellular pH to mitosis in the acellular slime mould Physarum polycephalum. J Cell Physiol 91:297, 1976.

Gerson DF: Intracellular pH and the mitotic cycle in Physarum and mammalian cells. In Jeter J, Cameron I, Padilla G, Zimmerman A (eds): "Cell Cycle Regulation." New York: Academic Press, 1977, p 105.

Gerson DF, Kiefer H: Intracellular pH and membrane potential changes following specific mitogenic stimulation of murine B and T lymphocytes. In Resch K (ed): "Mechanism of Lymphocyte Activation." Amsterdam: Elsevier (in press).

Gillies R, Deamer D: Intracellular pH, methods and applications. Curr Topics Bioenerg 9:63, 1979.

Gomes DJS: Fluorescence studies. I. Variations of fluorescence intensity of methylumbelliferone solutions at various pH values and concentrations. Rev Port Farmacia 21:54, 1971.

Grünhagen HH, Witt HT: Umbelliferone as indicator for pH changes in one turn-over. Z Naturforsch 256:373, 1970.

Heiple JM, Taylor DL: Intracellular pH in single motile cells. J Cell Biol 86:885, 1980.

Horan PK, Wheeless LL: Quantitative single cell analysis and sorting. Science 198:149, 1977.

Kano K, Fendler JH: Pyranine as a sensitive pH probe for liposome interiors and surfaces. Biochim Biophys Acta 509:289, 1978.

Kiefer H, Blume A, Kaback H: Membrane potential changes during mitogenic stimulation of mouse spleen lymphocytes. Proc Natl Acad Sci USA 77:220, 1980.

Nakashima M, Sousa JA, Clapp RC: Spectroscopic species of 4-methylumbelliferone in water and in ethanol. Nature Physical Sci 235:16, 1972.

Ohkuma S, Poole B: Fluorescence probe measurement of the intralysosomal pH in living cells and the perturbation of pH by various agents. Proc Natl Acad Sci USA 75:3327, 1978.

Roos A, Boron WF: Intracellular pH. Physiol Rev 61:296, 1981.

Rotman B: Changes in the membrane permeability of human leukocytes measured by fluorochromasia in a rapid flow fluorometer. In Thaer AA, Sernetz M (eds): "Fluorescence Techniques in Cell Biology." New York: Springer-Verlag, 1973, p 255.

Schreier M: In vitro immunization of dissociated murine spleen cell. In Lefkovits I, Pernis B (eds): "Immunological Methods." New York: Academic Press, 1979, p 327.

Valet G, Raffael A, Moroden L, Wünsch E, Ruhenstroth-Bauer G: Fast intracellular pH determination in single cells by flow-cytometry. Naturwissenschaffen 68:265, 1981.

Visser JWM, Jongeling AAM, Tanke HJ: Intracellular pH-determination by fluorescence measurements. J Histochem Cytochem 27:32, 1979.

Intracellular pH: Its Measurement, Regulation, and
Utilization in Cellular Functions, pages 135-160
© 1982 Alan R. Liss, Inc., 150 Fifth Avenue, New York, NY 10011

The Use of Fluorescent Amines for the Measurement of pH_i: Applications in Liposomes, Gastric Microsomes, and Sea Urchin Gametes

Hon Cheung Lee, John G. Forte, and David Epel
Department of Physiology, Medical School, University of Minnesota, Minneapolis, Minnesota 55455 (H.C.L.), Department of Physiology—Anatomy, University of California, Berkeley, California 94720 (J.G.F.), and Hopkins Marine Station, Stanford University, Pacific Grove, California 93950 (D.E.)

I.	Introduction	136
II.	Methods	136
	A. Liposome Preparation	136
	B. Preparation of Gastric Microsomes	137
	C. Isolation of Sea Urchin Gametes	137
	D. Fluorescence Measurements	137
	E. Fluorescence Microscopy	137
	F. Flow Dialysis	137
	G. Assessment of Sperm Motility	138
	H. Scoring of Acrosome Reaction	138
	I. Media	138
III.	Results and Discussion	138
	A. Probes That Exhibit Decrease in Quantum Yield in the Presence of a pH Gradient	138
	1. The principle; 2. Quantitative verification; 3. Applications to the study of H^+ transport in gastric microsomes; 4. Applications to the study of sea urchin sperm motility initiation	138
	B. Probes That Exhibit pH-Dependent Shift in Fluorescence Spectra	149
	1. The principle; 2. Quantitative verification; 3. Biological application	150
	C. Probes That Exhibit Concentration Dependent Shift in Fluorescence Spectra	154
IV.	Conclusions	157
V.	References	158

I. INTRODUCTION

The fundamental role of ions in regulating biological functions has received increasing recognition. Of particular interest is the role of H^+; in virtually all biological reactions that have been studied, the dependency on H^+ concentration has been demonstrated. It is therefore not surprising that biological systems would utilize H^+ to control and regulate their functions. With the recent advances in technology, such as NMR, microelectrodes, and fluorescence microscopy, the internal pH (pH_i) of an increasing number of systems has been measured and shown to have changes that correlate with their functions. Many of these techniques and their applications to a wide variety of systems are described in this volume.

In this chapter, we shall be concerned with the use of fluorescent amines for the measurement of intravesicular pH. This approach is based on the general principle of weak base (or acid) distribution as described in detail in a preceding article [Roos and Keifer, this volume] and can be summarized in simple terms as follows: At any given pH, the charged and uncharged form of a base (or an acid) is in equilibrium with the H^+ concentration. If the uncharged form is freely permeable and the charged form is not, the uncharged probes pass through the membrane to become charged and trapped in the more acidic (basic) internal environment. This process continues until the internal charged and uncharged probes attain equilibrium with the pH_i. The resultant effect is the concentration of probes into the internal space in accordance to the existing pH gradient. This concentrating effect can be utilized in conjunction with various spectral properties of fluorescent probes for the development of simple methods to monitor pH_i.

We shall describe three types of fluorescent probes that we have successfully employed in various systems (liposomes, gastric microsomes, and sea urchin gametes). They are probes that exhibit 1) decrease in quantum yield in the presence of a pH gradient, 2) pH-dependent, and 3) concentration-dependent shifts in their fluorescence spectra. We shall present quantitative verification for the use of these probes with liposomes as a model system and then apply them to biological membrane vesicles. Finally, sea urchin gametes will be used to illustrate the cellular applications.

II. METHODS

A. Liposome Preparation

Unless indicated otherwise, liposomes were prepared as follows. Crude asolectin from Sigma was partially purified by acetone extraction [Kagawa and Racker, 1971] under N_2. Partially purified asolectin (approximately 50 mg) was dried by a stream of N_2 to evaporate the chloroform; 10 ml of 20 mM glutamate–NaOH buffer at various acidic pH values (2.5–5.0) was added to the dried lipid, and the resulting suspension was sonicated in the cold under N_2 for 10 minutes using a Branson Sonifier (model W350). The pulsed mode of 50% duty cycle was used, and the output power was set at about 75W. All solutions were prepared without Cl⁻—eg, pH adjustment of buffers was carried out with H_2SO_4 and NaOH. This is because the fluorescence of one of the probes, quinine, is strongly quenched by Cl^-.

B. Preparation of Gastric Microsomes

Gastric microsomal vesicles were isolated from porcine fundic mucosal homogenates by differential and density-gradient centrifugation procedures as previously described [Lee and Forte, 1978].

C. Isolation of Sea Urchin Gametes

Shedding of gametes of S. purpuratus and L. pictus was induced by intracoelomic injection of 0.5 M KCl. Sperm were collected "dry" and stored undiluted at 4°C. Egg suspensions were kept at 16°C with constant stirring.

D. Fluorescence Measurements

A Perkin Elmer spectrofluorimeter was used (either model MPF 44A or 204A). All measurements were done at room temperature (21–23°C) except those involving sea urchin sperm, in which case a circulating water bath was used to maintain the cuvette at 19 °C. Wavelengths used for different probes were as follows: 400 → 450 nm (excitation → emission) for 9-aminoacridine; 493 → 530 nm for acridine orange; 425 → 505 nm for quinacrine (artebrin); 347 → 445 nm for the acidic peak of quinine and 340 → 380 for the alkaline peak; 350 → 480 nm for the acidic peak of acridine and 350 → 430 for the alkaline peak; 360 → 500 nm for o-hydroxycinnamic acid.

E. Fluorescence Microscopy

A Zeiss microscope equipped with epifluorescence was used. Micrographs were taken with Kodak Ektochrome 400 film with 10–20 seconds' exposure time and pushed process to 800 ASA.

F. Flow Dialysis

The amount of free dye can be measured by the flow dialysis method [Colowick and Womack, 1969; Ramos et al, 1976; Lee and Forte, 1978]. The apparatus consisted of two chambers separated by a dialysis membrane. The disk-shaped upper and lower chambers (19 mm diameter × 10 mm) both had a capacity of 2.8 ml, the contents of which were mixed continuously by two small magnetic stirring bars. The lower chamber was completely filled with buffer solution pumped through at a constant rate. The dialysate was either collected into fractions or directly fed into a flow-through cuvette monitored by a fluorimeter. After the steady state is reached, the free dye concentration in the

the steady state is reached, the free dye concentration in the upper chamber is directly proportional to the dye concentration in the lower chamber.

G. Assessment of Sperm Motility

Motility was assessed by examination of sperm samples (100 µl) on a serological ring slide using a Zeiss microscope at 40 × magnification with darkfield illumination. Each sample was recorded by a television camera on video tape. The records were viewed by three different observers who were unaware of the experimental conditions, and who were given a rating of 0–100% motility; a 0% corresponds to no movement at all, and 100% indicates that all sperm were motile. The three ratings were averaged and they rarely differed by more than 10%.

H. Scoring of Acrosome Reaction

Sperm samples were fixed with 6% formalin and put on copper grids coated with Formvar. They were then negatively stained with uranyl acetate (0.2%) and viewed with a Hitachi electron microscope. One hundred sperm were counted for each sample.

I. Media

Artificial seawater (ASW) contained 490 mM NaCl, 28 mM $MgSO_4$, 27 mM $MgCl_2$, 10 mM KCl, 10 mM $CaCl_2$, 2.5 mM $KHCO_3$ and 10 mM HEPES, pH 7.7. In Na^+-free artificial seawater (0NaSW), 490 mM choline chloride was used instead of NaCl. In high K^+-artificial seawater (KSW), 490 mM KCl was used instead of NaCl.

III. RESULTS AND DISCUSSION

A. Probes That Exhibit Decrease in Quantum Yield in the Presence of a pH Gradient

Included in this group are 9-aminoacridine (9AA), atebrin, and acridine orange (AO). The first two probes are probably the most widely used probes for monitoring the pH gradient formation in isolated cellular organelles. Atebrin was first discovered to show quenching of its fluorescence in the presence of energized chloroplasts [Kraayenhof, 1970]. This was later interpreted as being due to the formation of a pH gradient across the chloroplast membrane, and various other fluorescent amines (9AA included) were introduced as probes for the pH gradient [Schuldiner et al, 1972]. In the following, we shall use liposomes and gastric microsomes as two model systems to illustrate the principle of using these probes.

1. The principle. We shall first demonstrate that the quenching of the fluorescence is due to the uptake of the probe. A pH gradient can be artificially imposed across gastric microsomal vesicles by incubating them at pH 4.0 in the presence of a permeable organic acid (succinate). The external pH was then quickly raised to pH 8.0 by the addition of small aliquot of Tris base. This method is similar to that used in chloroplasts [Jagendorf and Uribe, 1966]. Figure 1A shows that the imposition of a pH gradient effected a large quenching of the fluorescence of 9AA, which can be reversed by the addition of valinomycin (a K^+-ionophore) and carbonylcyanide-m-chlorophenyl hydrazone (CCCP; a proton ionophore). In Figure 1B, the amount of free dye was directly measured by the flow dialysis method under the same conditions. The close correspondence

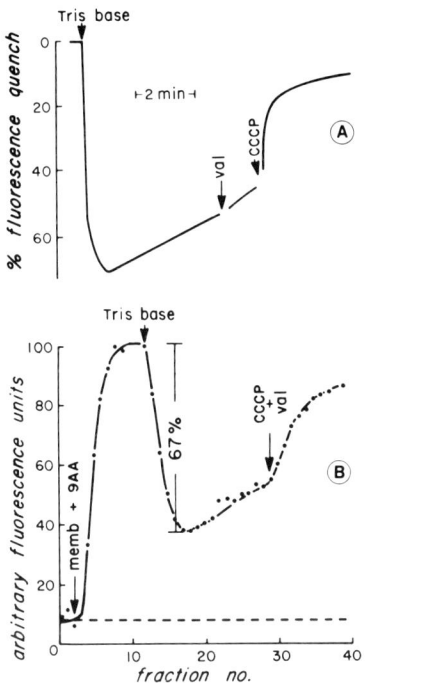

Fig. 1. pH-induced 9-aminoacridine uptake as measured by spectrofluorometric and flow dialysis methods. A) For the spectrofluorometric method, gastric microsomes (0.5 mg) were incubated in 2 ml medium containing 150 mM KCl, 10 μM 9-aminoacridine (9AA), and 10 mM succinate at pH 4.0 for 5 min. 25 mM Tris-base was added to bring the external pH to 8. Assuming that the vesicular interior equilibrated to about pH 4 in the presence of the low pH-succinate system, the addition of base bringing the external medium to pH 8 would thus impose a H^+ gradient of 4 pH units across the gastric vesicles. Valinomycin (val) and CCCP (2 μM each) were used to dissipate the H^+ gradient. B) For the flow dialysis measurement, the conditions in the upper chamber were exactly as described in (A). In the lower chamber, 150 mM KCl solution flowed through at a rate of 7.4 ml/min and collected every 0.5 min. The change induced by the acid-base shift was 67% of the 9-aminoacridine fluorescence, which agreed well with the 68% dye uptake measured by the spectrofluorometric method.

between the two independent measurements indicated that the quenching of the fluorescence is indeed a good measure of uptake of the probe.

We shall next consider the mechanism of the quenching. Various suggestions have been advanced: for example, screening by pigments in the membrane (chloroplasts), or interaction or binding to membrane components. Perhaps the most plausible explanation is the concentration-dependent self-quenching mechanism suggested by Deamer et al [1972]. According to this mechanism, the probe distributes across the membrane according to the existing pH gradient. The concentrating effect of the pH gradient would result in such a high internal probe concentration that energy transfer between molecules of the same species may cause a decrease in quantum yield. If this is true, then the extent of quenching at a given pH gradient should be strongly dependent on the total probe concentration. One could, presumably, lower the total concentration such that the internal concentration does not achieve the level where self-interaction is important and consequently eliminate the quenching phenomenon. Furthermore, under this condition, even though no quenching is observed, the probe should still be accumulated by the pH gradient.

To verify this postulate we prepared liposomes at pH 4.0 and diluted them into a buffer of pH 8.2, thus creating a pH gradient across the liposomes. The extent of quenching of 9AA fluorescence by these liposomes was measured at different concentrations of 9AA. As shown in Figure 2A, a 20-fold reduction in the total concentration of 9AA completely eliminated the pH gradient-induced quenching. However, when the flow dialysis method was used to measure the amount of free dye at the lowest 9AA concentration, a large proportion of the probes was still being taken up by the liposomes (Fig. 2B). The accumulated probes can be released by nigericin (K^+/H^+ exchanger), indicating that the uptake by the liposomes was due to the pH gradient.

Summarizing, we can conclude that this group of fluorescent amines behave as a weak base that distributes across the membrane in accordance with the existing pH gradient. At sufficiently high probe concentration, the pH gradient resulted in such a high internal probe accumulation that the fluorescence was quenched by self-interaction. This quenching can be so effective that it can be used as a direct measure of probe uptake.

2. Quantitative verification. If these fluorescent amines behave as a weak base and distribute in accordance to the pH gradient across the membrane, their distribution should conform with the following equation [Schuldiner et al, 1972; Deamer et al, 1972; Lee and Forte, 1978]:

$$pH_o - pH_i = \log(A_i/A_o) + \log(V_o/V_i) \tag{1}$$

where the subscripts i and o refer to internal and external quantities, A is the

pH$_i$ Measurement With Fluorescent Amines / 141

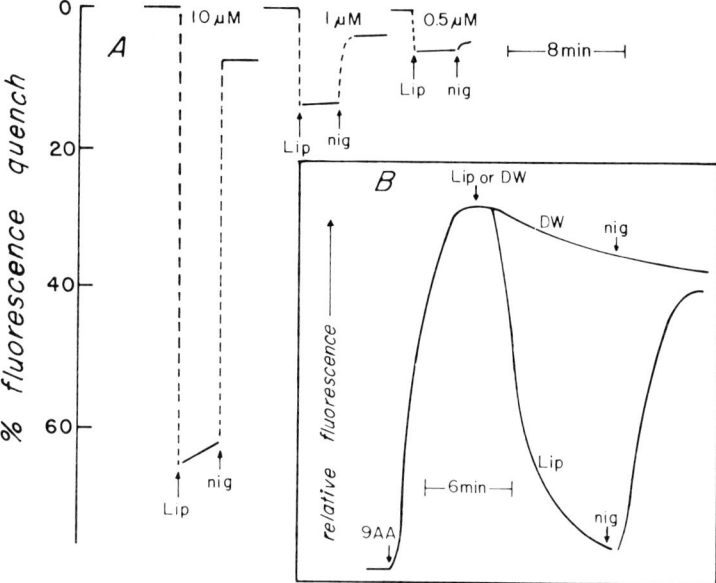

Fig. 2. The dependence of the ΔpH-induced fluorescence quenching on the total concentration of 9-aminoacridine (9AA). Liposomes were prepared by the sonication method as follows: Purified egg phosphatidyl choline (50 mg) in 1 ml of chloroform was dried down by a flow of N$_2$. Ten milliliters of 20 mM citrate buffer at pH 4.0 was added and the resulting mixture was sonicated for 50 min at 0–5°C as described in the Methods. The undispersed lipid was removed by centrifugation at 40,000g for 40 min at 0–5°C. A) A pH gradient was imposed across the liposomes by adding 100 μl of the liposome suspension (arrows labeled Lip) to 1.9 ml of 50 mM potassium borate buffer at pH 8.2 containing 0.5–10 μM of 9AA. The resulting fluorescence quenching was recorded and 0.5 μg/ml of nigericin (arrows labeled Nig) was added to discharge the pH gradient. B) The free 9AA concentration was monitored with the flow dialysis method. The upper chamber contained 1.9 ml of K-borate buffer at pH 8.2. The same buffer was passed through the lower chamber at a rate of 1.6 ml/min and fed into a flow-through cuvette. A small aliquot of 9AA was added (arrow labeled 9AA) to the upper chamber, giving a final concentration of 0.5 μM. The increase in the fluorescence intensity in the dialysate was monitored continuously. At the time indicated by the arrow labeled Lip, 100 μl of liposomes was added to the upper chamber. In a parallel experiment, the same volume of distilled water (DW) was added as control. At the end of the experiment the pH gradient was discharged by adding nigericin (0.5 μg/ml) to the upper chamber.

total amount of probe, and V is the volume. Since the percentage of fluorescence quenching (Q) is a direct measure of dye uptake, we can take the expression Q/(100−Q) as equivalent to A_i/A_o. The assumptions made in deriving this equation are: 1) The probe has only one ionizable group; 2) the uncharged form is freely permeable to the membrane; 3) the charged form is not permeable; and 4) the pK of the probe is much greater than either the internal or external pH.

According to equation 1, if the probe is behaving ideally, a plot of $\log(A_i/A_o)$ vs pH gradient should be a straight line with a slope of 1, and the intercept on the ordinate should be equal to $\log(V_o/V_i)$. A pH gradient of various magnitude was artificially imposed across gastric microsomes, and the resultant dye uptake was measured. The results were plotted in Figure 3. In the case of 9AA, the data approximated a straight line with a slope closely corresponding to the theoretical value of 1. From the intercept a vesicular volume was calculated as 1.6 µl/mg protein, which is consistent with the Cl space of 2 µl/mg protein measured by radioactive Cl equilibrium distribution [Lee and Forte, 1978].

In the case of acridine orange, also shown in Figure 3, the plot of $\log(A_i/A_o)$ vs the pH gradient also gives a straight line, but with a slope of 0.63 instead of 1 and an intercept giving a vesicular volume one to two orders of magnitude higher than 9AA. In spite of this complication, however, it is clear that the quenching of AO fluorescence is still directly related to the pH gradient.

We suggested that the binding of AO to some negative sites on the internal face of the membrane could explain this non-ideal behavior, and thus as the probe is accumulated by the pH gradient, and additional equilibrium between the internal dye and the binding sites has to be established [Lee and Forte, 1978]. Mass action law would then predict an increase in binding to these sites with increasing ΔpH, owing to higher internal dye concentration. The binding, therefore, does not invalidate the use of these probes if it is reversible; it does, however, present a complication to the quantification of the gradient. As long as one is content with qualitative changes (which in most cases are as important as the absolute value of the gradient), these probes are still very useful and highly sensitive tools for monitoring the pH gradient. We shall illustrate this point with two concrete applications: H^+ transport in gastric microsomes and in the initiation of sea urchin sperm motility.

3. Applications to the study of H⁺ transport in gastric microsomes.

The acid-secreting cells (oxyntic cells) of the gastric mucosa undergo dramatic morphological transformation when the tissue is stimulated to secrete acid. In the resting state, oxyntic cells are full of tubulovesicles which fuse with the plasma membrane and become long, extensive microvilli upon stimulation. These tubulovesicles, which can be isolated as microsomal vesicles from fundic gastric mucosa of several species, have been shown to possess K^+-stimulated, Mg^{+2}-dependent, ATPase activity [Forte et al, 1974]. Immunocytochemical and other studies have demonstrated that these membranes are largely derived from oxyntic cells of the gastric mucosa [Chang et al, 1977; Granser and Forte, 1973]. Furthermore, the gastric microsomal vesicles transport (accumulate) H^+ in the presence of K^+, Mg^{+2}, and ATP [Lee et al, 1974; Chang et al, 1977]. In a series of studies, we applied fluorescent amine probes to the analysis of the H^+-transport properties of these gastric microsomes [Lee et al, 1976, 1979; Forte

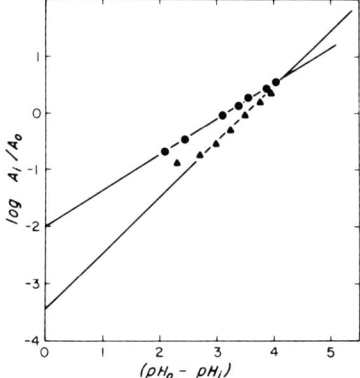

Fig. 3. Calibration curve for 9-aminoacridine and acridine orange uptake as a function of pH gradient. Data were plotted in accordance with equation 1 using the log A_i/A_o as the ordinate and the pH gradient on the abscissa, where A_o is the amount of dye in the bulk external medium and A_i is the amount of dye uptake. The ratio A_i/A_o is calculated as the ratio $Q/(100 - Q)$ where Q is the percentage of fluorescence quenched. Different pH gradients were imposed across the gastric microsomes as described in the legend of Figure 1. Dye uptake was measured by the percentage of fluorescence quenching. The dye concentration was 10 μM for both 9-aminoacridine (▲) and acridine orange (●); amount of gastric microsomes was 250 μg protein/ml for 9-aminoacridine and 50 μg protein/ml for acridine orange.

and Lee, 1977; Lee and Forte, 1978]. Some of the principal findings will be reviewed here.

The applicability of fluorescent amine probes to the gastric microsome system was first suggested by the empirical observation that they show spectral changes closely correlated with the H^+ movements as monitored by a pH electrode [Lee et al, 1976]. Both changes had a similar time course and were reversed by the combination of valinomycin and CCCP. Further analysis indicated the spectral shifts were also accompanied by a decrease in fluorescence, thus providing a higher sensitivity assay than the absorbance measurements. We then proceeded to calibrate the system by imposing various known pH gradients across the microsomal membranes and measure the corresponding fluorescence changes [Lee et al, 1976; Lee and Forte, 1978]. The results (Fig. 3) indicated that 9AA behaved ideally as a weak base and distributed in accordance to the established pH gradient, whereas AO showed binding in addition to accumulation by the ΔpH.

Active H^+ transport (as monitored by these fluorescent amines) in gastric microsomes required both K^+ and Mg^{+2} in addition to ATP. Moreover, K^+ was needed at some internal site as suggested by the requirement of preincubation in KCl medium to produce high transport rate [Lee et al, 1976]. This is consistent with the suggestions of Lee et al [1974] and Sachs et al, [1976] that the gastric ATPase is a K^+/H^+ exchange pump. Further support comes from the use of the K^+ ionophore, valinomycin, which dramatically stimulated both the ATPase and the H^+ transport activities [Forte and Lee, 1977; Lee and Forte, 1978]. Figure 4 shows this stimulatory effect of valinomycin as monitored by three different probes, 9AA, AO, and atebrine (quinacrine). Since valinomycin is an electrogenic ionophore, its facilitation of the entry of K^+ must be accompanied by an anion (Cl in this case). We confirmed this prediction by showing that valinomycin does increase the rate of passive Cl^- uptake by the vesicles [Lee and Forte, 1978]. Another consequence of this analysis is that if Cl^- is replaced by an impermeable anion, such as isethionate, then valinomycin should be ineffective in transporting K^+ to the internal sites and therefore should lose its stimulating effect on H^+ transport. This was indeed observed as Cl^- was partially replaced by isethionate; stimulation by valinomycin was reduced and, in fact, was virtually abolished when Cl^- was down to 7.5 mM [Lee and Forte, 1978].

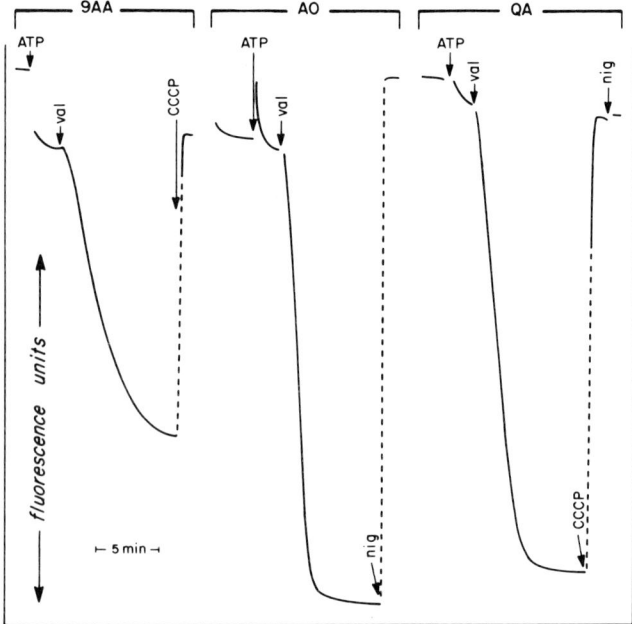

Fig. 4. Valinomycin (val) stimulation effect on fluorescence quenching of 9-aminoacridine (9AA), acridine orange (AO), and quinacrine (QA). Dye concentration was 10 μM in a medium containing 150 mM KCl, 2 mM $MgCl_2$, 10 mM Pipes, 0.1 mM EDTA, pH 7.0, with 80 μg/ml gastric microsomes. Reaction was started by adding 0.5 mM ATP. Valinomycin was added at 2.5 μM. Fluorescence quenching was reversed by either 2.5 μM CCCP or 2.5 μM nigericin (nig).

We can thus conclude that, in the presence of valinomycin, the net effect is the accumulation of HCl inside the gastric vesicles. Under the most favorable condition, a ΔpH of 4 to 4.5 pH units can be generated.

The microsomal ion transport activities have been correlated with the ATPase activity of the system and summarized into the pump-leak model schematized in Figure 5 [Lee et al, 1979]. The model consists of an electroneutral pump that transports one H^+ into the vesicle for each K^+ transported out, and the respective leak, or conductance pathways for H^+, K^+, and anions. A computer simulation of this model has been formulated using 1) Michaelis-Menten kinetics to describe the pump with the substrate being internal K^+, and 2) the well-established Nernst-Planck formulation to model passive pathways. Variations in the permeability coefficients of the contributing ionic species in the computer simulation rather nicely correlate with observed H^+ transport data obtained experimentally using ionophores or ion substitution to affect passive ionic fluxes [Lee et al, 1979].

4. Applications to the study of sea urchin sperm motility initiation. Application of these fluorescent probes to a cellular system such as sea urchin sperm represents a special challenge. The sperm is a multicompartment system and a large portion of the cell volume is occupied by the nucleus whose DNA contents may present many potential binding sites for these probes. However, as mentioned earlier, the presence of internal binding sites would make quantification difficult, but should not invalidate the qualitative response as long

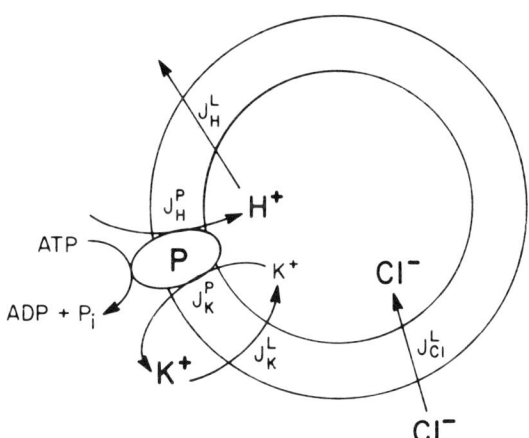

Fig. 5. Hypothetical scheme acccounting for ion movements across gastric microsomal vesicles. The J value are the ionic fluxes with the superscripts designating pump flux (P) or leak pathway (L). The model consists of an ATP-driven H^+/K^+ exchange pump and the passive leak pathways for the principal ions, K^+, H^+, and Cl^-.

as the binding is readily reversible. In fact, the binding phenomenon should amplify the affect of the pH gradient (eg, acridine orange in gastric microsomes as shown in Fig. 3) and thus for a given ΔpH, more probe will be taken up.

It also turns out that the sperm may present a lesser problem than many other cell types, since it is a highly differentiated cell having only a limited number of intracellular compartments. Of these, the only possible acidic compartment consists of acrosomal vesicles, which represent only a very small fraction of the total volume. Nonetheless, the total uptake of probe by the sperm should be treated as some average contributed by different compartments. With this reservation in mind, we proceeded to apply these probes to sea urchin sperm.

The two major events occurring during the activation of sea urchin sperm are initiation of motility and acrosome reaction. Sea urchin sperm are immotile in semen; motility is induced upon dilution of the semen into seawater, and is accompanied by acid release from the sperm. Both activation processes are Na^+-dependent and thus the sperm would remain immotile when semen is diluted into Na^+-free seawater. Addition of Na^+ would then initiate motility and the same amount of acid is released [Nishioka and Cross, 1978]. It is therefore of interest ot see if the internal pH is increased upon initiation of motility and, if so, whether it is the causal trigger.

The other activation event is the acrosome reaction, which involves the fusion of the acrosomal vesicle (containing molecules that will form attachment with the egg) with the plasma membrane and the formation of the acrosomal filament through the polymerization of actin. The acrosomal-reacted sperm will then have the attachment molecules exposed on the surface of the acrosomal filament, and the sperm is then ready for attachment and fusion with the egg. The acrosome reaction is naturally triggered by the interaction with a factor associated with the jelly layer of the egg. It can also be triggered artificially with Ca^{+2} ionophores [Collins and Epel, 1977]. Either way would result in activation of H^+ release and Ca^{+2} uptake by the sperm [Schackman et al, 1978]. It has been postulated that the release of H^+ would result in a corresponding cellular alkalinization, which is the prerequisite for the polymerization of actin, whereas the Ca^{+2} influx is responsible for the fusion event [Tilney et al, 1978]. The sea urchin sperm is therefore a very interesting system for the study of the mechanism of pH_i regulation of biological functions. We shall use 9AA for monitoring pH_i. Essentially the same results have been obtained using another structurally different amine, methylamine [Lee et al, 1980] (Lee et al, in preparation). This lends support for the validity of the results reported here since one would not expect that nonspecific effects, such as binding to cellular components, would be the same for two completely different probes.

Figure 6 shows the time course for the quenching of 9AA fluorescence after the addition of sperm to various artificial seawater solutions. The fluorescence change reached a steady level in about 10 minutes. To demonstrate that the

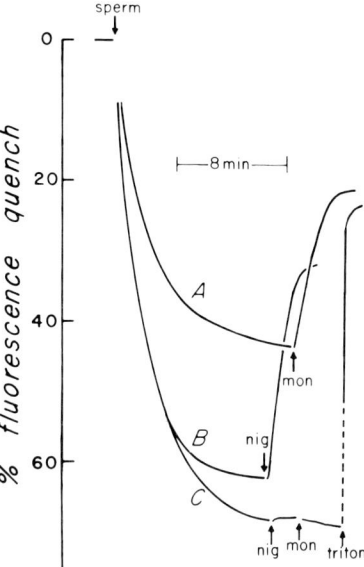

Fig. 6. Quenching of 9AA fluorescence by sea urchin sperm and subsequent reversal with various ionophores. A small aliquot (5 μl) of undiluted semen was added to 2.5 μM 9AA in 2 ml of various media: A) ASW, B) KSW, and C) ONaSW. The resulting fluorescence change was monitored continuously. The ionophores nigericin (nig; 2.5 μg/ml) and monensin (mon; 15 μg/ml) were added to discharge the pH gradient across the plasma membrane of the sperm. This resulted in partial recovery of the fluorescence intensity. A similar effect was also seen with the use of the detergent Triton X-100 (0.1 mg/ml).

observed fluorescence change was due to a pH gradient (internal acidic) across the sperm membrane, the pH gradient was discharged with selective ionophores. The Na^+:H^+ exchange ionophore, monensin, was used first. The high concentration of Na^+ (490 mM) in the artificial seawater should allow monensin to exchange internal H^+, thereby discharging the pH gradient. Addition of monensin to the sperm suspension resulted in partial recovery of the fluorescence as shown in curve A of Figure 6. Similarly, if the Na^+ in ASW is replaced by K^+, addition of the K^+:H^+ exchange ionophore, nigericin, should also discharge the pH gradient and lead to recovery of the fluorescence. This occurred, as shown in curve B of Figure 6. However, neither monensin nor nigericin affected a discharge of fluorescence when choline was used as the principal cation (without Na^+ or K^+), as shown in curve C of Figure 6. However, the nonselective permeablizing agent, triton, did allow fluorescence recovery in the choline-containing seawater. These results demonstrate the specificity of the action of the ionophore. It is noted that in all cases the fluorescence recovery was not complete.

This can be attributed to some irreversible binding to cellular components and therefore not responding to the pH gradient, or alternatively, it may be due to incomplete discharge of gradient in all these treatments.

The above results suggest that a large portion of the probe is sensing the pH gradient, and therefore the fluorescence quenching can be used as a qualitative index of pH_i. We shall then proceed to measure the changes in pH_i associated with sperm activation. Figure 7 shows that the dilution of semen into Na^+-free seawater containing 9-aminoacridine resulted in a dramatic quenching of fluorescence, indicative of an acidic pH_i. The sperm remained immobile under these conditions (0% motility at 1). Addition of 8 mM Na^+ activated motility (90%),

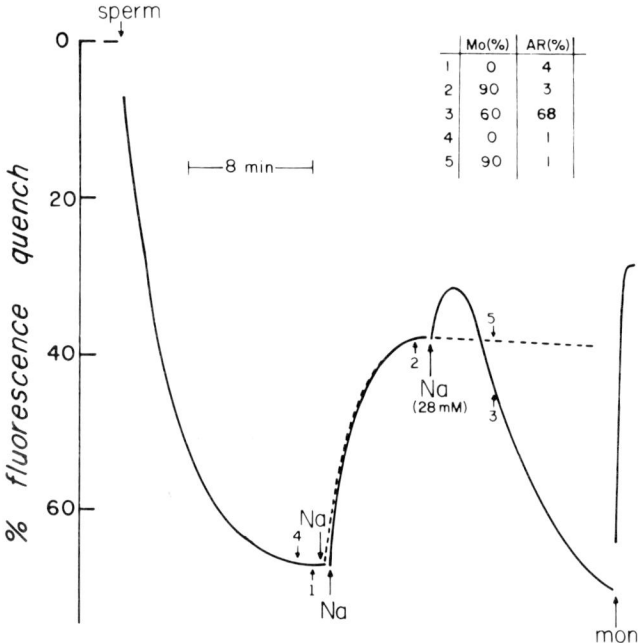

Fig. 7. Changes in 9AA fluorescence associated with Na^+ activation of sea urchin sperm. Sperm (5 μl semen) were suspended in 2 ml of 0NaSW containing 2.5 μM 9AA. As the fluorescence change approached a steady level, a small aliquot of ASW was added (final Na^+ concentration was 8 mM) to activate sperm motility. This resulted in a reversal of the fluorescence quenching (solid line), indicating alkalinization of pH_i. Further increase of Na^+ concentration to 36 mM induced a transient alkalinization and a prolonged reacidification. Addition of 20 μg/ml monensin partially reversed the quenching. In a parallel experiment (dashed curve) the Na^+ concentration was kept at the 8 mM level. At various times indicated by arrows (numerically labeled), small aliquots of samples were taken from the cuvette for assaying of sperm motility (Mo) and acrosome reaction (AR) as described in the Methods. The results are listed on the figure and suggest that the reacidification is associated with the acrosome reaction.

and at the same time caused a large release of the probe indicating that pH_i was alkalinized. This new pH_i remained stable for at least 20 minutes (the dashed curve). However, when more Na^+ was added (final concentration = 36 mM), there was a further but transient increase in pH_i. Measurements at this point indicated that 68% of the sperm had been acrosome-reacted and the sperm motility decreased as the pH_i reacidified. This phenomenon of acrosome reaction triggered by low Na^+ had been observed in another species of sea urchin [Schackman and Shapiro, 1981]. The results therefore demonstrate that a sustained alkalinization of pH_i is associated with initiation of motility, whereas a transient and smaller increase in pH_i resulted from acrosome reaction. An independent study by Schackman and Shapiro in yet another species of sea urchin confirmed the alkalinization of pH_i associated with the acrosome reaction (personal communication). Further studies indicate that the reacidification is due to a mitochondrial function since it can be inhibited by antimycin and is stimulated by uncouplers, such as valinomycin. This cellular acidification is the result of mitochondrial sequestration of the massive Ca^+ influx associated with the acrosome reaction [Schackman and Shapiro, 1981] (Lee et al, in preparation).

We shall next consider the question of whether the increase in pH_i causally triggers sperm motility. Previous study had demonstrated that NH_4Cl can activate motility in the absence of Na^+, presumably through the alkalinization of pH_i [Nichioka and Cross, 1978]. What remains to be shown is that NH_4Cl can indeed increase pH_i and that the change is not simply a cationic substitute for Na^+. Our approach was to block the action of Na^+ by high K^+, and use NH_4Cl to rescue the inhibited sperm. This is shown in Figure 8. When the sperm were diluted into ONaSW containing high K^+ (100 mM), they remained immobile (5% motility at point 1 of Fig. 8). Addition of Na^+ did not produce an alkalinization of pH_i, nor was motility increased (7% motility at point 2). However, further treatment of these inhibited sperm with 10 mM NH_4Cl resulted in an increase in both pH_i and motility (70% at point 3). NH_4Cl was effective independent of Na^+ since its addition in the absence of Na^+ also initiated motility (70% at point 4). These experiments therefore indicate that it is the increase in pH_i that is the trigger and not Na^+ per se. Further support comes from the use of nigericin in high K^+ medium; along with the increased pH_i there was an increase in motility (70% at point 5).

B. Probes That Exhibit pH-Dependent Shift in Fluorescence Spectra

Included in this group are two fluorescent amines, quinine and acridine. The fluorescence spectra of the two amines are pH-dependent. They emit fluorescence at higher intensity and at longer wavelength when they are in a more acidic environment. Since both probes behave similarly, we shall present a detailed characterization for only one of them, quinine. Liposomes will be used as a model system to illustrate the principle, and gastric microsomes will be used to demonstrate the biological applications.

1. The principle. The present method makes use of the pH-dependent shift of the emission maxima of the amines. As the probes distribute into the intravesicular space, they respond to the acidic environment and emit fluorescence at a longer wavelength than the external probes, which sense a more alkaline medium. By measuring both the decrease in the alkaline fluorescence peak and the enhancement of the acidic peak, direct determination of the internal

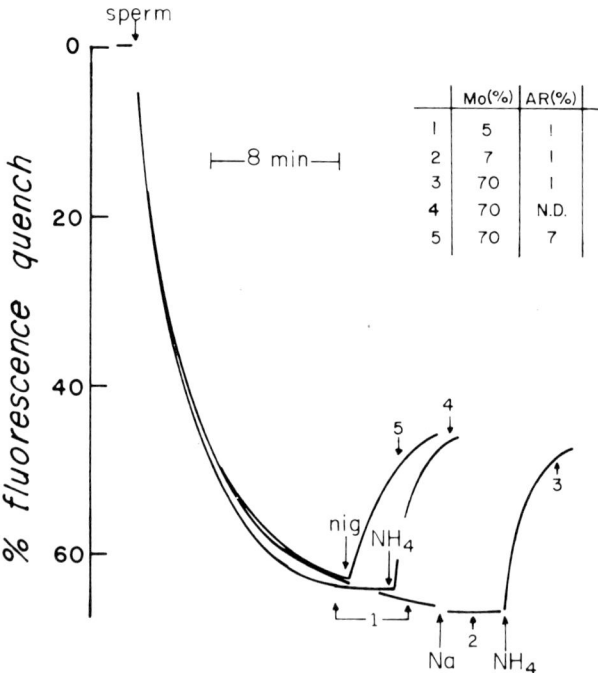

Fig. 8. Induction of sperm motility by ammonium and nigericin, and the associated change in pH_i. Sperm (5 μl semen) were suspended in 2 ml of 0NaSW containing high K^+ (100 mM). The pH_i was monitored qualitatively by the fluorescence changes of 9AA as described in the legend to Figure 7. Under this high K^+ condition, addition of 8 mM Na^+ (arrow labeled Na) did not activate motility (arrow labeled 2) or induce alkalinization of pH_i. On the other hand, treatment with 10 mM ammonium (arrow labeled NH_4) was able to alkalinize the pH_i and activate motility. The effect of ammonium is independent of Na^+ since its addition before Na^+ also produced the same results. Similarly, nigericin (1.8 μg/ml at the arrow labeled nig) alkalinized pH_i under the high K^+ condition and activated motility in the absence of Na^+. Sperm motility (Mo) and acrosome reaction (AR) were assayed at various times indicated by arrows (numerically labeled) by taking aliquots from the fluorescence cuvette, and were measured as described in the Methods.

pH can be obtained. A more detailed derivation of the method for using this shift in the spectrum to calculate intravesicular pH without the need of determining the vesicular volume is available elsewhere [Lee and Forte, 1980].

2. Quantitative verification. The above principle can be critically assessed by preparing liposomes of various pH_i and comparing the measured pH_i to that of the actual pH_i.

Figure 9 shows the emission spectrum of quinine at various pH values. It can be seen that the fluorescence intensity at 445 nm increases with decreasing pH, whereas that at 380 nm shows the reverse. Figure 10 shows the time course for the changes in fluorescence intensity of quinine at 445 nm (A, the acidic peak) and 380 nm (B, the basic peak) induced by the pH gradient. According to our

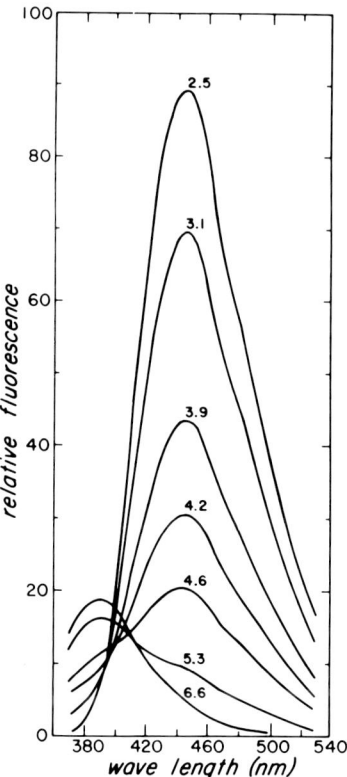

Fig. 9. Effect of pH on the fluorescence emission spectrum of quinine. Quinine (2.5 μM) was added to 2 ml of 20 mM citrate buffer at pH 2.5. The emission spectrum was recorded with the excitation wavelength set at 347 nm. Small aliquots of concentrated NaOH were added to titrate to the various pH values as indicated on the individual recorded spectra.

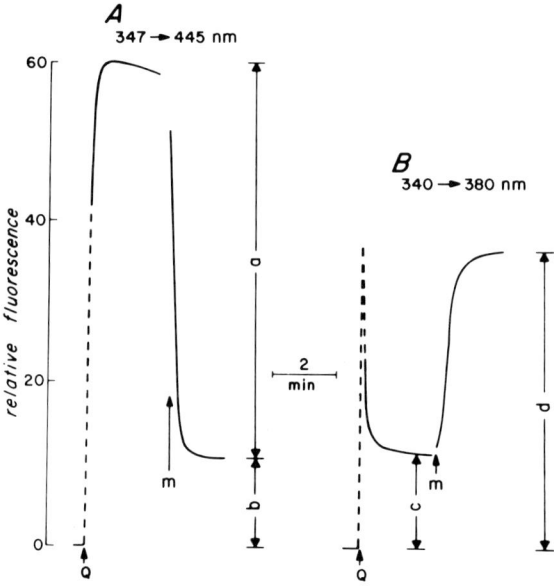

Fig. 10. Time course of the fluorescence changes induced by ΔpH. Internal acidic liposomes (pH_i = 2.5) were diluted 20-fold into a medium containing 25 mM citrate at pH 5.1 to create a ΔpH across the liposomes. Final liposome concentration was 0.25 mg/ml. A small aliquot of quinine was added to give a final concentration of 0.25 μM (indicated by the arrow labeled Q). The resulting increase in fluorescence intensity at 445 nm (excitation 347 nm) is shown in A; the concomitant decrease at 380 nm (excitation 340 nm) is shown in B. The respective reversal of fluorescence due to addition of 2.5 ng/ml of the Na^+-H^+ exchange ionophore monensin (arrow labeled m) is also shown.

previous analysis [Lee and Forte, 1980], the fraction of probe (r_o) remaining on the outside can be obtained from the ΔpH-induced decrease in fluorescence intensity at 380 nm. The fluorescence intensity in the presence of a pH gradient (c in Fig. 10B) is proportional to the amount of probe on the outside, and the intensity after the addition of the monensin used to discharge the pH gradient (d in Fig. 10B) should be proportional to total amount of probe. The fraction of the probe that remains on the outside (r_o) in the presence of a pH gradient is therefore the ratio of the two:

$$r_o = c/d \qquad (2)$$

The fraction that goes into the liposomes (r_i) should then be:

$$r_i = 1 - r_o \qquad (3)$$

To calculate the intravesicular pH we need to know the fluroescence intensity due to the intravesicular probe (f_i). This can be obtained from the difference between the total fluorescence at 445 nm and that contributed by the probe in the outside bulk medium. Thus:

$$f_i = (a + b) - br_o = a + b(1 - r_o) = a + br_i \qquad (4)$$

where a and b are the portions of the total fluorescence as defined in Figure 10A. Dividing f_i by r_i we get the total fluorescence (F_i) if all the probes were on the inside:

$$F_i = f_i/r_i = (a + br_i)/r_i \qquad (5)$$

By comparing F_i with a fluorescence vs pH curve constructed with the same concentration of probe, the intravesicular pH can thus be obtained directly. The accuracy of the pH thus determined requires that the fluorescence of the intravesicular probe respond only to internal pH and not be quenched by binding or any other factors. To evaluate this critically, we prepared liposomes of various internal pH while the external medium was fixed at pH 6.0. If the above reasoning is correct, and the intravesicular probe responds only to internal pH, then we should be able to calculate the exact internal pH of the liposomes from the fluorescence measurements. The results of such a calibration are shown in Figure 11 where the calculated internal pH is plotted against the actual pH at which the liposomes were prepared. As can be seen the measured values fall closely along a slope of 1, indicating excellent correlation.

In principle, the reasoning presented here should be equally applicable to the measurement of intravesicular pH of internally alkaline vesicles. However a fluorescent weak acid is needed instead of an amine. This fluorescent acid should have a pK_a lower than the internal pH. Similar to the amine probe, the undissociated form of the acid should be freely permeable to the membrane and be converted to the charged form when it senses the alkaline internal pH. If the charged form is impermeable then the resulting probe distribution should be in accord with the ΔpH across the membrane. We have shown [Lee and Forte, 1980] the applicability of this principle using o-hydroxycinnamic acid in phospholipid vesicles prepared with various alkaline interior environments (pH 8.5–10.5). Unfortunately, there is no pH-dependent spectral shift for o-hydroxycinnamic acid, and it was not possible to calculate the internal pH without an independent estimate of probe distribution. However, the material can be used as a qualitative probe of intravesicular alkalinization.

3. Biological application. The amine probes were tested using gastric microsomes. The time-course of fluorescence intensity for gastric microsomes

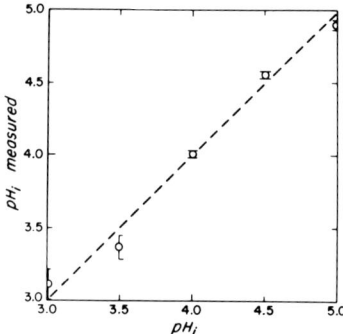

Fig. 11. Correlation between the internal pH of liposomes (pH_i) and the pH value measured by the present method (pH_i measured). Liposomes were prepared at various pH values in 20 mM glutamate according to the values represented on the horizontal axis. The liposomes were diluted 20-fold into a medium containing 25 mM citrate at pH 6.0 in order to produce the ΔpH. The final liposome concentration was 0.25 mg/ml. The measurement of internal pH (pH_i measured) was performed with 0.25 μM quinine as described in the text. The standard curve of fluorescence intensity vs pH was constructed with the same buffer (20 mM glutamate) as that used in preparing the liposomes. The indicated values are the mean of three separate measurements on the same liposome preparation with the bars representing the standard deviation. Perfect correlation between the predicted and measured values is represented by a 45° line (dashed line).

incubated with quinine is shown in Figure 12. Activation of the H^+-transport system by ATP produced a small increase in fluorescence intensity at 445 nm (Fig. 12A) and a concomitant decrease in intensity at 380 nm (Fig. 12B). Addition of valinomycin greatly stimulated the formation of the pH gradient and, therefore, also the fluorescence changes as shown by the recordings. Nigericin was used to discharge the ΔpH and reverse the fluorescence change. Using the data of Figure 12, we calculated the internal pH to be 3.8, whereas the external pH was fixed at pH 6.7, corresponding to a ΔpH of 2.9 pH units. An identical result was obtained using 9-aminoacridine as a ΔpH probe under the same conditions (data not shown).

C. Probes That Exhibit Concentration-Dependent Shift In Fluorescence Spectra

In this section, we shall discuss only one probe, acridine orange. As shown in section A, acridine orange shows concentration-dependent quenching of its monomer fluorescence (535 nm). However, as its concentration increases to such a high level that multimeric dye aggregates are formed, a new fluorescence emission peak appears at about 640 nm, although its quantum efficiency is much lower than the monomer fluorescence. The polymeric fluorescence can also be induced by binding to certain polyanions such as DNA or RNA. This latter

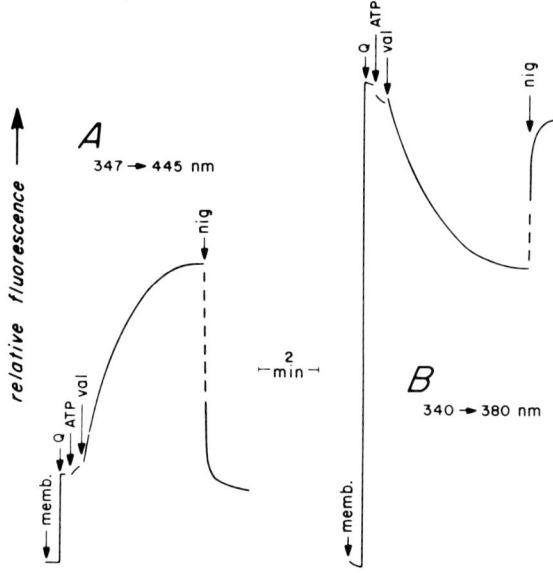

Fig. 12. Time course of fluorescence change of quinine induced by activation of H$^+$-transport system in gastric microsomes. Quinine (0.5 μM) was added (indicated by arrows labeled Q) into a medium containing 150 mM KNO$_3$, 1mM MgSO$_4$, 10 mM Pipes (pH 6.7), 0.05 mM EDTA, and 0.13 mg/ml gastric microsomes. The H$^+$-transporting system in these microsomes was activated by the addition of 0.5 mM ATP (arrows indicated by ATP) and further stimulated by 2.5 μM valinomycin (arrows labeled val). The resulting fluorescence enhancement at 445 nm and the concomitant decrease at 380 nm are shown in A and B, respectively. The H$^+$ gradient was dissipated by the addition of 0.5 μg/ml nigericin (arrows labeled nig).

process has certain structural requirements on the polyanions; namely, the adjacent sites on the polyanion have to be no farther apart than 10 Å such that the bound dyes are close enough to interact with each other [Stone, 1967; Massari and Pascolini, 1977].

The reasoning behind the use of acridine orange to visualize the intracellular acidic compartment is that one can adjust the total probe concentration such that the most acidic compartment would accumulate enough dye to produce the characteristic polymeric fluorescence. These regions would then appear red whereas the rest of the cell would be green (the monomer fluorescence). This approach had been used to visualize lysosomes in intact cells [de Duve et al, 1974; Zeitz et al, 1980]. In this section, we shall describe the applications of this method to sea urchin eggs.

Previous studies by a number of investigators [Johnson et al, 1976; Shen and Steinhardt, 1978, 1979] had demonstrated that fertilization of the sea urchin egg is accompanied by a Na-dependent acid release from the egg and alkalinization of the intracellular pH. This change in pH_i was postulated to be the triggering signal for the late changes (ie, protein and DNA synthesis) after fertilization [Vacquier and Brandiff, 1975; Epel et al, 1974]. Recently, the suggestion was advanced that the acid released by the egg is due to CO_2 [Gillies et al, 1980]. This raises the possibility that the alkalinization of the pH_i could be the result of the sequestration of the H^+ by some intracellular acid stores. Further implication of the existence of the intracellular acid stores as regulators of pH_i comes from the study of intracellular pH changes associated with fertilization in frog eggs [Webb and Nuccitelli, 1980]. In that system alkalinization of pH_i was observed, but not the release of H^+ to the external medium. It was therefore of interest to see if any intracellular compartments in sea urchin egg became acidified after fertilization.

Plate I (see page xxix) shows the acridine orange staining patterns of sea urchin eggs from two different species. Eggs from L. pictus were mixed with a minimum amount of sperm so that some eggs remained unfertilized. Acridine orange was then added to stain the eggs. This procedure would ensure that both fertilized and unfertilized eggs were exposed to the same amount of dye (fertilized eggs can be identified by the presence of the fertilization membrane). The unfertilized eggs appeared green, the characteristic fluorescence of the acridine orange monomers. Immediately after fertilization, red fluorescence (characteristic of acridine orange polymers) appeared at the cortical region, and reached a maximum intensity in 10–20 minutes. The correlation between the red fluorescence and fertilization was greater than 95%. At high magnification, the red fluorescence was seen to be originating from an abundance of granules ranging in size from 1–2 μm. At about 30 minutes after addition of sperm, progressive increase in red fluorescence in the central region of the cell was seen (Plate IA). At the clear streak stage (90 minutes after sperm addition), all the red fluorescence was centralized (Plate IB), indicating intracellular movements of these granules. The staining pattern at this latter stage suggests that these granules are associated or organized by the mitotic apparatus. At the two-cell stage (Plate IC), all the granules are in the perinuclear region. In the case of S. pupuratus eggs, the initial staining pattern (Plate ID) is the same as L. pictus. However, the centralization of the red fluorescence was not observed; instead, it gradually disappeared after 40 minutes.

The appearance of these granules can be triggered artificially by the Ca^{+2}-ionophore, A23187, which also activates the egg parenthogenically [Steinhardt and Epel, 1974; Steinhardt et al, 1974]. Two lines of evidence suggest that they are indeed acidic compartments: 1) The red fluorescence can be discharged by treatment of the egg with NH_4Cl, nigericin, or monensin, all of which are expected to dissipate the pH gradient across membranes; 2) when the cells were lysed gently, the granules were released and they were able to accumulate the dye and produce the red fluorescence. This latter accumulation, however, was pH gradient-sensitive, since acidification of the medium from pH 7.5 to 6.7 progressively decreased the red fluorescence.

The function of these granules is not clear at this point. Several possibilities are being investigated. 1) They may be involved in intracellular pH regulation. 2) Their association with the mitotic apparatus suggest that they may play a role in its formation. 3) They may serve some digestive functions in mobilizing nutrients from the yolk platelets.

IV. CONCLUSIONS

In this chapter we have presented several applications using fluorescent amines as probes for vesicular and cellular pH. A major advantage in the use of these probes is the simplicity. Equipment requirements are few, and the probes provide continuous indices of pH_i. We have demonstrated that in some simple systems, such as liposomes and gastric microsomes, quantitative results can be obtained. In cellular systems (eg, sea urchin sperm), complications arise because of compartmentation, and the existence of many potential binding sites makes quantification difficult. However, the presence of binding does not invalidate the use of these probes as a qualitative index of pH_i as long as the binding is reversible. In fact, we have shown that the presence of binding may increase the sensitivity of the method. Thus the use of these probes is still valid as a highly sensitive and convenient way for continuous monitoring of the relative changes in cellular pH_i.

We have discussed three major groups of fluorescent probes in the results section. The first group of probes that exhibit ΔpH-dependent quenching are probably the most widely used probes in a variety of subcellular organelles and liposomes [Schuldiner et al, 1972; Singh and Bragg, 1977; Rottenberg and Lee, 1975; Deamer et al, 1972; Pollard et al, 1976; Schuldiner and Fishkes, 1978; Hasan and Rosen, 1977; Casadio and Melandri, 1977]. Recently, we and others have extended their application to cellular systems [Lee et al, 1980; Meizel and Deamer, 1978] (Lee et al, in preparation; Shapiro BM, personal communication). The exact mechanism of quenching is not clearly understood. We presented evidence in support of the idea that quenching is due to concentration-dependent self-interaction of the accumulated probes. Therefore, in order to ensure complete quenching, one must use a relatively high concentration of probe. This may present significant perturbation of pH_i since the probe penetrates in its neutral form and becomes charged by accepting H^+ from inside. This would alter the pH_i if the internal probe concentration were to exceed the buffering power of the internal medium.

The second group of probes that exhibit pH-dependent spectral shifts can overcome some of these difficulties. These probes show red shifts in the emission maximum and increase in quantum yield in going from alkaline to acidic medium.

By measuring both the decrease in the alkaline peak and the increase in the acidic peak, the ΔpH-induced changes in the distribution and the fluorescence enhancement of the intravesicular probes can be calculated. The internal pH can then be obtained directly by comparing with a standard curve of fluorescence vs pH, and no measurement of volume of the H^+ space is necessary. Since this method does not depend on quenching, much lower probe concentration can be used. One of the probes described in this group, quinine, is a diamine, and thus its accumulation is much more sensitive to the pH gradient [Schuldiner et al, 1972]. A possible disadvantage of using quinine in biological systems is the extreme sensitivity to quenching by Cl^-, which makes it necessary to use Cl^--free solutions.

The use of the concentration-dependent spectral shift described for the third group of probes presents the exciting possibility of visualizing the intracellular acidic compartments in living cells. Thus the acridine orange-staining granules have been observed in a variety of living cells. In neuroblastoma cells [Zeitz et al, 1980], their distribution is dependent on the stage of differentiation and the condition of the intracellular microtubular system. In some cases, they have been identified with lysosomes [James-Kracke et al, 1979; Canonico and Bird, 1969; Allison and Young, 1969]. Finally, they have been used to show acidic compartments within gastric glands [Berglindh et al, 1980].

ACKNOWLEDGMENT

The work concerning the sea urchin gametes was done in the laboratory of Dr. David Epel. We wish to thank him for his support and valuable suggestions.

V. REFERENCES

Allison AC, Young MR: Vital staining and fluorescence microscopy of lysosomes. In Dingle JT, Feel HB (eds) "Lysosomes in Biology and Pathology." Amsterdam, London: North-Holland, 1969, Vol 2, p 600.

Berglindh T, Dibona DR, Ito S, Sachs G: Probes of parietal cell function Am J Physiol 238:G165, 1980.

Canonico PG, Bird JW: The use of acridine as a lysosomal marker in rat skeletal muscle. J Cell Biol 43:367, 1969.

Casadio R, Melandri BA: The behavior of 9-aminoacridine as an indicator of transmembrane pH difference in liposomes of natural phospholipids. J Bioenerg Biomemb 9:17, 1977.

Chang H, Saccomani G, Rabon E, Schakman R, Sacks G: Proton transport by gastric membrane vesicles. Biochim Biophys Acta 464:313, 1977.

Collins F, Epel D: The role of calcium ions in the acrosome reaction of sea urchin sperm. Exp Cell Res 106:211, 1977.

Colowick SP, Womack FC: Binding of diffusable molecules by macromolecules: Rapid measurement by rate of dialysis. J Biol Chem 244:774, 1969.

Deamer DW, Prince RC, Crofts AR: The response of fluorescent amines to pH gradient across liposomes membranes. Biochim Biophys Acta 274:323, 1972.

de Duve CH, de Baray TH, Poole B, Trout A, Tulkens P, van Hoof F: Lysosomotropic agents. Biochem Pharmacol 23:2495, 1974.

Epel D, Steinhardt RA, Humphreys T, Mazia D: An analysis of the partial metabolic derepression of sea urchin eggs by ammonia: The existence of independent pathways in the program of activiation at fertilization. Dev Biol 40:245, 1974.

Forte JG, Ganswer AL, Tanisawa A: The K^+-stimulated ATPase system of microsomal membranes from gastric oxyntic cells. Ann NY Acad Sci 242:255, 1974.

Forte JG, Lee HC: Gastric adenosine triphosphatases: A review of their possible role in HCl secretion. Gastroenterology 73:921, 1977.

Gillies RJ, Rosenberg M, Deamer DW: Inorganic carbonate release and pH changes during the activation sequence in sea urchin eggs. J Cell Biol 87:136a, 1980.

Granser AL, Forte JG: K^+-stimulated ATPase in purified microsomes of bullfrog oxyntic cells. Biochim Biophys Acta 307:169, 1973.

Hasan SM, Rosen BP: Energy transduction in Escherichia coli. Biochim Biophys Acta 459:225, 1977.

Jagendorf, AT, Uribe EG: ATP formation caused by acid-base transition of spinach chloroplasts. Proc Natl Acad Sci USA 55:170, 1966.

James-Kracke MR, Sloane BF, Shuman H, Somlyo AP: Lysosomal composition in culture vascular smooth muscle cells: Electron probe analysis. Proc Natl Acad Sci USA 76:6461, 1979.

Johnson JD, Epel D, Paul M: Intracellular pH and activation of sea urchin eggs after fertilization. Nature 262:661, 1976.

Kagawa Y, Racker E: Partial resolution of the enzymes catalyzing oxidative phosphorylation. XXV. Reconstitution of vesicles catalyzing ^{32}Pi-adenosine triphosphate exchange. J Biol Chem 246:5477, 1971.

Kraayenhof R: Quenching of uncoupler fluorescence in relation to the "energized state" in chloroplasts. Fed Eur Biochem Soc Lett 6:161, 1970.

Lee HC, Quintanilha AT, Forte JG: Energized gastric microsomal membrane vesicles. Biochem Biophys Res Commun 72:1179, 1976.

Lee HC, Forte JG: A study of H^+ transport in gastric microsomal vesicles using fluorescent probes. Biochim Biophys Acta 508:339, 1978.

Lee HC, Breitbart H, Berman M, Forte JG: Potassium-stimulated ATPase activity and hydrogen transport in gastric microsomal vesicles. Biochim Biophys Acta 553:107, 1979.

Lee HC, Forte JG: A novel method for measurement of intravescicular pH using fluorescent probes. Biochim Biophys Acta 601:152, 1980.

Lee HC, Schuldiner S, Johnson C, Epel D: Sperm motility initiation: Changes in intracellular pH, Ca and membrane potential. J Cell Biol 87:39a, 1980

Lee J, Simpson G, Scholes P: An ATPase from dog gastric mucosa: Changes of outer pH in suspensions of membrane vesicles accompanying ATP hydrolysis. Biochem Biophys Res Commun 60:825, 1974.

Massari S, Pascolini D: Phosphatidic acid distribution on the external surface of mixed vesicles. Biochemistry 16:1189, 1977.

Meizel S, Deamer DW: The pH of the hamster sperm acrosome. J Histochem Cytochem 26:98, 1978.

Nishioka D, Cross N: The role of external sodium in sea urchin fertilization. In Dirksen ER, Prescott D, Fox DF (eds): "Cell Reproduction." New York: Academic Press, 1978, p 403.

Pollard HB, Zinder O, Hoffman PG, Nikodejevic O: Regulation of the transmembrane potential of isolated granules by ATP, ATP analogs and external pH. J Biol Chem 251:4544, 1976.

Ramos S, Schuldiner S, Kaback HR: The electrochemical gradient of protons and its relationship to active transport in Escherichia coli membrane vesicles. Proc Natl Acad Sci USA 73:1892, 1976.

Roos A, Keifer DW: Theoretical considerations on the distribution of weak acids and bases and experimental results. This volume.

Rottenberg H, Lee CP: Energy dependent hydrogen ion accumulation in submitochondrial particles. Biochemistry 14:2675, 1975.

Sachs G, Chang HH, Rabon E, Schackman R, Lewin M, Saccomani G: A nonelectrogenic H^+ pump in plasma membranes of hog stomach. J Biol Chem 251:7690, 1976.

Schackman RW, Eddy EM, Shapiro BM: The acrosome reaction of Strongylocentrotus purpuratus sperm. Dev Biol 65:483, 1978.

Schackman RW, Shapiro BM: A partial sequence of ionic changes associated with the acrosome reaction of Strongylocentrotus purpuratus. Dev Biol 81:145, 1981.

Schuldiner S, Fishkes H: Sodium-proton antiport in isolated membrane vesicles of Escherichia coli. Biochemistry 17:706, 1978.

Schuldiner S, Rottenberg H, Avron M: Determination of ΔpH in chloroplasts. 2. Fluorescent amines as a probe for the determination of ΔpH of chloroplasts. Eur J Biochem 25:64, 1972.

Shen SS, Steinhardt RA: Direct measurement of intracellular pH during metabolic derepression of the sea urchin egg. Nature 272:253, 1978.

Shen SS, Steinhardt RA: Intracellular pH and the sodium requirement at fertilization. Nature 282:87, 1979.

Singh AP, Bragg PD: ATP-dependent proton translocation and quenching of 9-aminoacridine fluorescence in inside-out membrane vesicles of a cytochrom-deficient mutant of Escherichia coli. Biochim Biophys Acta 464:562, 1977.

Steinhardt RA, Epal D: Activation of sea urchin eggs by a calcium ionophore. Proc Natl Acad Sci USA 71: 1915, 1974.

Steinhardt RA, Epel D, Carroll ES, Yanaginiachi R: Is calcium ionophore a universal activator for unfertilized eggs? Nature 252:41, 1974.

Stone AL: Aggregation of cationic dyes on acid polysaccharides. II. Quantitative parameters of metachromasy. Biochim Biophys Acta 148:193, 1967.

Tilney LG, Kiehart DP, Sardet C, Tilney M: The polymerization of actin. IV. The role of Ca^{++} and H^+ in the assembly of actin and in membrane fusion in the acrosomal reaction of echinoderm sperm. J Cell Biol 77:536, 1978.

Vacquier VD, Brandiff B: DNA synthesis in unfertilized sea urchin eggs can be turned on and turned off by the addition and removal of procaine hydrochloride. Dev Biol 47:12, 1975.

Webb DJ, Nuccitelli R: The intracellular pH goes alkaline at fertilization in the Zenopus egg. J Cell Biol 87:137a, 1980.

Zeitz M, Lange K, Keller K, Herken H: Distribution of acridine orange accumulating particles in neuroblastoma cells during differentiation and their characterisation by subcellular fractionation and electron microscopy. Cell Mol Biol 25:305, 1980.

Intracellular pH Measurement Techniques: Their Advantages and Limitations

Richard Nuccitelli
Zoology Department, University of California, Davis, California 95616

I.	Introduction	161
II.	pH_i Measurement in Cell Populations	164
	A. Spatial Resolution	164
	B. Temporal Resolution	164
	C. Amount of Material Needed	165
	D. Advantages and Limitations of These Techniques	165
III.	pH_i Measurement in Single Cells	165
	A. pH-Sensitive Microelectrodes	166
	B. Fluorescein–Ovalbumin Injection	166
IV.	Summary of Discussion Following Technique Paper Presentations	166
	A. Weak Acid/Base Distribution	166
	B. ^{31}P-NMR for pH_i Measurement	167
	C. pH-Sensitive Glass Microelectrodes	168
	D. Optical Techniques for pH_i Measurement	168
V.	References	169

I. INTRODUCTION

Technical advances over the past decade have made several new intracellular pH (pH_i) measurement techniques available. These have been described in detail in the previous chapters in this volume, and here I will list a comparison of their advantages and limitations and summarize the discussions that followed each technique paper presentation at the meeting.

These pH_i measurement techniques fall into two broad categories depending on their applicability to a single cell or to populations of cells. For quick comparison, Table I lists significant features of each technique.

TABLE I. pH_i Measurement Techniques—Advantages and Limitations

Technique	Spatial resolution	Temporal resolution	Amount material needed	Advantages	Limitations
For cell populations					
Weak acid/base distribution					
A. Isotope-labeled	Some average of cytoplasm and organelles of many cells	Size-dependent: must measure equilibration time; seconds for organelles, 30 min for muscle fiber	0.1 μl cell volume	Simplicity of execution; useful for small cells and organelles	Only one measurement per aliquot possible unless dialysis technique used. Indicator may influence pH_i at high concentrations (mM)
B. Spectrally monitored (fluorescence or absorbance)	Same as A	Same as A. 9-Amino acridine has fast equil. time	0.05 μl cell volume (10^7 cells with 2 μm diameter)	Same as A plus pelleting unnecessary so cells can be studied further. Continuous monitoring possible	Absolute pH measurement unreliable (fluorescence quenching and unknown microenvironment)
^{31}P-NMR	Separate compartments at different pH may be detected as separate peaks	Concentration-dependent—typically a few seconds to 10 min	1 μmole of $^{31}P_i$; typically 10 ml cells in 20–50% suspension	Noninvasive ATP, PCr, and other PO_4 metabolite levels measured at same time	Large number of cells needed and fairly close packing required

Method	Resolution	Response time	Sample size	Advantages	Disadvantages
pH$_i$-dependent fluorescence-absorbance of carboxyfluorescein	Average over cytoplasm of many cells	Continuous indicator	0.05 μl cell volume (10^7 cells with 2 μm diameter)	Good for very small cells. Detects fast changes. Indicates cytoplasmic pH	Poor long-term (> 1 h) stability due to dye leakage; Inadequate esterase concentration in some cells
Fluorescence-activated cell sorter with coumarins	Average over many cells	Minutes	10^3–10^4 cells	Scans each cell separately and can sort	Requires permeant dyes, suspended cells needed, 1–50 μm diameter must soak cells in mM levels of dye
For single cells					
Recessed-tip pH microelectrode (Thomas)	1 μm	5–30 seconds	1 cell	Continuous record of pH$_i$ in single cell; easy calibration; long-term stability	Immobilized cell must withstand impalement with 2 electrodes or de Hemptinne's double-barreled electrode
Exposed-tip pH microelectrode (Hinke)	50 μm	1 second	1 large cell	Same as recessed-tip type	Same as recessed-tip type plus must be sure entire exposed tip is inside cell
Fluorescein–ovalbumin injection	Limited by beam width ≃ 10 μm	1 second	1 cell or less (1 phagosome)	Measure pH gradients in single cell, which may be motile. Indicates cytoplasmic pH	Must inject cells with label or stimulate endocytosis of it or use 6-carboxyfluoresceindiacetate technique

II. pH_i MEASUREMENT IN CELL POPULATIONS

A. Spatial Resolution

The weak acid/base distribution (isotope-labeled) and the fluorescence-activated cell sorter (FACS) techniques average over both cytoplasm and organelles of many cells and provide no spatial resolution of pH_i within single cells. However, fluorescent weak bases can be used to identify intracellular compartments with different pH values [see Meizel and Deamer, 1978; Lee and Forte, this volume]. Carboxyfluorescein is largely excluded from organelles and will therefore provide an indicator of cytoplasmic pH alone [see Thomas et al, and Simons et al, this volume]. ^{31}P-NMR will indicate pH_i by the spectral position of the inorganic phosphate (P_i) peak. If there are intracellular compartments at different pH values with adequate P_i in each, multiple P_i peaks will appear at the respective pH_i values. One could then gain spatial resolution if the sources of the separate P_i peaks could be identified as specific cytoplasmic organelles. This has been attempted for mitochondria [Cohen et al, 1978] and plant vacuoles [Roberts et al, 1980], for example. Some meeting discussion was directed to the difficulty of detecting phosphorous compounds in cellular mitochondria. Three possible explanations were presented: 1) Immobilization of mitochondrial compounds may spread out the peaks so much that they may become undetectable; 2) paramagnetics, such as manganese, might be present to broaden the signals; 3) P_i turnover, as it is incorporated into ATP, may be so rapid that free P_i is undetectable.

B. Temporal Resolution

The best temporal resolution is provided by carboxyfluorescein absorbance or fluorescence, which is a continuous indicator of cytoplasmic pH in a cuvette of cells. The other techniques all require a fixed sampling time that limits their temporal resolution. FACS must average 10^3–10^4 cells measured separately, thereby limiting its resolution to a few minutes. ^{31}P-NMR also signal-averages over many scans, which generally limits its resolution to several minutes. Events of shorter duration may be studied with NMR if they can be synchronized to occur at the start of each NMR pulse (see Gadian et al, and Gillies et al, this volume).

The temporal resolution of the weak acid/base distribution method using an isotope-labeled substance is limited by equilibration times which must be measured for the cell to be studied. This can be on the order of seconds for very small cells and organelles, or the order of minutes for large cells such as muscle fibers. Aliquots of cells must be removed from the population for each measurement. However, if the dialysis technique is used [Ramos and Kaback, 1977] with on-line mixing into scintillation fluid, a sample time resolution of 10–15 seconds could be possible. Spectrally monitored weak acid/base distribution permits continuous measurement and is better for the detection of transient pH_i changes.

C. Amount of Material Needed

All of these techniques require large numbers of cells. FACS scans each cell separately but must average the signals from 10^3–10^4 cells. ^{31}P-NMR can detect about 1 μmole of P_i, so the number of cells required depends on the amount of P_i per cell. Generally P_i is at millimolar levels in the cytoplasm so about 1 ml of cell volume is required. Typically, a 10-ml sample of cells in a 20–50% suspension is used. This is much more material than required by the other techniques. For example, isotope-labeled DMO requires only about 0.1 μl cell volume per aliquot, so it is 10,000 times more sensitive than ^{31}P-NMR in that respect. The fluorescence and absorbance techniques also have a relatively low cell requirement of roughly 0.1 μl or 10^7 cells 2 μm in diameter.

D. Advantages and Limitations of These Techniques

Most of the advantages and limitations of these population techniques are evident from the previous sections, and some are discussed in depth in an excellent review by Roos and Boron [1981]. The isotope-labeled weak acid/base technique is popular because of its simplicity of execution and its accuracy for determining the absolute pH_i value. However, this pH_i value will represent some *average* of the organelle and cytoplasmic pH_i. Only one measurement per aliquot is possible, and these cells are not normally used for subsequent observations. In that respect, the spectrally monitored weak acid or base has an advantage since continuous monitoring of pH_i is possible and the cells can be used for another experiment. On the other hand, absolute pH_i measurements have greater uncertainties with fluorescent probes than with isotope-labeled ones owing to fluorescence quenching and interactions with cytoplasmic molecules that can shift the pH_i calibration curve. Another limitation of the weak acid/base technique is that high concentrations of the probe (typically mM) can change pH_i. This can be contrasted with the ^{31}P-NMR technique, which does not introduce any probe at all but uses the naturally occurring $^{31}P_i$ as a pH_i indicator. It is the only technique listed that is totally noninvasive.

III. pH_i MEASUREMENT IN SINGLE CELLS

There are fewer techniques available for pH_i measurement in a single cell. In fact, only two are considered to be reliable indicators of cytoplasmic pH: the pH-sensitive glass microelectrode, and the spectrally monitored probes when coupled to large impermeant molecules such as ovalbumin to prevent leakage into organelles or out of cells.

A. pH-Sensitive Microelectrodes

The two most common forms of pH-sensitive microelectrodes are the Hinke exposed-tip pH microelectrode and the Thomas recessed-tip pH microelectrode (see R. C. Thomas, this volume, for sketches of both). The Hinke electrode is somewhat easier to make and has a faster response time, but is only useful in cells with the proper cylindrical geometry to assure that the entire exposed tip is inside the cell. In contrast, the recessed-tip design can be used with any cell shape, and its use is limited only by the cell's ability to withstand impalement with a pair of microelectrodes or a double-barreled microelectrode. The spatial resolution of this technique is determined by the electrode tip size, which is 1 μm or less, and the temporal resolution is dependent on the tip geometry and recess volume but is typically 5–30 seconds. The main advantages of this technique are the ease of calibration, accuracy of the absolute pH_i value, and the continuous record of pH_i provided by it with a long-term stability of several hours. Dr. de Hemptinne (this volume) has developed a double-barreled version of the recessed-tip pH microelectrode which is useful for pH_i measurement in small cells for which impalement with two electrodes is very difficult.

B. Fluorescein–Ovalbumin Injection

The most recently developed technique for single-cell pH_i measurement utilizes membrane-impermeant, spectrally monitored probes to indicate the local pH_i in a cellular region (see Heiple and Taylor, this volume). Once the probe is internalized, either by injection or endocytosis, continuous pH_i monitoring of the specifically labeled compartments is possible with a spatial resolution of about 10 μm, limited only by the aperture size and probe concentration. The temporal resolution is limited by the fluorescence filter change time, which is about 1 second on the present apparatus. This technique has several advantages over pH microelectrodes since it can be used with mobile cells, can detect pH_i gradients without electrode penetration injury, and can be used to measure specific organelle pH such as described for single phagosomes in this volume. The main limitation of this technique is the uncertainty concerning the absolute pH_i value which is common for all fluorescent probes. However, this problem could be solved by using a second pH_i measurement technique, such as the pH microelectrode, to calibrate the fluorescence in vivo. An exciting extension of this technique which is being worked on is pH_i imaging. By comparing videotapes taken at two different excitation wavelengths, a point–by–point ratio can be taken to give pH_i at each point across the cell.

IV. SUMMARY OF DISCUSSION FOLLOWING TECHNIQUE PAPER PRESENTATIONS

A. Weak Acid/Base Distribution

The discussion following Albert Roos's presentation centered on the limitations of the DMO technique. Its temporal resolution is restricted by equilibration time which in turn depends on the surface-to-volume ratio. Thus small cells will allow a much better temporal resolution than large cells with longer equilibration times. The importance of measuring this equilibration time for each cell to be studied was stressed. There is no spatial resolution with this technique since the uncharged form will permeate all cellular membranes. However, it will come into equilibrium with the charged form according to the pH gradient across each membrane, and Dr. Roos pointed out that some spatial resolution would be possible if an optical method were devised to localize the weak acid.

Another parameter which must be carefully chosen is the concentration of the weak acid or base to be used. This is because these same substances will change pH_i if applied at somewhat higher concentrations. When the possibility was raised that unstirred layers might interfere with equilibration, Dr. Roos mentioned some recent results indicating that this is not a problem with DMO since the same permeability was measured for both weakly and strongly buffered media. Dr. de Hemptinne commented that surface pH measurements usually indicate a slightly acid region at the cell surface which is probably due to local CO_2 release.

B. ^{31}P-NMR for pH_i Measurement

The ^{31}P-NMR technique was discussed in two papers—those of Drs. Gillies and Gadian. The discussion began with Dr. Moody questioning the accuracy of the technique for measuring true pH_i without interference from unknown factors. It was pointed out that Drs. Webb and Nuccitelli (this volume) had directly compared ^{31}P-NMR with pH microelectrode results on the same cell and had obtained very similar values for pH_i with both techniques. The question was restated by Dr. Roos, who asked if cytoplasm at pH 6 would give the same P_i chemical shift as extracellular fluid at pH 6. In reply it was stated that the only factor that significantly shifts the P_i chemical shift–pH calibration curve is ionic strength, because protein or divalent cations do not bind to P_i as they do to ATP, for example. Therefore, if the ionic strength were the same inside and out, one would measure the same pH_i from the P_i chemical shift. The ionic strength dependence is not very steep, however. For example, a large, 5-fold increase in ionic strength shifts the titration curve upward by only about 0.2 pH unit. One way to measure intracellular ionic strength would be to compare the chemical shift of methylphosphonate (if your cells take it up) with that of P_i. Both are pH_i-sensitive but only P_i is ionic strength-sensitive. Dr. Jacobus pointed out that the γ-ATP peak was considered for use as a pH_i indicator, since 97% of heart muscle ATP is cytoplasmic as opposed to only 70% of the total P_i. Unfortunately, ATP is found in three forms—Mg^{2+}-ATP, K^+-ATP, and free ATP—all of which have a different pH dependence, so one must know the relative amounts of each of these to calculate pH_i from the γ-ATP chemical shift.

Superconducting NMR machines are quite expensive so they are normally shared by many users as a departmental or institutional facility. However, this "pH_i meter" does much more than measure pH_i. Many other cellular components, such as the amount of ATP, NAD, and phosphocreatine, are measured at the same time—all noninvasively. This makes it an ideal tool for assessing the metabolic health of an organ or tissue as described by Bore et al, in this volume, and it can also be used to measure the kinetics of P_i exchange between molecules, such as PCr and ATP. The requirement for a high cell density creates problems that are solved by various methods of superfusion, including growing cells on hollow fibers up to tissue density.

C. pH-Sensitive Glass Microelectrodes

Presentations were made by Dr. R. C. Thomas who described his recessed-tip pH microelectrode and Dr. de Hemptinne who described the double-barreled modification of the recessed-tip pH microelectrode. Dr. Thomas followed an enthusiastic discussion of the NMR technique and began by commenting, "I'm going to go home and try to get an NMR machine." But he was quick to point out that for single-cell studies where pH, $[Na^+]_i$, and $[Cl^-]_i$ must be measured at the same time, the ion-sensitive microelectrode is hard to beat. The pH microelectrode is easily calibrated, has long-term stability over several hours, is inexpensive, and has excellent spatial and temporal resolution. The smallest cell to be successfully impaled is the guinea pig atrial cell (5 μm in diameter) using the double-barreled design. Drs. Steinhardt and Shen pointed out that some cells are harder to impale than others, and that the sea urchin egg is often activated by pH microelectrode impalement, particularly if the electrode had been used previously to impale an egg. Dr. Thomas's reply: "Well, you shouldn't work on those difficult preparations."

Dr. de Hemptinne pointed out that the reference half of his double-barreled electrode developed tip potentials more rapidly than conventional microelectrodes until he began adding 10 mM K^+ citrate and 10 mM EDTA to the 3M KCl filling solution. Apparently the calcium level in the electrode tip can affect the tip potential.

D. Optical Techniques for pH_i Measurement

Dr. Heiple summarized the main advantages of her fluorescein–ovalbumin fluorescence technique as being fast, reproducible to ± 0.02 pH unit, and capable of measuring pH_i in localized areas within single cells even while the cell is moving along. It is the only technique which allows you to "follow the pH changes within a single phagosome and within a single moving cell." Some limitations are imposed by the microinjection procedure, but it may be possible to load cells by liposome fusion to avoid impalement. All of the fluorescent probes share the problem of calibrating for absolute pH_i. It is important to make some attempts to calibrate these probes in situ as well as to compare them with other pH_i measurement techniques. Dr. Gerson pointed out the possibility that when you see dark and bright spots of fluorescence, they may represent microenvironment differences other than local pH differences.

The flow microfluorometry or FACS technique was next discussed by Dr. Gerson, who described the main difficulties as calibration and dye leakage. By using a very narrow nozzle opening, the amount of extracellular fluid in the beam path is minimized and the machine only measures when a cell is in the beam, so the influence of extracellular dye is minimized but is still a problem.

The carboxyfluorescein technique presented by Dr. Thomas requires intracellular esterase activity, which varies from cell to cell. Whereas most mammalian cells, with the exception of erythrocytes, will probably have this enzyme activity, bacteria have very little in general and "cells prepared by collagenase generally give you more problems than cells that you grow in tissue culture . . . and fed cells seem to retain the dye better than fasted cells." Dr. Simons pointed out that leakage of two types exists—slow leakage of hydrolyzed probe, and faster leakage of unhydrolyzed probe in cells with poor esterase activity. Unhydrolyzed probe is also hydrolyzed outside the cell and this nonenzymatic effect must be taken into account. Dr. J. A. Thomas felt that washing the cells after incubation should solve that problem, but Dr. Simons finds continuous leakage using continuous dialysis. Dr. Thomas also mentioned that whereas carboxyfluorescein mainly labels the cytoplasm, the more permeant fluorescein goes into the most basic compartment and so labels mitochondria very easily. It is a nice way to look at pH changes in a mitochondrial suspension.

ACKNOWLEDGMENT

I would like to thank Dr. Elizabeth Simons for helpful discussions concerning the limitations of some of these pH_i measurement techniques.

V. REFERENCES

Cohen SM, Ogawa S, Glynn P, Yamane T, Brown TR, Shulman RG: ^{31}P nuclear magnetic resonance studies of isolated rat liver cells. Nature 273:554, 1978.

Meizel S, Deamer DW: The pH of the hamster sperm acrosome. J Histochem Cytochem 26:98, 1978.

Ramos S, Kaback HR: pH-dependent changes in proton-substrate stoichiometries during active transport in Escherichia coli membrane vesicles. Biochemistry 16:4271, 1977.

Roberts JKM, Ray PM, Wade-Jardetsky N, Jardetsky D: Estimation of cytoplasmic and vacuolar pH in higher plant cells by ^{31}P NMR. Nature 283:870, 1980.

Roos A, Boron WF: Intracellular pH. Physiol Rev 61:296, 1981.

REGULATION OF INTRACELLULAR pH

Intracellular pH: Its Measurement, Regulation, and
Utilization in Cellular Functions, pages 173–187
© 1982 Alan R. Liss, Inc., 150 Fifth Avenue, New York, NY 10011

Proton Permeability in Biological and Model Membranes

David W. Deamer
Zoology Department, University of California, Davis, California 95616

I.	Introduction	173
II.	Methods and Potential Artifacts	176
III.	Proton Permeability of Model Membrane Systems	177
IV.	Discussion	182
	A. Proton Permeability of Biological Membranes	182
	B. Relative Cation Permeabilities of Model Membranes	183
	C. Mechanisms of Proton Flux Across Lipid Bilayers	184
V.	Summary	185
VI.	References	186

I. INTRODUCTION

Ion transport and ion concentration gradients have assumed a central role in our understanding of cell function. From studies of planar lipid membranes and liposome systems, it is clear that a major function of the lipid bilayer moiety of biological membranes is to act as a barrier to the free flux of ions. For instance, sodium and potassium ions have permeability coefficients in the range of 10^{-13} to 10^{-14} cm/sec across liposome membranes, whereas the permeability coefficient of these ions across typical biological membranes is in the range of 10^{-7} to 10^{-10} cm/sec. Most investigators agree that the higher permeability is a function of protein channels that permit ion flux across an essentially impermeable lipid bilayer.

Knowledge of sodium and potassium permeability was essential for understanding plasma membrane function, particularly that of the axon and sarcolemma. It has generally been assumed that proton permeability was similar to other monovalent cations, and that the cell maintained a relatively constant intracellular pH. Therefore proton permeability and pH gradients were largely overlooked as areas of research interest.

This changed dramatically with the realization that electrochemical proton gradients, as defined by the chemiosmotic theory [Mitchell, 1961, 1966] were central energy sources in the function of oxidative and photosynthetic coupling membranes. A number of new techniques were developed to measure both the kinetics of proton transport and magnitude of the resulting gradients, and these are now being applied to investigate broader questions related to intracellular pH. One of these questions concerns the proton permeability of biological membranes, since a complete understanding of intracellular pH gradients must include knowledge of the barrier to passive proton flux.

One of the early tests of the chemiosmotic hypothesis was to demonstrate quantitatively that coupling membranes provided a sufficient permeability barrier to protons. Mitchell and Moyle [1967] devised a method to estimate proton conductance that involved the addition of an acid or base to a mitochondrial suspension (Fig. 1). This produced a pH gradient across the membrane, and the kinetics of decay could be followed by a glass electrode. From the known buffering capacity of the mitochondria, it was possible to estimate proton conductance from such data. Mitchell calculated that proton conductance was in the range of that for sodium or potassium across other membranes, and that the mitochondrial membrane does provide a sufficient barrier to proton flux. However, it should be noted that the gradient decays with a half-time measured in minutes, much more rapidly than expected from knowledge of sodium and potassium permeabilities in other membrane systems.

This apparent paradox brings up a significant point frequently overlooked, and that is that a low conductance is not necessarily equivalent to a low permeability. Conductance is a measurement of ion flux, and depends on the concentration of the ionic species involved. It follows that if protons, sodium, and potassium have similar conductances in mitochondria, then the intrinsic permeability of the membrane to protons must be vastly greater, since proton concentration is 10^{-7} M, and sodium or potassium concentration is typically in the range of 10^{-1} M.

Another demonstration of a high apparent proton permeability in a biological membrane follows from the original demonstration by Neumann and Jagendorf [1964] that chloroplasts were capable of light driven proton transport. A typical result is shown in Figure 2. Upon illumination of a chloroplast suspension, the pH of the medium increases, reflecting the inward transport of protons. Rottenberg et al [1972] calculated from 9-aminoacridine distribution that a pH gradient of over 3 units was established under these conditions. Note that when the light is turned off, the gradient decays in less than a minute. The half-time of this decay is again much less than that of sodium or potassium gradients across other membranes, and one may conclude that chloroplasts, like mitochondria, have a relatively high proton permeability.

From these considerations it became apparent to us that the intrinsic permeability of natural membranes to protons may be quite different from that of other

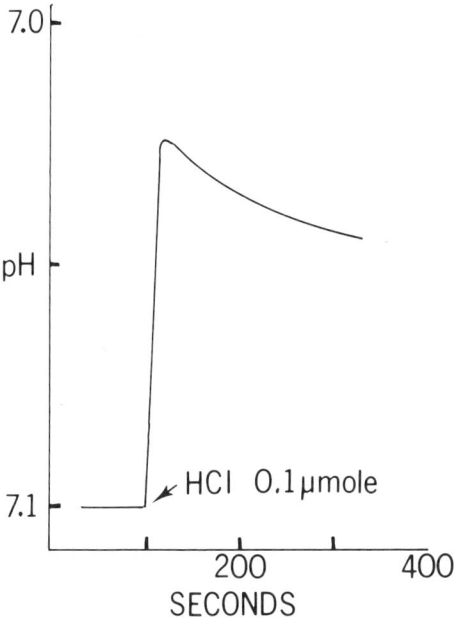

Fig. 1. Decay of pH gradients in a mitochondrial suspension. When a pulse of HCl was added, the external pH first decreased, then increased slowly over a few minutes as proton equivalents passively fluxed inward across the mitochondrial membranes. (Redrawn from Mitchell [1966].)

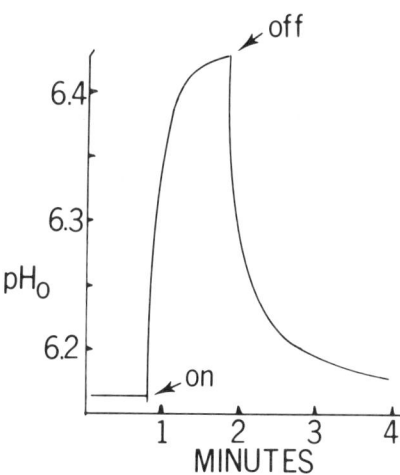

Fig. 2. Decay of pH gradients in a chloroplast suspension. When a chloroplast suspension is illuminated, the external pH increases as protons are actively transported inward. When illumination ends, the resulting pH gradient rapidly decays. Chloroplasts (20 µg chlorophyll per ml) were suspended in 0.1 M NaCl containing 20 µM phenazine methosulfate and illuminated with saturating light levels. (D.W. Deamer, unpublished results.)

monovalent ions. There are several possible explanations if this is the case. First, it has been known for years that pH gradients rapidly decay across erythrocyte membranes, and this has been accounted for by a protein channel that provides for bicarbonate–chloride exchange, thereby permitting a buffer anion to carry proton equivalents. Mitochondria and chloroplasts may have similar porter proteins that permit such movement [Mitchell, 1966] or perhaps nonspecific proton flux occurs through other transmembrane proteins. A second possibility is that cell membranes might contain small amounts of fatty acids or other unknown components that could act to enhance proton permeability. Finally, the lipid bilayer itself could have an unexpectedly high intrinsic permeability to protons.

This leads to a basic question related to proton flux across membranes and representing the main thrust of this chapter: What is the intrinsic proton permeability of the bilayer region of natural membranes, and how does it compare with the permeability of other monovalent cations?

The simplest way to approach this question experimentally is to use model systems such as liposomes and planar lipid membranes. Several laboratories have initiated such studies, and the rest of this chapter will review their findings to date.

II. METHODS AND POTENTIAL ARTIFACTS

Although most of the methods described earlier in this volume are potentially applicable for measuring proton flux across lipid bilayers, only a few have been applied to date. In liposome systems these include weak base distribution using 9-aminoacridine [Nichols et al, 1980] glass electrode methods [Nichols and Deamer, 1980; Nozaki and Tanford, 1981], the pH-sensitive fluorescent dye pyranine [Biegel and Gould, 1981], and a spin label method that monitors the membrane potential produced by proton flux [Cafiso and Hubbell, 1981]. One study with planar lipid membranes has also been directed toward establishing proton flux and permeability coefficients [Gutknecht and Walter, 1981]. In this section, each of these will be described in turn and the results compared.

It is first worthwhile to take note of some of the possible artifacts that can affect measurements of proton flux, and which must be controlled. First, a variety of contaminants are able to increase the general ionic permeability of the membrane. For instance, solvents such as chloroform may remain in the membrane from the initial lipid preparation procedures. Other possible contaminants include detergents such as cholate, Triton, and octyl glucoside that are used in the preparation of liposomes by detergent dialysis methods. Lipid hydrolysis and oxidation products may also increase bilayer permeability to ions. Oxidation damage has in fact been shown to increase potassium conductance of planar lipid membranes [Van Zutphen and Cornwell, 1973] and calcium conductance of liposomes [Serhan et al, 1981].

A second variety of artifact can be produced by contaminants that specifically promote the decay of proton gradients. For example, weak acids and weak bases may be present in small quantities. These compounds equilibrate across membranes with respect to pH and thereby discharge proton gradients (see Roos, this volume). Alternatively, trace contaminants such as fatty acids may have protonophoric properties.

In practice it is impossible to remove all trace contamination, and the best approach is to compare permeability in at least two different systems. If one obtains the same result, it is likely that trace contaminants associated with preparation methods or with the monitoring system are having negligible effects.

As noted earlier, unambiguous proton flux measurements become considerably more complicated in biological membranes, since the lipid bilayer contains small amounts of fatty acid and lysophosphatides that may enhance proton permeability. Furthermore, most biological membranes are penetrated by transmembrane peptide chains that may offer sites of ion flux. Since this chapter is directed toward describing the permeability barrier to protons in biological membranes, the approach will be to compare permeability in biological and model systems and to determine whether the model membrane permeability is in accord with results for biological membranes.

III. PROTON PERMEABILITY OF MODEL MEMBRANE SYSTEMS

The first attempt to measure proton permeability coefficients of lipid bilayers was by Nichols et al [1980] who utilized the 9-aminoacridine method in a liposome system to monitor decay of known pH gradients. Earlier measurements of proton fluxes have been carried out [Kornberg et al, 1972; Hopfer et al, 1968; Kano and Fendler, 1978] but were not directed at comparative or absolute measurements of the proton permeability. In the 9-aminoacridine study, pH gradients were established across liposome membranes composed of egg phosphatidylcholine plus 2 mol % egg phosphatidic acid, which served to prevent aggregation. The liposomes (large unilamellar vesicles, or LUV) were prepared by injecting a diethyl–ether solution of lipid into a sodium pyrophosphate buffer at low pH ranges, followed by Millipore and gel filtration. The Millipore filtration served to size the vesicles and remove aggregates, and the gel filtration removed residual ether and exchanged the external buffer for potassium sulfate. Aliquots of the liposomes were then injected into a potassium phosphate buffer at high pH, and the decay of the gradients ($\sim$ 3 pH units) was monitored by the resulting decrease in fluorescence quenching. Since the buffering capacity was known, and the lipid surface area could be calculated, it was possible to determine net proton–hydroxyl flux and the correlated permeability coefficient. In order to compare proton and sodium permeability, the flux of Na^{22} was measured in the same system.

A typical experiment is shown in Figure 3. Note that the half-time of decay of the buffered gradients is measured in minutes, comparable with the somewhat faster half-time of decay in mitochondria and chloroplasts that was shown earlier. If P_{net} for proton–hydroxyl flux is calculated, the values are in the range of

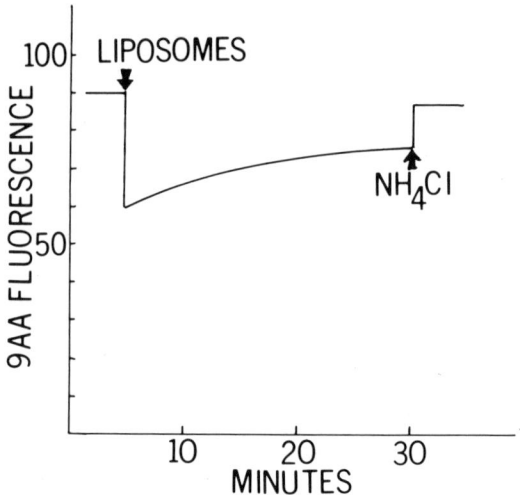

Fig. 3. Decay of pH gradients in liposomes. Liposomes containing pyrophosphate buffer at pH 5.8 were placed in a second pyrophosphate buffer at pH 8.6 containing 1 μM 9-aminoacridine (9-AA). Since the 9-AA is a weak base, it becomes concentrated in the acidic volume and undergoes a fluorescence quenching. As the pH gradient decays, 9-AA is released and quenching decreases. NH$_4$Cl addition causes the gradient to decay completely and the 9-AA fluorescence returns to its original level. (Redrawn from Nichols et al [1980].)

10^{-4} cm/sec. (The symbol P_{net} will be used in this chapter to indicate permeability coefficients in which both proton and hydroxyl flux may have participated. If an attempt has been made to obtain separate measurements, the symbols P_H or P_{OH} will be used.)

The result, if correct, reflects a remarkably high permeability of the lipid bilayer to proton equivalents. It was important to test the result in a second system, since the 9-aminoacridine method required large pH gradients that could have affected the result in some unknown manner. A glass electrode system similar to that described by Mitchell and Moyle [1967] was chosen since it could be used with gradients in the range of a few tenths of a pH unit. In this system, impermeant buffers were encapsulated in LUV liposomes as before, and these were placed in a lightly buffered medium. Sucrose was used to balance the internal and external osmotic strength.

A typical result is shown in Figure 4. Note that the buffered gradients again decay with a half-time measured in minutes, and that the kinetics of decay are similar in either direction. Addition of valinomycin in the presence of potassium had little effect, suggesting that the flux was not limited by counterion movement. The permeability coefficient calculated from the proton flux was in the range of 10^{-4} cm/sec. This value was obtained with several lipid mixtures.

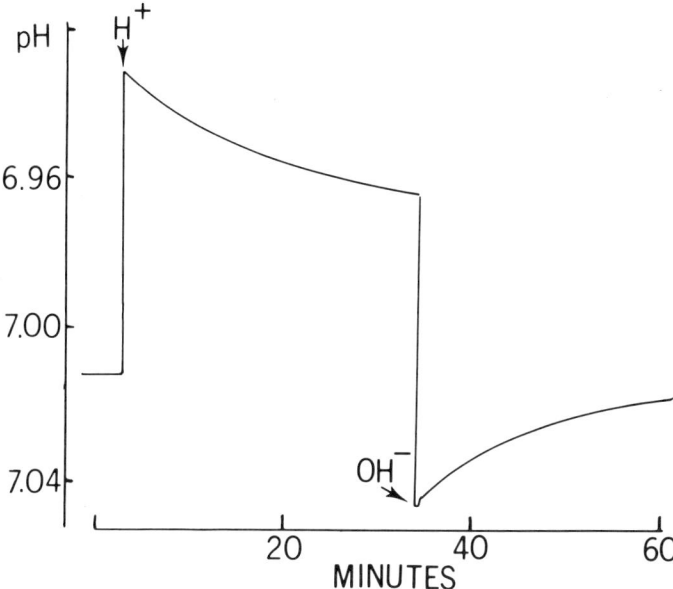

Fig. 4. Decay of gradients in a liposome system. Liposomes containing a concentrated buffer were placed in a relatively unbuffered solution. When the pH was displaced by acid or base addition, the rapid pH shift was followed by a slow return to an intermediate pH as proton equivalents passively crossed the membrane. (Redrawn from Nichols and Deamer [1980].)

An immediate question is whether the remarkably high permeability is an artifact of the preparation method. One may imagine that any of the possible artifacts described earlier could affect the flux and produce a high permeability value. Alternatively, phosphatidic acid may be increasing proton permeability, since Serhan et al [1981] have shown that phosphatidic acid greatly increases the permeability of calcium in liposome membranes.

We therefore extended our work to an entirely different liposome preparation system, using octyl glucoside and detergent dialysis [Mimms et al, 1981] to make LUV from egg phosphatidylcholine with and without admixtures of phosphatidic acid. Some comparative results are shown in Figure 5. Note that phosphatidic acid did moderately increase the proton flux, but the calculated permeability coefficients were still in the range of 10^{-4} cm/sec. Biegel and Gould [1981] obtained similar results in a sonicated liposome system, using dimyristoyl phosphatidylcholine with admixtures of phosphatidic acid. We conclude that neither the preparation method nor phosphatidic acid admixture explains the high proton permeability.

A number of other recent studies have been directed toward estimating proton permeability, with a nearly unprecedented disagreement about the actual value.

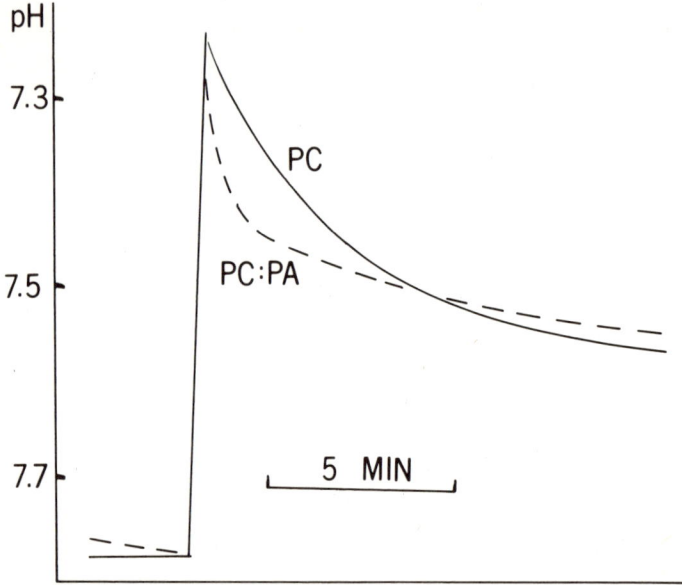

Fig. 5. Comparison of proton permeability of phosphatidylcholine liposomes with and without phosphatidic acid admixtures. Liposomes (8 mg per ml) were prepared by octylglucoside dialysis [see Mimms et al, 1981] in 50 mM Tricine buffer, followed by suspension in unbuffered sodium sulfate. Addition of small aliquots of 5 mM H_2SO_4 produced a rapid pH shift, followed by a slow return to an intermediate pH as proton equivalents crossed the liposome membranes. Phosphatidic acid admixture (PC:PA = 9:1) enhanced the proton permeability of the liposomes, but the calculated permeability coefficients were in the same range as liposomes prepared by the ether infusion method. (D.W. Deamer, unpublished results.)

Perhaps six orders of magnitude separates the highest and lowest estimates. These studies will be described here, then discussed with a view toward resolving some of the discrepancies.

Clement and Gould [1981] and Biegel and Gould [1981] used an entrapped pH-sensitive fluorescent dye to monitor rapid pH changes in the interior of small unilamellar vesicle (SUV) liposome preparations. This method has the advantage of reporting pH shifts in the millisecond range, but is not readily adapted to quantitative measurements of flux. The fluorescent dye pyranine (2.5 mM) was encapsulated during liposome preparation by sonication, the other components being 0.1 M KCl and 5 mM Tricine—MES buffers. Purified soybean phosphatide was the primary lipid used in these studies because it is commonly used in membrane reconstitution studies. However, the system was also tested with purified lipids such as dimyristoyl phosphatidylcholine. In a typical experiment, a pH gradient was produced in a stop-flow apparatus by mixing the liposomes with a second buffer of different pH, and the pH-dependent pyranine fluorescence shift gave an indication of the rate of internal pH change resulting from proton flux. In the absence of valinomycin, the decay had a half-time of a few minutes, and if valinomycin was present, the decay half-time was reduced to 300 msec. This result led the authors to conclude that the proton permeability was much greater than that of other cations. Potassium was the only counterion available, and in the absence of valinomycin, was rate limiting. If valinomycin was added so that potassium could move freely, the high proton permeability became apparent and the pH gradient decayed in less than a second. (The rapid decay relative to the LUV system described previously probably results from the low buffering capacity of the encapsulated medium and the small volume of the SUV preparation.)

These results offer independent confirmation that proton permeability is relatively high in liposome systems. Since sonication was involved, rather than ether infusion or detergent dialysis, the possibility of artifacts produced by trace contaminants is reduced. Furthermore, since the same result could be obtained with liposomes prepared from synthetic lipids containing saturated acyl chains, oxidation products could not have contributed to the result.

A considerably lower permeability was reported by Cafiso and Hubbell [1981], who took special care to reduce chloroform and fatty acid traces to minimal levels. In this study, the high relative permeability of protons was used to produce a membrane potential that could be detected by a spin label probe. From the kinetics of the potential buildup, it was possible to calculate proton flux and permeability, and the values obtained were in the range of 10^{-8} to 10^{-9} cm/sec. It is significant that these values are still 4–5 orders of magnitude greater than the permeability coefficients of sodium and potassium reported in SUV preparations using phosphatidylcholine.

Nozaki and Tanford [1981] have studied proton permeability in another LUV system. In this, liposomes were prepared by octyl glucoside dialysis at pH 4 and pH 10 in 40 mM aspartate buffer, then titrated to pH 7 to produce a pH gradient across the membrane. The rationale was that aspartate is a good buffer at pH 4 and 10, but buffers poorly at pH 7. A glass electrode could therefore monitor the external pH changes resulting from proton flux down the pH gradients. From the rate of pH change, and knowledge of the aspartate buffer capacity, permeability coefficients could be calculated for protons and hydroxyls at the low and high internal pH range, respectively.

In this system, the measured pH change occurred over a period of 40 hours, rather than the few minutes reported by other investigators. The apparent P_H was in the range of 10^{-9} cm/sec, and P_{OH} in the range of 10^{-7} cm/sec. The flux measured in the presence of nitrate was about 10 times that in the presence of chloride. Since this was in the same order as the dissociation constants of nitrate and chloride, the authors concluded that much of the flux, at least in the acid internal pH range, occurred as protonated anions. Using the data of Gutknecht and Walter [1981], the authors were able to subtract the apparent HCl contribution for the chloride result and concluded that the true P_H is in the same range (10^{-12} cm/sec) as that of other monovalent cations. They account for the high P_{OH} value by suggesting that hydrolysis products may have accumulated in the membranes during the extended dialysis required to remove the octyl glucoside used as a detergent, and that these increased the bilayer permeability.

Gutknecht and Walter [1981] have used planar lipid membranes to estimate proton permeability. Membranes were formed from several phospholipid species by brushing a decane–lipid solution over 1.8 mm² holes in a polyethylene partition separating two compartments. The phospholipid species included phosphatidycholine with and without cholesterol, and phosphatidylserine. One of the compartments was acidified (300 mM HCl) or neutral (50 mM NaCl), and proton flux was measured both as conductance and by a glass electrode in the second compartment that measured the pH shift produced as proton equivalents crossed the membrane. (In some experiments, the first compartment was also made basic, although the membranes were less stable under highly alkaline conditions.) Most of the flux occurred as HCl under these conditions, and the calculated HCl permeability was 2.9 cm/sec. No flux was observed when H_2SO_4 was substituted for HCl. The authors concluded that planar lipid membranes had very low conductance for protons and hydroxyl ions.

IV. DISCUSSION

Since there are so many variables among the published studies of proton permeability in lipid bilayer systems, it is not possible at this time to decide which represents the "true" value. (See Table I for summary.) Despite the variation in the results, it is still possible to compare the range of permeability values in the model membranes with those of biological membranes. Proton fluxes in the latter have been measured in mitochondria, sarcolemma, and sarcoplasmic reticulum, so it is possible to calculate approximate permeability coefficients in these systems. It is important to note that the net proton flux in the biological membranes may occur by mechanisms different from those in the model membrane system, and it is likely that the permeability of the biological membranes will be higher than that of lipid bilayers. However, if any biological membrane is much less permeable to protons than lipid bilayer membranes, it follows that the latter are not good models for studies of the barrier properties of natural systems.

A. Proton Permeability of Biological Membranes

Table II shows some fluxes and computed permeability coefficients of several biological membranes. Significantly, all of the values are greater than even the largest reported permeability coefficient from model systems. We can conclude that the lipid bilayer, even if much more permeable to protons than expected from studies of other ions, nonetheless offers a sufficient barrier to proton flux,

TABLE I. Summary of Proton Permeability Coefficients Measured in Lipid Bilayer Systems

System	Permeability coefficient (cm sec^{-1})	Reference
LUV, PC:PA 98:2 9-AA method	$P_{net} = 1.4 \pm 1.6 \times 10^{-4}$	Nichols et al, 1980
LUV, PC:PA 90:10 Glass electrode	$P_{net} = 3.2 \pm 1.1 \times 10^{-4}$	Nichols and Deamer, 1980
SUV, DMPC Pyranine fluorescence	$P_{net} = 3 \times 10^{-3}$	Biegel and Gould, 1981[a]
SUV, PC Membrane potential	$P_H \sim 10^{-8}-10^{-9}$	Cafiso and Hubbell, 1981
LUV, PC Glass electrode	$P_H = 3 \times 10^9$ $P_{OH} = 1 \times 10^{-7}$	Nozaki and Tanford, 1981[b]
Planar lipid membrane, PC-decane Conductance	$P_H = 3 \times 10^{-9}$ $P_{OH} = 4 \times 10^{-9}$	Gutknecht and Walter, 1981

[a]Calculated from data of Biegel and Gould using buffer capacity for Tricine and MES buffers, and assuming encapsulated volume of 0.5 liters per mole DMPC.
[b]This is a measured value in nitrate solution and may not represent an intrinsic value for proton flux. See Discussion.
Abbreviations: PC, phosphatidylcholine; PA, phosphatidic acid; DMPC, dimyristoyl phosphatidylcholine; SUV, small unilamellar vesicles; LUV, large unilamellar vesicles.

TABLE II. Approximate Proton Permeabilities of Several Biological Membranes

Membrane	Permeability coefficient (cm sec^{-1})	Reference
Mitochondria	$P_{net} = 10^{-3a}$	Mitchell and Moyle, 1967 Nichols et al, 1980
Frog muscle sarcolemma	$P_{net} = 10^{-3}$	Itzuzu, 1972
Sarcoplasmic reticulum	$P_{net} = 10^{-3}$	Meissner, 1981
Erythrocyte membrane	$P_{OH} = 2 \times 10^{-4}$ (pH 9)[b] $P_{OH} = 4 \times 10^{-1}$ (pH 4)	Crandell et al, 1971

[a]Calculated by Nichols et al from data of Mitchell and Moyle.
[b]Authors assumed that no proton flux occurred.

since biological membranes are even more permeable and yet are able to maintain transmembrane electrochemical proton gradients.

B. Relative Cation Permeabilities of Model Membranes

Although there is no consensus on absolute values of proton permeability in liposome systems, in all of the studies the measured permeability coefficients were orders of magnitude greater than those of sodium or potassium in similar systems. A possible exception is the result of Nozaki and Tanford [1981]. The measured permeability coefficient was in the range of that reported by others, but these authors subtracted an assumed flux of HCl from the measured flux and calculated that the actual permeability was similar to that of sodium in the same system. One difference in this study is that the authors did not test the effect of valinomycin-potassium on the system, so the proton flux may have been inhibited by the absence of counterion current. A second difference is that the authors did not report any observations in the first few seconds or minutes after establishing the gradients, which is the time interval used by other investigators.

As discussed earlier, Gutknecht and Walter [1981] did not find evidence for relatively high proton flux across planar lipid membranes. The essential point of this paper was that no anomolous conductance could be measured that might have reflected a high proton current, as would be expected if protons were in fact highly permeable. However, electrically neutral proton flux could be measured and was accounted for as HCl flux. One difference in this study in relation to the liposome systems is that the measurements were performed at very low pH ranges (300 mM HCl). This is low enough so that the lipid phosphate groups would be associated with protons. The resulting positively charged membrane might indeed be less permeable to a current of positive ions in the form of protons. A second difference is that the planar lipid membranes contain decane or tetradecane as solvent, and the effect of this on proton flux is unknown.

C. Mechanisms of Proton Flux Across Bilayers

We will close this discussion by speculating on mechanisms by which proton equivalents may cross lipid bilayers. It is a reasonable assumption that in the absence of specific carriers, ions cross lipid bilayers as hydrated species. Therefore we will first outline some of the concepts that have evolved to account for water flux across bilayers, since water and hydrated ions may share similar pathways.

Two approaches in considering water flux mechanisms have evolved. The first is that transient pores appear in the form of defects in the bilayer structure. Water, as well as solutes, may be able to enter such defects in the bilayer and thereby cross the membrane. This concept is supported by the finding that permeability in synthetic lipid bilayers is greatest near the phase transition temperature where such defects are at a maximum level [Papahadjopoulos et al, 1973]. Hauser et al [1973] have pointed out that if the defect is large enough, the process could not be considered to be equivalent to a true permeability. This remains an unresolved question, particularly for relatively impermeant substances such as sodium.

The second approach is that water flux occurs as diffusion of monomers. For instance, Traüble [1971] has proposed a molecular mechanism for water flux through bilayer structures in which it is envisaged that gauche-trans "kinks" produce defects in the fluid bilayer structure. Water monomers are able to enter the kinks and thereby diffuse with them across the membrane.

A mechanism involving some form of diffusion is supported by the demonstration of Finkelstein and Cass [1968] that one can calculate a permeability coefficient of water from knowledge of its solubility and diffusion coefficient in a bulk phase hydrocarbon system. The result is 6.4×10^{-3} cm/sec^{-1}, remarkably close to measured permeability coefficients in experimental bilayer systems. This is in part fortuitous, however, since water permeability of bilayers varies over a tenfold range depending on lipid composition. Furthermore, the solubility of water in the membrane hydrocarbon phase is unknown, and may vary markedly from its solubility in a bulk phase hydrocarbon.

We can now ask whether either of these alternative models helps in understanding the relatively high value of proton permeability obtained in the studies of liposomes that have been described here. Clearly, if water flux is entirely monomeric, then a hydrated proton is no different from any other hydrated cation that may attempt to find its way across the membrane. However, if some fraction of the water within the hydrophobic region of the bilayer is associated with other water molecules in a transient defect, an interesting possibility emerges. It is well known that protons have greater mobilities in aqueous solutions or ice compared with other cations. This is explained by the ability of protons to jump along hydrogen-bonded water molecules, rather than moving as discrete charges. Nagle and Morowitz [1978] have suggested the possibility that associated water may occur in the hydrocarbon region of bilayers, and Nichols and Deamer [1980] have proposed that the relatively high proton permeability of liposomes can be accounted for if protons are able to move through hydrogen-bonded strands or clusters of water. This concept will provide a useful working hypothesis in directing future research.

V. SUMMARY

The aim of this review chapter was to provide an estimate of the proton permeability barrier offered by the bilayer region in biological membranes. From the results presented, it is apparent that different model systems produce conflicting results for the proton permeability coefficient. The highest values were reported by Nichols and Deamer [1980] and are in the range of 10^{-4} cm/sec. A relatively high proton permeability was independently obtained by Biegel and Gould [1981]. However, in SUV liposomes composed of egg PC, Cafiso and Hubbell [1981] find P_H values in the range of 10^{-8} to 10^{-9} cm/sec. A similar value was obtained by Gutknecht and Walter [1981] in planar lipid membranes, and by Nozaki and Tanford [1981] in LUV preparations composed of egg PC. Both of the latter papers conclude that proton permeability is in the range of other monovalent cations, since they can account for much of the flux in their systems as nonelectrogenic flux of molecular HCl.

Despite these conflicting data, it is still possible to compare these values with the apparent proton flux and permeability of natural membranes. Flux measurements have been made for mitochondria, sarcolemma, and sarcoplasmic reticulum, and, when P_H is calculated from the data, the values are all in the range of 10^{-3} cm/sec, about 10 times greater than the highest estimate for proton permeability in lipid bilayer systems. We conclude that the lipid bilayer, even if it is much more permeable to protons than expected from the flux of other monovalent cations, is still a sufficient barrier to proton flux so that coupling membranes can function and transmembrane pH gradients can be maintained by cells.

VI. REFERENCES

Biegel CM, Gould JM: Kinetics of hydrogen ion diffusion across phospholipid vesicle membranes. Biochemistry 20:3474, 1981.

Cafiso DS, Hubbell WL: Spin label detection of electrogenic proton fluxes in phospholipid vesicles. Biophys J 33:114a, 1981.

Clement NR, Gould JM: Pyranine as a probe of internal aqueous hydrogen ion concentration in phospholipid vesicles. Biochemistry 20:1534, 1981.

Crandell ED, Klocke RA, Forster RE: Hydroxyl ion movement across the human erythrocyte membrane. J Gen Physiol 57:664, 1971.

Finkelstein A, Cass A: Permeability and electrical properties of thin lipid membranes. J Gen Physiol 52:145s, 1968.

Gutknecht J, Walter A: Transport of protons and hydrochloric acid through lipid bilayer membranes. Biochim Biophys Acta 641:183, 1981.

Hauser H, Oldani D, Phillips MC: Mechanism of ion escape from phosphatidylcholine and phosphatidylserine single bilayer vesicles. Biochemistry 12:4507, 1973.

Hopfer U, Lehninger AL, Thompson TE: Protonic conductance across phospholipid bilayer membranes induced by uncoupling agents for oxidative phosphorylation. Biochemistry 59:484, 1968.

Izutsu KT: Intracellular pH, H ion flux and H ion permeability coefficient in bullfrog toe muscle. J Physiol (Lond) 221:15, 1972.

Kano K, Fendler JH: Pyranine as a sensitive pH probe for liposome interiors and surfaces. Biochim Biophys Acta 509:289, 1978.

Kornberg RD, McNamee MG, McConnell HD: Measurement of transmembrane potentials in phospholipid vesicles. Proc Natl Acad Sci USA 69:1508, 1972.

Meissner G, Young RC: Proton permeability of sarcoplasmic reticulum vesicles. J Biol Chem 255:6814, 1980.

Mimms LT, Zampighi G, Nozaki Y, Tanford C, Reynolds JA: Phospholipid vesicle formation and transmembrane protein incorporation using octyl glucoside. Biochemistry 20:833, 1981.

Mitchell P: Coupling of phosphorylation to electron and hydrogen transfer by a chemiosmotic type of mechanism. Nature 191:144, 1961.

Mitchell P: Chemiosmotic coupling in oxidative and photosynthetic phosphorylation. Biol Rev 41:445, 1966.

Mitchell P, Moyle J: Acid–base titration across the membrane system of rat liver mitochondria. Biochem J 104:588, 1967.

Nagle JF, Morowitz HJ: Molecular mechanisms for proton transport in membranes Proc Natl Acad Sci USA 75:298, 1978.

Neumann J, Jagendorf AT: Light-induced pH changes related to phosphorylation by chloroplasts. Arch Biochem Biophys 107:109, 1964.

Nichols JW, Deamer DW: Net proton-hydroxyl permeability of large unilamellar liposomes measured by an acid–base titration technique Proc Natl Acad Sci USA 77:2038, 1980.

Nichols JW, Hill MW, Bangham AD, Deamer DW: Measurement of net proton-hydroxyl permeability of large unilamellar liposomes with the fluorescent pH probe aminoacridine. Biochim Biophys Acta 596:393, 1980.

Nozaki Y, Tanford C: Proton and hydroxyl ion permeability of phospholipid vesicles. Proc Natl Acad Sci USA 78:4324, 1981.

Papahadjopoulos D, Jacobson K, Nir S, Issac T: Phase transitions in phospholipid vesicles. Biochim Biophys Acta 311:330, 1973.

Rottenberg H, Grunwald T, Avron M: Determination of ΔpH in chloroplasts. Eur J Biochem 25:54, 1972.

Serhan C, Anderson P, Goodman E, Dunham P, Weissmann G: Phosphatidate and oxidized fatty acids are calcium ionophores J Biol Chem 256:2736, 1981.

Traüble H: The movement of water across lipid membranes: A molecular theory. J Membr Biol 4:193, 1971.

Van Zutphen H, Cornwell DG: Some studies on lipid peroxidation in monomolecular and bimolecular lipid films. J Membr Biol 13:79, 1973.

Snail Neuron Intracellular pH Regulation

Roger C. Thomas
Department of Physiology, University of Bristol, Bristol BS8 1TD, England

I.	Introduction	189
	A. Formal Introduction	189
	B. Informal Introduction	190
II.	Methods	190
	A. Experimental Setup	190
	B. Preparation	191
	C. Lighting	192
	D. Physiological Solutions	193
	E. Microelectrodes	193
	F. Calibration Solutions	194
	G. Electrical Arrangements	194
	H. Setting-up Procedure	194
	I. Stumbling Points	194
	J. Possible Mechanisms for pH_i Regulation	195
III.	Results	196
	A. Effect of pH_i Recovery on the Membrane Potential (E_m)	196
	B. Removal of External K^+	197
	C. Effect of Decreasing Internal Cl^-	197
	D. Effect of Removing Bicarbonate	198
	E. Effect of Removing External Na^+	199
	F. Effect of Metabolic Inhibitors on pH_i Recovery	200
IV.	Discussion	202
V.	References	204

I. INTRODUCTION

A. Formal Introduction

The biochemical importance of the H^+ ion in enzyme reaction rates, in determining membrane properties and even in generating ATP, is too well known to need description here. Random changes in intracellular pH (pH_i) are clearly biologically undesirable, and it is therefore theoretically probable that pH_i is regulated in some way, at least in higher animal cells.

In contrast to the pH_i regulating system, the sodium pump, first proposed in 1940, is now very well characterized as the mechanism keeping intracellular Na^+ low, and intracellular K^+ high, using ATP to drive these ions against their concentration gradients. Until recently almost nothing was known about any similar mechanism that might regulate pH_i. Indeed, thanks to experimental inadequacies there was some doubt that pH_i was regulated at all.

Some 5 years ago the situation began to improve. In St Louis Walter Boron and his co-workers began working on pH_i regulation by squid axons and barnacle muscles, and in Bristol I started working with recessed-tip microelectrodes on snail neurone pH_i. Of all the techniques for studying pH_i, the pH-sensitive electrode is clearly the best for studying pH_i regulation by single large cells. This is mainly because it is easy to calibrate and gives a continuous readout of pH_i from a cell that (often) appears to remain healthy for many hours.

In this chapter I will describe experiments designed to show which ions are involved in the recovery of snail neuron pH_i from an acid load, and what is the source of energy for the ion movements. The results illustrated are from recent new experiments that largely confirm earlier publications [Thomas, 1976a, b; 1977; 1978a]. I will not do more than mention work on other preparations, since an excellent review has just appeared [Roos and Boron, 1981].

B. Informal Introduction

I actually started to study pH_i regulation because I found myself with the ideal techniques at my fingertips. I first learned to inject salts into snail neurons and to record intracellular sodium, in order to better study the electrogenic Na pump [see Thomas, 1972]. My first Na^+-sensitive microelectrodes were very blunt, so I eventually developed the recessed-tip design, which was a great improvement. Urged by visitors to see if I could make recessed-tip pH microelectrodes, and having been given some pH glass on a visit to Corning, I did make some (very slow) pH-sensitive microelectrodes about 9 years ago. To prove they worked I eventually did a series of experiments on snail neuron pH_i [Thomas, 1974]. It seemed a rather boring subject, and appeared to be regulated only very slowly. But my standard snail Ringer of those days contained no bicarbonate. Slowly it dawned on me (and more quickly on Boron and De Weer [1976]) that normal pH_i regulation required bicarbonate, and that its mechanism was not only at that time quite unknown, but also very interesting.

II. METHODS

A. Experimental Setup

My basic procedure is to penetrate an exposed snail neurone with up to six microelectrodes. I then make the cell interior acid by exposing it to CO_2 or by iontophoretic injection of HCl, and observe the pH_i recovery. My experimental

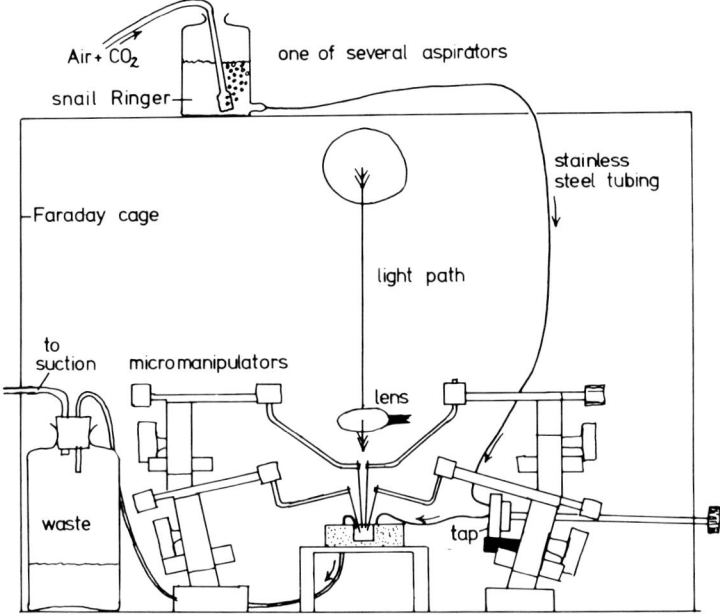

Fig. 1. Diagram of front view of experimental setup, excluding electrical apparatus. The preparation is viewed by a horizontally mounted dissecting microscope, which is not shown. The Faraday cage is 77 cm × 45 cm deep × 59 cm high, and has a hinged front that is not shown.

arrangement (excluding microscope and electrical apparatus) is illustrated in Figure 1. Six Prior micromanipulators are arranged around the experimental chamber, which is seen more clearly in Figure 2. The micromanipulators each hold one microelectrode within 15° of vertical.

B. Preparation

Snails (Helix aspersa) are collected locally, many coming from my own garden, and kept in a corner of my laboratory without food or water. I remove the whole circumesophageal ganglion (made up of nine connected ganglia) using forceps and scissors, and mount it on a plastic slide. I then use very fine scissors and forceps, and a dissecting microscope, to remove the outer connective tissue over the visceral and right pallial ganglia. I then put the plastic slide in the experimental bath shown in Figure 2, cover it with Ringer, and view it with a horizontally mounted binocular microscope. Finally I tear the remaining con-

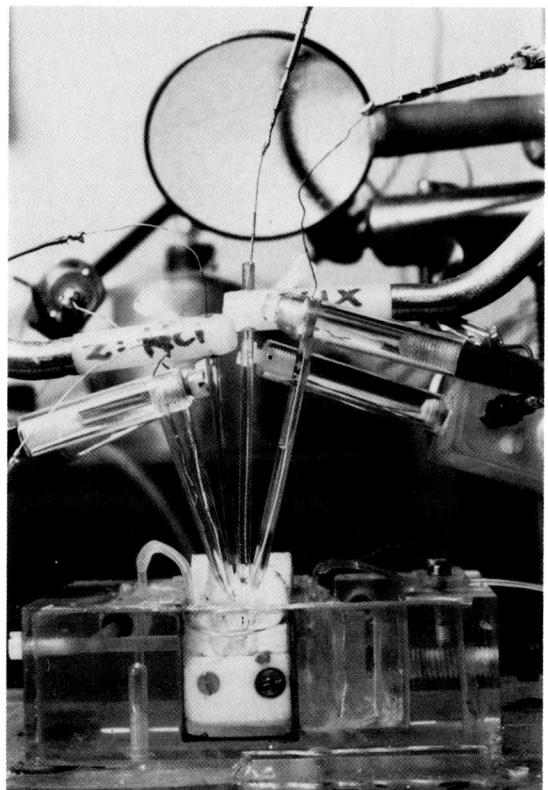

Fig. 2. Photograph of experimental chamber with six microelectrodes mounted above it. The width of the white plastic block on which the preparation is mounted is 17 mm. Ringer solution enters from the right and leaves via the inverted U-tube at the left.

nective tissue over the large neurons in the ganglia with a tungsten hook. If my favorite cell (the largest one at the rear of the right pallial ganglia) is not well exposed, I may try squirting Ringer at it from a syringe. If this cell appears dead, I repeat the dissection.

C. Lighting

The preparation is lit from above and behind by light from a microscope lamp outside the cage. It is focused onto the superficial neurons by a lens mounted on a small retort stand that can be moved manually to optimize the constrast. With care in focusing, excellent dark-field type illumination can be obtained.

D. Physiological Solutions

The preparation is continually superfused with a snail Ringer solution from one of several bottles on top of the Faraday cage enclosing my setup. Each bottle is connected by tubing to a 12-way tap [Partridge and Thomas, 1975]. Solutions containing CO_2 are run through stainless steel tubing to prevent loss of CO_2 between bottle and bath. From the tap the selected solution flows into the bath at a rate of 2–3 ml/min, and is sucked out by an inverted U-tube (see Fig. 2). Waste solution is collected in a bottle inside the Faraday cage to reduce electrical interference.

My normal snail Ringer is equilibrated with 2.5% CO_2 in air, has a pH of about 7.55, and contains the following salts (in mequiv/l) NaCl:80, KCl:4, $CaCl_2$:7, $MgCl_2$:5, $NaHCO_3$:20. (The precise pH of such a solution can be hard to measure with a commercial combination pH macroelectrode: I have found that combination electrodes with ceramic references tend to record different pHs in bicarbonate-buffered solutions than do separate reference and pH electrodes [see also Illingworth, 1981]. Gel-filled combination electrodes have also given trouble.) My standard CO_2-free snail Ringer has 20 mM HEPES, instead of $NaHCO_3$, and is adjusted to pH 7.5 by addition of NaOH.

The Na-free solution is made by replacing NaCl by n-methyl glucamine neutralized with NaOH, and replacing $NaHCO_3$ by 20 mM n-methyl glucamine. Several hours of bubbling with 2.5% CO_2 converts the latter to bicarbonate and the pH to about 7.5.

The Cl-free solution is made by substituting 1.15 mequiv of the gluconic acid salts for each mequiv of the chlorides. This factor is to allow for the lower activity of the gluconate solution.

E. Microelectrodes

Conventional microelectrodes for iontophoretic injections were made from filamented borosilicate glass and filled with 1 M KCl, HCl, or NaCl, or 0.02 M K vanadate, pH 9.9. They were then tested to see if they would carry at least 100 nA with their tips in snail Ringer. Resistances were usually about 10 MΩ for KCl and NaCl, 20 MΩ for HCl, and 40 MΩ for K vanadate.

Chloride, Na^+, and pH-sensitive microelectrodes were made using Corning 477315 liquid, NAS 11–18, and pH-sensitive glass as previously described [Thomas 1978b]. Their response was checked before, and calibrated after, each successful experiment.

Reference microelectrodes for recording the membrane potential were made by filling the tip of silanized micropipettes with a 2% solution of potassium tetrakis (p-chlorophenyl) borate in n-octanol [Thomas and Cohen, 1981].

F. Calibration Solutions

pH-sensitive microelectrodes were calibrated in CO_2-free pH 7.5 and pH 6.5 (buffered with PIPES) solutions. Na^+ and Cl^--sensitive microelectrodes were calibrated in a series of solutions in which K^+ was substituted for Na^+ and gluconate for Cl^-. The activity coefficient of 100 mM K gluconate was measured with a K^+-sensitive microelectrode and found to be 0.68 as compared with 0.78 for normal snail Ringer.

G. Electrical Arrangements

The potential from each recording microelectrode was led (via a unity-gain varactor-diode amplifier) to the inputs of one or more differential amplifiers and to an oscilloscope. The E_m amplifier amplified the difference between reference and bath electrodes (the latter is a Ag/AgCl wire in an Agar-Ringer bridge). The pH amplifier handled the difference between reference and pH microelectrodes, the aNa amplifier the reference and Na^+-sensitive microelectrodes, and the aCl the reference and Cl^--sensitive microelectrodes. The outputs of these amplifiers were recorded on a four-channel FM tape recorder (Racal Store 4) and a four-channel potentiometric pen recorder (Watanable multicorder).

The current passing microelectrodes were connected to a floating current clamp [Thomas, 1975], which permitted the passage of currents between two microelectrodes or between one and the bath. It also allowed the potential from any current microelectrode to be observed on the oscilloscope while the cell was being penetrated.

H. Setting-up Procedure

The first step is to ensure that all the required solutions are prepared and in bottles connected to the multiway tap, and that generous amounts of the CO_2:air mixture are bubbling through the appropriate solutions. Next a preparation is dissected. Then the microelectrodes are arranged with the tips above the preparation and their backs connected to the amplifiers or current clamp. After that the lid of the Faraday cage is closed, the pen recorder switched on, and the electrode response to a pH 6.5, low Cl^-, low Na^+ (Li^+ substituted) test solution is observed. If all electrodes respond as expected, the preparation is left, superfused with normal CO_2 snail Ringer, to equilibrate for at least 30 min.

I. Stumbling Points

1) The preparation needs to be continually superfused to prolong its life and prevent temperature and other changes when the solution is changed.

2) To minimize artifacts use aged polyethylene tubing to suck solution out of the bath. Adjust the suction or tubing orifice so that the solution level in the bath does not change.

3) It is surprisingly easy to get an incomplete penetration, even with a sharp microelectrode. Accurate intracellular measurements depend on all electrodes' being fully inserted and recording the same membrane potential. This should be checked by passing a hyperpolarizing current.

4) Injury to the cell not only causes a loss of membrane potential but also leads to a rise in internal Na^+, and Cl^- (and probably other ions) and a fall in pH. Minimizing the effects of cell penetration is a major problem, and there is no easy answer.

5) Leakage of salts such as KCl into the cell from a conventional microelectrode can be a serious problem, since resistances high enough to minimise leakage cause unstable electrodes (see Thomas & Cohen, 1981).

6) Preparations really are healthier in $CO_2 : HCO_3^-$ buffered solutions. Bubbling gas mixtures are a nuisance, but at least $NaHCO_3$ is cheaper than HEPES. Tris is not to be trusted.

J. Possible Mechanisms for pH_i Regulation

Before coming to the results of my experiments, I will describe various mechanisms by which pH_i might be regulated. An active extrusion of H^+ ions (or the equivalent uptake of OH^- or HCO_3^- ions) is required to maintain the normal value of pH_i. The mechanism by which the cell actively extrudes Na^+ ions is well known, and is shown in Figure 3 as the sodium pump. The pump consumes energy in the form of ATP, and extrudes 3 Na^+ ions in exchange for the uptake of 2K. It is thus electrogenic, and its operation tends to hyperpolarize the cell [Thomas, 1972].

In contrast, the pH_i-regulating system (as shown by the experiment illustrated in Fig. 4) is electroneutral. The simplest explanation for this is that it exchanges one cation for another, or one anion for another. The various possibilities are shown in Figure 3 under "pH_i-regulating systems." The first three—H:K exchange, Cl:OH exchange, and Cl^-:HCO_3^- exchange—would all require meta-

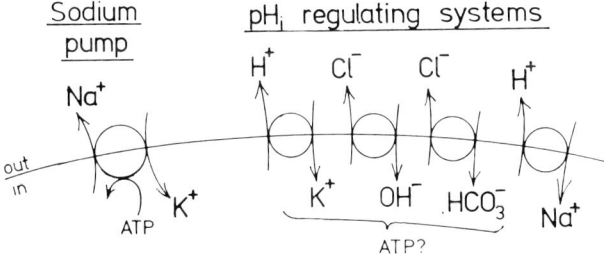

Fig. 3. Diagram showing the ionic mechanism of the sodium pump and various electroneutral models for the pH_i regulating system.

bolic energy, since the ions would be moving against their concentration gradients. On the other hand, Na^+-H^+ exchange would not require energy input, since there is a very large inward Na^+ gradient.

III. RESULTS

A. Effect of pH_i Recovery on the Membrane Potential (E_m)

The experiment shown in Figure 4 compares the effect on E_m of equal injections of NaCl and HCl. I will describe it in some detail.

At the start of the recording all microelectrodes had their tips poised above the cell, and were electrically stable. The first microelectrode to be lowered into the cell was the Na^+-sensitive one. Next I inserted the pH-sensitive microelectrode and then the reference liquid ion exchanger (RLIE) microelectrode. As can be seen, once conditions had stablized E_m was about 45 mV, pH_i about 7.4, and internal Na^+ activity (aNa_i) was about 6 mM. To test the perfusion system and pH_i microelectrode I then explosed the preparation to CO_2-free Ringer. This caused a large increase in the pH_i as HCO_3^- combined with H^+ ions and left the cells as CO_2 [see Boron and De Weer 1976; Thomas, 1976a].

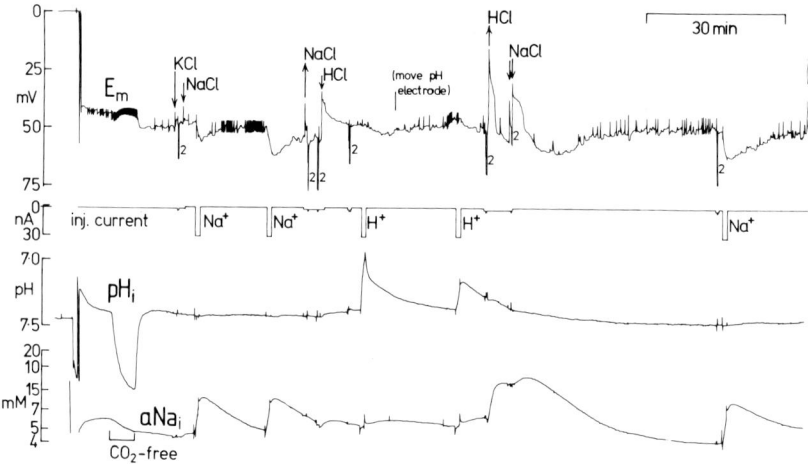

Fig. 4. Pen recording of an experiment showing the effect of NaCl and HCl injections on the membrane potential (E_m) (inside cell is negative, though not indicated in the convention used here), intracellular pH (pH_i), and intracellular Na activity (aNa_i) of a snail neurone. The injection current is also shown. The cell was spontaneously active, but the slow response of the pen recorder has largely eliminated the action potentials. Arrows above the E_m trace indicate the points where injection microelectrodes were inserted or withdrawn. Except where indicated, cell was in 2.5% CO_2 Ringer. The increases in E_m labeled "2" were caused by a 2nA hyperpolarizing current.

Having changed the superfusing solution back to CO_2 Ringer, I inserted a KCl-filled microelectrode into the cell. I then used it to pass a 2nA hyperpolarizing current to assess membrane resistance. Then I inserted a NaCl-filled electrode and switched on a 3nA current between these two electrodes to prevent leakage of Na^+ ions into the cell.

A few minutes later I reversed the current and increased it to 30nA for 1 minute to inject NaCl. This caused a rise in aNa$_i$, as shown on the bottom trace, corresponding to an increase in internal Na^+ of about 10 mequiv/l cell water. As the Na^+ was injected E_m increased, suggesting that the Na^+ pump was generating a current of about 1nA. As the Na^+ was pumped out, E_m returned to normal. A second injection was made, this causing a larger increase in E_m as the membrane resistance had apparently increased.

The NaCl electrode was then withdrawn and replaced with one filled with HCl. Two injections of HCl were then made, using the same charge as for the NaCl injections. (During the first pH$_i$ fell very rapidly, so the pH and HCl electrodes were moved apart before the next injection.) The effects of these two HCl injections on E_m were very small, showing that the pH$_i$ regulating system was not electrogenic. (The quantities of NaCl and HCl injected would have been very similar, and so were the rates of recovery of pH$_i$ and aNa$_i$. Thus if even only half the H^+ ions had been extruded uncoupled to other ion movements, E_m would certainly have been visibly affected.)

To show that this lack of change in E_m as pH$_i$ recovered was not due to cell damage, I then withdrew the HCl electrode, reinserted the NaCl electrode (causing a large rise in aNa$_i$ due to cell damage), and finally made another injection of NaCl. This result shows that the snail neuron pH$_i$-regulating system is electrically neutral.

B. Removal of External K^+

The first mechanism proposed in Figure 3 is the exchange of internal H^+ for external K^+. This should be blocked by K^+-free Ringer, as is the sodium pump. Although it is not illustrated here, K^+ removal has no effect on the pH$_i$ recovery from an HCl injection [Thomas, 1976c].

C. Effect of Decreasing Internal Cl^-

If pH$_i$ regulation involves the exchange of internal Cl^- for external OH^- ions, pH$_i$ recovery should inhibited by low internal Cl^-. An experiment to test this is shown in Figure 5. Rather than injecting HCl to decrease pH$_i$ (since it would raise Cl_i^-), this was done by applying CO_2 after a period in CO_2-free Ringer.

The experiment was started with the cell in the normal 2.5% CO_2 Ringer. Once the cell had been penetrated with Na^+, Cl^-, pH-sensitive, and reference microelectrodes, the external CO_2 was removed. This caused a rise in internal pH and aCl$_i$, and a fall in aNa$_i$. Some 20 minutes later the CO_2 was replaced.

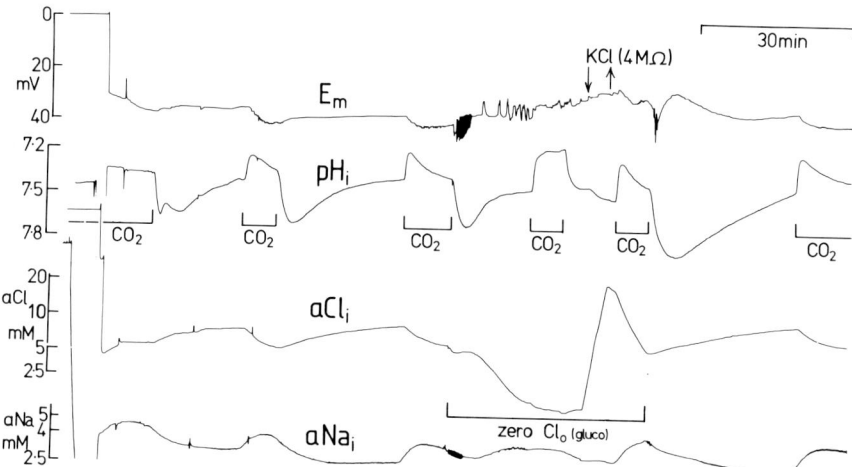

Fig. 5. The effect of external Cl removal on the response to 2.5% CO_2 application and removal of the E_m, pH_i, aCl_i, and aNa_i of a snail neurone. The arrows indicate where a 4-megohm KCl electrode was inserted and withdrawn.

Intracellular pH fell and then recovered. As it recovered, aCl_i fell and aNa_i increased. After a second removal and reapplication of CO_2, external Cl^- was removed. This led to a decrease in aCl_i. Once it had fallen below about 2 mM, CO_2 was again applied. This time pH_i fell as before, but did not return to normal until the CO_2 was removed.

To show that this blockage of pH_i recovery was not due to the lack of external Cl^-, I then increased aCl_i by penetrating the cell with a low resistance KCl microelectrode. Once aCl_i was well above normal I withdrew this electrode and applied CO_2. This time pH_i recovered. Thus normal pH_i recovery requires internal Cl^- (and is associated with a fall in aCl_i), strongly suggesting that normal pH_i regulation involves an efflux of Cl^- ions.

D. Effect of Removing Bicarbonate

If pH_i recovery involves anion exchange, it will require external hydroxyl or bicarbonate ions. The experiment shown in Figure 6 shows that in CO_2-free (and HCO_3^--free) Ringer not only is the intracellular buffering power much reduced [Thomas, 1976a] but pH_i recovery is very slow. Compare the rate of pH_i recovery after the fourth injection with that after the acidification occurring when CO_2 was reapplied. If allowance is made for the increased buffering power in CO_2, the relative rate of H^+ extrusion in CO_2 is even greater.

These last two experiments strongly favour $Cl^- HCO_3^-$ exchange as the mechanism for pH_i regulation in snail neurons. Such a mechanism was proposed for

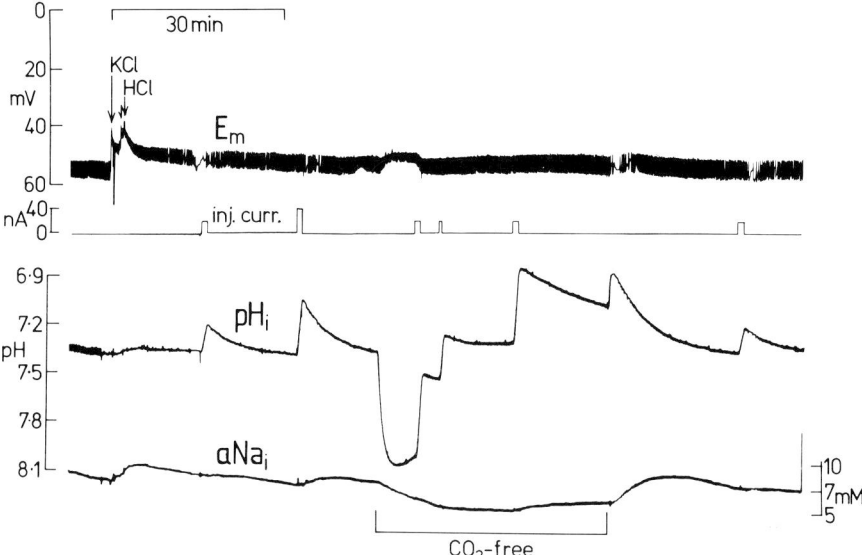

Fig. 6. The effect of removing CO_2 and bicarbonate on the pH_i and aNa_i reponse to HCl injections. Arrows indicate where the two injection electrodes were inserted. Except where indicated extracellular CO_2 was 2.5%.

squid axons by Russell and Boron [1976]. At the same time [Thomas, 1976b] I showed that snail neuron pH_i regulation was blocked by inhibitors of anion exchange. But before deciding that anion exchange was the whole story for snail neurons, I luckily tried removing external Na^+.

E. Effect of Removing External Na^+

In the experiment shown in Figure 7 I recorded the effects of four injections of HCl on both pH_i and aNa_i. The first two injections were made with the cell bathed in normal CO_2 Ringer. After each injection pH_i recovered normally. After these controls, I removed external Na^+ (causing a fall in internal Na^+) and made a third injection. This time pH_i did not recover until external Na^+ was replaced. Finally I applied the anion-exchange inhibitor SITS and made a fourth injection. Again there was no recovery of pH_i. Thus pH_i regulation appears equally inhibited by either removal of external Na^+ or by application of SITS.

If, as this result suggests, $Na^+:H^+$ exchange plays a part in pH_i regulation, there should be a rise in internal Na^+ as pH_i recovers. Inspection of the aNa_i record in Figures 4–7 shows that this does indeed occur, although not very clearly in Figure 4.

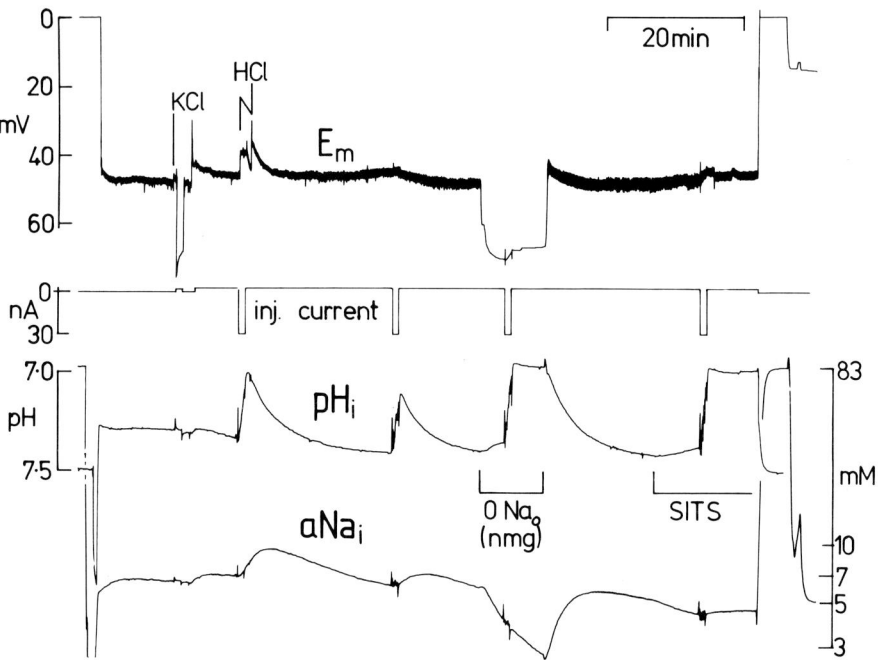

Fig. 7. The effect of removing external Na and of applying 10^{-5} SITS (4-acetamido-4'-isothiocyanato-stilbene-2,2'-disulphonic acid) on the pH_i and aNa_i response to HCl injection. Part of the calibration of the Na^+-sensitive microelectrode is shown at the end.

The results so far presented are thus in favor of both $Cl^-HCO_3^-$ *and* $Na^+:H^+$ exchange. Is there any requirement for ATP, as shown by Russell and Boron [1976] for squid axons?

F. Effect of Metabolic Inhibitors on pH_i Recovery

The uncoupling agent carbonyl cyanide m-chlorophenyl hydrazone (CCmP) lowers intracellular ATP levels by making mitochondria permeable to H^+ ions. When applied to snail neurons, as shown in Figure 8, it causes a decrease in pH_i and starts a steady increase in aNa_i. The rise in internal Na^+ is caused by inhibition of the Na^+ pump, which requires ATP. The pH_i recovery from HCl injections, however, is not much affected.

Perhaps some ATP remains under these conditions, enough to drive $Cl^-HCO_3^-$ exchange. If so, orthovanadate would be expected to have some effect, as it inhibits transport ATPases [Macara, 1980]. In the experiment shown

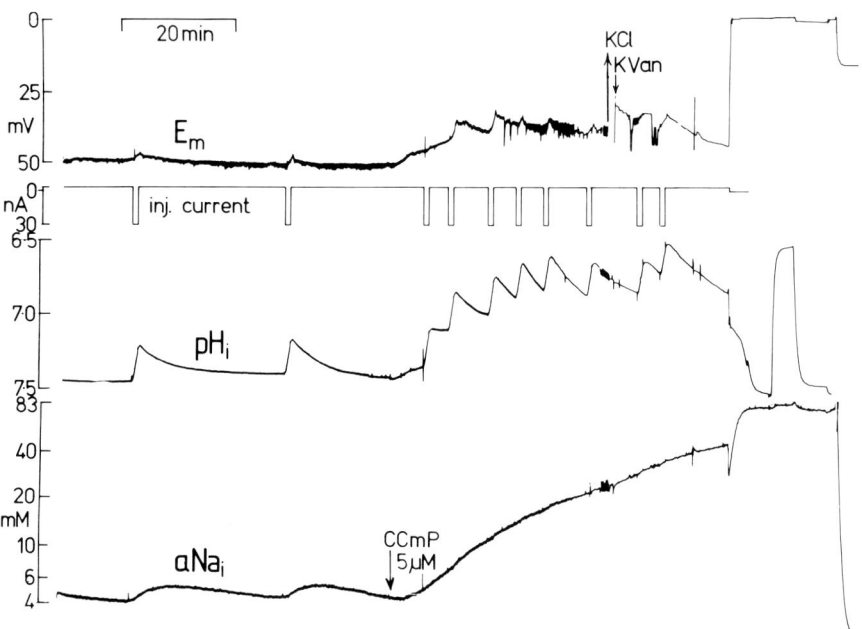

Fig. 8. The effect of the uncoupling agent CCmP on the E_m, pH_i, and aNa_i response to eight injections of HCl and two of vanadate. Calibration of the pH microelectrode is also shown.

in Figure 8, I withdrew the KCl electrode after the eighth HCl injection and replaced it with an electrode filled with K orthovanadate. Thus when I next passed the injection current, vanadate ions would be carried into the cell along with H^+ ions from the HCl electrode.

Two injections of vanadate ions had no apparent effect on pH_i recovery, reinforcing my earlier [Thomas, 1978a] suggestion that ATP plays no role in snail neuron pH_i regulation. Presumably energy is provided by the large inward Na^+ gradient.

To make sure that the pH_i recovery seen with CCmP was not due to passive H^+ movements across the cell membrane, I tested the effect of SITS in the experiment shown in Figure 9. As before, CCmP appeared to shift the pH_i baseline, but did not block pH_i recovery. SITS, however, did almost completely block pH_i recovery after the sixth and seventh HCl injections. There is no reason to suppose SITS affects any passive H^+ permeability of the cell membrane induced by CCmP, so this result shows that the pH_i recovery in CCmP takes place via the normal SITS-sensitive mechanism.

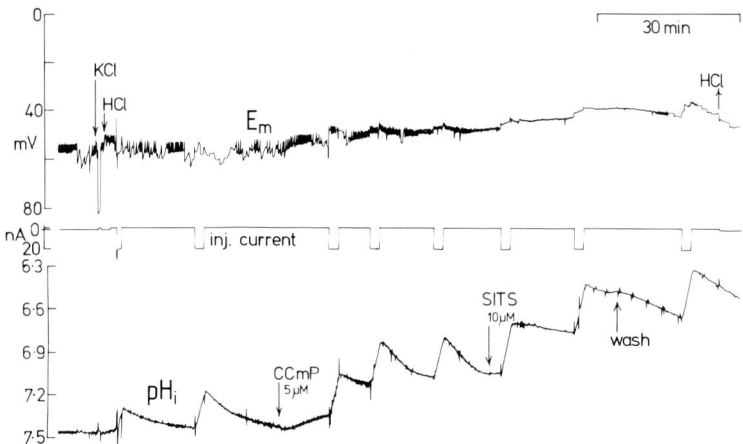

Fig. 9. The effect of CCmP and SITS on the pH$_i$ recovery from HCl injection. At the points indicated first CCmP and then SITS were added to the superfusion solution.

IV. DISCUSSION

None of the schemes shown in Figure 3 can explain all my results. If there were *separate* Cl$^-$:HCO$_3^-$ and Na$^+$:H$^+$ exchangers, then pH$_i$ recovery would not be completely blocked by any of the treatments described above. Furthermore, lowering internal Cl$^-$ would not prevent the rise in aNa$_i$ seen during the normal pH$_i$ recovery. The results confirm that pH$_i$ recovery from an acid load in snail neurons is almost completely blocked either by removal of external Na$^+$ or by lowering internal Cl$^-$. It appears to involve both a rise in internal Na$^+$ and a fall in internal Cl$^-$, and requires HCO$_3^-$. It is blocked by SITS but does not seem to require ATP. I therefore reaffirm my proposal [Thomas, 1977] that pH$_i$ in snail neurons is regulated by a membrane carrier exchanging external Na$^+$ and HCO$_3^-$ ions for internal H$^+$ and Cl$^-$ ions, as shown in Figure 10A. (Since I have no evidence that H$^+$ ions are involved, it is possible that a second HCO$_3^-$ ion enters the cell instead of the H$^+$ ion leaving.)

The energy available from the large inward Na$^+$ gradient is more than adequate to drive the other ion fluxes. Inspection of Figure 5, for example, shows that aNa$_i$ is about 4 mM, giving an inward electrochemical gradient (from the Nernst equation) of about 70 mV, to which can be added E$_m$. The other ions are all relatively close to equilibrium; E$_H$–E$_m$ and E$_{HCO_3}$–E$_m$ are both about 30 mV in favor of pH$_i$ decrease, and E$_{Cl}$–E$_m$ is close to zero. Indeed if there were no loss of energy in the operation of the pH$_i$-regulating system, and if each hypothetical site were perfectly selective for the ion it binds, the normal pH$_i$ should be more alkaline than it is. That pH$_i$ is as acid as it is is probably a question of selectivity.

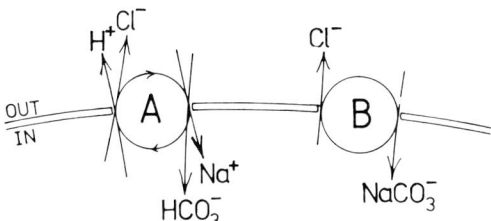

Fig. 10. Two models which could explain the ionic requirements for pH_i regulation in snail neurones. A was originally proposed by Thomas [1977], B by Becker and Duhm [1978].

Becker and Duhm [1978] have proposed that pH_i might be regulated by an exchange of internal Cl^- for an external ion pair consisting of one Na^+ ion linked to one CO_3^{2-} ion, as shown in Figure 10B. Such a scheme would beautifully explain the inhibition of pH_i regulation by SITS and its analogues. This system should run equally well with Li^+ instead of Na^+, but I have found that replacement of Na^+ by Li^+ outside the cell is almost as inhibitory as replacement by organic cations [Thomas, 1967c]. The ion pair has been considered much more thoroughly by Boron [1981] and Roos and Boron [1981], so I will not discuss it further.

Further evidence for a membrane carrier not requiring metabolic energy comes from preliminary experiments [Thomas, 1980] on the possible reversal of the snail neuron pH_i regulator. A large acid influx is seen when the cell is exposed to pH 6.5 Ringer, and this influx is blocked by removal of external Cl^- and by application of SITS.

The stoichiometry of pH_i regulation has not been carefully investigated in snail neurons [see Thomas, 1977]. In squid axons Boron and Russell have done a lot of work on ion influxes, and their results fit in with scheme A. Detailed discussion of their and other's results on other preparations is out of place here. Suffice it to say that in squid axons and barnacle muscle pH_i regulation appears similar to that in snail neurons. In mammalian muscle, however, Aickin and Thomas [1977; see also Roos and Boron, 1978] have obtained results suggesting that pH_i regulation is mainly achieved by $Na^+:H^+$ exchange, with a separate and smaller role for $Cl^-:HCO_3^-$. Moody [1981] finds a somewhat similar situation in crayfish neurons. For more details and discussion of the mechanisms proposed for other preparations, see Roos and Boron [1981].

Finally, a plea that the 4-ion carrier described above in Figure 10A should not be dismissed as too complicated to be believable. The Na^+ pump, after all, must have sites for three Na^+ ions, two K^+ ions, and one ATP, making six binding sites in all.

ACKNOWLEDGMENTS

I am most grateful to the MRC for money, to Michael Rickard for technical help, and to Mrs. Avril Lear for patience and skill far beyond that of a normal secretary.

V. REFERENCES

Aickin CC, Thomas RC: An investigation of the ionic mechanism of intracellular pH regulation in mouse soleus muscle fibres. J Physiol 273:295, 1977.

Becker BF, Duhm J: Evidence for anionic cation transport of lithium, sodium and potassium across the human erythrocyte membrane induced by divalent anions. J Physiol 282:149, 1978.

Boron WF, De Weer P: Intracellular pH transients in squid axons caused by CO_2, NH_3 and metabolic inhibitors. J Gen Physiol 67:91, 1976.

Boron WF, McCormick WC, Roos A: pH regulation in barnacle muscle fibers: Dependence on extracellular sodium and bicarbonate. Am J Physiol 240:C80, 1981.

Illingworth JA: A common source of error in pH measurements. Biochem J 195:259, 1981.

Macara IG: Vanadium—An element in search of a role. Trends Biochem Sci 5:912, 1980.

Moody WJ: The ionic mechanism of intracellular pH regulation in crayfish neurones. J Physiol 316:293, 1981.

Partridge LD, Thomas RC: A twelve-way rotary tap for changing physiological solutions. J Physiol 245:22P, 1975.

Roos A, Boron WF: Intracellular pH transients in rat diaphragm muscle measured with DMO. Am J Physiol 235:C49, 1978.

Roos A, Boron WF: Intracellular pH. Physiol Revs 61:296, 1981.

Russell JM, Boron WF: Role of chloride transport in regulation of intracellular pH. Nature 264:73, 1976.

Thomas RC: Electrogenic sodium pump in nerve and muscle cells. Physiol Revs 52:563, 1972.

Thomas RC: Intracellular pH of snail neurones measured with a new pH-sensitive glass microelectrode. J Physiol 238:159, 1974.

Thomas RC: A floating current clamp for intracellular injection of salts by interbarrel iontophoresis. J Physiol 245:20P, 1975.

Thomas RC: The effect of carbon dioxide on the intracellular pH and buffering power of snail neurones. J Physiol 255:715, 1976a.

Thomas RC: Ionic mechanism of the H pump in a snail neurone. Nature 262:54, 1976b.

Thomas RC: Comparison of the Na^+ and H^+ pumps in a snail neurone. J Physiol 263:212, 1976c.

Thomas RC: The role of bicarbonate, chloride and sodium ions in the regulation of intracellular pH in snail neurones. J Physiol 273:317, 1977.

Thomas RC: Comparison of the mechanisms controlling intracellular pH and sodium in snail neurones. Respir Physiol 33:63, 1978a.

Thomas RC: "Ion Sensitive Intracellular Microelectrodes." London: Academic Press, 1978b.

Thomas RC: Reversal of the pH_i-regulating system in a snail neuron. In Boulpaep EL (ed): "Current Topics in Membranes and Transport," Vol 13. New York: Academic Press, 1980, p 23.

Thomas RC, Cohen CJ: A liquid ion-exchanger alternative to KCl for filling intracellular reference microelectrodes. Pflugers Arch 390:96, 1981.

Regulation of Intracellular pH in Barnacle Muscle

Albert Roos and Walter F. Boron

Departments of Physiology and Biophysics, and of Anesthesiology, Washington University School of Medicine, 660 South Euclid Avenue, St. Louis, Missouri 63110 (A.R.), and Department of Physiology, Yale School of Medicine, New Haven, Connecticut 06510 (W.F.B.)

I.   Introduction 205
II. Results and Discussion 206
III. References 218

I. INTRODUCTION

The intracellular pH (pH_i) of most cells, including the muscle fiber of the giant barnacle (Balanus nubilus) is maintained at a steady-state value that is more alkaline than would be expected from passive H^+ distribution. The equilibrium relation between membrane voltage, V_m, and internal and external H^+ concentrations is given by the Nernst expression $V_m = 1,000 \, (RT/F)\ln((H^+)_o/(H^+)_i)$ where R and F are the gas constant and Faraday constant respectively, and V_m is membrane voltage in millivolts. Thus, at 22°C, $V_m = 59(pH_i - pH_o)$. That is, at a membrane voltage of -59 mV (a reasonable value for an isolated barnacle muscle fiber under experimental conditions), pH_i should be 1 unit less than pH_o. Actually, when the fiber is submerged in barnacle seawater (BSW) at pH 7.8 (which is about the pH of the animal's hemolymph), pH_i is 7.3 rather than 6.8.

This relatively alkaline pH_i is maintained in the face of acidifying influences (metabolism, H^+ influx, OH^- and HCO_3^- efflux) by a regulating mechanism located in the membrane, to which we have given the name "acid extrusion." This same mechanism responds to the imposition of acute acid loads by returning the pH_i toward normal. The following is a summary of some of the studies carried out in our laboratory on acid extrusion in the barnacle muscle fiber, a cross section of which is shown in Figure 1. This fiber is about 500 μm in

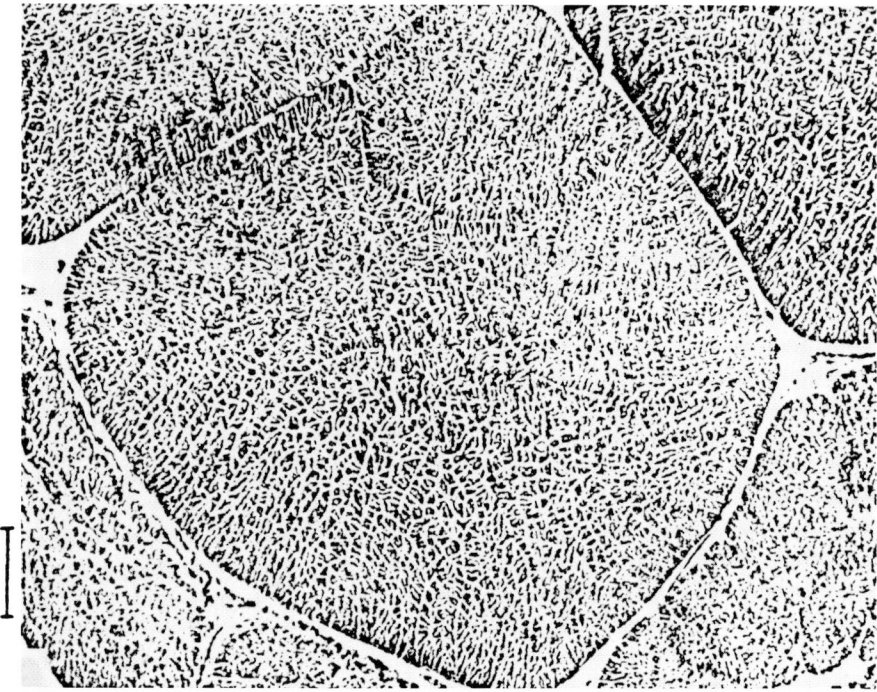

Fig. 1. Cross section through a muscle fiber of the giant barnacle (Balanus nubilus). The mark indicates 100 μm. Note the invaginations of the sarcolemma. [From Hoyle and Smyth, 1963.]

diameter; in general, fiber diameter ranges from 400 to more than 1,000 μm. Note the many invaginations of the sarcolemma, which probably testify to the large fiber's multifiber origin. However, both electrically and mechanically the fiber acts as a single unit. Figure 2 illustrates the intracellular pH-sensitive glass microelectrode used in this work. The electrode was developed by J.A.M. Hinke [Hinke, 1967].

II. RESULTS AND DISCUSSION

The technique of mounting a single muscle fiber and of inserting the pH-sensitive and reference electrodes has been described in detail [Boron, 1977], as have construction of the electrodes, calibration procedures, composition of the superfusates, and recording technique [Boron and De Weer, 1976]. We shall start with examining the course of pH_i observed when a fiber is exposed for a limited period to 5% CO_2 (pH_o 7.60) (Fig. 3). This is one of several ways in which an acid load can be imposed: Molecular CO_2 rapidly enters the fiber by

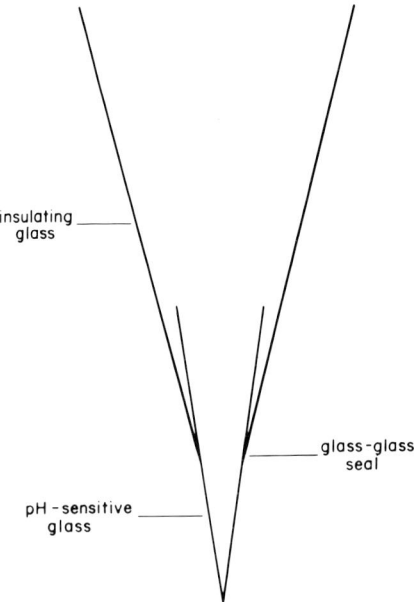

Fig. 2. Exposed tip microelectrode designed by Hinke [1967], used in the present studies. The entire pH-sensitive surface must be introduced into the cell.

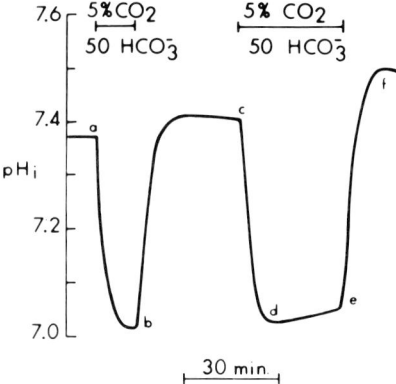

Fig. 3. Effect of short- and long-term CO_2 exposures on pH_i. Fiber ($V_m = -52$ mV) was first exposed to BSW containing 5% CO_2/50 mM HCO_3^- for about 12 min; pH_i fell rapidly and appeared to reach a steady value (point b). Removing CO_2 caused pH_i to overshoot its initial baseline value (7.37) by 0.04. During more lengthy (33 min), second CO_2 exposure, pH_i fell (cd) and then slowly rose (de). Returning fiber to CO_2-free BSW caused pH_i to overshoot its initial value (7.41) by an exaggerated amount, 0.09. [From Boron, 1977.]

diffusion, and then, after hydration, yields protons. The first CO_2 pulse was applied for only 12 minutes, the second for more than half an hour. Especially in the second pulse it can be seen that the pH_i, after its initial fall, slowly rises, beginning at point d, at which pH_i is about 7.05. The equilibrium potential for H^+ at this point is only about 59 (7.05–7.60) $\simeq$ -32 mV, whereas V_m = -49 mV. That is, there is an *inward* driving force on H^+ amounting to 17 mV, and—assuming CO_2 equilibration across the membrane—an outward driving force on HCO_3^- of the same magnitude. Therefore, the "plateau phase alkalinization" following point d must be the result of an active, energy-requiring process. The pH_i after CO_2 removal is seen to return to a value higher than the original one. This is the necessary consequence of the preceding removal of acid. Note that the post-CO_2 overshoot is greater than the total pH_i rise during the plateau. This difference is due to the greater internal buffering power due to the presence of HCO_3^-/CO_2 during the plateau. Figure 4 also shows two pH_i transients produced by CO_2. However, during the second, 0.5 mM SITS (4-acetamido-4'-isothiocyanostilbene-2,2'-disulfonic acid) was present in the superfusate. This compound is one of a series of disulfonic stilbenes that blocks carrier-mediated anion transport in red cells [Cabantchik et al, 1978; Ho and Guidotti, 1975; Knauf and Rothstein, 1971]. It can be seen from Figure 4 that SITS has blocked the plateau phase alkalinization, as well as the overshoot. This strongly suggests that HCO_3^- influx, rather than H^+ efflux, is responsible for the alkalinization. It is of considerable interest that, under favorable conditions, a plateau phase *acidification* can also be evoked (Fig. 5). Here, the acid load was imposed at a very low external pH, namely 6.3, which, as will be pointed out below, discourages acid extrusion. The SITS sensitivity of the plateau acidification (second pulse, Fig. 5) indicates that the acidification is not due to H^+ leaking in (even though there is a large inward driving force on H^+).

The CO_2 method of acid loading necessitates concomitant changes in the external environment. Because we wanted to analyze the ionic substrate of the acid extrusion process in more detail, we introduced an internal acidification procedure that leads to reduced pH_i while leaving the external medium unchanged. This method consists of pulsing the fiber with a NH_4Cl-containing solution, a procedure previously described [Boron and De Weer, 1976]. In brief, application of NH_4Cl-BSW first produces a rapid rise in pH_i because of the entry of NH_3, which then combines with protons. The pH_i then slowly decays owing to entry of NH_4^+, which yields protons. When the fiber is now returned to the original NH_4Cl-free BSW, both NH_3 and NH_4^+ leave in the form of NH_3, an excess of H^+ ions being left behind. Therefore, the pH_i value is now lower than before the pulse was applied, the degree of lowering being determined by external NH_4Cl concentration, external pH, and length of exposure. Figure 6 shows the pH_i course during and after the NH_4Cl pulse. The reduced pH_i returns only slowly towards normal when the fiber is in HEPES-BSW (pH 7.40). However,

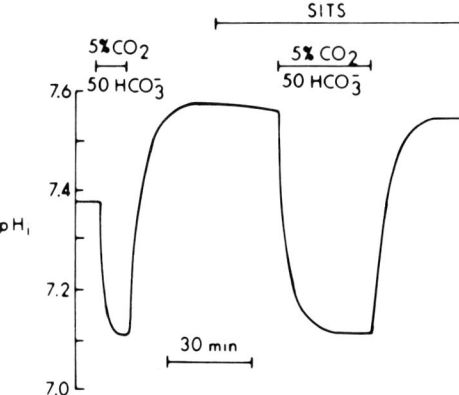

Fig. 4. Inhibitory effect of SITS on acid extrusion. Fiber ($V_m = -57$ mV) was first pulsed with 5% CO_2/50 mM HCO_3^-. After removal of CO_2, pH_i overshot the baseline (indicating that acid extrusion had taken place during CO_2 exposure) and then began a slow decline. Introduction of 0.5 mM SITS produced no immediate effect on pH_i but, when fiber was exposed to CO_2, SITS prevented a secondary alkalinization. When CO_2 was removed, pH_i did not overshoot initial pH_i. [From Boron, 1977.]

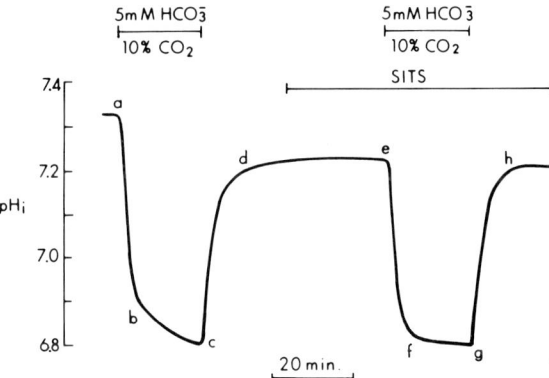

Fig. 5. SITS-sensitive acidification in pH_o 6.3 BSW. The fiber was first exposed to BSW containing 5 mM HCO_3^- and equilibrated with 10% CO_2 (pH_o 6.27). After the initial CO_2-induced fall in pH_i (ab), pH_i continued to fall (bc). Return to CO_2-free BSW caused pH_i to return to a value less than the original one. The fiber was then poisoned with 0.5 mM SITS for 25 min, and then once again exposed to 5 mM HCO_3^-/10% CO_2-BSW. This time, after its initial fall (ef), pH_i was nearly constant (fg), and, on removal of CO_2, nearly returned (gh) to its initial value. [From Boron et al, 1979.]

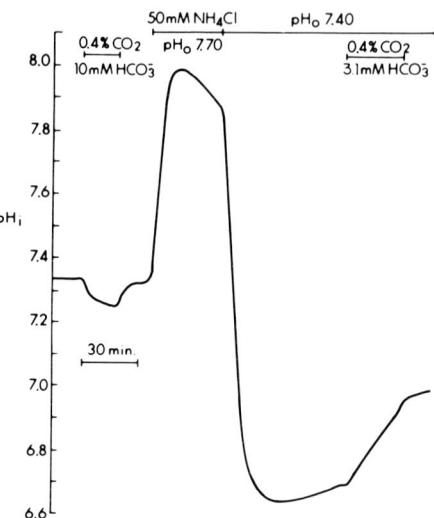

Fig. 6. Fiber (initial $V_m = -57$ mV) was first pulsed with 0.4% CO_2/10 mM HCO_3^-, to compare this acidification with one produced by same solution at lower baseline pH. Fiber was then pulsed with 50 mM NH_4Cl; this resulted in a rapid rise of pH_i, followed by a slower decay. Returning fiber to normal BSW (pH_o 7.40, 5 mM HEPES buffer) caused pH_i to fall to a very low value, and then to rise slowly. When this solution was replaced by one buffered with 0.4% CO_2/3.1 mM HCO_3^- (at same pH_o), rate of alkalinization was increased by more than 5-fold. Small ripple in pH_i tracing at outset of second CO_2 exposure represents competition between CO_2-induced acidification and the alkalinizing effect of "pump." [From Boron, 1977.]

it can be seen that, in a CO_2-HCO_3^--containing superfusate of the same pH_o, alkalinization occurs at a much brisker rate. This again points to HCO_3^- being the essential substrate. This is summarized in more detail in Figure 7, where the pH_i course is shown in five fibers during recovery from an acid load. The pH_o in all five was the same, but the bicarbonate concentrations differed. The pH_i was first lowered to 6.7 with the NH_4Cl prepulse technique (pulse not shown). It can be seen that 1) at a particular pH_i, the rate of pH_i recovery is greater, the higher the $(HCO_3^-)_o$; and 2) at a particular $(HCO_3^-)_o$, the rate of pH_i recovery is greater, the lower the pH_i.

When the effect of pH_o is examined (Fig. 8), we find that the higher the pH_o, the greater is the rate of alkalinization at a particular pH_i. It must be kept in mind that in these studies the pH_o changes were accomplished by changing $(HCO_3^-)_o$ in the same direction, CO_2 concentration being kept unchanged. Yet, even at constant $(HCO_3^-)_o$, an increase in pH_o enhances alkalinization (Fig. 9). It must be concluded that both pH_o and $(HCO_3^-)_o$ affect this process.

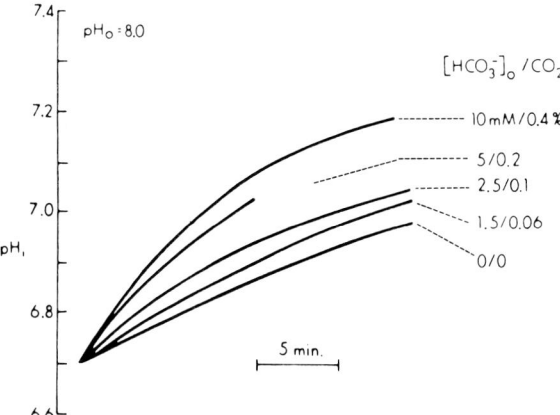

Fig. 7. Five experiments showing recovery of pH_i from acid loads at five different values of $(HCO_3^-)_o$. Five fibers were acid-loaded by exposing them to 50 mM NH$_4$Cl BSW, pH_o 7.7, followed by BSW pH_o 7.4. Courses of pH_i during this NH$_4$Cl prepulse are not shown; they resembled that in Figure 6. pH_i fell to below 6.7 in all five experiments, and subsequent recovery of pH_i is shown here in a composite drawing. Superfused BSW contained $(HCO_3^-)_o$ (in mM), percentage CO_2 as follows: 0,0; 1.5, 0.06; 2.5, 0.1; 5, 0.2; and 10, 0.4. By parallel changes in pCO_2 and $(HCO_3^-)_o$, a constant pH_o of 8.0 was maintained. [From Boron et al, 1981.]

The rate of acid extrusion (equivalent millimoles of protons removed per minute from a liter of myoplasm) can be obtained from records such as those shown in Figures 7 and 8 by multiplying, at each pH_i, the slope (dpH_i/dt) with the total intracellular buffering power, β_T, at that pH_i. This buffering power is the sum of the intrinsic non-CO_2-buffering power and that contributed to by the CO_2-HCO_3^- system, β_{CO_2}. The former has been measured by us over a wide pH_i range [Boron, 1977; Boron et al, 1979]. Since the $(CO_2)_i$ during the inscription of the curves remains nearly constant (equal to $(CO_2)_o$), $\beta_{CO_2} = 2.3\,(HCO_3^-)_i$, and can thus be calculated. The relationship between acid extrusion rate (millimoles l^{-1} min^{-1}), and pH_i at constant pCO_2 is given in Figure 10. It can be seen that, at each pH_o, there is a linear relationship between acid extrusion and pH_i.

We can conclude from the results presented thus far that acid extrusion is enhanced by reducing pH_i, by raising pH_o, and by raising $(HCO_3^-)_o$, and that it is accomplished by the uphill inward transport of bicarbonate ions. In a number of other cells [for references see Boron et al, 1978; Roos and Boron, 1981], this HCO_3^- entry is accomplished by an exchange of external HCO_3^- for internal Cl$^-$. A study was therefore undertaken of the relationship between HCO_3^- fluxes and Cl$^-$ fluxes in barnacle muscle under conditions of acid loading [Boron et al, 1978]. Both the external solution and, independently, the internal ionic envi-

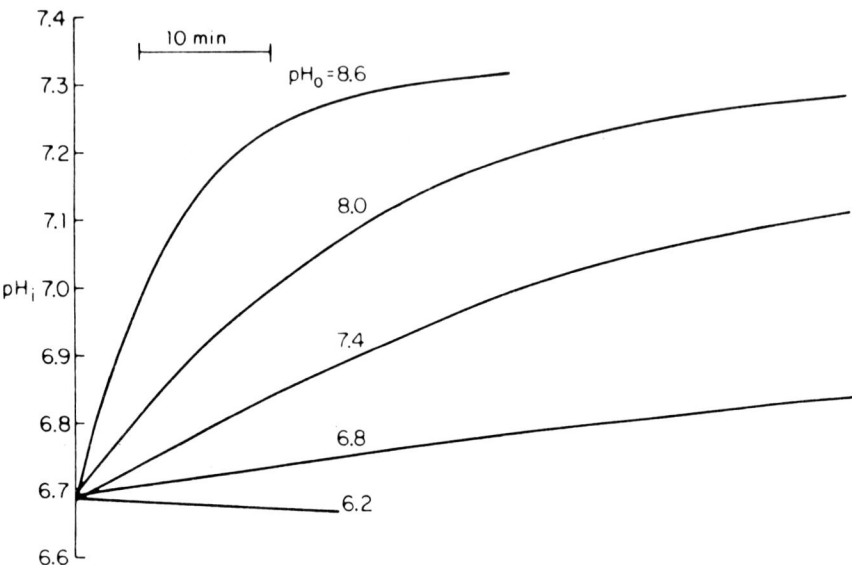

Fig. 8. Dependence of pH_i recovery upon pH_o (pCO_2 constant). Fibers were previously acid-loaded by prepulsing with 50 mM NH_4Cl-BSW, as in Figure 6. They were then exposed to BSW equilibrated with 0.4% CO_2 at pH_o 8.6, 8.0, 7.4, 6.8, or 6.2; $(HCO_3^-)_o$ was 40, 10, 2.5, 0.63, and 0.16 mM, respectively. Experiments are picked up at the point where pH_i is 6.7, and the subsequent recovery (or, in the case of pH_o 6.2, decay) is monitored. In the pH_o 7.4 experiment, pH_i eventually reached 7.3, but only after 2 h. [From Boron et al, 1979.]

Rates of pH_i recovery with $[HCO_3^-]_o$ = 2.5 mM following acid loading

pH_o	CO_2 %	Rate of pH_i Recovery	Rate of Acid Extrusion
6.8	1.6	4.6 ± 1.2 (4)	0.20 ± 0.05
7.4	0.4	15.3 ± 2.6 (4)	0.59 ± 0.10
8.0	0.1	30.3 ± 3.3 (4)	1.15 ± 0.13

Rates of pH_i recovery values are expressed in thousandths of a pH unit/min ± SE; number of fibers in parentheses. All data were obtained at pH_i 6.8. External solution contained 30 mM HEPES. Rates of acid extrusion values are expressed in $mmol \cdot l^{-1} \cdot min^{-1}$ ± SE.

Fig. 9. [From Boron et al, 1979.]

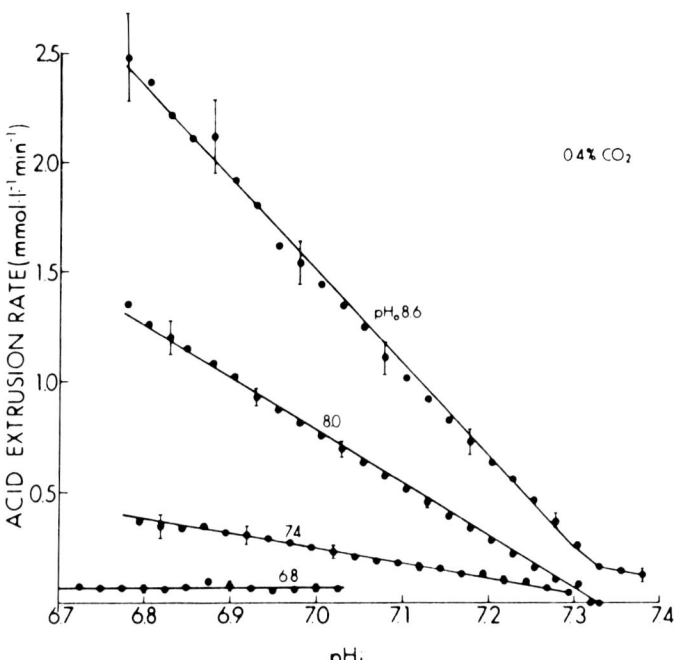

Fig. 10. Acid extrusion rate as a function of pH_i at 4 values of pH_o (pCO_2 constant). Summary of data from 63 experiments, in which pH_i was lowered by pretreating with NH_4Cl and then allowed to recover while the fiber was bathed in HCO_3^-/0.4% CO_2-BSW at pH_o 8.6, 8.0, 7.4, or 6.8. The curves at pH_o 7.4, 8.0, and 8.6 were obtained as follows: 1) At a given pH_o, the acid extrusion rate for each experiment was calculated at pH_i intervals of 0.025 by multiplying the rate of pH_i recovery by β_T, as described in text, and acid extrusion rate was then plotted against pH_i. 2) All curves at a given pH_o were normalized to the mean pH_i at which acid extrusion rate was zero; these pH_i values were 7.32 ± 0.025 (n = 7) at pH_o 7.4; 7.33 ± 0.023 (n = 11) at pH_o 8.0; and 7.38 ± 0.027 (n = 19) at pH_o 8.6. At pH_o 8.6 the acid extrusion rate never became zero; therefore, the pH_i to which these curves were normalized was obtained by extrapolation to zero acid extrusion rate of the linear portion of the curves. 3) All normalized data at a given pH_o were averaged at pH_i intervals of 0.025, and these averaged values are plotted in this figure. The pH_o 6.8 curve is a simple average of the data without normalization. Total number of experiments: 12 at pH_o 6.8, 12 at pH_o 7.4, 20 at pH_o 8.0, and 19 at pH_o 8.6. Each point represents the average of 2–7 experiments at pH_o 6.8, 3–11 at 7.4, 7–20 at 8.0, and 7–19 at 8.6. Vertical bars indicate standard error; lines represent linear regressions. [From Boron et al, 1979.]

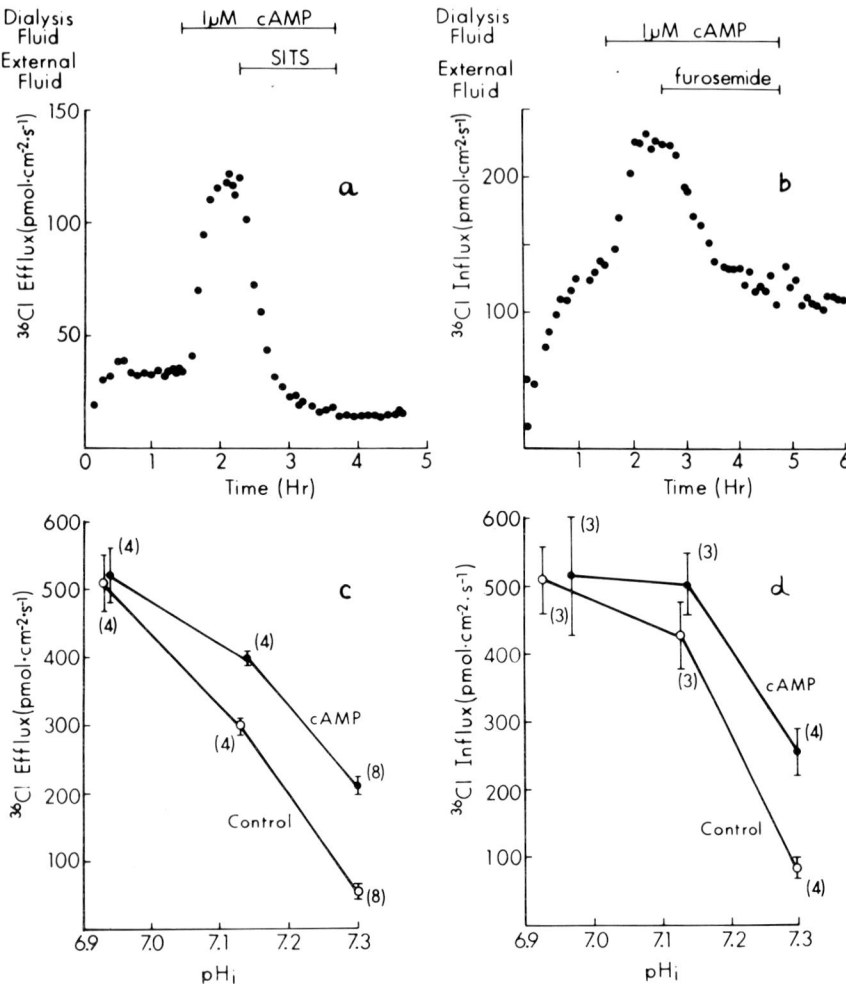

Fig. 11. a) Effect of cyclic AMP on Cl⁻ efflux. At time zero, flow of isotope-containing dialysis fluid, pH 7.3 (50 mM HEPES buffer) was begun. After Cl⁻ efflux had reached a steady state, 1 μM cAMP was added to the dialysis fluid, causing an increase in Cl⁻efflux. Subsequent addition of 0.5 mM SITS to the external fluid reduced Cl⁻efflux to slightly below control values. V_m was -53 mV at the outset and -49 mV at the end of the experiment. b) Effect of cyclic AMP on Cl⁻ influx. After Cl⁻ influx had reached a steady state, addition of 1 μM cAMP to the dialysis fluid caused an increase in Cl⁻ influx, which was then reversed by the presence of 0.6 mM furosemide in the external fluid. V_m was -51 mV and -47 mV at the beginning and end of the experiment, respectively. c, d) Effect of cyclic AMP on Cl⁻ efflux and influx at various values of pH_i. Experiments were similar to those above. Vertical bars represent 1 standard error; number of experiments is given in parentheses. Lines through points drawn by hand. Actual values of pH_i were verified with pH microelectrodes in separate experiments; 4–5 determinations of pH_i were made for each pH of the dialysis fluid; SE = ± 0.03–0.04. Fiber was assumed to be a cylinder. [From Boron et al, 1978.]

ronment could be varied, the latter by inserting through the fiber a porous dialysis capillary through which fluids of various composition were driven (dialysis technique of Brinley and Mullins [1967]). In one set of experiments, pH-sensitive and reference microelectrodes were placed alongside the dialysis tube. In another set, either Cl⁻ influx or efflux was measured with ^{36}Cl. The isotopic flux studies were performed by J.M. Russell and M.S. Brodwick (Department of Physiology and Biophyics, University of Texas Medical Branch, Galveston). Some of the results are plotted in Figure 11. The two lower panels of this figure show that both Cl⁻ influx and efflux are greatly enhanced when pH_i is reduced. Cyclic AMP (cyclic 3′,5′-adenosine monophosphate) stimulated both Cl⁻ influx and efflux, especially when the "basal" fluxes were low (relatively high pH_i). The two upper panels of Figure 11 show that either SITS or furosemide, a diuretic, blocks cyclic AMP stimulation. Lowering of pH_i not only enhances bidirectional Cl⁻ influxes, but also initiates acid extrusion (as has already been discussed), a process that is similarly stimulated by cyclic AMP (Figure 12). SITS and furosemide were found to block this stimulation also.

The common sensitivity of both the Cl⁻ fluxes and of acid extrusion (HCO_3^- influx) to cyclic AMP, SITS, and furosemide strongly suggests that all three are mediated by the same carrier. In fact, Russell (to be published) observed a requirement of Cl⁻ influx for internal Cl⁻, and we found [Boron et al, 1978] that if, at reduced pH_i, $(HCO_3^-)_o$ is raised at constant pH_o (7.8), Cl⁻ influx falls concomitantly with increase in acid extrusion.

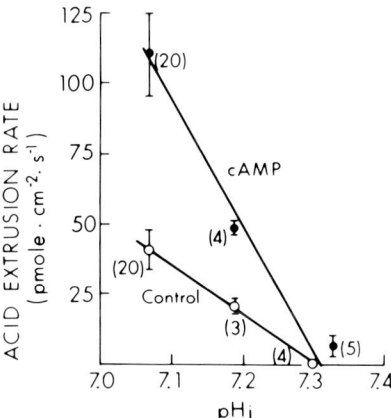

Fig. 12. Effect of cyclic AMP on acid extrusion rate at various values of pH_i. Values of acid extrusion rate were calculated from observed dpH/dt (during pH_i recovery) and measured intracellular buffering power, assuming that the fiber is a cylinder. Vertical bars indicate one standard error; number of experiments is given in parentheses. Lines though points represent least-squares fits. [From Boron et al, 1978.]

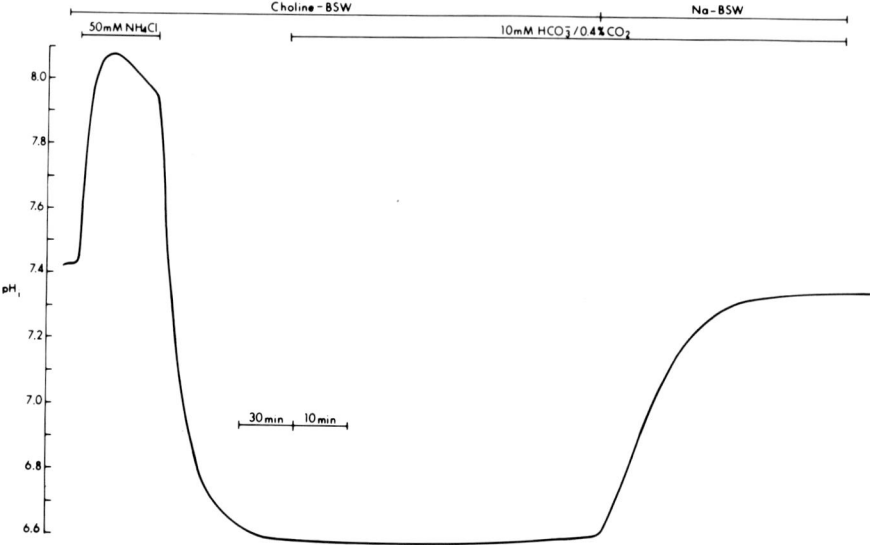

Fig. 13. Course of pH$_i$ in BSW-containing choline instead of Na$^+$. A fiber was exposed to BSW pH$_o$ 7.7 in which choline and 50 mM NH$_4^+$ replaced Na$^+$, followed by choline-BSW, pH$_o$ 7.4, after which choline-BSW containing 10 mM HCO$_3^-$ (added as KHCO$_3$) and 0.4% CO$_2$, pH$_o$ 8.0 was superfused for over 1 h. No recovery of pH$_i$ took place. Na$^+$ was introduced into superfusate in BSW containing 440 NaCl, 10 mM HCO$_3^-$, 0.4% CO$_2$, pH$_o$ 8.0. During this exposure to Na$^+$, pH$_i$ recovery proceeded normally. [From Boron et al, 1981.]

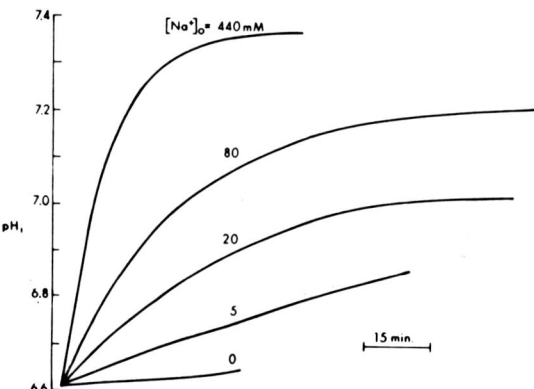

Fig. 14. Five experiments showing recovery of pH$_i$ from acid loads at five different values of (Na$^+$)$_o$. Each of five fibers was first acid-loaded by exposure to 50 mM NH$_4$Cl in choline-BSW, followed by a solution without NH$_4$Cl and containing 0–440 mM Na$^+$. The course of pH$_i$ during exposure to these two solutions is not shown; it resembles that in Figure 13. pH$_i$ fell to below 6.6 in all five experiments. Composite drawing above shows subsequent recovery of pH$_i$ during exposure to a third solution buffered with 10 mM HCO$_3^-$ and 0.4% CO$_2$, pH$_o$ 8.0. (Na$^+$)$_o$ of the second and third solutions was the same for a particular experiment. [From Boron et al, 1981.]

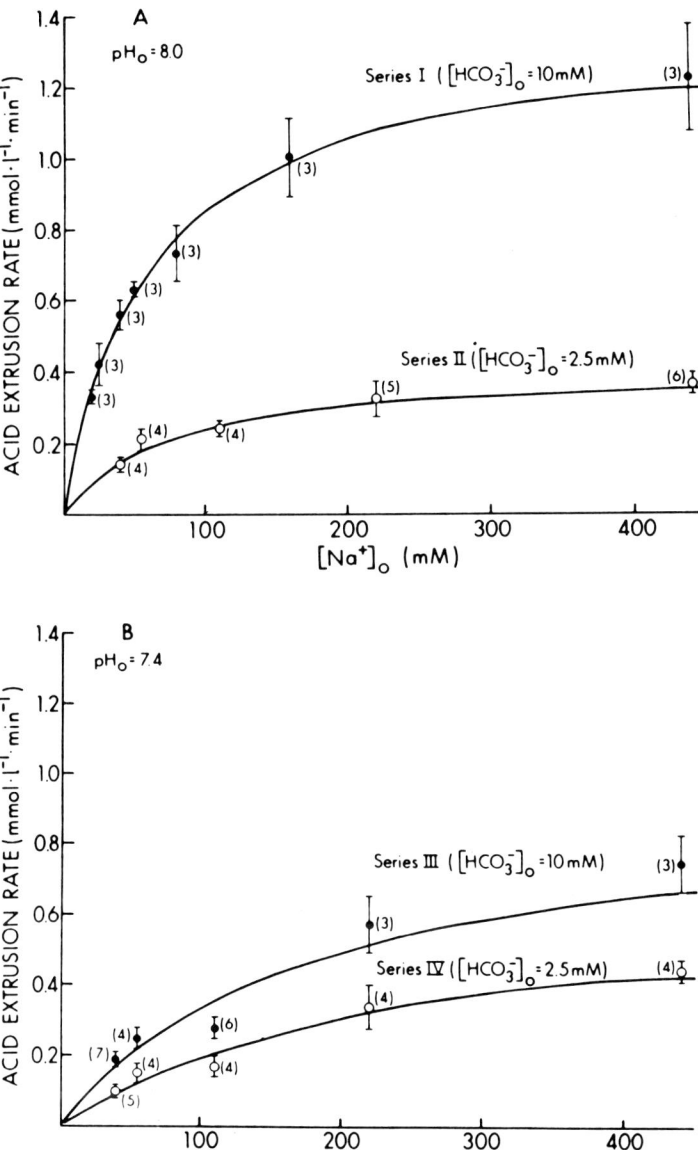

Fig. 15. Acid extrusion rate as a function of $(Na^+)_o$. Summary of data from 88 experiments in which pH_i was lowered by the NH_4Cl prepulse technique in choline-BSW, after which recovery of pH_i was observed during superfusion with BSW containing various $(Na^+)_o$ at A) 10 mM HCO_3^-, 0.4% CO_2, pH_o 8.0 (series I); 2.5 mM HC14, 0.1% CO_2, pH_o 8.0 (series II); and B) 10 mM HC14, 1.6% CO_2, pH_o 7.4 (series III); 2.5 mM HC14, 0.4% CO_2, pH_o 7.4 (series IV). Each point represents the average of 3–7 experiments. Vertical bars indicate SE. Curves representing simple Michaelis-Menten kinetics were obtained by computer using a nonlinear least-squares method, and assuming log-normal distributions. [From Boron et al, 1981.]

Finally, we must examine the role of sodium in acid extrusion. Figure 13 shows that, even though conditions for acid extrusion after the NH$_4$Cl prepulse are favorable (low pH$_i$, high pH$_o$, HCO$_3^-$ and Cl$^-$ present in medium and inside the fiber), pH$_i$ recovery cannot take place in the absence of Na$^+$. Lithium can partially substitute for it. The quantitative relationship between (Na$^+$)$_o$ and pH$_i$ recovery rate is illustrated in Figure 14, which shows the time course of pH$_i$ of five representative fibers. Each fiber was exposed to a different (Na$^+$)$_o$; (HCO$_3^-$)$_o$ and pH$_o$ were the same. The slopes of curves such as these can be converted into acid extrusion rates as described above. These rates, plotted against (Na$^+$)$_o$, are shown in Figure 15 in which all the measurements were made at pH$_i$ 6.8. The computed fitted curves represent simple Michaelis-Menten kinetics, assuming log-normal distributions. Similar curve fitting was applied to two sets of studies (not illustrated) in which acid extrusion was measured as function of (HCO$_3^-$)$_o$ (pH$_o$ either 7.4 or 8.0; normal (Na$^+$)$_o$). The kinetic parameters, V$_{max}$ and K$_m$, derived from these six curves allowed us to *rule out* at least two transport models: the ion pair model in which the ion pair NaCO$_3^-$ would be transported (in exchange for Cl$^-$), and the model in which Na$^+$ and HCO$_3^-$ would bind randomly to the carrier with the binding of HCO$_3^-$ competitively inhibited by H$^+$. A number of other models are compatible with our observations, such as random binding of Na$^+$ and HCO$_3^-$ or of Na$^+$ and CO$_3^{2-}$ to the carrier. Experiments involving different combinations of variables and constants are needed to allow a more decisive choice.

In the above, we have attempted to give a synopsis of the complex ionic interactions that, in barnacle muscle, regulate one of its important parameters, the intracellular pH.

ACKNOWLEDGMENTS

David W. Keifer and Wayne C. McCormick made significant contributions to some of the studies on which this paper is based. The work has been supported by Public Health Service grants HL-00082 to A. Roos and GM-06499 (National Research Service Award) to W.F. Boron. A. Roos is the recipient of National Institutes of Health research career award HR-19608.

III. REFERENCES

Boron WF: Intracellular pH transients in giant barnacle muscle fibers. Am J Physiol 233 (Cell Physiol 2):C61–C73, 1977.

Boron WF, De Weer P: Intracellular pH transients in squid giant axons caused by CO$_2$, NH$_3$ and metabolic inhibitors. J Gen Physiol 67:91–112, 1976.

Boron WF, McCormick WC, Roos A: pH regulation in barnacle muscle fibers: Dependence on intracellular and extracellular pH. Am J Physiol 237 (Cell Physiol 6):C185–C193, 1979.

Boron WF, McCormick WC, Roos A: pH regulation in barnacle muscle fibers: Dependence on extracellular sodium and bicarbonate. Am J Physiol 240 (Cell Physiol 9): C80–C89, 1981.

Boron WF, Russell JM, Brodwick MS, Keifer DW, Roos A: Influence of cyclic AMP on intracellular pH regulation and chloride fluxes in barnacle muscle fibers. Nature (Lond) 276:511–513, 1978.

Brinley FJ Jr, Mullins LJ: Sodium extrusion by internally dialyzed squid axons. J Gen Physiol 68:2303–2331, 1967.

Cabantchik ZI, Knauf PA, Rothstein A: The anion transport system of the red blood cell. The role of membrane protein evaluated by the use of 'probes.' Biochim Biophys Acta 515:239–302, 1978.

Hinke JAM: Cation-selective microelectrodes for intracellular use. In Eisenman G (ed): "Glass Electrodes for Hydrogen and Other Cations. Principles and Practice." New York: Dekker, 1967, pp 464–477.

Ho MK, Guidotti G: A membrane protein from human eythrocytes involved in anion exchange. J Biol Chem 250:675–683, 1975.

Hoyle G, Smyth T Jr: Neuromuscular physiology of giant muscle fibers of a barnacle, Balanus nubilus Darwin. Comp Biochem Physiol 10:291–314, 1963.

Knauf PA, Rothstein A: Chemical modification of membranes. I. Effects of sulfhydryl and amino reactive reagents on anion and cation permeability of the human red blood cell. J Gen Physiol 58:190–210, 1971.

Roos AR, Boron WF: Intracellular pH. Physiol Rev 61: 296–434, 1981.

Intracellular pH Regulation in Squid Giant Axons

John M. Russell and Walter F. Boron

Department of Physiology and Biophysics, University of Texas Medical Branch, Galveston, Texas 77550 (J.M.R.), and Department of Physiology, Yale University of School of Medicine, New Haven, Connecticut 06510 (W.F.B.)

I.	Introduction	221
II.	Methods	222
	A. General	222
	B. Solutions	222
	C. Measurement of pH_i and Net, Equivalent HCO_3^- Flux	223
	1. Definitions; 2. Apparatus; 3. Calculation of net, equivalent HCO_3^- flux	223
	D. Measurement of Na^+ and Cl^- Fluxes	226
	1. General; 2. ^{36}Cl and ^{22}Na efflux experiments; 3. ^{36}Cl and ^{22}Na influx experiments	226
III.	Summary of Previous Findings	227
IV.	Results	233
V.	Discussion	236
VI.	References	237

I. INTRODUCTION

From the point of view of acid-base homeostasis, the primary concern of the cell is to extrude acid as rapidly as it accumulates intracellularly. Mechanisms by which acid accumulates within the cell include 1) passive ion fluxes (eg, H^+ influx and HCO_3^- efflux); 2) metabolic production of acid; and 3) in certain circumstances, the release of acid from intracellular stores (eg, from mitochondria, in exchange for Ca^{++}). If the cell is to maintain a stable internal pH (pH_i), it must actively "extrude acid" from the cell (ie, transport acid out of and/or base into the cell) as rapidly as acid accumulates from the aforementioned three sources. In this chapter we shall examine some aspects of the ion transport mechanism responsible for pH_i regulation in squid giant axons.

II. METHODS

A. General

These experiments were performed at the Marine Biological Laboratory, Woods Hole, Massachusetts, using giant axons of the squid Loligo pealei. The internal dialysis technique (see Fig. 1) of Brinley and Mullins [1967] was used to control intracellular composition and to introduce and collect radioisotopic tracers. This involved cannulating an isolated, cleaned axon (500–600 μm diameter) at both ends, and introducing a length of cellulose acetate dialysis tubing (140 μm diameter). The central region of the latter is hydrolyzed (this is indicated by broken lines in Fig. 1), rendering it permeable to solutes of molecular weight less than about 1,000. Dialysis fluid (DF) was delivered to the dialysis tubing from a 250-μl syringe, driven by a Sage syringe pump. To avoid damage to the axon membrane, the osmolarities of the artificial seawater (ASW) and DF were closely controlled, with the former being about 20 mOsm higher than the latter. In the isotope efflux and pH_i experiments, axons were continuously superfused with ASW. CO_2-containing solutions were delivered via gas-tight syringes and CO_2-impermeable tubing. Temperature, controlled by a circulating water bath connected to the jacketed chamber, was 16°C in the stoichiometry study, and 22°C in other experiments.

B. Solutions

The standard ASW had the following composition (in mM): 425 Na^+, 12 K^+, 10 Ca^{++}, 50 Mg^{++}, 530 Cl^-, 12 HCO_3^-, 15 $EPPS^-$, and 15 neutral EPPS. The solutions were buffered to pH 8.00 both with 0.5% CO_2 and 30 mM 4-(2-hydroxyethyl)-1-piperazine-propane sulfonic acid (EPPS). Such dual buffering had the advantage of holding extracellular buffering power constant under the conditions of low $(HCO_3^-)_o$ used in some of the kinetics experiments. In those kinetic studies in which we studied the dependence of acid extrusion on $(Na^+)_o$ and $(HCO_3^-)_o$, axons were acid loaded by pretreating them with ASW containing 50 mM NH_4^+ (NH_4^+ replacing Na^+), containing no HCO_3^- (Cl^- replacing HCO_3^-), and buffered to pH 7.7. The NH_4^+ was washed out by exposing the axons to an NH_4^+-free, nominally HCO_3^--free ASW. The recovery of pH_i was then monitored as the axons were exposed to the standard ASW in which either $(HCO_3^-)_o$ or $(Na^+)_o$ had been reduced. The external HCO_3^- or Na^+ was replaced by Cl^- or choline, respectively. The choline used in these experiments was washed in activated charcoal and recrystallized from isopropyl alcohol [Boron et al, 1981]. In a few experiments, either Li^+ or N-methyl-D-glucammonium substituted for Na^+.

The standard dialysis fluid had the following composition (in mM): 58 Na^+, 350 K^+, 7 Mg^{++}, 150 Cl^-, 264 glutamate, 210 taurine, 10 HEPES, 4 ATP, 1 EGTA, 0.5 phenol red. It was titrated to pH 6.5 or 7.3 with KOH or glutamic acid. In the Cl^- kinetics experiments, Cl^- was replaced by glutamate in the DF.

C. Measurement of pH_i and Net, Equivalent HCO_3^- Flux

1. Definitions. Since one cannot use a pH-sensitive microelectrode to distinguish the movement of HCO_3^- or $CO_3^=$ in one direction from the movement of H^+ in the opposite direction, we shall refer to the movement of HCO_3^-, $CO_3^=$ and/or H^+ as "net, equivalent HCO_3^- flux." "Net" is used because the pH electrode cannot detect the results of unidirectional fluxes, but only net H^+ activity changes. By equivalent HCO_3^- flux is meant the flux of HCO_3^- plus one-half the flux (if any) of $CO_3^=$ in the same direction, plus the flux (if any) of H^+ in the opposite direction.

2. Apparatus. Our approach in these experiments was to acid load the axons (ie, lower pH_i), and then to monitor the subsequent recovery of pH_i with a pH-sensitive microelectrode. In the Na^+ and HCO_3^- kinetic studies, the axons were acid-loaded by pretreating with NH_4^+. In the internal Cl^- kinetic studies, and in the stoichiometry experiments, the axons were acid-loaded by lowering the pH of the DF to 6.5. A schematic representation of the apparatus in these dialysis experiments is given in Figure 1, top. The pH and voltage-measuring electrodes enter through opposite end cannulae, and lie alongside the dialysis tubing. In the NH_4^+ experiments, the situation was similar, except that the dialysis tubing was not present. The pH-sensitive electrodes of Hinke's design (see Fig. 2) were ideal for use with cannulated axons. The details for constructing these electrodes have been described in detail elsewhere [Hinke, 1967]. Briefly, lead glass tubing (which is pH-insensitive) is pulled on a microelectrode puller having a very long excursion. The shank of this pipette is maintained at an outer diameter of about 125 μ or less for the terminal 3 cm (the length inserted into the cannula and axon). The broken off end of this pipette is then formed into a tapered tip on a microforge, and the tip is cleanly broken off at a point where the inner diameter is about 50 μm. pH-sensitive glass tubing is drawn out by hand, and a short length is introduced into the large end of the lead glass pipette and advanced until it protrudes through, and finally jams in, the latter's open tip. Lastly, a microforge is used to seal the pH-sensitive and lead glass capillaries together, and to form the protruding pH glass into a spear-shaped tip about 100 μm long. The electrode is filled with 0.1 M HCl, and connected to a high-impedance electrometer by means of a Ag/AgCl half cell.

The internal reference (voltage-measuring) electrodes were similar to the pH-sensitive one except that their tapered tips were broken off at a much smaller diameter, and left open. In the Cl^- kinetics study, these electrodes were filled with 2 M K^+ glutamate to avoid altering the Cl^- gradient. (Such an electrode has a junction potential of ~ -15 mV when in ASW or DF; this offset is stable,

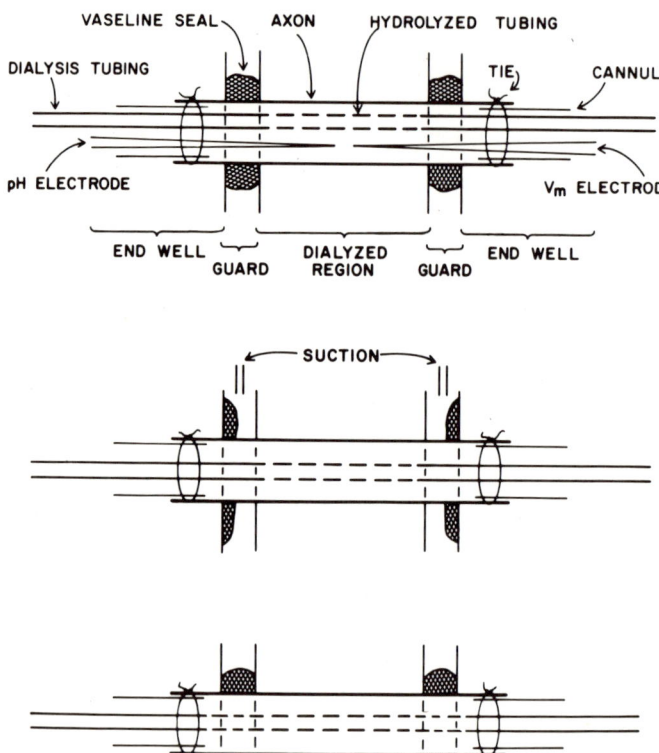

Fig. 1. Schematic representation of dialysis apparatus. Top: The apparatus as it is employed in experiments in which dialysis is used to lower pH_i, and electrodes are used to monitor the pH_i recovery. The diameter of the axon is 500–600 μm, that of the shanks of the electrodes is ≤ 125 μm, and that of the dialysis tubing is 140 μm. The length of the dialyzed region is ~18 mm. Middle: The apparatus as it is is used in isotope efflux experiments. Since isotope leaving the dialysis tube can diffuse laterally within the axons. Suction is applied to the medial aspect of the guard regions to sweep away any isotope that crosses that portion of the axon's membrane contained in the guard region. This ensures that the isotope collected by the ASW superfusing the axon's central or dialyzed region reflects only that isotope actually crossing that portion of the axon's membrane. Bottom: The apparatus as it is used in isotope influx experiments. Vaseline seals prevent the isotope from entering the guard regions. The dialysis tubing is permeable well into the guard regions to collect any incoming isotope that diffuses laterally within the axon.

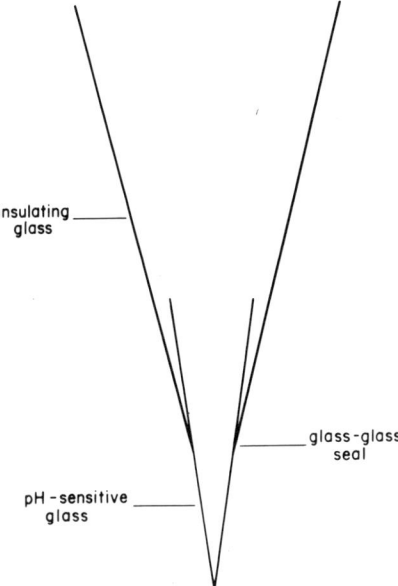

Fig. 2. Schematic diagram of protruding-tip pH electrode, designed by Hinke [1967]. The insulating glass is Corning 0120 lead glass. The pH-sensitive glass is Corning 0150. The diameter at the glass–glass seal was 50–70 μm in these experiments, and the diameter of the electrode shank was ≤ 125 μm for the terminal 3–4 cm. [From Boron, 1980. Reproduced by permission of Academic Press.]

and an appropriate correction is easily made.) In the other experiments, this electrode was filled with 1 M KCl. Connection to an electrometer was made via a calomel half cell. A similar electrode was placed in the bath and connected to an electrometer. The bath was grounded through a platinum wire. The voltage difference between the pH electrode and the internal reference electrode signals is the voltage due solely to pH_i; this difference was obtained electronically and plotted on a strip chart recorder. The difference between the internal and external reference electrodes is the membrane potential (V_m), and was similarly obtained and plotted.

3. Calculation of net, equivalent HCO_3^- flux.

After cells were acid-loaded, pH_i was allowed to recover in the test solution. The net, equivalent HCO_3^- flux was taken as the product of the rate of pH_i recovery (dpH_i/dt), the intracellular buffering power (β), and the volume-to-surface ratio. The value of dpH_i/dt is the slope of the pH_i change, and was read directly off the strip chart recording. The volume-to-surface ratio is one-fourth the axon's diameter, assuming the axon to be a circular cylinder. β is defined as dB/dpH where dB is the amount of strong base that must be added to the axoplasm to raise pH_i by dpH. This was determined as follows. The axon was first acid-loaded by dialysis and exposed to the same solutions as it would have been in an ordinary experiment, except that SITS was present in the ASW to block activity of the pH_i-regulating mechanism. The axon was then exposed to ASW containing 0.2 mM of NH_4^+ at pH_o 7.75. Such a solution contains a small amount of NH_3, which enters the axon and there combines with H^+ to form NH_4^+. For each NH_4^+ so formed, one H^+ has been abstracted from cellular buffers. Thus, the amount of strong base added to the axoplasm (dB in the above definition of β) was simply $d(NH_4^+)_i$. We allowed the entry of NH_3 to continue until pH_i reached a new steady level, about 0.10–0.15 higher than the initial one. At this point, the amount of strong base added to the cell was $\Delta(NH_4^+)_i$, which is the same as $(NH_4^+)_i$ since the initial $(NH_4^+)_i$ was zero. $(NH_4^+)_i$ was calculated from pH_i assuming that $(NH_3)_i = (NH_3)_o$. ΔpH_o could have been taken as the difference between the initial and final pH_i. However, this value is somewhat artificially reduced, owing to the passive entry of NH_4^+ (see Boron [1977] as well as the discussion of Fig. 4, below). Therefore, we took as ΔpH_i the change in pH_i upon removal of external NH_4^+. The average β ($\Delta(NH_4^+)_i/\Delta pH_i$) in eight experiments was 11.2 ± 1.1 mM.

D. Measurement of Na^+ and Cl^- Fluxes

1. General. The unidirectional fluxes of Na^+ and Cl^- measured with ^{36}Cl and ^{22}Na. ^{36}Cl experiments are complicated by the relatively low specific activity of this isotope. However, with long sample times (collecting ASW or DF for as long as 6–8 min per sample) and long counting times (~25 min, or enough for a 3% counting error), adequate counts were obtained. The Cl isotope (as Na ^{36}Cl for influxes and K ^{36}Cl for effluxes) was ashed to remove organic impurities. The ^{22}Na experiments were straightforward; in order to minimize background Na^+ fluxes, ouabain (10^{-5} M) and TTX (10^{-7} M) were present in all external solutions (including those for the ^{36}Cl and pH microelectrode stoichiometry experiments). Both ^{36}Cl and ^{22}Na were counted in a liquid scintillation counter, using a Triton X-100 cocktail.

2. ^{36}Cl and ^{22}Na efflux experiments. The appropriate isotope was added to the DF. The isotope leaves the dialysis tubing in the hydrolyzed central region (Fig. 1, middle), equilibrates with axoplasmic Cl^- or Na^+, and then leaves the axon, entering the central region of the chamber. There it is swept away by ASW superfusing the axon at ~2 ml/min, to be gathered by an automated fraction collector. Effluxes were calculated from the radioactivity of the sample, the specific activity of the isotope in the DF, the surface area of the axon in the chamber's central region, and the collection time. Corrections were made for variations in quenching due to alterations in ASW composition. It is crucial, of course, that the flow of DF and ASW from the syringe drivers and peristaltic pumps be stable. Care was also taken not to collect isotope that had left the portion of the axon enclosed by the two guard regions, since the composition of the axoplasm in these areas is not well controlled. This potential contamination was avoided by using a second set of syringe drivers to withdraw (and discard) ASW from each guard region at the rate of ~50 μl/min (for a total of 5% of ASW flow). Flow of isotope into the stagnant end wells was prevented by Vaseline seals in the lateral portion of each guard region.

3. ^{36}Cl and ^{22}Na influx experiments. The isotope was added to the ASW, which, for the sake of economy, was stagnant. Isotope enters the portion of the axon contained in the chamber's central region, whereupon it diffuses into the dialysis tubing. (Note that the dialysis tubing is hydrolyzed into the guard regions to maximize isotope uptake.) Collection of the isotope was achieved by periodically (at 2–5-minute intervals) washing the free end of the dialysis tubing with distilled water, and collecting the rinse in a scintillation vial. Influx of isotope through the guard region portions of the axon membrane was prevented by filling the guard regions with Vaseline. To maintain a stable CO_2 tension in CO_2-containing ASWs, a stream of the proper gas mixture, preequilibrated with ASW, was passed over the axon.

III. SUMMARY OF PREVIOUS FINDINGS

The pH_i-regulating mechanism of squid giant axons appears to be very similar to that operative in snail neurons (see Thomas, this volume) and barnacle muscle (see Roos and Boron, this volume). There are, however, a few minor differences among the transport systems in the three preparations; these will be pointed out below. The first evidence that pH_i is actively regulated comes from experiments in which both pH_i and membrane potential (V_m) were measured. In squid giant axons [Caldwell, 1958; Spyropoulos, 1960; Boron and De Weer, 1976a], pH_i is about 7.3 and V_m about -60 mV. At an external pH (pH_o) of ~7.8, the pH_i at which H^+ and HCO_3^- would be in equilibrium is ~6.8. Thus, in the steady state there is a substantial electrochemical gradient favoring the influx of H^+ and efflux of HCO_3^-. If the gradual decline of pH_i is to be prevented, an active, acid-extruding process must exist. More direct evidence for such an acid-extruding mechanism came in an experiment in which a squid axon was exposed to CO_2 and HCO_3^- at constant pH_o [Boron and De Weer, 1976a]. As shown in Figure 3, application of ASW containing 5% CO_2 and 50 mM HCO_3^- ($pH_o = 7.7$) causes an abrupt fall in pH_i, as measured with a pH-sensitive microelectrode. This is due to the influx of CO_2, the intracellular hydration to form H_2CO_3, and the subsequent dissociation to H^+ and HCO_3^-. As $(CO_2)_i$ approaches $(CO_2)_o$, the net influx of CO_2 slows and would eventually halt if no other transport were occurring. Instead, we find that pH_i slowly recovers during a period termed the "plateau phase." This gradual intracellular alkalinization cannot be accounted for by passive ion fluxes (ie, H^+ influx, HCO_3^- efflux), which would serve only to lower pH_i further, and must therefore reflect the activity of an active, acid-extruding transport system. When CO_2 is eventually removed from the axon,

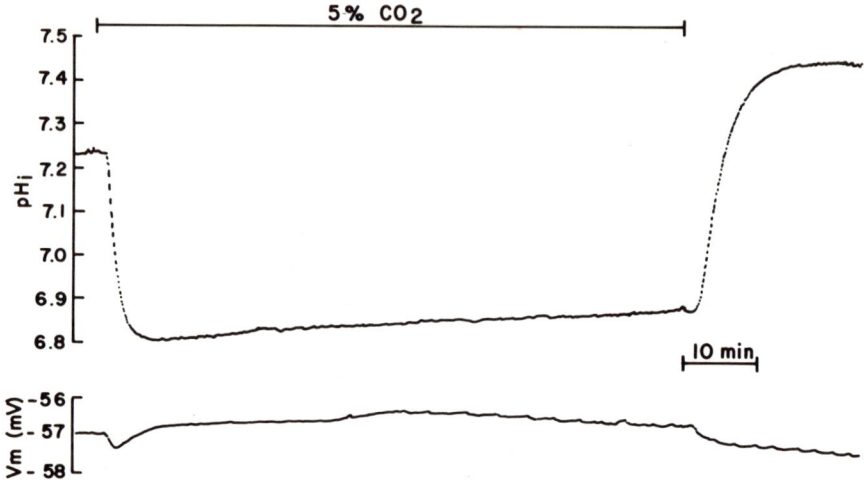

Fig. 3. Effect of CO_2 on pH_i and membrane potential (V_m) in a squid giant axon. pH_o was 7.70 throughout the experiment. At the indicated time, the CO_2-free HEPES buffer was replaced by 5% CO_2/50 mM HCO_3^-. Note that after its initial CO_2-induced fall, pH_i recovers during the long exposure to CO_2. When CO_2 is removed, pH_i overshoots its initial level. Both the pH_i recovery and the overshoot are manifestations of the active extrusion of acid. [From Boron and De Weer, 1976a. Reproduced by permission of the Rockerfeller University Press.]

pH_i rapidly rises to a value substantially above the initial one, as intracellular HCO_3^- combines with H^+ and eventually leaves the cell as CO_2. The overshoot of pH_i beyond its initial level reflects the net loss of intracellular acid that had occurred as the result of acid extrusion during the plateau phase. (Note that the magnitude of the overshoot is much larger than that of the plateau-phase alkalinization. This occurs because the total intracellular buffering power of the axoplasm after removal of CO_2 is merely that of the axon's intrinsic buffers. During the plateau phase, however, this is augmented by the buffering power of intracellular CO_2/HCO_3.) [See Roos and Boron, 1981.]

The primary method for studying pH_i regulation is to acid-load cells and observe the subsequent pH_i recovery, as during the plateau phase of Figure 3. Maneuvers which slow the pH_i recovery from an acid load are thus inferred to have inhibited the pH_i-regulating mechanism. The general properties of pH_i regulation were elucidated piecemeal on squid axons, snail neurons, and barnacle muscle. The first clue to the ionic mechanism of the transport process came in experiments in which pH_i recovery was studied in both the presence and absence of HCO_3^- at constant pH_i and pH_o [Boron and DeWeer, 1976b], as illustrated in Figure 4. In this experiment the axon was acid loaded using the NH_4^+ prepulse technique [Boron and DeWeer, 1976a]. The axon was exposed to ASW in which 50 mM Na^+ was replaced by NH_4^+, while pH_o was held constant at 7.7. Since an NH_4^+-containing solution has a small amount of NH_3, this treatment causes

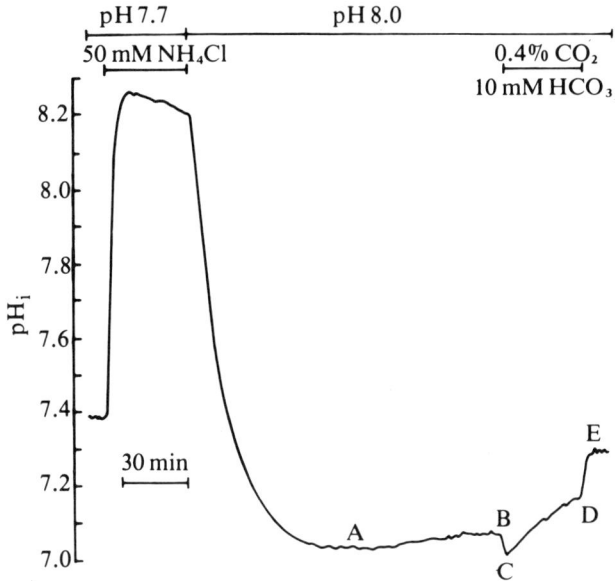

Fig. 4. The dependence of acid extrusion on HCO_3^- in a squid giant axon. The axon was acid loaded by a 30-minute application and then removal of NH_4^+ (see text for details). After washout of the NH_4^+, pH_i has fallen to nearly 7.0, reflecting an acid load. pH_i, however, recovers very slowly (A–B) in the nominal absence of HCO_3^- (pH_o = 8.00). When 10 mM HCO_3^-/0.4% CO_2 was added at constant pH_o, pH_i initially fell owing to the influx of CO_2 (B–C), and then recovered at an accelerated rate (C–D). Return to a nominally HCO_3^--free buffer caused a rapid rise in pH_i (D–E), owing to efflux of CO_2. [From Boron and De Weer, 1976b. Reproduced by permission of MacMillan Journals, Ltd.]

a rapid rise in pH_i as the neutral weak base enters the cell and combines with H^+ to form NH_4^+. This intracellular alkalinization would be expected to continue until $(NH_3)_i$ equalled $(NH_3)_o$, at which time the net influx of NH_3 would cease, and pH_i would level off. Instead, we find that after its initial rise, pH_i begins a slow decline. This plateau-phase acidification is due to the passive influx of NH_4^+, some of which dissociates to form H^+ and NH_3. The subsequent removal of external NH_4^+/NH_3 causes internal NH_4^+ to give up its H^+ and exit as NH_3. pH_i thus falls far below its initial level, the degree of undershoot reflecting the net accumulation of acid (ie, NH_4^+) during the previous plateau phase. (Note that the magnitude of the undershoot is appreciably larger than that of the plateau-phase acidification. This occurs because the total intracellular buffering power after removal of NH_4^+ is merely the buffering power of the axon's intrinsic buffers, whereas this is augmented by the considerable buffering power of the intracellular NH_4^+/NH_3 during the plateau phase.) This whole procedure of first applying and then removing external NH_4^+/NH_3 is merely a ploy for acid-loading the cells; indeed, it is formally equivalent to injecting HCl into the axon. Note that after the acid load (see point A), the axon was bathed in nominally HCO_3^--free ASW (pH_o = 8.0), and that the recovery of pH_i (segment A–B) was very slow. The addition of a CO_2/HCO_3^- buffer, however, caused pH_i to recover at a substantially greater rate (C–D); the small, initial fall in pH_i (B–C) upon application of the HCO_3^- ASW merely reflects the influx of CO_2. These results thus strongly suggest that the pH_i-regulating mechanism implements pH_i recovery by transporting HCO_3^- into the cell. At about the same time as the above study on squid axons, Thomas [1976a] observed that the recovery of pH_i from an acid load in snail neurons is faster in Ringer buffered with HCO_3^-. The HCO_3^--dependence of the pH_i-regulating mechanism has also been demonstrated in barnacle muscle [Boron, 1977].

The dependence of the pH_i recovery on HCO_3^- suggested that, for the sake of electroneutrality, another anion may be involved in the transport process, perhaps Cl^-. In the experiment of Figure 5 [Russell and Boron, 1976], which

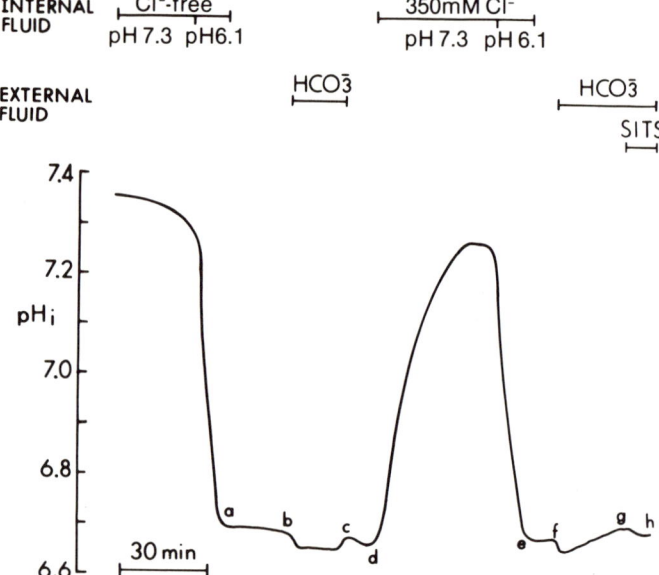

Fig. 5. Dependence of acid extrusion on internal Cl^- in a squid giant axon. The axon (V_m ~ −52 mV) was first dialyzed with a Cl^--free solution (glutamate replacing Cl^-) for ~40 minutes to greatly reduce $(Cl^-)_i$. During the final few minutes, the pH of the dialysis fluid (DF) was lowered to reduce pH_i. At point a, dialysis was halted returning control of pH_i to the axon. In the absence of internal Cl^- (a–d), pH_i did not recover from the acid load, even with the addition of HCO_3^- to the artificial seawater (ASW) at constant pH_o (8.00). Dialysis was then restarted with 350 mM Cl^-. After dialysis was halted, addition of HCO_3^- to the ASW caused pH_i to recover. This recovery, which reflects the active extrusion of acid, was blocked by 0.5 mM SITS added to the ASW. [From Russell and Boron, 1976. Reproduced by permission of MacMillan Journals, Ltd.]

was designed to test the dependence of the squid axon's pH_i-regulating mechanism on internal Cl^-, internal Cl^- was initially lowered by dialyzing the axon with a Cl-free fluid (glutamate replacing Cl^-). pH_i was simultaneously lowered by reducing the pH of the dialysis fluid (DF) to 6.1. At point a, dialysis was halted, thereby returning control of pH_i to the axon. Note that in the absence of Cl^- (a–d), there was no rise in pH_i, even when HCO_3^- was added to the ASW (b–c). This result thus supports the hypothesis that HCO_3^- uptake is accompanied by an efflux of Cl^-. Indeed, when the axon was then dialyzed with a Cl-containing solution (d–e), and dialysis subsequently halted (e–f), pH_i recovery was initiated by the addition of HCO_3^- (f–g). Interestingly, this pH_i recovery was blocked by the addition of 0.5 mM 4-acetamido-4'-isothiocyanostilbene-2,2'-disulfonate (SITS) to the ASW. SITS had previously been shown to block anion exchange in erythrocytes [Cabantchik and Rothstein, 1974], and also to block the pH_i regulation in snail neurons [Thomas, 1976b]. This effect of SITS on acid extrusion was subsequently confirmed on barnacle muscle [Boron, 1977]. The observation that internal Cl^- is required for acid extrusion in squid axons was subsequently confirmed on snail neurons [Thomas, 1977]. Cl^- is also involved in acid extrusion in barnacle muscle [Boron et al, 1978], although a demonstration that acid extrusion is absolutely dependent on internal Cl^- is yet to be attempted in this preparation.

The above data suggest that the pH_i-regulating mechanism of the squid axon (as well as of the snail and barnacle) might be a simple, SITS-sensitive, HCO_3-Cl exchanger. However, in 1977 Thomas found that the pH_i-regulating system in snail neurons requires external Na^+, in addition to external HCO_3^- and internal Cl^-. This observation indicated that the transporter is far more complex than previously suspected. Subsequent work on barnacle muscle [Boron et al, 1981] confirmed that the pH_i-regulating system in this preparation too requires Na^+, and further, that this dependence on external Na^+ apparently follows simple Michaelis-Menten kinetics (See Roos and Boron, this volume). The involvement of Na^+ led Thomas [1977] to propose the top model in Figure 6 for the ionic mechanism of the pH_i-regulating system in snail neurons. Note that the proposed system is electroneutral, mediating the uptake of Na^+ and HCO_3^- in exchange for H^+ and Cl^-. The second model of Figure 6 is very similar, except the efflux of H^+ is replaced by the influx of a second HCO_3^-. In the final model—the ion-pair hypothesis of Becker and Duhm [1978]—the uptake of a single $NaCO_3^-$ complex replaces the uptake of Na^+ plus two HCO_3^-. Note that the three models are equivalent thermodynamically, and could be distinguished only on kinetic grounds. So far, the only evidence we have discussed for these models is that the pH_i recovery following an acid load requires external HCO_3^- and Na^+ as well as internal Cl^-. However, there was also evidence that acid extrusion is accompanied by the efflux of Cl^- and the influx of Na^+. Russell and Boron [1976] used ^{36}Cl to demonstrate the Cl^- efflux in squid axons. Thomas, working on snail neurons, showed that the pH_i recovery (ie, acid extrusion) is accompanied

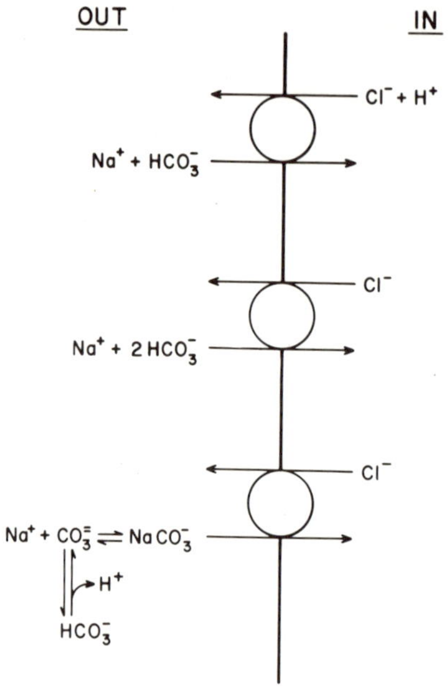

Fig. 6. Three models of acid extrusion. The top model [Thomas, 1977] has external Na^+ and HCO_3^- exchanging for internal H^+ and Cl^-. In the middle model, the efflux of H^+ is replaced by the influx of a second HCO_3^-. In the bottom model [Becker and Duhm, 1978], the $NaCO_3^-$ ion pair exchanges with Cl^-. Note that all models predict electroneutral transport with a stoichiometry of 1 Na^+ entering for each Cl^- leaving the cell, and for each pair of protons neutralized (by HCO_3^- or $CO_3^=$ influx and/or by H^+ influx) intracellularly. [From Roos and Boron, 1981. Reproduced by permission of the American Physiological Society.]

by a rise in intracellular Na^+ activity (a_i^{Na}) and a fall in intracellular Cl^- activity (a_i^{Cl}), both measured with ion-sensitive microelectrodes. In addition, Russell [1978] showed that a_i^{Cl} falls during periods of presumed acid extrusion in Aplysia neurons. Although these observations suggested that acid extrusion is accompanied by a net influx of Na^+ and a net efflux of Cl^-, some questions remained. The microelectrode measurements always leave open the possibility that the observed changes in intracellular ion activities were complicated by interference from some other ion, a change in cell volume, or the ion's release from or sequestration by an intracellular site. Also, the aforementioned study of isotopic Cl^- efflux on squid axons did not include a measurement of Cl^- influx; without the latter one cannot be certain that acid extrusion is accompanied by a *net* efflux of Cl^-. What was obviously required was a study in which bidirectional radioisotope fluxes are used to determine the net Na^+ and Cl^- fluxes accompanying acid extrusion. A by-product of such an undertaking would be a reliable measurement of the stoichiometry of the pH_i-regulating system.

IV. RESULTS

The ideal preparation for studying isotopic fluxes and the stoichiometry of the pH_i-regulating system is the internally dialyzed squid giant axon. Internal dialysis gives one nearly total control of the intracellular environment. In addition, the dialysis procedure can be used to dispense isotope in efflux experiments, or to collect isotopes in influx experiments. Our first goal was to determine whether there is an increase in the rate at which Cl^- leaves the axon during periods of acid extrusion. We know from previous studies that acid extrusion occurs only under conditions of intracellular acid loading, and only in the presence of external Na^+ and HCO_3^-. Our approach in this experiment was therefore to dialyze an axon with a DF containing ^{36}Cl, and having a pH of 6.5. The external solution contained Na^+, but no HCO_3^-. Since HCO_3^- was absent, any Cl^- leaving the axon during this period should have been unassociated with the pH_i-regulating mechanism. Any *increase* in the Cl^- efflux upon addition of external HCO_3^- (at constant pH_o) can then be attributed to the pH_i-regulating system. Figure 7 illustrates such an experiment. Since this experiment was also intended to test the ATP dependence of the pH_i regulating system, the axon was poisoned with cyanide throughout. In addition, we dialyzed with an ATP-free DF during the first half of the experiment, so that the axoplasm had a very low ATP during this period. After the Cl^- efflux reached a steady level in the absence

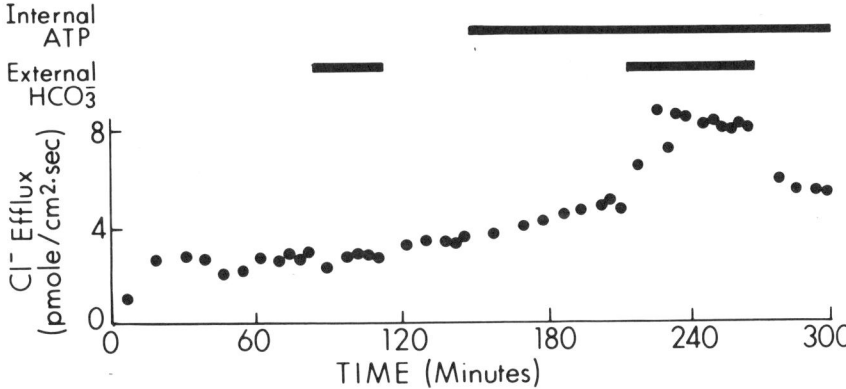

Fig. 7. Cl efflux from an internally dialyzed squid axon. The axon was poisoned with 2 mM cyanide (DF and ASW) throughout; in addition, the DF was ATP-free in the first half of the experiment. The DF was at pH 6.5 to stimulate the pH_i-regulating system. However, in the absence of ATP, addition of 10 mM HCO_3^- (pH_o constant at 8.0) had no effect on Cl^- efflux. When 4 mM ATP was added back to the DF, application of external HCO_3^- now caused Cl^- efflux to rise by 3–4 pmol·cm^{-2}·s^{-1}. This increment represents that portion of the Cl^- efflux that is mediated by the pH_i-regulating system. [From Russell and Boron, 1976. Reproduced by permission from MacMillan Journals, Ltd.]

of internal ATP and external HCO_3^-, HCO_3^- was added to the ASW. This produced no change in Cl^- efflux. During the second half of the experiment, however, when the DF contained 4 mM ATP, addition of HCO_3^- to the ASW caused the Cl^- efflux to rise by 3 to 4 $pmol \cdot cm^{-2} \cdot s^{-1}$. This experiment thus demonstrates that 1) ATP is required for activity of the pH_i-regulating system in squid axons, and 2) there is a component of Cl^- efflux associated with activity of the pH_i-regulating system. In a total of 15 similar experiments, the HCO_3^--dependent Cl^- efflux under conditions of intracellular acid loading was 3.9 ± 0.2 $pmol \cdot cm^{-2} \cdot s^{-1}$ (see Table I). The Cl flux measured in these experiments, of course, is a unidirectional efflux, and does not necessarily reflect a *net* Cl^- efflux. Therefore, under identical conditions, we also measured the HCO_3^--stimulated unidirectional Cl^- influx. In a total of 5 axons, the Cl^- influx was 0.1 ± 0.2 $pmol \cdot cm^{-2} \cdot s^{-1}$, not significantly different from zero. Thus, the net flux (see Table I) was -3.8 $pmol \cdot cm^{-2} \cdot s^{-1}$, the negative value denoting a net efflux. The same procedure can be used to obtain the net Na^+ flux associated with activity of the pH_i-regulating system. In a total of 13 axons, the mean, HCO_3^--stimulated Na^+ influx was 3.4 ± 0.4 $pmol \cdot cm^{-2} \cdot s^{-1}$. The mean, HCO_3^--stimulated Na^+ efflux was 0.0 ± 0.1 $pmol \cdot cm^{-2} \cdot s^{-1}$ in a total of six axons. These results, summarized in Table I, indicate that the *net* Na^+ flux mediated by the pH_i-regulating system was $+3.4$ $pmol \cdot cm^{-2} \cdot s^{-1}$, the positive value denoting a net influx. It is noteworthy that the "backflux" of ions on the pH_i-regulating system appears to be very small: Neither the unidirectional Na^+ efflux nor the Cl^- influx was significantly different from zero. This is in contrast to barnacle muscle where the pH_i-regulating system appears to mediate a significant Cl-Cl [Boron et al, 1978] and Na-Na exchange (Russell et al, unpublished).

Still required for a complete determination of the stoichiometry is a measure of the HCO_3^-, H^+, or $CO_3^=$ traversing the axon membrane. Isotopes of hydrogen or carbon cannot be used for this purpose inasmuch as the label on H^+ rapidly

TABLE I. Stoichiometry of the Squid Axon's pH_i-Regulating System*

Ion	Influx	Efflux	Net flux
Na^+	3.4 ± 0.4 (13)	0.0 ± 0.1 (6)	$+3.4$ $pmol \cdot cm^{-2} \cdot s^{-1}$
Cl^-	0.1 ± 0.2 (5)	3.9 ± 0.2 (15)	-3.8 $pmol \cdot cm^{-2} \cdot s^{-1}$
HCO_3^-	—	—	7.5 ± 0.6 $pmol \cdot cm^{-2} \cdot s^{-1}$ (15)

*Mean $\pm$ SE, with number of experiments given in parentheses. The HCO_3^- flux was calculated from the rate of pH_i recovery and is the net, equivalent HCO_3^- flux referred to in text. Standard conditions: T = 16°C, $(Na^+)_o$ = 425 mM, $[HCO_3^-]_o$ = 12 mM, pH_o = 8.0, $[Cl^-]_i$ = 150 mM, pH_i = 6.5.

exchanges with H_2O whereas that on HCO_3^- rapidly exchanges with CO_2. We are therefore left with computing the net flux from the rate of pH_i change, as measured with a pH-sensitive microelectrode. The approach is to dialyze an axon with pH 6.5 DF, just as in the isotope studies, until $pH_i = 6.5$. At this point dialysis is halted, returning control of pH_i to the axon. As long as no HCO_3^- is present in the ASW, pH_i will not recover, reflecting the requirement of the pH_i-regulating system for HCO_3^-. Addition of HCO_3^- (at constant pH_o), however, initiates pH_i recovery, which can be measured as the slope dpH_i/dt, where t is time. Subsequent removal of the HCO_3^- causes the pH_i recovery to halt. The approach of measuring dpH_i/dt cannot be used to distinguish between HCO_3^- or $CO_3^=$ flux, or H^+ flux in the opposite direction. We will therefore refer to a "net, equivalent HCO_3^- flux," realizing that the measured flux is a net one, and that it could reflect H^+, HCO_3^- and/or $CO_3^=$, transport. This flux is simply the product of dpH_i/dt, the intracellular buffering power (β), and the axon's volume-to-surface ratio (ρ):

$$J_{HCO_3}^{net} = (dpH_i/dt) \cdot \beta \cdot \rho$$

dpH_i/dt is read directly off the strip chart recording. β is determined empirically, using an exposure to ammonia as previously discussed [Boron, 1977]. ρ is one-fourth the axon's diameter. In a total of 16 such experiments, $J_{HCO_3}^{net}$ averaged 7.5 ± 0.6 pmol·cm^{-2}·s^{-1} (see Table I). These stoichiometry data are thus in excellent agreement with all three models of Figure 6: The pH_i-regulating system takes up one equivalent of Na^+ for each equivalent of Cl^- removed from the cell, and for each two equivalents of acid neutralized (whether by HCO_3^- or $CO_3^=$ influx, or by H^+ efflux) intracellularly.

For the sake of completeness, it is crucial to demonstrate that each of the observed fluxes has the properties expected of the axon's pH_i-regulating system: 1) dependence on intracellular acid loading, 2) dependence on internal ATP, 3) blockade by SITS, and 4) dependence on each of the cotransported ions. We found that the Cl^- efflux, the Na^+ influx and the net, equivalent HCO_3^- flux were all dependent on intracellular acid loading. That is, there was no HCO_3^--stimulated flux at a normal pH_i (ie, 7.3), whereas at low pH_i (ie, 6.5) there was. Similarly, we found that, under conditions of acid loading, external HCO_3^- failed to increase any of the three fluxes in the absence of internal ATP, or in the presence of external SITS. Finally, we showed that the HCO_3^--dependent Cl^- efflux requires external Na^+, that the HCO_3^--dependent Na^+ influx requires internal Cl^-, and that the net equivalent HCO_3^- flux requires both external Na^+ and internal Cl^-. These results strongly suggest that the measured fluxes truly reflect activity of the pH_i-regulating system.

A final project of ours has been to quantitate the dependence of the axon's pH_i-regulating system individually on external Na^+, external HCO_3^-, and internal

Cl⁻. This was done by acid-loading the axon and then observing the resultant recovery of pH_i. $J^{net}_{HCO_3}$ was then calculated from dpH_i/dt, as described above. In the case of the external Na^+ and external HCO_3^- studies, pH_i was lowered using an NH_4^+ pretreatment (see Fig. 4). In the case of the internal Cl⁻ study, pH_i was lowered and internal Cl⁻ was varied using the internal dialysis technique. In each study, $J^{net}_{HCO_3}$ was determined at a number of substrate concentrations, and the kinetic parameters calculated using a nonlinear least-squares curve fit to the Michaelis-Menten equation:

$$\frac{V}{V'_{max}} = \frac{S}{S + K'_m}$$

where V is the observed velocity (ie, $J^{net}_{HCO_3}$), V'_{max} is the apparent maximal velocity, S is the substrate concentration, and K'_m is the apparent substrate concentration at half maximal velocity. The standard conditions in these experiments were pH_i = 6.7, $(Cl^-)_i \cong 100$ mM, pH_o = 8.0, $(HCO_3^-)_o$ = 12 mM, $(Na^+)_o$ = 425–437 mM, and T = 22°C. The results of these experiments are summarized in Table II. It is interesting to note that the K'_m values for Na^+ and HCO_3^- are far lower than the likely concentrations of these ions in the extracellular fluid. K'_m (Cl⁻), however, is approximately the same as the normal $(Cl^-)_i$. It is also of interest that acid extrusion is completely blocked whether Na^+ is replaced by choline, N-methyl-D-glucammonium, or Li^+. This effect of Li^+ is in contrast to what is observed in barnacle muscle, where Li^+ supports acid extrusion about 70% as well as does Na^+ [Boron et al, 1981].

V. DISCUSSION

Two major areas of research hold promise in the future. The first is a detailed study of the transporter's energetics. It is clear, for example, that the axon's pH_i-regulating system requires internal ATP; it is not clear what role the ATP plays. ATP could be stoichiometrically hydrolyzed, and thereby help energize the transport of the ions. ATP could, however, play a catalytic role, being required, perhaps, for some crucial phosphorylation. If the latter is the case, then

TABLE II. Kinetics of the Squid Axon's pH_i-Regulating System*

Ion	n	K'_m	V'_{max}
External HCO_3^-	30	2.3 ± 0.2 mM	10.6 ± 0.4 pmol·cm⁻²·⁻¹
External Na^+	21	77 ± 13	10.3 ± 0.6
Internal Cl⁻	40	84 ± 15	19.6 ± 1.2

*Kinetic parameters obtained from nonlinear, least-squares curve fit. n, number of data points (measurements of net, equivalent HCO_3^- flux). V'_{max} is apparent maximal velocity. K'_m is substrate concentration at half V'_{max}. Standard conditions: $(Na^+)_o$ = 425–437 mM, $(HCO_3^-)_o$ = 12, pH_o = 8.0, $(Cl^-)_i \simeq 100$ mM, pH_i = 6.7.

the energy for transport must be derived from the Na^+ gradient ("gradient model"). One way of approaching this problem is to use the dialysis technique to set up ion gradients across the axon membrane, and then to determine whether the sum of gradients for Na^+, HCO_3^-, Cl^-, and H^+ uniquely determines the direction of net transport.

A second area for future work is a detailed kinetic analysis of the transporter. Our work has already shown that the transport process apparently follows simple Michaelis-Menten kinetics with respect to external Na^+, external HCO_3^-, and internal Cl^-. Each of these dependencies should now be examined as a function of pH_o, pH_i, and the other two parameters (ie, external Na^+ and HCO_3^- in the case of dependence on internal Cl^-). Only with this kind of approach can we hope to obtain more detailed insight into pH_i regulation.

VI. REFERENCES

Becker BF, Duhm J: Evidence for anionic cation transport of lithium, sodium and potassium across the human erythrocyte membrane induced by divalent anions. J Physiol (Lond) 282:149, 1978.

Boron WF: Intracellular pH transients in giant barnacle muscle fibers. Am J Physiol 233:C61, 1977.

Boron WF: Intracellular pH regulation. In Boulpaep EL (ed): "Current Topics in Membranes and Transport," Vol 13. "Cellular Mechanisms of Renal Tubular Ion Transport." New York: Academic Press, 1980, pp 3–22.

Boron WF, De Weer P: Intracellular pH transients in squid giant axons caused by CO_2, NH_3 and metabolic inhibitors. J Gen Physiol 67:91, 1976a.

Boron WF, De Weer P: Active proton transport stimulation by CO_2/HCO_3^-, blocked by cyanide. Nature 259:240, 1976b.

Boron WF, McCormick WC, Roos A: pH regulation in barnacle muscle fibers: Dependence on extracellular sodium and bicarbonate. Am J Physiol 240:C80, 1981.

Boron WF, Russell JM, Brodwick MS, Keifer DW, Roos A: Influence of cyclic AMP on intracellular pH regulation and chloride fluxes in barnacle muscle fibres. Nature 276:511, 1978.

Brinley FJ Jr, Mullins LJ: Sodium extrusion by internally dialyzed squid axons. J Gen Physiol 50:2303, 1967.

Cabantchik ZI, Rothstein A: Membrane proteins related to anion permeability of human red blood cells. J Mem Biol 15:207, 1974.

Caldwell PC: Studies on the internal pH of large muscle and nerve fibres. J Physiol (Lond) 142:22, 1958.

Hinke JAM: Cation-selective microelectrodes for intracellular use. In Eisenman G (ed): "Glass Electrodes for Hydrogen and Other Cations. Principles and Practice." New York: Marcel Dekker, 1967, pp 464–477.

Roos A, Boron WF: Intracellular pH. Physiol Rev 61:296, 1981.

Russell JM: Effects of ammonium and bicarbonate-CO_2 on intracellular chloride levels in Aplysia neurons. Biophys J 22:131, 1978.

Russell JM, Boron WF: Role of chloride transport in regulation of intracellular pH. Nature 264:73, 1976.

Spyropoulos CS: Cytoplasmic pH of nerve fibres. J Neurochem 5:185, 1960.

Thomas RC: The effect of carbon dioxide on the intracellular pH and buffering power of snail neurones. J Physiol (Lond) 255:715, 1976a.

Thomas RC: Ionic mechanism of the H^+ pump in a snail neurone. Nature 262:54, 1976b.

Thomas RC: The role of bicarbonate, chloride and sodium ions in the regulation of intracellular pH in small neurones. J Physiol (Lond) 273:317, 1977.

Chloride–Bicarbonate Exchange in the Sheep Cardiac Purkinje Fibre

R.D. Vaughan-Jones
University Laboratory of Pharmacology, Oxford University, South Parks Road, Oxford OX1 3QT, England

I.	Introduction	239
II.	Methods	240
	A. Preparation and Solutions	240
	B. Electrodes and Calibrations	241
III.	Results	242
	A. The Existence of $Cl^-–HCO_3^-$ Exchange in the Purkinje Fibre	242
	B. The Coupling Ratio	245
	C. Dependence on HCO_3^- and External Cl^-	246
	D. The Effects of Ammonia on $Cl^-–HCO_3^-$ Exchange	248
IV.	Discussion	250
V.	References	251

I. INTRODUCTION

In cardiac tissue the intracellular pH is maintained at a value of between 7.0 and 7.4. Furthermore direct measurements of pH_i in vitro using pH-sensitive microelectrodes have indicated that cardiac cells are capable of regulating their intracellular pH in a manner that, at least superficially, resembles that observed in many other excitable tissues. That is, the cells possess a system for removing excess acid from the cytoplasm. Hence at normal extracellular pH (~7.4), an experimentally induced intracellular acidosis is only transient, and pH_i returns to a more alkaline level within a few minutes. This was first observed by Ellis and Thomas [1976a] in sheep heart Purkinje fibres and also in ventricular muscle [Ellis and Thomas, 1976b]. Cytoplasmic buffers and intracellular organelles such as mitochondria will provide a means for initially removing some of the excess acid, but these mechanisms can have only a limited binding capacity. Hence in the longer term the acid must be extruded from the cell. This could take the form of an outward H^+ ion transport or an inward OH^- or HCO_3^- transport.

The ionic models of acid extrusion that have been proposed for other excitable cells offer a useful starting point when considering pH_i regulation in heart (see previous chapters). These appear as various combinations of Na^+–H^+ and Cl^-–HCO_3^- countertransport. In some cases there is a Na-dependent anion exchanger [Thomas, 1977; Boron et al, 1981; Moody, 1981], whereas in others the two types of countertransport appear to be independent of each other [Aickin and Thomas, 1977; see also Moody, 1981]. In sheep heart Purkinje fibres it has been found that the recovery of pH_i following removal of NH_4Cl (the "ammonia prepulse technique" described in previous chapters) is greatly slowed by the reduction of external Na levels and by the drug amiloride, an inhibitor of Na movements. Small transient changes of intracellular Na activity can also be measured during such recoveries and so a Na^+–H^+ countertransport system has been proposed to account for some of the acid extrusion [Deitmer and Ellis, 1980]. A role for Cl^- and HCO_3^- ions, however, has not so far been established. In this chapter I shall therefore consider evidence obtained using Cl^-- and pH-sensitive microelectrodes that a Cl^-–HCO_3^- countertransport exists, in the Purkinje fibre [Vaughan-Jones, 1979b] and that it is independent of Na^+–H^+ exchange. Beyond this, the situation may differ from that seen in other excitable cells. Preliminary experiments suggest that the anion exchanger plays at most only a minor role in the extrusion of acid. In contrast, under certain intracellular alkaline conditions, it effectively provides a route for acid *entry* into the cell (ie, a HCO_3^- efflux), which counterbalances the acid extrusion by other mechanisms. Two possibilities arise: (1) that under some conditions anion exchange is more effective at regulating pH_i in the face of an alkaline rather than an acid load, or (2) and perhaps more likely, that in heart, anion exchange has become specialised to maintain intracellular Cl^- levels rather than intracellular pH. The behaviour of anion exchange in heart is intriguing because it has always been assumed that both Cl^-–HCO_3^- and Na^+–H^+ exchange "switch off" when pH_i becomes alkaline [see Roos and Boron, 1981, p. 385].

II. METHODS

A. Preparation and Solutions

The experiments are performed on Purkinje fibres excised from the ventricular wall of a fresh sheep heart. These fibres are discrete bundles of fast-conducting tissue surrounded by a tough connective tissue sheath. A suitable preparation for experiments is about 2–5 mm long (a fibre can usually be trimmed without noticeable damage) and up to 0.5 mm in diameter. The sheep Purkinje fibre offers three major advantages over other types of cardiac preparation. First, although it is multicellular the individual cells are usually larger than those of cardiac muscle itself (~ 20 μ diameter or more compared to ~ 9 μ), so that prolonged microelectrode impalements become a little easier. Second, under most conditions the cells are well coupled electrically (length constant ~ 2 mm) so that two microelectrodes placed about 0.5 mm apart are likely to measure similar if not identical membrane potentials. Hence it is less imperative to insert the reference voltage electrode and the ion-sensitive electrode into the same cell in a bundle in order to measure an ion activity. Potential uniformity can often be checked by superfusing the preparation with a zero $(K)_o$ solution. This causes E_m to switch rapidly to a depolarised level and should produce no immediate ion-activity change if changes of E_m are measured equally by both electrodes. The internal activity of more than one ion can often be recorded simultaneously. For this to be of any use analytically one must suppose that the activity of an internal ion is homogeneous throughout the bundle, and there is now evidence that this may well be so, at least for the case of intracellular Na^+ [see Eisner et al, 1981, p 182]. The third advantage is that the tissue is only weakly contractile so that intracellular electrodes are less likely to be dislodged by movement of the fibre. Although it is possible to measure ion activities inside contracting fibres, the present experiments were performed on quiescent preparations to minimise the technical difficulties involved. It seems unlikely that the properties of a membrane ion-transport system will change in an active preparation, although intracellular pH may itself be a little different.

The fibres are pinned in a small superfusion chamber (volume ~ 0.2 ml). Solutions are selected by means of an eight-way rotary tap and are pregassed and superfused at 35–37°C ± 0.5°C. The composition of the normal Tyrode is: NaCl 118 mM; KCl, 4.5 mM; $CaCl_2$, 2.5 mM; $MgCl_2$, 1 mM; glucose, 11 mM; $NaHCO_3$ ~ 23 mM so that with nominally 5% CO_2 + 95% O_2 the pH is 7.40. Changes in ionic composition are made by isosmotic substitution. Cl^--free solutions contain the Na^+ and K^+ salts of glucuronic acid, 1 mM $MgSO_4$ and 12 mM Ca gluconate (an increased (Ca^{2+}) to compensate for Ca^{2+}-binding by glucuronate and gluconate). Nominally HCO_3^--free solutions contain equivalent amounts of NaHEPES titrated to the appropriate pH. When NH_4Cl is added, an equivalent amount of NaCl is removed.

B. Electrodes and Calibrations

Conventional microelectrodes are pulled from fibre-filled Pyrex tubing (o.d. 2 mm) and back-filled with either 3 M KCl or more usually 0.5 M K_2SO_4 + 5 mM KCl to minimise Cl leakage into the cells. Their resistance is typically 10–15 MΩ(3 M KCl). Membrane potential is measured relative to a 3 M KCl, 4% agar bridge positioned downstream from the preparation. The bath is also earthed via a second agar bridge.

pH-sensitive microelectrodes are of the recessed-tip, glass design described by Thomas [1974, 1978]. For efficient penetration of Purkinje fibres their tip diameters must be 0.5–1.0 μ. The exposed length of ion-sensitive glass is typically 15–30 μ. The electrodes are calibrated in HCO_3^--free Tyrode buffered to pH 7.40 with 20 mM NaHEPES and pH 6.4 with 20 mM NaADA (N-(2 acetamido)iminodiacetic acid). Acceptable electrodes respond to a pH change with a 95% response being achieved within at least 40 seconds. Cl^--sensitive microelectrodes [after Walker, 1971] are made using a drop of liquid Cl^--ion exchanger (Corning code 477315) incorporated into the tip of a microelectrode that has previously been dipped in a silane solution (15 seconds, in 2% tri-n-butyl chlorosilane in dry chloronapthalene) and baked for 24 hours at 200°C. The rest of the electrode is filled with 0.5 M KCl. An electrode is calibrated in solutions of NaCl using Na glucuronate to replace NaCl. Glucuronate is one of the few available Cl^- substitutes that is not sensed by the Cl^--electrode and so it is used in both calibrations and experiments. It has a pk_a of 3.2 but its undissociated form does not appear to penetrate membranes in quantities sufficient to produce an unwanted intracellular acidification [eg, Marrannes et al, 1980]. Indeed, as will be described, pH_i becomes alkaline in the presence of glucuronate.

Because Cl^-electrodes are not very sensitive to Cl^- over other simple anions, it is important to check that intracellular Cl^- activity measurements are not distorted by foreign anion interference. The simplest way of assessing this is to measure a^i_{Cl} whilst removing external Cl^-. Normal a^i_{Cl} is about 20 mM, and in Cl^--free media this declines to about 3 mM if the solutions are HCO_3^--free (bicarbonate adds about 1.5 mM of extra interference) [Vaughan-Jones, 1979a]. Hence the electrode mainly measures Cl^-. The residual level could be Cl^- trapped for some reason within the cells, or it could be the background level of interference produced by foreign anions. Since it is clearly low it has not been corrected for. It is unlikely to affect the analyses significantly [see Vaughan-Jones, 1979a, b].

The electrical arrangements are by now standard for experiments with ion-sensitive electrodes, and the reader is referred to Vaughan-Jones [1979b] and Thomas [1978] for more details.

III. RESULTS

A. The Existence of Cl^-–HCO_3^- Exchange in the Purkinje Fibre

The original evidence for Cl^--HCO_3^- exchange comes from measurements of pH_i and a^i_{Cl} whilst the extracellular levels of Cl^- are varied [Vaughan-Jones, 1979b]. Under these conditions there is a reciprocal relationship between intracellular Cl^- and intracellular pH. Figure 1 shows one such experiment. Here pH_i, a^i_{Cl} and membrane potential are recorded simultaneously. After an initial struggle whilst all three microelectrodes are inserted into the fibre, stable readings of intracellular activities are monitored. The important point to note is that neither

pH, Regulation in Purkinje Fibres / 243

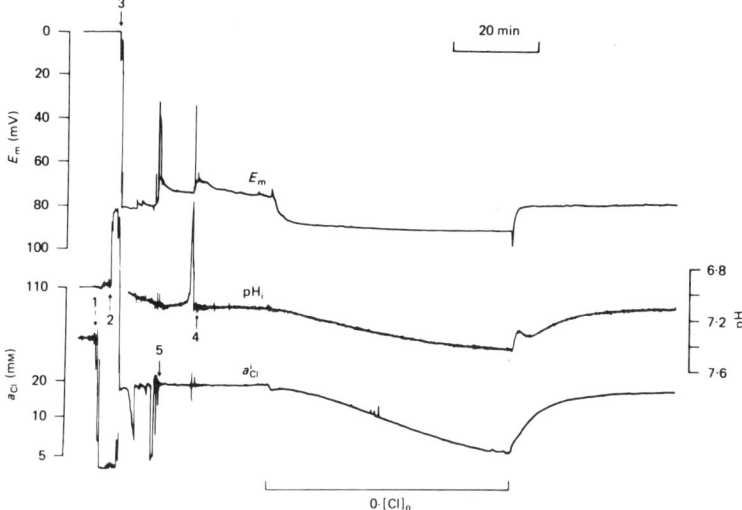

Fig. 1. Pen recording of an experiment to measure intracellular Cl activity, a^i_{Cl} (lower trace); intracellular pH, pH_i (middle trace); and membrane potential, E_m (upper trace) of a quiescent sheep cardiac Purkinje fibre. Initially all three electrodes are extracellular. At arrows 1 and 2 the pH and Cl⁻-electrodes electrodes are inserted. At arrow 3 the conventional microelectrode is inserted. This signal is electronically subtracted from the other two electrode signals, producing direct readings of activity. Stable recordings are achieved after arrows 4 and 5. The lower bar denotes a period in Cl-free solution (glucuronate-substituted) [from Vaughan-Jones, 1979b].

H⁺ nor Cl⁻ is passively distributed across the fibre membrane. Intracellular Cl⁻, a^i_{Cl}, is about four times higher than a passive level, so that the equilibrium potential for Cl⁻, E_{Cl}, is −45 mV, 30 mV positive to the resting potential, E_m. On the other hand, $(H^+)_i$ is lower than that predicted from a passive distribution. The H⁺ ion equilibrium potential, E_H, is −18.5 mV.

When external Cl⁻ is removed (glucuronate-substituted), Cl⁻ slowly leaves the fibre so that a^i_{Cl} declines. However, as a^i_{Cl} falls, pH_i *rises* at a similar rate. Hence in a Cl⁻-free solution pH_i is eventually about 0.3 units more alkaline. When external Cl⁻ is readmitted, Cl⁻ is rapidly reaccumulated inside the fibre whilst at the same time pH_i falls back to normal levels (the transient hump on the pH_i trace is probably artefactual since it was also measured in the bulk extracellular solution, it is most likely caused by a transient change of extracellular P_{CO_2}). The alkaline and acid shifts of pH_i upon removing and then readding external Cl⁻ are also seen with methylsulphate-substituted solutions.

Figure 2 shows that these effects on pH_i and a^i_{Cl} are inhibited by SITS (4-acetamido-4′-isothiocyanato-stilbene-2,2′-disilphonic acid) an inhibitor of anion movements. This suggests that intracellular Cl⁻ and pH are here interrelated by Cl⁻–HCO_3^- countertransport. Further evidence for this comes from the fact that

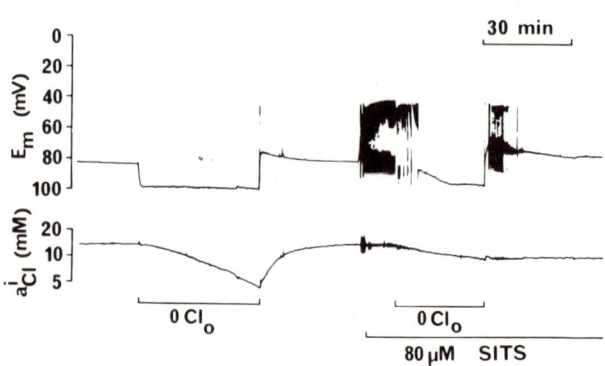

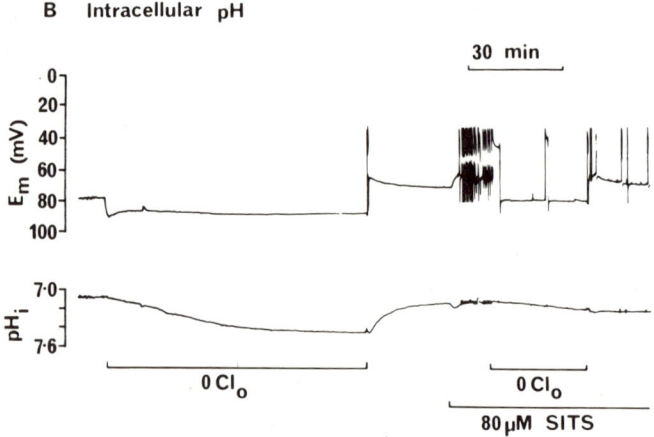

Fig. 2. Pen recordings of two experiments, A and B (separate fibres) showing the effects on E_m, a^i_{Cl} and pH_i of removing and re-adding external Cl and the effects of subsequently adding the anion-exchange inhibitor, SITS.

the washout and subsequent reaccumulation of intracellular Cl^- are both slowed in the nominal absence of HCO_3^- and CO_2. Figure 3 shows an experiment in which a fibre has been equilibrated in a Cl^--free solution buffered with HEPES + 100% O_2 to a pH_o of 7.4. Readmitting external Cl^- results in a very slow reaccumulation of intracellular Cl^- with a half-time of about 80 minutes. Later on in the experiment the reaccumulation is examined again, but this time the solutions are all buffered with 22 mM HCO_3^- + 5% CO_2 (95% O_2) at the same

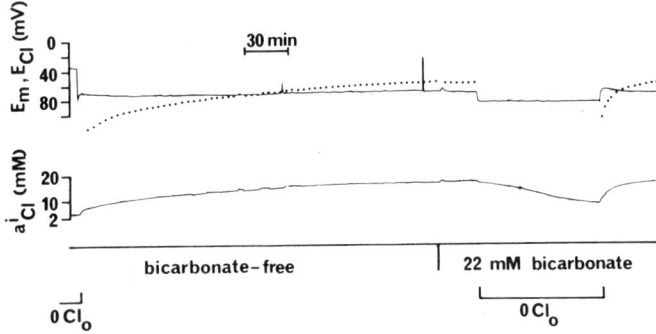

Fig. 3. When re-adding external Cl the reaccumulation of intracellular Cl is faster in the presence of bicarbonate (5% CO_2 + 95% O_2). External pH 7.4 throughout experiment. Bicarbonate-free solution buffered with 20 mM HEPES, 100% O_2. The filled circles show values of the Cl equilibrium potential, E_{Cl}, calculated from direct measurements of a^i_{Cl} and a^o_{Cl}. These have been superimposed on the E_m recording (upper trace). They demonstrate that E_{Cl} eventually becomes positive to E_m—ie, higher than passive levels of a^i_{Cl} are achieved.

pH_o, so that HCO_3^- will also be present *inside* the cell. The recovery of intracellular Cl^- is now much faster with a half-time of about 11 minutes. Similarly the washout of Cl^- is nearly twice as fast if HCO_3^--buffered solutions are used (not shown). It is interesting that removing HCO_3^- inhibits the reaccumulation of intracellular Cl^- (Fig. 3) by an amount that is comparable to the inhibition produced by SITS, in the presence of HCO_3^- (see Fig. 4).

The postulated link between Cl^- and HCO_3^- can perhaps be more clearly appreciated when the changes of a^i_{Cl} and pH_i measured in Figure 1 are replotted as in Figure 5A. Here changes of pH_i have been used to compute changes of intracellular HCO_3^- which can then be compared with the simultaneous changes of a^i_{Cl}. This emphasises the reciprocal relationship between internal Cl^- and HCO_3^- and shows that both levels change at very similar rates. There are many assumptions inherent in calculating intracellular HCO_3^- [see Vaughan-Jones, 1979b], the main ones being that the quantitative relationship between HCO_3^-, P_{CO_2} and pH is the same for cytoplasm as for extracellular Tyrode, and that the internal activity coefficient is 0.75. Nevertheless there is no compelling evidence to indicate that the approach does not give at least an approximate estimate of bicarbonate changes.

B. The Coupling Ratio

The rates of change of pH_i and a^i_{Cl} in Figure 1 can be used to compute net membrane fluxes of Cl^- and acid equivalents, in order to get some idea of the relative movements of the ions involved. For Cl^- this is quite easy, since there is little evidence of appreciable Cl^- buffering within the cell. For pH, however,

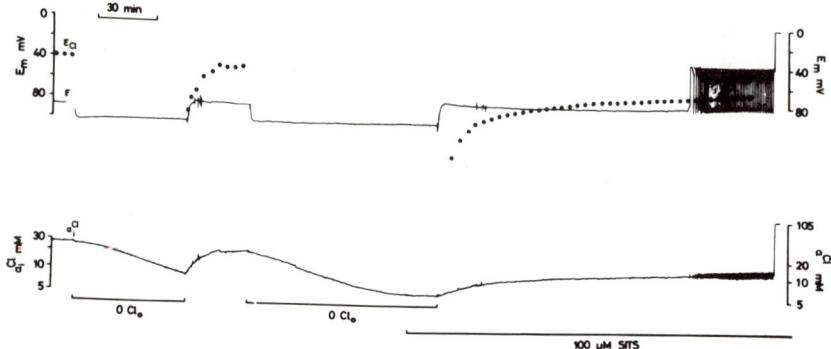

Fig. 4. The drug SITS inhibits the rapid reaccumulation of intracellular Cl. The filled circles denote calculated values of E_{Cl}, which have been superimposed on the E_m trace. Spontaneous slow oscillations of E_m occur toward the end of the experiment. These produce a thickening of the a^i_{Cl} trace since the Cl⁻-electrode's time constant is much slower than that of the voltage electrode. [From Vaughan-Jones, 1979b.]

the intracellular buffering power (including contributions from both CO_2 and non-CO_2 related buffers) must be allowed for, since this will damp down any changes of pH_i. Methods of estimating buffering power have been described [see Thomas, 1976; Boron, 1977]. The product of buffering power and the rate of change of pH_i gives an estimate of fluxes of acid equivalents in terms of an intracellular concentration change per unit time. When these are compared with the net Cl⁻ fluxes using data from Figure 1, then both fluxes are quite similar in size (Fig. 5B). On average, for four experiments, the ratio was 1·45 (acid ÷ Cl). Hence, if only Cl⁻ and HCO_3^- ions are involved then the movement of HCO_3^- will either equal or slightly exceed the counter movement of Cl⁻. However, simultaneous movements of H⁺ and OH⁻ cannot be excluded so that the apparent coupling ratio is only approximate. It does seem likely, however, that anion exchange is independent of Na⁺–H⁺ exchange in this tissue since removing external Na⁺, which produces a rapid fall of a^i_{Na} [Ellis, 1977], does not affect the reaccumulation of intracellular Cl⁻. Furthermore, Cl⁻ recovery is unaffected by amiloride and is not accompanied by changes of intracellular Na⁺ activity, measured using Na⁺-sensitive microelectrodes.

C. Dependence on HCO_3^- and External Cl⁻

It seems likely that the anion exchange system has a relatively high affinity for external Cl⁻ since most of it must be removed in order to produce a fall of a^i_{Cl}. The results of one such experiment are shown in Figure 6. In contrast, changes of external HCO_3^- at a constant external pH produce only small changes of a^i_{Cl}. This is evident in Figure 3 in which the addition of 22 mM $(HCO_3^-)_o$ at a pH_o of 7.4 produces virtually no steady-state change of a^i_{Cl}. Changing

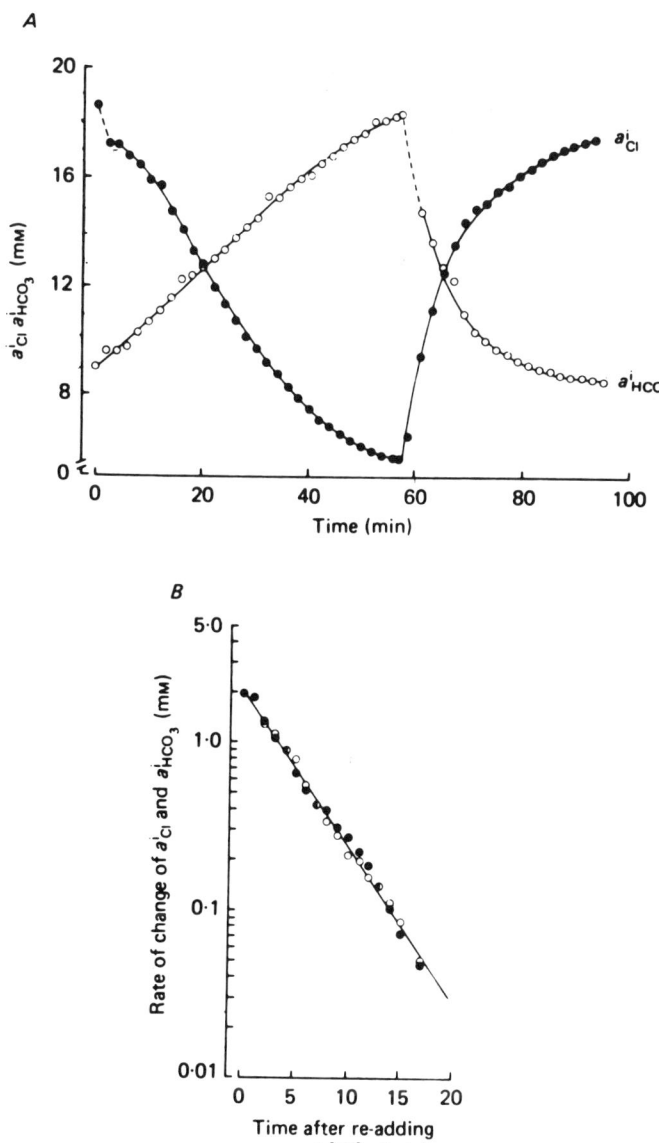

Fig. 5. A) The changes of a^i_{Cl} (filled circles) and the changes of intracellular bicarbonate activity (open circles; estimated from the measured changes of pH_i) that occur when external Cl is removed and then re-added. Data from Figure 1. B) The net membrane fluxes of Cl (filled circles) and of acid equivalents (open circles; denoted here tentatively as bicarbonate fluxes), estimated at various times after re-adding external Cl (expressed as mM/min). Data from the Cl reaccumulation phase shown in A. [From Vaughan-Jones, 1979b.]

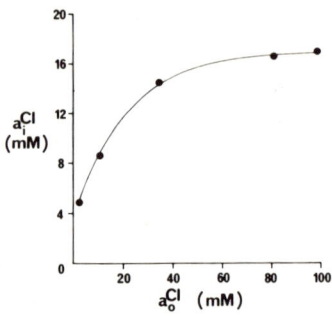

Fig. 6. Relationship between internal and external equilibrium levels of Cl. Data from a single fibre.

HCO_3^- can, however, affect the *rate* at which $Cl^--HCO_3^-$ exchange operates. The normally rapid reaccumulation of intracellular Cl^- is about 50% slower in the presence of 2 mM rather than 22 mM $(HCO_3^-)_o$ (pH$_o$ 7.4). However, this sort of experiment is difficult to analyse more precisely since it is clear that Cl^-–HCO_3^- exchange can operate in either direction depending on whether external Cl^- is high or low. Hence a change in the rate of Cl^- or HCO_3^- transport may reflect different degrees of activation of either an internal or an external site.

D. The Effects of Ammonia on $Cl^--HCO_3^-$ Exchange

The role of anion exchange in cardiac tissue is still uncertain. In this section I describe some preliminary experiments designed to test whether it is involved in the regulation of pH$_i$. The traditional approach is to induce an intracellular acidosis and then examine the subsequent recovery of pH$_i$, if any. The method used here is the "ammonia prepulse technique" [Aickin and Thomas, 1977; Boron and De Weer, 1976], which has been outlined in the previous chapters. It is illustrated in Figure 7. During the addition of 20 mM NH$_4$Cl, pH$_i$ initially becomes alkaline presumably because the small quantities of molecular NH$_3$ that exist at physiological pH$_o$ rapidly cross the membrane and chelate intracellular H$^+$ ions. After a few minutes, however, pH$_i$ begins to acidify, and it is notable that a^i_{Cl} also rises. Upon removing NH$_4$CL any intracellular NH$_4^+$ that has been formed is lost rapidly across the membrane as NH$_3$, leaving behind H$^+$ ions, hence acidifying the cell. There is then an exponential recovery of pH$_i$, presumably because of the extrusion of acid or its ionic equivalent. During this period a^i_{Cl} falls back to normal levels. Figure 7 also shows that a disulphonic stilbene, DIDS (4,4'diisothiocyanostilbene-2-2'-disulphonic acid) profoundly slows the secondary acidification that occurs in the presence of NH$_4$Cl. It also inhibits the rise of a^i_{Cl}. Consequently, the degree of "acid loading" is reduced so that the acidosis produced by removing NH$_4$Cl is much smaller than usual. Despite this,

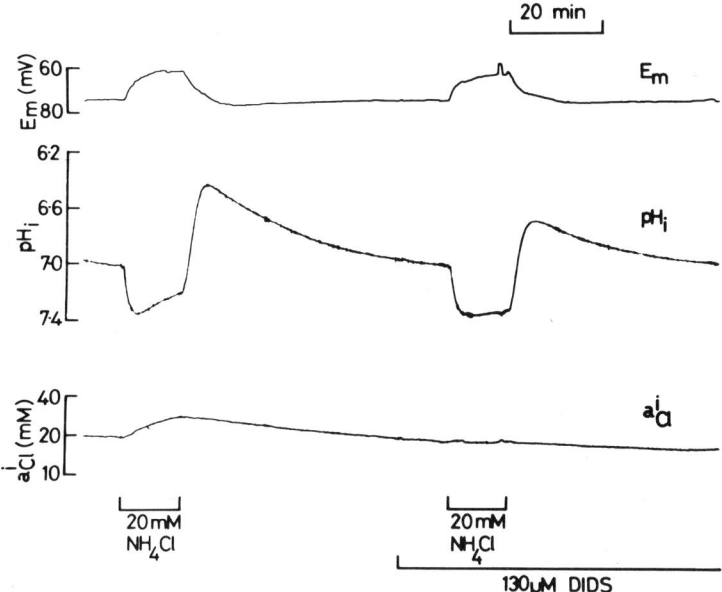

Fig. 7. Simultaneous measurement of pH_i, a^i_{Cl} and E_m. The effect of adding external NH_4Cl in the absence and then the presence of the drug DIDS. 10mM $(HCO_3^-)_o$, 3% CO_2 + 97% O_2, pH_o 7.26 [from Vaughan-Jones, 1981].

the rate constant of the subsequent recovery of pH_i in the presence of DIDS is the same as before.

The ability of NH_4Cl to produce a rise of a^i_{Cl} cannot be related to the depolarisation that it produces since similar depolarisations produced by raising external K^+ do not mimic the effect. Furthermore, if nominally HCO_3^--free, HEPES-buffered solutions are used, NH_4Cl produces little or no rise of a^i_{Cl}. It would appear then that during NH_4Cl application Cl^- enters the cell and HCO_3^- leaves. Much of the acidification during this period is caused by the Cl^-–HCO_3^- exchange mechanism. It is less clear whether the system subsequently operates in the reverse direction and assists acid extrusion during the recovery of pH_i following NH_4^+ removal. I shall return to this in the Discussion.

Finally, it is interesting to reexamine the action of amiloride upon the ammonia response. Unlike DIDS, it has no effect upon the Cl^- and pH_i changes that occur during NH_4Cl treatment, whereas it profoundly slows the subsequent recovery of pH_i following NH_4Cl removal (Fig. 8). The rate of fall of a^i_{Cl} during the recovery, however, is not greatly affected. Hence the acid-loading phase of NH_4^+ treatment is slowed by DIDS, whereas the acid-extruding phase following NH_4^+ removal is slowed by amiloride.

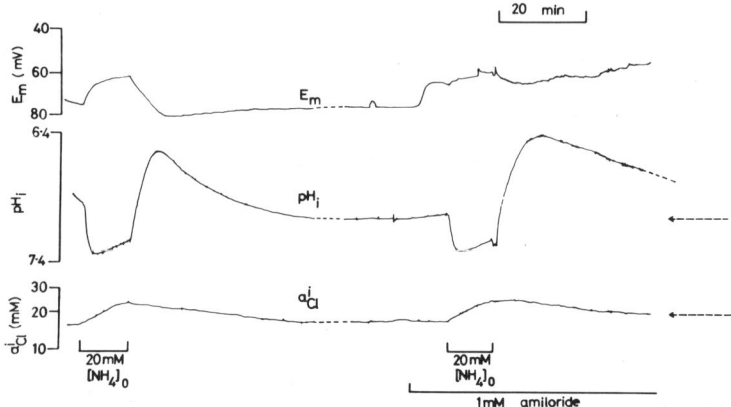

Fig. 8. The effect on pH_i, a^i_{Cl} and E_m of adding external NH_4Cl in the absence and presence of 1 mM amiloride. Dashed lines in the middle of the experiment denote a resting period of about 40 minutes. The arrows denote the final equilibrium levels of pH_i (upper arrow) and a^i_{Cl} (lower arrow) achieved in the presence of amiloride.

IV. DISCUSSION

The main findings described here are that a Cl^--HCO_3^- countertransport exists in the Purkinje fibre, and that it is independent of the Na^+-H^+ exchange described by Deitmer and Ellis [1980]. The system can operate in either direction across the membrane depending upon whether external Cl^- is low or high and it is stimulated by external NH_4^+. In some respects, then, the Purkinje fibre resembles mammalian skeletal muscle [Aickin and Thomas, 1977] and the crayfish giant neurone [Moody, 1981], where two separate exchange processes have been proposed. Here the similarity may end, for the apparent lack of effect of DIDS upon the recovery of pH_i following NH_4^+ removal (Fig. 7) suggests that the anion system in heart does not assist acid extrusion. A similar conclusion was reached by Ellis and Thomas [1976b]. The evidence is not conclusive, however, since a fall of a^i_{Cl} accompanies the recovery of pH_i, and this may indicate an extrusion of Cl^- in exchange for a HCO_3^- influx. Further experiments are required to see if this is blocked by DIDS but since Cl^- levels are high in cardiac tissue, such a decline could equally well be a passive efflux of Cl^- via some other membrane pathway.

Of greater interest, however, is the finding that much of the acid loading produced by NH_4Cl is mediated by anion exchange: Cl^- enters and HCO_3^- leaves the cell. This system could be switched on either by some peculiar property of NH_4^+ itself, or alternatively it may be activated by an alkaline shift of pH_i. Whatever the stimulus, the overall effect is to return pH_i toward normal resting levels in the face of an alkaline load. This is a novel finding since it has always been assumed that the intracellular acidification during NH_4^+ treatment represents passive NH_4^+ entry probably via K^+ channels [eg, Boron and de Weer, 1976] or perhaps partially on the Na^+–K^+ pump [Aickin and Thomas, 1977]. There is indeed evidence (unpublished) of a slow passive entry of NH_4^+ ions into the Purkinje fibre, probably via channels, but the intracellular acidification by this route is usually slower than that produced by Cl^-–HCO_3^- exchange. It will be interesting to see if the slow acidification in NH_4^+ found in other cells is also partly mediated by anion exchange. So far, it has always been assumed that this system switches off at an alkaline pH_i, or at least reverts to a Cl^-–Cl^- self-exchange [eg, Russell and Boron, 1976].

What, then, is the functional significance of anion exchange in cardiac tissue? Perhaps it helps to maintain a constant pH_i under abnormally alkaline conditions. This must be tested further, but experiments (unpublished) suggest that the alkaline shift of pH_i seen upon reducing P_{CO_2} at a constant pH_o is not subsequently influenced by the anion exchange. Furthermore, its apparent lack of contribution to acid extrusion must be examined more carefully under a greater variety of conditions. An alternative possibility is that Cl^-–HCO_3^- exchange has become specialised in heart to maintain a high intracellular Cl^-. The strict regulation of pH_i may therefore be delegated to other ion-transport systems. It is worth recalling that normal resting levels of a^i_{Cl} are achieved at the expense of acidifying the cell by about 0.3 pH units (Fig. 1). In this case the anion system, which in other cells extrudes Cl^- and takes up HCO_3^- may be effectively turned around in the Purkinje fibre. If this is so, the functional advantage may lie in the maintenance of a particular value of E_{Cl}. Perhaps it is related to a role of Cl^- ions in the generation of membrane currents. It has been suggested, for example, that Cl^- is involved in pacemaker currents in rabbit [eg, Seyama, 1979; Millar and Vaughan Williams, 1981], but such an idea is still rather controversial.

In conclusion, when external Cl^- is readmitted to a Cl^--depleted Purkinje fibre, or when external NH_4Cl is added, intracellular pH is initially alkaline, and anion exchange then provides an easy route for acid entry into the cell. This will act like an inward acid "leak" and will offset acid extrusion mediated by other mechanisms. The exchanger will therefore influence pH_i even if it is not necessarily a controller of pH_i. The acid extrusion mechanisms themselves still remain largely unknown, although Na^+–H^+ exchange most likely plays a role as evidenced from the inhibitory action of amiloride seen in Figure 8 [see also Deitmer and Ellis, 1980].

V. REFERENCES

Aickin CC, and Thomas RC: An investigation of the ionic mechanism of intracellular pH regulation in mouse soleus muscle fibres. J Physiol 273:295–316, 1977.

Boron WF: Intracellular pH transients in giant barnacle muscle fibres. Am J Physiol 233:C61–C73, 1977.

Boron WF, de Weer P: Intracellular pH transients in squid giant axons caused by CO_2, NH_3 and metabolic inhibitors. J Gen Physiol 67:91–112, 1976.

Boron WF, McCormick WC, Roos A: pH Regulation in barnacle muscle fibres: Dependence on extracellular sodium and bicarbonate. Am J Physiol 240:C80–C89, 1981.

Deitmer JW, Ellis D: Interactions between the regulation of the intracellular pH and sodium activity of sheep cardiac Purkinje fibres. J Physiol 304:471–488, 1980.

Eisner DA, Lederer WJ, Vaughan-Jones RD: The dependence of sodium pumping and tension on intracellular sodium activity in voltage-clamped sheep Purkinje fibres. J Physiol 317:163–187, 1981.

Ellis D: The effect of external cations and ouabain on the sodium activity in sheep heart Purkinje fibres. J Physiol 273:211–240, 1977.

Ellis D, Thomas RC: Microelectrode measurement of the intracellular pH of mammalian heart cells. Nature 262:224–225, 1976a.

Ellis D, Thomas RC: Direct measurement of the intracellular pH of mammalian cardiac muscle. J Physiol 262:755–771, 1976b.

Marrannes R, de Hemptinne A, Leusen I: pH Aspects of transient changes in conduction velocity in isolated heart fibres after partial replacement of chloride with organic anions. Pflugers Archiv, 389:199–209, 1981.

Millar JS, Vaughan Williams EM: Pacemaker selectivity: Influence on rabbit atria of ionic environment and of alinidine, a possible anion antagonist. Cardiovasc Res 15:335–350, 1981.

Moody WJ: The ionic mechanism of intracellular pH regulation in crayfish neurones. J Physiol 316:293–308, 1981.

Roos A, Boron WF: Intracellular pH. Physiol Rev 61:296–434, 1981.

Russell JM, Boron WF: Role of chloride transport in regulation of intracellular pH. Nature 264:73–74, 1976.

Seyama I: Characteristics of the anion channel in the sino-atrial node cell of the rabbit. J Physiol 294:447–460, 1979.

Thomas RC: Intracellular pH of snail neurones measured with a new pH-sensitive glass microelectrode. J Physiol 238:159–180, 1974.

Thomas RC: The role of bicarbonate, chloride and sodium ions in the regulation of intracellular pH in snail neurones. J Physiol 273:317–338, 1977.

Thomas RC: The effect of carbon dioxide on the intracellular pH and buffering power of snail neurones. J Physiol 255:715–735, 1976.

Thomas RC: Ion sensitive intracellular microelectrodes. How to make them and use them. New York: Academic Press, Biology Techniques Series, 1978.

Vaughan-Jones RD: Non-passive chloride distribution in mammalian heart muscle: Micro-electrode measurement of the intracellular chloride activity. J Physiol 295:83–109, 1979a.

Vaughan-Jones RD: Regulation of chloride in quiescent sheep heart Purkinje fibres studied using intracellular chloride and pH-sensitive microelectrodes. J Physiol 295:111–137, 1979b.

Vaughan-Jones RD: Ammonia stimulates chloride-bicarbonate exchange in sheep cardiac Purkinje fibres. J Physiol (in press), 1981.

Walker JL: Ion-specific liquid ion exchanger microelectrodes. Anal Chem 43:89–93A, 1971.

Hydrogen and Bicarbonate Transport by Salamander Proximal Tubule Cells

Walter F. Boron and Emile L. Boulpaep

Department of Physiology, Yale University School of Medicine, New Haven, Connecticut 06510

I.	Introduction	253
II.	Methods	254
	A. General	254
	B. Solutions	255
	C. Microelectrodes	256
III.	Results	260
	A. Control Studies	260
	B. Normal Values	260
	C. Mechanism of pH_i Regulation	260
	D. Basolateral HCO_3^- Transport	262
IV.	Discussion and Conclusions	266
V.	References	267

I. INTRODUCTION

The previous chapters of this volume have dealt with mechanisms of intracellular pH (pH_i) regulation in certain symmetrical cells: nerve, muscle, and Purkinje fibers. The problem of maintaining a stable pH_i in the face of chronic intracellular acid loading is shared by cells trapped within an epithelium. Failure to properly regulate pH_i in these cells would presumably lead to deranged cellular function. Epithelial cells, such as the amphibian renal proximal tubule cell with which this chapter is primarily concerned, are characterized by their asymmetry. One surface of the tubule cell faces one compartment (ie, the tubule lumen), and another surface, isolated from the first by tight junctions, faces a second compartment (ie, the blood). In the case of the amphibian proximal tubule cell, there is a substantial electrochemical gradient favoring the influx of H^+ into the cell and the efflux of HCO_3^- out of the cell across both the membrane facing the lumen ("luminal" membrane) and that facing the blood ("basolateral" membrane). In addition, these cells are probably occasionally subjected to periods of intracellular acid loading as a result of cellular metabolism or of the release of acid from intracellular stores (eg, mitochondria). Yet the renal cells thus far studied are able to maintain a relatively alkaline pH_i. This could be accomplished only if these cells possess an active pH_i-regulating mechanism, analogous to that operative in symmetrical cells.

In addition to the pH-related problems they share with symmetrical cells, certain epithelial cells have a problem unique to their physiologic role: They transport acid or base across themselves, from one side of the epithelium to the other. Considering the absolute necessity for intracellular acid–base homeostasis, one might suspect that transcellular transport of acid or base might be inextricably linked to the more fundamental process of pH_i regulation. This proposition will be the central focus of this chapter. Our experimental approach has been to employ the isolated, perfused tubule technique with proximal tubules of the tiger salamander. The isolated, perfused tubule method, devised by Burg and his colleagues in 1966, permits one to perfuse the lumen and superfuse the basolateral surface of a tubule that is completely isolated from the rest of the kidney. This approach enjoys two distinct advantages: (1) The tubule cells are not subjected to extraneous neural and humoral influences, and (2) one has nearly total control of the composition of the solutions on both luminal and basolateral surfaces of the cell. We chose the salamander proximal tubule because the cells are large enough to permit long-lasting impalements with ion-sensitive microelectrodes. Our data indicate that pH_i regulation in these cells is mediated by a symmetrically distributed Na–H exchanger. The potential for net acid secretion (blood to lumen) is conferred by a pathway for HCO_3^- that is confined to the basolateral membrane.

II. METHODS

A. General

Female tiger salamanders (Ambystoma tigrinum) were anesthetized in 0.1% tricaine. The kidneys were removed, and single proximal tubules were isolated with glomeruli intact. Lengths (700 to 1,000 μm) of early proximal tubule were dissected free of the rest of the nephron, and transferred to the chamber. The perfusion apparatus, similar to that originally described by Burg et al [1966], consisted of two assemblies of triple concentric pipettes (see Fig. 1). The tubule is surrounded by the outermost pipette and cannulated by the middle one. Since the space between these two pipettes is air-tight, application of vacuum draws the tubule up between the pipettes and causes it to jam at the outer pipette's constriction, forming a mechanically and electrically tight seal. Luminal perfusate is introduced through the innermost pipette (not shown) in the right-hand assembly at the rate of ~ 0.5 ml/min. The majority of this perfusate escapes between the innermost and middle pipettes; a very small amount (~ 20 nl/min) actually perfuses the tubule lumen. This perfusate is collected by applying suction to the innermost pipette of the left-hand assembly. The tubules were also continuously

CHAMBER

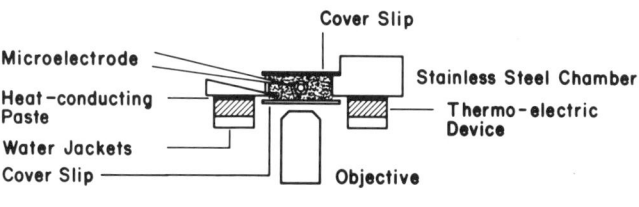

SIDE VIEW

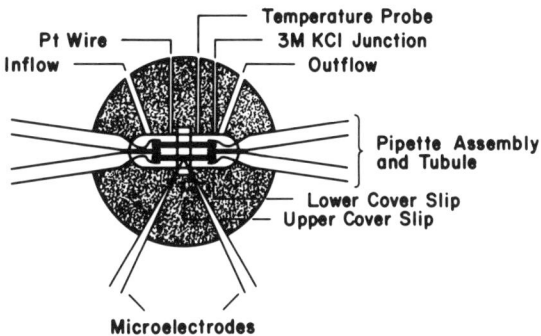

TOP VIEW

Fig. 1. Schematic diagram of apparatus. Top) Side view. The tubule rests on the lower coverslip, the upper surface of which is covered with a thin layer of hardened Sylgard 184. This Sylgard coating minimizes sliding of the tubule during microelectrode impalements. The upper coverslip is required only in experiments in which optical measurements (eg, absorbance) are made. Bottom) Top view. The tubule (thick parallel lines) is cannulated at each end by a pipette assembly consisting of three concentric pipettes. The outermost pipette surrounds the tubule. The middle pipette cannulates the lumen. The innermost pipette (not shown) delivers perfusate on the right-hand side, and collects it on the left-hand side. The tubule is continuously perfused and superfused.

superfused at about 15 chamber volumes per minute. The chamber was placed on the fixed stage of an inverted microscope; microelectrode impalements were made at a magnification of 400 ×.

B. Solutions

The standard Ringer's solution had the following composition (in mM): 90 NaCl, 2.5 KCl, 0.5 NaH_2PO_4, 1 $MgCl_2$, 2.2 glucose, 1.7 Ca lactate, 0.5 L-glutamine, 0.5 D,L-alanine, 0.05 D,L-Na glutamine, and 0.2 L-lysine HCl. In addition, the Ringer was buffered to pH 7.5 with either 10 mM $NaHCO_3^-$/1.5%

CO_2 in O_2 or 13.3 mM N-2-hydroxyethylpiperazine-N'-2-ethanesulfonic acid (HEPES) titrated with concentrated NaOH (6.7 mmol per liter Ringer). Various ion substitutions are described in the text.

C. Microelectrodes

We employed pH-sensitive microelectrodes (see Fig. 2, left) of the recessed-tip design of Thomas [1974]. They were fabricated from Corning 0150 pH-sensitive glass tubing (0.5 mm i.d., 1 mm o.d.) and Corning 1720 aluminosilicate glass tubing (1.1 mm i.d., 2.2 mm o.d.). Consult Thomas's monograph (and Thomas, this volume) for details [1978]. Based on measurements made at high magnification using oil immersion, we estimate that the electrode tips had outer diameters $\leq$ 0.5 μm. The electrodes' responses to solution changes had time constants of 10–20 seconds. The electrodes' slopes averaged $\sim$ 57 mV; resistances were 10^{11} to 10^{12} Ω.

The Na-sensitive microelectrodes were also of Thomas's [1969] recessed-tip design (similar to the pH-sensitive electrode of Fig. 2), and were fabricated from Corning NAS 11-18 Na-sensitive glass tubing (0.5 mm i.d., 1 mm o.d.) and aluminosilicate glass tubing (see above). These were filled with NaCl-saturated

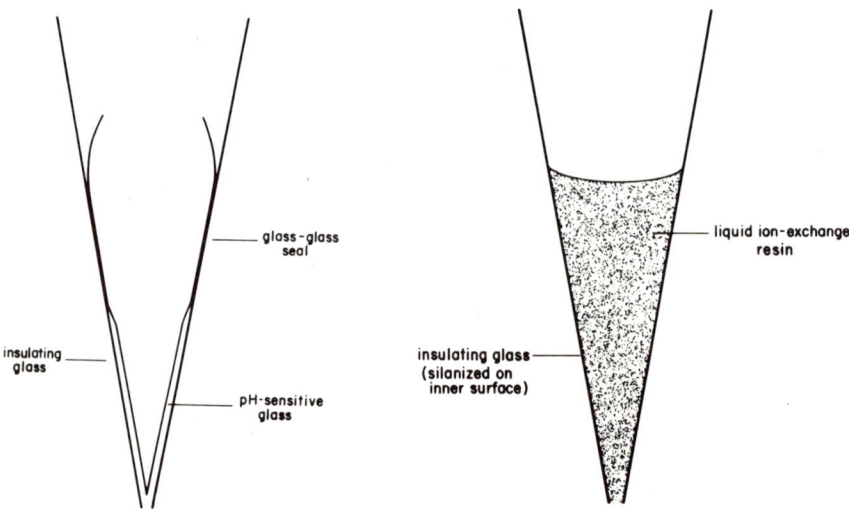

Fig. 2. Schematic diagram of ion-sensitive microelectrodes. Left) pH-sensitive electrode, of the recessed-tip design of Thomas [1974, 1978]. The Na^+-sensitive electrodes were of a similar design [Thomas, 1969, 1978], except the pH-sensitive glass was replaced by Na-sensitive glass. Right) Liquid–ion exchanger electrode. The aluminosilicate pipette was silanized as described in text. The electrode was then backfilled with Corning Cl exchanger (No. 477315).

dry methanol and stored with tips immersed in dry methanol to forestall hydration of the Na-sensitive glass membrane. Electrode slopes ranged from 56 to 60 mV, selectivities of Na^+ over K^+ from 100 to 450, and resistances from 10^{10} to 10^{11} Ω.

The Cl-sensitive microelectrodes were of the liquid–ion exchanger type (Fig. 2, right). We found the following silanization procedure optimal for producing functional electrodes even on days of relatively high humidity: Aluminosilicate micropipettes were dried for 2 hours at 200°C within an open-topped vessel of approximately 300 ml volume. Immediately following this, 10 μl of tri-n-butylchlorosilane was introduced into the container, which was then covered with a lid. After 2 minutes, the lid was removed and the silane vapor was vented from the vessel; the electrodes were baked for an additional 30 minutes at 200°C. The pipettes were finally allowed to cool in an evacuated dessicator over P_2O_5. The electrode tips were backfilled with Corning Cl exchanger (No. 477315). The electrodes had selectivities for Cl^- over HCO_3^- of 9 to 12, slopes of about 58 mV, and resistances of 10^9 to 10^{10} Ω.

Ling-Gerard microelectrodes were pulled from 1-mm "Omega Dot" borosilicate tubing (Frederick Haer, Brunswick, Maine). These were filled with 3 M KCl, and had resistances of 30–60 MΩ, and tip potentials < 5 mV.

The ion-sensitive microelectrode as well as a calomel electrode contacting the basolateral solution was connected to the inputs of a high impedance (10^{15} Ω) electrometer (WPI Model 223). The Ling-Gerard electrode and a calomel cell contacting the luminal solution were connected to the inputs of a medium impedance (10^{11} Ω) electrometer. The bath was grounded through a platinum wire. The voltage differences between various pairs of electrodes were obtained electronically and plotted on a four-channel strip chart recorder. We continuously recorded an intracellular ion activity (ion-sensitive vs Ling-Gerard electrodes), basolateral membrane potential (Ling-Gerard vs basoteral calomel electrodes), and transepithelial potential (luminal vs basolateral calomel electrodes). pH_i values were recorded according to the American convention (high pH_i values on top, low values at the bottom).

To facilitate microelectrode impalements, the tubule rested on a thin layer of hardened Sylgard 184, which covered the upper surface of the chamber's lower coverslip. This arrangement increases the success of microelectrode impalements because the tubule tends to stick to the Sylgard as the electrode advances [Sackin and Boulpaep, 1981]. To increase further the chances of a successful impalement, the electrodes were introduced into the cells by carrying the former on rapidly advancing piezoelectric devices (see Fig. 3), two kinds of which were employed. The first consisted of a 5.5-cm long stack of 20 piezoelectric crystal disks. The disks, each 6 mm in diameter and 2.5 mm thick, were made of ELB 6 (ELB Co., East Hartford, Connecticut) and were coated with nickel on the parallel faces. The disks were glued together with silver (ie, conductive) epoxy and wired

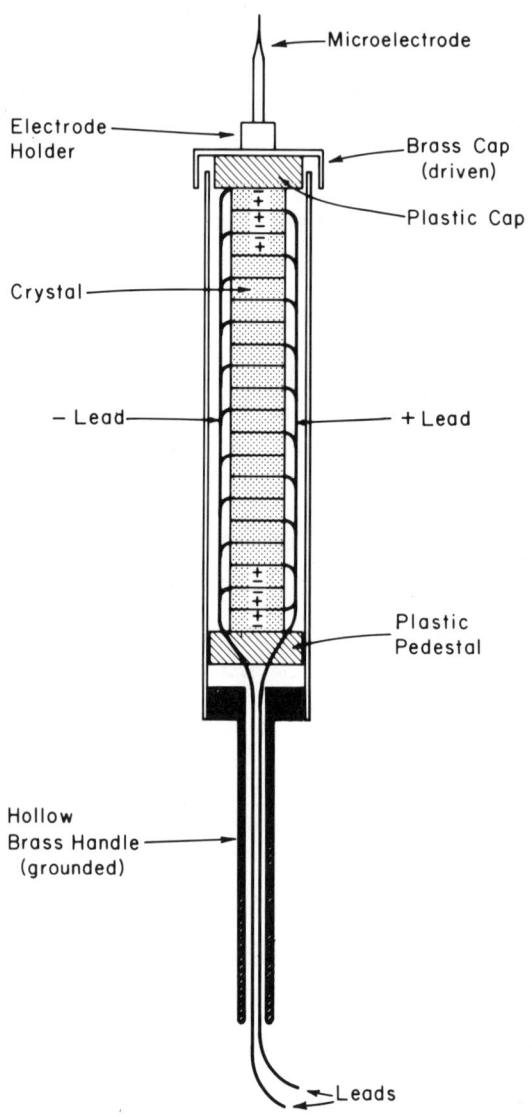

Fig. 3. Legend on next page.

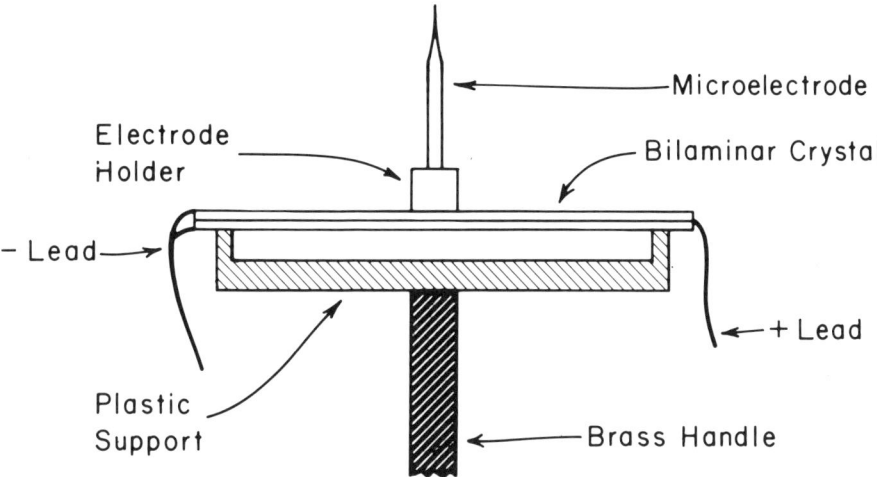

Fig. 3. Piezoelectric crystal assemblies. Left) Crystal stack (longitudinal section). Crystal disks (6 mm diameter, 2.5 mm thick) were glued together with conductive epoxy, surfaces of like polarity facing each other. Small tabs of brass shim stock (0.001 in. thick) were placed between crystals to provide electrical contact. Tabs of like polarity were folded together and the leads were strung through the hollowed-out handle of the assembly. The brass casing and the handle were both grounded. The brass cap was driven to the same voltage as the electrode. Right) Bender (cross section). The crystal is actually bilaminar (1 mm thickness), with the two (+) surfaces facing one another and contacting a center tab. When a positive voltage is placed across the crystal, the surface bows away from the handle (ie, advancing the electrode) in a cantilever action. A negative voltage causes the surface to bend toward the handle. When the voltage is applied in a square wave, the motion of the crystal has two phases, a rapid movement followed by a slow creep to the final position. Since this creep is undesirable, we applied only a brief pulse (30 msec at 70 V) so that the crystal rapidly advanced ~ 10 μm and then retracted, with no creep. Our first models employed a crystal with a circular (50 mM diameter) surface, later models a crystal with a rectangular surface (50 × 12 mm) for the sake of compactness.

in parallel. Applying 1,000 V across each of the disks caused the entire stack to shorten by ~ 10 μm. Discharging the crystals by connecting the leads caused the stack to expand rapidly to its initial length and thereby carry the electrode into the cell. Because such large voltages were employed the stack had to be carefully shielded. The stack's lateral and bottom surfaces were surrounded by brass tubing, which was grounded. The leads to the crystal stack were also surrounded by a grounded shield. The top of the crystal stack was covered by a brass cap that was "driven" to the same voltage as the ion-sensitive microelectrode. This output for the driven shield is obtained directly at the electrometer probe. The second piezoelectric device was a simple "bender," a bilaminar disk about 50 mm in diameter and 1 mm thick (LPZ-2H, Transducer Products, Inc.,

Goshen, Connecticut). When a square wave pulse of 75 V is passed across this crystal for about 30 msec, the center of the disk transiently advances in a drumhead action about 15 μm, carrying the electrode into the cell.

III. RESULTS

A. Control Studies

Ideally, both the ion-sensitive electrode and the Ling-Gerard electrode would be placed within the same cell. Since this is impractical with these relatively small cells, we opted for placing the electrodes in two separate cells. As a control we simultaneously impaled separate cells with a Ling-Gerard microelectrode and the aluminosilicate portion of an ion-sensitive microelectrode. Since the latter had an open tip and was filled with 3M KCl, it too measured basolateral membrane potential (V_{bl}). We found that the two electrodes measured the same V_{bl} (average difference: 0.8 ± 3.7 mV; n = 27) when the tubules were exposed to standard Ringer. When V_{bl} was altered by raising $(K^+)_{bl}$ to 12 mM or lowering it to 0 mM, the two electrodes measured virtually the same V_{bl}. This agreement may be due to excellent electrical coupling between cells, or to the nearly identical properties of the cells that make up the tubule.

B. Normal Values

When tubules were perfused and superfused with standard, HCO_3^--free Ringer (pH 7.50), the cells had an average pH_i of 7.43 ± 0.02 ($\pm$ SE; n = 49 tubules). V_{bl} was -56.7 ± 1.7 mV. Intracellular Na^+ and Cl^- were 24.0 ± 2.6 mM (n = 8) and 18.4 ± 2.4 mM (n = 9), respectively. When the Ringer was changed to CO_2/HCO_3^- Ringer at the same pH, pH_i fell by an average of 0.17 ± 0.02 to 7.29 ± 0.05 (n = 10). This difference is statistically significant (P = 0.00002).

C. Mechanism of pH_i Regulation

When a nerve or muscle cell is subjected to an acute intracellular acid load, pH_i recovers from the insult with approximately an exponential time course [Thomas, 1976; Boron et al, 1979]. This pH_i recovery cannot be accounted for by passive transport processes since the electrochemical gradient for H^+ favors a passive influx, and that for HCO_3^-, a passive efflux. To determine whether these tubule cells also actively regulate their pH_i we acid loaded the cells by the NH_4^+ prepulse technique. As described for squid axons (see Russell and Boron, this volume), exposing a cell to NH_4^+ causes a small net influx of this ion. When extracellular NH_4^+ is removed, nearly all this intracellular NH_4^+ dissociates into H^+ plus NH_3, and leaves the cell as the neutral weak base. As a result, protons are left behind and pH_i falls below the initial level; ie, the cells are acid-loaded. In our experiments with the kidney tubules (see Fig. 4), we replaced 20 mM

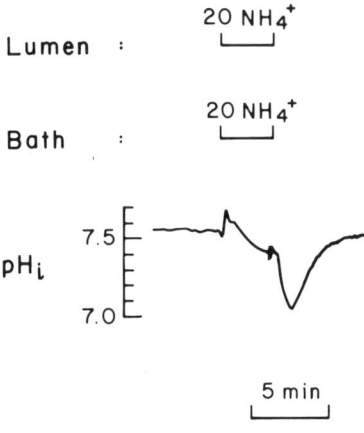

Fig. 4. Acid-loading a tubule with a pulse of NH_4^+. 20 mM Na^+ was replaced by an equivalent amount of NH_4^+ in both the lumen and bath; external pH was held constant at 7.5. Application of the NH_4^+ causes a rise in pH_i due to influx of NH_3, and then a fall in pH_i due to influx of NH_4^+. Removal of external NH_4^+ causes a sharp fall in pH_i as intracellular NH_4^+ releases H^+ to exit as NH_3. The cell then slowly recovers from this intracellular acid load, following an exponential time course.

Na^+ in both the luminal and basolateral solutions with NH_4^+, holding external pH constant at 7.50. In order that the pH_i changes not be complicated by transport of HCO_3^-, we performed these NH_4^+ experiments in the nominal absence of HCO_3^-. We found that the changes in pH_i produced by application and removal of NH_4^+ were very similar to those previously observed in symmetrical cells. Application of NH_4^+ caused an initial rise in pH_i of ~ 0.15 (due to entry of NH_3), followed by a slower fall of pH_i to, or slightly below, the initial value (due to entry of NH_4^+). This return of pH_i to the vicinity of the initial pH_i generally was complete in ~ 5 minutes. Removal of luminal and basolateral NH_4^+ caused pH_i to fall rapidly by 0.2 to 0.4, and then to recover more slowly to the initial value. This pH_i recovery followed an exponential time course, and generally had a rate constant of ~ 1.0 min^{-1}. Thus, the entire experiment requires about 10 minutes.

Considerable indirect evidence suggested that the proximal tubule cell possesses a Na–H exchanger at the luminal membrane [see Warnock and Rector, 1979]. In addition, Murer et al [1976] and Kinsella and Aronson [1980] have identified a Na–H exchanger in membrane vesicles derived from the dog kidney, primarily the brush border (luminal membranes) of proximal tubules. We therefore suspected that this pH_i recovery following an NH_4^+-induced acid load might be mediated by a luminal Na–H exchanger. Indeed, we found that when both luminal and basolateral Na$^+$ was replaced with bis-(2-hydroxyethyl) dimethyl ammonium (BDA), pH$_i$ failed to recover from an acid load. When 100 mM Na$^+$ was added back to the lumen only, pH$_i$ returned to its initial value following an exponential time course. However, the rate constant for this pH$_i$ recovery was only about half that observed in control experiments in which Na$^+$ was present in both luminal and basolateral solutions. We therefore performed a second experiment in which, after completely removing Na$^+$ from the system and acid-loading the cells, we returned Na$^+$ to the basolateral solution only. Once again, pH$_i$ recovered from the acid load at about half the normal rate. These results thus suggest that a Na-dependent pH$_i$-regulating system is present in both the luminal and basolateral membranes of these cells. If this pH$_i$-regulating system is truly a Na–H exchanger, then one would expect it to be inhibited by the diuretic amiloride, which inhibits the Na–H exchanger of mouse soleus muscle [Aickin and Thomas, 1977] and of brush border membrane vesicles [Kinsella and Aronson, 1980]. Accordingly, we acid-loaded tubule cells as before, but this time placed 2 mM amiloride in the basolateral and/or the luminal solution. Such a high level of the diuretic is required because 1) amiloride is competitive with Na$^+$; 2) the extracellular (Na$^+$) is much higher (ie, 100 mM) than the apparent K$_m$ for Na$^+$ (ie, ~ 5 mM; see Kinsella and Aronson [1980]); and 3) the apparent affinity for amiloride is low (ie, K$_I$ ≅ 15 μM; see Kinsella and Aronson, [1980]). We found that this dose of amiloride reduced the rate constant of pH$_i$ recovery by about 40% with the drug placed in either the lumen or bath, and by about 75% with the drug simultaneously in the lumen and bath. This result confirms our suspicion that a pH$_i$-regulating transporter is symmetrically distributed in these cells, and suggests that the transporter is a Na–H exchanger.

Our conclusion that pH$_i$ in these cells is regulated by a Na–H exchanger is supported by several other observations. In the first place, the recovery of pH$_i$ from an acid load is accompanied by a transient rise of intracellular Na$^+$ activity. In addition, several pieces of evidence rule out the involvement of a coupled Na/HCO$_3$/H/Cl transport system, such as is operative in squid axons, snail neurons, and barnacle muscle: In nominally HCO$_3^-$-free Ringer the pH$_i$ recovery in these tubule cells is neither blocked by SITS, nor blocked by removal of Cl$^-$, nor accompanied by a decrease of intracellular Cl$^-$ activity.

D. Basolateral HCO$_3^-$ Transport

The aforementioned experiments were carried out in nominally HCO$_3^-$-free Ringer, so as to minimize the influence of HCO$_3^-$ transport on the observed pH$_i$ changes. When we replaced this HCO$_3^-$-free Ringer (both on the luminal and basolateral sides) with a HCO$_3^-$-containing Ringer at the same pH, pH$_i$ rapidly declined by about 0.15. The sudden fall in pH$_i$ was expected, because of the influx of CO$_2$, which subsequently hydrates and dissociates into H$^+$ plus

HCO_3^-. However, given the existence of bilateral Na–H exchangers, we would have expected pH_i to recover from this acute acid load. Instead, pH_i remained at a depressed level. Thus, the application of CO_2 must produce a chronic, as well as an acute, intracellular acid load. It will be shown that this chronic acid load is imposed by the efflux of HCO_3^- (or an equivalent event) across the basolateral membrane. The sustained decrease in pH_i is just adequate to increase the Na–H exchange rate by an amount that balances the chronic, CO_2-induced acid load.

The most direct way of demonstrating basolateral HCO_3^- transport is to monitor pH_i while lowering *basolateral* (bl) (HCO_3^-). We found that when $(HCO_3^-)_{bl}$ is reduced from 10 to 2 mM at a constant CO_2 of 1.5% (ie, pH_{bl} falls from 7.5 to 6.8), pH_i rapidly falls by ~ 0.3 (see Fig. 5). Returning $(HCO_3^-)_{bl}$ to 10 mM causes pH_i to rapidly return to its initial level (time constant: ~ 2 min^{-1}). This recovery of pH_i upon restoration of basolateral HCO_3^- probably has two components: luminal and basolateral Na–H exchange, and basolateral HCO_3^- uptake. Indeed, the recovery rate can be reduced by half, but by only

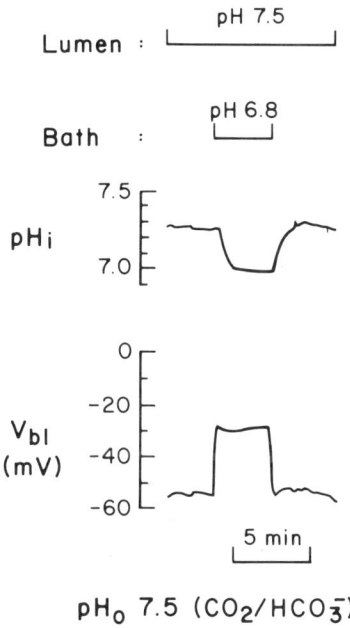

Fig. 5. Reduction of basolateral (HCO_3^-) at constant CO_2 tension. CO_2 was held at 1.5%, so that pH_{bl} fell from 7.5 to 6.8 as $(HCO_3^-)_{bl}$ was reduced from 10 mM to 2 mM. This caused a rapid, reversible fall in pH_i. Similarly lowering luminal pH had only a slight effect on pH_i.

half, if 2 mM amiloride is applied to both luminal and basolateral solutions. This reduction represents the majority of the Na–H exchange component. It is interesting to note that if *luminal* (HCO_3^-) is reduced to 2 mM, there is virtually no change in pH_i. Thus, it appears that there is a path for HCO_3^- transport at the basolateral, but not at the luminal, membranes of these cells. Furthermore, this basolateral HCO_3^- pathway is largely blocked by the stilbene derivative 4-acetamido-4′-isothiocyanostilbene-2,2′-disulfonate (SITS): Application of 0.5 mM SITS in the basolateral solution inhibits about 75% of the fall in pH_i brought on by reduction in $(HCO_3^-)_{bl}$.

What is the mechanism of this basolteral HCO_3^- transport? It is unlikely to be a simple diffusive pathway for HCO_3^-. The electrochemical gradient for HCO_3^- across the basolateral membrane is outwardly directed throughout the course of the above experiment: under control conditions, immediately after lowering $(HCO_3^-)_{bl}$, and also following the restoration of $(HCO_3^-)_{bl}$ to 10 mM. There is therfore no way that a passive movement of HCO_3^- could help account for the accelerated rise in pH_i upon restoring $(HCO_3^-)_{bl}$ to 10 mM. Another possibility is that HCO_3–Cl exchange accounts for the presumed movements of HCO_3^-. Unfortunately, the pH_i changes brought on by altering $(HCO_3^-)_{bl}$ are unaffected by removal of Cl^-, and are not accompanied by changes of intracellular Cl^- activity sufficiently large to account for the pH_i change. Also, since the Cl^- chemical gradient (about a fivefold gradient out to in) across the basolateral membrane is always larger than the HCO_3^- gradient, a simple, electroneutral HCO_3–Cl exchanger by itself could only mediate net HCO_3^- efflux, never (under the conditions of our experiments) net HCO_3^- influx.

Our search for a possible mechanism for basolateral HCO_3^- transport led us to the hypothesis that HCO_3^- transport may be linked to that of Na^+. When we removed Na^+ from the *luminal* solution (replacing Na^+ with BDA, TMA or N-methyl-D-glucammonium), there was very little effect on pH_i. When we removed Na^+ from the basolateral solution, however, pH_i rapidly declined by ~ 0.3 (see Fig. 6). When Na^+ was added back to the basolateral solution, pH_i recovered rapidly (rate constant: ~ 2 min^{-1}). Interestingly, removal of basolateral Na^+ also caused a transient depolarization of the basolateral membrane; readdition of the basolateral Na^+ produced a transient hyperpolarization, often reaching -110 mV or beyond. Similar, though smaller, transient changes in V_{bl} were observed in our experiments in which we varied $(HCO_3^-)_{bl}$. This immediately suggested that Na^+ and HCO_3^- might be crossing the basolateral membrane as a negatively charged complex, such as the ion pair $NaCO_3^-$. Thus, reducing either $(Na^+)_{bl}$ or $(HCO_3^-)_{bl}$ would lead to the basolateral efflux of what can be thought of as an anion, and thus to a depolarization. The depolarization would be short-lived, however, since transport of this "anion" would gradually slow. Returning either $(Na^+)_{bl}$ or $(HCO_3^-)_{bl}$ to its initial level would, of course, produce a transient hyperpolarization. This model makes several predictions:

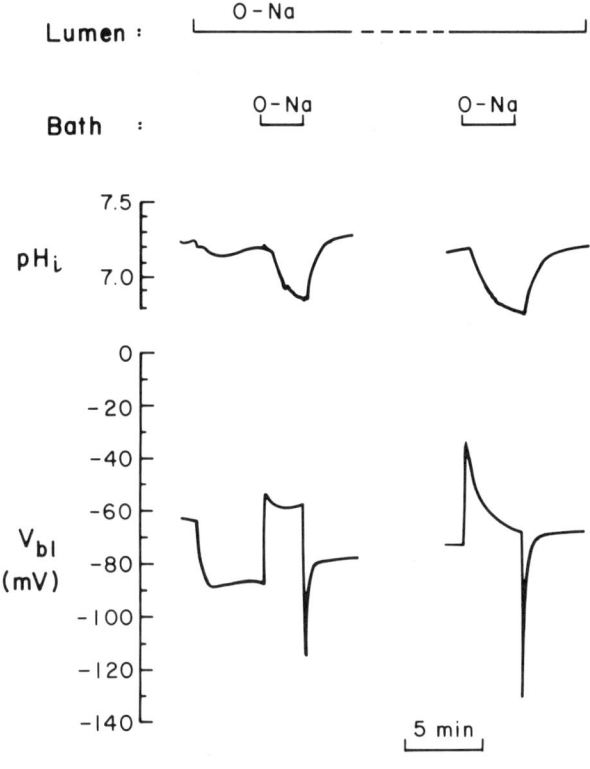

Fig. 6. Effect of removing basolateral Na$^+$ on pH$_i$ and basolateral membrane potential (V$_{bl}$). Luminal Na$^+$ was absent throughout. Also, CO$_2$ was 1.5%, (HCO$_3^-$) was 10 mM and pH was 7.5 in both the lumen and the bath, throughout. Removal of Na$^+$ (replaced by N-methyl-D-glucammonium) produced a rapid, reversible fall in pH$_i$. Note also that the removal of Na$^+$ is accompanied by a transient basolateral depolarization, whereas the readdition of Na$^+$ is accompanied by a transient basolateral hyperpolarization.

1) The effect of basolateral Na$^+$ removal on both pH$_i$ an V$_{bl}$ should be blocked by SITS. This was indeed observed. When tubules were pretreated with basolateral SITS, basolateral Na$^+$ removal had only slight effects on pH$_i$ and V$_{bl}$. When SITS was applied *after* removal of basolateral Na$^+$, subsequent re-addition of the Na$^+$ produced only a slow pH$_i$ recovery. The rate constant of this recovery, ~ 1 min^{-1}, is characteristic of Na–H exchange.

2) The effects on V$_{bl}$ of altering (HCO$_3^-$)$_{bl}$ should likewise be blocked by SITS. This was also observed.

3) The effects on pH_i and V_{bl} of altering $(Na^+)_{bl}$ should depend on the presence of HCO_3^-. We found that altering $(Na^+)_{bl}$ in the nominal absence of HCO_3^- did not produce the characteristic transient changes in V_{bl}. Also, the rate constant for pH_i recovery upon re-addition of Na^+ was only about half that occurring in the presence of HCO_3^-. That pH_i recovery remaining in the absence of HCO_3^- is probably identical to the Na–H exchange described above.

4) The effects on pH_i and V_{bl} of altering $(Na^+)_{bl}$ should be unaffected by removal of Cl^-. This prediction was also verified.

IV. DISCUSSION AND CONCLUSIONS

Our data indicate that salamander proximal tubule cells regulate their pH_i in much the same way as a symmetric cell. When the tubule cells are acutely acid-loaded, they respond by extruding acid from the cell. The mechanism of this acid extrusion is Na–H exchange, which apparently occurs at both the luminal and basolateral cell membranes. From the point of view of an epithelium, it is rather curious that these Na–H exchangers should be symmetrically distributed. Acid secretion into the tubule lumen would be most efficiently accomplished if all Na–H exchangers were located at the luminal membrane. One may speculate that these tubule cells evolved from a symmetrical cell with symmetrically distributed Na–H exchangers. When this ancestral cell eventually became part of an epithelium, pH_i regulation may have received a higher priority than transepithelial acid secretion, and therefore symmetrical Na–H exchangers were retained. However, net transport across an epithelium requires some asymmetry, and in the salamander proximal tubule this is conferred by the pathway for HCO_3^-, which is confined to the basolateral membrane. We suggest the following scheme for transepithelial acid secretion: Imagine a cell that initially has no HCO_3^- permeability, so that pH_i is rather high (eg, ~ 7.43) and the Na–H exchangers are nearly inactive. We now restore the normal HCO_3^- permeability to the basolateral membrane. HCO_3^- leaves the cell for the basolateral solution. As it does so, the intracellular equilibria $CO_2 + H_2O = H_2CO_3 = H^+ + HCO_3^-$ shift to the right, resulting in the formation of H^+ and more HCO_3^-. The fall in pH_i stimulates the Na–H exchangers. The increased rate of Na–H exchange on the basolateral membrane is, of course, of no benefit as far as acid secretion is concerned. This merely reflects a short-circuiting of a portion ($\sim 50\%$ in the case of the salamander) of the basolateral HCO_3^- efflux. That quantity of H^+ that is extruded across the luminal membrane, however, constitutes acid secretion. According to this interpretation, transepithelial acid secretion is a byproduct of pH_i regulation.

V. REFERENCES

Aickin CC, Thomas RC: An investigation of the ionic mechanism of intracellular pH regulation in mouse soleus muscle fibres. J Physiol (Lond) 273:295, 1977.

Boron WF, McCormick WC, Roos A: pH regulation in barnacle muscle fibers: Dependence on intracellular and extracellular pH. Am J Physiol 237:C185, 1979.

Burg M, Grantham J, Abramow M, Orloff J: Preparation and study of fragments of single rabbit nephrons. Am J Physiol 210:1293, 1966.

Kinsella JL, Aronson PS: Properties of the Na^+–H^+ exchanger in renal microvillus membrane vesicles. Am J Physiol 238:F461, 1980.

Murer H, Hopfer U, Kinne R: Sodium/proton antiport in brush-border-membrane vesicles isolated from rat small intestine and kidney. Biochem J 154:597, 1976.

Sackin HJ, Boulpaep EL: The isolated, perfused salamander proximal tubule: Methods, electrophysiology and transport. Am J Physiol 241:F39, 1981.

Thomas RC: Membrane current and intracellular Na changes in a snail neurone during extrusion of injected Na. J Physiol 201:495, 1969.

Thomas RC: Intracellular pH of snail neurones measured with a new pH-sensitive glass microelectrode. J Physiol (Lond) 238:159, 1974.

Thomas RC: The effect of carbon dioxide on the intracellular pH and buffering power of snail neurones. J Physiol (Lond) 255:715, 1976.

Thomas RC: "Ion-sensitive Intracellular Microelectrodes." How to Make and Use Them. London: Academic Press, 1978.

Warnock DG, Rector FC Jr: Proton secretion by the kidney. Ann Rev Physiol 41:197, 1979.

The Effect of External Ions on pH$_i$ in Sea Urchin Eggs

Sheldon S. Shen
Department of Zoology, Iowa State University, Ames, Iowa 50011

I.	Introduction	269
II.	Materials and Methods	271
	A. Experimental System	271
	B. Solutions	271
	C. Electrophysiological Measurements	272
	D. H$^+$-Selective Microelectrodes	272
	E. Problems in Measuring Intracellular pH	272
III.	Results	273
	A. E$_m$ and pH$_i$ of Unfertilized Eggs	273
	B. Effect of (H$^+$)$_o$ and (Cl$^-$)$_o$ on pH$_i$ in Unfertilized Eggs	274
	C. Effect of (Na$^+$)$_o$ on Cytoplasmic Alkalinization	276
	D. Effect of NH$_4$Cl on pH$_i$ of Unfertilized Eggs	278
IV.	Discussion	279
V.	References	281

I. INTRODUCTION

Fertilization or parthenogenetic activation of sea urchin eggs initiates a series of morphologic and metabolic changes. These have been divided into "early" and "late" events on the basis of time and by the fact that parthenogenetic activation with ammonia can initiate late responses independently of the early changes [Epel et al, 1974]. The experimental separation of the early from the late changes suggested there were at least two activating agents. The activating agent for the early events has been demonstrated to be a transient increase in internal calcium levels [Steinhardt and Epel, 1974; Chambers et al, 1974; Steinhardt et al, 1977]. The second activating agent, which is mimicked by ammonia, has been demonstrated to be an increase in intracellular pH [Johnson et al, 1976;

Lopo and Vacquier, 1977; Shen and Steinhardt, 1978]. Fertilization in natural sea water elicits a peak calcium transient 45–60 seconds after activation, lasting 2–3 minutes and cytoplasmic alkalinization begins 60–90 seconds after activation, stablizing 6–8 minutes after [Steinhardt et al, 1978]. The following observations have allowed separation of these two events; Ammonia treatment induced the late changes in the absence of early events [Steinhardt and Mazia, 1973]; divalent ionophore (A23187) treatment induced egg activation [Steinhardt and Epel, 1974; Chambers et al, 1974]; and external sodium was required for metabolic activation of the egg [Chambers, 1976].

Currently, the relationship between calcium and pH during fertilization is unknown; however, many experiments utilizing the above observations are able to separate the effects of a transient increase in internal calcium levels from an increase in cytoplasmic pH. These experiments suggest that cytoplasmic alkalinization is the pervasive change that initiates the subsequent late events. Treatment of eggs with ammonia in calcium-free artificial seawater does not cause a rise in internal calcium levels as detected by aequorin [Zucker et al, 1978]. Under this condition, they observed chromosome condensation and breakdown of the nuclear envelope. In the converse experiment, ionophore A23187 activation of eggs in sodium-free artificial seawater elicits elevated calcium levels without cytoplasmic alkalinization [Shen and Steinhardt, 1979]. Under this condition, acceleration of protein synthesis [Winkler et al, 1980] and development of new potassium conductance [Shen and Steinhardt, 1980] were not observed. Thus, the rise in internal pH mediates subsequent changes independently of the transient rise in internal calcium levels.

The idea of cytoplasmic alkalinization at fertilization of sea urchin eggs dates back many years [McClendon, 1910; Warburg, 1910]. Renewed interests were stimulated with the analysis of partial activation by ammonia [Steinhardt and Mazia, 1973]. Alkalinization of the cytoplasm at fertilization was first detected by cell homogenate studies [Johnson et al, 1976; Lopo and Vacquier, 1977]; however, numerous artifacts are possible with pH measurements of homogenates. Apart from the obvious difficulties created by the necessary time gap between homogenation and measurement of pH, it is well established that injury to cell contents releases acid and gives low pH values [Chambers and Chambers, 1961]. By utilizing the development of recessed-tip, pH-sensitive microelectrodes [Thomas 1974, 1976], the direct measurement of intact egg cytoplasmic pH was made. Measurements with microelectrodes demonstrated the postulated increase in pH_i at fertilization and at activation by weak bases [Shen and Steinhardt, 1978].

The increase in the internal pH has been shown to mediate a wide variety of events associated with the metabolic derepression of sea urchin eggs. Direct comparisons of pH_i measured with pH microelectrodes, and amino acid incorporation under the same conditions have shown that pH controls protein synthesis rate in a highly sensitive and reversible manner [Grainger et al, 1979]. Manipulations of pH_i with weak acids or bases have also shown that the subcellular location of glucose-6-PO_4-dehydrogenase can be regulated by pH_i [Aune and Epel, 1978], the development of new potassium conductance is dependent on cytoplasmic alkalization [Shen and Steinhardt, 1980], cytoplasmic pH changes may lead to alterations in protein phosphorylation [Keller et al, 1980], and the polymerization of egg cortex actin can be regulated by pH_i [Begg and Rebhum, 1979]. Since pH_i appears to be a pervasive change during fertilization that mediates a wide variety of cytoplasmic and nuclear events, it is important to consider the mechanisms regulating pH_i and its shifts during development.

Regulation of cell functions and development by pH_i may be a fundamental regulatory mechanism in biology. The importance of cytoplasmic pH has been demonstrated or at least implied in a variety of tissues, many of which are discussed within this volume. However, the mechanisms controlling pH_i are poorly understood. This has been largely for technical reasons, but the development of pH_i microelectrodes [Thomas, 1974; 1976] has stimulated progress in elucidation of the mechanisms regulating the pH_i of neurones and muscle fibers. However, these tissues are differentiated, nondividing cells with a vested interest in maintaining constant pH_i levels. The sea urchin egg offers a preparation for studying cytoplasmic pH regulation in an embryonic system, where the role of pH_i in regulating cellular activities during development has been documented.

II. MATERIALS AND METHODS

A. Experimental System

An important consideration in the selection of Lytechinus pictus was the availability of gametes year round. Gravid L. pictus can be constantly maintained with minimal difficulties in Instant Ocean tanks with Ocean 50 Seamix (Jungle Laboratories Corp) at 15°C and biweekly feedings of kelp. The shedding of gametes was induced by intracoelomic injection of 0.5 M KCl. The sperm was collected dry and stored at 4°C. The eggs were passed through silk mesh to remove the jelly, washed several times in Millipore-filtered artificial seawater, and stored at 16°C with constant gentle stirring.

B. Solutions

Artificial seawater (ASW) had the following composition in mM: 460, NaCl; 10, KCl; 11, $CaCl_2$; 29, $MgSO_4$; 27, $MgCl_2$; 2.5, $NaHCO_3$; pH 8. Ion replacement of Na^+ was with choline and $KHCO_3$ for $NaHCO_3$, replacement of K^+ or Ca^{2+} was with Na^+, and replacement of Cl^- was with isethionate and acetate. Ammonium chloride and 2,4-dinitrophenol (DNP, Sigma) were dissolved in the ASW, and the pH was readjusted to 8 just prior to use.

C. Electrophysiological Measurements

The sea urchin eggs were immobilized with polylysine-coated Petri dishes [Mazia et al, 1975]. The experiments were performed at 16–18°C with a refrigerated microscope stage. H^+-selective microelectrodes were made as described later. Membrane potential microelectrodes were pulled from 1.2 mm Omegadot tubing and when filled with 3 M KCl had resistances of 60–80 MΩ. Impalement of eggs with conventional microelectrodes was made by bringing the electrode against the egg, followed by briefly overtuning the negative capacitance compensation. This method of impalement unfortunately does not work for H^+-selective microelectrodes, whose entry into an egg required a sharp mechanical jolting of the manipulator. In order to determine that both microelectrodes were implanted in the egg with minimal damage, current pulses were passed through the KCl electrode periodically, and the corresponding voltage deflections were monitored by the pH microelectrode. Excessive membrane damage or vesiculation at the electrode tips resulted in a loss of electrical coupling. Membrane potentials were rocorded either with a Getting (model 5) or Biodyne (model AM-4) preamplifier. H^+-selective microelectrode potentials were recorded with a WP Instruments (model F-223) electrometer. The recorded potentials were fed into a Textronix 5111 oscilloscope and permanent records were made with either a Gould Brush 220 or Soltec 3314 chart recorder.

D. H^+-Selective Microelectrodes

Intracellular pH was measured with pH-sensitive microelectrodes with recessed tips [Thomas, 1974; 1976]. These electrodes were constructed with techniques very similar to those described by Thomas [1978], except the Corning 1720 aluminosilicate tubing was pulled with a single-stage pull. Since good visual observation was necessary for the success of constructing these electrodes, manipulation, sealing, and fusion were carried out on a microforge with a compound microscope fitted with a Leitz 32X long-working distance objective. The pH-sensitive electrodes were filled by heating in distilled water at 80–85°C for 2–3 hours. The distilled water was then displaced by backfilling with 0.1 M NaCl, 0.1 M citrate buffer, pH 6 solution. The filled electrodes were stored with their tips immersed in dichromate solution. The pH microelectrodes used gave a linear response of slope 56–59 mV per pH unit in 15 seconds or less when calibrated before and after each experiment. All pH_i measurements were made with unused electrodes and repeated at least twice with similar results. Experiments where either electrode had drifted more than ± 2 mV from the original baseline were discarded.

E. Problems in Measuring Intracellular pH

Certain special difficulties exist with measuring the intracellular pH of unfertilized sea urchin eggs, the foremost being the high input resistance of the sea urchin egg membrane [Jaffe and Robinson, 1978; Chambers and de Armendi, 1979; Shen and Steinhardt, 1980]. Recordings of the membrane potentials and resistances of unfertilized L. pictus eggs with a single KCl microelectrode have yielded values of -75 mV and 300 kΩ-cm^2 respectively (Shen and Steinhardt, unpublished observations). These eggs were capable of regenerative action potentials in response to outward current pulses. However, the additional impalement of the egg with a pH-sensitive microelectrode caused decreases of the membrane potential and resistance values to -12 mV and 40 kΩ-cm^2 respectively. These eggs have a nonregenerative response to applied current pulses. The lower resting potential and resistance may be due to a leakage artifact caused by the additional electrode impalement. In order to minimize both the leakage artifact and electrode activation of the egg, the outer 1,720 aluminosilicate electrodes were pulled with a resistance of 40–60 MΩ, when filled with 3 M KCl.

A frequent problem encountered with the pH-sensitive microelectrode was vesiculation at the electrode tip shortly after penetration. This could be visually observed, and resulted in a loss of electrical coupling with the conventional microelectrode. The frequency of vesiculation could be reduced by decreasing external calcium from 11 mM to 5.5 mM. Additional decreases of external calcium hindered membrane recovery from penetration. After a stable pH$_i$ value was attained, switching the bath to ASW had no effect. Another difficulty encountered with measuring pH$_i$ was that a pH-sensitive microelectrode could not be used repeatedly. For reasons that are unclear, an electrode previously used, caused spontaneous activation when utilized with other eggs. Finally, partial egg activation occasionally occurred. This could be detected visually by a localized elevation of the fertilization membrane and was also characterized by a slowly alkalizing cytoplasm. When partial egg activation occurred, the experiments were terminated.

III. RESULTS

A. E_m and pH$_i$ of Unfertilized Eggs

The measured pH$_i$ of unfertilized Lytechinus pictus eggs was 6.86 $\pm$ 0.01, and the recorded membrane potential (E_m) was -11.3 ± 0.5 mV (n = 65, mean $\pm$ SD). If protons were in passive equilibrium across the plasma membrane, the expected internal pH at this mean membrane potential and external pH (pH$_o$) of 8, would be 7.8. The measured value of 6.86 indicated an active transport of protons or its equivalent into the egg. However, due to the high input resistance of unfertilized sea urchin eggs [Jaffe and Robinson, 1978; Chambers and de Armendi, 1979; Shen and Steinhardt, 1980] and the additional impalement of a pH-sensitive microelectrode, the low resting potentials of these eggs may reflect a leakage artifact. If the pH$_i$ and E_m of unpenetrated eggs are near 6.9 and -65 mV respectively, protons would then be near passive equilibrium. Alternatively, the reported values of cytoplasmic pH of unfertilized eggs may be a consequence of impalement and may not reflect the true values.

Several lines of evidence suggest that the measured pH_i is close to the pH_i of intact eggs, despite the small resting potentials. Estimation of pH_i in unfertilized L. pictus eggs by the distribution of 5,5-dimethyl-2,4-oxazolidine-dione (DMO) was 6.86 [Johnson and Epel, 1981] and by ^{31}P-NMR was 7.02 (Winkler et al, personal communication). Preliminary experiments also indicate that the pH_i of unfertilized eggs, similar to neurons and muscle fibers, is insensitive to E_m. Clamping the E_m of unfertilized eggs more negative than -70 mV or more positive than $+15$ mV in excess of 15 minutes had little or no effect on pH_i (Shen and Steinhardt, unpublished observations). Furthermore, if pH_i is dependent upon E_m, I would expect to see a rapid rise in pH_i after penetration of the egg by the pH microelectrode. Such large alteration in pH_i has not been observed. Instead a slight alkalization, probably due to membrane "healing" [Aickin and Thomas, 1975], was observed. In 21 experiments, a comparison of pH_i was made immediately after entry of both electrodes and after stabilization of cytoplasmic pH. The pH_i increased slightly from 6.81 ± 0.02 to 6.89 ± 0.01 in 21.2 ± 1.9 minutes. Another possibility was that the pH electrode does not measure the same E_m as the KCl electrode. This possibility can be excluded since the pH electrode has a faster response time to changes of potential than pH levels. An estimation of the E_m measured by the pH electrode may be obtained from the near instantaneous change of the voltage recorded upon entry or withdrawn from the egg. The estimated E_m measured by the pH electrode was within ± 3 mV of the E_m measured by the KCl electrode in all cases. Although the unfertilized eggs with small resting potentials are probably damaged from the electrode penetrations, these eggs have been observed to fertilize and develop normally.

B. Effect of $(H^+)_o$ and $(Cl^-)_o$ on pH_i in Unfertilized Eggs

Previous investigations have examined the metabolic energy and external ion dependencies for maintaining the pH_i in unfertilized eggs. Both 2,4-dinitrophenol (DNP) at 0.5 mM and antimycin A at 2 μM had little effect on pH_i [Shen and Steinhardt, 1980]. Replacing sodium with choline and elevating external potassium to 100 mM did not alter pH_i significantly [Shen and Steinhardt, 1979] (Shen and Steinhardt, unpublished observations). However, the pH_i in unfertilized eggs was sensitive to changes of external proton or chloride concentrations.

If hydrogen ions are permeable across the plasma membrane of the egg, alteration of pH_o should affect pH_i. The initial experiments, as illustrated in Figure 1, examined the effects on pH_i in unfertilized eggs by large but brief acidic exposures. Changing the pH_o of ASW from 8 to 5.7 for 5 minutes caused a rapid fall in pH_i of 0.21 unit. Upon returning to ASW, pH_o 7.8, pH_i rose and overshot its original value to a new stable value of 7.1. The egg was now exposed to a second acidic perfusion of ASW, pH$_o$ 4.5 for 4 minutes, which caused another rapid drop in pH$_i$ of 0.26 unit. When the bath was replaced with ASW, pH$_o$ 8, the pH$_i$ rose and restabilized at 7.1. The second acidic pulse did not cause a further overshoot of pH$_i$. The egg was fertilized and the previously described alkalinization accompanying fertilization occurred, with the newly fertilized egg having a stable pH$_i$ of 7.24. During the time course of the recovery in pH$_i$ after the acidic pulses, the membrane hyperpolarized.

Figure 2 illustrates the effect of longer exposures to acidic or basic pH$_o$ on pH$_i$ in an unfertilized egg. Changing pH$_o$ from 8 to 6.5 for 15 minutes caused

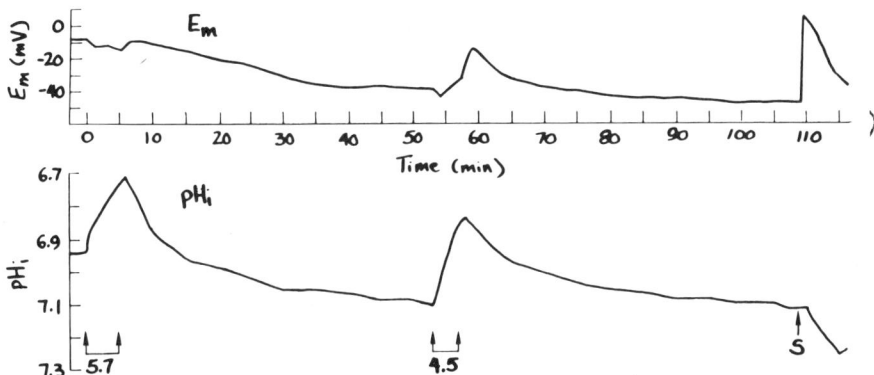

Fig. 1. An experiment to show the effect of brief pulses of acidic ASW on E$_m$ and pH$_i$ in unfertilized eggs. The initial acidic pulse caused pH$_i$ to overshoot its initial value and membrane hyperpolarization. Fertilization of the egg (denoted in all figures by an arrow with an "S") caused an alkalization of the cytoplasm, which was less than that accompanying fertilization of untreated eggs. However, the sum of pH$_i$ overshoot and increase with fertilization was equal to the magnitude normally expected for fertilization of untreated eggs.

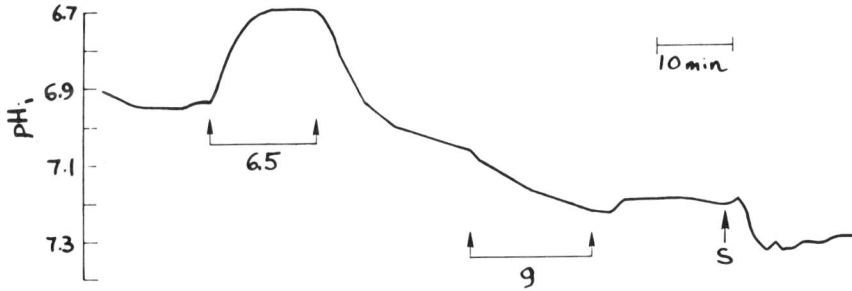

Fig. 2. An experiment to show the effect of longer exposures of altered pH$_o$ and pH$_i$ in unfertilized eggs.

a rapid fall in pH_i from 6.94 to 6.7, which stabilized after 6 minutes. No recovery in pH_i was observed during the succeeding 9 minutes. Upon returning to ASW, pH_o 8, pH_i rose and overshot its original value to a new stable value of 7.05. When pH_o was elevated to 9, pH_i increased. After a 15-minute exposure, the bath pH was returned to 8 and pH_i stabilized at 7.22. Upon fertilization, pH_i rose and stabilized at 7.3. It is worth noting that the increment of alkalinization accompanying fertilization of these eggs was not the usual magnitude [Shen and Steinhardt, 1978]. Instead, the sum of the alkalinization by altering pH_o and by fertilization was equal to the normal magnitude of cytoplasmic alkalinization with fertilization.

In addition to changes of pH_o, the pH_i in unfertilized eggs was sensitive to changes of external chloride concentration $((Cl^-)_o)$. In four experiments, where unfertilized eggs were exposed to Cl-free ASW, the pH_i increased to a mean of 7.23 ± 0.05. As seen in Figure 3, exposure of this unfertilized egg to Cl-free ASW caused an alkalinization of the cytoplasm from 6.92 to 7.28 over a 42-minute time course. Fertilization in Cl-free ASW resulted in an additional increase in pH_i, which was equal to the normal magnitude of cytoplasmic alkalinization with fertilization. In the example seen in Figure 4, pH_i rose from 7.21 to 7.63. Perfusion of the bath with ASW caused a fall in pH_i of the now fertilized egg, which suggested that the return to a normal chloride ion gradient drove protons into the egg.

C. Effect of $(Na^+)_o$ on Cytoplasmic Alkalinization

Chambers [1976] reported that small amounts of external sodium were required from 30 seconds to 10 minutes following fertilization for metabolic ac-

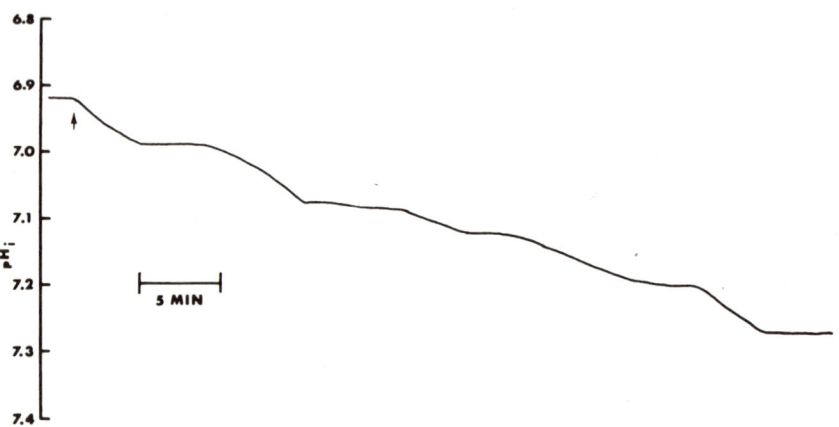

Fig. 3. An experiment to show the effect of Cl-free condition on pH_i in unfertilized eggs. At arrow, the bath was perfused with Cl-free ASW, pH_o 8.

tivation. This external sodium requirement has been demonstrated to be a prerequisite for cytoplasmic alkalization at fertilization [Johnson et al, 1976; Shen and Steinhardt, 1979]. However, the precise mechanism by which millimolar amounts of sodium permits this alkalization remains unclear. In order to elucidate the role of external sodium, it is of interest to know if the sodium requirement is a once in a lifetime event.

In a previous report [Shen and Steinhardt, 1980], exposure of fertilized eggs to DNP caused pH_i to fall to unfertilized levels. Upon the removal of DNP, the cytoplasm realkalinized. If the sodium requirement is a one-time event, this realkalinization should be sodium-independent. As seen in Figure 5, eggs were fertilized in 25 mM Na-ASW and the alkalinization at fertilization occurred. The eggs were then exposed to 0.5 mM DNP, which caused a brief fall and rise in pH_i, followed by a slow fall to unfertilized values. The bath was now perfused with Na-free ASW containing DNP, which had no significant effect on pH_i. Subsequently the DNP was washed out by perfusing with Na-free ASW, and no significant change in pH_i occurred. However, upon the addition of $(Na^+)_o$ to 80 mM, pH_i rose 0.17 unit. Thus, even two hours after insemination, the realkalinization of the cytoplasm required external sodium. These eggs were observed to develop normally to the blastula stage.

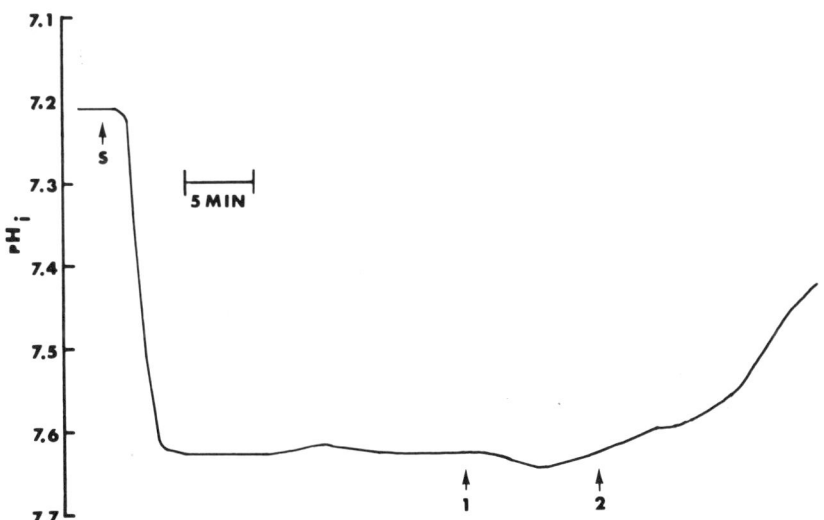

Fig. 4. Intracellular pH recorded during fertilization of an egg in Cl-free ASW, pH_o 8. At arrow 1, the bath was perfused with ASW containing 260 mM Cl$^-$, and at arrow 2 the perfusion was ASW.

D. Effect of NH₄Cl on pH$_i$ of Unfertilized Eggs

After exposure of unfertilized eggs to an acid load, pH$_i$ was observed to overshoot the original pH value of the egg (Figs. 1 and 2) [Grainger et al, 1979]. In contrast, after exposure to ammonia, pH$_i$ has been observed to undershoot the original pH$_i$ (Fig. 6A) [Shen and Steinhardt, 1978]. After exposure to ammonia, pH$_i$ stabilized at the pH$_i$ of the fertilized state, since fertilization of these

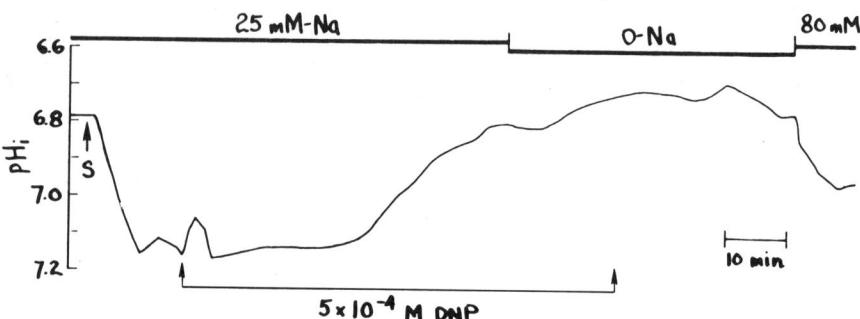

Fig. 5. An experiment to show that realkalinization of fertilized eggs required external Na$^+$. See text for further details.

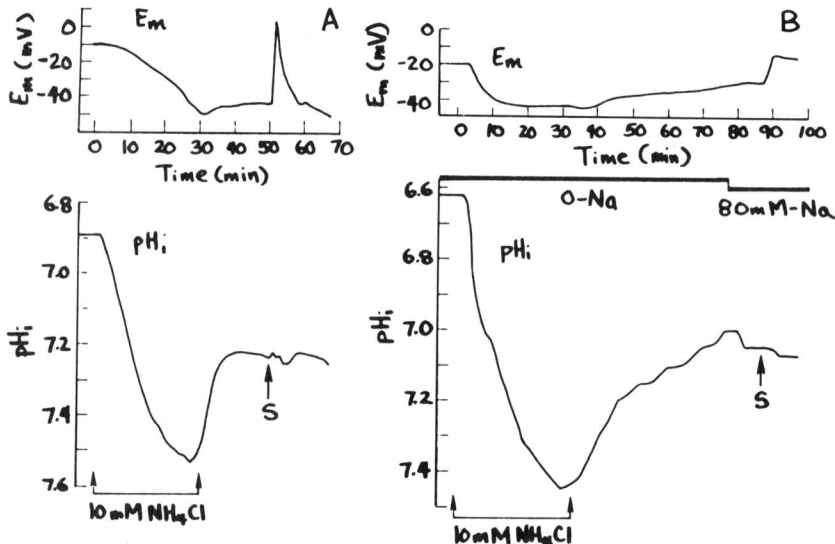

Fig. 6. The effect of ammonia on pH$_i$ in unfertilized eggs. A) A typical recording of E$_m$ and pH$_i$ in unfertilized eggs exposed to a pulse of 10 mM NH$_4$Cl. Upon the removal of ammonia, pH$_i$ fell and stabilized at the fertilized level, since fertilization of the egg did not cause significant changes of pH$_i$. B) The effect of ammonia on unfertilized eggs did not require external Na$^+$.

eggs did not cause further alkalinization (Fig. 6a). This observation suggested that the exposure to ammonia had activated mechanisms for regulating pH_i, which are normally associated with the fertilized egg. Since cytoplasmic alkalinization with fertilization requires sodium, the possibility that the effect of ammonia was also dependent on $(Na^+)_o$ was examined. As seen in Figure 6B, the pH_i of an unfertilized egg pulsed with 10 mM NH_4Cl in Na-free ASW also stablized at the fertilized level, which indicated that the effects of ammonia on the unfertilized egg does not have a sodium requirement.

Exposure of fertilized eggs to metabolic inhibitors caused a fall in pH_i to unfertilized levels [Shen and Steinhardt, 1980]. Similarly, exposure of unfertilized, but ammonia-pulsed, eggs to metabolic inhibitors caused pH_i to fall. In Figure 7A, exposure of an unfertilized egg in ASW with 0.5 mM DNP to 10 mM NH_4Cl elicited the expected rise in pH_i. However, upon the removal of the ammonia, the pH_i fell to its original baseline. Similarly, when 0.5 mM DNP was added to unfertilized eggs previously pulsed with 10 mM NH_4Cl, the cytoplasm acidified (Fig. 7B). Thus, the alkaline cytoplasm of unfertilized eggs previously exposed to ammonia was maintained by an energy-dependent, acid-extruding pump.

IV. DISCUSSION

Although electrode penetration of unfertilized L. pictus eggs caused membrane depolarization, there is agreement on the internal pH value of 6.9 determined by microelectrode, DMO and ^{31}P-NMR techniques. If the membrane potential of intact eggs is near -65 mV, then at pH_o 8, protons would be near electrochemical equilibrium. Interestingly, despite membrane depolarization with electrode impalement, the unfertilized egg maintained a pH_i of 6.9 for at least several

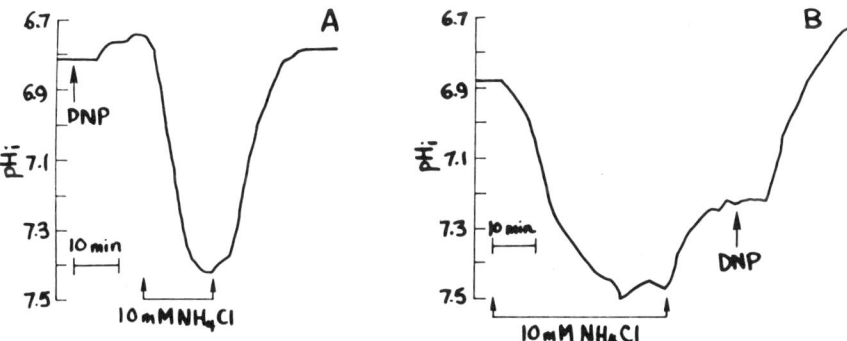

Fig. 7. The effect of DNP on pH_i in unfertilized eggs exposed to ammonia. A) In the presence of 0.5 mM DNP, the addition of 10 mM NH_4Cl caused pH_i to increase; however, when the ammonia was washed out, pH_i fell to the original unfertilized level. B) The addition of 0.5 mM DNP caused the elevated pH_i in an unfertilized, but ammonia-treated, egg to fall.

hours. Thus, for the mean recorded pH_i and E_m values of 6.86 and -11.3 mV, the unfertilized egg would have to exert an inward driving force on protons in excess of 50 mV in order to maintain its pH_i value. It is conceivable that the acidic pH_i could be maintained, despite membrane depolarization, if protons are impermeant. However, as seen in Figures 1 and 2, changes of pH_o caused rapid changes of pH_i. In addition, the pH_i of unfertilized eggs was insensitive to metabolic inhibitors [Shen and Steinhardt, 1980]. Thus, it appeared that maintenance of an acidic cytoplasm was dependent on an inward proton shuttle. Although changes of external sodium or potassium had little effect on pH_i, decreases of external chloride caused a rise in pH_i (Fig. 3). This effect suggested that the mechanism of pH_i regulation in unfertilized eggs is chloride-dependent. The exact nature remains to be elucidated.

After exposure to acidic pH_o or weak acid, the pH_i was observed to overshoot the original pH_i in the unfertilized egg. In comparison, after exposure to weak base, the pH_i was observed to undershoot the original pH_i. Perturbation of pH_i in unfertilized eggs apparently caused a shift toward the alkaline pH_i in fertilized eggs, which suggested activation of the metabolic energy-dependent, pH regulatory mechanism of fertilized eggs. This conclusion was supported by the observation that the increment of cytoplasmic alkalinization accompanying fertilization of eggs, which had been previously exposed to changes of pH_o, weak acid, or weak base, was not of the usual magnitude. Instead, the sum of the alkalinization caused by alteration of the external environment and by fertilization was equal to the normal magnitude accompanying fertilization. Furthermore maintenance of the alkaline pH_i in ammonia-pulsed eggs required metabolic energy, since the additional presence of DNP caused pH_i to fall to the unfertilized level (Fig. 7). But unlike cytoplasmic alkalinization associated with fertilization, the effect of ammonia treatment on pH_i regulation appeared to bypass the sodium dependent step (Fig. 6B).

Exposure of unfertilized sea urchin eggs to ammonia and other weak bases results in the activation of some of the events that normally follow fertilization, such as the acceleration of protein synthesis and the initiation of DNA synthesis [Epel et al, 1974; Vacquier and Brandriff, 1975; Grainger et al, 1979]. On the assumption that only the uncharged partner of the conjugate pair is permeant, Jacobs [1940] derived an equation for the steady-state transmembrane distribution of weak electrolytes. Depending upon parameters such as concentration, pH_o, pH_i, and intracellular buffering capacity, the addition of weak electrolytes should elicit changes in pH_i. This has been demonstrated for neurones and muscle fibers [Thomas, 1974; Aickin and Thomas, 1975, 1977; Boron and DeWeer, 1976; Ellis and Thomas, 1976; Hinke and Menard, 1976; Bolton and Vaughan-Jones, 1977; Boron, 1977]. However, from previous studies and the data presented here, the effects of ammonia on pH_i in unfertilized eggs have three distinct differences with those reported for nervous tissues. First, the pH_i change elicited by the addition of the weak base required a much longer time course than its equilibrium across the egg membrane [Winkler and Grainger, 1978]. Second, given enough time, different concentrations of ammonia caused similar increases in pH_i [Shen and Steinhardt, 1978]. Finally, when the ammonia was removed from the bath, there was only partial recovery of lower pH_i values, as discussed earlier. These different responses of the sea urchin egg to ammonia treatment may reflect changing regulation of internal pH associated with developmental activation in the egg.

ACKNOWLEDGMENTS

I would like to thank Dr. R.A. Steinhardt for his advice and the use of his facilities for some of the experiments. This investigation was supported by NSF grant PCM 80-03732.

V. REFERENCES

Aickin CC, Thomas RC: Micro-electrode measurement of the internal pH of crab muscle fibres. J Physiol (Lond) 252:803–815, 1975.

Aickin CC, Thomas RC: Micro-electrode measurement of the intracellular pH and buffering power of mouse soleus muscle fibres. J Physiol (Lond) 267:791–810, 1977.

Aune TM, Epel D: Increased intracellular pH shifts the subcellular location of G6PDH. J Cell Biol 79:164a, 1978.

Begg DA, Rebhun LL: PH regulates the polymerization of actin in the sea urchin egg cortex. J Cell Biol 83:241–248, 1979.

Bolton TB, Vaughan-Jones RD: Continuous direct measurement of intracellular chloride and pH in frog skeletal muscle. J Physiol (Lond) 270:810–833, 1977.

Boron WF: Intracellualr pH transients in giant barnacle muscle fibers. Am J Physiol 233:C61–C73, 1977.

Boron WF, DeWeer P: Intracellular pH transients in squid giant axons caused by CO_2, NH_3 and metabolic inhibitors. J Gen Physiol 67:91–112, 1976.

Chambers EL: Na is essential for activation of the inseminated sea urchin egg. J Exp Zool 197:149–154, 1976.

Chambers EL, deArmendi J: Membrane potential, action potential and activation potential of eggs of the sea urchin, Lytechinus variegatus. Exp Cell Res 122:203–218, 1979.

Chambers EL, Pressman BC, Rose B: The activation of sea urchin eggs by the divalent ionophore A23187 and X-537A. Biochem Biophys Res Commun 60:126–132, 1974.

Chambers R, Chambers EL: "Exploration into the Nature of the Living Cell." Cambridge: Harvard University Press, 1961, pp 141–174.

Ellis D, Thomas RC: Direct measurement of the intracellular pH of mammalian cardiac muscle. J Physiol (Lond) 262:755–771, 1976.

Epel D, Steinhardt R, Humphreys T, Mazia D: An analysis of the partial metabolic derepression of sea urchin eggs by ammonia: The existence of independent pathways. Dev Biol 40:245–255, 1974.

Grainger JL, Winkler MM, Shen SS, Steinhardt RA: Intracellular pH controls protein synthesis rate in the sea urchin egg and early embryo. Dev Biol 68:396–406, 1979.

Hinke JAM, Menard MR: Intracellular pH of single crustacean muscle fibres by the DMO and electrode methods during acid and alkaline conditions. J Physiol (Lond) 262:533–552, 1976.

Jacobs MH: Some aspects of cell permeability to weak electrolytes. Cold Spring Harbor Symp Quant

Biol 8:30–39, 1940.

Jaffe LA, Robinson KR: Membrane potential of the unfertilized sea urchin egg. Dev Biol 62:215–228, 1978.

Johnson CH, Epel D: Intracellular pH of sea urchin eggs measured by dimethyloxazolidinedione (DMO) method. J Cell Biol 89:284–291, 1981.

Johnson JD, Epel D, Paul M: Intracellular pH and activation of sea urchin eggs after fertilisation. Nature (Lond) 262:661–664, 1976.

Keller C, Gundersen G, Shapiro BM: Altered in vitro phosphorylation of specific proteins accompanies fertilization of Strongylocentrotus purpuratus eggs. Dev Biol 74:86–100, 1980.

Lopo A, Vacquier VD: The rise and fall of intracellular pH of sea urchin eggs after fertilisation. Nature (Lond) 269:590–592, 1977.

Mazia D, Schatten G, Sale W: Adhesion of cells to surface coated with polylysine. J Cell Biol 66:198–200, 1975.

McClendon JF: How could increase in permeability to electrolytes allow the development of the egg? Proc Soc Exp Biol Med 8:1–3, 1910.

Shen SS, Steinhardt RA: Direct measurement of intracellular pH during metabolic derepression of the sea urchin egg. Nature (Lond) 272:253–254, 1978.

Shen SS, Steinhardt RA: Intracellular pH and the sodium requirement at fertilisation. Nature (Lond) 282:590–592, 1979.

Shen SS, Steinhardt RA: Intracellular pH controls the development of new potassium conductance after fertilization of the sea urchin egg. Exp Cell Res 125:55–61, 1980.

Steinhardt RA, Epel D: Activation of sea-urchin eggs by a calcium ionophore. Proc Natl Acad Sci USA 71:1915–1919, 1974.

Steinhardt RA, Mazia D: Development of K^+-conductance and membrane potentials in unfertilised sea urchin eggs after exposure to NH_4OH. Nature (Lond) 241:400–401, 1973.

Steinhardt R, Zucker R, Schatten G: Intracellular calcium release at fertilization in the sea urchin egg. Dev. Biol 58:185–196, 1977.

Steinhardt RA, Shen SS, Zucker RS: Direct evidence for ionic messengers in the two phases of metabolic derepression at fertilization in the sea urchin egg. ICN–UCLA Symp Mol Cell Biol 12:415–424, 1978.

Thomas RC: Intracellular pH of snail neurones measured with a new pH-sensitive glass microelectrode. J Physiol (Lond) 238:159–180, 1974.

Thomas RC: Construction and properties of recessed-tip micro-electrodes for sodium and chloride ions and pH. In Kessler M, Clark LC, Lubbers DW, Silver IA, Simon W (eds): "Ion and Enzyme Electrodes in Biology and Medicine." Munich: Urban and Schwarzenberg, 1976, pp 141–148.

Thomas RC: "Ion-Sensitive Intracellular Microelectrodes." London: Academic Press, 1978, pp 24–40.

Vacquier VD, Brandriff B: DNA synthesis in unfertilized sea urchin eggs can be turned on and turned off by the addition and removal of procaine hydrochloride. Dev Biol 47:12–31, 1975.

Warburg O: Uber die oxydationen in lebenden zellen nach versuchen am seeigelei. Zeitschr Physiol Chem 66:305–340, 1910.

Winkler MM, Grainger JL: Mechanism of action of NH_4Cl and other weak bases in the activation of sea urchin egg. Nature (Lond) 273:536–538, 1978.

Winkler MM, Steinhardt RA, Grainger JL, Minning M: Dual ionic controls for the activation of protein synthesis at fertilization. Nature (Lond) 287:558–560, 1980.

Zucker RS, Steinhardt RA, Winkler MM: Intracellular calcium release and the mechanism of parthenogenetic activation of the sea urchin egg. Dev Biol 65:285–295, 1978.

Intracellular pH Regulation: A Summary of the Proposed Mechanisms and Meeting Discussion

William J. Moody, Jr.

Jerry Lewis Neuromuscular Research Center, University of California, Los Angeles, California 90024

I.	A Survey of pH_i-Regulating Mechanisms	283
II.	Summary of General Discussion of pH_i Regulation	285
III.	Cells in Which pH_i-Regulating Mechanisms Serve Additional Functions	286
IV.	Summary of Discussion on Renal Tubule Cells and Oocytes	287

A great deal of information has been obtained in the last several years about the mechanisms by which cells regulate pH_i. This information has been obtained from a variety of animals from arthropods to mammals, and in a wide range of tissues. Up to the present, investigators have relied almost exclusively on pH-sensitive microelectrodes to measure pH in these studies.

I. A SURVEY OF pH_i-REGULATING MECHANISMS

Virtually all cells maintain pH_i at a level somewhat alkaline to that predicted by an equilibrium distribution of H^+. The "average" cell might have a membrane potential of -60 mV, an extracellular pH of 7.5, and $pH_i = 7.25$. Were H^+ in electrochemical equilibrium across the membrane, pH_i would be 6.5. Thus to maintain pH_i the cell must extrude H^+ against a 45 mV driving force. The rate of H^+ extrusion, and thus the energy consumed by the cell per unit time in maintaining pH_i, depends on both the passive influx of H^+ (which as Dr. Deamer pointed out in his presentation, may be quite high) and the metabolic production of H^+ within the cell.

The pH_i-regulating mechanisms of virtually all cells studied to date share common features (an extensive review of pH_i regulation has recently been published: Roos and Boron, Physiol Rev 61:296, 1981): 1) They require the presence of Na^+ in the external medium, and transport Na^+ into the cell in exchange for the H^+ extruded; and 2) they involve the transport of HCO_3^- (almost always coupled with a Cl^- antiport) across the membrane. The major difference among cells discussed by the participants at this meeting was that of cells with a single

pH_i-regulating mechanism involving Na^+-dependent anion exchange *vs* cells that have two separate systems, one of which is independent of HCO_3^- transport but still requires Na^+ influx.

Three cells discussed with clear examples of single pH_i-regulating systems were the snail neuron, the squid giant axon, and the giant barnacle muscle fiber. In each of these, H^+ extrusion is completely blocked by either the anion exchange inhibitor SITS, or Na^+-free external solutions. Each system transports Na^+ into the cell and Cl^- out. In squid axons, the stoichiometry of the ion movements has been measured in tracer experiments as 1 Na^+ (in): 1 Cl^- (out): 2 H^+ (equivalents neutralized intracellularly). In each cell the exchange is electroneutral.

In contrast to these cells, other preparations appear to use two separate mechanisms to regulate pH_i. During the discussion of pH_i regulation, Dr. Moody presented an example of data from such a cell (see Moody, J Physiol 316:293, 1981). In experiments carried out on crayfish neurons, he showed that unlike in the above cells, SITS does not completely block H^+ extrusion. To insure that SITS was effectively reaching its site of action, he carried out the following experiment. H^+ was iontophoretically injected into crayfish neurons while pH_i was recorded. The injection was done with the cell exposed to HCO_3^--free Ringer. When pH_i recovery was about half complete, the HCO_3^- concentration of the external solution was increased to 25 mM. This caused a substantial increase in the rate of pH_i recovery, demonstrating the involvement of HCO_3^- transport. The cell was then returned to HCO_3^--free Ringer and the experiment repeated, with the cell exposed continuously to 100 μM SITS. In the presence of SITS, the rate of pH_i recovery following H^+ injection in HCO_3^--free Ringer was unaffected. However, in SITS, increasing the external HCO_3^- concentration to 25 mM did not increase the rate of pH_i recovery. This demonstrates that crayfish neurons have two pH_i regulating systems, only one of which is HCO_3^- dependent and SITS-sensitive. Both appear to require external sodium. Dr. Moody proposed that the SITS-sensitive component is identical to the Na^+-dependent anion exchange, which is the only pH_i-regulating mechanism in snail neurons, squid axons, and barnacle fibers. The additional system in crayfish neurons is most likely a pure Na^+-H^+ exchange. Other cells also appear to have two systems: Aickin and Thomas (J Physiol 273:295, 1977) obtained similar data with mouse soleus muscle fibers, although their interpretation was slightly different (see below).

Mammalian cardiac muscle also regulates pH_i using two separate systems, but, as Dr. Vaughan-Jones showed, with one important difference. In cardiac muscle, the SITS-insensitive Na^+-H^+ exchange serves to extrude H^+ in response to an acid load. The SITS-sensitive anion exchange, however, brings H^+ into the cell (by HCO_3^- extrusion) in response to *alkaline* loads. This is the only case in which an active recovery from intracellular alkalinization has been demonstrated. It is not yet clear why this type of response should be important in cardiac muscle. It remains possible that anion exchange in heart is primarily geared to maintain a high intracellular Cl⁻ level instead of performing the more familiar role of controlling pH_i.

II. SUMMARY OF GENERAL DISCUSSION OF pH_i REGULATION

Some attempt was made by the participants to find a rationale for the distribution of single vs dual mechanisms of pH_i regulation among cells. Snail neurons and squid axons are both molluscan nerve structures so it is not surprising that they should both have single systems. It is less clear, however, why a crustacean neuron should differ in this regard from crustacean muscle and molluscan neurons, and be more similar to mammalian muscle. Dr. Moody noted that cells with high intracellular Cl⁻ levels might tend to have a separate H^+ extrusion system not utilizing Cl⁻ efflux, so as not to operate against the need for active Cl⁻ influx. He provided a counterexample to his own suggestion, however, by noting that the squid giant axon maintains high internal Cl⁻ levels and has only one H^+ extrusion mechanism dependent on Cl⁻ efflux. No other possible reason for differences among cells was discussed.

Boron, Roos, Moody, and R.C. Thomas discussed the differences between the data of Aickin and Thomas in mouse muscle, and Moody in crayfish neurons. Moody had proposed that both pH_i regulating systems in crayfish neurons were Na^+-dependent, based on the fact that removal of Na^+ from the external solution appeared to block all pH_i regulation. In mouse muscle, Aickin and Thomas had proposed that the two systems were Na^+–H^+ exchange and Cl^--HCO_3^- exchange with no Na^+ dependence, as if the four-ion carrier hypothesized by Thomas for snail neurons had simply been split into two separate entities in mouse muscle. Thomas pointed out, however, that in removing Na^+ from the external Ringer in the mouse experiments, it was necessary to use Li^+, which is known to substitute partially for Na^+ in H^+ extrusion, because all of the inert cations caused the muscle to go into contracture. Therefore, they had not really addressed this particular point in their experiments, and had not intended to rule it out in making their proposal of pH_i regulation. It was concluded that the system in mouse muscle could in fact be identical to that in crayfish neurons.

Several other points about pH_i regulation in general were discussed: Dr. Deamer pointed out that if H^+ ions are actually the species transported by these pumps, then the affinity of the site for H^+ must be very high compared to the affinity of the Na^+ pump for Na^+, for example, since $[H^+]_i$ is only a few tenths micromolar. Dr. R.C. Thomas agreed, but noted that the rate of H^+ transport is comparable to the rate of Na^+ transport, since to change pH_i by 1 unit requires the extrusion of roughly 25 mM H^+ when the intracellular buffering power is considered.

The question of the energy source for H^+ extrusion was then raised. In the squid axon, Drs. Russell and Boron had shown that internal perfusion with ATP-free solution blocked pH_i regulation. However, a requirement for just the binding, not the hydrolysis, of ATP could not be ruled out. In snail neurons, Dr. R.C. Thomas had done the experiment of reducing ATP with metabolic poisons to levels sufficiently low to block the Na^+ pump, without affecting H^+ extrusion. In crayfish neurons, Dr. Moody had been able to show the H^+ extrusion could be blocked if the energy gradient against H^+ extrusion was made equal to that for Na^+ entry by lowering the external pH and increasing the internal Na^+ concentration. This occurred when sufficient ATP was still present to allow the Na^+ pump to operate normally So the energy source for H^+ extrusion is still not completely clear. In all preparations external Na^+ is required, and its electrochemical gradient could theoretically provide enough energy to drive all the other ion movements involved. But the squid axon experiments, which are probably the most direct, indicate that ATP is in some way involved in the process.

Can the pumps responsible for pH_i regulation be forced to work in reverse of the direction of their normal operation in the cell? Drs. de Hemptinne and R.C. Thomas briefly discussed their results on this issue. Thomas has demonstrated that at low external pH, the coupled uptake of H^+ and Cl^- occurs in snail neurons, apparently by the same carrier that normally extrudes these ions. However, the decrease in internal Na^+ concentration predicted by the sodium efflux expected in this situation could not be detected. In cardiac Purkinje fibers, however, de Hemptinne was able to record the decrease in $[Na^+]_i$ under these conditions. This discussion prompted Dr. Vaughan-Jones to comment that the reverse operation of these pumps makes the calculation of pump rates under various ionic conditions difficult, since the backflux rates are not accurately known and it is the one-way rates that are needed in the calculation. Roos and Boron agreed in principle, but argued that under almost all experimental conditions where the calculations are made, the backflux rates would be negligibly small.

III. CELLS IN WHICH pH_i REGULATING MECHANISMS SERVE ADDITIONAL FUNCTIONS

In at least two types of cells, the pH_i-regulating mechanisms serve functions in addition to merely maintaining pH_i. Renal tubule cells not only maintain pH_i, but also engage in the transcellular transport of H^+. Dr. Boron discussed how this could be accomplished by a rather simple variant on a symmetrical cell with separate pH_i-regulating systems. The kidney tubule cells appear to have Na^+–H^+ exchange systems symmetrically distributed, but separate Na^+-dependent HCO_3^- transport systems only at the basolateral membrane. With this distribution, an acid load via the basolateral membrane can result in the transcellular transport of H^+ merely by the normal operation of pH_i regulation in the cell.

Another cell in which the mechanisms of pH_i regulation merit special attention is the oocyte. In sea urchin and Xenopus oocytes, fertilization causes an apparent change in the pH_i-regulating system so that pH_i increases and is maintained at a new, higher level (see paper by Shen and Steinhardt, this volume). In sea urchin, at least, this pH_i change is clearly of physiological importance (see the article by Winkler in this volume). However, how—or even if—the unfertilized oocyte regulates its pH_i is not clear. In sea urchin oocytes, Drs. Shen and Steinhardt have shown that the resting pH_i is very close to that predicted by an equilibrium distribution of H^+. There is substantial evidence for the stimulation of Na^+–H^+ exchange at fertilization in that the pH_i increase requires external Na^+ and is accompanied by Na^+ influx. But as Shen and Steinhardt pointed out, things are not quite so simple, since the pH_i increase is not blocked by amiloride in normal sea water Na^+ levels (480 mM) and survives rather drastic reductions in external Na^+ levels, which might not be expected of a system whose rate of operation depends on the Na^+ gradient. It is clearly of great importance to understand the pH_i regulating system in oocytes, but the experiments of Shen indicate that this may be a difficult task indeed.

IV. SUMMARY OF DISCUSSION ON RENAL TUBULE CELLS AND OOCYTES

In the tubule cells, Dr. Boron proposed that the Na^+-dependent HCO_3^- transport at the basolateral membrane did not involve Cl^- efflux. Dr. Vaughan-Jones pointed out that if it did, the transport system would be more like that in other cells, and he questioned Boron's evidence on this point. The failure to record a decrease in internal Cl^- levels during H^+ extrusion, Vaughan-Jones asserted, could be due to a large passive Cl^- permeability at the basolateral membrane. Boron agreed, but cited both the electrogenicity of the transport and the lack of effect of changes in external Cl^- levels to support his original proposal.

The question of whether fertilizaton stimulates a true Na^+–H^+ exchange prompted a rather animated discussion among Drs. Shen, Steinhardt, Moody, and Boron. Shen and Steinhardt presented two observations that suggest caution in assuming that Na^+–H^+ exchange is involved: 1) The sensitivity of this process to amiloride is much lower than for Na^+–H^+ exchange in other systems; and 2) the reduction of external Na^+ levels from 480 mM to 5 mM had little effect on the pH_i increase at fertilization, and only when $[Na^+]_o$ was reduced below 5 mM was the alkalinization blocked. Shen noted that recording of internal Na^+ concentration during fertilization would be of value. Boron and Moody mentioned two possible explanations for the data: 1) The system might simply have a high affinity for Na^+, as is true in membrane vesicles from kidney tubules. This would explain the apparent lack of effect of reductions in external Na^+ levels, and the insensitivity to amiloride, which competes with Na^+ for a binding site. 2) When external Na^+ levels are reduced, internal Na^+ levels also fall, and thus the Na^+ gradient is disrupted less than one would expect. Under these conditions Na^+–H^+ exchange might appear to operate almost normally, even thought $[Na^+]_o$ had been greatly reduced.

Drs. Gillies and Deamer presented a series of experiments they have carried out suggesting a quite different source for acid extrusion from sea urchin eggs at fertilization. They found that the extracellular acidification produced by a suspension of eggs at fertilization behaves like a voltaile acid,—ei, CO_2 (J Cell Physiol, 108:115, 1981). Even in HCO_3^- and O_2-free seawater equilibrated with N_2, the fertilization acid is volatile. Confirming this, they find a 4.5 micromole/ml egg difference in acid labile CO_2 between fertilized and unfertilized eggs, which persists in eggs that have been ethanol-extracted and heated at 475°C for 8 hours. They suggest that the fertilization acid—and hence the rise in pH_i—is the result of the release of CO_2 from an inorganic carbonate compound in the egg. Gillies pointed out that the main drawback of this hypothesis is that it does not account for the external Na^+ dependence of the fertilization acid.

Dr. Steinhardt questioned some of the procedures in these experiments. He stated that sperm in the true absence of oxygen should not be able to fertilize. Hence, he thought that the seawater in Gillies and Deamer's experiments was probably not completely HCO_3^--free. Dr. Deamer expressed doubts about this objection, pointing out that the sperm were actually exposed to O_2-free seawater for only a very brief interval. Others suggested that perhaps the dilution of sperm into the seawater might have carried in some HCO_3^-.

In the general discussion, two technical points were raised about the experiments presented by Dr. Roos and Dr. Vaughan-Jones.

A. Roos presented data obtained in barnacle muscle concerning the relationship between external Na^+ concentration and the rate of acid extrusion. W. Moody asked what effect changes in $[Na^+]_i$ during the experiment would have on the plot of $[Na^+]_o$ vs pumping rate, since in many cells, $[Na^+]_i$ varies in a consistent manner with $[Na^+]_o$. Roos replied that at the start of the exposure to a particular $[Na^+]_o$, $[Na^+]_i$ was close to zero, and $[Na^+]_i$ probably did not increase very much up to the time at which the extrusion rates were being measured. The absence of internal Na^+ also blocked the reverse operation of the transport system, so that extrusion rates were probably very close to actual one-way transport rates (this in response to a question from R. Vaughan-Jones).

Following R. Vaughan-Jones's presentation on pH_i regulation in cardiac muscle, D. Spray asked whether changes in electrical coupling between fibers produced by pH_i changes might mean that the critical assumption of isopotentiality between voltage- and ion-sensitive microelectrodes (which must be placed in separate fibers in this preparation) might not be valid. R. Vaughan-Jones replied that one could easily test (and he had done so) that both electrodes record the same *changes* in membrane potential. He thought that this sort of problem might explain some transient artifacts seen on his records, but would not materially affect the overall results. (It is worth noting that double-barreled electrodes, as described in this volume by de Hemptinne, could circumvent this problem.)

At the end of the general discussion on pH_i regulation, two short presentations of new data were made: 1) A. de Hemptinne presented experiments using pH_i recordings to monitor the entry of organic acids into sheep heart Purkinje fibers. If one assues that the protonated forms of such acids cross the membrane and dissociate to generate free H^+ in the cell, then the rate of decrease of pH_i is proportional to the rate of entry of the acid. For a variety of acids (acetate, propionate, alpha-keto-butyrate, etc), the measured rate of pH_i change was well predicted by the product of the concentration of the uncharged form and its lipid–water partition ratio. A few acids, such as lactate and pyruvate, penetrated faster than expected. To test whether this might be due to a specific transport system for metabolically important acids, de Hemptinne did two experiments: He showed that (a) alpha-cyano-4-hydroxycinnamate, which blocks lactate and pyruvate transport into red cells and mitochondria, reduced the rate of lactate and pyruvate into Purkinje fibers; and (b) D-lactate permeated much less rapidly than L-lactate. 2) A. Roos discussed experiments that he and R.F. Abercrombie are doing in frog muscle fibers. As Vaughan-Jones had earlier shown, these fibers when exposed to an acid load regulate pH_i little if at all. Abercrombie and Roos have found that if the fibers are depolarized by whatever means, they acquire the ability to extrude H^+ after an acid load. This new H^+ extrusion is blocked by Na^+-free external solutions or amiloride. If contraction is blocked with hypertonic solutions, the depolarization itself triggers a large early transient acidification, in addition to stimulating pH_i regulation in response to a subsequent acid load. This spontaneous pH_i transient fall can be imitated with caffeine, which releases Ca^{2+} inside the fibers but does not cause depolarization. It is not at this point clear how depolarization either causes the pH_i transient or stimulates pH_i regulation in the fibers.

UTILIZATION OF pH_i IN THE CONTROL OF CELLULAR FUNCTIONS

Intracellular pH Changes Accompanying the Activation of Development in Frog Eggs: Comparison of pH Microelectrodes and ^{31}P-NMR Measurements

Dennis J. Webb and Richard Nuccitelli
Zoology Department, University of California, Davis, California 95616

I.	Introduction	294
II.	Materials and Methods	295
	A. Animals and Eggs for Microelectrode Experiments	295
	B. pH and Voltage Electrodes	296
	C. Animals and Eggs for ^{31}P-NMR Experiments	299
	D. Egg Extract Preparation	302
	E. Recording and Treatment of Spectra	302
III.	Results	304
	A. pH_i Changes Following Fertilization or Activation	304
	B. Evidence for Lack of Na^+–H^+ and Cl^-–HCO_3^- Exchange	306
	C. pH_i Change Independent of pH_o	309
	D. pH_i Oscillations Associated With the Cell Cycle	309
	E. Localized pH_i Changes Following Injury	312
	F. ^{31}P-NMR Spectra of Unfertilized Eggs	313
	G. ^{31}P-NMR Spectra of Fertilized Eggs	313
IV.	Discussion	316
	A. pH_i of the Unfertilized Egg	316
	B. Initial Transient Acidification	318
	C. Permanent pH_i Increase	319
	D. Nature of the Permanent pH_i Increase	320
	E. Manipulation of pH_i	320
	F. Cyclic pH_i Oscillations	321
	G. The Importance of the Permanent pH_i Increase for the Activation of Development	322
V.	References	322

I. INTRODUCTION

Fertilization activates the relatively quiescent, mature oocyte [Epel, 1979]. The most immediate and generally occurring events include changes in the membrane potential and Ca^{2+} activity. Recent studies with marine invertebrate eggs indicate that the increase in metabolic activity at fertilization can also involve intracellular pH (pH_i) changes that may be crucial to the successful activation of development. We have investigated pH_i of a freshwater vertebrate egg during fertilization and early cleavage using two completely different techniques.

The mature egg of the South African clawed frog Xenopus laevis is 1.3 mm in diameter with a brown pigmented animal hemisphere and a lighter vegetal hemisphere. In close apposition to the plasma membrane is the vitelline envelope which itself is surrounded by three jelly layers. When a sperm swims through the jelly and vitelline envelope to fuse with the egg's plasma membrane, a series of events is set into motion to activate development. The first event is the fertilization potential, a depolarization lasting for about 20 minutes, which appears to result from an increase in the chloride conductance leading to a net efflux of Cl^-. Grey et al [1981] have recently shown that the fertilization potential in Xenopus acts as a fast block to polyspermy in a similar fashion to that reported for the eggs of the sea urchin, marine worm, starfish, and frog [Jaffe, 1976; Gould-Somero et al, 1979; Miyazaki and Hirai, 1979; Cross and Elinson, 1980]. Within minutes of the onset of the fertilization potential in the teleost, Oryzias latipes, and the sea urchin, Lytechinus pictus, there appears to be a transitory increase in the Ca^{2+} activity that gives rise to the cortical reaction [Ridgway et al, 1977; Steinhardt et al, 1977]. This reaction involves fusion of the cortical granules with the plasma membrane beginning at the sperm entry site. Exocytosis of the granule contents into the perivitelline space is followed by the lifting off of the vitelline envelope from the plasma membrane, and a change in its structure to form the fertilization envelope. The latter is impenetrable to other sperm and acts as a permanent block to polyspermy. This process is complete about 4 to 6 minutes after sperm–egg fusion [Grey et al, 1981], and allows the egg to rotate freely within the fertilization envelope so that the vegetal hemisphere containing the dense yolk platelets is lowermost. Approximately 90 minutes after the fertilization potential begins, a furrow forms at the animal pole and the egg starts to cleave. The earliest cleavages occur synchronously and have a cycle of about 30 minutes [Hara et al, 1980]. Apart from cleavage these early events triggered by fertilization can also be initiated by activating the egg with a sharp needle or the ionophore, A23187.

In a number of marine invertebrates fertilization triggers an acid efflux [Johnson et al, 1976; Paul, 1975; Ii and Rebhun, 1979] which, in the sea urchin, is accompanied by a rise in pH_i [Johnson et al, 1976; Shen and Steinhardt, 1978]. This pH_i rise is dependent upon extracellular Na^+ and has been implicated in the dramatic increase in the rate of protein synthesis seen in sea urchin eggs at this time [Grainger et al, 1979; Winkler et al, 1980 and this volume]. In this chapter we report on the pH_i changes occurring at fertilization and early cleavage of a freshwater vertebrate Xenopus laevis using two different techniques [Nuccitelli et al, 1981; Webb and Nuccitelli, 1981]. Recessed-tip pH-sensitive glass microelectrodes as developed by Roger Thomas [1974] and ^{31}P-nuclear magnetic resonance (NMR) both reveal a similar permanent pH_i rise after fertilization or activation. The better time resolution of the pH microelectrodes allowed us to follow the time course of the pH_i rise and also revealed a small transient pH_i fall immediately following fertilization and a pH_i oscillation associated with the cleavage cycle. Considering that the pH microelectrode is inserted into a single egg while ^{31}P-NMR noninvasively scans about 1,000 eggs, our findings give us greater confidence in these results as well as in each of these techniques used as pH_i monitors. The pH_i changes appear to be unaffected by removal of extracellular Na^+ or Cl^- or by changes in extracellular pH over a wide range.

II. MATERIALS AND METHODS

A. Animals and Eggs for Microelectrode Experiments

Xenopus laevis females were induced to ovulate by injecting a primer dose of 100 IU of human chorionic gonadotropin (Sigma) followed 36–72 hours later by a booster dose of 1,000 IU. On occasions when previously ovulated females were allowed to rest less than 2 months, the booster dose alone was sufficient to induce the next ovulation. Frogs were kept at room temperature (22–24°C) in tap water. Mature eggs were squeezed out dry 7–10 hours after the booster injection. Typically, four eggs with their jelly coat intact were placed using forceps into a dry perfusion chamber to which they adhered. The jelly coat is essential for fertilization of Xenopus eggs. After a minute's delay the chamber was flooded with 2–3 ml of modified F_1 solution [Hollinger and Corton, 1980; Webb and Nuccitelli, 1981] and perfused either continuously or, as in the majority of experiments, intermittently. Modified F_1 solution (referred to as F_1 throughout) contained, in mM: NaCl 31.25, KCl 1.75, $CaCl_2$ 0.25, $MgCl_2$ 0.06, Na_2HPO_4 0.5, NaOH 1.9, $NaHCO_3$ 2.0, Tricine 10.0, pH 7.8. Solution pH was corrected using a combination pH electrode (model 39502, Beckman Instruments) calibrated with standard buffers pH 7.00 and pH 9.18 (part No. 3007, 3009, Beckman Instruments) and monitored on a model 3500 digital pH meter (Beckman Instruments). Illingworth [1981] recently pointed out that misleading pH measurements can be obtained using certain combination pH electrodes owing to differences in ionic strength between standard buffers and the unknown solution. Na^+-free solution was prepared by substituting potassium salts for all the sodium salts except where choline chloride replaced NaCl. In some experiments

PO_4^{2-} and HCO_3^- were left out altogether. Cl^--free solution was prepared by replacing $CaCl_2$ with $Ca(NO_3)_2$ and the other chloride salts with their sulphate equivalents. Amiloride (Merck), DIDS (4,4′-diisothiocyanostilbene-2,2′-disulfonic acid, Calbiochem) and 2,4-dinitrophenol (DNP, Sigma) were added directly to either the Na^+-free, Cl^--free, or F_1 solution. The DIDS solution was stored at 4°C in the dark. A23187 (Calbiochem) was made up as a stock solution in ethanol (1 mg/ml), 10 µl of which was added to 10 ml of the solution under test to give a final concentration of about 2 µM.

Testes were removed from decapitated male frogs and stored cold in 1.5 × OR-2 solution [Wallace et al, 1973], which contained in mM: NaCl 124, KCl 3.75, Na_2HPO_4 1.5, NaOH 5.7, HEPES 7.5, $CaCl_2$ 1.5, $MgCl_2$ 1.5, pH 7.8. These were normally used within 3 days. Fertilization of Xenopus eggs was achieved either by adding a small piece of testis or 2-3 drops of a concentrated sperm suspension to the chamber. A sperm suspension was made by blotting dry a small piece of testis before placing it in a few drops of F_1 solution for a couple of minutes. It was then blotted dry again and macerated in 2-3 drops of fresh F_1 solution.

B. pH and Voltage Electrodes

pH_i was measured using Thomas-type, recessed-tip, pH-sensitive glass microelectrodes [Thomas, 1974; see also chapter by Thomas in this book] and 3 M KCl-filled voltage microelectrodes. In construction of pH electrodes the outer pipettes were made from aluminosilicate glass (1.2 mm inner diameter) using a one-stage pull on a Narashige pipette puller (PE-2, Fig. 1) with zero solenoid setting. A six-turn coil (5.5 mm diameter) of nickel-chrome wire (1.3 mm diameter) was heated with about 12 amps. This gave tip diameters of less than 0.5 µm. pH glass capillaries of 2.5 cm length and 1.0 mm outer diameter (Microelectrodes Inc, also obtainable from Clark Electromedical Instruments) were pulled on a David Kopf Instruments pipette puller (Fig. 1). An 11-turn coil (4.5 mm diameter) of nickel-chrome wire (1.0 mm diameter) was heated with about 4 amps and the solenoid set at a value of about 4.5. The end of a steel tube (0.67 mm outer diameter), which was connected to a N_2 cylinder, was inserted into the completed pH pipettes and sealed on with wax. The extreme tips of the pH pipettes were then fused using a microforge with a heated filament of platinum wire (50 µm diameter) and then inserted into the aluminosilicate pipettes (Fig. 2) to within an intertip distance of 2-4 µm. About 1,300 lb/in² pressure was applied to the pH pipette during sealing with the outer pipette, leaving a 25-µm exposed length of pH glass beyond the seal. Completed pH glass pipettes sealed into outer pipettes were filled with deionized water containing a trace of detergent and heated overnight in a water bath at 80°C. Air bubbles were removed by a cat's whisker, and the bulk of the water in the pipettes was displaced with 0.1 M NaCl solution buffered to pH 6.0 with 0.1

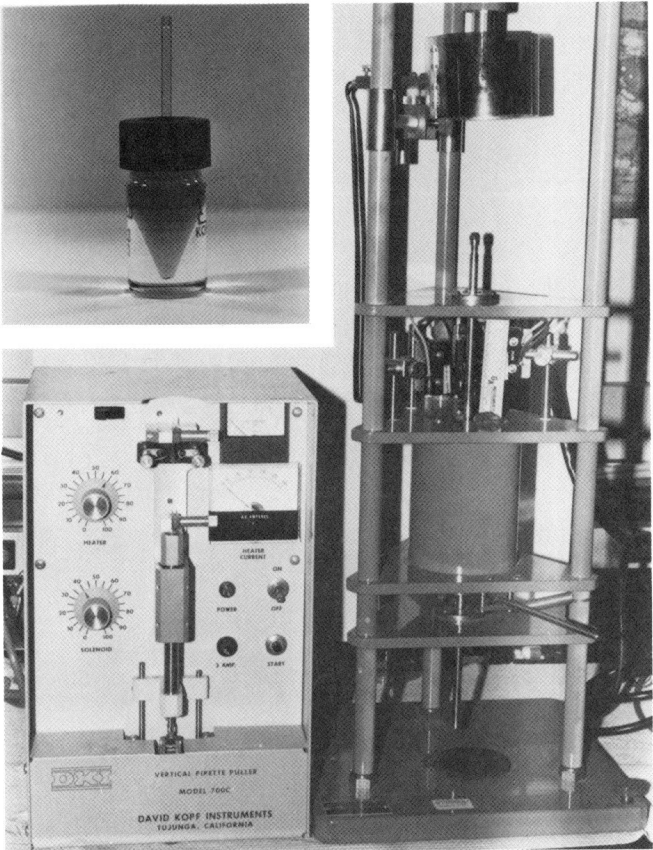

Fig. 1. Vertical pipette pullers used to make pH glass micropipettes (left: David Kopf) and outer glass micropipettes (right: Narashige). A pH microelectrode with its tip in chromic acid contained in a Kontes 2-ml microflex tube (upper left corner).

M Na citrate. Electrodes were stored with their tips in freshly prepared chromic acid in 2 ml capacity Kontes microflex tubes (Fig. 1) for at least 2 days before testing. Response times for a full unit change in pH varied between 6 and 30 seconds with a typical response of 55–59 mV, although some with a response as low as 50 mV were also used.

Voltage electrodes with low tip resistances of about 10 MΩ were chosen, since they were less susceptible to tip potential fluctuations while inserted in the egg. The pH and voltage electrodes were mounted on Narashige manipulators (MP-2) and connected via Ag/AgCl wire junctions to high input resistance op-

Fig. 2. The microforge with a pH glass micropipette mounted on steel tubing about to be inserted into the outer glass micropipette. A 50-μm thick platinum wire acts as a filament to apply heat locally during fusion of the two glass types while a pressure of about 1,300 lb/sq in. is applied to the inner pH micropipette from a N_2 cylinder.

erational amplifiers Analog Devices 311J and Biodyne AM4 (Santa Monica), respectively (Fig. 3). The differential of the two electrodes, to give the pH, and the voltage electrode response, were displayed on a four-channel chart recorder (Multicorder MC 66001 Watanabe). The chamber was grounded through an agar/ F_1 solution bridge and Ag/AgCl junction.

In general the egg was first impaled with the pH electrode by sharply tapping the electrode holder with a pair of forceps. The voltage electrode was then introduced by increasing the negative capacitance to oscillate the electrode tip. To avoid prick activation during electrode insertion, eggs were initially bathed in F_1 solution containing 5–10 mM of the anesthetic, chlorobutanol (1,1,1-trichloro-2-methyl-2 propanol, Sigma). This substance causes a slight but reversible decrease in pH_i, but once removed has no apparent effect on development. In some cases impalement of the egg without prick activation was possible in the absence of chlorobutanol (see Fig. 5B and 10A). Even though the pH and voltage electrodes were introduced into the egg in close proximity ($\sim$ 200 μm), at later divisions they may have ended up in different blastomeres. Fortunately, the blastomeres remain ionically coupled [Turin and Warner, 1980] and cleave synchronously for the first dozen or so divisions in the animal hemisphere [Hara and Tydeman, 1979].

C. Animals and Eggs for ^{31}P-NMR Experiments

Ten to twenty Xenopus laevis were injected for both morning and afternoon experiments so that a constant supply of eggs was available over an 8-hour period. Eggs were squeezed out of the females into modified F_1 solution minus the phosphate and inspected under a stereoscope for lysis and uneven pigmentation blotches. Only healthy-looking batches of eggs were used in the experiments, and each run required about 30 ml of eggs before dejellying, usually provided by 3–6 females. After each NMR experiment the eggs were again inspected visually and scored for pigmentation blotches and lysis. If any deterioration had occurred those NMR results were discounted. Fertilization was accomplished by mincing one-fourth of a testis in F_1 and adding it to eggs being agitated in a 6-inch Petri dish. Each dish was then scored for rotation before placing the eggs in the NMR tube (Fig. 4). The data presented here were taken from batches exhibiting greater than 80% rotation. These eggs proceeded to cleave normally in the NMR tube throughout our experiments and, when removed, looked identical to control embryos growing in a shallow petri dish. Ionophore activation was accomplished in a similar manner by adding a small amount of 1 mg/ml A23187 (Calbiochem) in ethanol to the petri dish so that the final A23187 concentration was 2 μM, and the final ethanol concentration was less than 0.1%.

Xenopus eggs are surrounded by three jelly layers that expand greatly upon hydration so that these 1.3-mm diameter eggs typically pack 4 mm center to center. In order to increase the packing density, the eggs were partially dejellied by combining 100 μl mercaptoethanol with 70 ml of F_1 plus 30 ml of eggs at pH 8.4. After 5–10 minutes of mild swirling, the packing density was approximately doubled, after which the mercaptoethanol was quickly washed off. Most

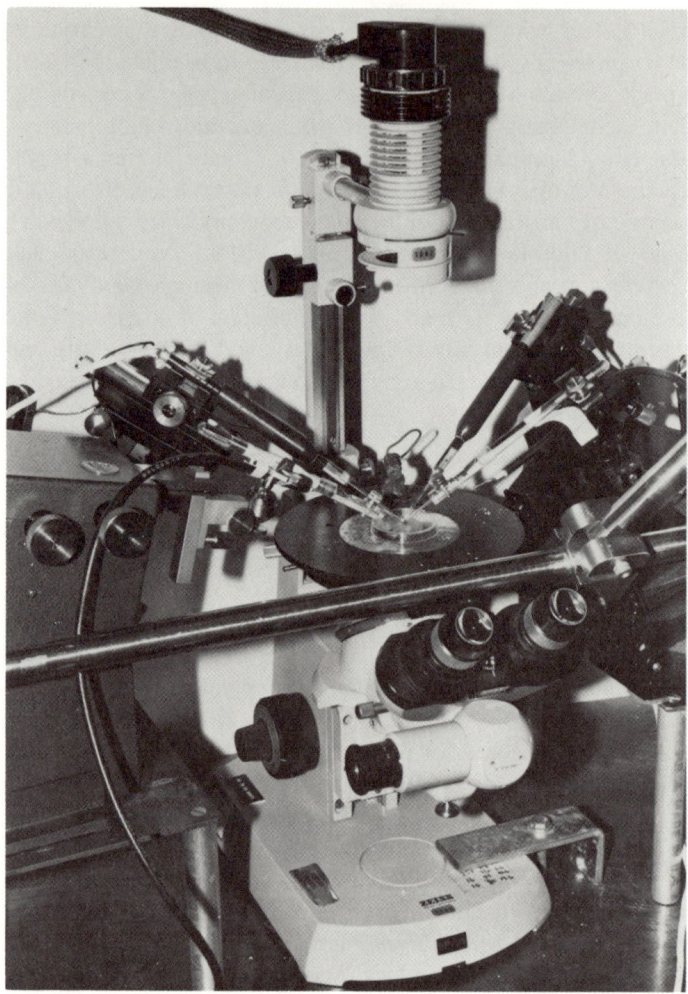

Fig. 3. The electrical setup.

eggs are still separated from their neighbors by jelly after this treatment. This is important because totally dejellied eggs pack so tightly that the perfusion of fluid around them is uneven and erratic. We used an 18-mm inner diameter sample tube and a 24-mm high spectrometer detector window. Eggs were partially dejellied to a 2-mm center-to-center packing density, providing spectra representing signal contributions from approximately 1,000 eggs.

Changes Accompanying Frog Egg Activation / 301

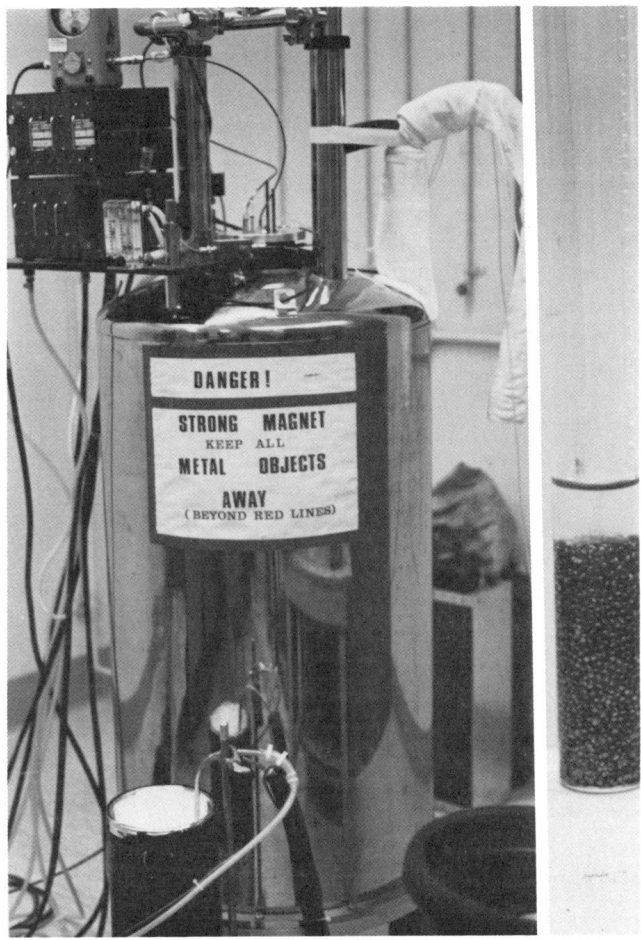

Fig. 4. The NMR machine containing the 20-mm diameter tube of frog eggs shown at right.

Using a peristaltic pump we continuously superfused the eggs by adding aerated F_1 at the bottom of the NMR tube while drawing off medium from the top of the tube. We typically used 15 ml of cells with a 15-ml column of fluid above them and superfused at 2–4 ml/min. This exchanged the 27 ml of extracellular fluid in 7–14 minutes. Xenopus eggs are ideal for this perfusion since, owing to their size and density, they quickly settle to the bottom of the tube and remain there undisturbed by the perfusion. This continuous flow also allowed us to change the medium and directly monitor extracellular pH while collecting spectra using a combination pH electrode (model 39505, Beckman Instruments).

D. Egg Extract Preparation

The egg extract solutions were prepared by first totally dejellying the eggs in 0.3% mercaptoethanol and washing them several times with an extract buffer described below. Next, they were gently lysed by drawing them through a 1-mm diameter syringe and then adding an equal volume of extract buffer. The extract buffer was designed to match closely the intracellular ionic environment when combined with an equal volume of lysed eggs. Both the total and free intracellular ion concentrations are listed in Table I, and substantial sequestering or binding of Mg^{2+} and Ca^{2+} is evident. To ensure similar low free Ca^{2+} and Mg^{2+} levels in the extract, the divalent chelators EDTA (ethylenedinitrilo-tetraacetic acid) or EGTA (ethyleneglycol-bis-β-amino-ethyl-ether-N,N'tetraacetic acid) were used to make two separate buffer solutions. We did not measure the total Ca^{2+} and Mg^{2+} released during our gentle lysing procedure, but would expect it to be lower than divalent cation store in the egg since some ions would be sequestered inside cellular organelles, and some would be bound to proteins and membranes. Extract solution number 1 used a final concentration of 1.5 mM EGTA in the buffer, which would hold the free Ca^{2+} down to low levels if the total Ca^{2+} was less than 1 mM. Solution 2 was designed to buffer both the Ca^{2+} and Mg^{2+} by using a final concentration of 5.5 mM EDTA. Using the stability constants of Martell and Smith [1974] (0.1 ionic strength corrected) assuming 5 mM total Mg^{2+} and 1 mM total Ca^{2+}, the free Mg^{2+} and Ca^{2+} levels would be 0.5 mM and 1.5 μM, respectively. Perchloric acid extracts were made by adding 3 ml of 50% $HClO_4$ to approximately 27 ml of eggs at 0°C for 10 minutes before titration back to pH 7.0 with KOH. After 12 hours at 4°C the solution was centrifuged for 2 hours at 10,000 rpm, and the supernatant was used for NMR.

E. Recording and Treatment of Spectra

For a general description of ^{31}P-NMR see Gadian et al, and Gillies et al, in this volume. ^{31}P-NMR spectra were recorded by us at 81 MHz on a Nicolet Magnetics NT 200 spectrometer operating in the Fourier transform mode. For the single-pulse experiments a 54° tipping pulse of 15 μsec was applied every 300 msec. All spectra in this paper have been plotted using 7 Hz line broadening. Three standards were used: 1) a sealed capillary containing 85% phosphoric acid was assigned 0 ppm chemical shift; 2) a second sealed capillary containing 1.6 M methylenediphosphonic acid (MdPA) was assigned 17.13 chemical shift; 3) 0.6 mM extracellular glycerophosphoryl choline (GPC) (Sigma grade I) was

TABLE I. Extract Buffer Solution Composition and Cytoplasmic Ion Concentrations in Xenopus Eggs

	Free cytoplasmic concentration		Total concentration in egg		Extract buffer 1	Extract buffer 2
	mM	Reference	mM	Reference	mM	mM
Na^+	19	De Laat et al, 1974	58	De Laat et al, 1974	12	19
K^+	52	De Laat et al, 1974	87	De Laat et al, 1974	52	52
Cl^-	54	De Laat et al, 1974	62	De Laat et al, 1974	53	28
Mg^{2+}	≤ 0.5	*	22	Colman and Gadian, 1976	0.5	—
Ca^{2+}	10^{-4}	Rink et al, 1980	2–7	Colman and Gadian, 1976	—	—
EGTA	—		—		3	—
EDTA	—		—		—	11
HEPES	—		—		—	10

*Sherwin Lee and Rick Steinhardt, personal communication.

assigned 0.49 ppm. It has been shown that the GPC chemical shift is independent of pH [Navon et al, 1977] and we confirmed this over the range of pH 7 to 8. All experiments were performed at 20°C in the absence of ^2H_2O. Therefore, field-frequency locking was not used, but our reference standard, GPC, was always present, and our spectrometer drift was negligible over these short experimental periods of less than 1 hour.

In order to pull the inorganic phosphate peak out from under the broad yolk-phosphate peak, we used two techniques: a two-pulse T_2 experiment [Campbell and Dobson, 1979], and a presaturation pulse on the yolk peak shoulder. The T_2 experiment discriminates between the two peaks on the basis of line width. The FID signals from broad lines such as the yolk phosphates decay more quickly than signals from narrow lines such as inorganic phosphate. Therefore, by applying a second pulse after most of the yolk contributions have died out, the refocused FID will contain mainly inorganic phosphate (P_i) contributions. Lines experiencing homonuclear decoupling such as ATP peaks are not refocused properly, and occur with anomalous phase in the transformed spectra [Campbell and Dobson, 1979]. This T_2 experiment resolved the P_i peak much better than convolution difference experiments.

The second technique is similar to the saturation transfer technique [Brown et al, 1977] and involves applying a presaturation pulse with a narrow frequency envelope centered on the yolk-phosphate peak shoulder. This will saturate the yolk peak so that it will not appear in the subsequent FID, but the inorganic phosphate peak will not be affected.

III. RESULTS

A. pH$_i$ Changes Following Fertilization or Activation

Fertilization results in changes in the membrane potential and pH$_i$ as shown in Figure 5A. The fertilization potential, a depolarization of about 10 mV, is followed approximately 2 minutes after its onset by a slight, transient pH$_i$ decrease averaging 0.03 ± 0.02 (SD, n = 8) pH unit. This is followed by a distinct, permanent increase averaging 0.31 ± 0.11 (SD, n = 7) pH unit beginning about 10 minutes after the start of the fertilization potential and completed about 1 hour later. We have summarized our pH$_i$ measurements in Table

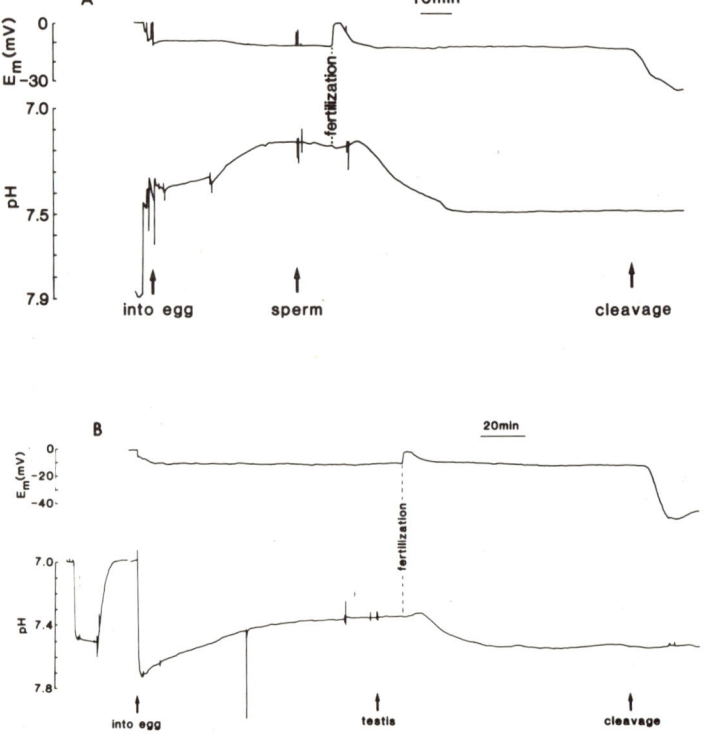

Fig. 5. Two records of intracellular pH and membrane potential (E$_m$) recorded in Xenopus eggs during fertilization. A) An unfertilized egg is impaled and sperm are added resulting in the typical fertilization response and cleavage. B) Calibrating solutions precede the impalement of the unfertilized egg in Na$^+$-free F$_1$ solution (K$^+$ substitute) at pH 7.0 in the absence of chlorobutanol. Within 2 min of electrode insertion at a pH$_i$ of about 7.7 the chamber was perfused with normal F$_1$ solution. A small piece of testis added about 2 h later led to successful fertilization and subsequent cleavage. Note that the pH scale is twice as large as that for the E$_m$ in Figures 5, 6A, and 9, but the same in all other figures.

TABLE II. Changes in the Membrane Potential and pH_i at Fertilization and Activation

	Resting potential (mV) Mean ± SD (N)	
	before activation	after activation[a]
	−10.1 ± 3.5 (38)	−10.1 ± 3.9 (42)

	Transient pH_i fall		
	Fertilized	Spontaneous activation	A23187 activation
Amplitude	0.03 ± 0.02 (8)	0.06 ± 0.05 (7)	0.08 ± 0.04 (6)
Onset time[b] (min)	2.6 ± 1.1 (8)	0.1 ± 1.3 (7)	−1.3 ± 2.0 (6)
Duration (min)	10.7 ± 4.6 (7)	9.6 ± 2.6 (7)	17.6 ± 7.9 (6)

	Permanent pH_i rise			
	Total Feb–Dec 1980	May–Dec 1980[c]	ΔpH_i	Time (min)
Unactivated	7.33 ± 0.11 (21)[d]	7.39 ± 0.11 (11)	—	—
Fertilized	7.67 ± 0.13 (10)[d]	7.69 ± 0.05 (6)	0.31 ± 0.11 (7)	67 ± 15 (6)
Activated[e]	7.64 ± 0.12 (28)	7.70 ± 0.10 (14)	0.30 ± 0.09 (15)	64 ± 14 (15)

Cyclical pH_i changes at cleavage		
ΔpH_i amplitude	Periodicity (min)	Onset time before cleavage (min)
0.03 ± 0.02 (38)	33 ± 6 (38)	24 ± 5 (9)

[a]Includes fertilized and artifically activated.
[b]Measured from the beginning of the fertilization potential.
[c]These results were obtained with improved technique and during the same period that our ^{31}P-NMR experiments were carried out.
[d]Range for unfertilized pH_i: 7.13 to 7.57, and fertilized or activated: 7.44 to 7.95.
[e]Includes all forms of artificial activation.

II. Egg rotation after activation might be expected to cause minor damage and thus affect the results presented here, but the pH_i changes were similar whether recorded in the animal or vegetal hemisphere. A further example of the membrane potential and pH_i changes seen in Xenopus eggs following electrode insertion and subsequent fertilization is shown in Figure 5B. Successful impalement was achieved in the absence of chlorobutanol in a solution in which K^+ replaced Na^+ and the extracellular pH (pH_o) had been lowered to 7.0. However, this success was probably due more to gentler insertion of the electrode rather than to the solution changes. A dramatic increase in the pH electrode response upon initial insertion was followed by a slow fall in pH_i at the beginning of which the egg bathing solution was replaced with normal F_1 solution at pH 7.8. The membrane potential hyperpolarized from −5 mV to −10 mV following the solution change probably as a result of the change in K$^+$ concentration. The mean resting potential of the unfertilized egg was -10.1 ± 3.5 mV (SD, n = 38) just prior to the activation potential, the same value as found approximately 1 hour after fertilization or activation, -10.1 ± 3.9 mV (SD, n = 42). The pH$_i$ eventually stabilized after an hour at about 7.4, and a small piece of testis was added to the chamber. Fertilization, 12 minutes later, gave rise to a small fertilization potential, a depolarization of about 10 mV, and a small transient acidification beginning about 2 minutes later. Beginning 10 minutes after the onset of the fertilization potential, the transient acidification was followed by a pronounced permanent increase to a level of about pH 7.7. Normal cleavage occurred about 100 minutes later accompanied by a membrane hyperpolarization of 40 mV. Similar pH$_i$ changes could also be elicited by the activation of eggs whether spontaneously, by pricking or upon application of the ionophore A23187. An example of the changes in membrane potential and pH$_i$ elicited by application of A23187 is shown in Figure 6A. The main difference between this response and that following fertilization is in the initial transient acidification. The ionophore-triggered acidification often begins before the onset of the activation potential and can have a larger amplitude as indicated in Table II. This suggests that A23187 can cause a pH$_i$ fall before the activation response.

B. Evidence for Lack of Na$^+$–H$^+$ and Cl$^-$–HCO$_3^-$ Exchange

In sea urchin eggs both the acid efflux and pH$_i$ rise at fertilization require some extracellular Na$^+$ [Johnson et al, 1976; Shen and Steinhardt, 1979]. Xenopus eggs, however, have no such requirement as shown in Figure 6B. Sperm are immotile in Na$^+$-free solution making fertilization impossible so the ionophore A23187 was used. Activation of the egg led to pH$_i$ changes similar to those seen with fertilization [Webb and Nuccitelli, 1981]. In Figure 6B electrode insertion was followed an hour later by replacement of the F$_1$ solution with Na$^+$-free F$_1$ solution (choline substitute). Fifteen minutes later this solution was exchanged for one containing 1 mM of the Na$^+$–H$^+$ exchange inhibitor amiloride. The activation potential of 20 mV some 40 minutes later upon addition of A23187 was followed by a dramatic rise in pH$_i$. The initial acidification was smaller in this particular example than usual for ionophore activation. However, the permanent pH$_i$ rise was very similar.

A number of pH$_i$ regulatory processes have been suggested to occur in certain tissues [Thomas 1976; various papers in this volume] including Na$^+$–H$^+$ and Cl$^-$–HCO$_3^-$ exchange. The lack of an effect of amiloride in Na$^+$-free solution would argue against Na$^+$–H$^+$ exchange, and in Figure 7 we demonstrate that Cl$^-$–HCO$_3^-$ exchange also appears to be absent. In Figure 7A the impaled egg has been perfused with Cl$^-$-free solution without affecting the typical pH$_i$ changes that eventually settle at about 7.3. Activation with A23187 triggers a large fertilization potential of almost 60 mV, which overshoots the zero potential by

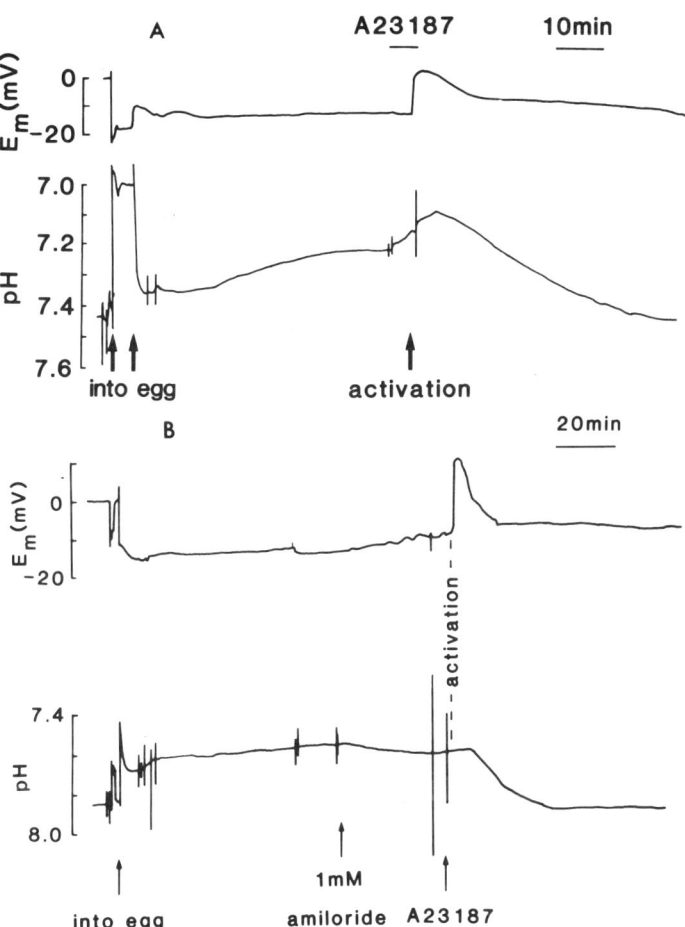

Fig. 6. Intracellular pH and membrane potential (E_m) recorded in Xenopus eggs during ionophore activation. A) Ca^{2+}–H^+ ionophore, A23187, activation. The voltage electrode was inserted first, followed by the pH electrode. In this example the chlorobutanol-containing F_1 solution was at pH 7.4, but following successful impalement was replaced with F_1 at pH 7.8. A23187 (Calbiochem), 2 μM, in F_1 solution was added for the time indicated by the bar. Ethanol, 0.1%, was also present but had no apparent effect when added alone. B) A Xenopus egg, ionophore-activated 60 min after perfusion with Na^+-free solution (choline substitute) and 45 min after addition of amiloride to the chamber.

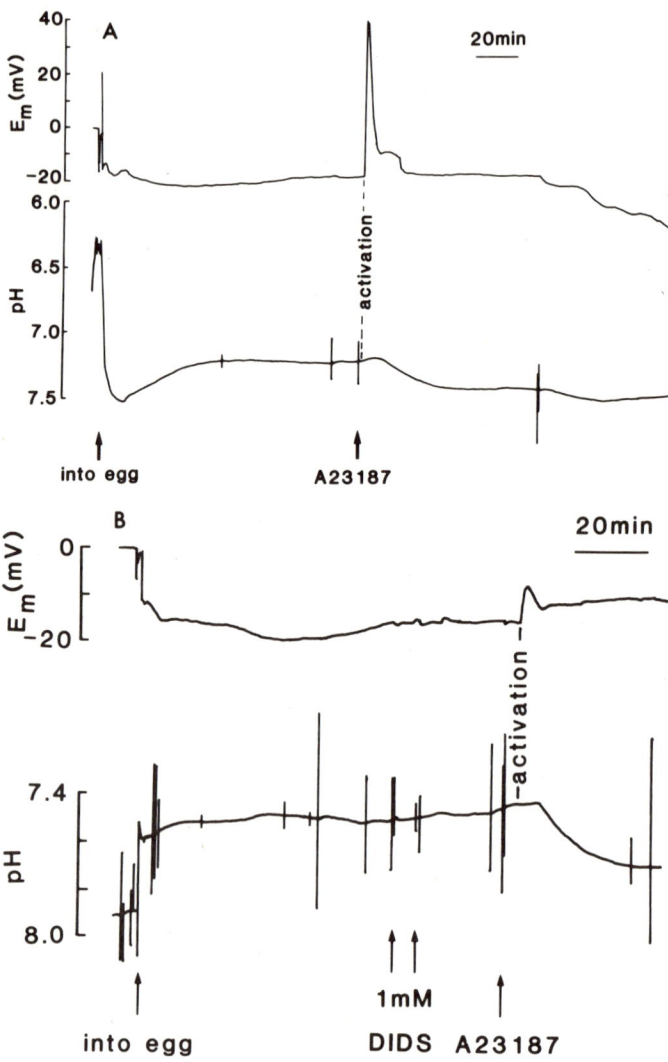

Fig. 7. Intracellular pH and membrane potential (E_m) recorded in Xenopus eggs during ionophore activation. A) The unfertilized egg is impaled in low pH_i (6.3) solution, which is then replaced with Cl^--free solution. Ionophore activation is elicited 2 h after egg impalement. B) The egg is impaled in F_1 solution (pH 7.9) and subsequently perfused with Cl^--free solution. DIDS in Cl^--free solution is added to the chamber an hour after electrode insertion and the egg was ionophore-activated 35 min later.

about 40 mV. This would be expected since the driving force for the Cl^- efflux has been increased by increasing the Cl^- concentration gradient. Nevertheless, a transient pH_i fall follows the onset of the fertilization potential and is followed itself by the permanent pH_i rise. There is also no obvious effect of the 60 mV change in the membrane potential on the pH_i. In Figure 7B the previous experiment is essentially repeated but this time in the presence of 1 mM DIDS, a Cl^-–HCO_3^- exchange inhibitor. The impaled egg is exposed to DIDS for 40 minutes in Cl^-–free solution before eventual activation with A23187. This time the fertilization potential is much reduced due to the blocking effect of DIDS on the Cl^- efflux. However, although the transient pH_i fall is virtually undetectable in this particular example, the permanent pH_i rise is unaffected.

C. pH_i Change Independent of pH_o

The pH_i of unfertilized and fertilized Xenopus eggs is unaffected by a wide range of changes in pH_o [Turin and Warner, 1980; Webb and Nuccitelli, 1981]. In Figure 8 the permanent pH_i rise is also shown to be unaffected by low pH_o. This egg was impaled in F_1 solution at pH_o 6.3, and 5 minutes later was perfused with pH 5.3. Fertilization becomes difficult at very low pH_o and so A23187 was added. The stabilized pH_i of about 7.5 in the unactivated egg rose to about 7.7 after activation. The activation potential appears to be prolonged somewhat (cf Fig. 6), a finding we frequently observed, suggesting a pH-sensitive ionic channel [Hutter and Warner, 1967]. pH_i was virtually unaffected by the later membrane potential fluctuations occurring around the time that cleavage would have been expected to occur if the egg had been fertilized.

Although we have shown that pH_i in Xenopus eggs is seemingly unaffected by drastic changes in pH_o, it can be altered by weak acids and bases [see Roos and Keifer, this volume]. The uncharged molecule is able to pass through the plasma membrane and either release or bind a proton. DNP is an example of a weak acid that acts as a protonophore, and in Figure 9 its effect on a spontaneously activated egg is shown. Spontaneous activation gives rise to the same pH_i changes as seen with fertilization or activation with A23187. If 1 mM DNP in F_1 solution (pH 7.8) is added to the chamber, the settled pH_i of about 7.7 decreases dramatically by almost 1 pH unit to about 6.9 over the next 40 minutes. It then slowly returns over the next hour or so to about 7.6 while still exposed to DNP. A similar response was obtained in two unactivated eggs.

D. pH_i Oscillations Associated With the Cell Cycle

Eggs impaled with pH and voltage microelectrodes and subsequently fertilized underwent normal cleavage for a number of divisions. It was possible to discern slight pH_i oscillations out of phase with the cleavage cycle but having a similar periodicity [Webb and Nuccitelli, 1981]. Whether impaled before or after fertilization, all eggs exhibited these cleavage-associated pH_i oscillations with a

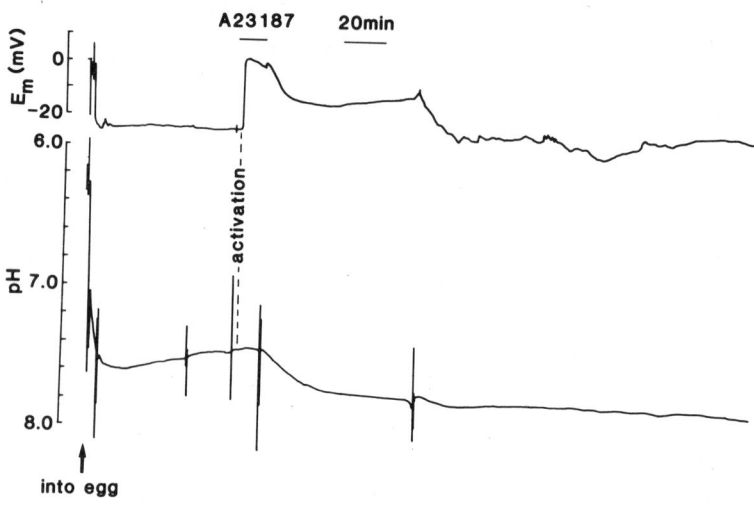

Fig. 8. Intracellular pH and membrane potential (E_m) recorded in a Xenopus egg activated at low extracellular pH. The egg was impaled in F_1 solution (pH 6.3), which was then replaced with F_1 solution (pH 5.3) about 5 min later.

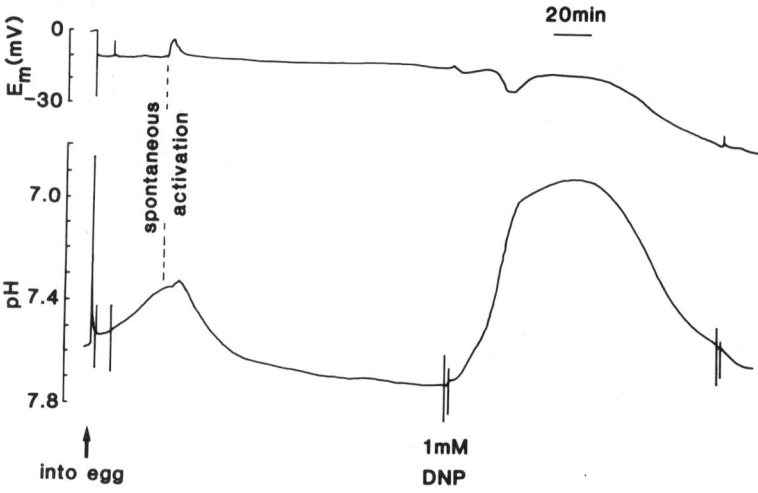

Fig. 9. Intracellular pH and membrane potential (E_m) recorded in a Xenopus egg showing the response to spontaneous activation and application of 1 mM DNP that was not washed off.

mean amplitude for 38 oscillations from 11 eggs of 0.03 ± 0.02 (SD) pH unit. The mean periodicity was 33 ± 6 minutes (SD, n = 38), and the mean onset time was 24 ± 5 minutes (SD, n = 9) before cleavage. Figure 10A shows an example of such a record during the first few cleavage divisions. This particular egg was impaled in the absence of chlorobutanol without prick activating. The

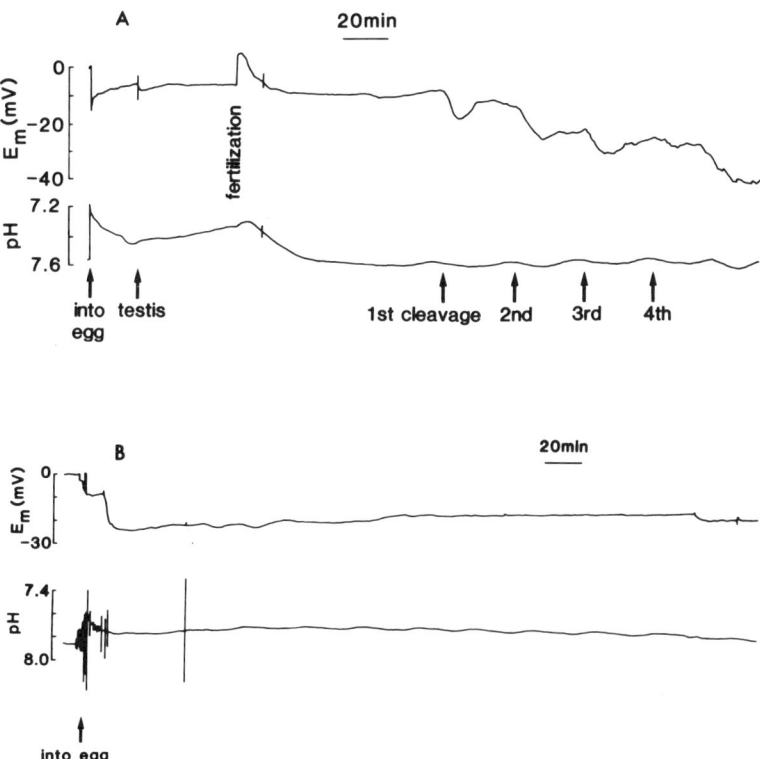

Fig. 10. A) Intracellular pH and membrane potential (E_m) recorded in a Xenopus egg during fertilization and early cleavage. This egg was impaled before flooding the chamber with F_1 solution in the absence of chlorobutanol. A fresh piece of testis was added to the chamber and fertilization ensued some 50 min later. The electrodes remained in the embryo up to the fifth cleavage division, and after electrode removal this embryo developed to the swimming tadpole stage. B) Intracellular pH and membrane potential (E_m) recorded in a fertilized, cleaving Xenopus egg. Ten percent F_1 solution replaced the normal F_1 solution about 10 min after impalement. Note the cyclical pH_i oscillations even in the absence of the usual cleavage-associated membrane potential hyperpolarizations.

usual pH_i changes can be seen following fertilization, but additionally it is possible to discern a series of small rhythmical pH_i oscillations. These oscillations precede each cleavage division as evidenced by the rhythmical membrane hyperpolarizations. The cyclical pH_i changes appear to be unaffected by removal of Na^+, Cl^-, or lowering of the ionic strength. In Figure 10B an already fertilized egg has been impaled by pH and voltage electrodes. The F_1 solution bathing the egg was then replaced with 10% F_1 resulting in a membrane hyperpolarization of almost 15 mV. The cleavage-associated membrane hyperpolarizations have been suppressed, whereas the pH_i oscillations remain discernible. Preliminary evidence indicates that ionophore- and prick-activated eggs also undergo rhythmical pH_i oscillations beginning at similar intervals after the activation potential and having a similar periodicity even though such eggs do not cleave.

E. Localized pH_i Changes Following Injury

In Figure 11 an unactivated egg has been impaled with two pH microelectrodes positioned at least 300 μm apart as well as with the voltage electrode (trace not shown). Chlorobutanol has been added to the F_1 solution to prevent prick ac-

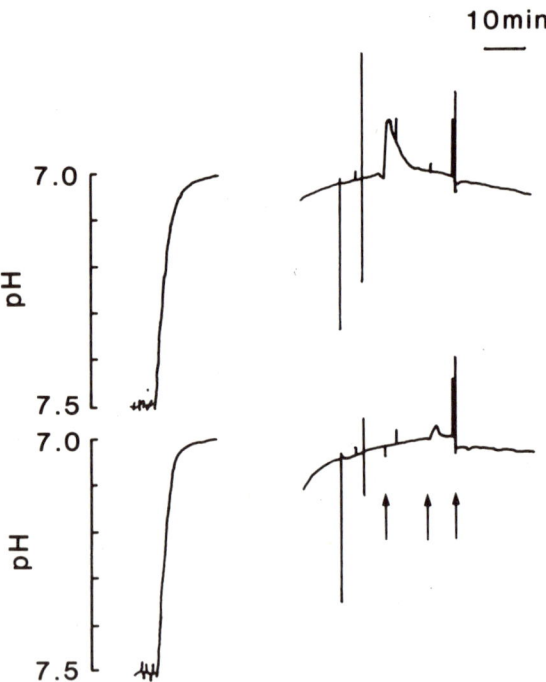

Fig. 11. Intracellular pH of an unfertilized Xenopus egg impaled with two pH microelectrodes separated by about 300 μm. Tapping of one electrode (first arrow, top trace) or the other (second arrow, bottom trace) produces a transient acidification localized at the site of the tapped electrode. Oscillations of the voltage electrode (trace not shown) by increasing the negative capacitance has negligible effect (third arrow). Chlorobutanol was present in the F_1 solution to prevent egg activation.

tivation, and has given rise to a slow transient acidification. Each electrode in turn was disturbed by tapping the back of the electrode holder. If electrode 1 was tapped, a large transient pH_i acidification occurred similar to that normally seen upon egg impalement. This recovered over the next 5–10 minutes. No pH_i change was registered by the second pH electrode. Similarly, a milder tap applied to electrode 2 revealed a small transient pH_i fall, whereas the pH_i detected by electrode 1 remained stable. Oscillation of the voltage electrode tip by increasing the negative capacitance produced similar voltage changes in both pH electrodes but no significant pH_i change. These observations suggest that a marked buffering capacity exists in the Xenopus egg as reported by Turin and Warner [1980] and the pH_i regulation can localize pH_i changes resulting from cortical injury.

F. ^{31}P-NMR Spectra of Unfertilized Eggs

A ^{31}P-NMR spectrum of unfertilized Xenopus eggs is shown in Figure 12A. This single-pulse experiment produced essentially the same spectrum as observed by Colman and Gadian [1976], with a large, broad yolk phosphoprotein peak obscuring the P_i peak. By using the two-pulse T_2 experiment, we were able to discriminate against broad line signals and pull out the P_i peak as shown in Figure 12B. This peak was identified as P_i by observing an increase in peak height when P_i was added to the extracellular medium adjusted to the same pH. It is also the only narrow line phosphate compound in that region of the egg extract spectrum. After studying six different batches of approximately 1,000 unfertilized, partially dejellied Xenopus eggs each, we found that the average P_i chemical shift was 2.63 ppm as listed in Table III. The T_2 spectra exhibit anomalous phasing for the ATP peaks, since the phosphates are closely coupled.

G. ^{31}P-NMR Spectra of Fertilized Eggs

Groups of fertilized eggs in which more than 80% of the eggs had rotated so that the pigmented animal hemisphere was uppermost were dejellied to the same extent as the unfertilized eggs had been, and their ^{31}P-NMR spectra were collected. Scoring for rotation and partial dejellying delayed the beginning of the NMR studies by 40 minutes, and 20–40 minutes of experiment time was needed

TABLE III. Summary of P_i Chemical Shifts from T_2 Experiments and Corresponding pH_i in Xenopus Eggs With pH Microelectrode Measurements Added for Comparison

Egg condition	P_i chemical shift (ppm)*	^{31}P-NMR pH_i	Microelectrode pH_i
Unfertilized	2.63 ± 0.04 (6)	7.42 ± 0.04 (6)	7.39 ± 0.11 (11)
Fertilized	2.81 ± 0.04 (5)	7.66 ± 0.06 (5)	7.69 ± 0.05 (6)
A23187-activated	2.80 ± 0.01 (3)	7.64 ± 0.02 (3)	7.70 ± 0.1 (14)

*These numbers represent the mean value obtained from the number of experiments in parentheses plus or minus the standard deviation. The spectra in each experiment were contributed by approximately 1,000 eggs.

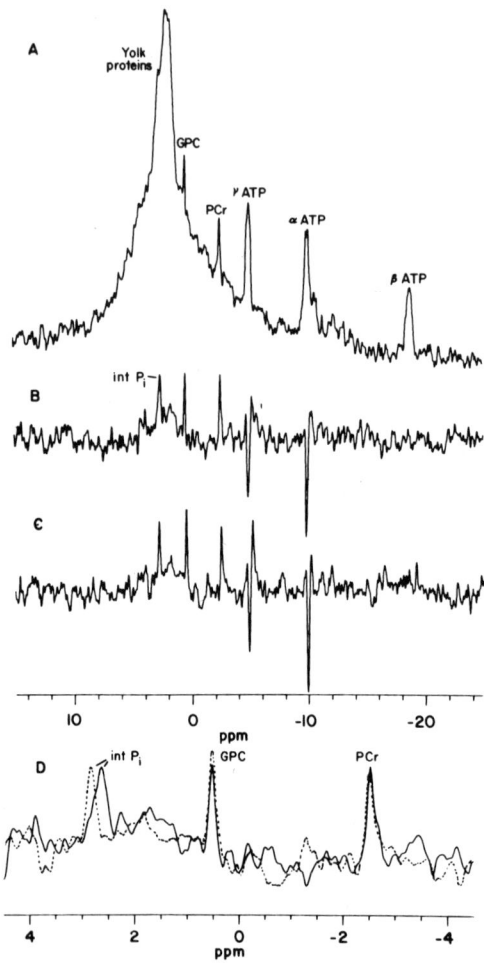

Fig. 12. ^{31}P-NMR spectra of approximately 1,000 Xenopus eggs before and after fertilization. A) Average of 1,000 single-pulse (54°) scans repeated every 300 msec on unfertilized eggs. The peaks were identified by adding those substances to egg extract solution. Peak assignments using GPC as a reference are (in ppm): yolk proteins, 2.18; GPC, 0.49; phosphocreatine (PCr), −2.53; γ-ATP, −5.02; α-ATP, −9.96; β-ATP, −18.65. B) Average of 2,000 two-pulse (90°–180° with 15 msec τ) scans repeated every 800 msec on the same eggs used in part A, above. Intracellular inorganic phosphate (int P_i) peak is now visible at 2.62 ppm. The other peak assignments are within 0.02 ppm of the values provided in part A. C) Average of 2,000 two-pulse (90°–180° with 15 msec τ) scans on fertilized eggs from the same group of females as in A and B. P_i peak is now visible at 2.82 ppm. Other peak assignments are within 0.02 ppm of the values provided in A except γ-ATP, −4.94; α-ATP, −10.09. D) Portions of Figure 12B (solid line) and 12C (dotted line) are overlaid on an expanded scale so that the 0.2 ppm shift of the P_i peak is clear.

to achieve a satisfactory signal to noise ratio. Since cleavage occurs between 60 and 90 minutes after fertilization, the eggs usually underwent first cleavage during our measurements. Figure 12C shows the spectrum from about 1,000 fertilized eggs originally taken from the same group as the eggs studied in Figure 12A and B. A distinct difference is revealed by this T_2 experiment, which is best illustrated by overlaying expanded versions of Figure 12B and C as shown in Figure 12D. While the GPC and PCr peaks coincide quite well, there is a distinct shift in the P_i peak of 0.2 ppm. This shift was observed in all five groups of fertilized eggs, and all three groups of A23187-activated eggs studied. The average P_i chemical shifts are listed in Table III.

In order to convert these P_i chemical shifts into pH_i indicators, we studied the P_i shift dependence on pH in the cell extract solutions described in the "Methods" section. The extract pH was adjusted and measured with a pH-sensitive glass electrode before and after determining the P_i chemical shift with NMR. Figure 13 shows this pH dependence using extract buffer 2, which chelates

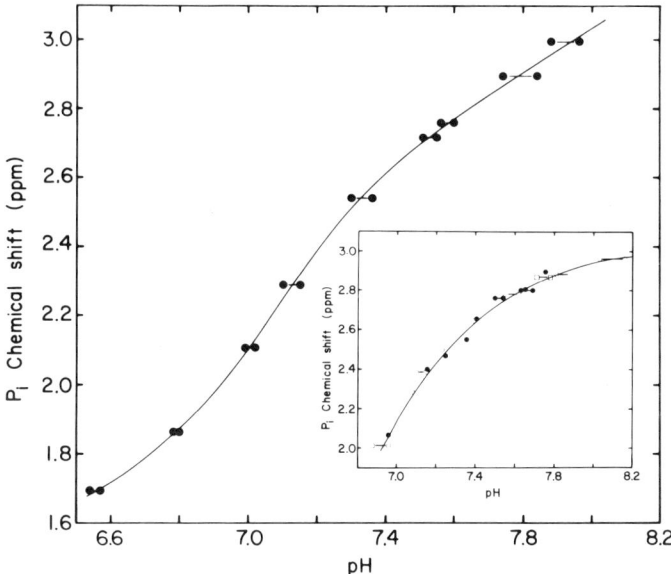

Fig. 13. Chemical shift of P_i as a function of pH in the frog egg extract solution made with buffer No. 2 at 20°C. A pH electrode was inserted into the NMR tube before and after the chemical shift measurement, and both values are plotted with a horizontal line drawn between them. The curve was drawn in by eye to fit the data points best. The inset figure is a plot of this same relationship in extract solution made with buffer No. 1 while varying temperature and protein conditions: Open circles represent data collected at 0°C on extract supernatant after centrifuging for 1 h at 12,000 g. Solid circles represent data collected at 20°C on supernatant of perchloric acid-precipitated extract after centrifuging for 2 h at 13,000 g. Squares represent data collected at 15°C on extract supernatant after centrifuging for 1 h at 12,000 g.

both Ca^{2+} and Mg^{2+} to physiological levels. The inset to Figure 13 shows the data taken with extract buffer 1 at two different temperatures and after perchloric acid precipitation of the protein. No significant temperature (0–20°C), protein, or Mg^{2+} (0.5–10 mM) dependence was found. The only way we could shift this curve was by making large ionic strength changes. When sea urchin egg extract solution [Winkler, this volume] with a 5-fold higher ionic strength was used, the curve was shifted approximately 0.2 ppm toward higher values. Thus, pH 7.4 corresponded to a P_i chemical shift of 2.62 in our Xenopus extract medium, but corresponded to a shift of 2.80 in sea urchin egg extract. Therefore, the cytoplasmic ionic strength must be known in order to generate accurate pH_i data. The pH_i values indicated by our P_i chemical shift measurements are listed in Table III. These measurements indicate that the fertilized egg's pH_i is 0.24 units more alkaline than that of the unfertilized egg. Similarly, eggs that had been activated with 2 µM of the $Ca^{2+}-H^+$ ionophore, A23187, showed an intracellular alkalinization of 0.22 pH units.

We have recently found that by applying a weak presaturation pulse at the frequency of the yolk phosphoprotein peak shoulder (1.7 ppm), we were able to eliminate this yolk peak from the subsequent FID and reveal the intracellular P_i peak. The main advantage of this method of eliminating the yolk peak over 1981] and first detected by us in single eggs using recessed-tip, pH-sensitive the T_2 is a better time resolution of 10 minutes. Figure 14 provides an example of this technique as applied to both unfertilized and activated eggs, and indicates a pH_i shift of 0.3 units between the two. We now plan to use this technique to study the time course of the pH_i change in order to compare it with the changes monitored with pH microelectrodes.

IV. DISCUSSION

These results represent the first measurement of pH_i during fertilization in a vertebrate egg. One of the most encouraging features of this study is the excellent agreement between the values for pH_i obtained in the Xenopus egg using two very different techniques. ^{31}P-NMR has allowed us to confirm the permanent rise in pH_i occurring at fertilization in a large number of eggs [Nuccitelli et al, microelectrodes [Webb and Nuccitelli, 1981]. The faster time resolution of the pH microelectrodes has enabled us to observe small, transient pH_i changes immediately following fertilization and during cleavage.

A. pH_i of the Unfertilized Egg

The pH_i of the unfertilized Xenopus egg is 7.4, which is more alkaline than that of the unfertilized sea urchin egg (6.9). The only previous report from Xenopus eggs at a similar stage indicated a pH_i of 7.7 in full-grown oocytes removed from the ovaries without their jelly coat [Lee and Steinhardt, 1981a]. In our study, we have investigated jelly-coated, mature oocytes squeezed from

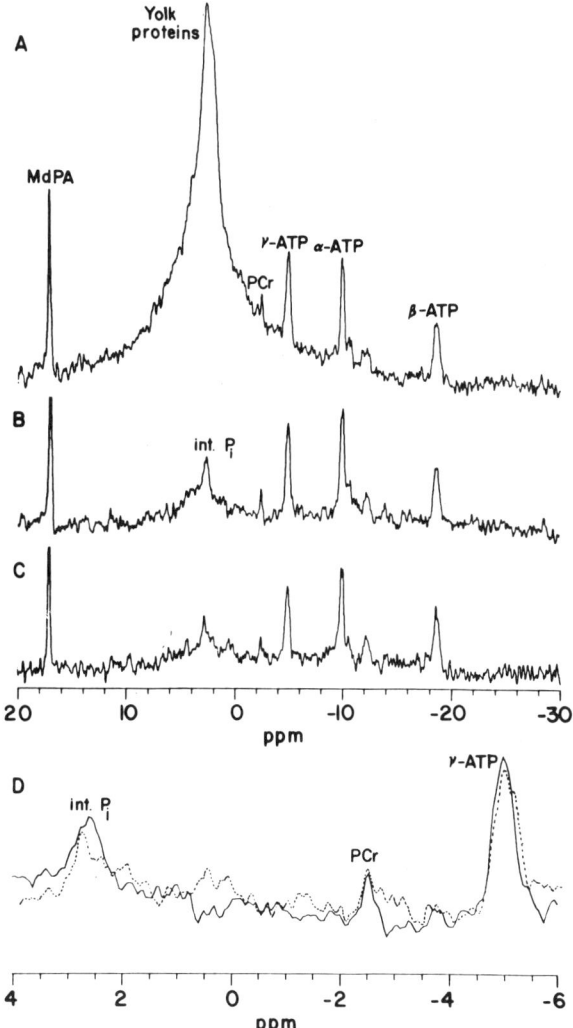

Fig. 14. ^{31}P-NMR spectra of approximately 1,000 Xenopus eggs before and after ionophore A23187 activation. A) Average of 1,000 single-pulse (54°) scans repeated every 300 msec on unfertilized eggs. Peak assignments using methylenediphosphonic acid (MdPA) as a reference are (in ppm): MdPA, 17.13; yolk proteins, 2.65; PCr, −2.51; γ-ATP, −5.01; α-ATP, −9.95; β-ATP, −18.84. B) Average of 1,000 two-pulse scans repeated every 500 msec on the same eggs as in A. The first pulse of each pair was a weak presaturation pulse on the yolk protein peak shoulder at 1.7 ppm, and the second pulse was a standard saturation pulse of 54°. Intracellular P_i peak is now visible at 2.59 ppm. The other peak assignments are within 0.03 ppm of the values in A. C) Average of 1,000 two-pulse scans repeated every 500 msec on A23187-activated eggs during a 9-min interval beginning 102 min after activation. The pulse pair was the same as used in B, and the P_i peak is now visible at 2.81 ppm. D) Portions of B (solid line) and C (dotted line) are overlaid on an expanded scale so that the 0.22 ppm shift of the P_i peak is clear.

the female frog. Prick activation often results from electrode insertion making these fertilization experiments difficult, but the application of 5 mM chlorobutanol prevented this. Several successful impalements in the absence of chlorobutanol showed that this substance was not the cause of the lower pH_i of the unfertilized egg (see Figs. 5B and 10A). The pH_i changes following impalement of the Xenopus egg include the expected injury-associated, transient acidification [Aickin and Thomas, 1975] but, in addition, a much slower acidification. This latter pH_i fall may reflect a recovery from alkalinization caused by localized activation at the site of pH electrode insertion. In Figure 11, injury induced in the already impaled unfertilized egg by tapping the pH electrode holder gave a transient acidification, with the pH_i returning to the level that existed prior to injury. The pH_i changes upon egg impalement were much less variable than any membrane depolarizations that occurred and, as in the sea urchin egg [Shen, this volume], there seemed to be no dependence of pH_i upon the membrane potential.

The Xenopus egg membrane has a high resistance [Kado et al, 1981] and an exceptionally low permeability to water and ions [De Laat et al, 1974]. Insertion of two electrodes might be expected to lead to depolarization of the membrane. Even eggs impaled with just voltage electrodes and having membrane potentials initially of -22 ± 4 mV (SD, n = 25), gradually depolarized to -17 ± 5 mV (SD, n = 23) 30 to 60 minutes later. A membrane potential of -8.65 ± 4.40 mV (SD) was reported for oocytes excised from Xenopus ovaries and matured in culture in the presence of progesterone [Kado et al, 1981]. Membrane potentials of a similar value were also found in other amphibian eggs including the toad, Bufo vulgaris (-12 mV) [Maéno, 1959], and the frog, Rana rugosa (-13 mV) [Ito, 1972], although Rana pipiens (-28 mV) [Cross and Elinson, 1980] was somewhat greater. Some of the variation in these values may be due to the different solutions used to bathe the eggs. It is worth noting that with a pH_o of 7.8, unfertilized pH_i of 7.4 and fertilized pH_i of 7.7, H^+ would be in equilibrium across the plasma membrane if the membrane potential were -24 mV and -6 mV, respectively. Nevertheless, pH_i remains unaffected by large changes in pH_o (Fig. 8) [Turin and Warner, 1980].

B. Initial Transient Acidification

The onset of the transient acidification 2–6 minutes after fertilization and its duration of about 11 minutes suggest an association with the rise in Ca^{2+} activity believed to occur at the time of the cortical reaction. Meech and Thomas [1977] showed that the injection of Ca^{2+} into snail neurones gives rise to an immediate decrease in pH_i that is proportional to the amount of Ca^{2+} injected. The transient nature of the acidification could be accounted for by sequestration of the released Ca^{2+} in exchange for H^+ by intracellular organelles, such as mitochondria, which are found in abundance in the cortex of Xenopus eggs [Grey et al, 1974].

The recovery of pH_i in the snail neurone following a 15-second injection of $CaCl_2$ takes approximately 10 minutes, which is very similar to the duration of the transient acidification in Xenopus eggs following fertilization, and also to the recovery time of the injury-related acidification following tapping of an already inserted pH electrode (Fig. 11). The variability in the size of the transient acidification and its onset time may be explained by localization of the Ca^{2+} release during the cortical reaction to the cortex and by the egg's strong buffering capacity [Turin and Warner, 1980], as well as by the distance the pH electrode tip is from the site of sperm entry. The cortical reaction in the 1.3-mm diameter Xenopus egg lasts for about 3.5 minutes [Hara and Tydeman, 1979; Grey et al, 1982], and probably does not begin until about 30 seconds after sperm–egg fusion [Gilkey et al, 1978; Steinhardt et al, 1977]. A wave of elevated free Ca^{2+} would be expected to cross the egg between 0.5 and 4 minutes after fertilization and reach the pH electrode halfway through this interval on average, ie, in 2.5 minutes. This is close to the average onset time of the transient acidification, and the standard deviation of 1.1 minutes indicates the range expected for such a mechanism. The source of the rise in Ca^{2+} activity at fertilization is unknown but is presumed to be derived from intracellular sites, since a similar rise occurs in eggs activated in Ca^{2+}-free EGTA solution [Gilkey et al, 1978].

Further support for an association with a rise in Ca^{2+} activity comes from the Ca^{2+}–H^+ ionophore, A23187, studies. In eggs activated with A23187, the transient acidification could begin *before* the activation potential and was larger in amplitude. Since the ionophore both carries Ca^{2+} into the egg and releases it from intracellular compartments, it may well raise the Ca^{2+} activity near the pH electrode very quickly before activation occurs.

C. Permanent pH_i Increase

The main pH_i change following fertilization is a permanent increase of 0.3 pH units. This is a slightly smaller increase than that seen in the sea urchin [Shen and Steinhardt, 1978] and occurs over a more alkaline range (pH 7.4–7.7). The onset time is also later in Xenopus and the time course much longer, which is not altogether surprising considering the greater size of this egg in comparison to that of the sea urchin. Impalement of a fertilized egg results in the injury-associated, transient acidification before the pH_i settles at about pH 7.7. This value is in close agreement with that previously obtained in fertilized Xenopus eggs [Turin and Warner, 1980; Lee and Steinhardt, 1981b] and also that found in Ambystoma [Spray et al, 1981]. These authors looked at dejellied embryos at a later stage of development. The large fertilization potential elicited in Cl⁻-free solution (Fig. 7A) and the hyperpolarizations at cleavage (Fig. 10A) are not associated with noticeable variations in pH_i, so that, as with the fertilized egg, pH_i is essentially unaffected by variations in the membrane potential. However, if the membrane potential of the unfertilized Xenopus egg is closer to -20 mV just before fertilization, then it would appear that fertilization results in a long-lasting depolarization to around -10 mV until the hyperpolarization begins at cleavage. This suggests a change in the membrane conductance, such as increased Cl^- or decreased K^+ conductance. In the sea urchin egg, fertilization has been shown to bring about an increased K^+ conductance [Shen and Steinhardt, 1980].

D. Nature of the Permanent pH_i Increase

Most of the contributions in this volume deal with either marine species or cells bathed by blood or hemolymph. Freshwater species such as Xenopus are in an environment where Na^+ and Cl^- are at much lower levels. The unreliable ionic environment of pond water and the exceptionally low membrane permeability of the Xenopus egg plasma membrane make ion exchange a less likely candidate for the pH_i change than in other systems. Unlike the sea urchin egg [Shen and Steinhardt, 1979], Xenopus has no extracellular Na^+ requirement for the pH_i increase at activation (Fig. 6B). Xenopus develops in an environment in which the Na^+ activity is normally much lower than the 20 mM intracellular Na^+ activity of the egg [De Laat et al, 1974; Slack et al, 1973], so that Na^+-H^+ exchange would require movement of Na^+ against its activity gradient. Even the development of a depolarizing fertilization potential in Xenopus relies on Cl^- efflux rather than Na^+ influx as in other species. Changes in extracellular Cl^- and pH_o also had some effect on pH_i of sea urchin eggs [Shen, this volume], which was not seen in Xenopus (Figs. 7 and 8). Although we have not ruled out exchange involving other ions, it is possible that the pH_i rise following fertilization is independent of extracellular ions, and that Xenopus relies instead upon an intracellular exchange system possibly involving membranous organelles such as mitochondria or endoplasmic reticulum. Our preliminary attempts at measuring an acid release concomitant with the permanent pH_i increase, as seen in certain marine species [Johnson et al, 1976; Paul, 1975; Ii and Rebhun, 1979], have so far proved unsuccessful.

E. Manipulation of pH_i

Although we have been unsuccessful at perturbing the permanent pH_i increase, our preliminary investigations with weak acids and bases have shown that the pH_i will respond to such treatments. Acetate, CO_2, and DNP all produce an acidification. The results with NH_3 are somewhat equivocal. The pH of F_1 solution containing 10 mM NH_4Cl or $(NH_4)_2 SO_4$ has to be raised to above 9.0 and close to its pK in order to achieve a pH_i increase. Even then the initial response is usually a drop in pH_i. This suggests that a marked permeability to the charged moiety (NH_4^+) exists in the Xenopus plasma membrane.

Amphibian eggs are capable of developing anaerobically up until the gastrula stage [Barth and Barth, 1954], and so the inhibition of ATP production by DNP due to uncoupling of oxidative phosphorylation will not, in itself, block this early development. Nevertheless, fertilized eggs so treated with DNP fail to develop further which may be due in part to the dramatic fall in pH_i (Fig. 9). Lee and Steinhardt [1981b] have shown that cleavage can be blocked in Xenopus by lowering pH_i to about 7.0. In the unfertilized sea urchin egg 0.5 mM DNP was reported to be without significant effect [Shen and Steinhardt, 1980; Shen, this volume], although causing a large acidification in the fertilized egg to around 6.8. The unfertilized egg pH_i is about 6.8 in the sea urchin, which is close to the pH_i to which Xenopus eggs are acidified by DNP.

F. Cyclic pH_i Oscillations

These have the same periodicity as the cleavage cycle but are out of phase with it, beginning about 24 minutes before the hyperpolarization associated with each cleavage. They also coincide very closely with the surface contraction waves reported by Hara et al [1980]. As in the case of the initial transient acidification following fertilization, these small cytoplasmic acidifications may indirectly reflect oscillations in the Ca^{2+} activity occurring during the cortical contraction waves. No fluctuations in Ca^{2+} activity have been detected through cleavage using Ca^{2+}-sensitive electrodes [Rink et al, 1980], but injection of Ca^{2+} beneath the plasma membrane or ionophore-induced release from intracellular stores will induce cortical contractions in frog eggs [Gingell, 1970; Schroeder and Strickland, 1974]. It therefore remains an appealing possibility that the surface contraction waves are accompanied by a Ca^{2+} increase that is beyond the spatial resolution of the Ca^{2+} electrode measurements. Slack et al [1973] reported an oscillating Na^+ activity in dejellied embryos that coincides with the surface contraction wave. However, the transient rise of 2–5 mM is in the wrong direction to support a Na^+–H^+ exchange mechanism at the plasma membrane, although it is consistent with a Na^+–Ca^{2+} exchange.

Another good candidate for the generation of these small pH_i oscillations is CO_2 production. The mitotic cycle of some dividing embryos including the frog, Rana platyrrhina, is associated with a rhythm in the rates of oxygen uptake and CO_2 output. In the frog embryo, the cycling of CO_2 output was observed from first cleavage onwards [Zeuthen and Hamburger, 1972], whereas in the embryos of two marine invertebrates, Urechis and Dendraster, this rhythm began after the first 5 or 6 divisions [Frydenberg and Zeuthen, 1960]. This observation suggested to the investigators a gradual fall in pH_i during mitosis and a gradual rise during interphase, which would agree with our findings in Xenopus.

Somewhat larger oscillations in pH_i during cell division have been reported in Physarum [Gerson and Burton, 1976; Steinhardt and Morisawa, this volume],

Tetrahymena [Gillies and Deamer, 1979; Gillies, this volume], and lymphocytes [Gerson, this volume] using other pH_i-measuring techniques. However, these larger changes are in the opposite direction, so they appear to be unrelated to the pH_i oscillations reported here. A recent study of sea urchin embryos did not detect significant changes in pH_i through the first two cleavage divisions [Johnson and Epel, 1981].

G. The Importance of the Permanent pH_i Increase for the Activation of Development

We do not yet have any evidence for a direct function for the pH_i rise in activation. However, there is a good temporal correlation with a doubling of the rate of protein synthesis between 10 and 45 minutes after activation, as indicated by the percentage of ribosomes present as polysomes [Woodland, 1974]. In contrast, measurements of pH_i during in vitro maturation of Xenopus oocytes do not support any direct relationship with the twofold increase in protein synthesis occurring at that time [Lee and Steinhardt, 1981a]. Therefore, direct measurements of protein synthesis rates during imposed changes in pH_i are needed to clarify this relationship.

Another observation related to this question is the insensitivity of Xenopus egg cleavage to pH_i above 7.3. Lee and Steinhardt [1981b] have shown that cleavage proceeds normally in embryos whose pH_i has been lowered to the unfertilized level of 7.3 beginning after the permanent pH_i increase. Cleavage is not blocked until the pH_i is lowered to about 7.0. Clearly, we must block the permanent pH_i increase entirely and observe the effect on activation.

ACKNOWLEDGMENTS

This investigation was supported by NSF grant PCM 78-26022 and a University of California Faculty Research grant to R.N., and NIH grant HD-04906 to Dr. Jerry Hedrick. The Nicolet magnetics NT-200 NMR spectrometer was purchased in part by an NSF grant to the Department of Chemistry, and was operated by Dr. Gerry Matson.

V. REFERENCES

Aickin CC, Thomas RC: Micro-electrode measurement of the internal pH of crab muscle fibres. J Physiol 252:803, 1975.

Barth LG, Barth LJ: "The Energetics of Development." New York: Columbia University Press, 1954, p 117.

Brown TR, Ugurbil K, Shulman RG: ^{31}P nuclear magnetic resonance measurements of ATPase kinetics in aerobic Escherichia coli cells. Proc Natl Acad Sci USA 74:5551, 1977.

Campbell ID, Dobson CM: The application of high resolution nuclear magnetic resonance to biological systems. Meth Biochem Anal 25:33, 47–48, 1979.

Colman A, Gadian DG: ^{31}P nuclear-magnetic-resonance studies on the developing embryos of Xenopus laevis. Eur J Biochem 61:387, 1976.

Cross NL, Elinson RP: A fast block to polyspermy in frogs mediated by changes in the membrane potential. Dev Biol 75:187, 1980.

De Laat SW, Buwalda RJA, Habets AMMC: Intracellular ionic distribution, cell membrane permeability and membrane potential of the Xenopus egg during first cleavage. Exp Cell Res 89:1, 1974.

Epel D: The triggering of development at fertilization. In Ebert J, Tokindo O (eds): "Mechanisms of Cell Change." New York: John Wiley and Sons, 1979, p 17.

Frydenberg O, Zeuthen E: Oxygen uptake and carbon dioxide output related to the mitotic rhythm in the cleaving eggs of Dendraster excentricus and Urechis caupo. Compt vend Lab Carlsberg 31:423, 1960.

Gerson DF, Burton AC: The relation of cycling of intracellular pH to mitosis in the acellular slime mould Physarum polycephalum. J Cell Physiol 91:297, 1976.

Gilkey JC, Jaffe LF, Ridgway EB, Reynolds GT: A free calcium wave traverses the activating egg of the medaka, Oryzias latipes. J Cell Biol 76:448, 1978.

Gillies RJ, Deamer DW: Intracellular pH changes during the cell cycle in Tetrahymena. J Cell Physiol 100:23, 1979.

Gingell D: Contractile responses at the surface of an amphibian egg. J Embryol Exp Morphol 23:583, 1970.

Gould-Somero M, Jaffe LA, Holland LZ: Electrically mediated fast polyspermy block in eggs of the marine worm, Urechis caupo. J Cell Biol 82:426, 1979.

Grainger JL, Winkler MM, Shen SS, Steinhardt RA: Intracellular pH controls protein synthesis rate in the sea urchin egg and early embryo. Dev Biol 68:396, 1979.

Grey RD, Wolf DP, Hedrick JL: Formation and structure of the fertilization envelope in Xenopus laevis. Dev Biol 36:44, 1974.

Grey RD, Bastiani MJ, Webb DJ, Schertel ER: An electrical block is required to prevent polyspermy in eggs fertilized by natural mating of Xenopus laevis. Dev Biol (in press), 1982.

Hara K, Tydeman P: Cinematographic observation of an "activation wave" (AW) on the locally inseminated egg of Xenopus laevis. Roux Arch Dev Biol 186:91, 1979.

Hara K, Tydeman P, Kirschner M: A cytoplasmic clock with the same period as the division cycle in Xenopus eggs. Proc Natl Acad Sci USA 77:462, 1980.

Hollinger TG, Corton GL: Artificial fertilization of gametes from the South African clawed frog, Xenopus laevis. Gamete Res 3:45, 1980.

Hutter OF, Warner AE: The pH sensitivity of the chloride conductance of frog skeletal muscle. J Physiol (Lond) 189:403, 1967.

Ii I, Rebhun LI: Acid release following activation of surf clam (Spisula solidissima) eggs. Dev Biol 72:195, 1979.

Illingworth JA: A common source of error in pH measurements. Biochem J 195:259, 1981.

Ito S: Effects of media of different ionic composition on the activation potential of anuran egg cells. Dev Growth Diff 14:217, 1972.

Jaffe LA: Fast block to polyspermy in sea urchin eggs is electrically mediated. Nature 261:68, 1976.

Johnson CH, Epel D: Intracellular pH of sea urchin eggs measured by the DMO method. J Cell Biol 89:284, 1981.

Johnson JD, Epel D, Paul M: Intracellular pH and activation of sea urchin eggs after fertilization. Nature 262:661, 1976.

Kado RT, Marcher K, Ozon R: Electrical membrane properties of the Xenopus laevis oocyte during progesterone-induced meiotic maturation. Dev Biol 84:471, 1981.

Lee SC, Steinhardt RA: pH changes associated with meiotic maturation in oocytes of Xenopus laevis. Dev Biol 85:358, 1981a.

Lee SC, Steinhardt RA: Observations on intracellular pH during cleavage of eggs of Xenopus laevis. J Cell Biol 91:(in press, November issue), 1981b.

Maéno T: Electrical characteristics and activation potential of Bufo eggs. J Gen Physiol 43:139, 1959.

Martell AE, Smith RM: "Critical Stability Constants," Vol 1. New York: Plenum, 1974, p 269.

Meech RW, Thomas RC: The effect of calcium injection on the intracellular sodium and pH of snail neurones. J Physiol (Lond) 265:867, 1977.

Miyazaki S, Hirai S: Fast polyspermy block and activation potentials: Correlated changes during oocyte maturation of a starfish. Dev Biol 70:327, 1979.

Navon G, Ogawa S, Shulman RG, Yamane T: High-resolution ^{31}P nuclear magnetic resonance studies of metabolism in aerobic Escherichia coli cells. Proc Natl Acad Sci USA 74:888, 1977.

Nuccitelli R, Webb DJ, Lagier ST, Matson GB: ^{31}P NMR reveals an increase in intracellular pH after fertilization in Xenopus eggs. Proc Natl Acad Sci USA 78:4421, 1981.

Paul M: Release of acid and changes in light-scattering properties following fertilization of Urechis caupo eggs. Dev Biol 43:299, 1975.

Ridgway EB, Gilkey JC, Jaffe LF: Free calcium increases explosively in activating medaka eggs. Proc Natl Acad Sci USA 74:623, 1977.

Rink RJ, Tsien RY, Warner AE: Free calcium in Xenopus embryos measured with ion-selective microelectrodes. Nature 283:658, 1980.

Schroeder TE, Strickland DL: Ionophore A23187, calcium and contractility in frog eggs. Exp Cell Res 83:139, 1974.

Shen SS, Steinhardt RA: Direct measurement of intracellular pH during metabolic derepression of the sea urchin egg. Nature 272:253, 1978.

Shen SS, Steinhardt RA: Intracellular pH and the sodium requirement at fertilization. Nature 282:87, 1979.

Shen SS, Steinhardt RA: Intracellular pH controls the development of new potassium conductance after fertilization of the sea urchin egg. Exp Cell Res 125:55, 1980.

Slack C, Warner AE, Warren RL: The distribution of sodium and potassium in amphibian embryos during early development. J Physiol (Lond) 232:297, 1973.

Spray D, Harris AL, Bennett MVL: Gap junctional conductance is a simple and sensitive function of intracellular pH. Science 211:712, 1981.

Steinhardt RA, Zucker R, Schatten G: Intracellular calcium release at fertilization in the sea urchin egg. Dev Biol 58:185, 1977.

Thomas RC: Intracellular pH of snail neurones measured with a new pH-sensitive glass microelectrode. J Physiol (Lond) 238:159, 1974.

Thomas RC: Ionic mechanisms of the H^+ pump in a snail neurone. Nature 262:54, 1976.

Turin L, Warner AE: Intracellular pH in early Xenopus embryos: Its effect on current flow between blastomeres. J Physiol 300:489, 1980.

Wallace RA, Jared DW, Dumont JN, Sega MW: Protein incorporation by isolated amphibian oocytes. III. Optimum incubation conditions. J Exp Zool 184:321, 1973.

Webb DJ, Nuccitelli R: Direct measurement of intracellular pH changes in Xenopus eggs at fertilization and cleavage. J Cell Biol 91:(in press, November issue), 1981.

Winkler MM, Steinhardt RA, Grainger JL, Minning L: Dual ionic controls for the activation of protein synthesis at fertilization. Nature 287:558, 1980.

Woodland HR: Changes in polysome content of developing Xenopus laevis embryos. Dev Biol 40:90, 1974.

Zeuthen E, Hamburger K: Mitotic cycles in oxygen uptake and carbon dioxide output in the cleaving frog egg. Biol Bull 143:699, 1972.

Intracellular pH: Its Measurement, Regulation, and
Utilization in Cellular Functions, pages 325-340
© 1982 Alan R. Liss, Inc., 150 Fifth Avenue, New York, NY 10011

Regulation of Protein Synthesis in Sea Urchin Eggs by Intracellular pH

Matthew M. Winkler
Department of Biological Chemistry, School of Medicine, University of California, Davis, California 95616

I.	Introduction	325
II.	Materials and Methods	326
	A. Preparation of the Cell-Free System	326
	B. Protein Synthesis in the Cell-Free System	327
	C. Solutions	327
III.	Results and Discussion	327
	A. Intracellular pH and Protein Synthesis	328
	B. Translational Efficiency	332
	C. Summary	339
IV.	References	339

I. INTRODUCTION

One of the most spectacular regulatory events in nature is the activation of metabolism that occurs after fertilization of the sea urchin egg. The unfertilized egg is a repressed cell with low rates of protein synthesis and no nuclear DNA synthesis. Fertilization results in a total alteration of this cell both structurally and metabolically. This includes changes in surface morphology and membrane permeability, a large increase in respiration, activation of various enzymes such as NAD kinase, the onset of DNA synthesis, and a 5–30-fold increase in the rate of protein synthesis [for a review see Epel, 1978]. There is also a release of intracellular Ca^{++} and an increase in the intracellular pH of about 0.4 units [Steinhardt et al, 1977; Shen and Steinhardt, 1978], both of which events have been shown to play major regulatory roles at fertilization.

In this work I will focus on the activation of protein synthesis and its control by intracellular pH (pH_i). I will describe experiments which 1) demonstrate the regulatory role of pH_i, 2) separate the effects of changes in intracellular Ca^{++} from those of pH_i, and 3) describe the use of a cell-free translation system to analyze more directly the mechanism by which intracellular pH regulates protein synthesis in the sea urchin egg.

The unfertilized egg is a cell primed for protein synthesis. It contains large amounts of apparently functional ribosomes, tRNAs, and other components of the protein synthetic machinery. It also contains stored maternal mRNA, which has been shown by the activation of protein synthesis in the absence of RNA synthesis in anucleate and actinomycin D-treated eggs [Denny and Tyler, 1964; Gross and Cousineau, 1963]. Thus potentially functional translational machinery and mRNA are present in unfertilized eggs, and yet protein synthesis is repressed. These observations and others have given rise to the "masked message hypothesis," which suggests that mRNA in the unfertilized egg is stored as a nontranslatable messenger ribonucleoprotein (mRNP) particle that is activated following fertilization. Further support for this hypothesis comes from measurements of translational efficiency in eggs and early embryos. Following fertilization there is a twofold increase in the peptide elongation rate and no increase in the number of ribosomes per polysome [Brandis and Raff, 1979; Hille and Albers, 1979; Humphreys, 1969]. Less than 1% of the ribosomes are present in polysomes in the unfertilized egg. This value increases to 10% by 30 minutes after fertilization and to 20% by 2 hours after fertilization [Humphreys, 1971]. Thus the major component in the increase in the rate of protein synthesis is the mobilization of stored maternal mRNA into polysomes. Although these findings are consistent with the masked message hypothesis, there are few data that directly implicate masked message in the control of protein synthesis.

II. MATERIALS AND METHODS

Measurement of protein synthesis and intracellular pH in intact eggs was as described [Grainger et al, 1979; Winkler et al, 1980].

A. Preparation of the Cell-Free System

The preparation of the cell-free translation system is described in Winkler and Steinhardt [1981] or by a modification described below. Eggs of the sea urchin Lytechinus pictus are shed by introduction of 0.55 M KCl to the coelomic cavity and dejellied by settling in seawater acidified to pH 4.5 with HCl. Eggs of the sea urchin Stronglyocentrotus purpuratus yield essentially no activity, and those of Arbacia punctulata yield a system with marginal activity. All further operations are performed on ice. Eggs are then washed four times in either the pH 6.9 or 7.4 media described below using a hand centrifuge. The fourth wash has 1 mg/ml soy bean trypsin inhibitor and 0.5 mg/ml reduced glutathione added. An equal volume of packed eggs and the last wash is homogenized with two strokes of a tight-fitting stainless steel dounce type of homogenizer. The homogenate is centrifuged at 25,000 RPM in an SW-56 rotor, and the supernatant is quickly frozen in small aliquots with liquid nitrogen.

B. Protein Synthesis in the Cell-Free System

Typically 180 μl aliquots of frozen system are thawed quickly and put on ice. Five microliters of a master mix containing 30 mM ATP, 4 mM GTP, 100 or 200 mM Arginine phosphate, and 34 mM MgCl adjusted to pH 7 is added. This master mix is stored at −70°C. Five microliters of labeled amino acid, typically ^{35}S-methionine and 10 μl of water or whatever a particular experiment requires is added. The system can still accommodate the addition of 10% volume of mRNA. It is not necessary to add extra amino acids with the nondialyzed system. Protein synthesis is initiated by incubating at 20°C. Incorporation generally continues for about 1 hour. The system shows the highest activity when it is prepared during the summer months, the normal breeding season for Lytechinus pictus. High levels of arginine phosphate yield the most active systems, but decrease the differences in the rate of protein synthesis seen between the two pH values.

C. Solutions

Media were so designed that only pH and gluconate would vary between the two pHs. Gluconate was used as a counter-ion instead of chloride. The composition of the pH 6.9 medium was 40 mM NaCl, 4.19 mM $CaCl_2$, 0.59 mM $MgCl_2$, 0.3 M glycine, 50 mM HEPES, 10 mM EGTA, 225 mM glycerol, and 115 mM K-gluconate. The composition of the pH 7.4 medium was 40 mM NaCl, 8.56 mM $CaCl_2$, 0.57 mM $MgCl_2$, 0.3 M glycine, 50 mM HEPES, 10 mM EGTA, 277 mM glycerol, and 100 mM K-gluconate. The pH was adjusted to 6.9 and 7.6 with KOH. This resulted in media that contained 150 mM K^+, 40 mM Na^+, approximately 50 mM Cl^-, 2 mM Mg^{++}, 5×10^{-7} M Ca^{++}, and 0.95 OsM. The differing ionic compositions at pH 6.9 and 7.4 were required because the calcium and magnesium binding constants for EGTA vary with pH, and because of the differing amounts of KOH required to achieve the desired pH. Eggs homogenized in the pH 6.9 media yield a homogenate with a pH of about 6.85, and eggs homogenized in the pH 7.4 media adjusted to pH 7.6 yield a homogenate with a pH of about 7.42.

III. RESULTS AND DISCUSSION

Fertilization results in a rapid increase in pH_i from 6.84 to 7.26 [Shen and Steinhardt, 1978]. There are a number of experimental results that indicate that this increase in pH_i plays a major role in controlling the increase in the rate of protein synthesis. NH_4Cl and other weak penetrating bases will increase pH_i, activate protein synthesis, and start the cell cycle [Shen and Steinhardt, 1978; Epel et al, 1974]. These bases raise pH_i by the action of the unchanged species diffusing into the egg and combining with a ^+H as the unchanged species comes to equilibrium with the changed species [Winkler and Grainger, 1978]. Many of the early events associated with Ca^{++} release are bypassed and the egg may still be fertilized [Epel et al, 1974]. Another line of evidence linking the increase in pH_i with the activation of protein synthesis is experiments in which the increase in pH_i is blocked by rapidly transferring eggs, 30 seconds after fertilization to Na^+ free seawater. Under these conditions, protein synthesis remains at the unfertilized level and development is arrested. If Na^+ is added back, there is an increase in pH_i followed by an increase in the rate of protein synthesis and the resumption of development [Johnson et al, 1976; Shen and Steinhardt, 1979; Winkler et al, 1980].

A. Intracellular pH and Protein Synthesis

Figure 1 shows measurements of intracellular pH and protein synthesis for unfertilized, NH_4Cl-activated, and fertilized eggs. In this experiment eggs were "preloaded" with labeled amino acid prior to the start of the experiment in order to circumvent artifacts due to changes in amino acid permeability that occur at fertilization. About 3 minutes after the increase in intracellular pH, the rate of protein synthesis increases in the fertilized eggs. In spite of the fact that NH_4Cl treatment results in an even larger pH increase, the rate of protein synthesis is less than that seen in fertilized eggs. The significance of this low rate of protein synthesis in NH_4Cl activated eggs will be discussed later.

Another line of evidence demonstrating the link between pH_i and protein synthesis is experiments in which the pH_i of fertilized eggs is artificially decreased by treatment with weak penetrating acids such as Na-acetate. Figure 2 shows the effect of 10 mM Na-acetate in SW pH 6.5 on pH_i and protein synthesis. The pH_i is decreased to the unfertilized level by a 25-minute acetate treatment. The "preload" method of measuring protein synthesis is not suitable for experiments lasting long enough to deplete the labeled amino acid significantly. In this experiment the rate of protein synthesis was measured with a pulse label technique that compensates for differences in amino acid uptake. Acetate treatment reduces the rate of protein synthesis, although not to levels so low as in the unfertilized egg. In addition the cell cycle is arrested. Once the eggs are returned to SW pH 8.0, protein synthesis quickly accelerates and the cell cycle resumes. Thus pH_i appears to act as more than just a trigger to signal the onset of protein synthesis, but can act in a reversible manner to increase and decrease the rate of protein synthesis. The lag between changes in pH_i and the response of the protein synthetic machinery suggests that the changes in intracellular pH are not effecting the protein synthetic machinery directly, the way pH effects an enzymatic optimum, but rather is acting as a component of a regulatory pathway.

The observation that NH_4Cl activation does not stimulate protein synthesis to levels so high as that seen in fertilized eggs suggests that other factors in addition to pH play a role in the activation of protein synthesis at fertilization. There is a large transient release of intracellular Ca^{++} immediately following

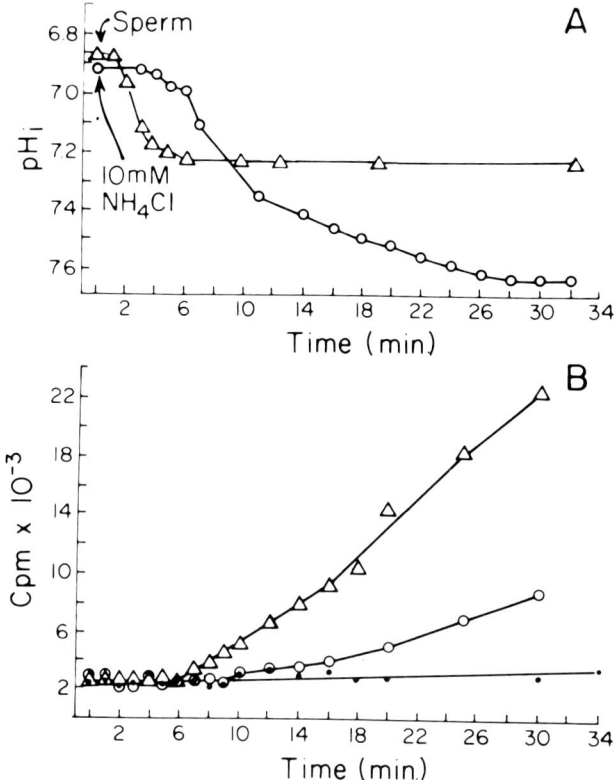

Fig. 1. A) The change in intracellular pH (pH_i) following fertilization or NH_4Cl activation. The eggs were either fertilized (△) or activated with 10 mM NH_4Cl, pH 8 (○), and the increase in pH_i was continuously monitored with H^+-sensitive microelectrodes. B) The increase in the rate of protein synthesis following fertilization of NH_4Cl activation. Eggs were preloaded and then placed in SW (unfertilized, ●), SW containing sperm (fertilized, △), or SW containing 10 mM NH_4Cl, pH 8 (NH_4Cl activated, ○). Samples were taken every 1 minute for the first 10 minutes and every 2 minutes thereafter.

fertilization that is a likely candidate for a regulatory role [Steinhardt et al, 1977]. Figure 3 shows the results of an experiment designed to separate the effects of pH and Ca^{++} on the rate of protein synthesis. Treatment of eggs with the Ca^{++} inophore A23187 in O-Na SW results in a Ca^{++} release, but no increase in pH_i (Fig. 3a). There is no stimulation of protein synthesis. Treatment of eggs with 10 mM NH_4Cl O-Ca^{++} SW (Fig. 3b) results in an increase in pH_i with no Ca^{++} release or influx. This results in a modest increase in the rate of protein synthesis. This increase is smaller than that observed after treatment with 10 mM NH_4Cl (Fig. 3c) in which there is both an increase in pH_i and a small Ca^{++}

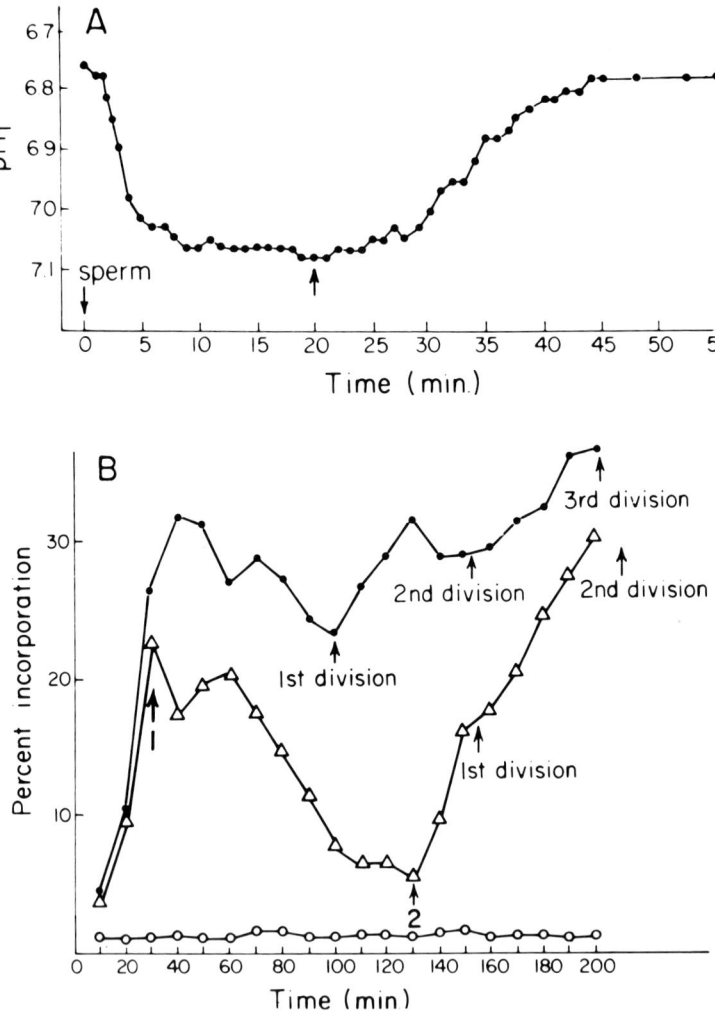

Fig. 2. A) The effect of Na-acetate on pH_i of fertilized eggs. Eggs were fertilized at t = 0, and at 20 minutes after fertilization (arrow) the egg was perfused with SW containing 10 mM Na-acetate, pH 6.5. B) The effect of Na-acetate on the rates of protein synthesis and cell division. Unfertilized control (○); fertilized control (●); fertilized with Na-acetate treatment as described below (△). Unfertilized eggs were added to SW containing sperm at t = 0; time points were taken every 10 minutes, and the rate of protein synthesis was determined by the pulse label technique. At 30 minutes after fertilization, one batch of eggs (△) was resuspended in SW containing 10 mM Na-acetate, pH 6.5 (arrow 1). At 130 minutes, this same batch of eggs was washed twice in SW, pH 8, and resuspended in SW at the original concentration (0.1%, arrow 2). Samples were also taken every 15 minutes, fixed in EtOH:acetic acid (3:1), stained in orcein, and scored for nuclear envelope breakdown, chromosome condensation, mitotic spindle formation, and cell division.

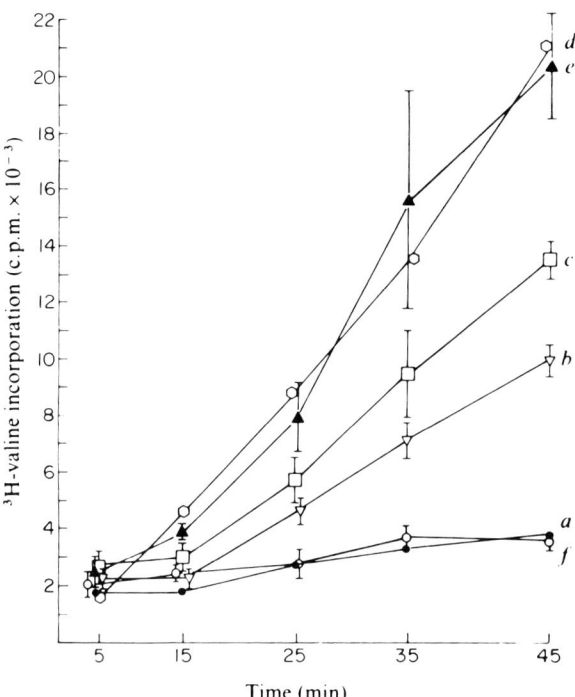

Fig. 3. Total ^{3}H-valine incorporation over a 45-minute period starting 5 minutes after activation in various ionic media. The mean incorporation of label into unfertilized eggs from four experiments was determined for each time point. The ratio of this mean to the unfertilized rate in each of these four experiments was used to normalize the data presented for the different ionic conditions. The plotted data points represent the mean of the normalized incorporation data. The value shown represents the amount of incorporation ± SEM for four repeats of this experiment testing five different conditions plus fertilization. The standard errors for fertilization have been omitted, and the time points slightly spread out for clarity. *a*) Eggs activated by 2.5 μM A23187 in O-Na$^+$ artificial sea water: intracellular Ca^{++} release without an intracellular pH rise. *b*) Eggs activated by 10 mM NH$_4$Cl in O-Ca^{++} artificial seawater: a rise in intracellular pH without a Ca^{++} release or influx. *c*) Eggs activated in 10 mM NH$_4$Cl in O-Na$^+$ artificial seawater: an intracellular pH rise with some Ca^{++} influx. *d*) Eggs activated in 10 mM NH$_4$Cl and 2.5 μM A23187 in O-Na$^+$ artificial seawater: both an intracellular pH rise and a intracellular Ca^{++} release. *e*) Eggs fertilized normally in natural seawater for comparison purposes. *f*) Unfertilized eggs in natural seawater. Artificial seawater: 484 mM NaCl, 10 mM KCl, 27 mM MgCl$_2$, 29 mM MgSO$_4$, 11 mM CaCl$_2$, 2.4 mM NaHCO$_3$, pH 8.0. O-Na$^+$ artificial seawater had choline Cl substituted for NaCl and KHCO$_3$ for NaHCO$_3$. O-Ca^{++} artificial seawater had 16 mM NaCl substituted for CaCl$_2$ and 2 mM EGTA added.

influx. Treatment of eggs with both Ca^{++} ionophore and 10 mM NH_4Cl in O-Na SW results in both an increase in pH_i and a Ca^{++} release (Fig. 3d). Protein synthesis is stimulated to levels comparable to those of the fertilized controls (Fig. 3e). The conclusions from these experiments are that a Ca^{++} release in the absence of a pH_i increase will not stimulate protein synthesis. A pH_i increase in the absence of a Ca^{++} release will yield a partial stimulation of the rate of protein synthesis. Both a pH_i increase and a Ca^{++} release are required, however, for protein synthesis to be stimulated to fertilized levels.

In order to study further the mechanism by which pH regulates protein synthesis, it became obvious that a cell-free translation system would be a large asset. It would allow the introduction of different components of the translation system such as exogenous mRNAs, and would provide conditions in which ionic parameters such as pH and Ca^{++} could be precisely varied at will. Our strategy in developing such a system was to use the most recent knowledge of the ionic conditions in the egg to develop media that would closely mimic the intracellular environment, including the use of Ca^{++}-Mg^{++}-EGTA buffers to maintain the extremely low levels of free Ca^{++} measured intracellularly, low Cl^- levels, and high levels of glycine and glycerol for osmotic balance. Levels of amino acid incorporation were high enough in our initial formulations to allow a process of sequential optimization for each component of the media. Figure 4 shows the incorporation of ^{3}H-valine into protein of homogenates of unfertilized eggs incubated at pH 6.9 and 7.4. These pH values are close to those of the unfertilized and fertilized eggs. Figure 5A shows the kinetics of incorporation at various pH values. Figure 5B shows the effect of pH on protein synthesis across a wide range of pH values. There is a sharp increase in the rate of protein synthesis starting at pH 6.9 with an optimum at pH 7.4. This corresponds closely to the pH values in vivo where protein synthesis is activated at fertilization.

B. Translational Efficiency

The effect of fertilization, in the intact egg, or an increase in pH in the cell-free system, on the protein synthetic machinery can be determined by measuring the translational efficiency before and after the stimulation of protein synthesis. For the purpose of this discussion there are three variables in question: The first is the speed at which a ribosome moves down the message, the second is the number of ribosomes per polysome, and the third is the total amount of mRNA in polysomes. Recent measurements of the peptide elongation rate in intact eggs demonstrate a doubling after fertilization [Brandis and Raff, 1978; Hille and Albers, 1979]. Figure 6 shows the measurement of elongation rate in the cell-free system by a very simple and direct method. Tobacco Mosaic Virus (TMV) mRNA is added to the cell-free system and, at 2-minute intervals, aliquots are removed and analyzed on SDS polyacrylamide gels with fluorography. The amount of time required for the first appearance of the completed TMV product

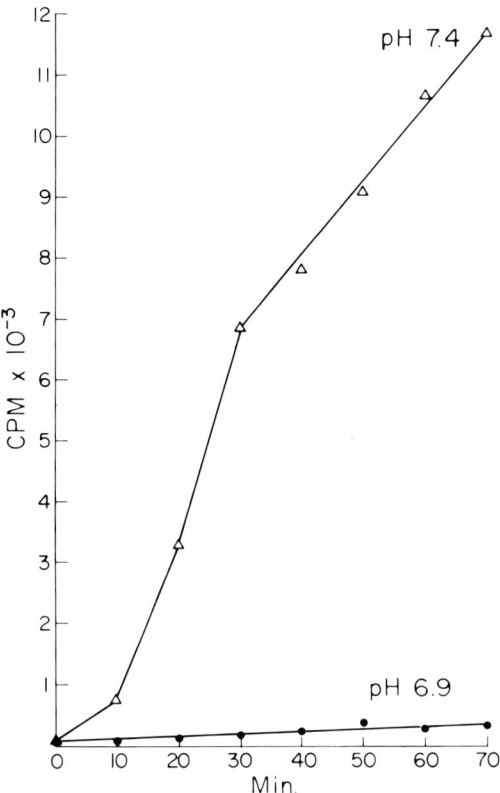

Fig. 4. Incorporation of ^{3}H-valine at pH 6.9 and 7.4 per 10 µl aliquot of the cell-free system prepared from unfertilized eggs and dialyzed to pH 6.9 and 7.4.

represents the maximal possible transit time. The cell-free system required 36 minutes at pH 6.9, and 18 minutes at pH 7.4 to synthesize the 110,000 MW TMV product. This finding suggests that the doubling of elongation rate at fertilization is due to the increase in pH_i and may represent a direct effect of pH on the protein synthetic machinery [for another view see Brandis and Raff, 1979].

Polysome size has been compared in the unfertilized and fertilized egg [Humphreys, 1969]. The absence of a significant change in size suggested that the initiation rate did not increase after fertilization. If the initiation rate increased, there should have been an increased number of ribosomes per polysome. (Note: This conclusion is dependent on the assumption that there does not exist a mechanism for maintaining a constant size for polysomes.)

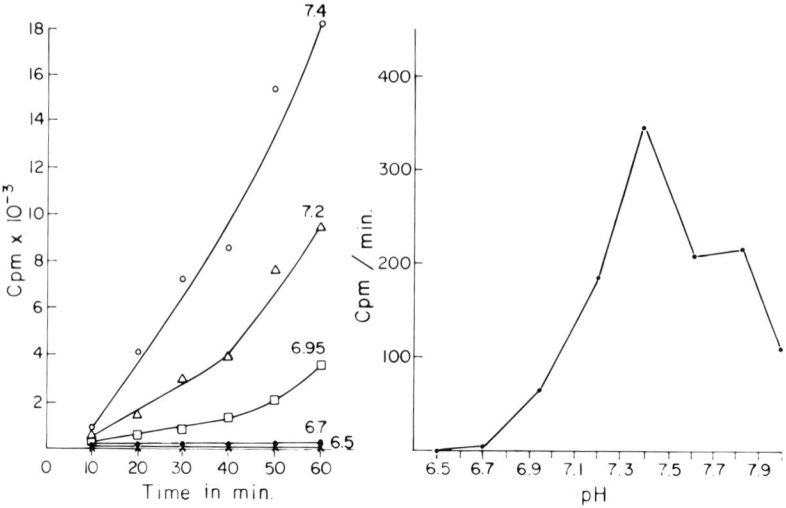

Fig. 5. Effect of pH on (^{3}H)valine incorporation. Left) (^{3}H)Valine incorporation per 10 μl aliquot of the cell-free system prepared from unfertilized eggs and dialyzed to the various pH values. The pH value is that of the dialysis medium at the completion of the experiment. For clarity, pH values above 7.4 have been omitted. Ca- and Mg-acetate were used instead of $CaCl_2$ and $MgCl_2$ so that the Cl^- concentration would not change as differing amounts of Ca and Mg were added. This is necessary because EGTA binding constants vary with pH. Differing amounts of glycerol were added to maintain 0.95 Osm and the pH was adjusted with Tris base. Right) The rate of (^{3}H)valine incorporation plotted against pH. This is the same data as at left except that values are given up to pH 8. The rate is the slope of a line determined by linear regression of six time points taken at 10-minute intervals.

Figure 7 shows a comparison of polysome size in the cell-free system at pH 6.9 and 7.4. The cell-free system at pH 6.9 was incubated with ^{14}C-valine and at pH 7.4 with ^{3}H-valine. The two fractions were combined and sedimented on the same sucrose gradient so that the profiles would be directly comparable. Thus in the cell-free system, as in the intact egg, polysome size remains constant even when the rate of protein synthesis is increased. In this particular experiment there was a 16.6-fold stimulation of the rate of protein synthesis. Since polysome size remains constant between the 2 pH values, only a 2-fold of the 16.6-fold stimulation is due to increased translational efficiency. The remaining 8.3-fold stimulation is presumably due to the mobilization of stored maternal mRNA in the cell-free system.

Development of the cell-free system makes it possible to test directly the hypothesis that the low rate of protein synthesis in the unfertilized egg is due

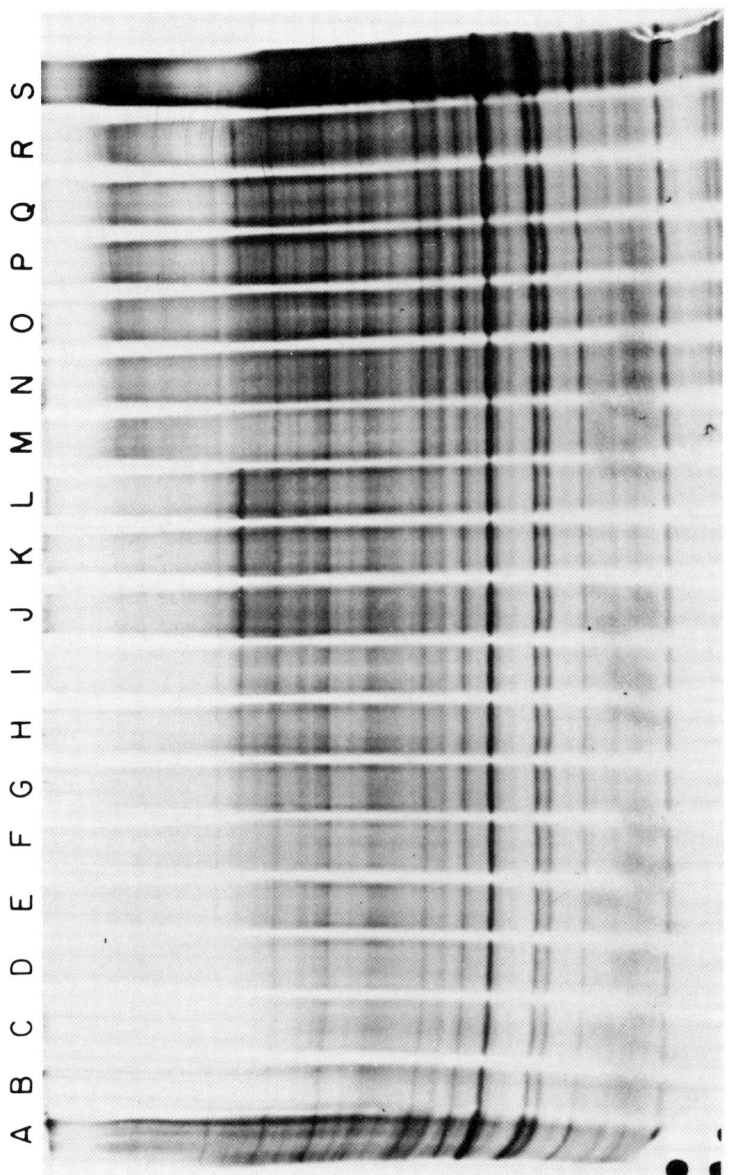

Fig. 6. Autoradiograph of a 10% SDS polyacrylamide gel displaying the newly synthesized products at 2-minute intervals after the addition of 25 µg/ml TMV mRNA. Slots A and S are controls without added TMV at pH 6.9 and 7.4. Slots B–L are pH 6.9 from 26 to 46 minutes after the addition of TMV mRNA. Slots M–R are pH 7.4 from 14 to 24 minutes after the addition of TMV mRNA.

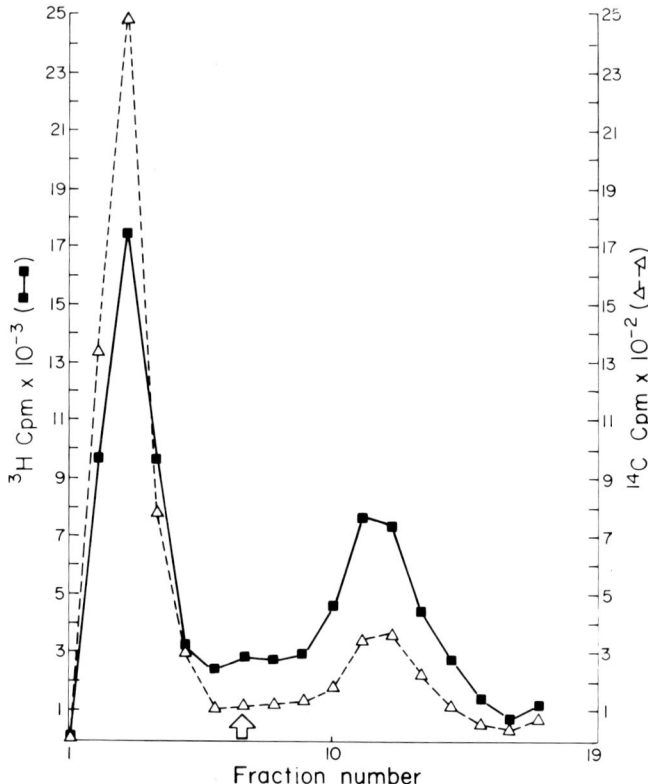

Fig. 7. Distribution of ^{14}C and ^{3}H-valine in nascent polypeptides on sucrose gradients from the cell-free system at pH 6.9 and 7.4. After a 30-minute incubation at either pH 6.9 or 7.4, the two fractions were combined and centrifuged on a 15–50% sucrose gradient for 150 minutes at 37,000 rpm in a Beckman SW-41 rotor at 5°C. Direction of sedimentation is from left to right. The arrow indicates the center of the monosome peak.

to the absence of translatable mRNA. This hypothesis, in its simplest form, predicts that the addition of exogenous mRNA to the cell-free system at pH 6.9 should stimulate protein synthesis to the levels seen at pH 7.4 (or to one-half the level if the difference in elongation rate is taken into account). Figure 8 shows the incorporation of ^{35}S-methionine as a function of added sea urchin histone and rabbit globin mRNA. The cell-free system displays a complex response to the adding of exogenous mRNA. Protein synthesis is stimulated at pH 6.9, but not to the levels seen at pH 7.4. Protein synthesis is also stimulated at pH 7.4. It thus appears that the factors that limit the rate of protein synthesis in the unfertilized egg are more than a simple absence of translatable mRNA.

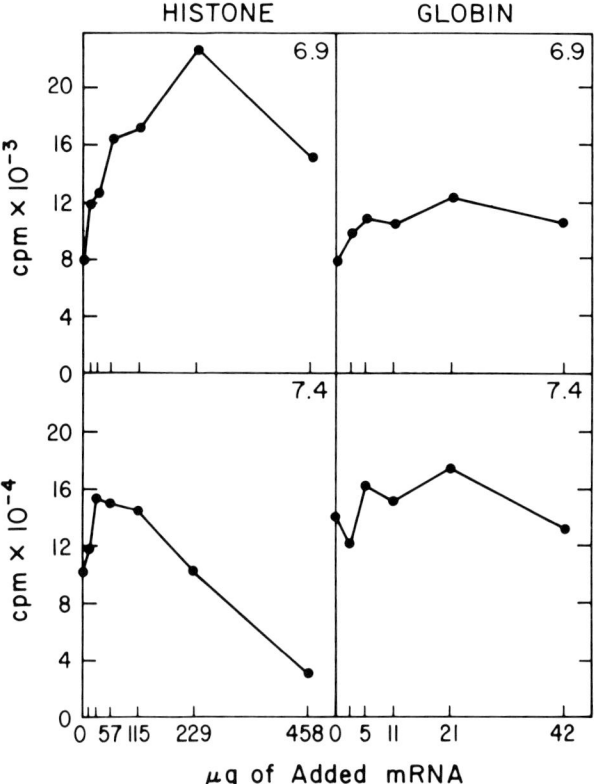

Fig. 8. The effect of exogenous sea urchin histone and globin mRNA on total incorporation of ^{35}S-methionine at pH 6.9 and 7.4. mRNA was added at the highest concentration indicated and then successively diluted by 2 with the cell-free system. Aliquots were fixed after 60 minutes of incubation.

Figure 9 shows a fluorograph of an SDS polyacrylamide gel analysis of the newly synthesized products with added histone mRNA. Several interesting observations emerge from careful scrutiny of this gel. One is that the added histone mRNA is translated more efficiently at pH 7.4 than at pH 6.9. In lanes G and N, where 14 µg/ml of histone mRNA was added, 12 times as much histone was synthesized at pH 7.4. This result suggests that added mRNA is subject to controls similar to those of endogenous mRNA. In addition, with the exception of lane I, the added mRNA does not compete with the endogenous mRNA. There is probably about 50 µg/ml of endogenous mRNA in the cell-free system of which only a small percentage would be active in polysomes. Histone mRNA was added to levels as high as 458 µg/ml. In a reticulocyte lysate the addition of this much mRNA would compete out the endogenous synthesis. In fact the added mRNA appears to stimulate the synthesis of endogenous proteins. Al-

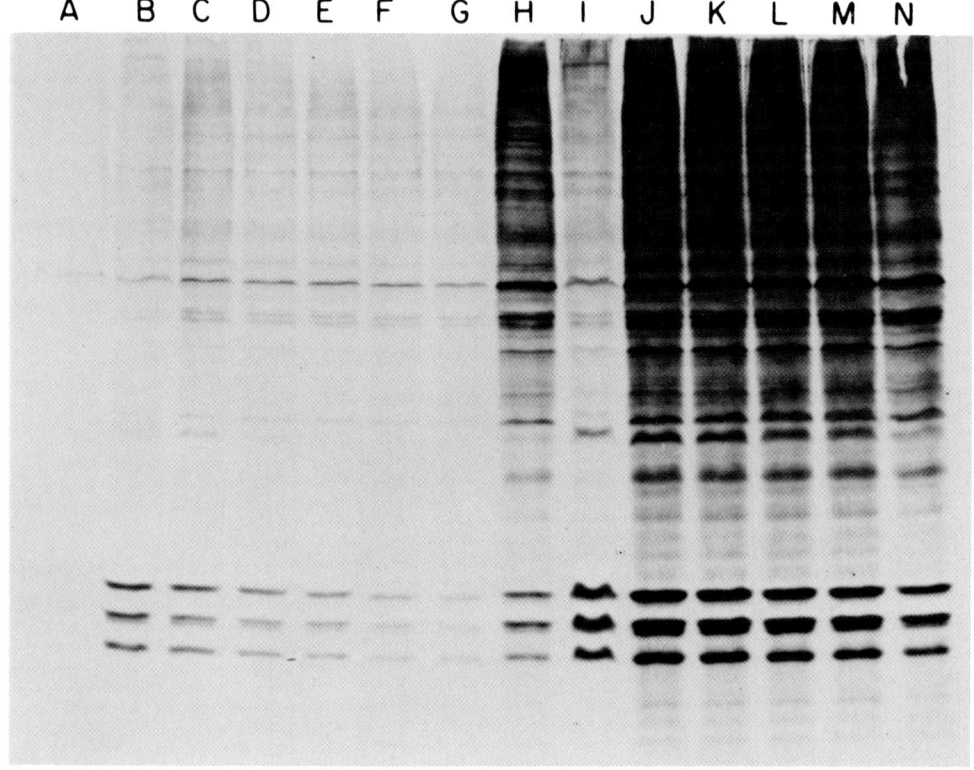

Fig. 9. Autoradiograph of a 17.5% SDS polyacrylamide gel displaying the newly synthesized products of the cell-free system with added sea urchin histone mRNA. Lanes A–G were incubated at pH 6.9 and lanes H–N at pH 7.4. Lanes A and H had no added histone mRNA. Lanes B and J had 458 μg/ml added histone mRNA. Each successive lane had 50% as much added histone mRNA. This gel shows the products of Figure 8.

though it would be premature to draw hard conclusions from these results, they are consistent with a simple model. This model suggests that there is an excess of a masking factor that would be able to bind to and inactivate most added mRNA. Increasing the pH would be expected to lead to a reduction in the mRNA binding affinity of the masking factor. A prediction of this model is that added mRNA would be in competition with the total pool of mRNA rather than that small portion in polysomes. This could explain why large amounts of added mRNA do not totally swamp out endogenous synthesis. Another prediction is that added mRNA would compete with endogenous mRNA for masking factor and should lead to the stimulation of endogenous proteins. This is seen in some instances. In addition, added mRNA would be expected to stimulate protein synthesis, especially at the higher pH value.

C. Summary

This work describes the apparent regulatory role of intracellular pH in the activation of protein synthesis at fertilization. Experimental manipulation of the pH_i of unfertilized and fertilized eggs is correlated with increases and decreases in the rate of protein synthesis. In addition, the use of various ion-substituted seawaters and parthenogenetic activating agents have enabled us to separate the effects of the increase in pH_i from those of the Ca^{++} release, which occurs simultaneously. The use of the cell-free translation system has enabled us to measure the effect of pH on the translational machinery directly. It appears that the increase in pH from 6.9 to 7.4 has a direct effect of doubling the peptide elongation rate, and also results in the translation of more mRNA. It also has the effect of allowing the more efficient translation of exogenous mRNAs. Although the basic controlling role of the increase in pH_i at fertilization is now beginning to be fairly well understood, the details of how pH_i interacts with the protein synthetic machinery are just beginning to be worked out.

ACKNOWLEDGMENTS

This work was supported by NSF grant PCM 8022833 and NIH grant PHS GM 22135-06 to John Hershey. M.M.W. is a fellow of the Jane Coffin Childs Memorial Fund for Medical Research. This investigation has been aided by a grant from the Jane Coffin Childs Memorial Fund for Medical Research.

IV. REFERENCES

Brandis JW, Raff RA: Translation of oogenetic mRNA in sea urchin eggs and early embryos. Demonstration of a change in translational efficiency following fertilization. Dev Biol 67:99, 1978.

Brandis JW, Raff RA: Elevation of protein synthesis is a complex response to fertilization. Nature (Lond) 278:467, 1979.

Denny PC, Tyler A: Activation of protein synthesis in non-nucleate fragments of sea urchin eggs. Biochem Biophys Res Commun 14:245, 1964.

Epel D: Mechanisms of activation of sperm and egg during fertilization of sea urchin gametes. In Moscona AA, Monroy A (eds): "Current Topics in Developmental Biology." New York: Academic Press, 1978, p 185.

Epel D, Steinhardt RA, Humphreys T, Mazia D: An analysis of the partial metabolic derepression of sea urchin eggs by ammonia: The existence of independent pathways. Dev Biol 40:245, 1974.

Grainger JL, Winkler MM, Shen SS, Steinhardt RA: Intracellular pH controls protein synthesis rate in the sea urchin egg and early embryo. Dev Biol 68:396, 1979.

Gross PR, Cousineau GH: Effects of actinomycin D on macromolecule synthesis and early development in sea urchin eggs. Biochem Biophys Res Commun 10:231, 1963.

Hille MB, Albers AA: Polypeptide chain elongation increases after fertilization of sea urchin eggs. Nature (Lond) 278:469, 1979.

Humphreys T: Efficiency of translation of messenger RNA before and after fertilization in sea urchins. Dev Biol 20:435, 1969.

Humphreys T: Measurement of messenger RNA entering polysomes upon fertilization of sea urchin eggs. Dev Biol 26:201, 1971.

Johnson JD, Epel D, Paul M: Intracellular pH and activation of sea urchin eggs after fertilization. Nature (Lond) 262:661, 1976.

Shen SS, Steinhardt RA: Direct measurement of intracellular pH during metabolic derepression at fertilization and ammonia activation of the sea urchin egg. Nature (Lond) 272:253, 1978.

Shen SS, Steinhardt RA: Intracellular pH and the sodium requirement at fertilization. Nature (Lond) 282:87, 1979.

Steinhardt RA, Zucker R, Schatten G: Intracellular calcium release at fertilization in the sea urchin egg. Dev Biol 58:185, 1977.

Winkler MM, Grainger JL: Mechanism of action of NH_4Cl and other weak bases in the activation of sea urchin eggs. Nature (Lond) 273:236, 1978.

Winkler MM, Steinhardt RA: Activation of protein synthesis in a sea urchin cell free system. Dev Biol 84:432, 1981.

Winkler MM, Steinhardt RA, Grainger JL, Minning L: Dual ionic controls for the activation of protein synthesis at fertilization. Nature (Lond) 287:558, 1980.

Intracellular pH and Proliferation in Yeast, Tetrahymena and Sea Urchin Eggs

Robert J. Gillies
Department of Molecular Biophysics and Biochemistry, Box 6666, Yale University, New Haven, Connecticut 06511

I.	Introduction	341
	A. Regulation of Proliferation	341
	B. Cell Cycle	342
	C. Intracellular pH	342
II.	Intracellular pH and Growth	343
	A. Tetrahymena	343
	B. Yeast	344
III.	Intracellular pH and the Cell Cycle	344
	A. Tetrahymena	344
	B. Yeast	347
IV.	Intracellular pH Manipulation	347
V.	Mediation of pH Changes in Sea Urchin Eggs Following Fertilization	353
VI.	Conclusions	358
VII.	References	358

I. INTRODUCTION

A. Regulation of Proliferation

Although investigated for over a century, the mechanisms regulating cellular metabolism and proliferation remain one of the greatest unknowns in biology today. Ever since Claude Bernard investigated the growth requirements of cells [Bernard, 1879], researchers have pursued the problem of cellular proliferation from physiological, biochemical, biophysical, and molecular biological perspectives.

Many physiological and biochemical parameters are known to affect cellular proliferation. The most noteworthy of these include growth factors [Sato and Ross, 1979], medium pH [Ceccarini and Eagle, 1971a, b], external calcium [Rasmussen et al, 1974], and external magnesium [Rubin, 1975], in addition to those compounds that supply essential molecules such as sugars, amino acids, phosphate, and the like. These "proliferation-stimulating factors" signal the cell to grow through the action of "second messengers" such as cyclic nucleotides [Pastan et al, 1975], intracellular calcium [Rasmussen and Goodman, 1977], intracellular Na$^+$ [Cameron et al, 1979], membrane potential [Cone and Tongier, 1973], and intracellular pH (pH_i). These "second messengers" stimulate cellular metabolism and proliferation intracellularly. Metabolic changes known to correlate with proliferation are protein phosphorylation [Nimmo and Cohen, 1977], tubulin [Rubin and Weiss, 1975] and actin [Wehland and Weber, 1980] polymerization, protein synthesis [Pardee, 1974], and glycolysis [Hassell et al, 1975; Rubin and Fodge, 1974]. Clearly, there is a need to integrate these diverse observations into more defined interrelationships.

B. Cell Cycle

One of the most successful models for investigating control of cellular proliferation is the cell division cycle. Within the last decade, concepts of the cell cycle have changed drastically. Most of this has come about through the efforts of Leland Hartwell and his colleagues [Hartwell et al, 1974], who have defined over 150 different temperature-sensitive cell division cycle (cdc) mutants of S. cerevisiae, representing 50 complementation groups. Each of these groups arrests growth at a specific point in the cell cycle when placed at the restrictive temperature. The relationship of these restriction points in yeast to proliferation cycles in other cell types remains to be established. At stationary phase under physiological conditions, most eukaryotic cells arrest growth at a point in G_1. This is termed the R point [Pardee, 1974] and is similar to mutation 28 in the *cdc* system. The most important observation concerning the R point is that it is sensitive to protein synthesis and that, once it is crossed, most cells are committed to replicate their DNA and divide.

C. Intracellular pH

The present chapter focuses on pH_i as a parameter of control of proliferation in yeast, Tetrahymena, and sea urchin eggs. Although the study of pH and cell growth is not new [Warburg et al, 1926], it has only been with the technological advances of the last few years or so that we have been able to probe the pH inside the cell. By way of introduction, I refer readers to comprehensive and recent reviews on regulation of pH_i, its interaction with other physiological parameters, and its relationship to control of cellular proliferation [Gillies, 1981; Roos and Boron, 1981]. This chapter will review my work which has centered on four basic questions: 1) Can pH$_i$ be regulated and, if so, does it correlate with cell proliferation? 2) Does pH$_i$ change throughout the cell cycle? 3) What are the metabolic and growth consequences of pH$_i$ manipulation? 4.) How are changes in pH$_i$ generated?

II. INTRACELLULAR pH AND GROWTH

There are many biochemical pathways that result in a net production of proton equivalents, resulting in a metabolic acid load to the cytosol. These include glycolysis, the TCA cycle, and hydrolysis of nucleoside triphosphates. In addition, at a medium pH at or below that of the cytoplasm, the electrochemical potential for protons across the plasma membrane also generates a cytosolic acid load. In order for a cell to regulate its pH$_i$ therefore, a mechanism must be expressed that involves extrusion of proton equivalents. the mechanisms of proton extrusion have been extensively studied in diverse systems and are the topic of other chapters in this volume. My research has focused on the phenomenon, rather than the mechanism, of pH$_i$ regulation. In these studies, I have induced acid or base loads by titrating the external pH. The resulting curve of intracellular vs extracellular pH gives a good indication of a cell's ability to regulate its pH$_i$ [Gillies and Deamer, 1979a]. Cell growth can be studied as a function of extracellular pH and correlations between pH$_i$ and growth can thus be drawn.

A. Tetrahymena

Tetrahymena are free-living ciliated protozoa, which are approximately 70 μm × 30 μm in size. Their growth characteristics have been extensively studied under a variety of conditions [Hill, 1972], and cultures can be easily synchronized by several methods. These characteristics make Tetrahymena an ideal system with which to investigate pH$_i$ as it relates to cell growth. For all of our studies on Tetrahymena, we have employed the DMO technique [Waddell and Butler, 1959]. Tetrahymena present unique problems when using this technique, owing to extensive cytoplasmic compartmentation. Our technique for resolving cytoplasmic pH out of the DMO data has previously been explained in another review [Gillies and Deamur, 1979b].

At 28°C, under conditions of varying external pH, tetrahymena (DN3) maintain their intracellular pH at 7.15 ± 0.05 within the external pH range 5.75–7.40. [Gillies and Deamer, 1979a]. These data are identical whether the cells are grown at the pH indicated or are titrated to that pH just prior to the determination. The ability of the cells to regulate their intracellular pH falls off rapidly at an external pH below 5.75 or above 7.4. In addition, doubling time is most rapid between external pH of 5.75–7.40. Growth rate falls off rapidly at pH values outside this range, coincident with the loss of intracellular pH regulation. There are two conclusions one can draw from the above observations. First, cell growth rate is not affected by the presence or absence of a pH gradient across the plasma membrane, since the cells grow equally well at pH_{ex} 6.2 ($\Delta pH = 1.0$) and pH_{ex} 7.2 ($\Delta pH = 0$). Second, there appears to be a sharp pH_{in} optimum for growth. This is illustrated in Figure 1, and shows that maximum growth rate correlates with a pH_{in} of 7.20 and falls off rapidly at pH_{in} below 6.90. It should be reiterated here that no causality is implied in these data. Maximum growth rate might well be a function of optimal intracellular pH. Alternatively, the possibility that maintenance of optimal pH is a function of optimal growth rate is equally likely, given the data. This question will be discussed further in section IV of this chapter.

B. Yeast

Another organism with which we have worked is the brewer's yeast, Saccharomyces cerevisiae (NCYC 239). Much work has been done on this system with regard to the biochemistry and physiology of growth regulation [Rose and Harrison, 1971]. Another major advantage of yeast is that is can easily be investigated using nuclear magnetic resonance spectoscopy (NMR). The NMR technique as it applies to pH_i determinations in yeast is reviewed elsewhere in this volume (see Gillies et al; Gadian et al).

Owing in large part to the ease and rapidity of pH_i determinations using NMR, we have been able to investigate pH_i under a variety of conditions. As shown in Table I, the pH_i of yeast is dependent on the presence of glucose, the state of oxygenation, and the growth history of the cultures. At 25°C, under aerobic conditions in the presence of glucose, glucose-repressed yeast maintain their pH_i at 7.10 ± 0.05 within the external pH range 5.0–6.5. (Fig. 2). As with Tetrahymena, this pH range also corresponds with optimal growth rate (inset, Fig. 2).

III. INTRACELLULAR pH AND THE CELL CYCLE

Intracellular pH was first postulated to change during the cell cycle in 1960 [Frydenberg and Zeuthen, 1960]. However, demonstration of this phenomenon was not forthcoming until 1976, when Gerson and Burton showed fluctuations of pH_i during the cell cycle of Physarum [Gerson and Burton, 1976]. In this system, the pH_i maximum occurred just prior to mitosis, and the minumum occurred during S phase. The amplitude of the fluctuations in this system was approximately 0.4 pH units. However, since Physarum does not express a G_1 period, it remained unknown whether the alkalinization of pH_i correlated to mitosis or to the initiation of DNA synthesis.

A. Tetrahymena

I first investigated this question in Tetrahymena. As mentioned previously, these cells are easily synchronized to a high degree by a variety of methods. The first synchronization technique used was the heat-shock method [Zeuthen,

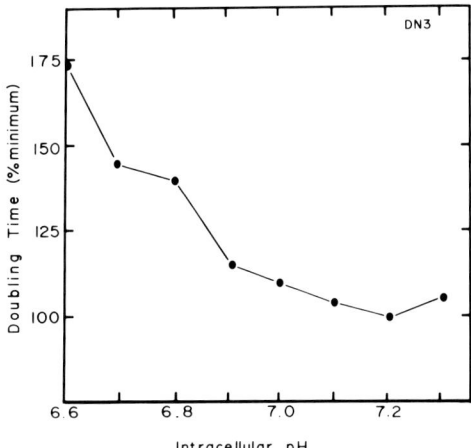

Fig. 1. Growth Rate and Intracellular pH of T. Pyriformis (DN3). Cells were grown at various values of external pH (range pH = 4.0 to pH = 9.0) in YEPD media containing 40 mM Zwitterionic buffer (TES, MES, HEPES). Growth was followed for 24 hours, at which time intracellular pH was determined by DMO technique. Raw data from Gillies and Deamer [1979a].

TABLE I. Intracellular pH of Saccharomyces cerevisiae Under Various Conditions

Condition		With glucose	Without glucose
Glucose-	Aerobic	7.30	7.28
repressed	Anaerobic	7.30	7.25
Glucose-	Aerobic		7.52
derepressed	Anaerobic		7.23
Raffinose	Aerobic	7.35	7.34
grown	Anaerobic	7.19	7.21
Acetate	Aerobic	7.45	7.50
grown	Anaerobic	7.03	7.15

pH^{EX} = 6.00. Intracellular pH of S. cerevisiae was estimated from the chemical shift of intracellular orthophosphate as determined by ^{31}p-NMR. Cells were grown in YEP medium containing 50 mM MES and either 3% dextrose, 2% raffinose, or 2% sodium acetate with bipthalate buffer. Glucose-derepressed cells were harvested in mid-stationary phase (after glucose depletion), whereas all other samples were harvested in mid-log phase. Raffinose grown cells are expected to be intermediate between glucose-repressed and derepressed (acetate grown). From den Hollander et al [in press].

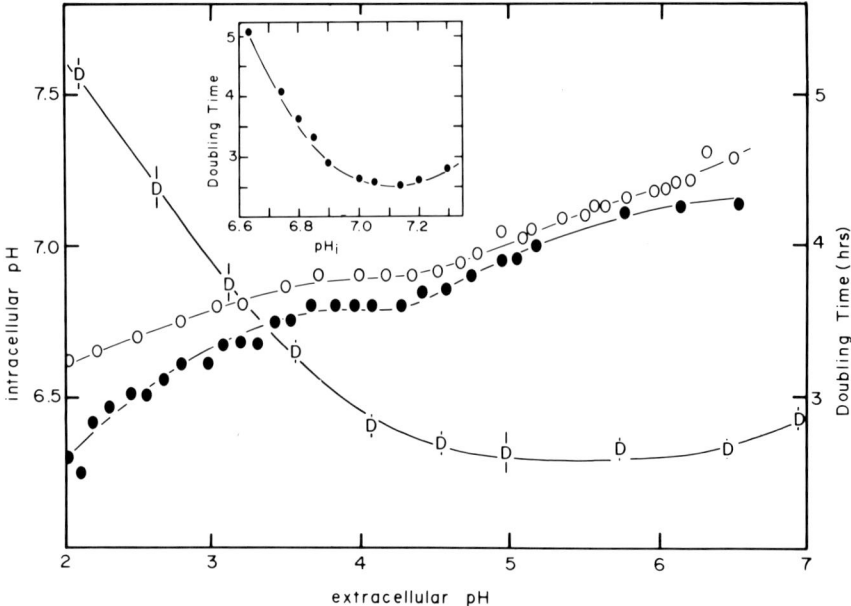

Fig. 2. Intracellular pH and doubling time as a function of extracellular pH in S cerevisiae (A364a). Intracellular pH was estimated from chemical shift of intracellular orthophosphate using ^{31}P-NMR spectroscopy on a 10% (v/v) suspension of yeast in YEPD containing 20 mM each MES, TES, and Tricine (pH = 7.4). Suspension was infused with 0.5N HCl containing 20% dextrose at a rate of 1 ml hr^{-1}. Data were collected in 5-minute blocks of 2930 FIDs resulting from 57° tipping pulses with a repetition rate of 0.1028 second. ● = Anaerobic, ○ = aerobic. Doubling time(s) was determined densitometrically in YEPD containing 20 mM each MES, TES, and tricine on parallel aerobic cultures inoculated at various values of external pH. Error bars represent standard deviations of duplicate cultures. Inset indicates correlation of growth rate with intracellular pH.

1971]. This technique synchronizes cells for many generations both with respect to cytokinesis and DNA synthesis. DNA synthesis proceeds immediately after cytokinesis, indicating the lack of a G_1 period. In this system, there are two alkaline pH transients of about 0.3 pH units above a baseline of pH 7.25 [Gillies and Deamer, 1979a]. These transients take 30 minutes apiece (0.15 cell cycles), and occur just prior to and after either cytokinesis or DNA synthesis. Both the magnitude and the timing of the first transient are consistent with the results of Gerson and Burton. To resolve whether the pH transients were correlated with DNA synthesis or cytokinesis, cells were also synchronized by starvation/refeeding [Cameron and Jeter, 1970]. This procedure synchronizes cells to a high degree with respect to DNA synthesis, comparable to the heat-shock method. However, cytokinetic synchrony is greatly reduced and occurs much later than

DNA synthesis. Cultures synchronized by staravation/refeeding also exhibit two alkaline pH_i transients of approximately 0.3 pH units above a baseline of pH 7.3. These transients have similar kinetics as, are of the same magnitude as, and bear the same temporal relationship to DNA replication as the shifts seen previously using the heat-shock method. Comparison of the data from the two synchronization procedures indicates that the pH shifts are related to DNA replication and are not affected by the time, duration, or magnitude of cytokinesis.

B. Yeast

I have also begun work on the cell division cycle of S. cerevisiae. Although the yeast are not as easy to synchronize as are Tetrahymena, they offer certain other advantages. First, they are very amenable to investigation by NMR, allowing for noninvasive studies of metabolic regulation coincident with pH studies. In addition, as mentioned earlier, there are many ts *cdc* mutants available that make possible more detailed biochemical studies of the nature of cell division cycle restrictive points.

My first studies centered on using populations of yeast synchronized by a complex starvation/refeeding scheme [Gillies et al, 1981]. Since these experiments involved using the NMR technique, conditions had to be met that allowed for cell synchrony at the high densities required by NMR. As shown in Figure 3, under proper conditions of aeration and feeding, cultures synchronously replicated their DNA and initiated budding at densities 2 orders of magnitude higher than previously reported. As also shown in Figure 3, synchronous populations of *S. cerevisiae* exhibit an alkaline pH transient of approximately 0.3 pH units above an initial pH_i of 7.0. This transient occurs immediately after feeding glucose and does not depend on whether or not the cultures were preoxygenated. The response of asynchronous cells to glucose is a permanent rise in pH_i from 7.0 to 7.1 (data not shown). Some interesting observations arise from these data. First, the pH transient in synchronous cultures results in an intracellular alkalinization to pH 7.3, regardless of the starting pH_i. This indicates that the transient is not controlled by release of hydroxyl equivalents, but instead seems to be regulated to reach a certain critical value of pH_i. Second, even though the pH transient in this system is much slower than that seen in Tetrahymena, it is similar in that it occurs entirely before the onset of DNA synthesis. Mitosis in this system begins at about 130 minutes which is over 1 hour after the pH_i transient. These data suggest that the transient in yeast pH_i is temporally related to DNA synthesis and not to mitosis.

IV. INTRACELLULAR pH MANIPULATION

As shown in the previous sections, intracellular pH correlates highly with cellular proliferation and with the initiation of DNA synthesis. These observations, however, do not address the question of whether pH_i plays any sort of regulatory role in these processes. The best studied system regarding control of cell metabolism and growth by pH_i is sea urchin egg activation. In this system, it has convincingly been shown that weak base parthenogenetic activation results from the imposed increase in pH_i which, in turn, increases K^+ conductance, increases protein synthesis, and initiates the replication of DNA [see Epel, 1978, for review]. In the following study, we have used similar logic to ask whether pH_i might play a role in Tetrahymena growth regulation.

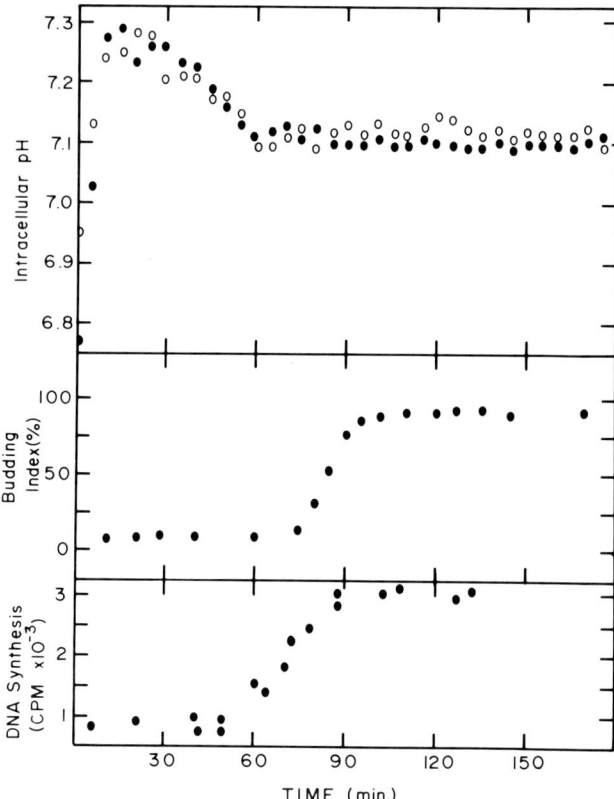

Fig. 3. Intracellular pH, budding index, and DNA synthesis cultures of S. cerevisiae. Conditions were identical to those reported previously [Gillies et al, 1981]. Intracellular pH was determined on cultures either preoxygenated (O) or not (●) for 30 minutes prior to addition of glucose, which initiates cell cycle progression. Asynchronous cultures did not express transient as seen in synchronous cultures (data not shown).

The end product of sugar metabolism in nonfermentative eukaryotes is CO_2. CO_2 is a weak acid in aqueous solutions and should, in sufficient quantities, act to lower pH_i. In addition, protein catabolism in Tetrahymena results in the production of ammonia which, as a weak base, could act to raise pH_i. We therefore asked whether accumulation of these waste products occurred in sufficient quantities to affect pH_i, cell proliferation, or both.

In Figure 4, it is shown that both ammonia and CO_2 accumulate in surprisingly high concentrations in cultures grown aerobically on proteose peptone, and glucose. In a culture with a saturation density of 7.5×10^5 cells·ml^{-1}, CO_2 con-

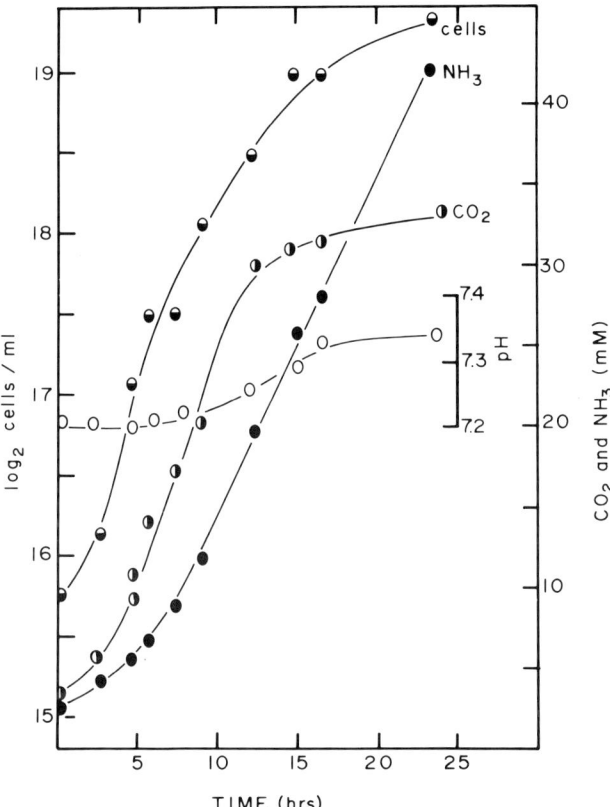

Fig. 4. Cell density, CO_2 concentration, NH_3 concentration, and pH in culture of T. pyriformis (GL) grown with shaking (50 rpm) at 23/C in YEPD media containing 40 mM HEPES. Cell number was determined using an electronic particle counter (celloscope). Ammonia and CO_2 content was determined using a soluble gas analyzer as described previously [Gillies and Deamer, 1981a].

centration reaches a steady-state level near 30 mM coincident with cessation of cell growth. Ammonia concentration, on the other hand, increases steadily beyond 40 mM by mid-stationary phase.

We then asked whether these concentrations of ammonia or CO_2 were sufficient to affect cell growth. However, since CO_2 is volatile, maintenance of controlled, long-term, steady-state concentrations above equilibrium is technically difficult. Because of this, we used a nonvolatile week acid to mimic the pH_i effects of CO_2. An ideal choice for this is DMO, since its pKa is almost identical to that of CO_2 and because it has been thoroughly tested and found to be metabolically inert [Waddell and Bates, 1969]. We have previously shown that both DMO and ammonia inhibit growth of Tetrahymena [Gillies and Deamer, 1981a]. The weak acid is much more potent, presumably because of its increased distribution into the relatively more alkaline intracellular volume. At an external pH of 7.2, the effect of ammonia on cell growth is essentially linear up to 100 mM, whereas DMO virtually arrests growth at concentrations above 20 mM. Comparison of these observations with the data of Figure 4 suggests some interesting correlations. First, medium ammonia concentration at the onset of stationary phase is approximately 25 mM under our conditions. At this concentration, ammonia has only a slight effect on the growth rate. In contrast, medium CO_2 concentration at the onset of stationary phase is approximately 30 mM. This medium concentration of a similar weak acid (DMO) results in a virtual inhibition of growth. These observations indicate that accumulation of medium CO_2 occurs in sufficient quantities to arrest cell growth, and that the effect of CO_2 is probably through its properties as a weak acid.

Are these concentrations of DMO sufficient to affect the intracellular pH? The data presented in Figure 5 indicate that they are. As shown, DMO acidifies pH_i, and its effect is clearly expressed in a pH-dependent fashion. At an external pH of 7.20, 20 mM DMO lowers the pH_i to 6.9, whereas at an external pH of 6.8, only 6 mM is necessary to achieve the same effect. Figure 6 illustrates the effect of DMO on culture doubling time at these same values for external pH. These data suggest a strong effect of pH_i on cell proliferation. Once again, if we plot cell doubling time as a function of pH_i, we see a sharp pH_i optimum for growth (Fig. 7). These data are qualitatively similar to those shown in Figure 1. The major difference between the two sets of data is that the earlier study was simply correlative, whereas the latter employed artificially imposed pH_i values. Taken together, these observations strongly indicate that pH_i can actually regulate cellular proliferation and that, under physiological conditions, stationary phase can be induced by accumulation of metabolically produced weak acids, which act to lower pH_i.

We have tested this hypothesis further by subjecting cultures to treatments, which resulted in attenuating the accumulation of medium CO_2 during culture growth. The first treatment was to culture cells in the presence of dialysis sacs

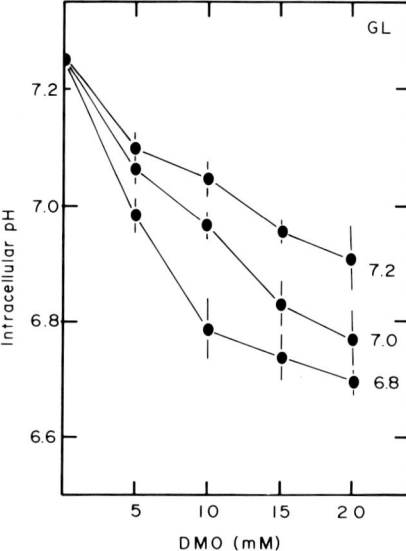

Fig. 5. Intracellular pH of T. pyriformis (GL) as a function of DMO concentration at various values of external pH. Cultures of Tetrahymena containing 10^5 cells ml^{-1} in YEPD buffered with 20 mM HEPES was adjusted to the pH indicated. Varying amounts of 1 M DMO in 95% EtOH was added to 5 ml of the cultures in triplicate and incubated for 5 minutes at 23/C prior to intracellular pH determination [Gillies and Deamer, 1981a]. In controlled experiments ethanol was shown not to affect intracellular pH at 2% (v/v).

containing 20 mM NaCl. The dialysis volume represents a "sink" for metabolic waste products, yet does not affect the total amount of food available on a per-cell basis. As shown in Figure 8, CO_2 accumulation rate decreased linearly with respect to increased volume of the dialysis sacs. The final density of cultures at stationary phase was increased over 4-fold in this study. In other studies, this treatment has led to over 6-fold increases in culture density at stationary phase [Gillies and Deamer, 1981a]. Comparison of this data with those presented in the previous paragraph indicate that the increase in culture density was a result of the lower steady-state levels of CO_2 in the medium. As a further test of this hypothesis, cells were grown to stationary phase and either refed with concentrated media, refed with a complete media change, or left untreated [Gillies and Deamer, 1981a]. The untreated controls began to die off as expected. Cultures fed with concentrated media did not change in density. This treatment provides food without affecting media CO_2 concentrations. In contrast, cultures refed with a complete media change were induced to grow further. These data demonstrate that the grwoth induction was not achieved by refeeding, but by removal of metabolic byproducts.

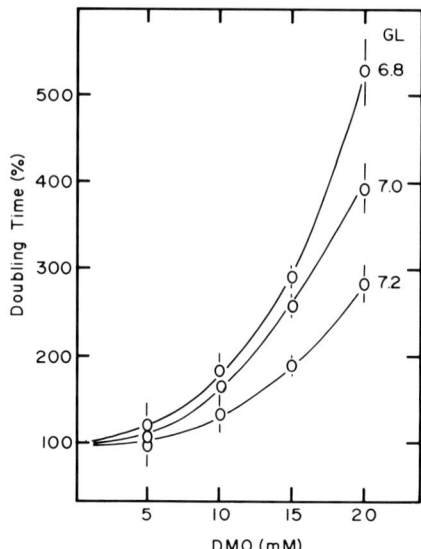

Fig. 6. Growth rate of T. pyriformis cultures as a function of DMO concentration at various values of external pH. Tetrahymena were inoculted in YEPD media containing 40 mM HEPES adjusted to pH indicated. Growth was followed using an electronic cell counter for 6 hours, at which time varying amounts of 1 M DMO in 95% EtOH were added. Growth was followed for an additional 24 hours. 2% EtOH was previously determined not to affect cell grwoth rate.

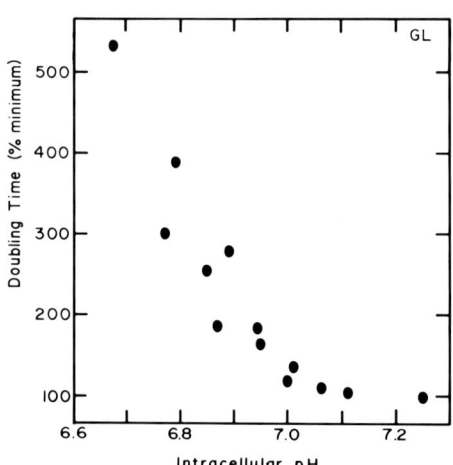

Fig. 7. Doubling time as a function of intracellular pH in T. pyriformis (GL). Data were obtained from experiments illustrated in Figures 5 and 6.

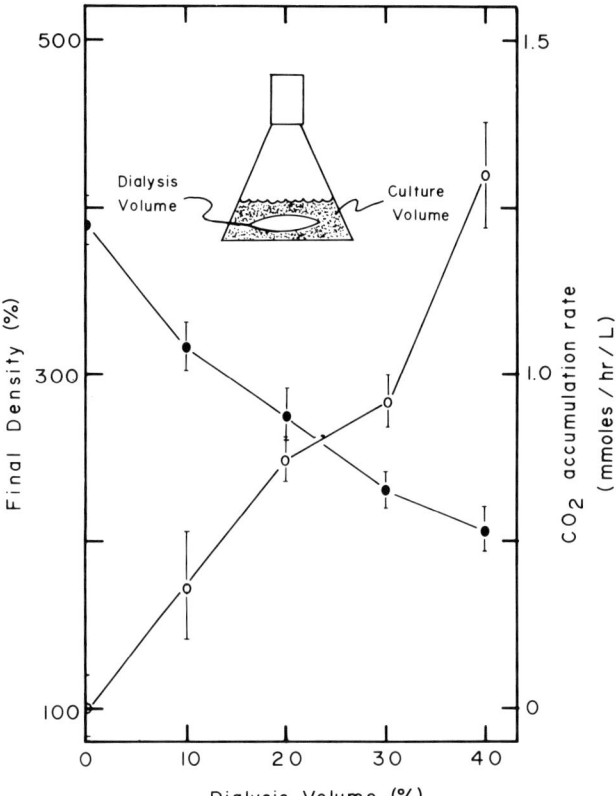

Fig. 8. CO_2 accumulation rate (●) and final density (○) of T. pyriformis cultured in presence of dialysis sacs containing 20 mM NaCl (inset). Final density was determined at mid-stationary phase and is expressed as percentage increase over control values (no dialysis volume). CO_2 content was determined at different times with soluble gas analyzer as previously described [Gillies and Deamer, 1981a].

V. MEDIATION OF pH CHANGES IN SEA URCHIN EGGS FOLLOWING FERTILIZATION

As indicated previously, many systems have been shown to exhibit alkalinization in response to either external stimuli or time during the cell division cycle. The final question posed in this communication deals with the nature of such changes in pH_i.

One of the best studied systems with regard to intracellular alkalinization is the activation of sea urchin eggs. In this system, fertilization or partheogenetic activation precipitates a permanent rise in pH_i 6.7 to 7.2 [Shen and Steinhardt, 1978]. At the same time, a stoichiometrically equivalent fall in extracellular pH is also seen. The nature of this fertilization acid has been controversial since its discovery more than 50 years ago [Runnstrom, 1933].

Our studies were initiated with the observation that the pH of seawater, lowered by acid release, slowly returns to prefertilization values [Gillies and Deamer, 1981b]. Covering the suspension with mineral oil prevented this reequilibration. Two plausible explanations for these observations are that either the released acid is interacting with some volatile component of the seawater buffering system or that the fertilization acid is itself volatile. To test this, eggs were fertilized in the presence of CO_2-free artificial seawater. Once acid release was complete, eggs were removed by centrifugation and nitrogen gas was bubbled through the sea water to hasten equilibration. As shown in Figure 9, the pH of the seawater once again reequilibrated to a more alkaline value. This behavior is strikingly similar to that of CO_2 (volatile acid) and is quite different from that of HCl (nonvolatile acid). These observations strongly suggest that the fertilization acid is itself volatile, and that it might possible be CO_2.

To test this possibility, the pH and CO_2 content of seawater was analyzed at times after fertilization in either open or closed systems. The data are presented in Figure 10 and indicate a strong correlation between CO_2 content and release of H^+-equivalents from the eggs. The correlation between CO_2 and H^+-equivalents is linear with a coefficient of 97%. The slope of 0.83 CO_2-eq/H^+-eq indicates that virtually all of the fertilization acid can be accounted for by the release of CO_2 from the eggs.

We next investigated possible sources of CO_2. One may imagine a biochemical process such as an enzyme-catalyzed decarboxylation reaction. Alternatively, CO_2 could be released from an organic or inorganic carbonate compound. To choose between these alternatives, cells were fixed and dehydrated with absolute ethanol, and acid-labile CO_2 content was determined manometrically. If the CO_2 is produced via decarboxylation, we would expect to see no difference in the content in unfertilized or fertilized eggs. As shown in Table II, there is a difference of about 4 μmoles per ml of packed eggs, which is similar to the amount of CO_2 released as fertilization acid. This suggests that CO_2 is produced from an endogenous compound that can release carbon dioxide nonenzymatically upon fertilization. These observations are in agreement with those of Runnstrom [1933], who demonstrated that release of fertilization acid was independent of metabolism.

To determine whether the CO_2 might be derived from an organic or inorganic carbonate, ethanol extracted eggs were incinerated at 475°C for 8 hours. This procedure completely oxidizes the organic compounds present in the ethanol

pH, and Cell Proliferation / 355

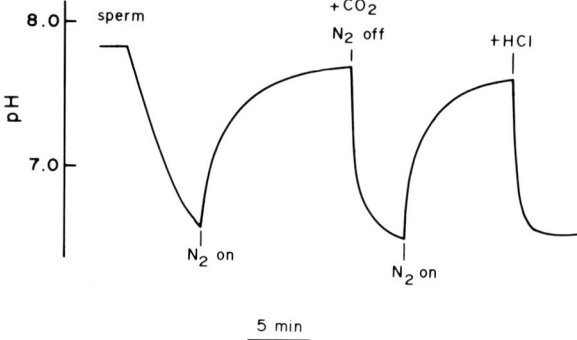

Fig. 9. The pH change caused by fertilization in bicarbonate-free artificial seawater was compared with similar pH shifts induced by CO_2 or HCl additions. The seawater was flushed with a gentle stream of nitrogen bubbles, which served to stir the suspension and to increase the rate at which a volatile acid might be driven off. Note that the pH shift caused by fertilization is reversible, and resembles that produced by CO_2 addition, whereas HCl produces a permanent shift.

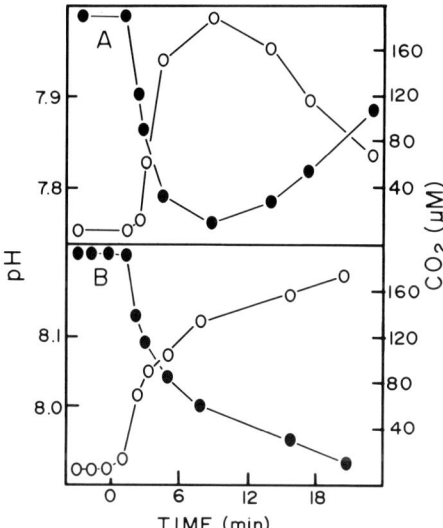

Fig. 10. Extracellular pH (●) and CO_2 concentration (○) in seawater following fertilization. Conditions as in Figure 1, with the upper figure showing the open system and the lower the closed system (mineral oil overlay). The CO_2 concentration is expressed as μmoles per liter above unfertilized values.

TABLE II. Acid-Labile CO_2 and Calcium Content of Extracted and Ashed Sea Urchin Eggs (μmole/ml packed eggs)

	Acid-labile CO_2		Calcium
Condition	Ethanol extracted	Ashed	Ashed
Unfertilized	17.1 ± 2.9	4.7 ± 1.9	3.7 ± 0.2
Fertilized	12.9 ± 1.5	0.6 ± 0.5	4.2 ± 0.3
Difference	4.2	4.1	−0.5
P	<0.025	<0.005	NS

CO_2 and calcium contents of sea urchin eggs were measured after ethanol dehydration or incineration. Ethanol extracts and ash were prepared as described in the text. Calcium was determined by atomic absorption spectroscopy. CO_2 was determined manometrically in a Gilson respirometer as gas volume released upon addition of 0.1 N HCl. All concentrations are expressed as micromoles per ml of packed eggs. Numbers in parentheses refer to the number of replicates, and P values were obtained fromm Student's t-distribution for the difference in CO_2 or calcium content of the unfertilized and fertilized samples. NS = not significantly different.

powder, and the resultant ash was analyzed for acid-labile CO_2. As seen in the table, the 4 μmole per ml difference persisted in the ash. This observation suggests that the CO_2 is bound in an inorganic form. Since the amount of calcium in the ash is similar to the amount of CO_2 released, one possibility is calcium carbonate.

Calcium carbonates are known to occur in diverse cell types [Pautard, 1970], and it is not unreasonable to suggest that similar deposits might occur in sea urchin eggs. Afzelius [1976] examined unfertilized eggs that had been fixed under conditions designed to minimize solution of calcium deposits. Electron-dense structures were observed just beneath the plasma membrane of the eggs, and Afzelius proposed that these may represent calcium deposits associated with sperm binding sites. It was estimated that the deposits and associated micropapillae compose about 1–2% of the egg surface area. Cardassis et al [1978] utilized osmium antimonate stain followed by electron microprobe analysis to localize calcium in eggs from two species of urchins. The microprobe provided evidence for calcium in densely stained deposits just beneath the plasma membrane.

If calcium carbonate deposits are present in unfertilized sea urchin eggs, we expect to visualize them by TEM as electron-dense structures in unstained sections. As shown in Figure 11A, occasional deposits are observed near the cortical surface, and, in some eggs, the particles formed clear rows. In post-stained eggs, the particles could be visualized in more detail (Fig. 11B) and resemble those described by Afzelius. These observations are consistent with the notion that

pH, and Cell Proliferation / 357

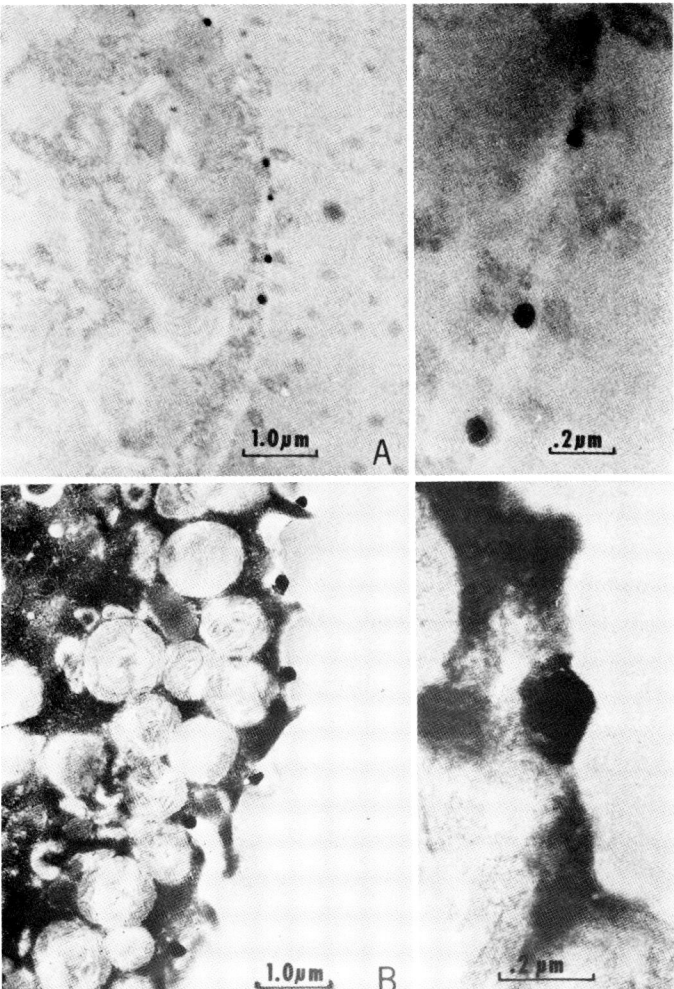

Fig. 11. Unfertilized sea urchin eggs were fixed 3 hours in 2% paraformaldehyde–2.5% glutaraldehyde buffered with 0.1 M cacodylate and saturated with calcium carbonate (0.15 mM). The fixed eggs were dehydrated in acetone and embedded in Epon-Araldite. Sections (100 nm) were examined unsustained (5A) and following unranyl acetate–lead citrate poststaining (5B). Rows of electron-dense particles were observed in the cortical region. The experiment was repeated twice with similar results.

some fraction of the calcium in sea urchin eggs is present as a mineral deposit in the cortical region. As a working hypothesis, we propose that a store of calcium carbonate is exposed to the cytoplasm upon fertilization. The mechanism for this step is unclear, but it may involve membrane depolarization and sodium influx. The calcium carbonate dissolves and free calcium and carbonate are released. The calcium may then react with the cortical granules, and the carbonate combines with protons to form bicarbonate and carbonic acid, thereby increasing pH_i. The carbonic acid (or carbon dioxide) crosses the plasma membrane to appear externally as fertilization acid.

VI. CONCLUSIONS

Many cations, such as Na^+, K^+, Ca^{++}, and Mg^{++} are known to be important in the regulation of cellular metabolism and proliferation. Until recently, the role of protons as "second messengers" in cellular growth control has been virtually unexplored. The data presented in this chapter indicate that changes in pH_i do occur during the life cycle of some cells, that there is an optimum pH_i for growth, and that acidification of pH_i will inhibit growth. The mechanisms by which pH_i affects cell growth and metabolism are not yet understood. However, it is clear that protons must be included in the "ionic network," which is important in the regulation of proliferation.

VII. REFERENCES

Afzelius BA: Micropappillae in sea urchin eggs. Dev Growth Diff 18:195–198, 1976.

Bernard C: "Le Cahier Rouge." (Translated by Hoff H, Guillemin L, Guillemin R, 1967.) Cambridge, Massachusetts: Schenkman Pub, 1879.

Cameron IL, Jeter JR: Synchronization of the cell cycle of Tetrahymena by starvation and refeeding. J Protozool 17:429–431, 1970.

Cameron IL, Smith NKR, Pool TB, Sparks RL: Intracellular sodium concentration as related to mitogenesis and to oncogenesis. J Cell Biol 83:7a, 1979.

Cardasis CA, Schuell H, Herman L: Ultrastructural localization of calcium in unfertilized sea urchin eggs. J Cell Sci 77:101–115, 1978.

Ceccarini C, Eagle H: Induction and reversal of contact inhibition of growth by pH modification. Nature New Biol 233:271–273, 1971a.

Ceccarini C, Eagle H: pH as a determinant of cellular growth and contact inhibition. Proc Natl Acad Sci USA 68:229–233, 1971b.

Cone CD, Tongier M: Contact inhibition of division: Involvement of the electrical transmembrane potential. J Cell Physiol 82:373–386, 1973.

Epel D: Mechanisms of activation of sperm and egg during fertilization of sea urchin gametes. Curr Top Dev Biol 12:186–246, 1978.

Frydenberg O, Zeuthen E: Oxygen uptake and carbon dioxide output related to the mitotic rhythm in the cleaving eggs of Dendraster excentricus and Urichis caupo. Compt Rend Lab Carlsberg 31:423–455, 1960.

Gerson DF, Burton AC: The relation of cycling of intracellular pH to mitosis in the acellular slime mould Physarium polycephalum. J Cell Physiol 91:297–304, 1976.

Gillies RJ, Deamer DW: Intracellular pH changes during the cell cycle in Tetrahymena. J Cell Physiol 100:23–32, 1979a.

Gillies RJ, Deamer DW: Intracellular pH: Methods and applications. Curr Top Bioenerg 9:63–89, 1979b.

Gillies RJ, Deamer DW: Endogenously produced CO_2 induces stationary phase in Tetrahymena cultures. J Cell Physiol 105:221–225, 1981a.

Gillies RJ, Deamer DW: CO_2 accompanies release of fertilization acid from sea urchin eggs. J Cell Physiol 108:37–43, 1981b.

Gillies RJ: Intracellular pH and growth control in eukaryotic cells. In Cameron I, Pool TB (eds): "The Transformed Cell." New York and London: Academic Press, 1981, pp 347–395.

Gillies RJ, Ugurbil R, den Hollander JA, Shulman RG: ^{31}P NMR studies of intracellular pH and phosphate metabolism during cell division cycle of S. cerevisiae. Proc Natl Acad Sci USA 78:2125–2129, 1981.

Hartwell LH, Culotti J, Pringle JR, Reid BJ: Genetic control of the cell division cycle in yeast. Science 183:46–51, 1974.

Hassell JA, Colby C, Romano AH: The effect of serum on the transport and phosphorylation of 2-deoxyglucose by untransformed and transformed mouse 3T3 cells. J Cell Physiol 86:37–46, 1975.

Hill D: "The Biochemistry and Physiology of Tetrahymena." New York: Academic Press, 1972.

Nimmo HG, Cohen P: Hormonal control of protein phosphorylation. Adv Cycl Nucleotide Res 8:146–266, 1977.

Pardee A: A restriction point for control of normal animal cell proliferation. Proc Natl Acad Sci USA 71:286–290, 1974.

Pastan I, Johnson GS, Anderson WB: Role of cyclic nucleotides in growth control. Ann Rev Biochem 44:491–524, 1975.

Pautard FGE: Calcification in unicellular organisms. In Schraer H (ed): "Biological Calcification." New York: Appleton-Century-Crofts, 1970, pp 105–201.

Rasmussen H, Bordier P, Kurokawa K, Nagata N, Ogata E: Hormonal control of skeletal and mineral homeostasis. Am J Med 56:751–763, 1974.

Rasmussen H, Goodman DBP: Relationships between calcium and cyclic nucleotides in cell activation. Physiol Rev 57:422–509, 1977.

Roos A, Boron WF: Intracellular pH. Physiol Rev (in press), 1981.

Rose AH, Harrison JS: "The Yeasts." London and New York: Academic Press, 1971.

Rubin H: Central role for magnesium in coordinate control of metabolism in animal cells. Proc Natl Acad Sci USA 72:3551–3555, 1975.

Rubin H, Fodge D: Interrelationships of glycolysis, sugar transport and initiation of DNA synthesis in chick embryo cells. In Clarkson B, Baserga R (eds): "Control of Proliferation in Animal Cells." New York: Cold Spring Harbor Labs, 1974, pp 801–816.

Rubin RW, Weiss GD: Direct biochemical measurements of microtubule assembly and disassembly in Chinese hamster ovary cells. J Cell Biol 64:42–53, 1975.

Runnstroem J: Zur Kenntnis der Stoffwechselvorgaenge bei der Entwicklungserregung des Seeigeleies. Biochem Z 258:257–279, 1933.

Sato GH, Ross R: Hormones and cell culture. In Clarkson B, Baserga R (eds): "Control of Proliferation in Animal Cells." New York: Cold Spring Harbor Lab, 1979, pp 27–35.

Shen S, Steinhardt RA: Direct measurement of intracellular pH during metabolic depression of the sea urchin egg. Nature 272:253–254, 1978.

Waddell WJ, Butler TC: Calculation of intracellular pH from the distribution of DMO. J Clin Invest 38:720–729, 1959.

Waddell WJ, Bates RG: Intracellular pH. Physiol Rev 49:285–336, 1969.

Warburg O: "Ueber den Stoffwechsel der Tumoren." Berlin: Springer, 1926.

Wehland J, Weber K: Distribution of fluorescently labeled actin and tropomyosin after microinjection in living tissue culture cells as observed with TV image intensification. Exp Cell Res 127:397–408, 1980.

Zeuthen E: Synchrony in Tetrahymena by heat shocks spaced a normal cell generation apart. Exp Cell Res 68:49–60, 1971.

Changes in Intracellular pH of Physarum plasmodium During the Cell Cycle and in Response to Starvation

R.A. Steinhardt and M. Morisawa

Department of Zoology, University of California, Berkeley, California 94720 (R.A.S.), and Ocean Research Institute, University of Tokyo, 15-1-1-Chome, Minamidai, Nakano-Ku, Tokyo 164, Japan (M.M.)

I.	Introduction	361
II.	Materials and Methods	362
	A. Plasmodia	362
	B. Microelectrode Recordings	363
	C. Solutions	363
III.	Results	364
	A. Choice of Recording Method	364
	B. Intracellular pH and the Cell Cycle	366
	C. Delay of Mitosis by Lowering Intracellular pH	366
	D. Decrease of Intracellular pH and the Delay of Mitosis During Starvation	369
	E. Increase of Intracellular pH With Refeeding	370
	F. Morphological Changes During Starvation and Sodium Acetate Treatment	371
IV.	Discussion	372
V.	References	373

I. INTRODUCTION

Gerson and Burton [1977] reported that there was cycling of intracellular pH in the plasmodium of the acellular slime mold Physarum polycephalum with the same period as the mitotic cycle and with a peak intracellular pH at mitosis. The sharp rise in pH_i; just before mitosis suggested a possible role in the synchronization of nuclear divisions in this acellular organism.

This seems an especially attractive hypothesis since it would fit with evidence favoring the synthesis of specific "initiator protein(s)," which are synthesized less than 1 hour before the onset of prophase [Sachsenmaier et al, 1972] and with the evidence linking the regulation of intracellular pH to the control of protein synthesis in other systems [Steinhardt and Winkler, 1979; Winkler et al, 1980; Grainger et al, 1979].

However, Gerson and Burton had used the rather blunt antimony pH microelectrodes that were commercially available and had reported unusually low values for intracellular pH (5.8–6.6) and had found much lower membrane potentials than previous workers [Miller et al, 1968]. Since Physarum polycephalum is grown in highly acidic medium [Daniel and Baldwin, 1964] it seemed that Gerson and Burton's results might have been affected by various degrees of leak of extracellular medium into the plasmodium during recordings of intracellular pH due to damage during the electrode penetrations. Gerson and Burton had also used a fluorescent method in measuring pH_i that had given qualitative support to their electrode measurements of cycling pH_i with a maximum near mitosis. Because their results suggested pH_i might be used in the regulation of the cell cycle, we were interested in trying to confirm their results by another method and to extend them by manipulating intracellular pH while observing the timing of mitosis.

By the use of very fine recessed-tip microelectrodes of the Thomas type, we were able to confirm that there is a cycle of intracellular pH corresponding to the period of the cell cycle. However the cycle of pH_i ranges between 7.0 and 7.5, and shows an additional feature of a slight dip about 2 hours before mitosis. Furthermore, we were able to show that whereas the intracellular pH must be at the higher levels during a critical period just before mitosis, mitosis itself can proceed while the plasmodium is at the lower values. We extended these studies to the process of differentiation induced by starvation, and found that intracellular pH does fall to very low levels after a period of starvation and that refeeding can reverse this process. Additionally, artificial manipulations that lower intracellular pH in the complete medium will give rise to a morphology that closely resembles the first stages of starvation-induced differentiation.

II. MATERIALS AND METHODS

A. Plasmodia

Microplasmodia of Physarum polycephalum were grown in agitated suspension culture on a rotary shaker at 130–140 rpm in semidefined growth medium (nutrient medium) at room temperature. They were later spread onto dry filter paper over stainless steel mesh. Two and one-half to three hours after spreading, the microplasmodia fuse with each other to form an acellular plasmodium. At this time growth medium is added to the cultures. All procedures for preparing and culturing the surface plasmodia were performed under aseptic conditions at 26°C [Daniel and Balwin, 1964]. Fused plasmodia entered first mitosis approximately 6–8 hours after spreading on filter paper. They divide in synchrony approximately 12–14 hours after spreading. Starved plasmodia were derived from surface cultures by incubation for the appropriate time in starvation medium. Repair was induced by transferring the starved plasmodia to nutrient medium.

Small unstained pieces of plasmodia were squashed under coverslips and observed at 1,350 × using oil immersion. The optics employed were phase contrast with a green filter.

B. Microelectrode Recordings

We measured intracellular pH with Thomas type recessed-tip pH-sensitive microelectrodes [Thomas, 1976]. The pH microelectrodes used in these experiments gave a linear response of slope 57–60 mv/pH unit and had less than a 3-mv drift when calibrated before and after each experiment in standard solution at pH 4.6. Measurement of resting potentials was made with conventional electrodes filled with 3M KCl and 10 mM EGTA with a resistance of 40–60 megaohms. The pH potential and the resting membrane potential were measured differentially between intracellular microelectrodes. To obtain the intracellular pH, the resting membrane potential was subtracted from the pH potential.

We used two methods to penetrate the cells with the microelectrodes. In the first method (full immersion), conventional and pH-sensitive microelectrodes penetrated cells 2–3 cm in diameter which were totally immersed in nonnutrient medium. As shown in Figures 1A and 1B, a decrease in potential occurred with small interval of 2–4 minutes at points a, b, and c, probably owing to leakage of the medium into the cell. In the second method (thin layer), two electrodes penetrated plasmodia directly as grown (on filter paper on the stainless steel mesh in either medium) immediately after calibrating the conventional and pH-sensitive microelectrodes in the standard medium at pH 4.6 and growth medium. Both media (nutrient and nonnutrient) gave identical pH readings. Following the intracellular recording, the two electrodes were again calibrated in standard pH 4.6 and growth media. Since better penetrations and healing yielding more stable potentials result from the thin layer technique (fig. 1c and d), subsequent pH measurements (Figs. 3–8) were performed by this method.

C. Solutions

The composition of semidefined growth medium (nutrient medium): Bactotrytone 10g/liter, yeast extract 3g/liter, dextrose 9g/liter, citric acid 3.6g/liter, $Mg_2SO_47H_2O$ 0.5 g/liter, KH_2PO_4 2g/liter, $CaCl_22H_2O$ 0.6g/liter. pH was adjusted to 4.6 with 30% KOH, final K^+ slightly above 40 mM. After autoclaving the solution, autoclaved hematine solution was added to a final level of 0.005% [Daniel and Baldwin, 1964]. Starvation medium, nonnutrient medium, and pH standard solutions have the same composition, except that tryptone, yeast extract, dextrose, and hematine are eliminated. Sodium acetate solution was made by adding the appropriate amount of 1 M sodium acetate to growth medium.

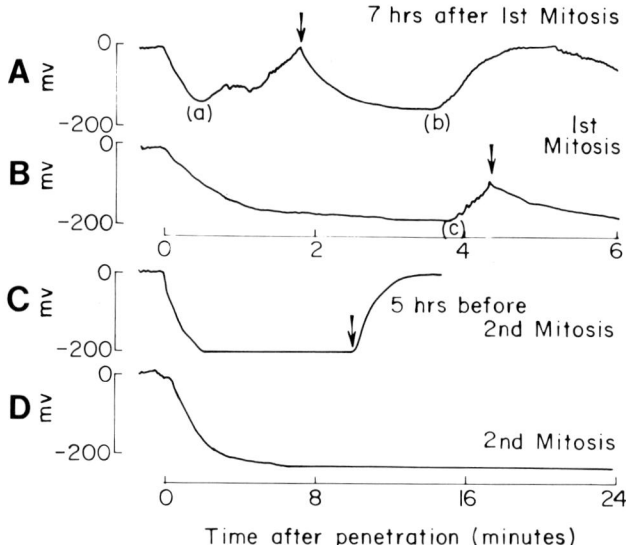

Fig. 1. Potentials recorded by pH microelectrodes inserted into macroplasmodia of Physarum polycephalum comparing full immersion and thin-layer recording conditions (see Methods). (A) Potential measured in the nonnutrient medium at 7 hours after first mitosis (full immersion). Maximum potential was 150 mv. (B) Potential measured as (A) at first mitosis (full immersion). Maximum was 170 mv. Arrow indicates the point where the electrode was inserted again. (C) Potential measured in Physarum on the filter paper at 5 hours before second mitosis (thin layer). Maximum was 200 mv. (D) Potential at second mitosis (thin layer). Maximum was 230 mv. Estimates of intracellular pH are obtained by subtracting the membrane potential from these records.

The Physarum polycephalum culture was a gift from Dr. W.M. LeSturgeon, Department of Molecular Biology, Vanderbilt University, Nashville, Tennessee.

III. RESULTS

A. Choice of Recording Method

Since an accurate measurement of intracellular pH with microelectrodes depends on a reliable method to implant both a conventional and a pH microelectrode, we experimented with different methods of recording. We first tried penetrations with growing plasmodia fully immersed in the nonnutrient medium. Although the -30 mV potentials obtained with the conventional microelectrode were higher than the -20 mV potentials found by Gerson and Burton [1977], they were lower than the -45 mV potentials reported by Miller et al [1968]. As is shown in Table I we could obtain better recordings of membrane potentials (-46 mV) by either making the recording solution more isotonic or mounting the plasmodia in the standard manner on top of the stainless steel mesh where

TABLE I. Membrane Potential in Plasmodia of Physarum polycephalum (Mean ± SE)

	n	Resting potential
Full immersion		
Growing plasmodia in nonnutrient or growth media	65	−30 ± 0.8 mV
Growing plasmodia in growth medium plus 120 mM NaCl	10	−46 ± 2.4 mV
Thin layer		
Growing plasmodia in growth or nonnutrient media	40	−46 ± 0.9 mV
Starved plasmodia in nonnutrient medium	32	−50 ± 1.5 mV
Starved plasmodia repaired by over 1 hour in growth medium	19	−40 ± 2.3 mV
Sodium acetate treated plasmodia in growth medium + acetate	15	−23 ± 1.8 mV
Acetate treated plasmodia repaired by return to growth medium	8	−38 ± 2.2 mV

the medium just soaks through the filter paper, forming a thin layer over the plasmodium. The thin-layer technique gave similar results for potential when the plasmodium was freshly placed in nutrient or nonnutrient medium but would develop higher inside negative potentials with several hours of starvation (Table I).

pH_i recordings were also more reliable with the thin-layer technique, and it was selected as the standard method. Figure 1 compares recordings made with the full immersion and thin-layer techniques. The pH electrode reports both the membrane potential and the potential due to proton concentration. Recordings (A and B) made with the full immersion technique were not as stable nor as high in magnitude as those made with the thin-layer technique (C and D). This result implied to us that even with our much finer microelectrodes there was some leak due to bad penetrations if the plasmodium was fully immersed. Since in both preparations fluid is accessible to the site of penetrations, we believe the better results with the thin-layer method are probably due to less movement of the plasmodium, and therefore less damage during microelectrode implantation, although they could equally be due to an increase in osmolarity at the surface when the dish is opened for a few minutes during recording, and evaporation occurs.

Figures 2 and 3 compare results obtained by both methods for a complete cell cycle centered on the second mitosis (20 hours after spreading). Figure 2 closely approximates the results obtained by Gerson and Burton [1977], showing a similar shape with a peak at mitosis, and large standard errors. However, these

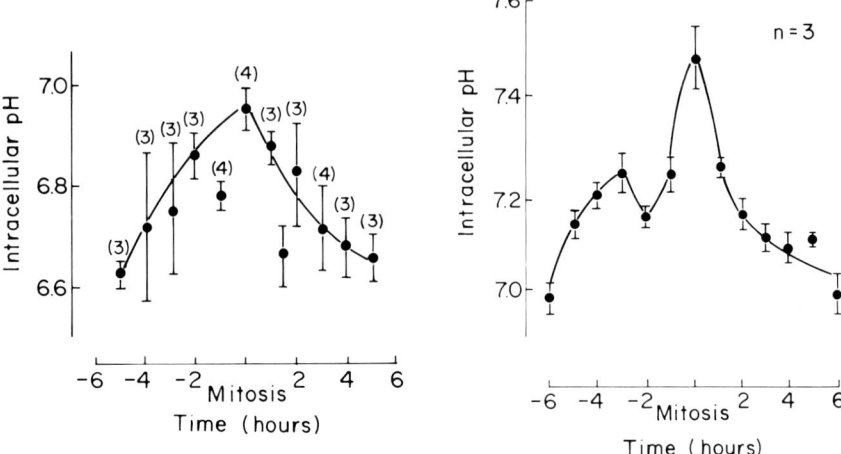

Fig. 2. Changes in intracellular pH during the mitotic cycle in plasmodia of Physarum polycephalum (Means + SE). pH was measured in plasmodia that had just been fully immersed in nonnutrient medium. The data, measured between first and second mitosis of three different plasmodia, were pooled with respect to the time of mitosis. Experimental number is indicated in parentheses. These conditions are closest to Gerson and Burton [1977].

Fig. 3. Intracellular pH around second mitosis measured on filter paper (N = 3, means + SE).

results differ in one important respect, in that the entire curve is shifted 0.6 pH units more alkaline. This difference we attribute to the use of much smaller tipped pH microelectrodes, and less leak after penetration, even under conditions of full immersion.

B. Intracellular pH and the Cell Cycle

Figure 3, showing the results obtained by the thin-layer method, represents the best approximation to what the actual pH_i is during an entire cell cycle, again centered on the second mitosis. The chief feature of these results is that the entire curve is 1.0 pH units more alkaline than that reported by Gerson and Burton, ranging from pH 7.0 to a maximum of 7.5 at mitosis. An additional feature is more clearly revealed, a plateau or slight dip at 2 hours before mitosis, which interrupts the steady increase from interphase to mitosis.

C. Delay of Mitosis by Lowering Intracellular pH

By manipulating composition of the medium we were able to lower pH_i and interrupt progress through the cell cycle. Figure 4 shows an experiment in which the growth medium is changed every hour with various concentrations of sodium

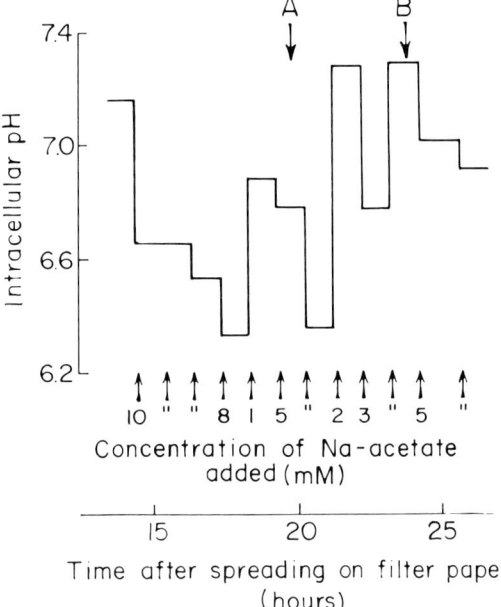

Fig. 4. Delay of second mitosis in the sodium acetate solution. Growth medium containing appropriate concentration of sodium acetate (pH 4.6) added at 14.5 hours after spreading the microplasmodial suspension culture on filter paper. Media containing sodium acetate was changed every hour. Normal second mitosis (control) occurred at point A (20 hours after spreading), and delayed mitosis in the sodium acetate medium (24 hours after spreading, at point B) occurred accompanied by increased intracellular pH with decreasing concentrations of sodium acetate.

acetate added while progress through the cell cycle is monitored along with measurements of pH_i. Acetate can penetrate cells in the uncharged form and release protons in the cytoplasm, a technique we have previously used to study the relation of pH to the control of protein synthesis in the activation at fertilization in sea urchin eggs [Grainger et al, 1979]. At A on Figure 4 the control plasmodium underwent mitosis, whereas the experimental plasmodium whose pH has kept below 7.0 did not. As the concentration of acetate is lowered, the pH_i rises and then is lowered again by a higher concentration. Eventually the plasmodium overcomes the influence of 3 mM acetate and completes mitosis. In this experiment a total of 2 hours above 7.3 (in two separate periods) is enough to go into mitosis. In other experiments it is possible to define more sharply the requirement for the higher pH to 1 hour at close to pH 7.4., just before the normal time of mitosis.

Figure 5 is an example of such an experiment in which the driving force to elevate pH_i close to the time of normal mitosis is strong enough to overcome

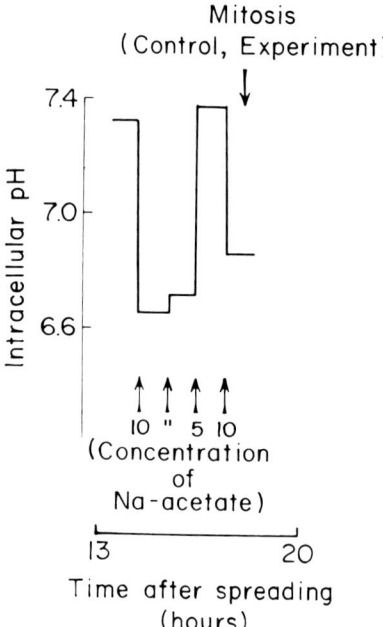

Fig. 5. The effect of short periods of treatment with sodium acetate on the intracellular pH and mitosis. Experimental conditions were the same as those of Figure 4. Treatment with sodium acetate started 14.5 hours after spreading, just 3.5 hours before mitosis. Both mitosis in normal growth medium and sodium acetate medium occurred at the same time (18 hours after spreading, see arrow).

the influence of 5 mM acetate, and mitosis occurs at the same time as the control. In this experiment 1 hour at pH 7.4 just before the normal time of mitosis is enough to enter and complete mitosis at the same time as the control untreated plasmodium. This result also illustrates two other important points. 1) that acetate is not acting to inhibit the initiation of mitosis by some other effect than its action on pH_i since the initiation of mitosis is taking place in a high concentration of sodium acetate (5 mM), which is not lowering the pH_i enough this close to mitosis to stop the initiation process; 2) even when the pH_i is dropped below 7.0 by the application of 10 mM sodium acetate, the actual process of mitosis itself can proceed once the initiation is complete. Thus if the timing is right, just a brief period of 7.4 pH_i can complete the process to initiate mitosis, which itself does not require the alkaline pH_i values.

That timing is important can be seen by experiments in which pH_i is kept below 7.0 right through the normal time of mitosis and only released 4 hours or more later. Mitosis occurred 6–7 hours after transfer back to the acetate free growth medium and 4 hours or more after the pH had recovered to normal levels (Fig. 6). In this experiment it is impossible to tell if the extra time is needed to

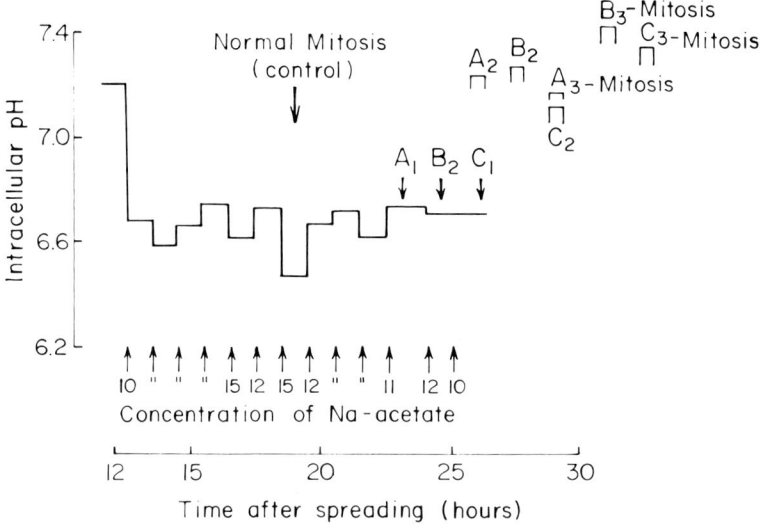

Fig. 6. The effect of sodium acetate on intracellular pH and mitosis. Experimental conditions were the same as for Figures 4 and 5, except that the treatment with sodium acetate started 12 hours after spreading. Normal second mitosis (control) occurred at 19 hours after spreading (arrow). Three plasmodia, grown on the same filter paper, were transferred from sodium acetate medium to normal medium at the times indicated by A_1, B_1, and C_1. pH was measured at the times indicated by A_2, A_3, B_2, B_3, C_2, and C_3, respectively. Mitosis occurred 6–7 hours after transfer to growth medium (A_3, B_3, and C_3).

get above pH 7.3 because of the interrupted cycle of regulating mechanisms for intracellular pH, or if the length of the delay implies the requirement to get another biochemical cycle back in phase with pH.

D. Decrease of Intracellular pH and the Delay of Mitosis During Starvation

Starvation experiments provide evidence that points to an absolute requirement for the correct pH in the timing of mitosis. When growing plasmodia are transferred to nonnutrient medium, the plasmodia spread out and form a network of veins [Guttes et al, 1961]. Under these conditions the nuclei of the starved plasmodium undergo an asynchronous abnormal mitosis as they begin the process of becoming dormant microsclerotia [Guttes et al, 1961; Rusch, 1970]. Figure 7 shows the pH_i recorded in three cases of starved plasmodia. The pH_i begins to drop within 2–3 hours after the transfer to non-nutrient medium. This drop is independent of the timing of the normal cycle of pH. In experiments 1 and 2, plasmodia were transferred to the nonnutrient medium 16 hours after the original spreading. Control plasmodia spread at the same time as experiment 1

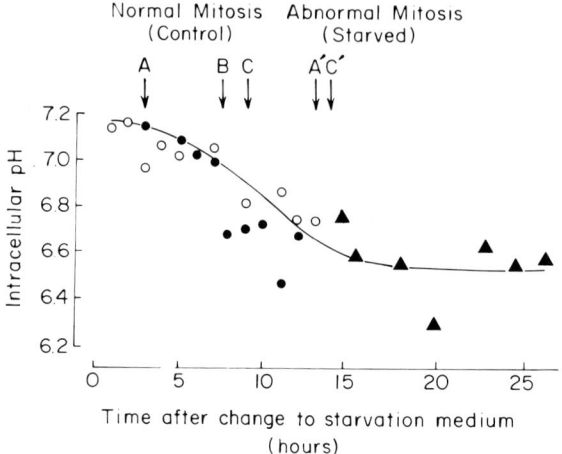

Fig. 7. Changes in intracellular pH during starvation of plasmodia. Medium changed from growth medium to starvation medium 16 hours (experiments 1 and 2) and 11 hours (experiment 3) after spreading. Normal second mitoses (controls), indicated by arrows, occurred 18.5 (A), 23.5 (B), and 20 (C) hours after spreading in experiments 1 (A), 2 (B), and 3 (C), respectively. Abnormal mitoses of starved Physarum in experiments 1 and 3 occurred 29.5 (A′) and 25.0 (C′) hours after spreading (see Guttes et al [1961] for description of abnormal "mitosis").

underwent mitosis at 18.5 hours (A) after spreading, whereas those controls associated with the spreading in experiment 2 underwent mitosis at 22.5 hours (B). In experiment 3 of this type, the plasmodium was transferred to the nonnutrient medium 11 hours after spreading on the filter paper with the mitosis in the controls occuring at 20 hours after spreading (C). In all cases the plasmodia transferred to nonnutrient medium remained at stable pH_i values for about 2 hours and then the pH_i began to decline gradually. The decline became steep at about 8 hours after the start of starvation, and pH_i reaches a low stable value of approximately 6.6 at 12–15 hours after transfer. Figure 7 also shows the relative timing of the less synchronous abnormal mitosis of the starved plasmodium for experiments 1 (A′) and 3 (C′). In these experiments a block of normal mitosis and the start of a differentiation process follows the drop in pH_i induced by starvation, which tends to confirm our interpretation of the acetate experiments.

E. Increase of Intracellular pH With Refeeding

By retransferring the starved plasmodium back to the complete nutrient medium, pH_i can recover to normal values, cycling resumes, and mitosis occurs only after the pH has attained values above 7.3 for a sufficient period of time (Fig. 8). Period from spreading to transfer to nonnutrient medium was 11 hours

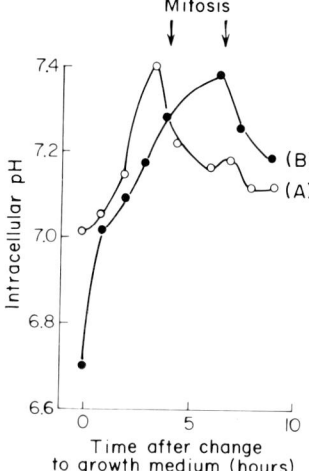

Fig. 8. Repair of the intracellular pH and mitosis after transfer from starvation medium to growth medium. (A) and (B): At 11 and 14 hours after spreading on filter paper, medium was changed from growth to starvation. The plasmodia were then incubated 9 and 13 hours followed by a change back to growth medium. The intracellular pH was measured in (A) and (B). The time of mitosis is indicated by an arrow.

for A and 14 hours for B. Period of starvation before transfer back to complete medium was 9 hours for A and 13 hours for B. Plasmodium A recovers faster with refeeding, attaining pH 7.4 within 3.5 hours, and enters mitosis by 4 hours after refeeding. Plasmodium B, which was starved longer, takes longer to raise its intracellular pH, and mitosis is delayed until after a period of pH 7.3. The starting intracellular pH of 7.02 after 9 hours of starvation for plasmodium A, and the starting pH of 6.7 after 13 hours of starvation for plasmodium B are in good agreement with the previous starvation experiments shown in Figure 7.

F. Morphological Changes During Starvation and Sodium Acetate Treatment

If starvation results in a lowering of pH_i, we reasoned that the acetate treatment might trigger some of the morphological signs of the differentiation process associated with the formation of dormant microsclerotia, even if the plasmodium is kept in the complete medium with the addition of 10 mM sodium acetate at 17 hours. Within 10 hours after the start of both starvation and acetate treatment, one sees the characteristic "fanlike extension consisting of an advancing front of a solid layer of protoplasm followed by an arboreal network of loosely scattered anastomosing strand," as described by Guttes et al [1961]. Although the appearance induced by these various periods of starvation is different in some respects from that induced by artificially lowering the pH, the major features of a fanlike extension at the tips of the scattered strands are fairly well reproduced even though the acetate plasmodia are supplied by a complete growth medium.

IV. DISCUSSION

Our results confirm a cycle of intracellular pH with the cell cycle in Physarum and demonstrate that the cycling exists in a reasonable pH_i range within which normal biochemistry can operate. Furthermore, our experiments demonstrate that the peak values obtained at the time of mitosis are essential for entry into mitosis, although not necessary for mitosis itself. We do not think, however, that these results extend generally to other systems, since other work from this laboratory has shown that pH_i does not regulate mitosis and cytokinesis in the early Xenopus embryo; nor does clamping at high pH_i values interfere with normal chromosome cycling or cell division in the sea urchin embryo [Lee and Steinhardt, 1981; Grainger et al, 1979].

Instead, the value of Physarum as an experimental system lies in two new directions. The results with pH_i and the entry to mitosis point to a pH_i-sensitive step in the preparation of nuclear material for mitosis, which, whereas it may be unique to this organism, may help unravel higher order chromosomal processing. In addition, the use of low pH_i to inhibit metabolism in unfavorable conditions such as starvation may be a more general response acting via the storage of mRNA by a "masking" mechanism as discussed below.

There are three lines of evidence for a pH-sensitive step prior to mitosis. First, there is the cycle itself, with peak values of pH 7.4–7.5 just within 1 hour of mitosis. Second, there is the delay of mitosis induced by lowering pH_i with sodium acetate. This effect of acetate is correlated with the actual lowering of pH_i rather than some other unknown side effect as seen in the experiments typified by Figures 4 and 5. It was not the concentration of acetate but the timing and amount of higher alkaline values that determined the entry into mitosis. Finally, there are the effects of starvation and refeeding on pH_i and mitosis. Starvation lowers the pH_i and interrupts the cell cycle blocking mitosis. Entry into mitosis upon refeeding depends on the timing of the recovery of pH_i to the pH 7.4 range. The timing of the pH-sensitive step just 1 hour before mitosis indicates that there is a higher-order process well past the histone H1 phosphorylation that Bradbury and Matthews have hypothesized as a necessary prelude to chromosome condensation [Bradbury et al, 1974].

It is not the study of the control of the cell cycle that directed our attention to Physarum, but rather Jeffery's work on the translational regulation of protein synthesis during starvation and refeeding [Jeffery, 1979]. His studies indicated that the process of dormancy involves the repression of functional mRNA at the level of initiation. This suggested similarities with the derepression of protein synthesis in sea urchin eggs that was known to be dependent on the rise in pH_i

at fertilization [Winkler and Steinhardt, 1981; Winkler, this volume]. Combining our results on pH_i with Jeffery's work on message makes it clear that Physarum should allow experimental analysis of the mechanisms of both repression (by lower pH_i) and derepression (by increasing pH_i). In fact, evidence is rapidly accumulating that dormancy in general is accompanied by low pH_i and activation by increases in pH_i. Since the first suggestion that a pH_i rise was involved in the derepression of sea urchin eggs [Steinhardt and Mazia, 1973], many other truly dormant cell types have been found to utilize pH_i in a similar fashion. The most recent examples include the germination of bacterial and yeast spores [Setlow and Setlow, 1980; Barton et al, 1980].

It is a great temptation to extend the interpretation of these results and suggest that the cycle of pH_i fits the requirements for the circadian clock for the mitotic oscillator in Physarum; however, I think this will have to wait for an extensive series of fusion experiments where the clock is advanced and pH_i monitored. Clearly, the involvement of protein synthesis in the control of circadian clocks [Jacklet, 1980] and the dependence of mitosis on the accumulation of cytoplasmic factors [Sachsenmaier et al, 1972] suggests a pathway by which ph_i may act in this system.

Finally, I would like to offer some thoughts as to why Physarum may be using pH_i as its primary and most important regulator both during the cell cycle and during dormancy. Physarum is adapted to growth in the highly acidic conditions in soil. To keep its pH_i in the range for active biosynthesis requires an enormous amount of energy. By relaxing its regulation of pH_i to the times it is actually needed rather than maintaining it at all times at alkaline values would be very much favored energetically. From this point of view the nuclear synchrony we marvel at is really a by-product rather than a principal adaptation itself.

ACKNOWLEDGMENTS

We thank Janet Alderton for the construction of the pH microelectrodes and the National Science Foundation for support (PCM 79-04744 to R.A.S.).

V. REFERENCES

Barton JK, Den Hollander JA, Lee TM, MacLaughlin A, Shulman RG: Measurement of the internal pH of yeast spores by ^{31}P nuclear magnetic resonance. Proc Natl Acad Sci USA 77:2470, 1980.

Bradbury EM, Inglis RJ, Matthews HR, Langan TA: Molecular basis of control of mitotic cell division in eukaryotes. Nature 249:553, 1974.

Daniel JW, Baldwin HH: Methods of culture for plasmodial myxomcetes. In Prescott DM (ed): "Methods in Cell Physiology," Vol 1. New York: Academic Press, 1964, p 9.

Gerson DF, Burton AC: The relation of cycling of intracellular pH to mitosis in the acellular slime mould Physarum polycephalum. J Cell Physiol 91:297, 1977.

Grainger JL, Winkler MM, Shen SS, Steinhardt RA: Intracellular pH controls protein synthesis in the sea urchin egg and early embryo. Dev Biol 68:396, 1979.

Guttes E, Guttes S, Rusch HP: Morphological observations on growth and differentiation of Physarum polycephalum grown in pure culture. Dev Biol 3:588, 1961.

Jacklet JW: Protein synthesis requirements of the Aplysia circadian clock. J Exp Biol 85:33, 1980.

Jeffery WR: Translational regulation of polysome formation during dormancy of Physarum polycephalum. J Bacteriol 140:490, 1979.

Lee SC, Steinhardt RA: Observations on intracellular pH during cleavage of eggs of Xenopus laevis. J Cell Biol (in press), 1981.

Miller DM, Anderson JD, Abbott BC: Potentials and ion exchange in the slime mold plasmodia. Comp Biochem Physiol 27:633, 1968.

Rusch HP: Some biochemical events in the life cycle of Physarum polycephalum. In Prescott DM, Goldstein L, McConkey E (eds): "Advances in Cell Biology," Vol 1. New York: Appleton-Century-Crofts, 1970, p 297.

Sachsenmaier W, Remy U, Plattner-Schobel R: Initiation of synchronous mitosis in Physarum polycephalum. Exp Cell Res 73:41, 1972.

Setlow B, Setlow P: Measurements of the pH within dormant and germinated bacterial spores. Proc Natl Acad Sci USA 77:2476, 1980.

Steinhardt RA, Mazia D: Development of K^+ conductance and membrane potentials in unfertilized sea urchin eggs. Nature 241:400, 1973.

Steinhardt RA, Winkler MM: The ionic hypothesis of cell activation at fertilization. In Kaplan JG (ed): "The Molecular Basis of Immune Cell Function." New York: Elsevier/North-Holland, 1979, p 11.

Thomas RC: "Ion-Sensitive Intracellular Microelectrodes." New York: Academic Press, 1978.

Winkler MM, Steinhardt RA: Activation of protein synthesis in a sea urchin cell-free system. Dev Biol 84:432, 1981.

Winkler MM, Steinhardt RA, Grainger JL, MinningL L: Dual ionic controls for the activation of protein synthesis at fertilization. Nature 287:558, 1980.

The Relation Between Intracellular pH and DNA Synthesis Rate in Proliferating Lymphocytes

Donald F. Gerson
Basel Institute for Immunology, 487 Grenzacherstrasse, CH-4005 Basel, Switzerland

I.	Introduction	375
II.	Methods	376
	A. Spleen Lymphocytes	376
	B. Measurement Techniques	376
III.	Results and Discussion	378
	A. Intracellular pH	378
	B. Correlation of pH_i and DNA Synthesis	380
IV.	Conclusions	382
V.	References	382

I. INTRODUCTION

Resting lymphocytes can be stimulated to increase in size and begin several rounds of cell division by treatment with concanavalin A (Con A) or bacterial lipopolysaccharide (LPS). Populations of murine spleen lymphocytes which have been exposed to either mitogen show a burst of protein and RNA synthesis, and reach a maximal rate of DNA synthesis in approximately 48 hours. The spleen lymphocyte population consists of B lymphocytes sensitive to LPS, and T lymphocytes, sensitive to Con A. We have found that in both subpopulations there is a transient rise in intracellular pH (pH_i) of 0.2 pH units by 6 hours poststimulation [Gerson and Kiefer, 1981]. There is also a second rise in pH_i that parallels the rate of DNA synthesis, and reaches a slightly higher maximum at 48 hours.

The fertilization of eggs presents a similar situation. Fertilization is also followed by increased rates of protein, RNA, and DNA synthesis. In sea urchin eggs, the first period of DNA synthesis occurs approximately 35 minutes after fertilization. There is a concomitant rise in pH_i by that time [Steinhardt and Winkler, 1979]. Because of the much more rapid kinetics of the activation process in sea urchin eggs, it is not possible to determine whether there are separate pH_i rises associated with stimulation and DNA synthesis, as in the mitogen stimulated lymphocyte [Johnson and Epel, 1981].

A change in the average cytoplasmic pH of a resting cell can be expected to have many general effects, rather than a small number of specific ones. Steinhardt and Winkler [1979] have shown that the rapid protein synthesis that follows fertilization of sea urchin eggs is facilitated by pH increases similar to those observed after fertilization. A key enzyme in DNA synthesis is DNA polymerase. Hobart and Infante [1978] have found that DNA polymerase β is synthesized rapidly during the early burst of protein synthesis occurring in the first 45 minutes after sea urchin egg fertilization.

The pH optimum of DNA polymerases is generally quite high (Table I). The pH optimum for sea urchin DNA polymerase β is pH 9.0. This is above physiological pH, as are most of the pH optima for this family of enzymes. All show increasing activity at increasing pH from ~7.0 to ~8.0, which encompasses the usual physiological pH_i range. This most probably is related to the rise in the free energy of hydrolysis of ATP and other triphosphates with increasing pH [Alberty, 1968]. In the synchronous coenocytic slime mold Physarum polycephalum, intracellular pH is at its highest point during the part of the cell cycle when DNA synthesis begins [Gerson and Burton, 1976; Gerson, 1977]. As in sea urchins and many other species, the pH optimum for DNA polymerase (pH 7.7) in Physarum is much higher than the average intracellular pH [Baer and Schiebel, 1978]. RNA polymerases also increase in activity from pH 7 to pH 8 [Chakraborty et al, 1973; Modak and Srinivasan, 1973]. Thus we see that three major biosynthetic processes (protein, RNA, and DNA synthesis) are all stimulated by increased pH in the physiological range. It is propitious that in resting cells stimulated to begin macromolecular biosynthesis, a physiological response is also initiated that results in an increase of the intracellular pH.

II. METHODS

A. Spleen Lymphocytes

Mouse (Balb/c) spleen lymphocytes were obtained by macerating the organ with a stainless steel mesh, separating the cells from particles of tissue by sedimentation, and removing red cells by NH_4Cl lysis [Schreier, 1979]. These were cultured at an initial cell density of 1×10^6 cells/ml in Iscove's Modified Dulbecco's medium (Gibco 785220) plus 10% fetal calf serum and 10 mM each of HEPES, PIPES, and TRICINE (Sigma, St Louis). Spleen cells were stimulated with either Con A (5 μg/ml, Pharmacia) or LPS (50 μg/ml).

B. Measurement Techniques

Intracellular pH was determined by the ^{14}C-DMO technique. ^{14}C-5,5-dimethyl-2,4-oxazolidinedione (New England Nuclear) was equilibrated with the cells at pH 7.3 for 30 minutes, which is approximately twice the minimum time required for equilibration. Intracellular water volume was measured with ^3H_2O

TABLE I. pH Optima of DNA Polymerases

Cell type	Polymerase type	pH optimum	Reference
HeLa + herpes simplex	—	8.0	Weissbach et al, 1973
Human, KB cells	C	9.5	Sedwick et al, 1975
Rat + Kilham virus	—	8.7	Salzman and McKerlie, 1975a
Rat nephroma	I	8.9	Salzman and McKerlie, 1975b
Rat nephroma	II	8.9	Salzman and McKerlie, 1975b
Rat nephroma	1	8.9	Salzman and McKerlie, 1975b
Rat nephroma	2	7.1	Salzman and McKerlie, 1975b
Rat nephroma	3	8.9	Salzman and McKerlie, 1975b
Rat nephroma	4	7.7	Salzman and McKerlie, 1975b
HeLa	C	8.0	Chiu et al, 1975
HeLa	N-I	9.0	Chiu et al, 1975
HeLa	N-II	8.0	Chiu et al, 1975
Calf thymus	—	7.2	Bollum et al, 1974
Calf thymus	γ	9.0	Bollum et al, 1974
Calf thymus	chromatin	9.0	Chang, 1974
Human lymphocyte	α	7.2	Goodman et al, 1978
Human lymphocyte	β	8.5	Goodman et al, 1978
Human lymphocyte + EBV	induced	9.5	Goodman et al, 1978
Human lymphocyte + EBV	associated	8.0	Goodman et al, 1978
Sea urchin	embryo	8.0	Loeb, 1969
Sea urchin	β	9.0	Hobart and Infarte, 1978
Xenopus	oocyte	7.9	Brown and Tocchini-Valentini, 1974
Drosophila	embryo	8.5	Karkas et al, 1975
Physarum	cytoplasmic	7.7	Baer and Schiebel, 1978
Saccharomyces	I	8.2	Chang, 1977
Saccharomyces	II	7.6	Chang, 1977
B. subtilis	I	8.2	Gass and Cozzarelli, 1974
B. subtilis	II	8.2	Gass and Cozzarelli, 1974
M. lutens	—	7.8	Hamilton, 1974
E. Coli	III	7.2	Livingston et al, 1975
Rauscher leukemia virus	—	7.8	Modak and Marcus, 1977
T4	—	8.2	Lo and Bessman, 1976

and corrected for extracellular water space with ^{14}C-Inulin. Paired triplets were used for each determination, one triplet with ^{14}C-DMO and ^{3}H$_2$O, and the other with ^{14}C-Inulin and ^{3}H$_2$O. The cells were separated from the incubation medium by centrifugation through a layer of silicone oil with a Beckman microfuge, as described for TPP uptake measurements by Kiefer et al [1980]. Dead cells were measured in each sample by hemocytometer counts of trypan blue admitting cells. Separate experiments demonstrated that trypan blue admitting cells had a pH$_i$ equal to the pH of the medium. Calculation of pH$_i$ from these data was performed by the standard method [see Gillies and Deamer, 1979], with the additional correction for dead cells.

The rate of DNA synthesis was determined from the amount of ^{3}H-thymidine incorporation into water-insoluble, glass fiber adhering material after a 1-hour incubation period. Cells were collected with a Hiller cell filtration device (O. Hiller, Madison). All results are expressed as ^{3}H-thymidine cpm/h–10^6 live cells, and thus are a measure of the average rate of DNA synthesis over a 1-hour period.

III. RESULTS AND DISCUSSION

A. Intracellular pH

Resting spleen lymphocytes begin substantial ^{3}H-thymidine incorporation rates 24 hours after exposure to mitogen (Fig. 1). The rate of DNA synthesis increases to a maximum at 48 hours for Con A-stimulated spleen cell populations, and at 54 hours for LPS-stimulated populations, then subsides to low values by 72–80 hours. In the late phases of the growth response to mitogen, there are substantial numbers of dead cells. However, the results reported here are calculated for only the live proportion of the population.

Intracellular pH was determined by the ^{14}C-DMO method at various times after mitogen addition. A biphasic response of pH_i to either mitogen was observed: from 0 to 6 hours intracellular pH rose from pH 7.18 to a maximum of pH 7.4; from 6 to 12 hours it fell back to the original level; from 12 to 48 hours, pH_i increased from pH 7.2 to pH 7.45; and from ~48 to 72 hours it fell once again.

The pH_i of resting mouse spleen lymphocytes was found to be pH 7.18 ± 0.03. The intracellular pH depends on the extracellular pH, which was always pH 7.3 for the measurements reported here. The value we have obtained compares favorably with previous measurements of similar cells. Zieve et al [1967] determined the intracellular pH of the human lymphocyte, and found it to be 7.31 under physiological conditions. Ahmed and Baron [1971] measured the intracellular pH of human leucocytes and reported it to be pH 7.104 ± 0.115 at an extracellular pH of 7.4 and a pCO_2 of 40 mmHg. Levin et al [1976] also measured the intracellular pH of human leucocytes and obtained 7.11 ± 0.02. More recently, Deutsch et al [1979] determined the intracellular pH of human peripheral lymphocytes, and obtained pH 7.25 in pH 7.4 media using the DMO technique.

The early time course of pH_i following mitogenic stimulation is preceded by a sharp depolarization of the membrane potential [Kiefer et al, 1980]. These changes are coincident with changes in ion fluxes across the cell membrane [Kaplan, 1978; Kaplan and Owens, 1980]. A similar pattern is seen in the first few minutes after the entry of sperm into sea urchin eggs [Epel, 1978, 1980]. In both lymphocytes and sea urchin eggs, this sequence is followed by increased rates of synthesis of protein, RNA, and DNA, and, ultimately, by cell division.

pH$_i$ and DNA Synthesis in Lymphocytes / 379

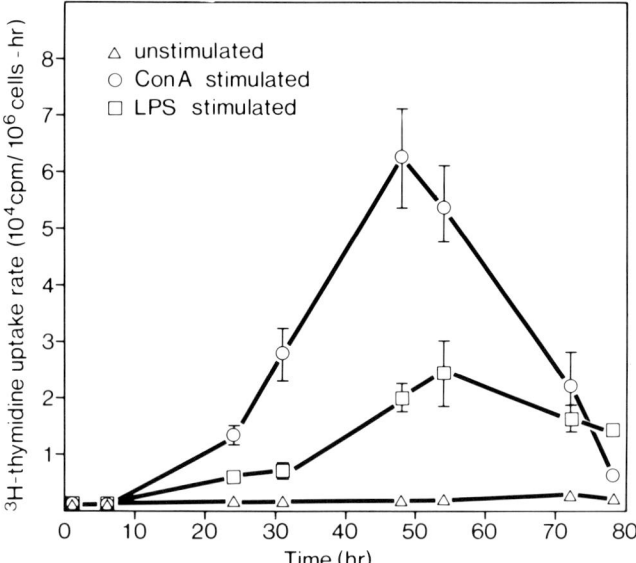

Fig. 1. The time course of the rate of DNA synthesis (^{3}H-thymidine uptake rate) following the mitogenic stimulation of spleen lymphocytes.

In some other systems, stimulation results in changes of both the membrane potential and the intracellular pH, but does not lead to cell division. Examples of this pattern include the response of platelets to thrombin [Sepersky et al, 1980], and the response of muscle cells to insulin [Moore, 1973; Moore, 1979]. In these cases, an activation has occurred in response to stimulus, but it is not followed by division. In the light of our observation of a biphasic pattern of pH changes following lymphocyte activation, it is possible that increased pH$_i$ can be independently associated with either activation responses or the initiation of cellular growth and division.

Lymphocyte activation appears to be a 2-step process. Mitogen initially activates a subset of the lymphocyte population, while simultaneously inducing other, as yet unidentified, cells to produce growth factors. These growth factors are then absorbed by the activated subset, which subsequently grow [Larsson and Coutinho, 1979; Larsson et al, 1980]. The time required to activate the responding subsets is approximately 4–6 hours. This coincides with the first phase of increasing intracellular pH. The growth kinetics of the responsive population, in the presence of growth factor, is similar to the pattern seen in Figure 1. It thus appears that the two phases of increasing pH$_i$ correspond temporally to the two phases of lymphocyte activation.

B. Correlation of pH_i and DNA Synthesis

Figure 2 gives the correlation between intracellular pH and DNA synthesis rate for various times after stimulation. The figure includes data only from the time growth begins to the time senescence begins (24–72 hours), ie, the second phase of activation. Populations stimulated with either mitogen increase in both intracellular pH and the rate of DNA synthesis during this period, whereas unstimulated cells remain at a low pH and do not have significant rates of DNA synthesis. In growing populations, the rate of DNA synthesis is strongly correlated with intracellular pH.

This correlation could result either from independent but temporally associated charges in pH_i and DNA synthesis, or from interdependent physiological processes. To help distinguish between these possibilities, a rapidly dividing population of Con A-stimulated cells (48 hours) was incubated in media of various pH from 6.8 to 7.5. The cells were then transferred to pH 7.3 medium (to avoid differential effects of pH on uptake processes) and pH_i and ^{3}H-thymidine incorporation rate were measured. The correlation between pH_i and ^{3}H-thymidine incorporation rate in these cells is given in Table IIA. Manipulation of pH_i influences the rate of DNA synthesis in much the same way as is seen to occur naturally during the growth response to mitogens. This favors the conclusion that these two physiological processes are interdependent. Inhibition of DNA synthesis with hydroxyurea or excess thymidine drastically reduces the rate of

TABLE II. Interrelations of pH_i and DNA Synthesis Rate

A. The effect of intracellular pH on DNA synthesis rate*

Intracellular pH	^{3}H-Thymidine 10^3 cpm/10^6 cells–h
7.32	50.0
7.29	47.0
7.15	30.0
6.95	5.0

B. The effect of DNA synthesis inhibitors on intracellular pH*

Treatment	^{3}H-thymidine 10^3 cpm/ 10^6 cells–h	Intracellular pH
Control	1.7	6.9
Con A	46.8	7.3
Con A + 1 mM Hydroxyurea	0.3	7.3
Con A + 5 mM Hydroxyurea	0.2	7.3
Con A + 2.5 mM Thymidine	0.2	7.3
Con A + 7.5 mM Thymidine	0.2	7.3

*Spleen lymphocytes from Balb/c mice were exposed to Con A, as described in the Methods section, incubated for ~48 hours, and then subjected to altered extracellular pH (A) or inhibitors (B) for 2 hours.

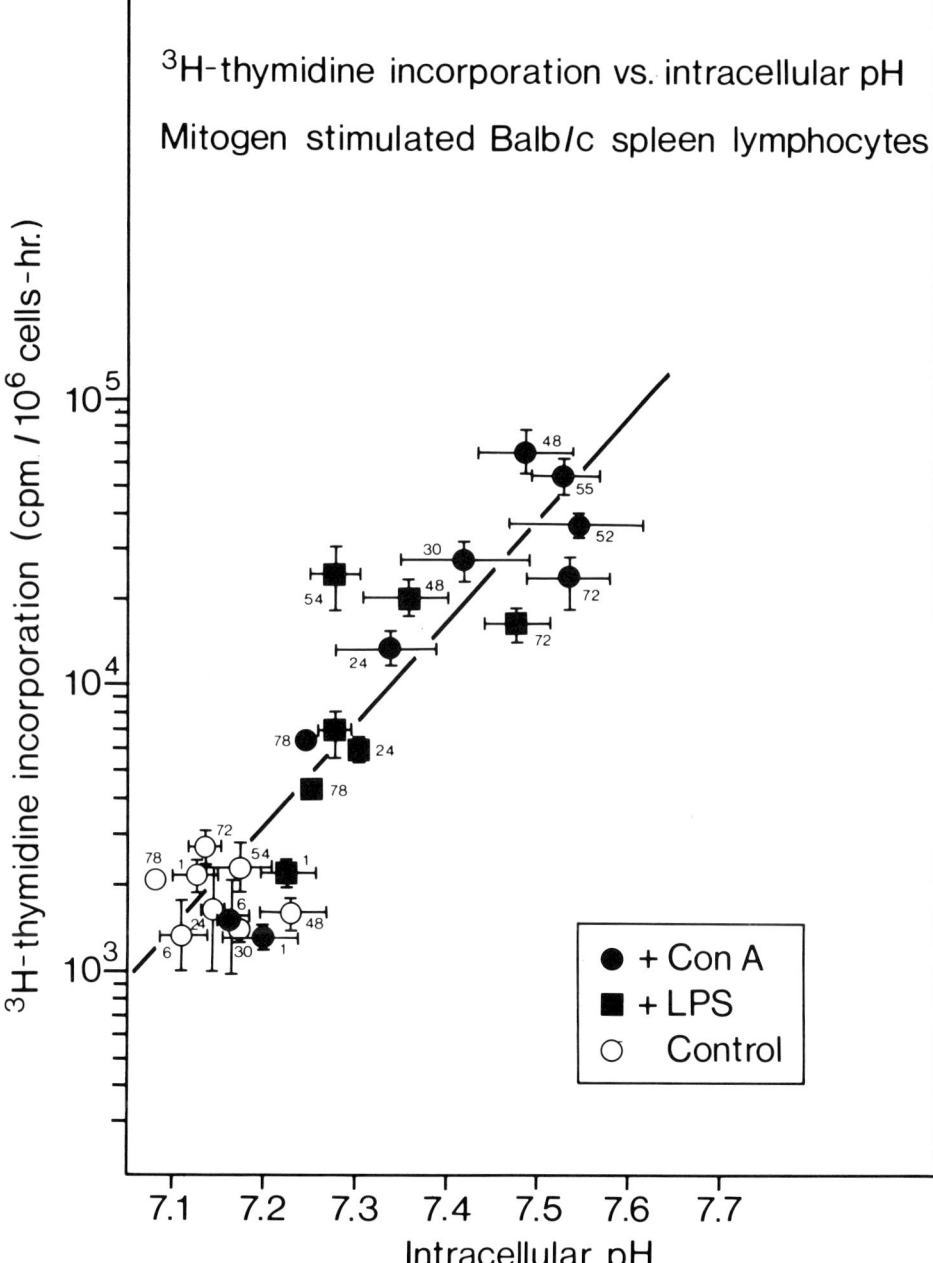

Fig. 2. The relation between ^{3}H-thymidine incorporation rate and intracellular pH over the time period from 24 to 78 hours following the mitogenic stimulation of spleen lymphocytes.

DNA synthesis, but does not affect the intracellular pH, as is seen in Table IIB. This further supports the hypothesis that increased intracellular pH stimulates the rate of DNA synthesis in growing cells.

Whereas increased pH_i facilitates DNA synthesis, and is part of the activation sequence, it alone is not the signal that results in the transition from a resting to an active state. Numerous attempts to increase the intracellular pH of resting lymphocytes in the absence of mitogen failed to result in the initiation of growth.

IV. CONCLUSIONS

We have found that intracellular pH increases in mitogen-stimulated lymphocytes, and that the pattern of the pH_i variation is temporally related to the two phases of lymphocyte stimulation. The initial rise in pH_i appears to be part of an activation sequence common to many systems. The late rise in pH_i appears to be a physiological response that facilitates the increased protein, RNA, and DNA synthesis required for growth and cell division. During the growth phase, the rate of DNA synthesis is strongly correlated with intracellular pH.

ACKNOWLEDGMENTS

Mr. Wolfgang Eufe provided expert technical assistance in all aspects of this work. The advice and encouragement of Dr. H. Kiefer and Dr. C. Steinberg are greatly appreciated.

V. REFERENCES

Ahmed S, Baron D: Intracellular pH measured in isolated normal human leucocytes. Clin Sci 40:487, 1971.

Alberty R: Effect of pH and metal ion concentration on the equilibrium hydrolysis of ATP to ADP. J Biol Chem 243:1337, 1968.

Baer A, Schiebel W: DNA polymerase from Physarum polycephalum. Eur J Biochem 86:77, 1978.

Bollum F, Chang L, Tsiapalis C, Dorson J: Nucleotide polymerizing enzymes from calf thymus gland. Meth Enzymol 29:70, 1974.

Brown R, Tocchini-Valentini G: The preparation of RNA directed DNA polymerase from ovaries of Xenopus laevis. Meth Enzymol 29:173, 1974.

Chakraborty P, Sarkav P, Huang H, Maitra U: Studies on T3 induced RNA polymerase. J Biol Chem 248:6637, 1973.

Chang L: Purification and properties of low molecular weight DNA polymerase from mammalian cells. Meth Enzymol 29:81, 1974.

Chang L: DNA polymerases from baker's yeast. J Biol Chem 252:1873, 1977.

Chiu R, Baril E: Nuclear DNA polymerases and the HeLa cell cycle. J Biol Chem 250:7951, 1975.

Deutsch C, Holian A, Holian S, Daniele R, Wilson D: Transmembrane electrical and pH gradients across human erythrocytes and human peripheral lymphocytes. J Cell Physiol 99:79, 1979.

Epel D: Regulation of cell activity at fertilization by intracellular Ca^{++} and intracellular pH. in Lerner R, and Bergsma D (eds): "The Molecular Basis of Cell–Cell Interactions." New York: Alan R Liss, 1978, p 377.

Epel D: Ionic triggers in the fertilization of sea urchin eggs. Ann NY Acad Sci 339:74, 1980.

Gass K, Cozzarelli N: B. subtilis DNA polymerases. Meth Enzymol 29:27, 1974.

Gerson D, Burton AC: The relation of cycling of intracellular pH to mitosis in the acellular slime mould Physarum polycephalum. J Cell Physiol 91:297, 1976.

Gerson D: Intracellular pH and the mitotic cycle in Physarum and mammalian cells. In Jeter J,

Cameron I, Padilla G, Zimmerman A (eds): "Cell Cycle Regulation." New York: Academic Press, 1977, p 105.
Gerson D, Kiefer H: Intracellular pH and membrane potential changes following specific mitogenic stimulation of murine B and T lymphocytes. In Resch K (ed): "Mechanism of Lymphocyte Activation." Amsterdam: Elsevier (in press).
Gillies R, Deamer D: Intracellular pH, methods and applications. Curr Top Bioenerget 9:63, 1979.
Goodman S, Prezyna C, Benz W: Two Epstein-Barr virus-associated DNA polymerase activities. J Biol Chem 253:8617, 1978.
Hamilton L: The purification of the DNA polymerase from M. luteus. Meth Enzymol 29:38, 1974.
Hobart P, Infante A: A low molecular weight DNA polymerase β in the sea urchin Strongylocentrous purpuratus. J Biol Chem 253:8229, 1978.
Johnson C, Epel D: Intracellular pH of sea urchin eggs measured by the DMO method. J Cell Biol 89:284, 1981.
Kaplan J: Membrane cation transport and the control of proliferation of mammalian cells. Ann Rev Physiol 40:19, 1978.
Kaplan J, Owens T: Activation of lymphocytes of man and mouse, monovalent cation fluxes. Ann NY Acad Sci 339:191, 1980.
Karkas J, Margulies L, Charagaff E: A DNA polymerase from embryos of D. melanogaster. J Biol Chem 250:8057, 1975.
Kiefer, H, Blume A, Kaback H: Membrane potential changes during mitogenic stimulation of mouse spleen lymphocytes. Proc Natl Acad Sci USA 77:2200, 1980.
Larsson E, Coutinho A: The role of mitogenic lectins in T cell triggering. Nature 280:239, 1979.
Larsson E, Iscove N, Coutinho A: Two distinct factors are required for induction of T-cell growth. Nature 283:664, 1980.
Levin G, Collinson P, Baron D: The intracellular pH of human leucocytes in response to acid–base changes in vitro. Clin Sci Mol Med 50:293, 1976.
Livingston D, Hinkle D, Richardson C: DNA polymerase III of E. coli. J Biol Chem 250:461, 1975.
Lo K, Bessman M: An antimutator DNA polymerase. J Biol Chem 251:2475, 1976.
Loeb L: Purification and properties of DNA polymerase from nuclei of sea urchin embryos. J Biol Chem 244:1672, 1969.
Modak M, Srinivasan P: Purification and properties of a RNA primer independent polyriboadenylate polymerase from E. coli. J Biol Chem 248:6904, 1973.
Modak M, Marcus S: Purification and properties of Rauscher leukemia virus DNA polymerase. J Biol Chem 252:11, 1977.
Moore R: Effect of insulin upon the sodium pump in frog skeletal muscle. J Physiol 232:23, 1973.
Moore R: Elevation of intracellular pH by insulin in frog skeletal muscle. Biochem Biophys Res Commun 91:900, 1979.
Salzman L, McKerlie L: Characterization of the DNA polymerase associated with Kilham rat virus. J Biol Chem 250:5583, 1975a.
Salzman L, McKerlie L: Nuclear and cytoplasmic DNA polymerases from rat nephroma cells. J Biol Chem 250:5589, 1975b.
Schreier M: In vitro immunization of dissociated murine spleen cells. In Lefkovits I, Pernis B (eds): "Immunological Methods." New York: Academic Press, 1979, p 327.
Sedwick W, Wong T, Korn D: Cytoplasmic DNA polymerase. J Biol Chem 250:7045, 1975.
Sepersky S, Simons E: Inhibition of platelet membrane potential changes by metabolic inhibitors. Fed Proc 39:1893, 1980.
Steinhardt R, Winkler M: The ionic hypothesis of cell activation at fertilization. In Kaplan J (ed): "The Molecular Basis of Immune Cell Function." New York: Elsevier, 1979, p 11.
Weissbach A, Hong S, Aucher J, Muller R: Characterization of herpes simplx virus induced DNA polymerase. J Biol Chem 248:6270, 1973.
Zieve P, Haghshenass M, Krevans J: Intracellular pH of the human lymphocyte. Am J Physiol 212:1099, 1967.

The Role of Intracellular pH in Insulin Action

Richard D. Moore, Mark L. Fidelman, Jeffrey C. Hansen, and John N. Otis

Biophysics Laboratory, State University of New York, Plattsburgh, New York 12901 (R.D.M., J.C.H., J.N.O.), and Department of Physiology, Medical College of Virginia, Virginia Commonwealth University, Box 551, MCV Station, Richmond, Virginia 23298 (M.L.F.)

I.	Introduction	386
	A. Model for Insulin Transduction	387
	B. Effect of Insulin on $Na^+:H^+$ Exchange	388
	C. Effect of Insulin on Glycolysis	389
II.	Review of Methodologies	390
	A. Ringer Solutions	390
	1. Ringer A (used to identify mechanism responsible for ΔpH_i); 2. Ringer B (used to analyze the action of insulin upon glycolysis)	390
	B. Procedures	391
	1. For experiments to identify the mechanism of insulin effect upon pH_i; 2. For analysis of insulin action upon glycolysis; 3. Determination of intracellular pH; 4. Determination of intracellular Na^+	391
	C. Materials	394
III.	Review of Experimental Results	394
	A. Effect of Insulin on $Na^+:H^+$ Exchange	394
	1. Confirmation of prediction 1; 2. Confirmation of prediction 2; 3. Confirmation of prediction 3; 4. Confirmation of prediction 4; 5. Conclusions	395
	B. Mechanism of Insulin Action on Glycolysis	398
	1. Confirmation of prediction 5; 2. Confirmation of predictions 6 and 7; 3. Confirmation of prediction 8; 4. Confirmation of prediction 9; 5. Confirmation of prediction 10; 6. Conclusions	399

IV. Discussion 406
  A. The Insulin Transduction System Mediates the Acute Action of Insulin Upon Glycolysis 406
    1. Na:H exchange must play a role in mediating insulin action upon glycolysis; 2. ΔpH_i is most likely the intracellular signal in the acute insulin effect on glycolysis 406
  B. Blocking Cellular Effects Due to the Insulin Transduction System 407
  C. Possible Role of the Insulin Transduction System in Regulation of Other Cellular Functions 408
    1. Effect of insulin on glycogen synthesis; 2. Effect of insulin on DNA synthesis and mitogenesis 408
  D. Possible Nonionic Mechanism for Mediation of Insulin Action ... 409
    1. Redundancy 409
  E. General Implications for Cell Function...................... 409
  F. Implications for Studies of Intracellular Enzymes 411
  G. The Proton as an Intracellular Signal 411
V. References 412

I. INTRODUCTION

In addition to its classic action upon glucose transport, insulin stimulates many intracellular processes including glycolysis, protein synthesis, RNA synthesis, and DNA synthesis. Until recently, no concrete model existed to account for these actions of insulin. Most investigators are now convinced that the effects of insulin are not mediated by changes in cyclic nucleotide levels [Czech, 1977]. The hypothesis has been advanced that Ca^{++} may be the intracellular signal, or mediator, for insulin [Schudt et al, 1976; Clausen, 1976; Clausen et al, 1975; Kissebah et al, 1975]. However, the evidence for this is still inferential since this signal would be manifested by changes in the intracellular thermodynamic activity of Ca^{++}, and the intracellular activity of this cation has not yet been followed during insulin action.

More recently, it has been reported that addition of insulin to membrane fragments results in the release of a factor, probably a peptide, that can affect the activities of some enzymes known to be affected by insulin [Jarett and Seals, 1979; Larner et al, 1979].

The clue that monovalent cations might be involved in insulin action was provided by the very first studies of the effect of this hormone upon the intact animal when it was observed that insulin not only lowers blood glucose levels, but decreases blood K^+ as well [Harrop and Benedict, 1923; Briggs et al, 1924]. It is now well established [Kamminga et al, 1950] that this is due to stimulation of K^+ uptake, largely by muscle, and that this is secondary to stimulation by insulin [Moore, 1973; Gavryck et al, 1975; Clausen and Kohn, 1977; Kitasato et al, 1980a, b, c] of the mechanism that pumps Na^+ out of and K^+ into cells. This mechanism, sometimes called the sodium-potassium pump, is identical to

(Na^+ + K^+)–ATPase [Racker and Fisher, 1975], an enzyme system that is activated by these two monovalent cations, which requires Mg^{++}, and which is referred to here as the Na pump. However, during the half century after the initial demonstration that insulin affects blood K^+ levels, studies of the effects of this hormone have focused almost exclusively upon organic metabolism. Even today, when most investigators speak of "insulin effect," they refer to the effects upon organic metabolism such as glucose transport and consumption, synthesis of macromolecules, and antilipolysis.

The suggestion that the Na pump is involved in mediating insulin action [Moore, 1965, 1973] was further supported by the finding that the activation by insulin of glycogen synthetase in intact cells is blocked by ouabain, a specific inhibitor of the Na pump [Blatt et al, 1972; Horn et al, 1973]. Also, in rat diaphragm, the stimulation of glycolysis by insulin is blocked by ouabain [Clausen, 1966]. In his review [1976], Hers stated his opinion that ions are the "main missing link" in understanding the mode of action of insulin.

In 1973, Moore had suggested that insulin might increase intracellular pH, pH_i, and thereby regulate intracellular events such as glycolysis [Gavryck et al, 1975]. The extreme sensitivity of phosphofructokinase, the rate limiting enzyme of glycolysis, to very small changes in pH [Trivedi and Danforth, 1966] is consistent with the suggestion that a change in pH_i, ΔpH_i, might represent the intracellular signal for insulin action upon this metabolic system. The prediction that insulin increases pH_i has been confirmed in frog sartorius using both the weak acid DMO [Moore, 1979, 1981a; Moore et al, 1979] and the noninvasive technique of ^{31}P-NMR [Moore and Gupta, 1980].

There is considerable evidence that the effect of insulin upon pH_i is due to stimulation of the Na:H exchange system in the plasma membrane [Moore, 1981].

A. Model for Insulin Transduction

Based upon the fact that insulin stimulates both the Na pump and Na:H exchange, a model has been proposed [Moore, 1981a] to account for many of the intracellular actions of insulin. In this model, called the "insulin transduction system" (see Fig. 1), the binding of insulin to its receptors on the plasma membrane sends a signal, presumably intramembrane, to stimulate both the Na:H exchange system and the Na pump in the membrane. Na^+ entering the cell through the Na:H exchange mechanism drives H^+ outward (or OH^- inward), raising pH_i [Moore, 1981a]. Because of the simultaneous stimulation of the Na pump by insulin, this Na^+ does not accumulate in the cell, but is extruded [Moore, 1973], thus maintaining the thermodynamic gradient required to drive Na:H exchange. Therefore, the primary result of stimulation of the insulin transduction system is an elevation of intracellular pH. The operating hypothesis is that this change in pH_i represents the intracellular signal whereby the information

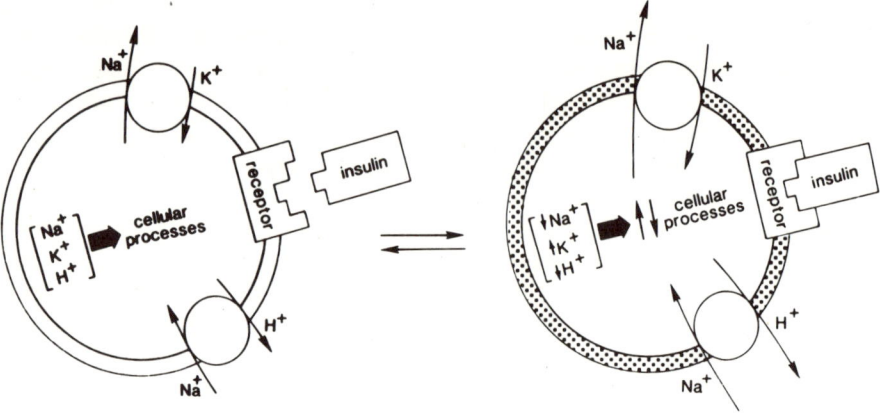

Fig. 1. The model for the insulin transduction system to mediate the acute action of insulin upon glycolysis and perhaps mediate this hormone's effect upon other cell functions such as DNA synthesis. This system depends on the integrity of the intact cell. In this model, the intracellular signal for the action of insulin is a change in the ionic atmosphere, especially intracellular pH, of the cell interior. When insulin binds to its receptor on the plasma membrane, it sends a signal, presumably within the membrane, which simultaneously stimulates (presumably by decreasing the activation energy) the Na:H exchange system and increases the activity of the sodium pump. Stimulation of the Na:H exchange system allows Na^+ moving down its own free-energy gradient into the cell to provide energy to move protons outward, against their free energy gradient, with a resulting decrease in intracellular proton activity. The simultaneous stimulation of the sodium pump not only prevents Na^+ from increasing inside the cell, but increases intracellular K^+. Figure reproduced from Moore [1981a].

that insulin has bound to receptors on the cell surface is communicated to intracellular systems. This model describes a biophysical system that depends on the Na^+ free-energy gradient across the plasma membrane and thus is a property only of the intact cell.

B. Effect of Insulin on $Na^+:H^+$ Exchange

The Na:H exchange mechanism operates by using the energy made available by Na^+ moving down its free gradient, ΔG_{Na}, to drive the proton up its own free energy gradient, ΔG_H. Therefore, the average free energy change, $<\Delta G>_{Na:H}$, for an Na:H exchange mechanism that couples Na^+ influx to proton efflux is the sum of the free energy required to transport n Na^+ ions inward and that required to transport m H^+ ions outward:

$$<\Delta G>_{Na:H} = n\Delta G_{Na} + (-m\Delta G_H)$$

or

$$<\Delta G>_{Na:H} = n\left[eV_m + kT \ln \frac{\gamma_{Na,i}[Na^+]_i}{\gamma_{Na,o}[Na^+]_o}\right] - m\left[eV_m + kT \ln \frac{\alpha_{H,i}}{\alpha_{H,o}}\right] \quad (1)$$

where e = the protonic charge, k = Boltzmann constant, T = absolute temperature, V_m = membrane potential, $\gamma_{Na,i}$ ($\gamma_{Na,o}$) = intracellular (extracellular) Na^+ activity coefficient, $[Na^+]_i$ ($[Na^+]_o$) = intracellular (extracellular) Na^+ concentration, and $\alpha_{H,i}$ ($\alpha_{H,o}$) = intracellular (extracellular) H^+ activity. This mechanism will have sufficient energy available from ΔG_{Na} to transport protons outward when $<\Delta G>_{Na:H} < 0$. If $[Na^+]_o$ is decreased sufficiently to null the average free energy, ie, $<\Delta G>_{Na:H} = 0$, this system should not transport protons. If $[Na^+]_o$ is decreased below this null point so that $<\Delta G>_{Na:H} > 0$, activation of the system should transport protons inward. Under physiological conditions the following relations are observed: $\Delta G_{Na} < 0$, $\Delta G_H < 0$, and $<\Delta G>_{Na:H} < 0$.

Therefore, if the increase in pH_i caused by insulin is due to stimulation of Na:H exchange, it follows that 1) The elevation of pH_i should be associated with an increased influx of Na^+. In the presence of sufficient ouabain to inhibit the stimulation of the Na pump by insulin [Gavryck et al, 1975], this increased Na^+ influx would result in an increase in intracellular Na^+ which should be correlated with the increase in pH_i produced by insulin. 2) By equation 1, that value of $[Na^+]_o$ can be calculated (assuming n/m = 1) for which $<\Delta G>_{Na:H} = 0$, and at that value of $[Na^+]_o$, insulin should have no effect upon pH_i. 3) Lowering $[Na^+]_o$ still further reverses the sign of $<\Delta G>_{Na:H}$ and thus reverses the direction of Na:H exchange. Therefore, removing extracellular Na^+ should convert the action of insulin from an increase to a decrease in pH_i. 4) If the diuretic drug amiloride does block Na:H exchange as previously reported [Aickin and Thomas, 1977; Johnson et al, 1976], this drug should block all the above effects of insulin.

Each of these predictions has been verified [Moore, 1979, 1981a; Moore et al, 1979].

C. Effect of Insulin on Glycolysis

The stimulation of glycolysis by insulin can occur independently of glucose transport and apparently is due to increased activity of phosphofructokinase, PFK [Ozand and Narahara, 1964; Beitner and Kalant, 1971]. The extreme sensitivity of PFK to small changes in pH is consistent with the hypothesis that changes in pH_i may be the intracellular signal for insulin.

In light of the above, the hypothesis that ΔpH_i, mediated by the insulin transduction system (Na:H exchange and the Na pump), represents the signal whereby insulin affects glycolysis leads directly to the following six predictions (numbered sequentially with the four above): 5) Blocking the effect of insulin upon pH_i by amiloride would also block the effect of this hormone upon glycolysis. 6) Blocking the effect of insulin upon pH_i by lowering $[Na^+]_o$ would also block the effect of this hormone upon glycolysis. 7) Since Na:H exchange is a function of both the Na^+ free-energy gradient, ΔG_{Na}, and the H^+ free-energy gradient, ΔG_H, the effect of insulin upon glycolysis should be a function of both ΔG_{Na} and ΔG_H. Therefore when ΔG_H varies only slightly, as in the experiments below, a reversal of ΔG_{Na} of sufficient magnitude to reverse the sign of $<\Delta G>_{Na:H}$ should reverse Na:H exchange and therefore result in an inhibition of glycolysis by insulin. 8) As ΔG_{Na} is varied, there should be a consistent correlation in both magnitude and direction between the effect of insulin upon pH_i and the effect of the hormone upon glycolysis. 9) Changing pH_i by a means other than insulin (eg, by varying CO_2) should mimic the action of insulin upon glycolysis. 10) *a*. Stimulation of glycolysis either by insulin or by increasing pH_i secondary to a decrease in CO_2 should be associated with activation of PFK. *b*. Inhibition of glycolysis either by insulin in a Na-free Ringer or by increasing CO_2 to decrease pH_i should be associated with a decrease in the activity of this rate-limiting enzyme.

These predictions were tested using anaerobic lactate production to measure the flux through the glycolytic pathway and the weak acid DMO to measure pH_i. Predictions 1–7 and 9–10 have been confirmed and partial support provided for prediction 8.

II. REVIEW OF METHODOLOGIES

Paired sartorius muscles from the frog Rana pipiens were used throughout. Muscles weighing less than 100 mg and usually 40–60 mg were removed from healthy frogs that had been stored at 22°C and force-fed liver twice a week. After passing microscopic checks for damage, the muscles were mounted at rest length on platinum frames [Sjodin and Henderson, 1964]. In all experiments one muscle of a pair from the same frog was used as the control and the other used as the experimental.

A. Ringer Solutions

1. Ringer A (used to identify mechanism responsible for ΔpH_i). To avoid any complications that might be due to $Cl^-:HCO_3^-$ exchange [Russell, 1978; Thomas, 1977; Russell and Boron, 1976], $H_2PO_4^-/HPO_4^{2-}$ was chosen as the only buffer. Therefore, the Na-Ringer contained only 104 mM Na^+, 2.5 mM K^+, 2 mM Ca^{2+}, and 1.6 mM HPO_4^{2-}; the remaining anion was Cl^-. Na-free Ringer was prepared by substituting 104 mM Li^+, 104 mM choline, or 75 mM Mg^{2+} for Na^+; 50 µM d-tubocurarine was added to the choline Ringer to prevent twitching. All Ringer was glucose-free and was titrated to a pH of 7.40 ± 0.03. When ouabain was used, its concentration was 1 mM. When amiloride was used, its concentration was 0.5 mM. In the two series of experiments where lactate production was determined using this Ringer, 10 µM l-epinephrine bitartrate was added to increase glycogen breakdown thus ensuring sufficient substrate for glycolysis [Ozand and Narahara, 1964].

2. Ringer B (used to analyze the action of insulin upon glycolysis).
The composition of this Ringer, which was also glucose-free, was 104 mM Na^+, 2.2 mM K^+, 2.0 mM Ca^{2+}, 1.2 mM Mg^{2+}, 1.6 mM HPO_4^{2-}, 1.9 mM SO_4^{2-}, 76.4 mM Cl^-, and 30.0 mM HCO_3^- at pH 7.4 ± 0.03 in 5% CO_2. When $[Na^+]_o$ was reduced, an equal concentration of choline was used as the Na^+ substitute and 50 μM tubocurarine was added to prevent twitching.

In order to rule out complications in interpretation due to the effect of insulin upon glucose transport, glucose (and any other) substrate was absent from the Ringer in all experiments. Therefore, 0.5 μM l-epinephrine bitartarate was used to stimulate the rate of glycogen breakdown and thus ensure sufficient substrate for PFK to be the rate limiting condition of the glycolytic reaction [Ozand and Narahara, 1964]. Ascorbate, 1 mM, was added to inhibit auto-oxidation of the epinephrine.

B. Procedures

1. For experiments to identify the mechanism of insulin effect upon pH_i.
After dissection, the muscles were kept in Na-Ringer (A) at room temperature for 2–3 hours before each experiment.

In those experiments using ouabain or amiloride, the dissected muscles were first placed in Ringer containing the drug for a 20-minute equilibration with the extracellular space prior to addition of insulin to the experimental muscle. The drug was present in the Ringer of all subsequent steps. In those experiments using Na-free Ringer, muscles were first placed in the Na-free Ringer for 1 minute followed by 10 minutes in a second tube to clear most (>95 %) of the Na^+ from the extracellular space.

For all subsequent steps (except where explicitly indicated), all solutions in which the experimental muscles were placed contained 250 mU/ml insulin. The control muscles were treated identically except for the absence of insulin. The experimental muscles were equilibrated with insulin in Ringer for 20 minutes before placing them, and their paired controls, in Ringer containing [^{14}C]5,5-dimethyloxazolidine-2,4-dione (DMO) and [^{3}H]sucrose for 90 minutes to prepare both muscles for determination of pH_i by the method of Waddell and Butler [1959] as described below.

In those experiments using Na-Ringer, immediately after the isotope washout period, the muscles were placed in Na-free, K-free, Ringer containing ouabain [Moore and Rabovsky, 1979] without insulin for an additional 20 minutes to clear the extracellular space of Na^+. The composition of this Ringer was identical to Ringer A except for the addition of 10^{-4} M ouabain and the replacement of both Na^+ and K^+ by an osmotic equivalent (82 mM) of Mg^{2+}. Muscles were then blotted and weighed. Dry weights were determined by reweighing after drying the muscles for 14 hours at 105°C. The dried muscles were then ashed, and intracellular Na^+ was assayed by flame emission as described below.

Except where indicated, experiments were aerobic. The anaerobic experiments were conducted in 100% N_2 in glove bags. Lactate production was determined using lactic dehydrogenase (E.C. 1.1.1.27) to convert lactic acid and NAD^+ to pyruvic acid and NADH as described below. All results are presented as the mean ± the standard error.

2. For analysis of insulin action upon glycolysis. After dissection, muscles were placed overnight at 4°C in separate stoppered tubes each containing 25 ml of Ringer B previously bubbled with 5% CO_2 in O_2.

Experiments were conducted under anaerobic conditions (5% CO_2 in N_2) in a glove bag in order to prevent pyruvate from entering the Krebs cycle, thus ensuring that lactate production accurately reflected the rate of glycolysis. Since glycogen was the only source of substrate for glycolysis, muscle glycogen content was periodically determined by using the amyloglucosidase enzymatic technique of Keppler and Decker [1974].

All experiments were conducted at constant temperature between 20.5°C and 22.5°C. After overnight storage, muscles were allowed to equilibrate anaerobically in Ringer containing 104 mM Na^+ for 30 minutes and were then preincubated for 40 minutes in the appropriate Ringer, with the experimental muscles exposed also to 400 mU insulin/ml. In those experiments in which the action of insulin upon glycolysis was measured in the presence of DMO, 1 mM unlabeled DMO was added during the equilibration period in order to insure that any depression in pH_i due to the presence of the weak acid [Iles and Cohen, 1974] had reached the steady state prior to measurement of lactate production. In the insulin-free experiments, where pH was changed by varying CO_2, control and experimental muscles were run in separate glove bags at appropriate CO_2 values. A Union Carbide series 150 two-tube gas proportioner with a ± 5% accuracy was used to mix CO_2 and N_2.

Muscles were then incubated in 2.5 ml of the appropriate Ringer for 90 minutes. The muscles were then blotted on filter paper, and wet weights were determined. They were then rapidly frozen by plunging them into isopentane cooled to near its freezing point (-170°C) with liquid N_2. Next, the muscles were pulverized with a mortar and pestle previously chilled with dry ice. Lactate was extracted from the muscles using a modified perchloric acid extraction technique [Karpatkin et al, 1964] in which 1 ml of 1 N perchloric acid was added to the pulverized muscle and ground to a slurry for 3 minutes. The perchlorate ion was precipitated with 0.7 ml of a potassium bicarbonate buffer (1.3 M K_2CO_3, 2.3 M $KHCO_3$, pH 9.2). Activated charcoal, 15 mg, was added to the ground slurry prior to the precipitation of perchlorate in order to eliminate material from the muscle extract that caused an interference peak at 290 nm, without affecting the peak due to lactate. This step is important because failure to remove the interference could produce substantial overestimates of lactate production.

Lactate was measured in both the muscle extract and in the efflux tubes by a modified enzymatic technique of Hohorst [1965] using lactic dehydrogenase (E.C. 1.1.1.27) to convert lactic acid and NAD^+ to pyruvic acid and NADH, with the former trapped by hydrazine. The absorbance at 340 nm due to newly formed NADH became a measure of the lactic acid originally present. The change in lactate production (in 90 minutes) is the total of the lactate in the experimental muscle plus that accumulated in its incubation Ringer, minus the same total for the control. In computing the rate of lactate production in the control muscles, it was assumed that the 40-minute preincubation period is sufficient for their lactate content to reach a steady state [Karpatkin et al, 1964]. The effect of any error in this assumption is small, since the lactate content of the muscle was only about 20% of that accumulated in the Ringer over 90 minutes. Therefore, the control glycolytic rate was obtained from the lactate accumulated in 2.5 ml of Ringer during the 90-minute incubation period. d-Fructose-6-phosphate (F-6-P) in the muscle extract was determined by the method of Lang and Michal [1974] in which glucose-6-phosphate dehydrogenase was first added to convert glucose-6-phosphate and $NADP^+$ into 6-phosphogluconolactone, NADPH and H^+, with the 6-phosphogluconolactone changing spontaneously into 6-phosphogluconate. Subsequent addition of phosphoglucose isomerase converts F-6-P into G-6-P and the additional increase in NADPH is determined at 340 nm.

3. Determination of intracellular pH.

Steady-state intracellular pH was determined by measuring the distribution of the weak acid 5,5-dimethyloxazolidine-2,4-dione (DMO). After the 40-minute preincubation period, both experimental and control muscles were loaded with [^{14}C]DMO (12.5 μCi/2.5 ml loading solution and [^{3}H]sucrose (4 μCi/2.5 ml loading solution) for 90 minutes, a time sufficient for both the effect of insulin and the equilibration of labeled compound to reach a steady state. Sufficient unlabeled DMO was added to the loading solution to bring the total concentration of DMO to 1 mM in order to minimize effects of intracellular binding. The external pH, pH_o, of the loading solution for each muscle was determined before and after the 90-minute period. The [^{14}C]DMO and [^{3}H]sucrose were then washed out of the muscle over a period of 140 minutes by transfer through a series of three tubes containing Ringer identical to the loading solution except for the absence of radioactive DMO and sucrose. This procedure removes more than 99% of this [^{14}C]DMO [Graves and Moore, 1976]. The wet weight of the muscle was determined, and the muscles were dried overnight at 105°C and reweighed to determine water content. Aliquots of Ringer from the washout tubes were dissolved in Aquasol II (New England Nuclear) and counted in a liquid scintillation counter to an accuracy of 3% (SD). The channels ratio technique was used to discriminate between the two isotopes. Back addition of the counts gave the amounts of [^{14}C]DMO and [^{3}H]sucrose present at the beginning of washout. The extracellular space (ECS) determined by [^{3}H]sucrose was used to calculate the intracellular [^{14}C]DMO in the muscle before washout. pH_i was calculated by substituting the experimentally determined values into the following equation [Waddell and Butler, 1959]:

$$pH_i = pK_a + \log_{10}\left[\frac{[DMO]_i}{[DMO]_o}(1 + 10^{pH_o - pK_a}) - 1\right] \quad (2)$$

The value used in these experiments for the pK_a of DMO was 6.28.

Except for Figure 6, all results are presented as the mean ± the standard error (SE).

4. Determination of intracellular Na^+. Internal Na^+ content of muscles was determined by first clearing the extracellular space of Na^+ by placing the muscle for 16 minutes in a Na-free, K-free Ringer [Moore and Rabovsky, 1979] containing 8.20 mM Mg^{2+}, 2.0 mM Ca^{2+}, 1.0 mM Tris (tris (hydroxymethyl) aminomethane), and 1.0 mM ouabain at pH 7.4. Muscles were then blotted on filter paper, weighed, dried at 110°C overnight, and reweighed to determine wet and dry weights. Muscles were prepared for flame emission analysis by ashing at 800°C for 10 hours and dissolving the ash in 0.1 N HNO_3 for 30 minutes. The value for the extracellular space was obtained with [^{3}H]sucrose from muscles of similar size used in the pH experiments.

C. Materials

Insulin was porcine sodium insulin (0.00% Zn), 25.9 U/mg or 26.5 U/mg, and was a gift from Eli Lilly and Company (Indianapolis). Ouabain, d-tubocurarine, and 1-epinephrine bitartrate were obtained from Sigma Chemical Co. (St. Louis). [^{14}C]DMO and [^{3}H] sucrose were obtained from New England Nuclear (Boston). Amiloride was a gift from Merck, Sharp and Dohme Research Laboratories (West Point, Pennsylvania).

III. REVIEW OF EXPERIMENTAL RESULTS

A. Effect of Insulin on $Na^+:H^+$ Exchange

Although inhibitors such as ouabain and amiloride may be useful in the identification of ion transport systems, in the final analysis such identification must be based upon the demonstration that all functional consequences of the transport system exist. For Na:H exchange, these were listed in the Introduction as the necessity to demonstrate that 1) the H^+ efflux is associated with a Na^+

influx; 2) when $<\Delta G>_{\text{Na:H}} = 0$, H^+ flux is blocked; 3) when $<\Delta G>_{\text{Na:H}} > 0$, H^+ flux is inward.

In these experiments, in order to eliminate any possible complications from HCO_3^-:Cl^- exchange, Ringer A was used except where indicated.

1. Confirmation of prediction 1. If the elevation of pH_i is driven by an influx of Na^+, this influx will produce an increase in Na_i^+ that should be proportional to the H^+ pumped out of the cell and therefore correlated with the increase in pH_i. However, this will not be observed unless the Na pump is inhibited to prevent this increment in Na_i^+ from being immediately pumped back out of the cell. Ouabain, 1 mM, inhibits the Na pump even in the presence of insulin [Gavryck et al, 1975].

To test whether ouabain itself affects pH_i, a series of experiments was performed in which 1 mM ouabain was added to one muscle of each pair. In 14 such pairs, the average difference in pH_i between the ouabain-treated muscle and its paired control, -0.019 ± 0.017, was not significant ($P > 0.25$) [Moore, 1981a]. This is the same result that Roos [1975] obtained for the effect of ouabain upon rat skeletal muscle.

In the presence of ouabain, insulin still produces a significant ($P < 0.01$) increase in pH_i [Moore, 1981a]. In these same muscles, insulin produced an average increase of 11.7 ± 2.3 mmole Na^+/kg muscle in intracellular Na_i^+ ($P < 0.005$) compared to the paired controls (see legend to Fig. 2 for discussion). As illustrated in Figure 2, the change in pH_i produced by insulin in each muscle is positively correlated ($r = 0.689$, $P < 0.01$) with the elevation in Na_i^+ produced by the hormone in the same muscle. Consistent with these results is the observation by Clausen and Kohn [1977] that in the presence of 10^{-3} M ouabain, insulin produced a significant 30–40% increase in $^{22}Na^+$ influx in rat soleus muscle.

2. Confirmation of prediction 2. Equation 1 may be used to calculate that value of $[Na^+]_o$ that would zero $<\Delta G>_{\text{Na:H}}$. In the absence of definitive information about the ratio of Na^+ to H^+ (n/m) carried by the putative Na:H exchange system, we assumed values of 2:3, 1:1, and 3:2. For $pH_o = 7.4$ and $[Na^+]_i = 7-8$ mM, decreasing $[Na^+]_o$ to about 6.8 mM should null the free energy for Na:H exchange.* In Ringer A (which contains bicarbonate), decreasing $[Na^+]_o$ to this calculated value (a 15-fold reduction) completely ($P > 0.5$) blocks the effect of insulin upon pH_i [Moore et al, 1979].

---

*The membrane potential, V_m, was taken to be -90 mV [Moore and Rabovsky, 1979]. Na^+ activities were estimated by using activity coefficients of 0.8 for extracellular Na^+ [Mullins and Noda, 1963] and 0.67 for intracellular Na^+ (based upon the intracellular activity determined by Lee and Armstrong [1974] and our own determination of $[Na^+]_i$ under the same experimental conditions).

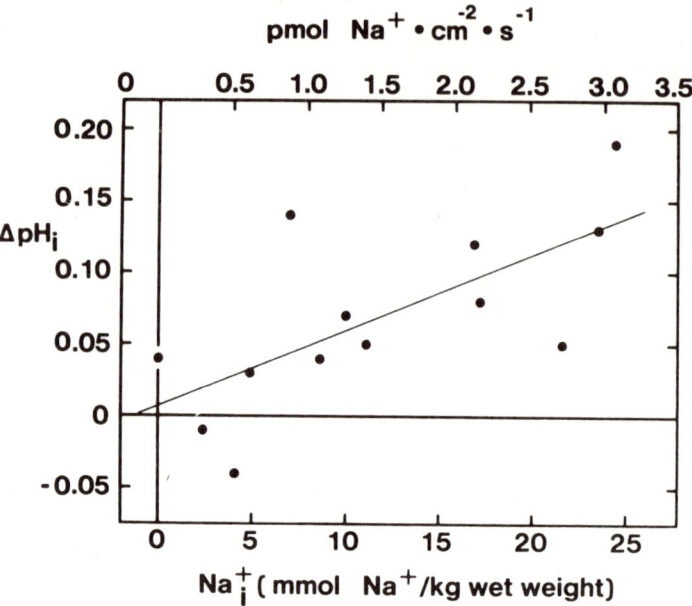

Fig. 2. Change in pH_i versus elevation of Na_i^+ produced by insulin in Ringer containing 1 mM ouabain. The increase in Na_i^+ is in mmole/kg muscle wet weight (per 250 min) on the bottom abscissa. Using a figure of 552 cm²/g muscle wet weight, this is converted to net Na^+ influx in $pmole \cdot cm^{-2} \cdot sec^{-1}$ on the top abscissa. The plotted line is a linear regression of the data. Figure reproduced from Moore [1981a].

3. Confirmation of prediction 3. If the efflux of H^+ is driven by the free-energy gradient for Na^+ across the plasma membrane, the effect of insulin on H^+ transport should be reversed in a Na-free (actually about 0.12 mM Na^+ [see Moore, 1981a]) Ringer where $<\Delta G>_{Na:H} > 0$. When either Mg^{2+} or choline is used to replace the Na^+ in the Ringer, the effect of insulin upon pH_i is converted to a statistically significant *decrease* (P < 0.005) [Moore, 1981a]. This decrease in pH_i produced by insulin in Mg^{2+} Ringer has been confirmed using the noninvasive method of ^{31}P-NMR to measure pH_i [Moore and Gupta, 1980]. To rule out the possibility that the decrease in pH_i might have been due to insulin stimulation of glycolysis with a resulting increase in metabolic production of protons, the effect of insulin was determined in Mg-Ringer under identical experimental conditions, except for a 100% N_2 atmosphere, and anaerobic lactate production was observed to *decrease* by 49% ± 13% [Moore, 1981a]. It is therefore unlikely the decrease in pH_i reflects metabolic events, but most probably reflects an influx of protons into the cell.

In contrast to the above, when Na^+ in the Ringer is replaced by Li^+, insulin still produces an increase in pH_i, which is significantly greater (P < 0.05) than the elevation observed in Na-Ringer. This indicates that Li$^+$ can substitute for Na$^+$ in the Na:H exchange [Moore, 1981a].

In the above experiments, the conclusion that in Na-Ringer $<\Delta G>_{Na:H} < 0$ was confirmed by application of Equation 1 to measured values of [Na$^+$]$_i$ and pH$_i$ and assuming the stoichiometry m/n = 1. Likewise, for the experiments conducted in Ringer where Na$^+$ was replaced by either choline or Mg^{2+}, Equation 1 (again assuming m/n = 1) was used to confirm that $<\Delta G>_{Na:H}$ actually was > 0. Since equation 1 is derived for proton efflux, a value of $<\Delta G>_{Na:H} > 0$ indicates that *reverse* operation of the system—ie, Na$^+$-driven proton *influx* is thermodynamically downhill, or spontaneous. Thus, the direction of change in pH$_i$ produced by insulin is consistent with the direction of ΔG_{Na} as would be expected of a process driven by Na:H exchange. Of equal importance, the magnitude of ΔG_{Na} is sufficient to move H$^+$ against its energy gradient, as is indicated by the fact that $<\Delta G>_{Na:H}$ always has the proper sign for the observed flux of H$^+$.

It is possible that all of the above results could be due to the operation of a Na$^+$–CO$_3^{2-}$ cotransport system [Funder et al, 1978]. Although in the experiments supporting points 1 and 3, HCO$_3^-$ was not present in the Ringer, some bicarbonate, and therefore CO$_3^{2-}$, would be present due to atmospheric and metabolic CO$_2$. To eliminate these sources of CO$_3^{2-}$, the effect of insulin in Ringer A (containing 104 mM Na$^+$) was determined in the presence of a 100% N^2 atmosphere. Even in the virtual absence of CO$_2$, where only trace amounts of CO$_3^{2-}$ should be present, insulin still produces a significant increase in pH$_i$ thus ruling out the possibility that the effect is mediated by a Na$^+$-CO$_3^{2-}$ cotransport system [Moore, 1981a]. A similar elevation of pH$_i$ in a N$_2$ atmosphere was also observed using a Ringer of the same composition (as Ringer A) except for the addition of 1.2 mM Mg^{2+}, 1.9 mM SO$_4^{2-}$, and 1 mM ascorbate [Moore et al, 1979].

The above evidence provides rather convincing confirmation of the predictions resulting from the hypothesis that the change in pH$_i$ produced by insulin is due to activation of Na:H exchange.

4. Confirmation of prediction 4. If amiloride does block Na:H exchange, it should block the effect of insulin upon both H$^+$ efflux and the associated Na$^+$ influx and upon the decrease in pH$_i$ produced by insulin in Na-free Ringer.

The diuretic drug amiloride (3,5-diamino-6-chloropyrazinoyl-guanidine) has no effect upon the Na pump [Bentley, 1968], but it prevents recovery of pH$_i$ in acid loaded mouse soleus muscle [Aickin and Thomas, 1977] and blocks a Na$^+$-dependent elevation in pH$_i$ during activation of sea urchin eggs after fertilization [Johnson, et al, 1976], suggesting it blocks Na:H exchange.

In the presence of 5% CO$_2$/30 mM HCO$_3^-$, 0.5 mM amiloride does block the effect of insulin upon pH$_i$ [Moore et al, 1979]. In Ringer lacking CO$_2$/HCO$_3^-$

Ringer A), amiloride still blocks the action of insulin upon pH_i. In an identical Ringer lacking amiloride, insulin significantly ($P < 0.001$) increases pH_i by 0.096 ± 0.016 [Moore, 1979].

Since amiloride blocks the increase in pH_i produced by insulin, it follows that if this increase is due to activation of an Na:H exchange mechanism it should also block the increase in Na_i^+ observed in Na-Ringer. As explained above, the Na pump must be inhibited to demonstrate this. When both muscles of a pair are exposed to 1 mM ouabain and 0.5 mM amiloride for 20 minutes before addition of insulin to the experimental muscle, the increase in Na_i^+ upon exposure of the muscles to the hormone for 90 minutes in the presence of both ouabain and amiloride is essentially zero ($P > 0.5$), thus confirming this prediction [Moore, 1981a].

If the system responsible for the action of insulin upon pH_i is Na:H exchange, and if amiloride blocks Na:H exchange, the insulin-induced decrease in pH_i observed in Na-free Ringer should also be blocked by this drug. This prediction is confirmed by the demonstration that in Ringer A, where Na^+ is replaced by osmotically equivalent amounts of Mg^{2+}, 0.5 mM amiloride inhibits the decrease in pH_i produced by insulin [Moore, 1981a].

5. Conclusions. From the above, it seems reasonable to conclude that insulin stimulates the Na:H exchange system in the plasma membrane, that amiloride does block this transport system as reported previously [Aickin and Thomas, 1977; Johnson et al, 1976], and that amiloride blocks the stimulation by insulin of this transport system.

B. Mechanism of Insulin Action on Glycolysis

Is the Na:H exchange system part of the insulin transduction system? That is, is this part of the mechanism whereby insulin affects cell function such as glycolysis? Specifically, is the action of insulin upon glycolysis dependent upon ΔG_{Na} and is the immediate signal the change in pH_i?

In the initial experiments designed to test these questions, we used Ringer A in order to minimize the number of possible ionic species that could possibly be involved in transport. In particular, the buffer system CO_2/HCO_3^- was absent from the Ringer in order to avoid any complications in interpretation arising from relatively permeant buffers [Boron and Roos, 1976; Boron and DeWeer, 1976]. In this less complicated Ringer, although still producing a significant elevation of pH_i, insulin has no effect on glycolysis [Moore et al, 1979]. Shaw and Stadie [1957, 1959] had previously found that absence of CO_2/HCO_3^- blocked the effect of insulin action on glycolysis.

When CO_2/HCO_3^- is restored (Ringer B), insulin not only produces an elevation of pH_i that is not significantly different from that in Ringer A, but also produces a significant stimulation of glycolysis that averaged 43% [Moore et al,

1979]. The reason for the requirement of CO_2/HCO_3^- is not certain, but a possible clue is provided by the observation that in the Ringer (Ringer A) lacking this buffer system, even without insulin, anaerobic lactate production is very near the maximum value observed when insulin is added in the presence of the CO_2/HCO_3^- system. Furthermore, in the presence (Ringer B) of CO_2/HCO_3^-, pH_i of the controls was 6.98, whereas in the absence of CO_2/HCO_3^-, pH_i of the controls is considerably higher, 7.27. This suggests that in the absence of CO_2/HCO_3^- the higher value of pH_i may result in glycolysis operating near its V_{max}. As in all other experiments, in order to rule out any effect due to stimulation of glucose transport by insulin, all Ringer was glucose-free with substrate for the glycolytic pathway being provided by epinephrine stimulation of glycogen breakdown. Because of the necessity for CO_2/HCO_3^-, Ringer B was used in all experiments discussed subsequently.

1. Confirmation of prediction 5. The addition of 5×10^{-4} M amiloride to Ringer B blocks not only the effect of insulin upon pH_i, but also upon glycolysis [Moore et al, 1979] as predicted by the hypothesis that changes in pH_i mediated by Na:H exchange represent the intracellular signal for insulin upon glycolysis.

2. Confirmation of predictions 6 and 7. The action of insulin upon pH_i is due to stimulation of Na:H exchange and therefore depends upon $<\Delta G>_{Na:H}$ [Moore, 1981a]. Therefore, since the direction of change in pH_i is determined by the sign of $<\Delta G>_{Na:H}$, a test of prediction 6 is that the effect of insulin upon glycolytic flux should be reversed as log $[Na^+]_o$, a component of the Na^+ free-energy gradient, is lowered sufficiently to reverse the sign of $<\Delta G>_{Na:H}$. Since removal of extracellular Na^+ will, by equation 1, reverse the sign of $<\Delta G>_{Na:H}$ and thus reverse the Na:H exchange, Na-free Ringer should result in an inhibition of glycolysis by insulin.

Variations in $[Na^+]_o$ have been shown to have little effect on the membrane potential, V_m, in frog sartorius muscle [Mullins and Noda, 1963]. Since in the present experiments pH_o is held constant, and since $[Na^+]_i$ and pH_i have relatively small variations (see legend to Fig. 3), it follows from equation 1 that changes in $<\Delta G>_{Na:H}$ are determined primarily by extracellular Na^+:

$$<\Delta G>_{Na:H} = A - B \log [Na^+]_o \qquad (3)$$

where A and B are positive and are approximately constant in the present experimental conditions. Although the magnitude of ΔpH_i, and therefore of the change in glycolytic flux, is not expected to be determined by $<\Delta G>_{Na:H}$ alone, this free energy should be one of the variables determining ΔpH_i.

When Na-free Ringer is used, the effect of insulin upon glycolysis is, as predicted, *reversed,* being converted to a 52% inhibition [Fidelman et al, in

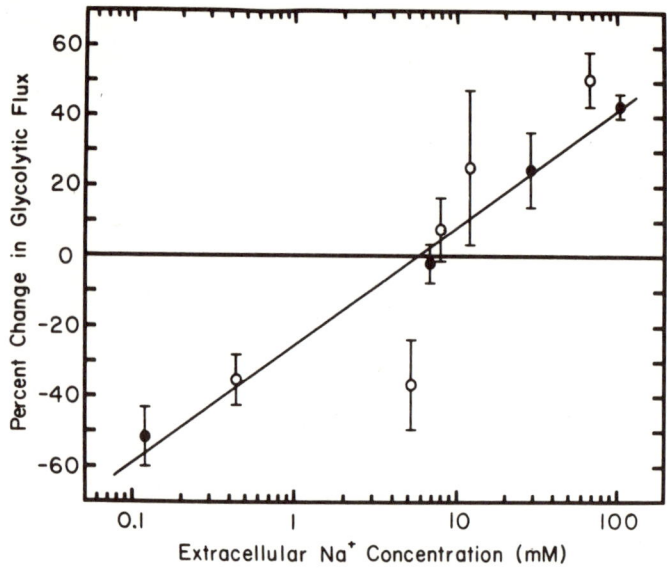

Fig. 3. The effect of lowering $[Na^+]_o$ on the action of insulin upon glycolytic flux, as reflected by anaerobic lactate production, in frog sartorius. Percent change in glycolytic flux due to insulin in plotted versus log $[Na^+]_o$. Solid circles: mean ± SE for 7 to 14 determinations. Open circles: mean ± SE for 3 or 4 determinations. The mean glycolytic flux for all controls was 23.1 ± 1.2 nmol lactate min^{-1}/g wet wt. For the Na-free Ringer, a value of $[Na^+]_o = 0.12$ mM was used (see text). Varying $[Na^+]_o$ does have some effect upon control values of $[Na^+]_i$ and pH_i. However, since control values of $[Na^+]_i$ and $\alpha_{H,i}$ varied by at most a factor of 2, their effect on $<\Delta G>_{Na:H}$ was extremely small (see equation 1) compared to that of $[Na^+]_o$, which was varied by almost three orders of magnitude. A least-squares linear regression was performed on the 60 actual data points (r = 0.823, P < 0.001; slope > 0, P < 0.001). Semilogarithmic scale. Figure reproduced from Fidelman et al [in press].

press]. Since the hypothesis is that $<\Delta G>_{Na:H}$ is one of the variables determining the effect of insulin on glycolysis, equation 3 suggests plotting the effect of insulin upon glycolysis versus log $[Na^+]_o$. In the "Na-free" Ringer, some Na^+ (about 0.04 mM) is present owing to contaminants in the reagent grade chemicals, and an additional amount (about 0.17 mM) was added by Na^+ diffusing from the muscles during the experiment. Therefore, a time-averaged value of 0.12 mM was used for $[Na^+]_o$ in this analysis. As illustrated in Figure 3, although a linear relationship is not theoretically predicted, the data indicate a reasonably linear correlation of percent change in glycolytic flux with log $[Na^+]_o$. The intercept at $[Na^+]_o = 5.8$ mM represents the value at which $<\Delta G>_{Na:H}$ should equal zero in the hypothesis tested here, and is in excellent agreement with the range, approximately 6.8 mM, predicted by equation 1.

At this point it may be considered established that the Na^+ free-energy gradient, ΔG_{Na}, provides the driving force for insulin action on glycolysis. According to the hypothesis tested, the exact effect is determined by $<\Delta G>_{Na:H}$ and this is supported by the observation that the experimental estimate of the null point at about $[Na^+]_o = 5.8$ mM agrees well with the theoretical prediction (based upon application of equation 1 to experimental determination of $[Na^+]_i$ and pH_i) of about 6.8 mM. However, although these results support the above hypothesis and also suggest that ΔpH_i is the connecting link between insulin stimulation of Na:H exchange and the hormone's effect upon glycolysis, it does not prove that ΔpH_i is the immediate signal whereby insulin affects this metabolic system. Before pH_i may be considered the immediate signal, confirmation of the remaining three predictions is required.

3. Confirmation of prediction 8. The effect of insulin upon pH_i should parallel its effect on glycolysis. As illustrated in Figure 4, at three different levels of $[Na^+]_o$, 104, 28.7, and 6.8 mM, the effects of insulin upon both pH_i and glycolysis remain parallel to each other [Fidelman et al, in press]. Since pH_i was measured with DMO, 1 mM DMO was also added to the Ringer when the

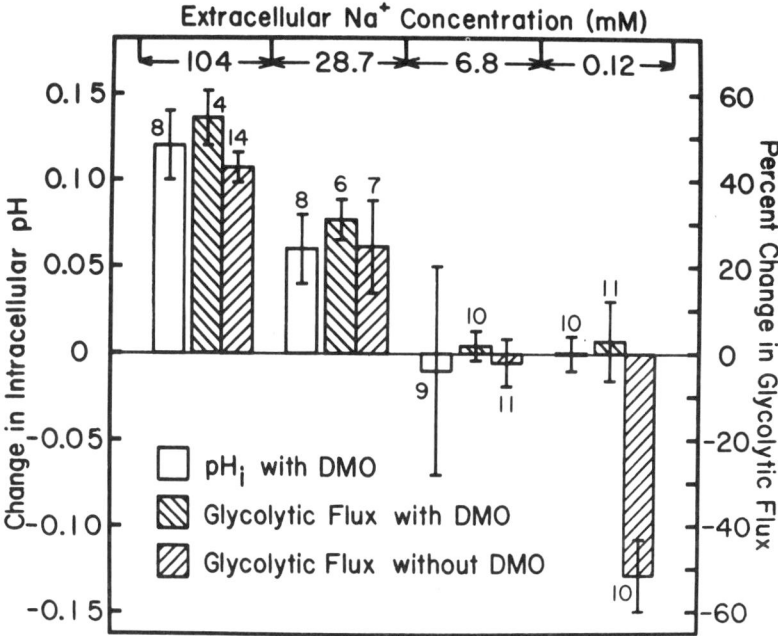

Fig. 4. The effect of lowering $[Na^+]_o$ on the change in internal pH produced by insulin and upon the changes in glycolytic flux due to insulin in frog sartorius. Unshaded columns indicate the change in internal pH produced by insulin, as measured by [^{14}C]DMO. Shaded columns indicate the percent change in glycolytic flux due to insulin. Each column represents the mean ± SE. The number of determinations for each mean is listed near the SE bar. Figure reproduced from Fidelman et al [in press].

effect of insulin upon glycolytic flux was determined, thus ensuring identical experimental conditions when comparing the effect of insulin upon pH_i to the effect upon glycolysis. In the absence of DMO, when $[Na^+]_o$ is decreased to 0.12 mM, insulin produces a 52% inhibition of glycolysis. It is therefore surprising that in Ringer B containing this same level of Na^+, as well as DMO, both glycolysis and pH_i were unaffected by insulin [Fidelman et al, in press]. Moreover, when DMO is present in choline-Ringer, which has this same low level of $[Na^+]_o$ (about 0.12 mM) but which lacks CO_2/HCO_3^-, SO_4^{2-}, epinephrine, and ascorbate, insulin does produce a decrease in pH_i [Moore, 1981a]. Also, when Mg^{2+} is used to replace Na^+ in the Ringer, insulin again produces a decrease in pH_i, whether measured by DMO [Moore, 1981a] or by ^{31}P-NMR [Moore and Gupta, 1980]. Therefore, the only common condition present when insulin fails to decrease pH_i in Na^+-free Ringer seems to be the presence of both DMO and the Ringer components: CO_2/HCO_3^-, SO_4^{2-}, epinephrine, and ascorbate. This suggests that somehow one or more of these components together with DMO act to block Na:H exchange. This evidently occurs only when $<\Delta G>_{Na:H}$ is reversed as indicated by the fact that at both 28.7 and 104 mM, where $<\Delta G>_{Na:H}$ is not reversed, insulin stimulates glycolysis whether or not DMO is present [Fidelman et al, in press]. That the presence of DMO in Ringer containing these two concentrations of Na^+ did not prevent insulin from stimulating glycolysis indicates that DMO is not somehow acting *directly* upon the glycolytic system.

From these results, the simplest conclusion is that in choline Ringer containing the above components, the presence of the nonphysiological anion DMO⁻ blocks the decrease in pH_i otherwise produced by insulin. It is possible that the mechanism for this might involve interference of the anionic form of DMO with the $HCO_3^-:Cl^-$ exchange system. This system may be coupled to the Na:H exchange system in snail neurons [Thomas, 1977], although probably not in rat diaphragm muscle [Roos and Boron, 1978] or in mouse soleus muscle [Aickin and Thomas, 1977].

In any case, the results are in every case consistent with the hypothesis that the effect of insulin upon glycolysis is mediated by that hormone's effect upon pH_i since, under identical experimental conditions, the two effects always parallel each other.

4. Confirmation of prediction 9. If ΔpH_i represents the signal between the action of insulin upon Na:H exchange at the plasma membrane and upon glycolysis, changes in pH_i produced by any other means should produce the same effect upon glycolysis as does the ΔpH_i produced by insulin. To demonstrate this, advantage was taken of the fact that in frog skeletal muscle, at constant extracellular HCO_3^-, pH_i can be altered by changing the level of CO_2 [Bolton and Vaughan-Jones, 1977]. In order to mimic the elevation of pH_i by insulin, CO_2 was reduced from 5% to a calculated 2.7% in Ringer B containing 104 mM Na^+. This increased pH_i by 0.16 ± 0.02 (SE) and increased glycolytic flux by about 55% [Fidelman et al, in press]. To mimic the effect of insulin in Na^+-free Ringer, CO_2 was increased from 5% to 10% in the same Na-free Ringer used in the insulin experiments, with the result that pH_i was decreased by 0.10 ± 0.02 and glycolytic flux was inhibited by 48%. Neither of these effects in the two Ringers is significantly different from that produced by insulin. A plot of ΔpH_i versus the associated change in glycolytic flux (see Fig. 5) demonstrates that the relationship between these two variables is independent of whether the effects are produced by insulin or by the purely physical-chemical procedure of

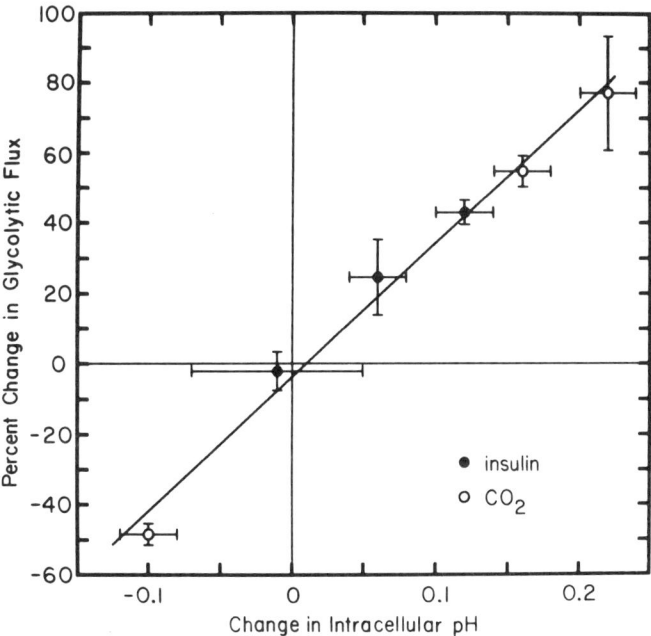

Fig. 5. The relationship between the change in intracellular pH and the percent change in glycolytic flux, as reflected by percent change in anaerobic lactate production, produced in frog sartorius under identical experimental conditions. Solid circles: changes due to insulin at 5% CO_2 and (from left to right) $[Na^+]_o$ = 6.8 mM, 28.7 mM, or 104 mM. Open circles: changes produced (in the absence of insulin) by CO_2 levels of (from left to right) 10%, 2.7%, or 2%, compared to paired controls at 5% CO_2. $[Na^+]_o$ = 104 mM at 2% and 2.7% CO_2 and, in order to mimic the effect of insulin when the Na^+ gradient is reversed, $[Na^+]_o$ = 0.12 mM at 10% CO_2. Each point represents the means ± SEs. The plotted line is a least-squares fit to the points. Figure reproduced from Fidelman et al [in press].

varying CO_2 levels. The results clearly demonstrate that ΔpH_i produced by insulin is of sufficient magnitude to account for the entire effect of this hormone upon glycolysis, at least within the 130 minutes during which insulin was acting.

5. Confirmation of prediction 10. Since insulin stimulation of glycolysis evidently coincides with increased activity of PFK [Ozand and Narahara, 1964; Beitner and Kalant, 1971], the hypothesis that pH_i mediates the action of insulin upon glycolysis requires that 1) stimulation of glycolysis either by insulin or by increasing pH_i secondary to a decrease in CO_2 should be associated with activation of PFK; and 2) inhibition of glycolysis either by insulin in Na-free Ringer or by using an increase in CO_2 to decrease pH_i should be associated with a decrease in the activity of PFK.

For a nonequilibrium, rate-limiting process such as that catalyzed by PFK [Rolleston and Newsholme, 1967; Williamson, 1970] an inverse change in the amount of substrate (F-6-P in the present case) and the flux through the enzyme (represented by anaerobic lactate production in the experiments discussed here) will unambiguously indicate a change in the activity of the enzyme [Rolleston, 1972]. Accordingly, when the change in glycolytic flux is plotted against the change in F-6-P (see Fig. 6), data in the upper left-hand quadrant will unambiguously represent activation of PFK, and data in the lower right-hand quadrant will represent inhibition of this enzyme.

In each of nine experiments where insulin is added to Ringer containing 104 mM Na^+, the data points fall within the upper left-hand quadrant, confirming that insulin activates PFK. When CO_2 is decreased to produce an elevation of pH_i of the same magnitude as that produced by insulin, all ten individual points fall within the upper left-hand quadrant similarly indicating activation of PFK [Fidelman et al, in press].

When glycolysis is inhibited by addition of insulin to Na-free Ringer, all nine data points fall within the lower right-hand quadrant, confirming the expectation that PFK was inhibited. When CO_2 is increased to produce an elevation of pH_i of the same magnitude as that produced by insulin in the Na-free Ringer, 10 of 12 individual data points fall within the lower right-hand quadrant, again indicating inhibition of PFK [Fidelman et al, in press].

6. Conclusions. From the above, it is clear that both the magnitude and the direction of the effect of insulin upon glycolysis can be determined by merely varying the concentration of extracellular Na^+. More specifically, the effect is most likely determined by $<\Delta G>_{Na:H}$ and is mediated by Na:H exchange. The evidence also supports the view that the immediate signal for glycolysis is ΔpH_i and that this acts primarily (although not necessarily exclusively) on the rate-limiting enzyme of glycolysis, PFK.

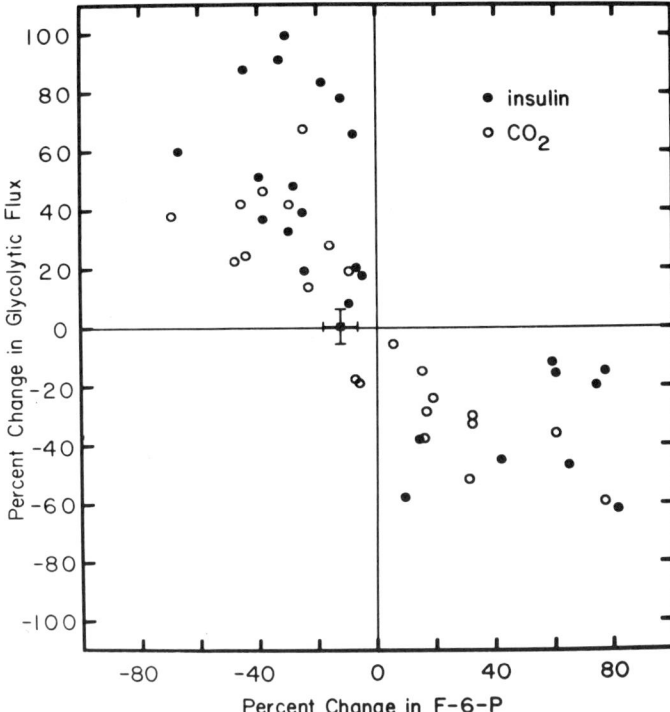

Fig. 6. The change in activity of phosphofructokinase, PFK, associated with changes in glycolytic flux due to either insulin or to changes in CO_2 levels. The percent change in glycolytic flux, as reflected by percent change in anaerobic lactate production, is plotted versus the percent change in the level of fructose-6-phosphate, F-6-P, in the muscle. In these studies, since the muscle extract was used for determination of F-6-P, only the lactate accumulated in the Ringer during the 90-minute experimental period was used to indicate glycolytic flux. This was justified by an analysis of the data used in Figures 3, 4, and 5, which indicated that ignoring the changed level of lactate within the muscle introduces only a few percent error into the estimate of the percent change in glycolytic flux and in no case changes the *direction* of the estimate, which is all that is important in the present analysis.

If PFK is activated, the data points should lie in the upper left quadrant, whereas if PFK is inhibited, data points should lie in the lower right quadrant. If the model is correct, data points should lie only in these two quadrants. Of 51 data points, two fail this criterion by a small amount.

Glycolysis was stimulated either by addition of insulin to Ringer containing the usual 104 mM Na^+, or by decreasing the CO_2 level from 5% to 2.7% in the same Ringer. Glycolysis was inhibited either by addition of insulin to Na-free Ringer, or by increasing the CO_2 level from 5% to 10% in Na-free Ringer. Three data points due to insulin and one due to CO_2 are omitted from the lower right quadrant because their F-6-P levels were off scale (115–240%).

In eight experiments, insulin was added to Ringer containing a level of Na^+, 6.8 mM, that should bring $<\Delta G>_{Na:H}$ near zero. In these experiments, as predicted by the model, there was no effect of insulin on glycolytic flux, and the change in F-6-P was not significant ($P > 0.05$) as indicated by the square and SE bars. Figure reproduced from Fidelman et al [in press].

IV. DISCUSSION

A. The Insulin Transduction System Mediates the Acute Action of Insulin Upon Glycolysis

1. Na:H exchange must play a role in mediating insulin action upon glycolysis. In view of the studies reviewed here it seems well established that insulin increases pH_i and that this is almost certainly due to stimulation of Na:H exchange. Moreover, it is clear that the Na^+ concentration, or activity, component of $<\Delta G>_{Na:H}$ can determine not only the magnitude but also the *direction* of the acute action of insulin upon glycolysis.

An especially interesting feature of this model is that it predicts a means to reverse the direction of insulin's effect upon glycolysis. The confirmation of the prediction that merely changing the concentration of Na^+ *outside* the cell to 0.12 mM will *reverse* the action of insulin on glycolysis provides especially powerful support for the model and strongly implies a direct, functional relationship between the Na^+ free-energy gradient and the intracellular signal that mediates the effect of insulin upon glycolysis. This single consideration makes it difficult to argue against the thesis that ionic phenomena involving the free-energy gradient for Na^+ play the predominant role in mediating the acute action of insulin upon glycolysis. If other factors are involved in mediating the acute effect of insulin upon this metabolic system, the ionic mechanism discussed here must be overriding their effects.

2. ΔpH_i is most likely the intracellular signal in the acute insulin effect on glycolysis. The idea that pH_i can affect glycolysis in animal cells is not new. Glycolysis is increased by relatively small changes in pH in cell-free extracts from rat diaphragm [Ui, 1966] and guinea pig leukocytes [Halperin et al, 1969] and in isolated rat heart [Scheuer and Berry, 1967], and is decreased by depressions of pH_i in intact human red blood cells [Tomoda et al, 1977] and in in vivo rat brain [Folbergrova et al, 1972]. PFK, which is the rate-limiting enzyme [Karpatkin et al, 1964] of glycolysis, is extremely sensitive to changes in pH. Trivedi and Danforth [1966] found that PFK isolated from frog skeletal muscle can be maximally activated (see Fig. 7) by pH elevations of only 0.1 to 0.2 units—ie, a 20–37% decrease in H^+ activity.

The stimulation by insulin of glycolysis, independent of glucose transport, in frog skeletal muscle [Ozand and Narahara, 1964] and in rat hemidiaphragm [Beitner and Kalant, 1971] had previously been concluded to be due to activation of PFK.

Four main considerations support the hypothesis that ΔpH_i is the intracellular signal for the acute effect of insulin upon glycolysis: 1) In frog skeletal muscle, the dependence of glycolysis upon ΔpH_i whether produced by insulin or by

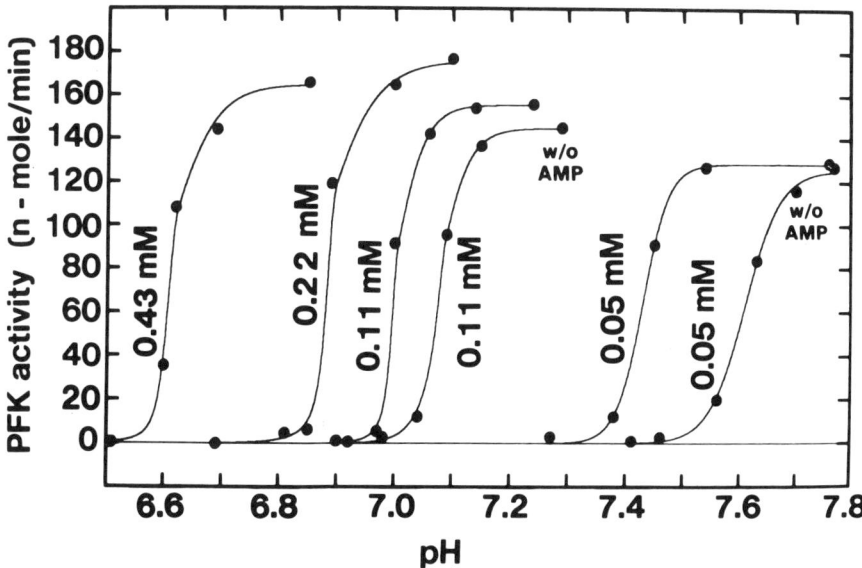

Fig. 7. The effect of pH on phosphofructokinase activity at various concentrations of fructose-6-phosphate in the presence or absence of 0.1 mM 5'-AMP. The data are taken from Figures 1 and 2 in Trivedi and Danforth [1966], with permission of W.H. Danforth.

changes in the level of CO_2 is clearly demonstrated by the studies reviewed here (see Fig. 5). 2) The ΔpH_i produced by insulin is due to activation of Na:H exchange, which couples H^+ efflux to Na^+ influx to obtain the necessary free energy from the Na^+ gradient. 3) The direction of the effect of insulin not only on glycolysis, but also on the activity of PFK, is determined by the direction of the Na^+ concentration gradient. This effect is mimicked by CO_2 induced changes in pH_i, consistent with the predictions of the model. 4) When both parameters are determined under identical experimental conditions, in each case changes in glycolysis are proportional to changes in pH_i.

B. Blocking Cellular Effects Due to the Insulin Transduction System

By consideration of the physical factors involved, the procedures that can block the operation of the insulin transduction system to increase pH_i can be divided into three classes: 1) Those that block the Na:H exchange system directly. An example is the use of amiloride, a drug that specifically inhibits this system and thus blocks the change in pH_i due to insulin [Moore, 1981a]. 2) Those that eliminate the driving force for the Na:H exchange system by manipulating relevant parameters so that $<\Delta G>_{Na:H}$ becomes zero. This may be done experimentally by *a)* sufficiently decreasing $[Na^+]_o$. The Na^+ substitute cannot be Li^+, which is a good substitute for Na^+ in this system [Moore, 1981a]; or *b)* elevating $[Na^+]_i$ by decreasing activity of the Na pump. This may be accomplished by using a specific inhibitor of the Na pump, such as ouabain, or by decreasing the ratio of ATP/ADP sufficiently to inhibit the active extrusion of Na^+ [Kennedy and DeWeer, 1976]. 3) Those that shunt the proton gradient with ionophores, such as monensin [Pressman, 1976], that greatly increase the proton conductance of the plasma membrane.

C. Possible Role of the Insulin Transduction System in Regulation of Other Cellular Functions

The studies from our laboratory reviewed here provide substantial evidence that the effect of insulin upon glycolysis is mediated by the insulin transduction system, although they do not directly relate this signal system to other effects of insulin. Results from several other laboratories, however, when interpreted in terms of this model, suggest that the action of insulin upon additional cellular processes may also be mediated, at least in part, by the insulin transduction system.

1. Effect of insulin on glycogen synthesis. It is quite possible that the insulin transduction system is also involved in mediating the effect of insulin on glycogen synthetase. Concentrations of ouabain sufficient to inhibit the sodium pump block activation of glycogen synthetase by insulin [Blatt et al, 1972; Horn et al, 1973]. Inhibition of the Na pump would, given sufficient time, inhibit the insulin transduction system by dissipating the Na^+ free-energy gradient, thereby removing the driving force for Na:H exchange.

2. Effect of insulin on DNA synthesis and mitogenesis. In tissue culture, insulin is required for cell division. Although doubt has been expressed about the physiological significance of this action of insulin, concentrations between 30 and 70 picomolar, well within the physiological range, stimulate cultured rat hepatoma cells to enter S phase and begin division [Koontz and Iwahashi, 1981]. A large number of results from several laboratories are consistent with the hypothesis that the insulin transduction system is part of the trigger mechanism for mitogenesis [Moore, 1981b], and that this system mediates the action of insulin upon RNA and DNA synthesis.

Gerson and Burton [1977] have suggested that changes in intracellular pH may trigger mitosis, and have reported that pH_i does rise during mitosis of synchronized cultures of slime mold consistent with the model proposed here. More recently, Gillies et al [1981] have used ^{31}P-NMR to demonstrate that pH_i rises during cell division.

Considerable experimental evidence suggests that the rise in pH_i during mitosis is due to Na:H exchange. For example, either amiloride or lowering $[Na^+]_o$ to 20 mM inhibits serum stimulated DNA synthesis in 3T3 cells [Rozengurt and

Mendoza, 1980]. Moreover, inhibition of the Na pump by ouabain in 3T3 cells blocks initiation of DNA synthesis by serum [Rozengurt and Heppel, 1975]. In human lymphocytes, ouabain inhibits the initiation by phytohemagglutinin or concanavalin A of DNA, RNA, and protein synthesis [Szamel et al, 1980]. In terms of the model proposed here, these effects of ouabain are not unexpected since inhibition of the Na pump should, especially in small cells, result in a rapid deterioration of the Na^+ free-energy gradient across the plasma membrane. As this free-energy gradient collapses, the driving force for Na:H exchange will be dissipated, resulting in a failure of this system to increase pH_i. These results suggest that the insulin transduction system is also involved in the action of several mitogens.

Carboxylic ionophores, such as monesin and nigericin, exchange H^+ for Na^+ or K^+ across the plasma membrane, thus collapsing both H^+ and monovalent cation gradients [Pressman, 1976]. Both monensin and nigericin block the initiation of DNA synthesis in cultured hepatocytes by a mixture of insulin, glucagon, and epidermal growth factor [Koch and Leffert, 1979]. Agents such as tetrodotoxin, which block components of Na^+ flux unrelated to the insulin transduction system, failed to block the initiation of DNA synthesis.

D. Possible Nonionic Mechanism for Mediation of Insulin Action

Addition of insulin to membrane fragments results in the release of a factor that can stimulate pyruvate dehydrogenase activity [Jarett and Seals, 1979], activate a phosphoprotein phosphatase that converts glycogen synthetase to its active form, and inhibit the cyclic AMP-dependent protein kinase [Larner et al, 1979]. Insulin also generates a peptide in diabetic animals that increases levels of phosphofructokinase by inhibiting the breakdown of this enzyme [Dunaway et al, 1978], although this has not yet been demonstrated in broken cell preparations. Whether or not the peptide released in vitro from membrane fragments is specific to the in vivo action of insulin has not yet been established, and may be open to question because the peptide is also released spontaneously [Seals and Czech, 1980, 1981].

1. Redundancy. It may be taken as an axiom that any essential regulatory mechanism will utilize redundancy. This is true of physiological systems such as the cardiovascular regulatory system. It should therefore not be surprising if, in addition to ΔpH_i, a peptide also functions as an intracellular signal for insulin. Indeed, the evidence cited here suggests that activation of glycogen synthetase may be mediated by both signal systems.

E. General Implications for Cell Function

In addition to playing a role in insulin action and being implicated in triggering cell division, Na:H exchange and the resulting increase in pH_i are also involved in activation of the egg during fertilization [Johnson et al, 1976]. All three of these processes are characterized by increased synthesis of macromolecules and cell growth. Taken together, this suggests that an elevated pH_i, initiated and presumably maintained by Na:H exchange, is required for the cell to enter the biophysical state optimal for growth.

In the case of insulin, this state is characterized by an "energization," or increased energy level, of the cell. It is common knowledge that insulin causes an increase in energy storage in the form of glycogen and fat. However, in addition to storing energy in macromolecules, insulin also increases the energy stored in biophysical mechanisms. For example, insulin energizes the plasma membrane in that it increases both the "concentration" [Moore, 1973; Clausen and Kohn, 1977] and the electrical components [Zierler, 1957, 1959; Moore and Rabovsky, 1979] of the free energy for Na^+, ΔG_{Na}, across the membrane.

Furthermore, even though insulin may not increase intracellular levels of ATP [Walaas et al, 1973], an insulin-induced increase in pH_i would result in an increase in the free energy, ΔG_{ATP}, available from hydrolysis of ATP since ΔG_{ATP} is a function of the proton activity:

$$\Delta G_{ATP} = \Delta G°_{ATP} - RT \ln \left[\frac{\alpha_{MgATP}}{\alpha_{MgADP} \cdot \alpha_{Pi} \cdot \alpha_H} \right] \quad (4)$$

Since pH is defined as the negative logarithm of the hydrogen ion activity, α_H:

$$pH \equiv - \log_{10} \alpha_H$$

the free energy available from ATP hydrolysis in the cell may be written explicitly as a function of the intracellular pH:

$$\Delta G_{ATP} = \Delta G°_{ATP} - RT \ln \left[\frac{\alpha_{MgATP}}{\alpha_{MgADP} \cdot \alpha_{Pi}} \right] - 2.3026 \, RT \cdot pH_i \quad (5)$$

In textbooks, the term $2.3026 \, RT \cdot pH$ is incorporated into the standard free energy, $\Delta G°_{ATP}$, thus leading to the statement that this term is pH-dependent. However, when the equation is expressed in its more complete form, as above, $\Delta G°_{ATP}$ becomes more nearly a true constant. From equation 5 it is apparent that at 37°C, an elevation of pH_i typical of insulin, say from 7.15 to 7.3, would result in an additional 0.212 kcal/mole available from ATP hydrolysis. Although this is only about 2% of the total available, some processes have energy thresholds, and in these even a 2% increase could be critical. Moreover, in disease states, such as diabetes, where the initial pH_i may be lower than normal, the increase due to insulin in available energy could be substantially greater.

In any case, much of the increase due to insulin in energy stored in the cell is not in the form of chemical bonds, but in biophysical parameters such as electric fields and activity gradients. Precisely how the increase in the energy state of the cell plays a role in other aspects of insulin action is not yet clear. However, Zierler and Rogus [1980] have proposed a related concept that the increased electrical field across the plasma membrane mediates the action of insulin upon glucose permeability.

F. Implications for Studies of Intracellular Enzymes

The thesis that pH_i is a normal regulatory factor in the cell requires that there be a pattern to the effects of pH on intracellular enzymes. It is feasible that the pH profiles of these enzymes could rule out the possibility that ΔpH_i produces coordinated changes in cell function. For example, if each enzyme in a linear metabolic pathway had pH profiles of alternatively positive and negative slopes in the physiological range, pH activation of one enzyme would be associated with deactivation of the next, and so on. This would require that pH_i be an invariant constant, rather than a regulatory factor as demonstrated in this publication.

It should be cautioned that pH profiles per se are not sufficient to identify possible control points for pH_i. Because of the complex nature of allosteric enzymes, pH may not always affect the activity of an enzyme directly, but rather may affect the interaction of another allosteric effector with that enzyme. In such cases, the enzyme may be insensitive to changes in pH unless the other factor is present. As an example of this type of interaction, α-glucose-1,6-diphosphate inhibition of hexokinase II is greatly increased as the pH increases above about 7 [Rose and Warms, 1975]. Also, activation of muscle pyruvate kinase by fructose-1,6-diphosphate is highly dependent upon pH, approaching the maximum as pH increases to about 7.4 [Gregory and Ainsworth, 1981].

G. The Proton as an Intracellular Signal

In the view presented here [also see Moore, 1981b], the effect of insulin upon many intracellular processes is mediated at least in part by a system that depends upon thermodynamic gradients. This system is thus quite different from the usual purely molecular system in which the intracellular signal, or "second messenger," is a discrete molecule that can therefore be isolated from the cell. As illustrated in Figure 1, a thermodynamic transduction system, of the type described here, exists *only* when the cell is *physically intact,* and therefore cannot be isolated by the in vitro *"reductionist"* procedures used in classic molecular biological investigations. It is clear that factors other than this model are involved since, although insulin activates this model in skeletal muscle, it does not stimulate DNA synthesis in this same tissue. The insulin transduction system may therefore be viewed as a signal system that activates intracellular systems (which actually carry out processes such as DNA synthesis), many of which are tissue-specific.

Another difference is that in this ionic, or thermodynamic, mechanism the signal is not carried by discrete molecules (for which the name "second messenger" aptly applies), but is conducted by a change in the intracellular ionic atmosphere, which provides more a pervasive influence than a limited number of individual molecules. In this connection, the proton is an especially appropriate choice for intracellular signaling since, because of its high mobility in water—particularly structured water—a change in the activity of protons in the cytoplasm of the cell would be rapidly communicated throughout the intracellular compartment. In view of the relatively short distances inside the cell, this might not seem significant. However, since in many tissues rapid diffusion of ions from one cell to the next occurs through cell–cell junctions [Loewenstein et al, 1978], whole collections of cells, such as the thyroid follicle, behave as one giant cell to ions such as the proton. This topological fact plus the fact that the mobility of the proton exceeds that of any other ion or molecule makes the proton an ideal choice for rapid signaling throughout the intracellular environment.

In view of the profound differences in physical behavior of protons and discrete molecules, it would perhaps be wise to confine the term "second messenger" to molecules. Molecular signals represent discrete signals analogous to letters, whereas the proton is a less discrete factor producing a more pervasive, environmental effect.

In view of the fact that the proton plays a central role in the chemiosmotic aspect of ATP synthesis, the choice of this same physical element as a cellular regulatory factor has interesting implications. Using the same currency for both energy transduction and information transduction obviously provides opportunities for close coordination of these two critical phenomena.

Although the mechanism whereby changes in intracellular proton activity (ΔpH_i) regulate many intracellular events remains unclear, in perhaps most of these cases it may involve allosteric activation of an enzyme—for example, phosphofructokinase. The suggestion that the proton can act as an allosteric effector has an historical precedent in the effect of pH upon the oxygen-binding curve of hemoglobin, ie, the Bohr effect.

ACKNOWLEDGMENTS

This work was supported by the National Institutes of Health under grants AM-17531 and AM-21059.

V. REFERENCES

Aickin CC, Thomas RC: An investigation of the ionic mechanisms of intracellular pH regulation in mouse soleus muscle fibres. J Physiol (Lond) 273:295, 1977.
Beitner R, Kalant N: Stimulation of glycolysis by insulin. J Biol Chem 246:500, 1971.

Bentley PJ: Amiloride: A potent inhibitor of sodium transport across the toad bladder. J Physiol (Lond) 195:317, 1968.

Blatt LM, McVerry PH, Kim K: Regulation of hepatic glycogen synthetase of Rana catesbeiana: Inhibition of the action of insulin by ouabain. J Biol Chem 247:6651, 1972.

Bolton TB, Vaughan-Jones RD: Continuous direct measurement of intracellular chloride and pH in frog skeletal muscle. J Physiol (Lond) 270:801, 1977.

Boron WF, DeWeer R: Intracellular pH transients in squid giant axons caused by carbon dioxide, ammonia and metabolic inhibitors. J Gen Physiol 67:91, 1976.

Boron WF, Roos A: Comparison of microelectrode, DMO, and methylamine methods for measuring intracellular pH. Am J Physiol 244:799, 1976.

Briggs AP, Koechig I, Doisy EA, Weber CJ: Some changes in the composition of blood due to the injection of insulin. J Biol Chem 58:721, 1924.

Clausen T: The relationship between the transport of glucose and cations across cell membranes in isolated tissues. II. Effects of K^+-free medium, ouabain and insulin upon the fate of glucose in rat diaphragm. Biochim Biophys Acta 120:361, 1966.

Clausen T, Elbrink J, Dahl-Hansen AB: The relationship between the transport of glucose and cations across cell membranes in isolated tissues. IX. The role of cellular calcium in the activation of the glucose transport system in rat soleus muscle. Biochim Biophys Acta 375:292, 1975.

Clausen T: The role of ions in the control of intermediary metabolism. In Kessler M, Clark LL, Lubbers DW, Silver I, Simon W (eds): "Ion and Enzyme Electrodes in Biology and Medicine." Munich: Urban and Schwarzenberg, 1976, p 237.

Clausen T, Kohn PG: The effect of insulin on the trnasport of sodium and potassium in rat soleus muscle. J Physiol (Lond) 265:19, 1977.

Czech MP: Molecular basis of insulin action. Annu Rev Biochem 46:359, 1977.

Dunaway GA, Leung GL-Y, Thrasher JR, Cooper MD: Turnover of hepatic phosphofructokinase in normal and diabetic rats. J Biol Chem 253:7460, 1978.

Fidelman ML, Seeholzer SH, Walsh KB, Moore RD: Intracellular pH mediates action of insulin upon glycolysis in frog skeletal muscle. Am J Physiol, Cell Physiol (in press).

Folbergrova J, MacMillan V, Siesjo BK: The effect of moderate and marked hypercapnia upon the energy state and upon the cytoplasmic $NADH/NAD^+$ ratio of the rat brain. J Neurochem 19:2497, 1972.

Funder J, Tosteson DC, Wieth JO: Effects of bicarbonate on lithium transport in human red cells. J Gen Physiol 71:721, 1978.

Gavryck WA, Moore RD, Thompson RC: Effect of insulin upon membrane-bound $(Na^+ + K^+)$ –ATPase extracted from frog skeletal muscle. J Physiol (Lond) 252:43, 1975.

Gerson DF, Burton AC: The relation of cycling of intracellular pH to mitosis in the acellular slime mould physarum polycephalum. J Cell Physiol 91:297, 1977.

Gillies RJ, Ugurbil K, DenHollander JA, Shulman RG: Phosphorus-31 NMR studies of intracellular pH and phosphate metabolism during cell division cycle of saccharomyces cerevisiae. Proc Natl Acad Sci USA 78:2125, 1981.

Graves B, Moore RD: A modification of the DMO technique for determination of intracellular pH in frog sartorius. Biophys J 16:200a, 1976.

Gregory RB, Ainsworth S: The regulatory properties of rabbit muscle pyruvate kinase. The effect of pH. Biochem J 195:745, 1981.

Halperin ML, Connors HP, Relman, AS, Karnovsky ML: Factors that control the effect of glycolysis in leukocytes. J Biol Chem 244:384, 1969.

Harrop GA, Benedict EM: The role of phosphate and potassium in carbohydrate metabolism following insulin administration. Proc Soc Exp Biol Med 20:430, 1923.

Hers HG: The control of glycogen metabolism in the liver. Annu Rev Biochem 45:167, 1976.

Hohorst HJ: L–(+) lactate determination with lactate dehydrogenase and DPN. In Bergmeyer HU (ed): "Methods of Enzymatic Analysis." New York: Academic, 1965, p 266.

Horn RS, Walaas O, Walaas E: The influence of sodium, potassium and lithium on the response of glycogen synthetase I to insulin and epinephrine in the isolated rat diaphragm. Biochim Biophys Acta 313:296, 1973.

Iles RA, Cohen RD: The effect of varying the amount of unlabelled 5,5-dimethyloxazolidine-2,4-dione (DMO) in the measurement of rat hepatic intracellular pH using (^{14}C)DMO. Clin Sci Mol Med 46:277, 1974.

Jarett L, Seals JR: Pyruvate dehydrogenase activation in adipocyte mitochondria by an insulin-generated mediator from muscle. Science 206:1407, 1979.

Johnson JD, Epel D, Paul M: Intracellular pH and activation of sea urchin eggs after fertilization. Nature (Lond) 262:661, 1976.

Kamminga CE, Willebrands AF, Groen J, Blickman JR: Effect of insulin on the potassium and inorganic phosphate content of the medium in experiments with isolated rat diaphragms. Science 111:30, 1950.

Karpatkin S, Helmreich E, Cori CF: Regulation of glycolysis in muscle. II. Effect of stimulation and epinephrine in isolated frog skeletal muscle. J Biol Chem 239:3139, 1964.

Kennedy BG, DeWeer P: Strophanthidin-sensitive sodium fluxes in metabolically poisoned frog skeletal muscle. J Gen Physiol 68:405, 1976.

Keppler D, Decker D: Glycogen determination with amyloglucosidase. In Bergmeyer HU (ed): "Methods of Enzymatic Analysis." New York: Academic, 1974, vol 3, p 1127.

Kissebah AH, Hope-Gill H, Vydelingum N, Tullock BR, Clarke PV, Fraser TR: Mode of insulin action. Lancet 1:144, 1975.

Kitasato H, Sato S, Murayama K, Nishio K: The interaction between the effects of insulin and ouabain on the activity of Na transport system in frog skeletal muscle. Jpn J Physiol 30:115, 1980a.

Kitasato H, Sato S, Marunaka Y, Murayama K, Nishio K: Effects of ouabain on Na efflux in high internal Na and insulin-preincubated muscles. Jpn J Physiol 30:591, 1980b.

Kitasato H, Sato S, Marunaka Y, Murayama K, Nishio K: Apparent affinity changes induced by insulin of Na–K transport system in frog skeletal muscle. Jpn J Physiol 30:603, 1980c.

Koch KS, Leffert HL: Increased sodium ion influx is necessary to initiate rat hepatocyte proliferation. Cell 18:153, 1979.

Koontz JW, Iwahashi M: Insulin as a potent, specific growth factor in a rat hepatoma cell line. Science 211:947, 1981.

Lang G, Michal G: D-glucose-6-phosphate and D-fructose-6-phosphate. in Bergmeyer HU (ed): "Methods of Enzymatic Analysis." New York: Academic, 1974, vol 3, p 1238.

Larner J, Galasko G, Cheng K, DePaoli-Roach AA, Huang L, Daggy P, Kellogg J: Generation by insulin of a chemical mediator that controls protein phosphorylation and dephosphorylation. Science 206:1048, 1979.

Lee CO, Armstrong WM: State and distribution of potassium and sodium ions in frog skeletal muscle. J Membr Biol 15:331, 1974.

Loewenstein WR, Kanno Y, Socolar SJ: The cell-to-cell channel. Fed Proc 37:2645, 1978.

Moore RD: The ionic effects of insulin. Abstr Biophys Soc FA12, 1965.

Moore RD: Effect of insulin upon the sodium pump in frog skeletal muscle. J Physiol (Lond) 232:23, 1973.

Moore RD: Elevation of intracellular pH by insulin in frog skeletal muscle. Biochem Biophys Res Commun 91:900, 1979.

Moore RD, Fidelman ML, Seeholzer SH: Correlation between insulin action upon glycolysis and change in intracellular pH. Biochem Biophys Res Commun 91:905, 1979.

Moore RD, Rabovsky JL: Mechanism of insulin action on resting membrane potential of frog skeletal muscle. Am J Physiol 236:C249, 1979.

Moore RD, Gupta RD: Effect of insulin on intracellular pH as observed by ^{31}P NMR spectroscopy. Int J Quan Chem Quan Biol Symp 7:83, 1980.

Moore RD: Stimulation of Na:H exchange by insulin. Biophys J 33:203, 1981a.

Moore RD: The insulin transduction system: A biophysical model for mitogenesis. Int J Quan Chem Quan Biol Symp 8:(in press) 1981b.

Mullins LJ, Noda K: The influence of sodium-free solutions on the membrane potential of frog muscle fibers. J Gen Physiol 47:117, 1963.

Ozand P, Narahara HT: Regulation of glycolysis in muscle. III. Influence of insulin, epinephrine, and contraction on phosphofructokinase activity in frog skeletal muscle. J Biol Chem 239:3146, 1964.

Pressman BC: Biological applications of ionophores. Ann Rev Biochem 45:323, 1976.

Racker R, Fisher LW: Reconstitution of an ATP-dependent sodium pump with an ATPase from electric eel and pure phospholipids. Biochem Biophys Res Commun 67:1144, 1975.

Rolleston FS, Newsholem EA: Control of glycolysis in cerebral cortex slices. Biochem J 104:524, 1967.

Rolleston FS: A theoretical background to the use of measured concentration of intermediates in study of the control of intermediary metabolism. Curr Top Cell Reg 5:47, 1972.

Roos A: Intracellular pH and distribution of weak acids across cell membranes. A study of D- and L-lactate and of DMO in rat diaphragm. J Physiol (Lond) 249:1, 1975.

Roos A, Boron WF: Intracellular pH transients in rat diaphragm muscle measured with DMO. Am J Physiol 235:C49, 1978.

Rose IA, Warms JVB: pH dependence of the α-glucose 1,6-diphosphate inhibition of hexokinase II. Arch Biochem Biophys 171:678, 1975.

Rozengurt E, Mendoza S: Monovalent ion fluxes and the control of cell proliferation in cultured fibroblasts. Ann NY Acad Sci 339:175, 1980.

Rozengurt E, Heppel LA: Serum rapidly stimulates ouabain-sensitive ^{86}Rb$^+$ influx in quiescent 3T3 cells. Proc Natl Acad Sci USA 72:4492, 1975.

Russell JM, Boron WF: Role of chloride transport in regulation of intracellular pH. Nature (Lond) 264:73, 1976.

Russell JM: Effects of ammonium and bicarbonate–CO_2 on intracellular chloride levels in aplysia neurons. Biophys J 22:131, 1978.

Scheuer J, Berry MN: Effect of alkalosis on glycolysis in the isolated rat heart. Am J Physiol 213:1143, 1967.

Schudt C, Gaertner U, Petts D: Insulin action on glucose transport and calcium fluxes in developing muscle cells in vitro. Eur J Biochem 68:103, 1976.

Seals JR, Czech MP: Evidence that insulin activates an intrinsic plasma membrane protease in generating a secondary chemical mediator. J Biol Chem 255:6529, 1980.

Seals JR, Czech MP: Characterization of a pyruvate dehydrogenase activator relased by adipocyte plasma membranes in response to insulin. J Biol Chem 256:2894, 1981.

Shaw WN, Stadie WC: Coexistence of insulin-responsive and insulin-non-responsive glycolytic systems in rat diaphragm. J Biol Chem 227:115, 1957.

Shaw WN, Stadie WC: Two identical Embden-Meyerhof enzyme systems in normal rat diaphragms differing in cytological location and response to insulin. J Biol Chem 234,2491, 1959.

Sjodin RA, Henderson EB: Tracer and non-tracer potassium fluxes in frog sartorius muscle and the kinetics of net potassium movement. J Gen Physiol 47:605, 1964.

Szamel M, Somogyi J, Csukas I, Solymosy F: Effect of ouabain on macromolecular synthesis during the cell cycle in mitogen-stimulated human lymphocytes. Biochim Biophys Acta 633:347, 1980.

Thomas RC: The role of bicarbonate, chloride and sodium ions in the regulation of intracellular pH in snail neurone. J Physiol (Lond) 273:317, 1977.

Tomoda A, Tsuda-Hirota S, Minakami S: Glycolysis of red cells suspended in solutions of impermeable solutes. J Biochem (Tokyo) 81:697, 1977.

Trivedi B, Danforth WH: Effect of pH on the kinetics of frog muscle phosphofructokinase. J Biol Chem 241:4110, 1966.

Ui M: A role of phosphofructokinase in pH-dependent regulation of glycolysis. Biochim Biophys Acta 124:310, 1966.

Waddell WJ, Butler TC: Calculation of intracellular pH from the distribution of 5,5-dimethyl-2,4-oxazolidinedione (DMO). Application to skeletal muscle of the dog. J Clin Invest 38:720, 1959.

Walaas O, Walaas E, Gronnerod O: Hormonal regulation of cyclic-AMP-dependent protein kinase of rat diaphragm by epiniphrine and insulin. Eur J Biochem 40:465, 1973.

Williamson JR: General features of metabolic control as applied to the erythrocyte. Adv Exp Med Biol 6:117, 1970.

Zierler KL: Increase in resting membrane potential of skeletal muscle produced by insulin. Science 126:1067, 1957.

Zierler KL: Effect of insulin on membrane potential and potassium content of rat muscle. Am J Physiol 197:515, 1959.

Zierler KL, Rogus EM: Hyperpolarization as a mediator of insulin action increased muscle glucose uptake induced electrically. Am J Physiol 239:E21, 1980.

Cellular Dormancy and the Scope of pH_i-Mediated Metabolic Regulation

William B. Busa

Zoology Department, University of California, Davis, California 95616

I.	Introduction	417
II.	Three Cryptobiotic Systems Demonstrating Large pH_i Changes	418
	A. Bacteria	418
	B. Yeast	419
	C. Crustaceans	419
III.	Determination of pH_i in Artemia Cysts	419
	A. The Organism	419
	B. Materials and Methods	420
	1. Origin and treatment of cysts; 2. Superfusion; 3. Spectroscopy	420
IV.	Results and Discussion	422
V.	Conclusions	424
VI.	References	425

I. INTRODUCTION

Although the last decade has witnessed impressive advances in the biologist's ability to determine and manipulate intracellular pH (pH_i), our understanding of the role(s) of the hydrated proton as a regulator of cellular metabolism is yet in its infancy. At present, the best example of the scope of pH_i-mediated metabolic regulation remains the first—the fertilized sea urchin egg (see Winkler, this volume), in which the pH_i increase of about 0.5 unit following fertilization [Shen and Steinhardt, 1978] has been implicated in the increases in protein synthesis [Grainger et al, 1979], mRNA polyadenylation [Wilt and Mazia, 1974], membrane K^+ conductance [Shen and Steinhardt, 1980], and microvillar actin polymerization [Begg and Rebhun, 1979], which accompany derepression and the initiation of embryogenesis. Similar (though generally smaller) pH_i transients in the eggs of higher phyla [see eg, Nuccitelli et al., 1981; Webb and Nuccitelli, this volume] suggest the generality of this phenomenon.

Questions regarding the scope of pH_i-mediated metabolic regulation necessarily involve some consideration of the *magnitude* of the pH_i excursions involved, and this in part explains the popularity of the urchin egg system, which exhibits one of the larger pH_i changes observed under nontraumatic conditions. It is intuitive that a large pH_i change will effect a broader range of cellular processes than a small change can—if only because the large change will be "felt" by a greater number of ionizable groups within the cell. Numerous examples of enzyme systems that might be regulated by large ($\sim$ 1 unit) but not by small ($\sim$ 0.1 unit) pH_i excursions may be offered by way of illustration. One such, the "PEP branchpoint" of invertebrate anaerobic metabolism [Hochachka and Somero, 1973] derives its potential for pH_i-mediated regulation from the lack of overlap of the pH activity profiles of two key enzymes of glycolysis: phosphoenolpyruvate carboxykinase (PEPCK) and pyruvate kinase (PK). In vitro, PEPCK is activated whereas PK is inactivated below about pH 6.5; in vivo this is expected to help switch carbon flow from pyruvate to oxaloacetate during anaerobiosis. The pH_i change presumed to accomplish this has not been observed, to my knowledge, but the in vitro pH activity profiles would seem to require it to be about as large as that observed in the sea urchin zygote.

Clearly, what is needed in order to investigate the potential scope of pH_i as a cellular regulator are reliable examples of large pH_i excursions determined under conditions nontraumatic to the cell. In what follows, I shall mean by a "large" change one of about 1 pH unit. This rather arbitrary definition will severely restrict the number of examples available; those that remain, however, will be seen to share an uncommon metabolic adaptation with some interesting implications for the roles of pH_i.

II. THREE CRYPTOBIOTIC SYSTEMS DEMONSTRATING LARGE pH_i CHANGES

We have already seen that, in the unfertilized sea urchin egg, dormancy is associated with a depressed pH_i. A dormant stage, in the broadest sense of the term, is a feature common to some portion of the life cycles of many organisms. In the most extreme cases, dormant organisms may be nearly or even completely ametabolic—a state Kielin [1959] has termed "cryptobiosis," literally "hidden life." Examples of such cryptobiotic systems include plant seeds, bacteria, such micrometazoa as nematodes and tardigrades, fungal spores, and certain crustacean embryos. Three of these also provide us with examples of large pH_i increases during transitions between the profound dormancy of cryptobiosis and the active metabolism of development.

A. Bacteria

Using the methylamine technique, Setlow and Setlow [1980] have demonstrated a 1-unit pH_i increase upon germination in two species of Bacillus spores, and the acidic pH_i of the dormant spore ($\sim$ 6.3) has been confirmed via ^{31}P nuclear magnetic resonance (^{31}P-NMR) by Barton et al [1980]. As the former point out, these observations may help explain the absence of glycolytic activity from the dormant spore, since phosphoglycerate mutase (the key enzyme in spore glycolysis) would be about 90% less active at pH 6.3 than at pH 7.3. In view of the fact that some spore enzymes display their greatest heat resistance in vitro at pH 6.5 [Sadoff, 1970], the acidic pH_i of dormant spores may also play a role in stabilizing protein structure.

B. Yeast

In a companion paper to that of the Setlows, Barton et al [1980] have demonstrated via ^{31}P-NMR (see Gadian, this volume) that the pH_i of dormant Pichia pastoralis spores is distinctly acidic, about 6. The pH_i excursion accompanying germination was not quantified in this study, but more recent observations of pH_i in vegetative Pichia cells suggest it spans about one unit (personal communication, J.A. den Hollander). The NMR linewidths observed indicate that the interior of the yeast spore is liquid, thus apparently ruling out dehydration of the spore interior as a dormancy-regulating mechanism under these conditions. These authors, too, speculate that an acidic pH_i may play some role in the spore's dormancy, citing the decreased glycolytic rate of E. coli at low pH_i [Ugurbil et al, 1978].

C. Crustaceans

My collegues and I have recently employed the ^{31}P-NMR technique to estimate pH_i during both dormancy and development in gastrula-stage embryos of the brine shrimp, Artemia. Briefly, we have observed an alkalinization of more than one unit accompanying the transition between dormancy and development. Since this transition is freely reversible in Artemia (see below), we have also been able to study the complementary acidification of pH_i accompanying the *initiation* of dormancy. The detailed report of our findings is in preparation at the time of publication of this volume [Busa et al, 1982]; below I shall focus mainly on the rationale and methodology we have employed.

III. DETERMINATION OF pH_i IN ARTEMIA CYSTS

A. The Organism

In many respects, the Artemia embryo is a model cryptobiotic system [see Clegg and Conte, 1980, for review]. The gastrula-stage embryo, encysted in a spherical shell about 0.1 mm in radius, can be completely desiccated and thereafter stored under vacuum or at ultralow temperatures—conditions under which concerted metabolic activity cannot occur. Even fully hydrated cysts can remain dormant, yet viable, for at least 5 months in the absence of oxygen [Dutrieu and Chrestia-Blanchine, 1966], resuming or halting development as O_2 is supplied or withdrawn. It is this characteristic in particular that renders the Artemia cyst of value in the study of large pH_i changes, particularly in conjunction with a noninvasive pH_i-determining technique such as ^{31}P-NMR. Such other characteristics as the cysts' relatively low respiratory rate and spherical shape render them very suitable subjects for the NMR technique, since it is possible to attain high sample densities within the spectrometer probe while providing all the necessary conditions for metabolism and development by suitable superfusion.

B. Materials and Methods

1. Origin and treatment of cysts.
Artemia cysts are available commercially, desiccated and vacuum-packed, from locations around the world. A difficulty common to any study such as this is their pronounced variability among locations, among suppliers, and even among batches from the same supplier; it is therefore important to begin a series of investigations with a single, sufficiently large batch of cysts on hand. The cysts used here originated from the Great Salt Lake, Utah.

The outermost layer of the cyst shell, the chorion, contains iron in the form of hematin and is often heavily contaminated with microorganisms. Prior to spectroscopy, therefore, the cysts are dechorionated essentially according to the method of Sorgeloos et al [1977]. This involves prehydration at 0°C (to prevent initiation of development) and incubation in 50% household bleach (about 1.25% NaOCl) on ice until microscopic examination reveals the complete removal of the deeply pigmented chorion. Finally, the cysts are extensively washed and equilibrated against the superfusion buffer to be used, either on ice or in nominally O_2-free (N_2-saturated) buffer.

2. Superfusion.
For in vivo spectroscopy, cysts are packed under gravity in the superfusion apparatus illustrated in Figure 1. The active volume of the tube ($\sim$ 6 ml) contains roughly 8×10^5 embryos and 2.4 ml buffer. O_2-saturated superfusion at a rate of 8 ml/min supports hatching of "free-swimming" nauplii along the length of the cyst column, and control experiments have shown that cysts incubated under 100% O_2 display hatching kinetics essentially indistinguishable from those of cysts incubated in air.

The inner layer of the cyst shell (which survives dechorionation) is essentially impermeable to all solutes of low volatility [Clegg and Conte, 1980]. This has two beneficial consequences for the spectroscopist. First, the principal (if not sole) metabolic waste product to be removed by superfusion is CO_2. Second, the shell's impermeability allows development to occur over a wide range of buffer compositions, ionic strengths, and pHs (see below).

Dormant cysts hydrate passively to levels dictated by the activity of water in the buffer [Clegg, 1978]; it is therefore good practice to equalize (at least roughly) the osmolalities of all buffers. Our basic superfusion medium is 0.46 M glycerol; when pH buffering is required the glycerol concentration is reduced to achieve a similar osmolality, as determined by vapor pressure osmometry.

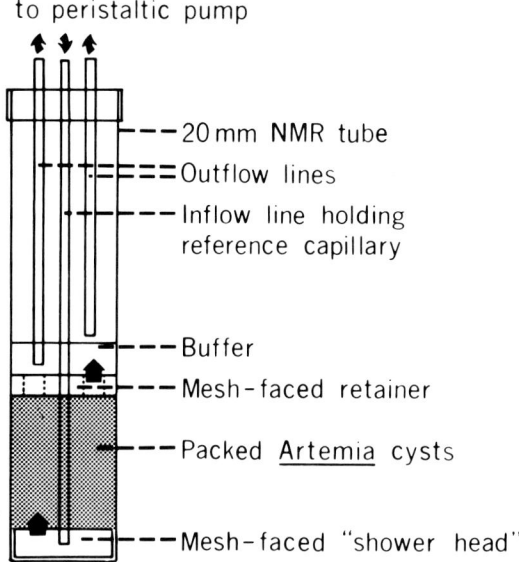

Fig. 1. Superfusion apparatus for in vivo ^{31}P-NMR spectroscopy of Artemia cysts. The central glass inflow line holds a fine, sealed capillary containing 1.6 M methylenediphosphonic acid, and delivers buffer to a glass and nylon mesh "shower head." A column of cysts 3 cm high rests atop the shower head and is retained at its upper surface by a Plexiglass and nylon mesh disk, machined to fit closely both the i.d. of the NMR tube and the o.d. of the inflow line. Buffer is supplied at a rate of 8 ml min^{-1} via peristaltic pump from a reservoir vigorously bubbled with either 100% N_2 or O_2.

3. Spectroscopy. ^{31}P-NMR spectra of whole cysts, cyst homogenates, and perchloric acid extracts were collected on a Nicolet NT-200 spectrometer in the Fourier transform mode at 81 MHz, using a pulse angle of 60°, a sweep width of 8,000 Hz, 1W proton decoupling, and a controlled probe temperature of 23°C. No D_2O lock was used. The external reference was 1.6 M methylenediphosphonic acid (trisodium salt, Sigma), capillaries of which were calibrated against 85% H_3PO_4. Spectra were prepared by summing 200–400 free induction decays collected at repetition times of 0.5–1.25 seconds. Routinely, ≤ 200-second temporal resolution is possible.

Peak identification was accomplished by addition of standards to perchloric acid extracts prepared essentially by the method of Stocco et al [1972]. pH$_i$ was determined from the chemical shift of the inorganic phosphate (P$_i$) resonance of in vivo spectra. The calibration curve relating P$_i$ chemical shift to pH was prepared using whole cyst homogenates, a compromise necessitated by the absence of data regarding ionic strength inside the cyst. Homogenates were prepared with a chilled mortar and pestle. pH was adjusted with NaOH or H_2SO_4, and determined with a combination electrode immediately before and again just after spectroscopy. The apparent pK_a of P_i thus determined is 6.7, and the curve fitting the NMR titration data is of the form

$$pH = 6.7 + \log \frac{\delta - 0.57}{3.08 - \delta}$$

where δ is the observed P_i chemical shift in ppm (relative to 85% H_3PO_4).

IV. RESULTS AND DISCUSSION

Our initial studies, utilizing well-buffered superfusion media, have shown that the pH_i of anaerobic dormant Artemia cysts is independent of buffer pH (pH_o) over the range pH_o 4–8. The independence of pH_i probably extends over a greater range of pH_o as well, since Horne [1966] observed that incubation at pH_o 1.8 had no apparent effect on cyst hatchability. This is probably a result of the previously mentioned impermeability of the cyst shell to solutes such as Cl^-, which might otherwise serve as counterions to support the transport of H^+ across the shell. The impermeability of the shell may thus permit maintenance of a pH_i different from the pH of the medium at no energetic expense to the dormant embryo.

The ^{31}P-NMR spectrum of anaerobic dormant cysts (Fig. 2A) reveals a prominent P_i resonance, the chemical shift of which varies between 1.8 and 1.3 p.p.m. This represents an anaerobic pH_i ranging between 6.7 and 6.3, the latter value being achieved only after about 12 hours of anaerobic incubation at room temperature. Just downfield (to the left) of the P_i resonance a broad phosphomonoester peak (probably composed of sugar phosphate resonances) is observed. Finally, a poorly resolved group of unidentified resonances occurs between about -10 and -12 ppm.

O_2-saturated superfusion of previously dormant cysts gives rise to a complex redistribution of intracellular phosphates as visualized by ^{31}P-NMR (Fig. 2B). The three resonances of nucleoside di- and triphosphates (NDP and NTP), undetectable in the dormant cysts, first appear shortly after the initiation of aerobiosis and increase in intensity with prolonged superfusion. Both the P_i and sugar phosphate peaks of Figure 2A rapidly decrease in intensity during the first minutes of aerobiosis. The shrinking P_i resonance is not observed to change its chemical shift significantly, but the decline of the sugar phosphate resonances reveals a narrower peak at about 3 ppm arising from P_i at a pH of ≥ 7.9. This identification is based in part on spectra of perchloric acid extracts of aerobic cysts, which at

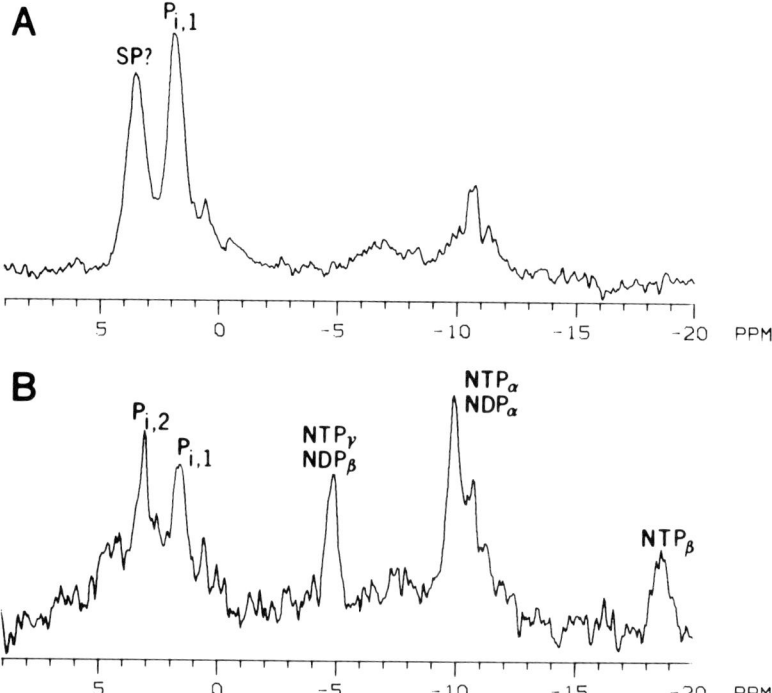

Fig. 2. ^{31}P-NMR spectra of dormant and developing Artemia embryos. A) Dormant embryos superfused with nominally O_2-free (N_2-saturated) 0.46 M glycerol (sum of 400 free induction decays, each 1.25 seconds). B) Developing embryos superfused 8 hours with O_2-saturated 0.46 M glycerol (400 free induction decays, each 0.5 seconds). Abbreviations: SP, phosphomonoester peak probably composed of sugar phosphate resonances; $P_{i,1}$, acid inorganic phosphate resonance; $P_{i,2}$, alkaline inorganic phosphate resonance; NTP_α, NDP_α, etc. overlapping nucleoside di- and triphosphate resonances. All chemical shifts are expressed relative to 85% H_3PO_4. No peaks (other than the methylenediphosphonic acid reference) occur outside the displayed region.

pH 8 display a single P_i resonance at about 3 ppm and no peak corresponding to the upfield (more acidic) P_i resonance in Figure 2B. I shall refer to these as the alkaline and acid P_i resonances, respectively.

The identification of the 3-ppm resonance in aerobic spectra as P_i at an alkaline pH is substantiated by its behavior during the reinduction of anaerobic dormancy. When superfusion is switched from O_2-saturated to nominally O_2-free buffer the alkaline P_i resonance responds by increasing in intensity and simultaneously shifting upfield, finally merging with the acid P_i resonance. Roughly coincident with this the three NTP/NDP resonances vanish, and the sugar phosphate resonances grow and shift upfield. The net result is a spectrum indistinguishable from that of Figure 2A.

The small acid P_i resonance visible in aerobic spectra such as Figure 2B has been tentatively identified as arising from the nonviable fraction of the sample, which amounts to about 35% of this batch of cysts.

The observations outlined above indicate that ≥ 1 unit pH_i excursions accompany the reversible transitions between dormancy and development in Artemia embryos. The *absolute* pH_i values we have determined must be considered tentative, however, since there is little reason to expect the ionic strength of the homogenates used to prepare our calibration curve to identically reflect in vivo ionic strength. The rather alkaline pH_i of aerobic embryos (≥ 7.9) further complicates quantification of aerobic pH_i. Nevertheless, the P_i chemical shift *changes* reported here (directly visualized in the aerobic-to-anaerobic transition) clearly reflect pH_i changes traversing at least half of the $HPO_4^{2-}/H_2PO_4^-$ titration curve.

V. CONCLUSIONS

It seems clear that pH_i changes of the magnitude observed in bacterial and fungal spores—and now in Artemia—are likely to affect numerous intracellular enzyme activities, and recent observations of the pH sensitivity of assembly of microtubules [Regula et al, 1981], microfilaments [Begg and Rebhun, 1979], and gap junction crystalline arrays [Peracchia and Peracchia, 1980] suggest that cytoskeletal and membrane proteins may also be affected. The significance of these effects in regard to the establishment, maintenance, and termination of dormancy remains unknown. Nevertheless, the occurrence of such pronounced pH_i changes in all three of the cryptobiotic systems so far examined—particularly in light of what is now known regarding the role of the pH_i increase in the fertilized sea urchin egg—makes a functional role for pH_i in cryptobiotic dormancy an intriguing possibility.

What is the source of the proton equivalents produced during the induction of anaerobic dormancy in Artemia? The cysts possess very low lactate dehydrogenase activity, and thus do not produce lactate during anaerobiosis [Ewing and Clegg, 1969]. Other metabolic acids (eg, succinate) might be involved, and this issue is currently under investigation. Multiple proton sources may in fact be involved, since a variety of metabolic pathways either produce or consume proton equivalents [Gillies, 1981]. Important examples include NTP hydrolysis and oxidative phosphorylation, respectively. During steady-state metabolism such linked pathways probably cancel each others' effects on pH_i to a great extent [Vághy, 1979]; during metabolic rundown, however, NTP hydrolysis without rephosphorylation will acidify pH_i, as will reduction of the NAD/NADH couple. Indeed, the prominent role played by protons in all aspects of energy metabolism suggests that H^+ activity may well be a sensitive indicator of metabolic status, through which a variety of intracellular processes can be coordinated to function in the manner most appropriate to the metabolic status of the cell.

A potential candidate for this type of pH_i-mediated metabolic regulation in the Artemia cyst was described by Warner et al [1974]. GTP:GTP guanylyltransferase reversibly catalyzes the synthesis of P^1,P^4-diguanosine-5′-tetraphosphate (GP_4G) from GTP in cyst yolk platelets [see Warner, 1980, for review]. The purified enzyme displays an acidic pH optimum ($\sim$ 6) and almost no activity at pH 8. This, as Warner observed, is a rather unusual pH profile for a nucleoside triphosphate enzyme, since these usually require the fully ionized substrate for maximum activity. GP_4G, sequestered in yolk platelets, constitutes the principal high-energy phosphate-bond store of the cyst and is slowly consumed during anaerobiosis [Stocco et al, 1972]. Whether its disappearance represents concerted (albeit low-level) energy metabolism or merely adventitious chemistry remains unclear, but the possibility that the depressed pH_i of anaerobic dormancy might activate alternative energy-producing pathways such as this to meet the minimal "housekeeping" requirements for the maintenance of biological integrity during dormancy clearly deserves further study.

The potential for pH_i-mediated regulation of the cell cycle and proliferation has recently excited great interest [see Gillies, 1981, for review]. Questions regarding the role of pH_i in the regulation of dormancy clearly address the same concerns, although from an opposing point of view; thus dormant cells such as the sea urchin egg can provide very useful experimental systems in which to assess the scope of pH_i-mediated regulation. As a multicellular system exhibiting the largest pH_i excursion yet reported, Artemia may someday prove especially useful in this regard.

VI. REFERENCES

Barton JK, den Hollander JA, Lee TM, MacLaughlin A, Shulman RG: Measurement of the internal pH of yeast spores by ^{31}P nuclear magnetic resonance. Proc Natl Acad Sci USA 77:2470–2473, 1980.

Begg DA, Rebhun LI: pH regulates the polymerization of actin in the sea urchin cortex. J Cell Biol 83:241–248, 1979.

Busa WB, Crowe JH, Matson GB: Intracellular pH and the metabolic status of dormant and developing Artemia embryos. Manuscript in preparation, 1982.

Clegg JS: Interrelationships between water and cellular metabolism in Artemia cysts. VIII. Sorption isotherms and derived thermodynamic quantities. J Cell Physiol 94:123–138, 1978.

Clegg JS, Conte FP: A review of the cellular and developmental biology of Artemia. In Persoone G, Sorgeloos P, Roels O, Jaspers E (ed): "The Brine Shrimp, Artemia," Vol 2. Wetteren, Belgium: Universa Press, 1980, p 11.

Dutrieu J, Chrestia-Blanchine D: Résistance des oeufs durables hydratés d'Artemia salina à l'anoxie. Compt Rend Acad Sci Paris D263:998–1000, 1966.

Ewing RD, Clegg JS: Lactate dehydrogenase activity and anaerobic metabolism during embryonic development in Artemia salina. Comp Biochem Physiol 31:297–307, 1969.

Gillies RJ: Intracellular pH and growth control in eukaryotic cells. In Cameron I, Poole TB (eds): "The Transformed Cell." New York: Academic Press, 1981, p 347.

Grainger JL, Winkler MM, Shen SS, Steinhardt RA: Intracellular pH controls protein synthesis rate in the sea urchin egg and early embryo. Dev Biol 68:396–406, 1979.

Hochachka PW, Somero GN: "Strategies of Biochemical Adaptation." Philadelphia: Saunders, 1973, pp 49–50.

Horne FR: The effect of digestive enzymes on the hatchability of Artemia salina eggs. Trans Am Microsc Soc 85:271–274, 1966.

Keilin D: The problem of anabiosis or latent life: History and current concept. Proc R Soc Lond B150:149–191, 1959.

Nuccitelli R, Webb DJ, Lagier ST, Matson GB: ^{31}P NMR reveals an increase in intracellular pH after fertilization in Xenopus eggs. Proc Natl Acad Sci USA 78:4421–4425, 1981.

Peracchia C, Peracchia LL: Gap junction dynamics: Reversible effects of hydrogen ions. J Cell Biol 87:719–727, 1980.

Regula CS, Pfeiffer JR, Berlin RD: Microtubule assembly and disassembly at alkaline pH. J Cell Biol 89:45–53, 1981.

Sadoff HL: Heat resistance of spore enzymes. J Appl Bacteriol 33:130–140, 1970.

Setlow B, Setlow P: Measurements of the pH within dormant and germinated bacterial spores. Proc Natl Acad Sci USA 77:2474–2476, 1980.

Shen SS, Steinhardt RA: Direct measurement of intracellular pH during metabolic derepression of the sea urchin egg. Nature 272:253–254, 1978.

Shen SS, Steinhardt RA: Intracellular pH controls the development of new potassium conductance after fertilization of the sea urchin egg. Exp Cell Res 125:55–61, 1980.

Sorgeloos P, Bossuyt E, Lavina E, Balza-Mesa M, Persoone G: Decapsulation of Artemia cysts: A simple technique for the improvement of the use of brine shrimp in aquaculture. Aquaculture 12:311–316, 1977.

Stocco DM, Beers PC, Warner AH: Effects of anoxia on nucleotide metabolism in encysted embryos of the brine shrimp. Dev Biol 27:479–493, 1972.

Ugurbil K, Rottenberg H, Glynn P, Shulman RG: ^{31}P nuclear magnetic resonance studies of bioenergetics and glycolysis in anaerobic E. coli. Proc Natl Acad Sci USA 75:2244–2248, 1978.

Vághy PL: Role of mitochondrial oxidative phosphorylation in the maintenance of intracellular pH. J Mol Cell Cardiol 11:933–940, 1979.

Warner AH: The biosynthesis, metabolism and function of dinucleoside polyphosphates in Artemia embryos: A compendium. In Persoone G, Sorgeloos P, Roels O, Jaspers E (eds): "The Brine Shrimp, Artemia," Vol 2. Wetteren, Belgium: Universa Press, 1980, p 105.

Warner AH, Beers PC, Huang FL: Biosynthesis of the diguanosine nucleotides. I. Purification and properties of an enzyme from yolk platelets of brine shrimp embryos. Can J Biochem 52:231–240, 1974.

Wilt FH, Mazia D: The stimulation of cytoplasmic polyadenylation in sea urchin embryos by ammonia. Dev Biol 37:422–424, 1974.

The Effect of Decreased Intracellular pH on the Electrical Properties of Invertebrate Muscle Fibers and Oocytes

William J. Moody, Jr.
Jerry Lewis Neuromuscular Research Center, University of California, Los Angeles, California 90024

I.	Introduction	427
II.	Methods	429
	A. pH Microelectrode Construction	429
	B. The Experimental Setup	430
III.	Results	430
	A. Effect of Low pH_i on Ca^{2+} Spike Generation in Crayfish Slow Muscle Fibers	430
	B. Effect of pH_i on Inwardly Rectifying K^+ Currents in Starfish Oocytes	437
IV.	Discussion	441
V.	References	442

I. INTRODUCTION

There are two principal reasons behind examining the effects of changes in intracellular pH (pH_i) on the electrical properties of excitable cells. First, by knowing in detail the relationship between pH_i and the behavior of a single type of membrane ionic channel, you can make certain statements about the role of titratable chemical groups in the ion permeation mechanism of that channel [Hille, 1968, discusses the effects of external pH in this regard]. Second, you might want to know whether a change in pH_i is responsible for a change in the electrical properties of an excitable cell under some physiologically relevant circumstance, such as during a particular developmental stage or in a mature cell during the action of some hormone.

For the person driven by the first motivation, the most direct control over the intracellular ionic environment of the cell to be studied is the most desirable approach. The recent development of techniques for the control of the ionic environment on both sides of the membrane [Lee et al, 1980; Horn and Patlak, 1980; Byerly and Hagiwara, 1981; see also Spray, Harris, and Bennett, this volume] has given new life to researchers grown weary of giant cephalopod axons and barnacle muscle fibers for this purpose. For those interested in the possible role of the intracellular ionic environment as a modulator of cell electrical properties, there are additional experimental goals. In addition to demonstrating that a change in pH_i, for example, has a particular electrophysiological effect, it would be nice to be able to show that a similar pH_i change occurs in the cell in the specific physiological situation under investigation. Clearly, this requires the cell to be as intact as practicable, with the cytoplasmic ionic environment under the cell's, not your, control. Furthermore, it is crucial to understand the response of the intact cell's homeostatic mechanisms to transient perturbations of the internal ionic milieu. For example, it has been shown that when neurons and muscle fibers are subjected to a cytoplasmic acidification, one of the recovery mechanisms is a membrane exchange system that extrudes H^+ and takes up Na^+ [see, for example, Thomas, 1977; Moody, 1981]. Because of this exchange, cells subjected to a cytoplasmic acid load—either experimentally or under physiological circumstances—will take up Na^+. Experiments investigating pH_i effects must therefore also consider effects of increased intracellular Na^+ concentration. (It is worth mentioning some numbers here. Cells can extrude H^+ very rapidly; for example, crayfish neurons can return cytoplasmic pH to normal after a 0.6–1.0 unit acid load in 5–10 minutes [Moody, 1981]. During this time, internal Na^+ concentration can increase by up to 40–50 mM. Even under conditions of direct internal perfusion, the intracellular concentration changes produced by these rapid fluxes may not be adequately prevented.)

In the experiments I report in this chapter, I have used ion-sensitive microelectrodes to record pH_i (and in some instances $[Na^+]_i$), and have used either direct injection of H^+ into nonperfused cells, or changes in Ringer composition to alter pH_i. I will describe two sets of experiments concerning the effect of decreased pH_i on membrane electrical properties. The first concerns the generation of Ca^{2+}-dependent action potentials in crayfish muscle fibers at normal and low pH_i. The second describes a block of inward K^+ currents in starfish oocytes by internal H^+. In the crayfish experiments, much of the potential physiological significance rests in the well-documented generality of Ca action potential properties in a variety of cells [Hagiwara and Byerly, 1981], and their participation in such important phenomena as secretion. In the case of the oocyte, however, there is some reason to believe that in this cell itself both pH_i and inward K^+ currents change at certain stages of early development, and the future direction of these experiments is to determine whether there is a causal connection between the two.

II. METHODS

Both the crayfish slow flexor muscle and starfish oocyte preparations have been described in detail elsewhere [Moody, 1978; Hagiwara et al, 1975]. The construction of pH-sensitive microelectrodes has also been described at length in the fine book by R.C. Thomas [1978]. I will use this section to emphasize modifications of the pH electrode techniques described by Thomas that I have found useful (I should emphasize that some of these were developed by Thomas himself since the publication of his volume), and to discuss points of experimental technique that I have found to be troublesome to newcomers to pH microelectrode experiments.

(The following section is best preceded by a reading of the relevant chapters in the Thomas book.)

A. pH Microelectrode Construction

1) A necessity in the construction of pH electrodes is to have an outer pipette that has a short shank-to-tip distance, and a relatively steep taper near the tip. With conventional microelectrode pullers, this shape almost invariably gives a fairly blunt tip. I have recently, after an initial period of distaste, become an addict of the Brown-Flaming air-jet electrode puller (Sutter Instrument Co., San Francisco) for pulling outer pipettes. With this puller, very nice tips can be made on very blunt-looking electrodes. I use a single operation pull, rather than the two-stage operation described by Thomas. Oddly, the finer tip does not seem to slow the electrode response appreciably. The electrodes are unquestionably better than those made on conventional pullers at penetrating crayfish muscle and neurons, and starfish eggs, the three preparations on which I have tested them. Since I have worked on large cells, I have made no effort to make these electrodes sharper. I feel confident that a determined effort with this puller would yield quite fine electrodes for reasonably small cells, still keeping 90% response times under one minute.

2) High N_2 pressure is essential for good seals and small recess volumes. Use steel tubing to connect the N^2 tank to the inner pipette, and apply at least 500 lbs/in^2 and preferably 1,000–1,200 lbs/in^2 to the inner pipette. Although the force generated by such pressures over the small surface area of the 30-gauge tubing making the connection to the inner pipette is only a few pounds, appropriate precautions are advisable: anchor the tubing well; pressurize the system and then close if off from the tank so that the volume of escapable N_2 is small; wear safety goggles if you don't wear glasses normally. For fun, I shoot the used inners by melting the wax seal over a small flame while pointing the tubing at a styrofoam target across the room (or perhaps at a scientific competitor).

3) Work in the microforge at 400 × magnification; less, and you can't see well enough. If you don't mind never making a successful pH electrode, use a commercial microforge with dissecting scope optics.

4) Fill the electrodes by heating at 60–70°C overnight, not by boiling. If you use aluminoscilicate outers, put a few drops of detergent into the water you inject into the electrodes, but not in the water in which you immerse them, since this glass is very hydrophobic.

5) Number all your inners, outers, and finished electrodes, and record all the appropriate dimensions and calibration data in a notebook. This will save your making the same mistake twice.

6) If you are just starting, read the book by Thomas and get a demonstration from someone experienced in the construction if you have any trouble.

B. The Experimental Setup

The differences between a conventional electrophysiological setup and one for recording with pH microelectrodes revolve principally around obtaining noise-free recordings from these high-impedance electrodes. I have found that the following are sources of trouble for many people who are using pH microelectrodes for the first time.

1) Shielding. Shield the preparation with a solid metal Faraday cage with a front that can be lowered during the recording. Static noise from air currents, movement in the room, or solution turbulence are the problems most commonly encountered. Do not shield the cable from the pH electrode to the follower amplifier—unless the shield is "driven" by the follower output signal.

2) Solution flow. It is virtually essential to have continuous solution flow over the preparation. You should be able to change solutions rapidly, without interupting the flow, changing the rate of flow, or changing the level of the bath. The importance of this cannot be overemphasized.

3) Record the pH electrode signal with the membrane potential signal subtracted from it directly on a slow speed chart recorder. You learn a lot about your preparation by just watching this record during the experiment. You also learn a lot about artifacts. Record the membrane potential alone simultaneously on another channel of the recorder.

4) Take time to let the electrode stabilize before starting the experiment, and time to let the cell recover from penetration. I normally allow a minimum of 10 minutes and often up to 30 minutes after penetration before doing anything, unless the penetration is exceptionally good.

III. RESULTS

A. Effect of Low pH_i on Ca^{2+} Spike Generation in Crayfish Slow Muscle Fibers

Ca^{2+}-dependent action potentials are of great interest to electrophysiologists because they are involved in important cellular phenomena such as neurosecretion, and because they seem to be quite labile in amplitude and waveform both during development and in mature cells [Spitzer and Baccaglini, 1976; Shapiro et al, 1980; see also Hagiwara and Byerly, 1981, for a general review]. The modulation of Ca^{2+} action potential waveform by a change in the intracellular concentration of H^+ or other ions would seem to be a likely mechanism by which a cell could exert control over processes triggered by the voltage-dependent entry of Ca^{2+} ions.

Crayfish slow muscle fibers, like most crustacean slow fibers, do not generate action potentials [Kennedy and Takeda, 1965; Atwood, 1976], although they clearly possess the requisite membrane channels to generate Ca^{2+} action potentials, since, in the presence of agents which block K^+ currents, such as tetraethylammonium, all-or-none Ca^{2+} action potentials can be elicited [see eg, Moody, 1980]. It appears that the inward Ca^{2+} current in these cells is short-circuited by an outward K^+ current—also triggered by depolarization. This insures the lack of all-or-none electrical behavior in fibers which must contract in a graded fashion, while at the same time retaining the voltage-dependent entry of Ca^{2+} that seems to be required for contraction [Atwater et al, 1974]. Since the balance of K^+ and Ca^{2+} currents, and hence the size of the action potential, is rather strictly related to the contractile function over a wide range of crustacean fiber types within a single animal [Atwood, 1976], this seems a good preparation in which to study the sensitivity of the Ca^{2+} spike mechanism to intracellular ionic events. Below, I will describe experiments that show that the generation of Ca^{2+} action potentials in these fibers is greatly facilitated when pH_i is decreased.

The principal result—that all-or-none Ca^{2+} action potentials can be elicited in these fibers when pH_i is decreased—is shown in Figure 1. I will describe this figure in some detail, concentrating on experimental procedures and problems that may be faced by others intending to do this sort of work. The figure is a direct photograph of a chart record of membrane potential (V_m) and pH_i. The upper trace displays V_m, referenced to an extracellular microelectrode; the lower trace shows pH_i, the output of the pH electrode with the output of the V_m electrode subtracted from it. A third electrode has been inserted into the fiber in order to inject current and examine electrical responses; response to depolarizing current pulses are shown at the bottom of the figure at four different times during the experiment.

At the start of the experiment, the pH_i, V_m, and current passing electrodes are inserted into the fiber and during the first 10 minutes, a steady record of the resting pH_i is obtained (7.17 in this fiber). At point 1, two tests are done on the cell: 1) A long hyperpolarizing current pulse is injected through the third electrode, to insure that both V_m and pH_i electrodes record the same change in

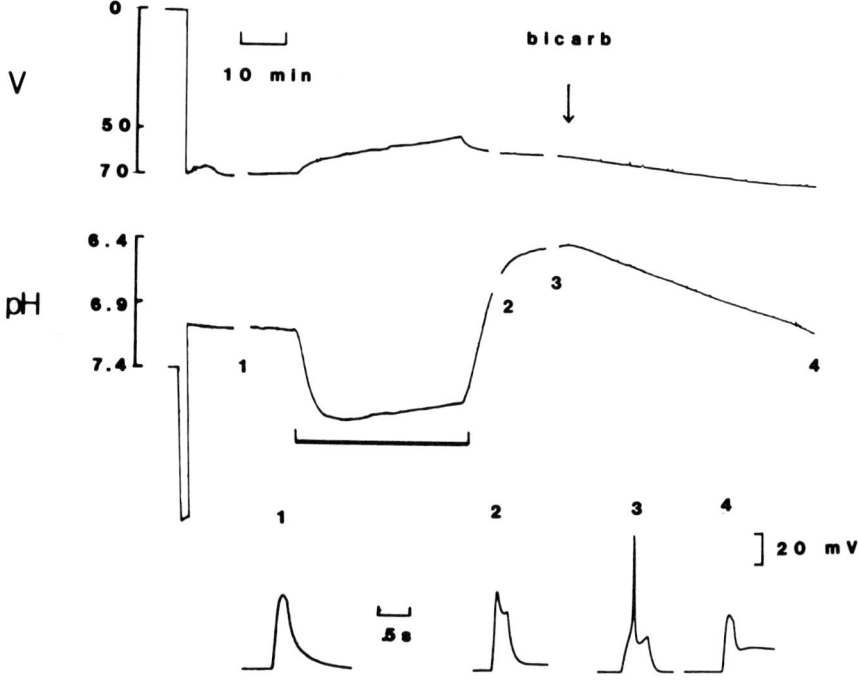

Fig. 1. Experiment showing the effect of decreased pH_i, produced by ammonium loading, on depolarizing responses in crayfish slow muscle. The bar below the pH_i trace marks the time during which 50 mM NH_4Cl was present in the Ringer. The gaps in the pH_i record indicate the times at which depolarizing responses, assessed by current injection, were examined. These responses are shown below the pH_i record. At "bicarb," 10 mM HCO_3^- was added to the Ringer.

potential; and 2) a series of short depolarizing current pulses is delivered to test the electrical response of the fiber.

No action potentials are elicited in these fibers by depolarization at normal pH_i, as shown in record #1 in Fig. 1 (see also Fig. 2B7). I used the "NH_4Cl acid loading" technique to decrease the pH_i, because the direct injection of H^+ from a microelectrode is very difficult in muscle fibers. The NH_4Cl technique has been described in detail by Boron and De Weer [1976]. Essentially, when a cell is exposed to Ringer containing 10–50 mM NH_4Cl, pH_i increases and the cell becomes loaded with NH_4^+ ions. Upon removal of NH_4Cl Ringer, NH_4^+ is "trapped" inside the fiber by the negative membrane potential, and exits by dissociating into uncharged NH_3 and H^+. This injects H^+ into the fiber, causing pH_i to fall below its pre-NH_4Cl value after NH_4Cl Ringer is removed. In Figure 1, NH_4Cl is present in the Ringer during the time marked by the bar below the

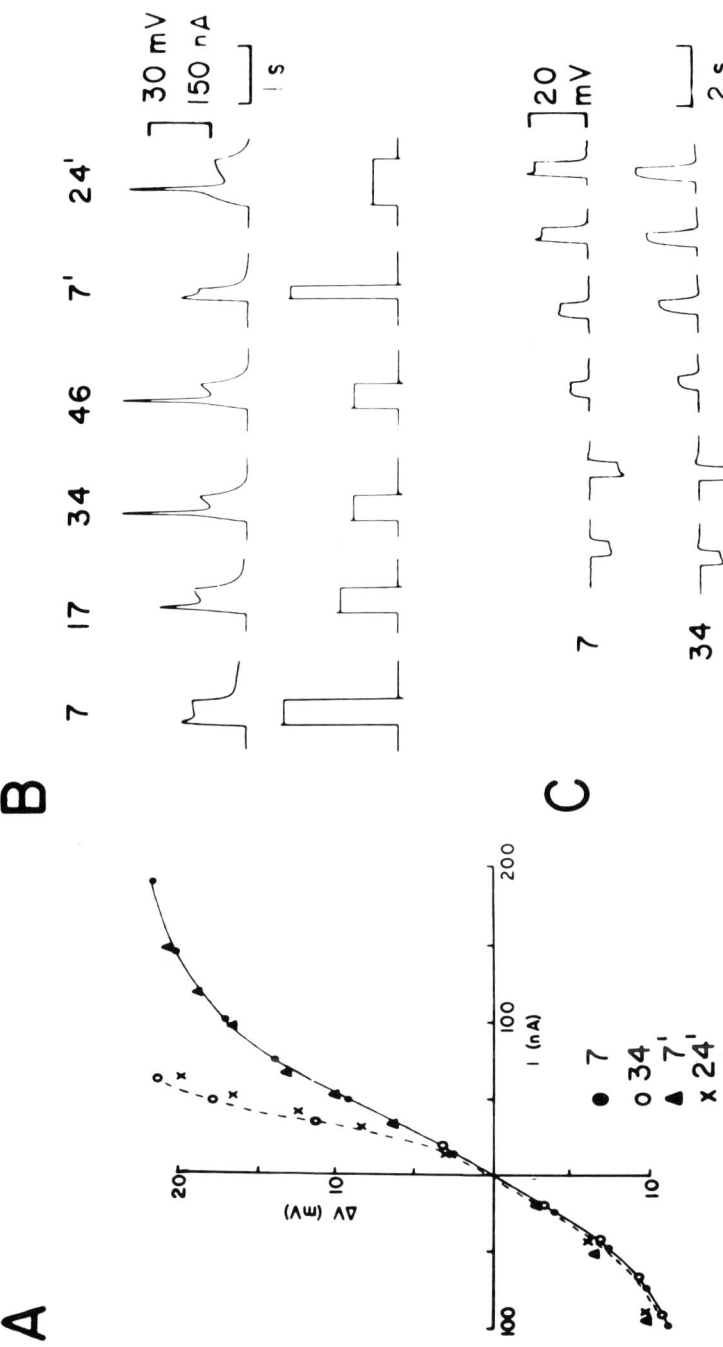

Fig. 2. Current–voltage relations (current clamp) and depolarizing responses in a fiber exposed to two NH$_4$Cl loads (see text). In B, the numbers indicate times in minutes after the *washout* of NH$_4$Cl after the first (7, 17, 34, and 46) and second (7', 24') exposures. C shows records from which two of the I–V plots in A were taken.

pH$_i$ record. You can see the large acidification recorded in the cytoplasm upon removal of NH$_4$Cl Ringer.

At two times after NH$_4$Cl removal (#2, pH$_i$ = 6.7, and #3, pH$_i$ = 6.4), electrical responses were again tested. The gradual increase in excitability is apparent, culminating in the ability to generate an all-or-none action potential at pH$_i$ = 6.4 (see also Fig. 2B7–46). The ability to generate all-or-none action potentials at pH$_i$ values below 6.5 was found in virtually every fiber tested. The action potentials are Ca-mediated by the criteria of block by cobalt or manganese ions, and the lack of effect of TTX or Na$^+$-free Ringer [Hagiwara and Byerly, 1981]. Note that these action potentials at low pH$_i$ are recorded in normal external Ringer, since the internal acidification occurs upon removal of NH$_4$Cl.

Having described the basic phenomenon, I turn to several important questions one would like to ask about this rather dramatic effect of low pH$_i$: 1) Is the effect reversible? 2) Can an effect of NH$_4^+$ inside the fiber on action potential generation be ruled out as the mechanisms of the apparent pH$_i$ effect? 3) Is the effect specific for the action potential mechanism, or is a general membrane resistance change responsible?

1) The reversibility of the low pH$_i$ effect is tested in the experiment of Figure 1, and also in Figure 2 (see below). After record 3 in Figure 1 was taken, 10 mM HCO$_3^-$ was added to the Ringer, which accelerates H$^+$ extrusion from these fibers [Boron, 1977]. After 35 minutes, pH$_i$ had recovered to near its normal value, and the control electrical properties returned. So when the internal acidification is reversed, the Ca^{2+} action potentials disappear.

2) The NH$_4$Cl loading procedure results in the accumulation of NH$_4^+$ ions within the fiber, as well as decreased pH$_i$. The fibers would also tend to lose K$^+$ in gaining NH$_4^+$. Three observations argue that decreased pH$_i$ and not "side effects" of NH$_4$Cl treatment are responsible for the change in excitability.

a) The action potentials persist for long times after NH$_4$Cl washout. Figure 2 shows a full action potential recorded 46 minutes after NH$_4$Cl washout. Although the kinetics of NH$_4^+$ washout from the cytoplasm vary among fibers, in every case where pH$_i$ was recorded, dpH$_i$/dt became zero at 25 minutes or less after return to normal Ringer. Since most NH$_4^+$ leaves the fiber as NH$_3$, d[H$^+$]$_i$/dt is approximately equal to $-$d[NH$_4^+$]$_i$/dt, so that when the peak acidification is reached, NH$_4^+$ should have been completely washed out of the fiber. Thus, any effects of internal ammonium should disappear by 25 minutes after washout of NH$_4$Cl from the Ringer. Therefore, the persistence of action potentials for 45 minutes to 1 hour after NH$_4$Cl removal argues against an effect of intracellular NH$_4^+$ or low intracellular [K$^+$].

b) Action potentials disappear transiently following exposure of a spiking fiber to NH$_4$Cl Ringer a second time. In a fiber whose pH$_i$ has been previously decreased, a second exposure to NH$_4$Cl Ringer will increase pH$_i$ to well above the normal value [see Moody, 1980, Fig. 1]. Just after washout of NH$_4$Cl, pH$_i$

will be high, but will soon decrease again to the value before the second NH_4Cl exposure. Any electrophysiological effects of internal NH_4^+ should be greater just after the second washout of NH_4Cl, whereas effects of low pH_i should transiently disappear, and then reappear as pH_i falls. Figure 2B shows that the latter is the case. Sample responses of a fiber to depolarizing current pulses are shown at various times following the removal of NH_4Cl after the first (7, 17, 34, 46) and second (7', 24') exposures. Note that the action potential disappears immediately after the second washout of NH_4Cl and reappears 17 minutes later.

c) Action potentials are also seen when pH_i is lowered by other means. Low pH Ringer buffered with acetate or equilibrated with 50–100% CO_2 have the same effect on fiber excitability. The results with CO_2 are shown in Figure 3A and C. Exposure of the fiber to pH 6.0 (PIPES) Ringer had little effect on either pH_i or the depolarizing response. pH 6.0 (CO_2) Ringer, however, caused pH_i to fall rapidly to 6.42 and action potentials to be elicited by depolarization. Low-frequency repetitive spiking in a fiber in CO_2 Ringer is shown in Figure 3C.

3) As I mentioned above, the wave form and amplitude of Ca^{2+}-dependent action potentials are in great part determined by the balance of voltage-dependent inward Ca^{2+} and outward K^+ currents. Internal H^+ could act on the voltage-dependent conductances to alter the balance in favor of net inward current, or it could increase the input resistance of the fiber by blocking non-voltage-dependent channels so that the small net inward current normally present could become regenerative. The current–voltage relations in Figures 2A, 2C, and 3B show that the former hypothesis is the correct one. Near the resting potential (origin) and for more negative potentials, the normal and low pH_i curves superimpose. In the depolarizing quadrant, they diverge, indicating that internal H^+ acts on some conductance activated by depolarization. This result is obtained when either NH_4Cl loading (Fig. 2) or CO_2 (Fig. 3) is used to decrease pH_i. Which of the conductances activated by depolarization are affected? Evidence presented in the full account of this work [Moody, 1980] indicates that H^+ acts to block the delayed outward K^+ currents in these cells, thus allowing the inward Ca^{2+} to become regenerative.

To summarize: When pH_i is decreased in crayfish slow muscle fibers, the delayed outward K^+ conductance is decreased, allowing the inward Ca^{2+} current to become regenerative, producing all-or-none action potentials in these normally relatively inexcitable cells.

What about the possibility that an increase in intracellular Na^+ activity caused by low pH_i (see above) is responsible for the change in spike generation? Two observations make this unlikely. When pH_i is lowered, Na^+ activity in the fibers does in fact increase, although the increases I have recorded are smaller and slower than those in crayfish neurons [Moody, 1981, Fig. 3]. However, a similar increase in $[Na^+]_i$ produced by blocking the Na pump with K^+-free Ringer [Thomas, 1972] does not alter the depolarizing responses of the fibers. Fur-

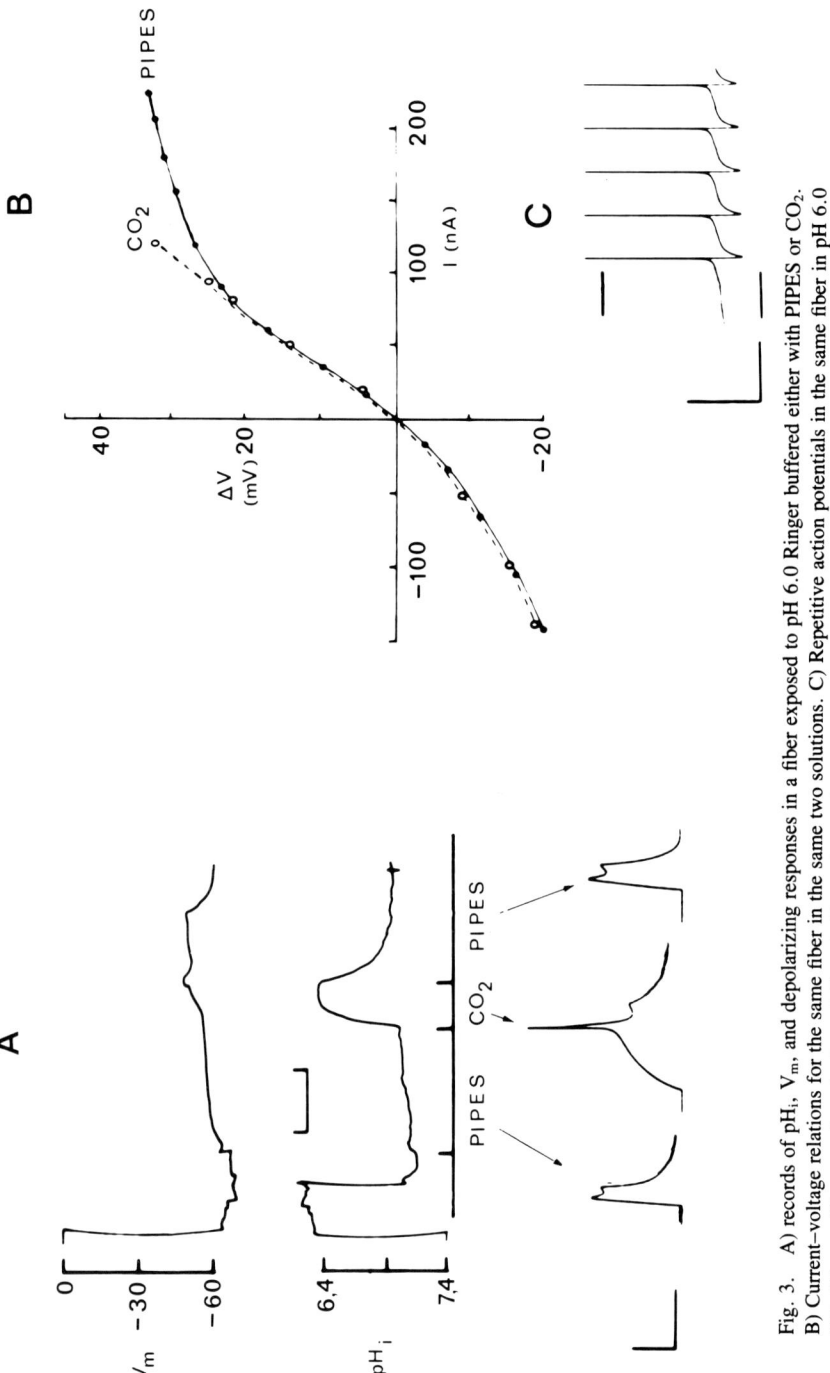

Fig. 3. A) records of pH_i, V_m, and depolarizing responses in a fiber exposed to pH 6.0 Ringer buffered either with PIPES or CO_2. B) Current–voltage relations for the same fiber in the same two solutions. C) Repetitive action potentials in the same fiber in pH 6.0 CO_2 Ringer. The two short horizontal lines indicate the resting potential in the absence of applied current, and the 0 mV point. Calibrations for the depolarizing response records; 20 mV, 0.5 sec (left), and 20 mV, 2 sec (right). The horizontal bar on the pH_i record indicates 10 min.

thermore, the disappearance of action potentials after H^+ extrusion is stimulated by HCO_3^- (Fig. 1) also argues against a causal role for increased $[Na^+]_i$, since stimulation of H^+ extrusion should increase $[Na^+]_i$ further.

To conclude this section, I will briefly describe a situation in which a change in energy metabolism in these fibers causes pH_i to decrease sufficiently to allow the generation of full action potentials. Several years ago I discovered that crayfish slow fibers treated with metabolic inhibitors or made anoxic acquire the ability to generate all-or-none action potentials [Moody, 1978]. This is an electrophysiological effect of anoxia quite unlike others reported in neurons and muscle fibers. I have quite recently been able to record pH_i during anoxia, and it does in fact decrease by an amount sufficient to cause the appearance of action potentials.

B. Effect of pH_i on Inwardly Rectifying K^+ Currents in Starfish Oocytes

The immature oocyte of the starfish (Mediaster aequalis) has several properties of great interest to anyone who wishes to study the possible modulation of membrane electrical properties by internal ions during early development. The oocyte is large (1 mm diameter) and can be impaled by at least three microelectrodes without appreciable loss of its resting potential. The egg displays a variety of interesting electrical properties, including a Ca^{2+}-mediated action potential that involves two separate inward currents, and an inward K^+ current that is activated by membrane hyperpolarization (inward, or anomalous, rectification [Hagiwara and Takahashi, 1974; Hagiwara et al, 1975]). As I will discuss below, these currents undergo a series of interesting changes during certain stages of early development.

Since the detailed properties of the ionic conductances in the oocyte can be studied under voltage clamp conditions, and since intracellular ionic activities can be measured with ion-sensitive microelectrodes, the starfish oocyte is a very advantageous preparation in which to study the questions outlined at the beginning of this chapter.

I have begun my investigations by studying the sensitivity of inwardly rectifying potassium currents to decreased intracellular pH. Inward rectification is the term applied to the large inward (approximately 1 microampere) K^+ currents that are activated when the oocyte membrane is hyperpolarized; inward rectification has been studied extensively in the starfish oocyte [Hagiwara and Takahashi, 1974; Hagiwara et al, 1978; Hagiwara and Yoshii, 1979; see also Hagiwara and Jaffe, 1979, for review].

For each experiment, an oocyte was impaled with three microelectrodes: voltage recording, current passing, and pH-sensitive. With three electrodes in place, resting potentials were -69 to -71 mV, not significantly different from single electrode impalements. Cells normally displayed a small acid injury on impalement, which recovered within 20–30 minutes. The mean resting pH_i value for 14 oocytes was 7.09 ± 0.08 unit. This is almost 0.5 units more alkaline than the value of 6.61 predicted for an equilibrium distribution of H^+ across the membrane given $V_m = -70$ mV. Thus, both in terms of absolute pH_i and its value relative to an equilibrium distribution of H^+, this oocyte behaves like any nerve or muscle cell.

The experimental strategy was quite straightforward. pH_i was decreased by exposing the oocytes to low pH Ringer buffered with the membrane-permeant weak acid acetate. The undissociated form (acetic acid) crosses the membrane readily and dissociates intracellularly, releasing H^+ and lowering pH_i [see Sharp and Thomas, 1981]. Control experiments for the effect of low external pH alone were done using the membrane-impermeant buffer biphthalate. For each value of pH_i inwardly rectifying currents were recorded under voltage-clamp conditions.

One of the experiments is shown in Figure 4. The record begins 15 minutes after penetration. The cell is in 10 K^+ Ringer and the resting pH_i was 7.08. After the K^+ concentration of the external solution was increased to 25 mM, the first series of voltage-clamp pulses was delivered (+10, −10, −20, and −30 mV from resting potential). Then the pH of the external Ringer was lowered to 5.0 with biphthalate. This had almost no effect of pH_i, and the inward rectifier currents were unchanged (record 2; Fig. 4). After a brief recovery period in normal external solution, the Ringer pH was again lowered to 5.0, but this time with acetate as the buffer. This caused pH_i to fall rapidly to 5.9, and the inwardly rectifying currents were virtually completely blocked. Return to normal Ringer caused pH_i to recover (actually pH_i became more alkaline than initially), and the currents returned to within 2% of their control values (record 4, Fig. 4). A technical note on this experiment: 20 mM biphthalate was used in the low external pH control solution, but only 10 mM acetate in the solution for lowering pH_i. This was done to equalize the buffer strengths of the two solutions, so that one could not argue that the effect was actually one of low external pH, but biphthalate simply was not as effective as acetate at changing external pH.

Knowing that at $pH_i = 5.9$, inward rectification is completely blocked, the next experiment was to determine the relation between pH_i and the current amplitude for intermediate pH_i values, to see whether this relation has the general shape of a titration curve. The experiments were similar to that in Figure 4, except that a series of acetate-buffered ASWs of pH values between 7.8 and 5.0 were sequentially washed into the experimental chamber. At each new value of pH_i, the inward rectifier currents were measured under voltage-clamp conditions. In three cells, four-point titration curves were determined in this way, and in several others, two or three pH_i values were examined. In each case the entire sequence of solutions was reversed to return pH_i in several steps to its normal value, and in each cell the relationship between pH_i and current amplitude showed

pH, Effects on Muscle and Egg Electrophysiology / 439

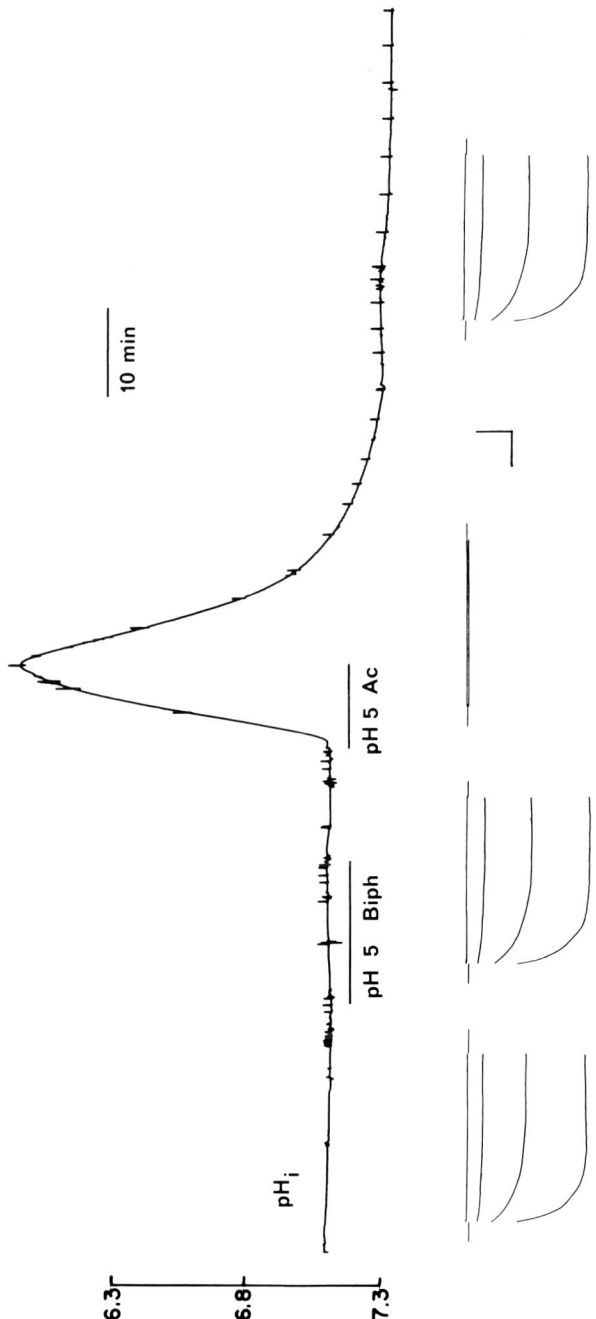

Fig. 4. Experiment showing the effects of pH 5.0 Ringer buffered with either biphthalate or acetate on pH_i and inward rectifier currents in a starfish oocyte. Calibrations for the current records: 0.2 microamperes, 0.2 sec. V_m was held under voltage-clamp throughout.

little if any hysteresis. The results were quantitatively very similar among cells. The pooled data are plotted in Figure 5, as normalized current amplitude versus pH_i. The points fall fairly well along the sigmoid shape of a standard titration curve, indicating that the block of the channel can probably be explained by the protonation of some site associated with the channel. The continuous line in Figure 5 is drawn according to the equation:

$$I/I_{pH = 7.23} = K^3/(K^3 + [H^+]^3)$$

where $K = 10^{-6.26}$. This equation represents the prediction of a model in which three H^+ ions bind to a site with a pK of 6.26 in order to block the channel. The theoretical curve fits the points almost well enough to make one believe the model. At least the equation represents a good empirical description of the data.

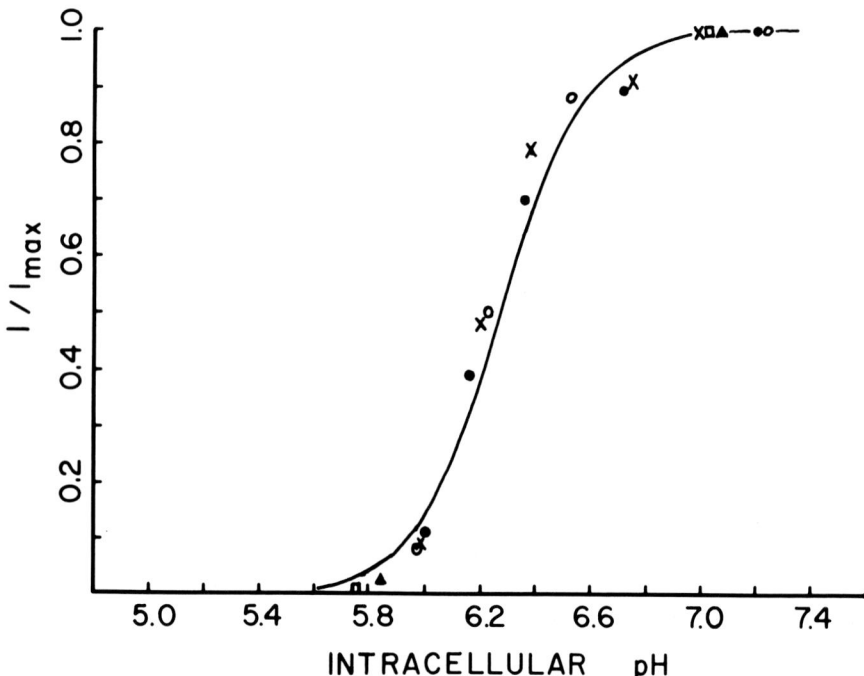

Fig. 5. Titration curve for the pH_i effect on inward rectification. Data points for five experiments are shown, using different symbols. See text.

The sensitivity of inward rectification to changes in the concentrations of intracellular ions is of particular interest in oocytes. In Mediaster, as well as in other species of starfish, inward rectification decreases substantially during in vitro maturation of the oocytes by 1-methyladenine [Miyazaki et al, 1975] (Lansman, personal communication). Inward calcium and outward potassium currents elicited by depolarization also change during maturation [Miyazaki et al, 1975; Miyazaki and Hirai, 1979; Miyazaki, 1979]. The effect of these changes appears to be to increase the amplitude of the fertilization potential so that an effective electrical block to polyspermy [Jaffe, 1976] develops in the mature egg. Both the active transport and the intracellular concentration of ions are affected by the maturation process in various oocytes [Moreau et al, 1978; Lee and Steinhardt, 1981]. It is possible that changes in pH_i or internal Na^+ activity during maturation cause the emergence of electrical properties characteristic of the mature starfish oocyte.

IV. DISCUSSION

In this chapter, I have described two series of experiments concerning the effects of decreased pH_i on the membrane electrophysiology of excitable cells. In both crayfish muscle and starfish oocytes, the mode of action of H^+ appears to be a blockade of K^+ permeability. Similar actions of H^+ on both inward and delayed outward K^+ currents have been reported in vertebrate muscle fibers [Blatz, 1980] and in the squid giant axon [Wanke et al, 1979].

It is likely that the facilitation of action potential generation caused by decreased pH_i is more pronounced in cells which generate Ca-dependent action potentials than in those with Na^+ spikes, even if in both types of cells, H^+ acts to decrease delayed rectification; Calcium currents, in contrast to Na^+ currents, overlap substantially with voltage-dependent outward K^+ currents [Hagiwara et al, 1969]. The peak amplitude of Ca^+ action potentials is therefore much more sensitive to changes in the amplitude of outward currents than the peak amplitude of the Na^+-dependent action potential. It would obviously be interesting to see whether decreased pH_i affects the generation of calcium action potentials in other preparations, especially ones in which the physiological role of voltage-dependent calcium entry is well established.

A block of inwardly rectifying K^+ currents by internal H^+, such as I have described here for oocytes, has also recently been reported in frog skeletal muscle fibers [Blatz, 1980]. This adds to the list of similarities between the channel for inward rectification in these two preparations [see Hagiwara et al, 1978, for example]. As discussed above, however, such as effect of a change in the intracellular ionic composition takes on special significance in oocytes, because of the known changes in electrical properties of these cells during hormone-induced maturation and early development, and the likelihood that these changes are accompanied by changes in the activities of several intracellular ions. By combining the voltage-clamp technique with the use of ion-sensitive microelectrodes in the starfish oocytes, it should be possible to generate and test a series of specific hypotheses concerning the role of various intracellular ionic activity changes in causing alterations in the properties of ionic conductances that occur during maturation and following fertilization.

ACKNOWLEDGMENTS

The experiments described in this chapter were carried out in the laboratory of Professor S. Hagiwara. They were supported by grants from the NIH and the Muscular Dystrophy Association of America to S. Hagiwara, and by postdoctoral fellowships from the MDA and the Helen Hay Whitney Foundation to the author. Figure 2 is reprinted from Moody [1980] with the permission of the Journal of Physiology.

V. REFERENCES

Atwater I, Rojas E, Vergara JV: Calcium influxes and tension development in perfused single barnacle muscle fibers under membrane potential control. J Physiol 243:523, 1974.

Atwood HL: Organization and synaptic physiology of crustacean neuromuscular systems. Prog Neurobiol 7:291, 1976.

Blatz AL: Chemical modifiers and low internal pH block inward rectifier K channels. Fed Proc 39:2073, 1980.

Boron WF: Intracellular pH transients an giant barnacle muscle fibers. Am J Physiol 233:C61, 1977.

Boron WF, De Weer P: Intracellular pH transients in squid giant axons caused by CO_2, NH_3, and metabolic inhibitors. J Gen Physiol 67:91, 1976.

Byerly WL, Hagiwara S: Calcium currents in internally-perfused nerve cell bodies of Limnea stagnalis. J Physiol (in press).

Hagiwara S, Hayashi H, Takahashi K: Ca^{2+} and K^+ currents of the membrane of barnacle muscle fibre in relation to the Ca^{2+} spike. J Physiol 205:115, 1969.

Hagiwara S, Takahashi K: The anomalous rectification and cation selectivity of the membrane of a starfish egg cell. J Membr Biol 18:61, 1974.

Hagiwara S, Ozawa S, Sand O: Voltage clamp analysis of two inward current mechanisms in the egg cell membrane of a starfish. J Gen Physiol 65:617, 1975.

Hagiwara S, Miyazaki S-I, Moody W, Patlak J: Blocking effects of barium and hydrogen ions on the potassium current during anomalous rectification in the starfish egg. J Physiol 279:167, 1978.

Hagiwara S, Yoshii M: Effects of internal potassium and sodium on the anomalous rectification of the starfish egg as examined by internal perfusion. J Physiol 292:251, 1979.

Hagiwara S, Jaffe L: Electrical properties of egg cell membranes. Ann Rev Biophys Bioeng 8:385, 1979.

Hagiwara S, Byerly WL: Calcium channel. Ann Rev Neurosci 4:69, 1981.

Hille B: Charges and potentials at the nerve surface: Divalent ions and pH. J Gen Physiol 51:221, 1968.

Horn R, Patlak J: Single channel currents from excised parches of muscle membrane. Proc Natl Acad Sci USA 77:6930, 1980.

Jaffe L: Fast block to polyspermy in sea urchin eggs is electrically mediated. Nature 261:68, 1976.

Kennedy D, Takeda K: Reflex control of abdominal flexor muscles in crayfish. II. The tonic system. J Exp Biol 43:229, 1965.

Lee KS, Akaike N, Brown AM: The suction pipette method for internal perfusion and voltage clamp of small excitable cells. J Neurosci Meth 2:51, 1980.

Lee S, Steinhardt RA: pH changes associated with meiotic maturation in oocytes of Xenopus laevis. Dev Biol 85 (in press), 1981.

Miyazaki S-I, Ohmori H, Sasaki S: Potassium rectifications of the starfish oocyte membrane and their changes during oocyte maturation. J Physiol 246:55, 1975.

Miyazaki S-I, Hirai S: Fast polyspermy block and activation potential. Correlated changes during oocyte maturation of a starfish. Dev Biol 70:327, 1979.

Miyazaki S-I: Fast polyspermy block and activation potential. Electrophysiological bases for their changes during oocyte maturation of a starfish. Dev Biol 70:341, 1979.

Moody WJ: Gradual increase in the electrical excitability of crayfish slow muscle fibers produced by anoxia or uncouplers of oxidative phosphorylation. J Comp Physiol 125A:327, 1978.

Moody WJ: Appearance of calcium action potentials in crayfish slow muscle fibres under conditions of low intracellular pH. J Physiol 302:335, 1980.

Moody WJ: The ionic mechanism of intracellular pH regulation in crayfish neurones. J Physiol 316:293–308, 1981.

Moreau M, Guerrier P, Doree M: Hormonal control of meiosis reinitiation in starfish oocytes. Exp Cell Res 115:245, 1978.

Shapiro E, Castellucci VF, Kandel ER: Presynaptic inhibition in Aplysia involves a decrease in the Ca^{2+} current of the presynaptic neuron. Proc Natl Acad Sci USA 77:1185, 1980.

Sharp AP, Thomas RC: The effects of chloride substitution on intracellular pH in crab muscle. J Physiol 312:71, 1981.

Spitzer NC, Baccaglini P: Development of the action potential in embryonic amphibian neurons in vivo. Brain Res 107:610, 1976.

Thomas RC: Intracellular Na^+ activity and the Na^+ pump in snail neurones. J Physiol 220:55, 1972.

Thomas RC: The role of bicarbonate, chloride, and sodium ions in the regulation of intracellular pH in snail neurones. J Physiol 238:317, 1977.

Thomas RC: "Ion Sensitive Intracellular Microelectrodes." London: Academic Press, 1978.

Wanke E, Carbone E, Testa PL: K^+ conductance modified by a titratable group accessible to protons from the intracellular side of the squid axon membrane. Biophys J 26:319, 1979.

ns, pages 445-461
Comparison of pH and Calcium Dependence of Gap Junctional Conductance

D.C. Spray, A.L. Harris, and M.V.L. Bennett

Division of Cellular Neurobiology, Department of Neuroscience, Albert Einstein College of Medicine, Bronx, New York 10461, and Marine Biological Laboratory, Woods Hole, Massachusetts, 02543

I.	Introduction	445
II.	Techniques	448
	A. Measurements of Gap Junctional Conductance	448
	B. Intracellular pH Measurement	450
III.	Results	450
	A. pH_i Electrode Measurements	452
	B. Does the pH_i Change Affect Intracellular Ca^{2+}?	454
	C. Increases in H_i^+ and Ca_i^{2+} Uncouple Independently	455
IV.	Discussion	457
	A. Mechanism of Gating by Hydrogen Ions	457
	B. Conclusions	458
V.	References	458

I. INTRODUCTION

The gap junction is composed of channels that span the membranes of coupled cells to allow the direct transfer of ions without access to extracellular space [cf Bennett and Goodenough, 1978]. In thin section, the gap junction is recognizable as a generally linear apposition of two membranes that have a characteristic septilaminar appearance (Fig. 1A). The corresponding freeze fracture image of the gap junction is usually a discoid array of large (8–10 nm) intramembrane particles cleaving with the P-face in all nonarthropod phyla (Fig. 1B). A central depression is often imaginable in the particle centers; this central depression is believed to represent the lumen of the channel. From sixfold

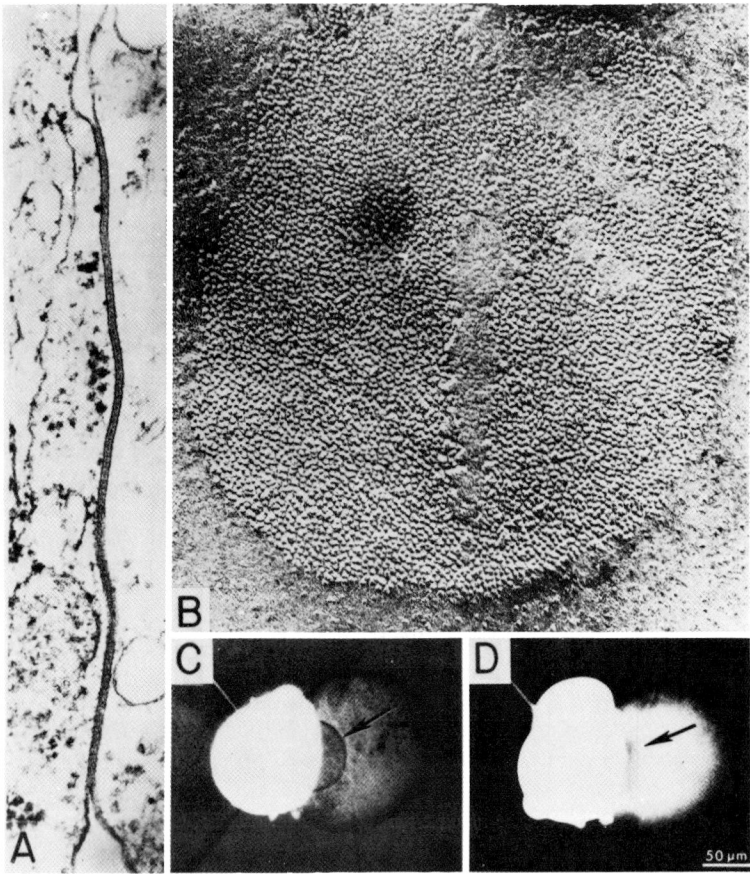

Fig. 1. Intercellular junctions in early amphibian embryos and intercellular movement of dye. A) Thin section of a large gap junction between axolotl blastomeres. The unit membranes in the apposed cells approach to within about 2 nm in the junctional region; a septilaminar contact is seen except where the membranes are sectioned obliquely. B) Freeze-fracture replica of a large junction. The intramembrane particles in the P face are mostly irregularly packed; pits are seen in the region of E face but are poorly resolved. A and B are at the same magnification, the region of apposition in A extending 1.2 μM. From Hanna et al [1980]. C, D) Passage of dye between coupled blastomeres. C) Mixed transmitted light and fluorescence micrograph of two coupled cells penetrated by four electrodes, one containing the fluorescent dye Lucifer yellow. A lens-shaped cavity occurs between them (arrow). D) Fluorescence micrograph showing spread of Lucifer yellow in the same pair from the injected cell on the left. Dye is excluded from the space between the cells, indicating that the dye had not moved by way of extracellular space. From Bennett et al [1978b].

symmetry evident in negatively stained preparations and x-ray crystallographic studies, it is commonly accepted that each hemichannel spanning a simple membrane is a hexamer of identical subunits [Makowski et al, 1977; Unwin and Zampighi, 1980]. Biochemical studies of isolated hepatic gap junctions have identified a 25–26 kilodalton protein that presumably represents the monomer [cf Hertzberg and Gilula, 1979; Finbow et al, 1980] and sequencing studies are underway [Nicholson et al, 1980].

An obvious function of gap junctions in electrically excitable tissues is to allow currents to flow from one cell to another through the intercellular channels. These electrotonic synapses can allow rapid transmission along a pathway and synchronization of activity within a neuronal population [cf Bennett, 1977; Spira et al, 1980]. The role played by gap junctions in inexcitable tissues is unclear, but presumably involves the exchange of informational ions and small molecules or metabolites between cells (cf Fig. 1C, D). Consequent physiological functions may include roles in tissue growth and differentiation in embryonic and transformed tissues [cf Bennett et al, 1981; Loewenstein, 1979] or the diffusional exchange of nutrients.

The process by which membrane channels are opened and closed is termed gating. Channel gating by voltage and by binding of various externally applied neurotransmitter substances or intracellular messengers, are phenomena central to contemporary neurobiology and cell biology. While most research on gating by chemical signals has centered on how the binding of externally applied substances can open or close channels, there is evidence that conductance of at least some channels is gated by the concentrations of certain ions at the cytoplasmic face of the channel [Takahashi and Yoshii, 1978; Marty, 1981; Spray et al, 1981a].

Until very recently the role that gap junctions might play in the control of physiological processes was limited by the belief that junctional conductance was unmodifiable except insofar as gap junctions could form and disappear. This view was changed by the demonstrations that intracelluar calcium injection [Rose and Loewenstein, 1975] and exposure to CO_2 [Turin and Warner, 1978] could rapidly and reversibly uncouple cells. Although the usefulness of coupling measurements to determine junctional conductance changes is limited (see Techniques), these studies prompted further exploration of the effects of various agents on gating of gap junction channels and therefore on intercellular communication. We present quantitative evidence that gap junctional conductance is gated by either cytoplasmic pH or calcium (Ca) as well as (in amphibia) by transjunctional voltage [Spray et al, 1981b]. Whereas small changes in pH produce profound changes in junctional conductance (g_j), similar changes in g_j by Ca ions require almost millimolar concentrations and thus seem to be nonphysiological. We also provide evidence that distinct regions of the gap junction molecule mediate pH and voltage dependence.

II. TECHNIQUES

Late cleavage stage embryos of the killifish (Fundulus) and the axolotl (Ambystoma) were dissociated into pairs or into single cells which were then reassociated as pairs. Cells were treated with up to 0.1% colchicine to inhibit mitosis in appropriate physiological saline (double strength Holtfreter solution for Fundulus; and either single strength Holtfreter or Nui-Twitty solution for Ambystoma, to which was added 5 mM HEPES buffer). Sodium salts of weak organic acids were substituted for NaCl in some experiments.

A. Measurements of Gap Junctional Conductance

Each cell of an isolated pair was impaled with two 5-10 M ohm KCl-filled microelectrodes to allow simultaneous current injection and potential measurement (Fig. 2A). With current pulses passed alternately in the two cells and potentials of both cells measured, the component resistances of the equivalent circuit can be unequivocally solved by means of the pi-tee transform [Bennett, 1966]. With current injected into cell 1,

$$r_{11} = \frac{V_1}{I_1} = \frac{r_1(r_2 + r_j)}{r_1 + r_2 + r_j}$$

and

$$r_{21} = \frac{V_2}{I_1} = \frac{r_2 r_j}{r_1 + r_2 + r_j}$$

Similar equations can be written for voltages resulting from current injected into cell 2, resulting in four equations that can be solved for the three unknown resistances. The coupling coefficient, k, which is widely used as an index of coupling (and by implication, junctional resistance), is V_2/V_1 with current applied in cell 1. Thus,

$$k = \frac{V_2}{V_1} = r_2/(r_2 + r_j)$$

Even in the absence of a change in g_j the coupling coefficient would decrease by any treatment that decreases nonjunctional resistance. [cf Spira et al, 1980]. The usefulness of the coupling coefficient as a measurement of g_j is therefore severely limited.

A second method by which junctional conductance can be measured with four microelectrodes in the two-cell system is the dual voltage clamp technique [Spray et al, 1979, 1981b] (Fig. 2B). Each cell is clamped to its resting potential by an independent voltage clamp circuit. Voltage steps are imposed in one cell, and junctional current (I_j) is measured as the current supplied by the other clamp to

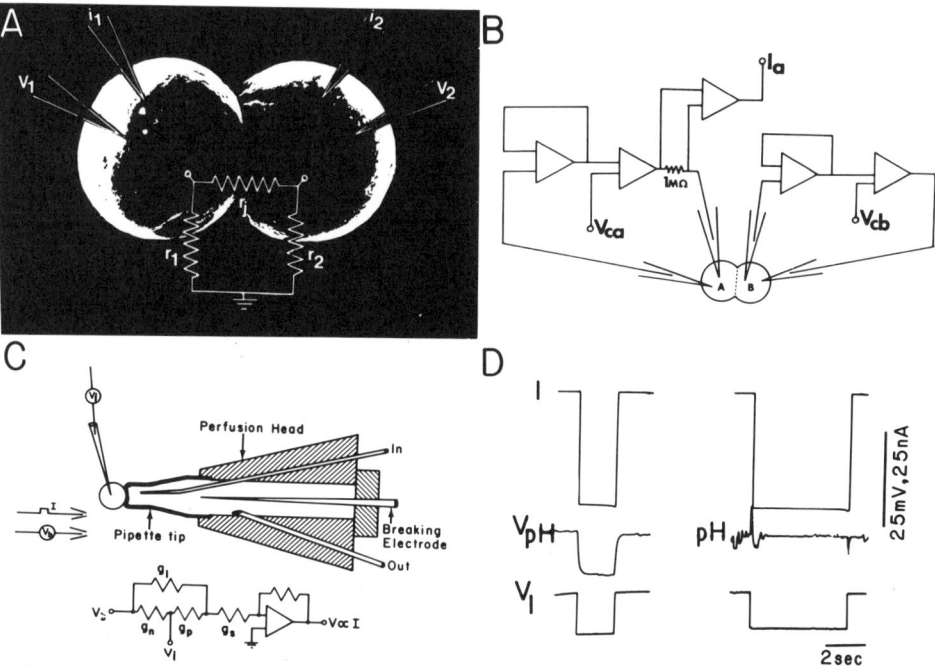

Fig. 2. Experimental paradigms for measurement of junctional conductance and intracellular pH. A) Scanning micrograph of a pair of teleost blastomeres with diagrammatic representation of microelectrodes and equivalent circuit for determining junctional conductance (micrograph by E.A. Morales). B) Schematic diagram of double voltage clamp electronics. Each cell of the coupled pair is placed in a separate voltage clamp circuit and held at its resting potential. Voltage steps are delivered to cell B. Any current flowing via the junctions from the pulsed cell into cell A is exactly matched by current of the opposite polarity supplied by the voltage clamp of cell A to hold its potential constant. The transjunctional current (I_a) divided by the transjunctional driving voltage gives a direct measure of junctional conductance (g_j) [from Spray et al, 1981b]. C) Apparatus and equivalent circuit for perfusion experiments. The perfusion head included four small glass tubes extending to within 50 μM of the pipette tip through which different solutions could be passed (one of which is shown, labeled In). One cell was sucked into the pipette and its membrane disrupted spontaneously or with the breaking electrode. To evaluate patch conductance, the intact cell was penetrated with a voltage microelectrode (V_i). Two calomel electrodes in the bath were used to measure voltage (V_b) and deliver current (I), which was measured in the outlet pathway (Out) with a virtual ground current probe. The ratio of the conductances of the patch to nonpatch membranes is given by $g_p/g_n = (V_b - V_i)/(V_i - V_s)$ where $V_s = I/g_s$ is the voltage drop across the series conductance of the perfusion pipette, g_s. The ratio g_p/g_n is independent of the leakage conductance, g_l, and is a valid measure of g_p when g_n is constant [from Spray et al, 1982]. D. A rigorous criterion was established for intracellular placement of the pH microelectrode. The pH (V_{pH}) and voltage (V_i) microelectrodes in the same cell were required to record the same steady-state voltage displacements in response to a long current pulse delivered through a third electrode in the same cell. This criterion is satisfied by the impalement on the left. Although the time constant of the high resistance pH microelectrode is much longer than that of the voltage electrode, the steady state voltages recorded are comparable. Thus, differential recording between V_{pH} and V_i (trace on the right labeled pH) yields a flat record between the transients, which are due to dissimilar time constants of the pH and voltage electrodes.

hold that cell's voltage constant. That current, divided by the voltage step, gives the junctional conductance.

The third measure of junctional conductance used in these studies involves a refinement of the internal dialysis technique, which has been used on single cells in a number of laboratories. Our apparatus was designed by Dr. John Lisman and Jeff Stern at Brandeis, and differed from previous designs in that the inflow tubing could be positioned to within 30–50 μ of the pipette tip (Fig. 2C). This placement allowed rapid change of perfusion solutions at the pipette tip. One cell of a coupled pair of Fundulus blastomeres was sucked into the 25 μ pipette orifice, and the heat-polished tip sealed onto the second cell. The first cell was then mechanically disrupted and its cytoplasm washed away. This procedure resulted in the cytoplasmic aspect of the junctional membrane being directly exposed to solutions within the pipette. The equivalent circuit for this arrangement is drawn in Figure 2C. The conductance of interest, that of the patch of membrane within the pipette tip, could be measured as a ratio of patch conductance (g_p) to that of the membranes outside the patch (g_n) without regard to the leakage conductance. (Spray et al, 1982).

Internal perfusion solutions were strongly buffered with regard to Ca^{2+} and H^+ ions containing 5 mM HEPES, 10 mM EGTA or nitrilotriacetic acid, 1 mM free Mg, and 100 mM KCl. Calcium and hydrogen ion concentrations cited were those concentrations measured in bulk solution with glass electrodes.

B. Intracellular pH Measurement

In many experiments a Thomas-type pH microelectrode was inserted into one cell of the coupled pair of blastomeres in order to record intracellular pH (cf Thomas, this volume). To reduce response times the tips of the outer electrodes were often either beveled or broken to about 2 μ prior to insertion and sealing of the pH-sensitive glass inside. Rigorous criteria for intracellular placement of the reference and pH electrodes were established: superimposable steady-state voltage responses to 10-second current pulses in the same cell (Fig. 2D). The pH signal was either recorded differentially with respect to the reference electrode or by itself in voltage clamp experiments in which the pH electrode was placed within the cell whose voltage was never changed.

III. RESULTS

Exposure of coupled cells to saline equilibrated with 100% CO_2 results in a marked decrease in the spread of current from one cell to another. This phenomenon has been observed in cells of pancreatic acini [Iwatsuki and Petersen, 1980], insect salivary gland [Rose and Rick, 1978], mammalian heart [Weingart and Reber, 1979; de Hemptinne et al, this volume; DeMello, 1980], crayfish nerve cord and gastropod ganglia [Giaume et al, 1980], and embryos of amphibia, fish, and squid [Turin and Warner, 1978, 1980; Bennett et al, 1978a; Spray et al, 1980, 1981a and unpublished]. With pairs of cells the changes in the resistances of junctional and nonjunctional membranes that underlie CO_2-produced uncoupling are readily analyzed. In the experiment of Figure 3A, the cells are initially coupled with a coupling coefficient of about 0.9 (Fig. 3A, left panel). Within a minute after exposure to saline equilibrated with CO_2, currents of similar magnitudes produce larger voltages in the injected cell and less spread to the other cell (coupling coefficient less than 0.1, Fig. 3A, right panel). Application of the pi-tee transform to the measured input and transfer resistances in the two directions allows calculation of the junctional and nonjunctional conductances during uncoupling. Exposure to propionate (Fig. 3B) rapidly and reversibly decreases g_j to less than 0.5% of its initial value. Similar profound reductions in g_j are seen when the cells are exposed to solutions in which chloride is replaced with acetate or lactate. Although reduction in g_j is seen even when these weak acids are applied at pH 7.8 [Spray et al, 1980], they are progressively more potent when the pH of the solution is reduced toward 6.8. This increased

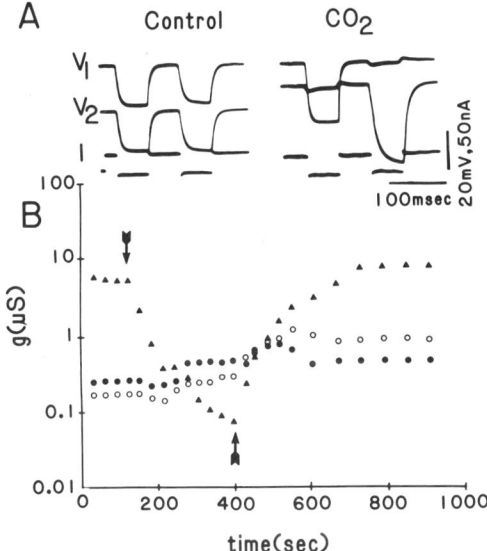

Fig. 3. Decrease in coupling between Fundulus blastomeres in response to CO_2 and propionate. A) Uncoupling of blastomers by CO_2. The voltages in a pair of blastomeres (V_i and V_2) were recorded while current pulses were passed first in cell 1 and then in cell 2. Following CO_2 exposure, the input resistances of both cells increased and the transfer resistance decreased. B) Propionate replacement for chloride also uncouples Fundulus blastomeres. The time course of changes in junctional (▲) and nonjunctional (○, ●) conductances of the two cells were calculated from the pi-tee transform. During propionate exposure (between arrows), g_j decreased to less than 0.1% of its original value and then recovered during rinsing. Conductances of nonjunctional membranes of the two cells changed transiently at each change of solution and showed an overall drift upwards.

effectiveness is consistent with the notion that these weak acids exert their effect by crossing the membrane in the undissociated form and then dissociating to liberate protons intracellularly [cf Sharp and Thomas, 1981]. Although chloride substitutes were shown to uncouple crayfish septate axon 10 years ago [Asada and Bennett, 1971], it only recently became apparent that this effect was due to decreased intracellular pH as shown directly by Giaume et al [1980]. Strong acids or impermeant buffers do not affect g_j even after 10 minutes exposure at pH 6 [cf Spray et al, 1981a].

A. pH_i Electrode Measurements

In order to test the hypothesis that CO_2 and other weak acids exert their effects through cytoplasmic acidification, we penetrated one cell of the pair with a Thomas-type pH microelectrode and repeated the experiments. In the experiments of Figure 4, pairs of short current pulses are applied alternately to the cells at 2-second intervals. Brief application of saline containing CO_2 acidifies the cytoplasm (H^+ activity increased) and uncouples the cells. Cytoplasmic pH and intercellular coupling return toward normal when the cells are rinsed with normal saline. Calculated conductances of junctional and nonjunctional membranes corresponding to the time course of the experiment appear in the lower portion of the figure. As pH falls to about 6.8, g_j decreases to less than 0.1% control, and recoveries of g_j and pH_i are parallel. Similar records have been obtained in response to the other weak acids mentioned above.

Depolarization of the two cells often accompanies treatment with weak acids. To control for an effect of depolarization on g_j, we voltage clamped both cells to a common holding potential while measuring pH_i of one cell and found that CO_2 still reversibly depressed g_j (Fig. 5, inset).

A summary of the relation found between pH_i and g_j is shown in Figure 5. Solid black triangles represent data from an experiment partly shown in Figure 4; those triangles with apices downward were obtained while pH_i decreased, upward oriented triangles are for increasing pH_i during recovery. Data from experiments with lactate superimpose on the data of Figure 5, as do voltage clamp measurements with CO_2 and lactate.

The observed relation between g_j and pH_i is well fit by a form of the Hill equation.

$$g_j = K^n/(K^n + [H]^n)$$

where K is about 50 nM (pK ~ 7.3) and n, the Hill coefficient, is between 4 and 5.

The Hill coefficient represents the minimal number of highly cooperative titratable sites that would result in the calculated curve; however, there may be

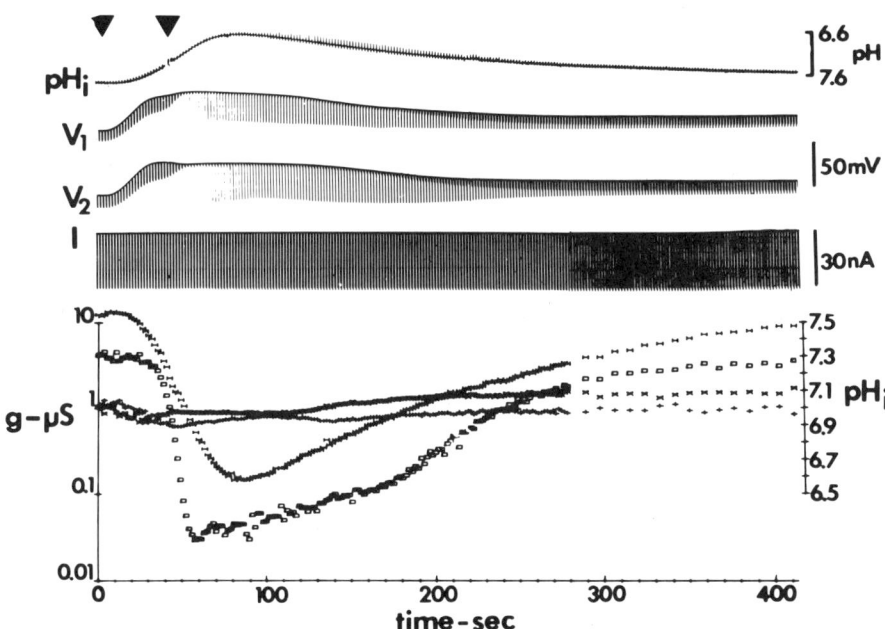

Fig. 4. Effect of CO_2 exposure on internal pH and electrotonic coupling between two axolotl blastomeres. Current pulses (I) are alternately passed into the two cells whose potentials are measured (V_1 and V_2). Brief application of saline equilibrated with 100% CO_2 (between arrowheads) causes cytoplasmic pH to decrease (uppermost trace, H^+ activity increasing upwards) and decreases the spread of current to the second cell. Washing the cells with CO_2-free saline restores pH_i and intercellular coupling. Conductances of junctional (square symbols) and nonjunctional (x, +) membranes calculated from these data are shown below. Junctional conductance is initially about 5 μS and begins to decrease as pH_i (H symbols) reaches about 7.4; g_j reaches a minimum value of about 0.02 μS at the minimum pH_i and then slowly recovers [from Spray et al, 1981a].

more sites if they are less cooperative, and these data are not inconsistent with the notion of one binding site per channel subunit. Normal pH_i of late cleavage stage amphibian and fish embryos is between 7.5 and 7.8 [Spray et al, 1981a; Turin and Warner, 1980; Webb and Nuccitelli, this volume], just above the steep region over which g_j is strongly affected by pH_i. There are data to suggest that gap junctional conductance may vary over a similar pH_i range in vertebrate heart, although the steepness of the relation is unknown [Weingart and Reber, 1979; Weingart et al, 1981]. Normal pH_i is about 7.2 in cardiac tissue, and CO_2 decreases pH_i and increases internal longitudinal resistance whereas alkanization with NH_4 decreases longitudinal resistance.

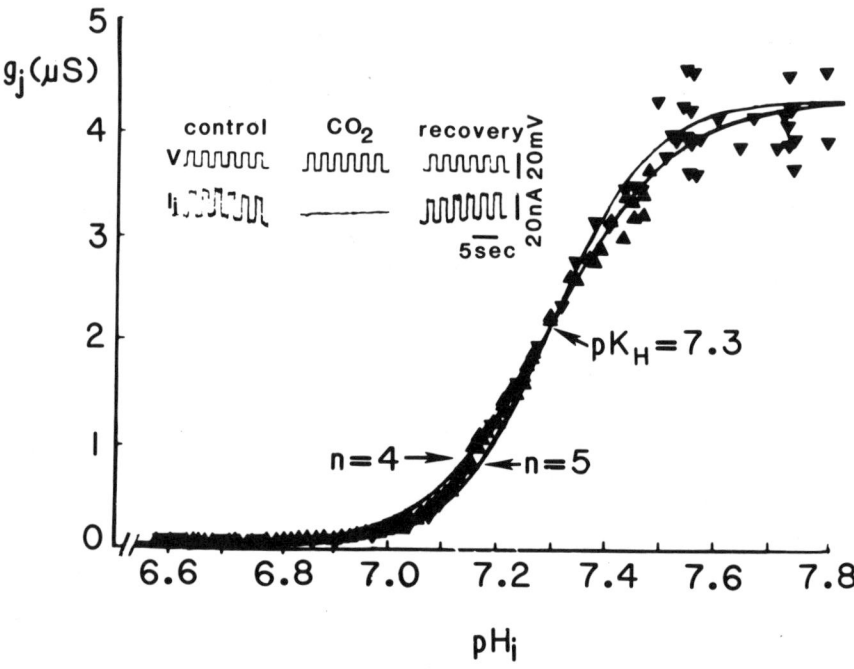

Fig. 5. Sensitivity of junctional conductance to intracellular H ions. The conductance data of Figure 4 are plotted as a function of pH_i in the main portion of this figure. Triangles with apices down represent initial values and values during cytoplasmic acidification. Triangles with apices up are values during recovery. The points fall along Hill curves with $pK_H = 7.3$ and n between 4 and 5. There is no evidence of hysteresis between falling and rising pH_i. The data are consistent with H ions acting directly on a channel macromolecule. [from Spray et al, 1981a.] Inset: The effect of CO_2 on g_j is not due to depolarization since g_j is reduced during CO_2 exposure when the two cells are clamped to a common holding potential and one cell receives test pulses (upper traces). The current on the lower traces that clamps the other cell at the holding potential is a measure of g_j.

B. Does the pH_i Change Affect Intracellular Ca^{2+}?

The rapidity and reversibility of the dependence of g_j on pH_i independent of the weak acid used and the lack of hysteresis in this relation with falling and recovery of pH_i suggests that the effect is direct rather than through a cytoplasmic intermediate. Since calcium ions have been shown to uncouple some cells [see below] and since Ca/H exchange is a common mode of cytoplasmic buffering [cf Meech and Thomas, 1977], we examined whether the brief CO_2 exposures used in our experiments elevated intracellular ionized calcium. In experiments with Dr. Joel Brown, the photoprotein aequorin was injected into one cell of a pair and luminescence was monitored with a dry ice-cooled photomultiplier tube. Despite the high sensitivity of the system, which could detect aequorin luminescence of the resting cell above background, brief CO_2 exposure which reduced g_j by more than 99% led either to no detectable aequorin glow or a slight glow too late to account for the change in g_j [Bennett et al, 1978a]. A similar conclusion was reached using Ca-sensitive microelectrodes to measure Ca activity in amphibian embryos [Rink et al, 1980], and in cardiac Purkinje fibers CO_2-induced acidification actually reduces cytoplasmic Ca activity [Hess and Weingart, 1980].

C. Increases in H_i^+ and Ca_i^{2+} Uncouple Independently

Cells of several tissues have been shown to uncouple when the intracellular concentration of ionized calcium is sufficiently elevated. Calcium injection through intracellular microelectrodes uncouples canine cardiac Purkinje fibers [DeMello, 1975], molluscan neurons [Baux et al, 1978], insect salivary gland cells [cf Rose and Loewenstein, 1975], and cleavage stage fish embryos [Spray et al, 1982]. Treatment of coupled cells with the calcium ionophore A23187 also uncouples [Rose and Loewenstein, 1975], although arthropod tissues are much more susceptible to this treatment than are those of vertebrates [Gilula and Epstein, 1976; Flagg-Newton and Loewenstein, 1979]. Other treatments expected to elevate intracellular calcium also uncouple (eg ouabain and dihydroouabain: Weingart [1977], Dahl and Isenberg [1980]; dinitrophenol: Rose and Loewenstein [1975]; anoxia: Wojtczak [1979]; injury: Asada and Bennett [1971].

The question has largely remained whether Ca could uncouple independent of a pH decrease. We developed a technique whereby one cytoplasmic face of the junctional membrane was perfused with known concentrations of H^+ and Ca ions in order to quantify the relation between g_j and Ca_i [Stern et al, 1980; Spray et al, 1982; Fig. 2C]. With Ca held constant at pCa 7, pH changes from 7.8 to 6.8 still changed patch conductance. When these data were normalized to maximum and minimum values at the pH extremes, the values of patch conductance at intermediate pH were near that value expected from experiments on intact cells (Fig. 6, inset).

With pH maintained at 7.8, free calcium concentrations as high as 0.10 mM (pCa 4.0) did not affect patch conductance (Fig. 6). Above this level of free calcium, patch conductance decreased substantially to about 25% of control at 1.0 mM, the highest concentration used in these experiments. Calculations fit to the data showed a pK of about 3.3 (K_{Ca} about 0.5 mM and Hill coefficient (n) of about 2 to 3). Both the apparent dissociation constant and the slope are very similar to those found for low affinity calcium binding to isolated guinea pig and ox ventricular gap junctions using spectroscopic methods [Nishiye et al, 1980].

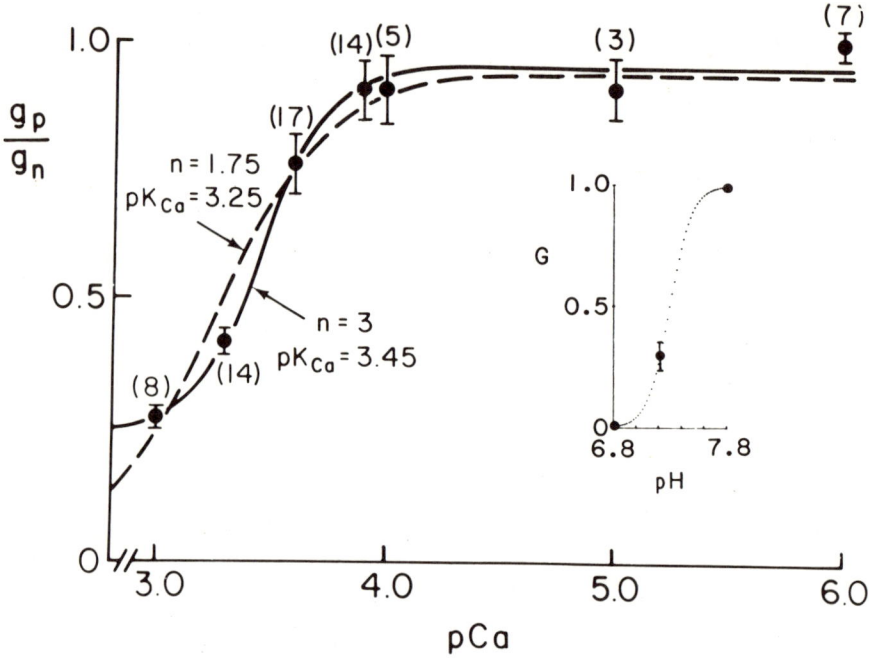

Fig. 6. Sensitivity of junctional conductance to intracellular Ca ions. From cell pairs in which one cytoplasmic aspect of the junctions was perfused with a solution buffered to pH 7.8 and various pCa levels. The ordinate g_p/g_n is a normalized measure of junctional conductance; standard errors and number of experiments are indicated. The curves are Hill plots fit by eye, assuming that the minimum value of g_p/g_n was either zero or just below that at pCa 3.0 (1 mM). Junctional conductance is very insensitive to Ca as compared to H ions. The inset shows experiments in which pCa was held at 7.0 whereas pH was changed between 7.8 and 6.8. The ordinate G represents a normalized value of conductance. The smooth curve is the Hill plot for $pK_H = 7.3$ and $n = 4.5$ obtained from intact cell pairs (see Fig. 5). The junctional conductance at pH 7.2 (error bar for six trials) falls close to the curve, indicating that the perfused preparation has the same pH sensitivity. [from Spray et al, 1982.]

Since pH_i affects junctional conductance about twice as steeply as does calcium near their pKs, two H ions or one Ca ion could be acting on a common site, H of course having a much higher affinity. Measurement of patch conductance using calcium concentrations between 0.1 and 1 mM over a range of pH values may allow us to determine whether or not the ions compete for common binding sites.

IV. DISCUSSION

A. Mechanism of Gating by Hydrogen Ions

Although there is as yet little direct evidence, we hypothesize that gap junction channels fluctuate between open and closed states, as do other channels in biological membranes, this distribution being affected by ligand binding to the channel molecule [cf Neher and Stevens, 1977]. Thus the distribution of channels in the two states would shift toward closed with decreasing pH_i. The mechanism by which a few hydrogen ions act to close the large gap junction channel presumably involves a conformational change in the channel protein. Junctional conductance in amphibian blastomeres is also gated by transjunctional voltage, the interaction of intrinsic protein dipoles with the field perhaps mediating a conformational change [Spray et al, 1981b; Harris et al, 1981]. Whether or not these two gating mechanisms involve the same portion of the gap junction molecule would have implications for the substructure of the gap junction channel. Physiological evidence suggests that the two gating mechanisms are distinct: The voltage dependence of the conductance that remains as pH_i is reduced is unchanged. This behavior would be observed if low pH_i simply closed a gate in series with the voltage-sensitive gate. Pharmacological evidence is even more dramatic: Weak concentrations of glutaraldehyde (5–10 μM), which reduce g_j by 40–50%, completely abolish the ordinary pH dependence. Voltage dependence is unaffected. Formaldehyde in higher concentration (10 mM) reduces g_j by a similar amount; the remaining conductance has a reduced voltage dependence with little change in pH dependence [Spray et al, 1981c, d].

A number of anesthetic agents reduce junctional conductance. In blastomeres tricaine at 10^{-4}–10^{-5} M and benzocaine at 2–5×10^{-3} M reduce junctional conductance, whereas procaine, tetracaine, and lidocaine (at 10^{-2} M) do not [Spray et al, 1981d]. Octanol reduces junctional conductance by half at concentrations of about 5×10^{-5} M in blastomeres [Spray et al, 1981d] and 10^{-3} M in crayfish septate axon [Johnston et al, 1980]. Heptanol and lower alcohols are much less effective in crayfish. No local anesthetic used has shown obvious effects on voltage or pH dependence of junctional conductance; these compounds may therefore act either on separate binding sites or by a mechanism different from H and Ca ions or transjunctional voltage.

Whether a morphological basis for gating of gap junctional conductance will be found at the electron microscopic level is still conjectural. There are a number of reports of fine structural changes with conductance state of the junction [cf Raviola et al, 1981; Page and Shibata, 1981]. Highly regular structure may or may not be associated with low conductance, and even though regularization of structure may follow treatments that reduce junctional conductance, it is apparently a secondary and slow process [cf Hanna et al, 1978, 1981].

B. Conclusions

The data summarized in this chapter indicate that H and Ca can independently decrease junctional conductance, although H is about 10,000 times as effective as Ca (apparent pKs 7.3 and 3.3, respectively). The profoundly different apparent affinities of the gap junction channel for H and Ca ions imply that these ions may play quite different roles in the regulation of intracellular communication.

Since the normal intracellular pH for cleavage stage Fundulus and amphibian embryos is about 7.5–7.7 [Turin and Warner, 1980; Spray et al, 1981a; Webb and Nuccitelli, this volume] and normal pH_i for other cell types can be even lower [cf Hess and Weingart, 1980; Aicken and Thomas, 1977; other authors, this volume], it seems reasonable that small deviations of pH_i from its normal level can sensitively modulate junctional conductance. Thus pH_i changes accompanying ischemia [cf Wojtczak, 1979; Jacobus et al, this volume], glucose stimulation of insulin release by pancreatic islet cells [cf Pace et al, this volume], and the progressive alkalinization during early embryonic development [cf Shen, and Webb and Nuccitelli, this volume] may change the properties of the tissues by changing the extent to which the cells communicate.

The high level of intracellular calcium ions required to affect junctional conductance presumably excludes a role for Ca ions in the normal physiological functioning of gap junctions. However, cytoplasmic calcium concentration may approach millimolar levels under pathological conditions such as cell death or membrane disruption. The normally high levels of extracellular calcium may thus provide the necessary stimulus for a severely diseased or damaged cell to isolate itself from its coupled neighbors and thereby preserve tissue integrity.

An implication of series arrangement of voltage and pH gates on gap junction channels is that these two gating mechanisms may multiply their effects on conductance of the channels. If voltage-dependent junctional conductance serves to isolate cell populations during development, the sensitivity of this process, and thus the effect, may be enhanced by even a small decrease in pH_i. Now that dyes are available that partition into cells depending on the gradient between external and internal pH [cf Heiple and Taylor, this volume], pH changes as boundaries form can be sought during later development and these pH_i changes may be correlated with embryonic fate.

ACKNOWLEDGMENTS

Research supported by NIH grants HD-02428, NS-12627, and NS-07512. Much of the work was done while DCS was a McKnight Scholar in Neuroscience and ALH was the recipient of an NIH Research Service Award.

V. REFERENCES

Aicken CF, Thomas RC: An investigation of the ionic mechanism of intracellular pH regulation in mouse soleus muscle fibres. J Physiol 273:295–316, 1977.

Asada Y, Bennett MVL: Experimental alteration of coupling resistance at an electrotonic synapse. J Cell Biol 49:159–172, 1971.
Baux G, Simonneau M, Tauc L, Segundo J: Uncoupling of electrotonic synapses by calcium. Proc Natl Acad Sci USA 75:4577–4581, 1978.
Bennett MVL: Physiology of electrotonic junctions. Ann NY Acad Sci 137:509–539, 1966.
Bennett MVL: Electrical transmission: A functional analysis and comparison to chemical transmission. In: Kandel E (ed): "Cellular Biology of Neurons," Vol 1, Section 1, "Handbook of Physiology of the Nervous System," Baltimore: Williams and Wilkins, 1977, pp 357–416.
Bennett MVL, Brown JE, Harris AL, Spray DC: Electrotonic junctions between Fundulus blastomeres: Reversible block by low intracellular pH. Biol Bull 155:442, 1978a.
Bennett MVL, Spira ME, Spray DC: Permeability of gap junctions between embryonic cells of Fundulus: A reevaluation. Dev Biol 65:114–128, 1978b.
Bennett, MVL, Goodenough DA: Gap junctions, electrotonic coupling, and intercellular communication. Neurosci Res Prog Bull 16:373–486, 1978.
Bennett MVL, Spray DC, Harris AL: Electrical coupling in development. Am Zool 21:413–427, 1981.
Dahl G, Isenberg G: Decoupling of heart muscle cells: Correlation with increased cytoplasmic calcium activity and with changes of nexus ultrastructure. J Membr Biol 53:63–75, 1980.
DeMello WC: Effect of intracellular injection of calcium and strontium on cell communication in heart. J Physiol 250:231–245, 1975.
DeMello WC: Influence of intracellular injection of H^+ on the electrical coupling in cardiac Purkinje fibres. Cell Biol Int Rep 4:51–58, 1980.
Finbow M, Yancey SB, Johnson R, Revel JP: Independent lines of evidence suggesting a major gap junctional protein with a molecular weight of 26,000. Proc Natl Acad Sci USA 77:970–974, 1980.
Flagg-Newton J, Loewenstein WR: Experimental depression of junctional membrane permeability in mammalian cell culture. A study with tracer molecules in the 300 to 800 dalton range. J Membr Biol 50:65–100, 1979.
Giaume C, Spira ME, Korn H: Uncoupling of invertebrate electrotonic synapses by carbon dioxide. Neurosci Lett 17:197–199, 1980.
Gilula NB, Epstein M: Cell-to-cell communication, gap junctions and calcium. Symp Soc Exp Biol 30:257–272, 1976.
Hanna RB, Spray DC, Model PG, Harris AL, Bennett MVL: Ultrastructure and physiology of gap junctions of an amphibian embryo, effects of CO_2. Biol Bull 155:442, 1978.
Hanna RB, Model PG, Spray DC, Harris AL, Bennett MVL: Gap junctions in an early amphibian embryo. Am J Anat 158:111–114, 1980.
Hanna RB, Reese TS, Ornberg RL, Spray DC, Bennett MVL: Fresh-frozen gap junctions: Structural detail in coupled and uncoupled states. J Cell Biol 91:125a, 1981.
Harris AL, Spray DC, Bennett MVL: Kinetic properties of a voltage dependent junctional conductance. J Gen Physiol 77:95–117, 1981.
Hertzberg EL, Gilula NB: Isolation and characterization of gap junctions from rat liver. J Biol Chem 254:2138–2147, 1979.
Hess R, Weingart R: Intracellular free calcium modified by pH_i in sheep cardiac Purkinje fibres. J Physiol 307:60P–61P, 1980.
Iwatsuki N, Petersen OH: Pancreatic acinar cells: The effect of carbon dioxide, ammonium chloride, and acetylcholine on intercellular communication. J Physiol 291:317–326, 1979.
Johnston MF, Simon SA, Ramon F: Interaction of anesthetics with electrical synpases. Nature 286:498–500, 1980.
Loewenstein WR: Junctional intercellular communication and the control of growth. Biochim Biophys Acta 560:1–65, 1979.

Makowski L, Caspar DLD, Phillips WC, Goodenough DA: Gap junction structure. II. Analysis of the x-ray diffraction data. J Cell Biol 74:629–645, 1977.

Marty A: Ca-dependent K channels with large unitary conductance in chromaffin cell membranes. Nature 291:497–500, 1981.

Meech RW, Thomas RC: The effect of calcium injection on the intracellular sodium and pH of snail neurones. J Physiol 265:867–879, 1977.

Neher E, Stevens CF: Conductance fluctuations and ionic pores in membranes. Ann Rev Biophys Bioeng 6:345–372, 1977.

Nicolson BJ, Hunkapiller MW, Hood LE, Revel JP: Partial sequencing of the gap junctional protein from rat lens and liver J Cell Biol 87:200a, 1980.

Nishiye H, Ishida A, Machima H: Ca binding of isolated nexus membranes related to intercellular uncoupling. Jpn J Physiol 30:131–136, 1980.

Page E, Shibata Y: Permeable junctions between cardic cells. Ann Rev Physiol 43:431–441, 1981.

Raviola E, Goodenough DG, Raviola G: Structure of rapidly frozen gap junctions. J Cell Biol 87:273–279, 1981.

Rink TJ, Tsien RY, Warner AE: Free calcium in Xenopus embryos measured with ion-selective microelectrodes. Nature 283:658–660, 1980.

Rose B, Loewenstein WR: Permeability of cell junction depends on local cytoplasmic calcium activity. Nature 254:250–252, 1975.

Rose B, Rick R: Intracellular pH, intracellular free Ca, and junctional cell–cell coupling. J Membr Biol 44:377–415, 1978.

Sharp A, Thomas R: The effects of chloride substitution on intracellular pH in crab muscle. J Physiol 312:71–80, 1981.

Spira ME, Spray DC, Bennett MVL: Synaptic organization of expansion motoneurons of Navanax inermis. Brain Res 195:241–269, 1980.

Spray DC, Harris AL, Bennett MVL: Voltage dependence of junctional conductance in early amphibian embryos. Science 204:432–434, 1979.

Spray DC, Bennett MVL, Harris AL: Conductance of gap junctions is highly sensitive to cytoplasmic pH. Soc Neurosci Abstr 6:96, 1980.

Spray DC, Harris AL, Bennett MVL: Gap junctional conductance is a simple and sensitive function of intracellular pH. Science 211:712–715, 1981a.

Spray DC, Harris AL, Bennett MVL: Equilibrium properties of a voltage dependent junctional conductance. J Gen Physiol 77:77–93, 1981b.

Spray DC, Harris AL, Bennett MVL: Glutaraldehyde differentially affects gap junctional conductance and its pH and voltage dependence. Biophys J 33:108a, 1981c.

Spray DC, Harris AL, White RL, Bennett MVL: Pharmacologically distinct sites mediate voltage and pH dependence of gap junctional conductance. VII Int Biophys Congr Abstr p 155, 1981d.

Spray DC, Stern JH, Harris AL, Bennett MVL: Gap junctional conductance: Comparison of sensitivities to H and Ca ions. Proc Natl Acad Sci USA (in press), 1982.

Stern JH, Spray DC, Harris AL, Bennett MVL: Gap junctions: Quantitative comparison of reduction in conductance by H and by Ca ions in a perfused preparation. Biol Bull 159:493, 1980.

Takahashi K, Yoshii M: Effects of internal free calcium upon the sodium and calcium channels in the tunicate egg analyzed by the internal perfusion technique. J Physiol 279:519–549, 1978.

Turin L, Warner AE: Carbon dioxide reversibly abolishes ionic communication between cells of early amphibian embryo. Nature 270:56–57, 1978.

Turin L, Warner AE: Intracellular pH in early Xenopus embryos: Its effect on current flow between blastomeres. J Physiol 300:489–505, 1980.

Unwin PNT, Zampighi G: Structure of the junction between communicating cells. Nature 283:545–549, 1980.

Weingart R, Hess P, Reber WR: Influence of intracellular pH on cell-to-cell coupling in sheep Purkinje fibers. In Hoffman B, Lieberman M, Paes de Carvalho A (eds): "International Symposium on Normal and Abnormal Conduction in the Heart." Amsterdam: Elsevier (in press), 1981.

Weingart R, Reber WR: Influence of internal pH on r_i of Purkinje fibres from mammalian heart. Experientia 35:929, 1979.

Wojtczak J: Contractures and increase in internal longitudinal resistance of cow ventricular muscle induced by hypoxia. Circ Res 44:88–95, 1979.

Stimulus Response Coupling in Human Platelets: Thrombin-Induced Changes in pH_i

Elizabeth R. Simons, David B. Schwartz, and Nancy E. Norman

Department of Biochemistry, Boston University School of Medicine, Boston, Massachusetts 02118

I.	Introduction	463
II.	Materials	464
	A. Buffers	465
	B. Platelets	465
III.	Fluorometric Methods	465
	A. 9-Aminoacridine	466
	B. 6-Carboxyfluorescein Diacetate	471
	C. Dimethyl 6-Carboxyfluorescein Diacetate	474
	D. General Comment	476
IV.	Results	477
V.	Discussion	478
VI.	References	480

I. INTRODUCTION

Like many other secretory cells, the response of human platelets to a specific stimulus (eg, thrombin) can be represented by a series of steps. These can be loosely defined as, sequentially:

1) Binding of stimulus to its specific membrane receptor.
2) Membrane response
   a) altered cation permeability
   b) activation of certain membrane-bound enzymes
3) Cytoplasmic coupling factors
   a) altered internal cation concentrations
   b) altered internal cyclic nucleotide levels
4) Cytoplasmic response
   a) activation of metabolic cycles
   b) activation of contractile system
   c) activation of organellar systems

The membrane response (2a) of stimulated platelets [Bramhall et al, 1976; Horne and Simons, 1978; Larsen et al, 1979], like that of activated granulocytes [Korchak and Weissman, 1978, 1980; Whitin et al, 1980, 1981; Cohen et al, 1981], of target cells for peptide hormones [Zieler and Rogus, 1981], of eggs undergoing activation or fertilization [Steinhardt and Mazia, 1973; Shen and Steinhardt, 1978; Webb and Nuccitelli, this volume; Winkler and Grainger, 1978; Finkel and Wolf, 1980], and of amoebae stimulated with chemotactic factor [Heiple and Taylor, 1981] is expressed by an immediate change in the transmembrane potential. In those cells for which concurrent measurements of pH_i have been reported, a concomitant change in pH_i accompanies the depolarization or hyperpolarization (usually alkalinization and acidification, respectively), as does a change in $(Na^+)_i$ and in $(Ca^{++})_i$. Subsequent events (eg, secretion, acceleration of DNA or RNA synthesis, germinal vessel breakdown) depend upon the type of cell studied. These consequences of stimulation have, for several types of cells, been discussed by others in this volume.

Our laboratory has reported [Horne and Simons, 1978, 1979; Larsen et al, 1979] that a rapid stimulus dose-dependent depolarization accompanies thrombin stimulation of washed human platelets. These observations have now been extended [Schwartz et al, 1980; Horne et al, 1981] to show that, like secretory cells whose stimulus response sequence is described above, platelets exhibit a stimulus-induced increase in pH_i that is dependent upon the thrombin dose. The increase has been followed by two different techniques which utilize the fluorescence of 9-aminoacridine's distribution between platelets and suspending medium, on the one hand, and that of internalized 6-carboxyfluorescein, on the other hand, as continuous monitors of internal pH. The dose dependence of thrombin-stimulated platelet membrane depolarization, internal alkalinization, and serotonin secretion is identical. The methods we describe here allow continuous monitoring of platelet response to thrombin stimulation and corroborate the previously postulated [Friedman and Detwiler, 1975] stimulus response coupling in the interaction of platelets with thrombin.

II. MATERIALS

Bovine serum albumin (BSA), adenosine diphosphate (ADP), and 9-aminoacridine (9-AA) were purchased from Sigma Chemical Company, St Louis. 6-Carboxyfluorescein diacetate (6-CF diA) and dimethyl-6-carboxyfluorescein diacetate (diMe 6-CF diA) were obtained from Molecular Probes, Inc, Plano, Texas. Silicone oil was purchased from Contour Chemical Company, Woburn, Massachusetts. Tetraphenyl phosphonium bromide (Ph_4P^+Br) was purchased from Alpha Division, Ventron, Danvers, Massachusetts. (^3H)-Ph_4P^+Br was the kind gift of Dr. W.R. VanPelt of Hoffman-LaRoche, Inc. 3,3'-Dipropylthiodicarbocyanine (diS-C_3-(5)) was the kind gift of Dr. A. Waggoner, Amherst,

Massachusetts, and nigericin the kind gift of Dr. Nils Bang of Eli Lilly Inc. All other chemicals were of reagent grade, and were purchased from Fisher Scientific Co. Bovine α-thrombin was prepared by Dr. N.E. Larsen from Parke-Davis topical thrombin by the procedure of Lundblad et al [1975].

A. Buffers

Buffers included 0.01 M phosphate (KH_2PO_4/K_2HPO_4), pH 5.75–8.70; modified Tyrode's solution, pH 7.35: (0.14 M NaCl, 11.9 mM $NaHCO_3$, 0.1% D-glucose, 2.7 mM KCl, 1 mM $MgCl_2 \cdot 6H_2O$, 0% BSA); or HEPES (N-2-2-hydroxyethyl-1-piperazine-N'-2-ethane sulfonic acid), pH 7.0 and 7.35: (3.3 mM NaH_2PO_4, 2.7 mM KCl, 137.0 mM NaCl, 5.5 mM D-glucose, 3.8 mM HEPES, 0.98 mM $MgCl_2 \cdot 6H_2O$). The Tyrode's solution was used and stored under a 5% CO_2 atmosphere and kept covered to maintain pH. For some calibration experiments the buffers were prepared from K^+ rather than Na^+ salts. They are then referred to as, for example, K^+-HEPES buffer in this chapter.

B. Platelets

Freshly drawn blood, anticoagulated with citrate, was treated as previously described [Horne and Simons, 1978, 1979]. Washed platelets were prepared by gel filtration over Sepharose 2B in either modified Tyrode's or HEPES buffer containing 0.15 U apyrase/ml. Because binding to plasma proteins interferes with the pH studies, albumin was omitted from the buffers. The platelet response to thrombin, as measured by diS-C_3-(5) fluorescence increase and/or by platelet aggregometry, was utilized to verify that treatment of platelets with probes described in this chapter did not affect their stimulus response.

Permission to draw blood by venipuncture from normal human volunteers was obtained from the Institutional Review Board for Human Research of the Boston University Medical Center.

III. FLUOROMETRIC METHODS

All fluorescence measurements were performed on either a Perkin-Elmer MPF-2a or 650/10 spectrofluorometer. Each instrument is supplied by the company with a thermostatable cell holder, holding 4 or 1 cuvette(s), respectively. Since the system to be studied is heterogeneous (ie, a suspension), continuous stirring is necessary. While Perkin-Elmer now has some stirred thermostatable cell holders available, our experience is with the homemade variety.

In the MPF-2a, which has so massive a rotating turret that magnetic stirring becomes impossible, we have mounted a stirring motor, equipped with a small Teflon paddle, above the cuvette. The length of shaft plus paddle must be designed so that no light path interference occurs. Under such conditions, the minimum volume required in a 1-cm light path square cuvette is 2 ml. Whereas our current motor is mounted via a thumb screw onto a mounting on the central turret, we have used an unmounted motor, simply held in place on the cuvette by means of a rubber stopper, with equal success. For multiple-sample experiments, easy demounting of the stirring system is very desirable. The motor should be rated at 3 to 6 V DC and should be operable at low speed ($<$ 500 to max 2,000 rpm) for long time periods. We have had the greatest success with small ($\sim$ 2 cm diameter, 2–3 cm long) motors from Japanese toy airplanes operated from a 6-V battery. Although small high speed motors are available, (eg, 20 V) they are not satisfactory; at low running speeds the motor bearings burn out, and at higher speeds the cells are whipped and fragmented.

In the 650 series of Perkin-Elmer fluorometers, a magnetic stirring device can be mounted under the cuvette holder. We have again used a small, low-speed motor whose shaft, protruding through the air supply hole from the open compartment space, is topped by a strong magnet. The cuvette is then stirred with a Teflon-coated 1½ × 8 mm magnet stirring bar without impinging on the light path. With this stirring system, measurements in the 650/10 can be performed on anaerobic as well as aerobic systems. The minimum sample volume is 0.7 ml.

The combination of small sample volume, adaptability to study of samples in sealed cuvettes, and availability of zero offset and of higher sensitivity make the 650/10 an easier fluorometer to use for pH gradient change measurements. Using either instrument, the excitation and emission wavelengths are selected, and fluorescence emission is monitored and recorded continuously. All additions to the cuvette are made remotely via Hamilton microliter syringe and capillary tubing.

The basis upon which utilization of weak base distribution, on the one hand, and of fluorescein esters, on the other hand, is utilized for pH_i measurements is given in other chapters of this book. The following shall thus deal exclusively with the specific advantages and disadvantages of these two types of probes as they relate to the rapid stimulus response of human platelets. These are summarized in Table I.

A. 9-Aminoacridine

Since the first report by Deamer et al [1972] of the applicability of this weak base to intraliposomal pH measurements, this probe has been utilized to follow pH_i changes in mitochondria [Schuldiner and Avron, 1971; Schuldiner et al, 1972; Pick and Avron, 1976], in secretory granules [Salama et al, 1980; Johnson et al, 1978], in bacterial chromatophores [Casadio et al, 1974], in chloroplasts [Chow and Hope, 1976], and in sperm cells [Gillies and Deamer, 1979]. An overall review of recent applications is contained in Gillies and Deamer [1979], as well as elsewhere in this volume.

The applicability of any probe to a given cellular system must be verified. The following experiments were performed in order to provide such verification.

1) Since the intensity of fluorescence is expected to be a measure of the concentration of the probe, the Beer-Lambert law must be obeyed. This has been shown to be true in other buffer systems [Casadio et al, 1974; Deamer et al, 1972] and also holds true in HEPES (Fig. 1). All subsequent experiments were performed at final concentrations of 4 µM 9-aminoacridine, well within the linear range.

2) In order to utilize the detected fluorescence as a quantitative measure of the distribution of probe between cell and medium, it must be independent of pH over the region studied. We find this to be true for 4 µM probe (Table II) between pH 6.6 and 7.8, a region far enough below its pK to ensure that it is essentially fully protonated. Thus a change in detectable fluorescence at a constant

TABLE I. Advantages and Disadvantages of Methods for Evaluation of Platelet ΔpH

Technique	Utilizable pH range	Advantages	Disadvantages
9-Aminoacridine	6.9 → 8.0	Rapid response; little manipulation; $F_{9aa\ in} \approx 0$	50% in organelles; ΔF proportional but not equal to ΔpH; no absolute pH_i
6-Carboxyfluorescein diacetate	< 6.0 → 7.3	Rapid response; $F_{6cfdia} \approx 0$; cytoplasmic; can calculate pH_i	Hydrolysis in platelets slow; correction for external hydrolysis and leakage necessary; pH range limited
Dimethyl-6-carboxyfluorescein diacetate	< 6.0 → 8.0	Rapid response; $F_{dim6cfdia} \approx 0$; can calculate pH_i; broad pH range	Hydrolysis in platelets even slower; correction for external hydrolysis and leakage necessary

TABLE II. Fluorescence of 4×10^{-6} M 9-Aminoacridine (λ_{exc} = 400 nm; λ_{em} = 456 nm)

pH	Fluorescence arbitrary units
6.8	30.6
7.0	31.4
7.2	29.7
7.4	31.7
7.8	30.1

total 9-aminoacridine concentration (4 μM) in the pH range indicated is attributable to a change in the degree of quenching, as previously demonstrated for other systems [Deamer et al, 1972; Casadio et al, 1974].

3) In order to convert such a change in fluorescence into a change in internal pH (at constant, well-buffered external pH) a calibration curve must be generated, and the actual pH gradient between the inside and the outside of the platelet compared to that calculated by the equations of Deamer [Deamer et al, 1972]. Platelets washed in ordinary (ie, Na^+) HEPES buffer were hence diluted to a final concentration of 5.5×10^7/ml K^+-HEPES buffer at a $(K^+)_{out}$ corresponding to 118 mM, the $(K^+)_i$ of the resting platelet, so that no K^+ concentration gradient existed. The addition of nigericin to 2 μM then collapsed the pH gradient as well [Thomas et al, 1979] and pH_{in} became equal to pH_{out}. The actual pH gradient change is hence known and can be compared with the apparent $\delta(\Delta pH)$, calculated as described by Deamer et al [1972].

If one can assume that virtually all of the base is protonated up to 7.8 (cf above) and that the remaining unprotonated portion permeates the membrane freely, then

$$(A)_T = (AH^+)_{out} + (AH^+)_{in} + (A) \simeq (AH^+)_{out} + (AH^+)_{in} \qquad (1)$$

and

$$\Delta pH = \log \frac{(AH^+)_{in}}{(AH^+)_{out}} + \log \frac{V_{out}}{V_{in}} \qquad (2)$$

where V_{out} and V_{in} are, respectively, the external and intracellular volume in which the probe is distributed. If one can assume that the ratio V_{out}/V_{in} does not change significantly upon stimulation of a cell, one can calculate $\delta(\Delta pH) = \Delta pH_{after} - \Delta pH_{before}$.

Since $AH^+_{in} = (A)_T - (AH^+)_{out}$, the equation becomes

$$\delta(\Delta pH) = \log \left\{ \left[\frac{(A_T) - (AH^+)_{out}}{(AH^+)_{out}} \right]_{\text{after stimulation}} \div \left[\frac{(A)_T - (AH^+)_{out}}{(AH^+)_{out}} \right]_{\text{before stimulation}} \right\} \qquad (3)$$

which, since fluorescence of 9-aminoacridine was shown to be proportional to its concentration, can be expressed as:

$$\delta(\Delta pH) = \log \left[\frac{F_T - F_a}{F_a} \times \frac{F_b}{F_T - F_b} \right] \qquad (4)$$

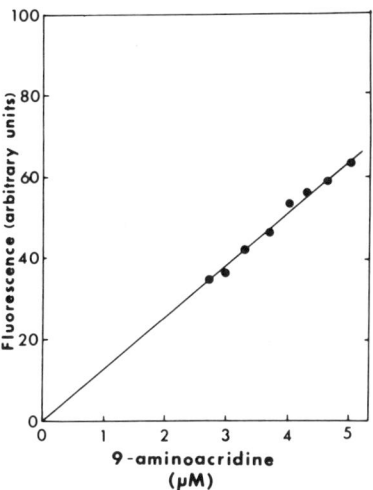

Fig. 1. Fluorescence of 9-aminoacridine in HEPES buffers of varying pH.

where the subscripts T, a, and b refer respectively to total, after, and before stimulation.

Comparison between the actual imposed ΔpH and the $\delta(\Delta pH)$ calculated as described above yielded a linear calibration above pH 7 with a slope of 0.3 (Fig. 2); below pH 7 there is apparently too little unprotonated probe to partition. That is, the calculated change in pH gradient by the equation given above is approximately one-third of the actual change between pH 7 and 8, which reflects the fact that part of the probe is associated with organelles and does not partition according to cytoplasmic versus external pH. (We have not attempted to go to high pH values, and the upper limit of linearity is thus not known.) Since the relationship is linear, the calculated pH can be used to measure a change in pH gradient—ie, a change in pH_{in} since pH_{out} is buffered and hence invariant—but not to evaluate the absolute change without reference to the calibration curve.

4) In order to verify the relative amount of 9-aminoacridine in the cytoplasm, platelets were diluted to 5.5×10^7 per ml HEPES buffer in the fluorometer cuvette and maintained at 37°C. 9-Aminoacridine was then added via microsyringe to a final concentration of 4 μM (NB: the final concentration of ethanol, the probe stock solution solvent, was less than 1% and, in control experiments, evoked no change in platelet properties). As shown in Table III for a representative experiment, approximately 13% of the fluorescence exhibited by the probe in the absence of platelets is quenched in their presence owing to inter-

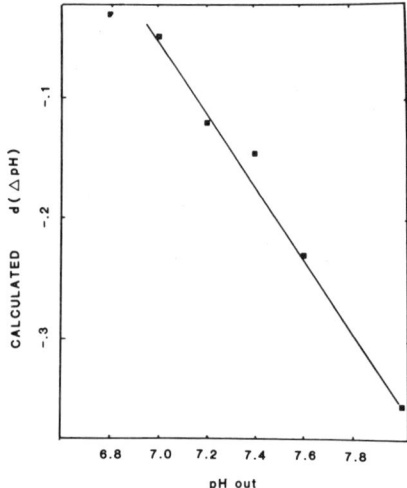

Fig. 2. Apparent $\delta(\Delta pH)$ calculated from 9-aminoacridine redistribution upon addition of nigericin to 5.5×10^7 platelets/ml K^+-HEPES buffer

nalization of the probe. Since platelets have many organelles and an extensive canalicular system, the internalized probe is probably partitioned into many or all of these as well as into the cytoplasm. This may be verified by treating the platelets with 0.08 mM digitonin, an agent that has been found [Akkerman et al, 1980] to release only the cytoplasmic contents of platelets at this concentration. As indicated in Table III, roughly half (57%) of the quenched fluorescence was released by this digitonin treatment. The remainder is presumably bound to or internalized within the organelles and/or the canalicular system. A 2 × saturating dose of thrombin [Larsen et al, 1979], 0.05 U/ml, led to only half the reduction in quenching obtained with digitonin. Addition of the latter after thrombin released the remainder, bringing the final quenching to the same level, ie, 50%, attained by digitonin treatment alone. Stimulation with more thrombin or more digitonin does not release the remaining platelet-associated 9-aminoacridine. It is possible to conclude from these data that therefore thrombin stimulation increases the pH_i of the platelet cytoplasm. It has been reported [Salama et al, 1980; Johnson et al, 1978] that 9-aminoacridine accumulates in secretory granules. Our findings with digitonin are not in agreement with these reports: Thrombin releases no more probe than does digitonin, even though thrombin causes the granules to secrete their contents into the external medium, and should therefore release additional probe into the medium.

TABLE III. 9-Aminoacridine Partition Into and Release From Platelets[a]

	F arbitrary units[b]	ΔF	% Released (5.6 ΔF)/5.6
Buffer only	46.6		
5.5 × 10⁷ Platelets	41.6	5.6	
Platelets + 0.05 U/ml thrombin	42.4	4.2	25
Platelets + 0.08 mM digitonin	44.2	2.4	57
Platelets + thrombin	42.4	4.2	25
Platelets + thrombin + digitonin	44.2	2.4	57

[a]Representative experiment. For an actual thrombin dose-response curve, zero suppression and scale expansion would be used to expand ΔF to full scale on the recorder.
[b]λ_{exc} = 400 nm, λ_{em} = 456 nm.

In order to increase the sensitivity of our thrombin dose-dependence measurements, the controls on the 650/10 were used to suppress the resting platelet suspension's fluorescence. The sensitivity scale factor allowing display of the thrombin-stimulated fluorescence increase (decrease in quenching) over a major portion of the recorder chart was then selected.

B. 6-Carboxyfluorescein Diacetate

In order to compare the pH_i changes observed via 9-aminoacridine with values obtained by means of a different probe, and in addition, to obtain values of absolute pH_i rather than of relative pH_i as given by the 9-aminoacridine technique, we have used the trapped probe method of Thomas et al [1979] by which a pH-sensitive indicator is generated in situ in the cell.

1) The diester of an indicator, 6-carboxyfluorescein diacetate, was allowed to incubate with platelets washed at pH 7.0 instead of 7.35 in HEPES buffer at 37°C for 15 minutes. The stock solution of probe in dimethylsulfoxide was sufficiently concentrated so that the final concentration of dimethylsulfoxide in the platelet suspension was kept below 0.1%, a concentration that has no effect on platelet response. Although Thomas et al found better internalization of the probe at lower pH, the inability of the platelet to maintain function if exposed to such low pH values made such facilitation impossible.

The incubated platelets were then rewashed over Sepharose 2B in pH 7.35 HEPES as usual, removing all noninternalized 6-carboxyfluorescein diacetate as well as any external already hydrolyzed 6-carboxyfluorescein. The diester hydrolyzed within the platelet by cytoplasmic esterases is considerably less membrane-permeable and remains inside the platelet. It should be noted that these 6-carboxyfluorescein-loaded platelets are fully responsive to platelet stimuli.

The amount of fluorescent probe in the external medium, negligibly small immediately after the second column separation, slowly increases for two reasons: There is some leakage of hydrolyzed 6-carboxyfluorescein from the inside to the outside, and there is also constant redistribution between the platelet and the external medium of unhydrolyzed 6-carboxyfluorescein diester. The latter undergoes spontaneous nonenzymatic hydrolysis in the medium. Whereas continuous dialysis of the labeled platelets decreases the amount of external hydrolyzed probe and of diester, it also enhances the redistribution of the latter. Thus, although we have used and continue to use this method on some occasions, correction for extracellular probe concentration must be made and is not trivial. This correction can be made by centrifuging an aliquot of the 5.5×10^7/ml platelet suspension immediately before each run and subtracting the fluorescence of the supernatant, ie, of external indicator, from the total fluorescence observed.

2) As shown in Thomas's chapter in this volume, the absorbance spectrum of 6-carboxyfluorescein shows a maximum at 492 nm and an isosbestic point at 464 nm [Thomas et al, 1979], whereas the diester does not absorb in this region. It has also been demonstrated that the emission spectrum of 6-carboxyfluorescein shows a maximum at 518 nm for excitation wavelengths of either 492 or 464 nm [Thomas et al, 1979]. Use of a ratio of fluorescence at two excitation wavelengths (eg, 492: 464) and the same emission wavelengths corrects for any differences in light path or in number of cells [Heiple and Taylor, 1981; Thomas et al, 1979]. We have used this approach to calculate a calibration curve for 6-carboxyfluorescein in Hepes buffer. (Fig. 3). It should be noted that excitation of 464 nm yields approximately half the emission signal generated by excitation of 492 nm. While the ratio calculation described above is utilized in our final calibration, it is not necessary to utilize it with a cell suspension containing a constant number of cells in a cuvette of constant light path, since these corrections then cancel out. Hence, for rapid kinetic (cf. below) observation during the course of the experiment, measurement of fluorescence emitted at 518 nm upon excitation at 464 nm has been omitted.

Correction for external probe (ie, hydrolyzed diester) is necessary. By using the isosbestic point, whose absorbance is independent of pH, as one of the excitation wavelengths we can obtain an inner control of, for example, the amount of this external hydrolyzed probe. Even when this is relatively constant, however, as is true for each individual run in the series discussed below, one must correct for the external probe's contribution to the observed fluorescence, and therefore one must calculate the net fluorescence increase ratio,

$$\frac{I_{492} \text{ (suspension } - \text{ supernatant)}}{I_{464} \text{ (suspension } - \text{ supernatant)}} \tag{5}$$

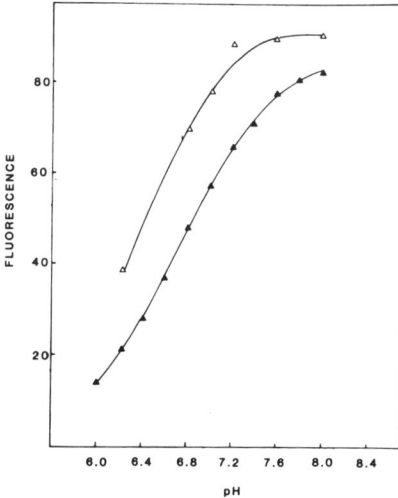

Fig. 3. Fluorescence (arbitrary units) of 6-carboxyfluorescein (△) (λ_{exc} = 492, λ_{em} = 518 nm) and of 4′,5′dimethyl-6-carboxyfluorescein (▲) (λ_{exc} = 507 nm, λ_{em} = 535 nm) in K^+-HEPES buffer.

for the calibration curve with cells. Such a calibration curve was obtained by incubating 6-carboxyfluorescein-loaded platelets for less than 5 minutes in K^+-HEPES buffer at 37°C, and then adding nigericin. In the presence of this ionophore under such conditions, $(K^+)_{in}/(K^+)_{out} = (H^+)_{in}/(H^+)_{out} = 1$, and pH_{in} becomes equal to the buffered pH_{out} as previously described for 9-aminoacridine calibration. The calibration for relative fluorescence of 6-carboxyfluorescein as a function of pH was the same with or without the presence of platelets [Horne et at, 1981].

3) As with all probes, it is necessary to verify the location of the hydrolyzed 6-carboxyfluorescein. As shown in Table IV, when probe-loaded resting platelets are centrifuged, ~ 63% of the probe remains in the supernatant. This value varies from 20%, immediately after a platelet preparation has been rechromatographed through Sepharose 2B and reequilibrated to 37°C, to 80% after 4½–5 hours without continuous dialysis. Since we have a rise in fluorescence intensity for excitation of the suspension at 464 nm, this represents an actual increase in the total quantity of hydrolyzed diester. Thus the sizable increase in external probe concentration is attributable not only to leakage of 6-carboxyfluorescein

TABLE IV. Localization of Intraplatelet 6-Carboxyfluorescein

Sample	Fluorescence arbitrary units		% Internal 6CF
	Suspension	Supernatant	
Platelets alone[a]	114	72.4	36.5
Platelets + 0.08 mM digitonin	128[b]	124	3.1
Platelets + 0.08 mM digitonin + 0.05 U + 0.05 U/ml thrombin		124	
Platelets + 0.05 U/ml thrombin + 0.08 mM digitonin	134.6[b]	124	7.9

[a]5.5×10^7 Platelets/ml HEPES buffer, pH 7.35.
[b]In the presence of digitonin the fluorescence intensities are enhanced.

from the platelets but also to nonenzymatic hydrolysis of leaked diester. Since platelets are relatively short lived ex vivo and can support neither low temperatures nor low pH without impairment of function, there is no way currently obvious to us in which this high external fluorescence can be avoided. It increases very slowly and hence can be considered constant for any single run (a period of less than 5 minutes), allowing a simple correction to be made for supernatant fluorescence. That is, we have routinely followed fluorescence continuously only for $\lambda_{exc} = 492$, $\lambda_{em} = 518$ nm, and have then divided the difference between suspension and supernatant by that for $\lambda_{exc} = 464$ for interpolation on the calibration curve (via equation 5).

C. Dimethyl 6-Carboxyfluorescein Diacetate

In an attempt to extend the utilizable pH indicating range of 6-carboxyfluorescein toward pH 8, we asked Molecular Probes Inc to design and prepare a modified ester of the probe. Although we have not yet had as much experience with 4′,5′dimethyl-6-carboxyfluorescein diacetate as with the Thomas et al probe, it may be useful for some applications (Table I).

1) The excitation spectrum of dimethyl-6-carboxyfluorescein (or of base hydrolyzed diacetate) is shown in Figure 4 ($\lambda_{em} = 535$). Unlike the unmethylated precursor, no true isobestic point was detectable.* As discussed above for 6-carboxyfluorescein, this is not a deterrent in our system where concentration and light path length are constant, but should be taken into account by anyone planning single cell applications (cf Heiple and Taylor, this volume, for discussion of such applications). The pH calibration curve for the dimethyl derivative (Fig. 3) indicates that the pK has indeed been moved to pH 6.8 and that the probe, in all other respects, behaves like its 6-carboxyfluorescein precursor.

*While the ratio of fluorescence intensity for any two excitation wavelengths can still be used, the absence of an isosbestic point makes the correction for a net total increase in probe (ie, increase in hydrolyzed dimethyl 6-carboxyfluorescein), on both sides of the membrane, more difficult.

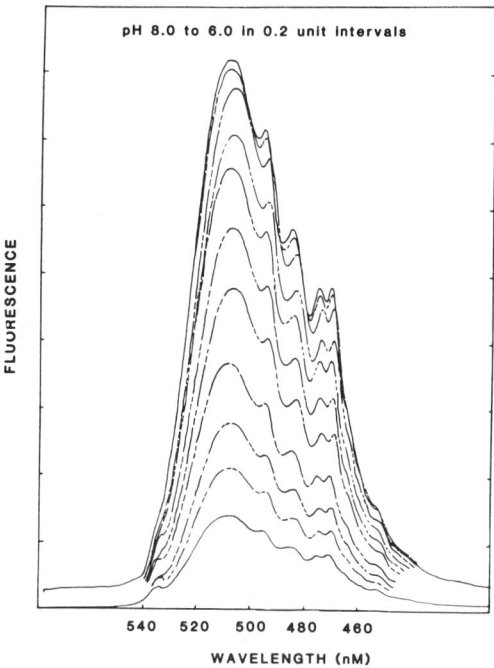

Fig. 4. Excitation spectra of 4',5'dimethyl-6-carboxyfluorescein (arbitrary units) as a function of K^+-HEPES buffer pH from 6.0 to 8.0 in 0.2 unit intervals (λ_{em} = 535 nm).

2) The cellular uptake of the dimethyl compound by platelets is worse than that of 6-carboxyfluorescein diacetate, and the correction for extracellular probe higher. The reason, as shown in Figure 5, is apparently the much lower rate of enzymic hydrolysis (but not of base hydrolysis), presumably attributable to steric hindrance. Thus the rate of hydrolysis of the dimethyl-6-carboxyfluorescein diacetate by an aliquot of platelet sonicate at pH 7 is, for example, less that 30% of that exhibited by the same concentration of 6-carboxyfluorescein diacetate under the same conditions. Since most of this hydrolysis is accomplished by granular enzymes not expected to be present in the cytoplasm, the situation in the whole platelet is exacerbated, as a higher concentration of uncleaved diester remains available. As yet these problems remain unsolved. It should be noted, however, that the technique of pH_i measurement by in situ probes is currently being adapted to bacteria as well as other cell types. In the former, the complicating factor of a high pH_i makes it advisable to use the broader ranged dimethyl 6-carboxyfluorescein (Simons et al, to be published).

D. General Comment

This Methods section has described in some detail our adaptation of 9-aminoacridine and of 6-carboxyfluorescein fluorescence measurements as indicators of changes in pH_i of washed human platelets. This description has included several verification and calibration experiments for each probe. These are valid *only* for the particular cell and conditions employed. It is hence mandatory that controls, verifications, and calibrations similar to those described here be performed for each new system to assure that 1) the observed parameter, change in parameter, or quantity calculated from it, is a linear function of the actual ΔpH; 2) an adequate quantity of any distributive probe is present so that the distribution is limited by the desired property (eg, ΔpH) and not by the amount of probe available and its diffusion constant; 3) the localization of the probe corresponds to a compartment in which stimulation evokes a quantitative response; and 4) correction for leakage or background fluorescence is adequate.

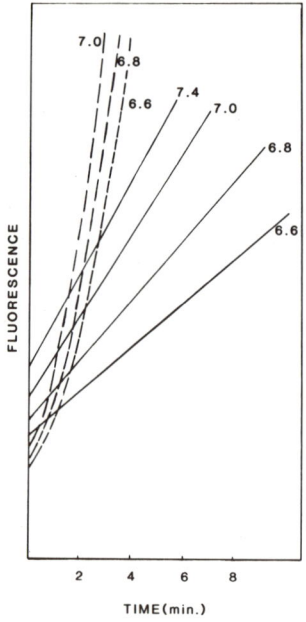

Fig. 5. Rate of fluorescence intensity change (ie, of hydrolysis) of 6-carboxyfluorescein diacetate (------) and of 4′,5′dimethyl 6-carboxyfluorescein diacetate (—) by equal quantities of the soluble portion of platelet sonicates. pH of K^+-HEPES buffer as indicated.

IV. RESULTS

We have observed and reported elsewhere [Horne and Simons, 1978] that thrombin stimulation of washed human platelets leads to a dose-dependent depolarization of the membrane that parallels that of serotonin secretion by these stimulated platelets. For each parameter, $\Delta\Psi$ being measured either by $diSC_3(5)$ fluorescence or by Ph_4P^+Br distribution while secretion is evaluated by ^3H-serotonin release, there is an asymptotic effect at 0.025 U/ml, corresponding to $\sim$ 4.5 nM α-thrombin.

We have now generated a curve depicting thrombin dose versus relative decrease in 9-aminoacridine fluorescence quenching by continuously recording this decrease as exhibited by a constant number (5.5×10^7/ml) of platelets stimulated by remote addition (via microsyringe) of varying amounts of α-thrombin. Zero suppression of the $t \leq 0$ signal was used in order to enlarge the recorded change in fluorescence upon stimulation. The response increased with increasing thrombin and attained a constant value at or above 0.025 U/ml ($\sim$ 4.5 nM). The response at 30 seconds, when it is still linear, was converted into a $\delta(\Delta pH)$ value by means of equation 4, and plotted as a function of thrombin concentration (Fig. 6). By interpolation on the 9-aminoacridine calibration curve (Fig. 2), the calculated maximal $\delta(\Delta pH)$ of 0.134 ± 0.001 (for ten experiments) corresponds to a true rise in pH_i of 0.27.

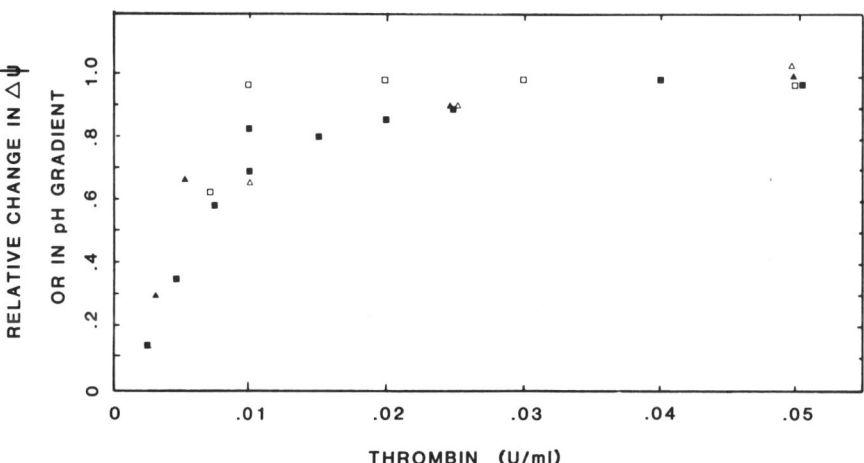

Fig. 6. Relative change in $\Delta\Psi$ or ΔpH as a function of thrombin dose (normalized so that asymptote (at thrombin = 0.025 U/ml) $\equiv$ 1.0). Relative $\Delta\Psi$ as measured by Ph_4P^+Br, △, or by $diS-C_3(5)$, ▲; relative ΔpH_i as measured by 6-carboxyfluorescein, □, or by 9-aminoacridine, ■.

We have also prepared a thrombin dose-dependence curve for 6-carboxyfluorescein-labeled platelets. In order to develop thrombin dose-response data, the fluorescence of the stirred, thermostated platelet suspension was again suppressed electrically on the fluorometer, and thrombin was added remotely as previously described. The results of a typical series are indicated in Figure 6. The maximal change in fluorescence upon thrombin stimulation is observed at 0.025 U/ml. It corresponds to an increase of 0.3 in pH_i, according to the calibration curve (Fig. 2). One can also use the nigericin experiments at high $(K^+)_{out}$, and can show thereby that pH_i of the resting platelet is 7.01 ± 0.02. Thus the observed ΔpH_i for a saturating thrombin dose rises from 7.01 to 7.30 when the exterior is at pH 7.35. This 88% alkalinization is of the same order of magnitude as the thrombin-induced depolarization (71%). The possible significance of this parallelism will be discussed in the next section.

We have previously shown [Horne and Simons, 1978b] that the observed thrombin induced membrane potential changes involve altered passive Na^+ transport since they could be inhibited by 10^{-4} and 10^{-3}M amiloride, a specific inhibitor of such transport. As shown in Table V, the thrombin-induced changes in pH_i are inhibited to the same extent as $\Delta\Psi$, but the secretion of serotonin—one of the last steps in the platelet response—is much less perturbed. (For example, 10^{-4}M amiloride leads to complete inhibition of $\Delta\Psi$ and ΔpH_i but only 46% inhibition of serotonin secretion). Thus thrombin induced alkalinization of the platelet cytoplasm and depolarization of its membrane both involve passive Na^+ transport. These findings are compatible with, but do not prove the existence of, the activation of a Na^+-H^+ antiport in the platelet membrane by α-thrombin stimulation.

V. DISCUSSION

The methods described in this paper and in others from this laboratory [Horne and Simons, 1978, 1979; Larsen et al, 1979; Larsen and Simons, 1981; Whitin et al, 1980, 1981; Horne et al, 1981; Cohen et al, 1981] have allowed us to investigate some of the first steps in the stimulus-coupled response of a cell as described in the introduction.

By a technique not described in this paper, we have verified step 1, the location of binding of thrombin to its specific platelet receptor, and have isolated that receptor [Larsen and Simons, 1981] on the human platelet, a "typical" secretory cell. The binding is rapid since we arrested it at 30 seconds in the above isolation and obtained complete binding [Larsen and Simons, 1981]. We have followed the membrane response via step 2a, by proving that depolarization accompanies binding of α-thrombin but not the equally strong binding of enzymatically inactive tosyllysylchloromethylketone-treated thrombin [Larsen et al, 1979]. This corroborates previous reports [Tollefsen and Majerus, 1976;

TABLE V. Effect of Amiloride on the Platelet Response to a Sub-saturating Dose (0.01 U/ml) of α-Thrombin

Concentration of amiloride	% of control response		
	$\Delta\psi$	ΔpH_i	(^{3}H)5HT release
Control	100	100	100
10^{-4}M	48	41	67
10^{-3}M	0	3	46

Martin et al, 1975; Ganguly and Sonnichsen, 1976; Workman et al, 1977; Tam and Detwiler, 1978, 1980] that thrombin binding to platelets does not require a free active site whereas activation of the platelets does. The depolarization is very rapid, detectable within the time limit of our current experiments (< 5 seconds), and has reached its maximum (approximately 71% depolarization, from approximately -52 to -15 mv) within 60 to 90 seconds [Horne and Simons, 1978, 1979; Larsen et al, 1979; Horne et al, 1981].

We have not addressed the study of step 2b, the activation of certain membrane-bound enzymes such as phospholipases, but this facet has recently been reviewed by others [Rittenhouse-Simmons, 1981]. We have summarized here and elsewhere [Schwartz et al, 1980; Horne et al, 1981] our observations relating to the next step in secretory cell response, 3a—the alteration of certain intracellular cation concentrations. As shown (Fig. 6), $(H^+)_{in}$ changes to the same relative extent and, to the accuracy of our time measurements, along the same time frame as the membrane potential (Fig. 6). That is, the intraplatelet pH, which we find to be 7.01, begins to rise immediately upon stimulation with thrombin, and reaches a value of 7.30 when a saturating dose (> 0.025 U/ml or 4.5 nM) is used. As with the membrane potential, this change, corresponding to ~ 71% of the initial pH gradient across the membrane, is complete within 60 to 90 seconds [Schwartz et al, 1980; Horne et al, 1981]. It is therefore impossible at this time to delineate the sequence of $\Delta\Psi$ and ΔpH changes—ie, to determine whether they occur sequentially or simultaneously. The latter could, for example, be the result of a membrane protein's altered conformation and a consequent Na^+-H^+ antiport opening, a hypothesis compatible with our finding that amiloride blocks both changes equally [Horne and Simons, 1979; Horne et al, 1981]. The involvement of cytoplasmic, rather than organellar, (H^+) changes is clear, but neither the mechanism by which they are initiated nor their time course with respect to the potential has yet been determined. The relative changes in other cation concentrations (eg, $(Ca^{++})_i$ and in cyclic nucleotides (steps 3a and b) will not be addressed here.

Although the thrombin activation of platelet metabolic processes, step 4a, has been well documented and reviewed [Fukami et al, 1976; Holmsen, 1977], its time course in relation to membrane potential and pH gradient changes is only now being explored [Sepersky and Simons, 1980]. The activation of organellar systems (step 4c), leading to secretion (often known as "release" in the platelet literature), is one of the end results of platelet stimulation [Jerushalmy and Zucker, 1966; Holmsen, 1976, 1977]. We have already reported elsewhere that the thrombin dose-dependence of platelet secretion parallels that of membrane depolarization [Horne and Simons, 1979; Larsen et al, 1979] and transmembrane pH gradient decrease [Schwartz et al, 1980; Horne et al, 1981].

The advantage of the techniques we have reviewed here, approaches that allow correlation of secretory cell stimulation with their sequential response, is that they allow dissection of that response into steps, and that they appear to be generally applicable. Thus, having started with platelets, we now have shown that granulocytes can be investigated by similar approaches [Whitin et al, 1980, 1981; Cohen et al, 1981]. The pH measurements we describe here and in other publications from this laboratory [Schwartz et al, 1980; Horne et al, 1981] are continuous, rapid, and simple, and require few platelets, as do the membrane potential measurements. These techniques are widely applicable to the study of coupled stimulus responses of other cells, as well as of perturbation of these responses by disease, [Whitin et al, 1980, 1981; Cohen et al, 1981; Lew et al, 1981] or by drugs [Schwartz et al, 1980], provided that the requisite controls and calibrations are performed for each cell.

ACKNOWLEDGMENTS

We thank Drs. John Whitin, Lucienne Letellier, and Emanuel Shechter for valuable discussions. We are grateful for Dr. Alan Waggoner's continuing interest and supply of di-S-C_3-(5). The partial support of NIH grants HL 15335 and HL 16357 is gratefully acknowledged.

VI. REFERENCES

Akkerman JWN, Ebberink, RHM, Lips, JPM, Christiaens, GCML: Rapid separation of cytosol and particle fraction of human platelets by digitonin-induced cell damage. Br J Haematol 44:291, 1980.

Bramhall JS, Morgan JI, Perris AD, Britten AZ: The use of a fluorescent probe to monitor alterations in trans-membrane potential in single cell suspensions. Biochem Biophys Res Commun 72:654, 1976.

Casadio R, Baccarini-Melandri A, Melandri BA: On the determination of the transmembrane pH difference in bacterial chromatophores using 9-aminoacridine. Eur J Biochem 47:121, 1974.

Chow WS, Hope AB: Light-induced pH gradients in isolated spinach chloroplasts. Aust J Plant Physiol 3:141, 1976.

Cohen HJ, Newburger PE, Chovaniec ME, Whitin JC, Simons ER: Opsonized zymosan-stimulated granulocytes—Activation and activity of the superoxide generating system. Blood (in press), 1981.

Deamer DW, Prince RC, Crofts AR: The response of fluorescent amines to pH gradients across liposome membranes. Biochim Biophys Acta 274:323, 1972.

Finkel T, Wolf DP: Membrane potential pH and activation of surf clam oocytes. Gamete Res 3:299, 1980.

Friedman F, Detwiler TC: Stimulus-secretion coupling in platelets. Effects of drugs on secretion of adenosine 5' triphosphate. Biochemistry 14:1315, 1975.

Fukami MH, Holmsen H, Salganicoff L: Adenine nucleotide metabolism of blood platelets. Biochim Biophys Acta 444:633, 1976.

Ganguly P, Sonnichsen WJ: Binding of thrombin to human platelets and its possible significance. Br J Haematol 34:291, 1976.

Gillies RJ, Deamer DW: Intracellular pH: Methods and applications. Curr Topics Bioenerg 9:63, 1979.

Heiple JM, Taylor DL: Intracellular pH in single motile cells. J Cell Biol 86:885, 1981.

Holmsen H: Classification and possible mechanisms of action of some drugs that inhibit platelet aggregation. Ser Haematol 8:50, 1976.

Holmsen H: Platelet energy metabolism in relation to function. In Mills DCB, Pareti FI (eds): "Platelets and Thrombosis," London: Academic Press, 1977, p 45.

Horne WC, Simons ER: Probes of transmembrane potentials in platelets: Changes in cyanine dye fluorescence in response to aggregation stimuli. Blood 51:741, 1978.

Horne WC, Simons ER: Effects of Amiloride on the response of human platelets to bovine thrombin. Thrombos Res 13:599, 1979.

Horne WC, Norman NE, Schwartz DB, Simons ER: Changes in cytoplasmic pH and in membrane potential in thrombin stimulated human platelets. Eur J Biochem (in press), 1981.

Jerushalmy Z, Zucker MB: Some effects of fibrinogen degradation products (FDP) on blood platelets. Thromb Diath Haemorrh 15:413, 1966.

Johnson RG, Carty SE, Fingerhood BJ, Scarpa A: FEBS Letters 120:75, 1980.

Johnson RG, Scarpa A, Salganicoff L: The internal pH of isolated serotonin containing granules of pig platelets. J Biol Chem 253:7061, 1978.

Korchak HM, Weissmann G: Changes in membrane potential of human granulocytes antecede the metabolic responses to surface stimulation. Proc Natl Acad Sci USA 75:3818, 1978.

Korchak HM, Weissman G: Stimulus-response coupling in the human neutrophil. Biochim Biophys Acta 601:180, 1980.

Larsen NE, Horne WC, Simons ER: Platelet interaction with active and TLCK-inactivated thrombin. Biochem Biophys Res Commun 87:403, 1979.

Larsen NE, Simons ER: Preparation and application of a photoreactive thrombin analogue. Biochemistry 20:4141, 1981.

Lundblad R, Uhteg LC, Vogel CN, Kingdon HS, Mann K: Preparation and partial characterization of two forms of bovine thrombin. Biochem Biophys Res Commun 66:482, 1975.

Martin BM, Feinman RD, Detwiler TC: Platelet stimulation by thrombin and other proteases. Biochemistry 14:1308, 1975.

Pick U, Avron CM: A method for measuring the internal pH in illuminated chloroplasts based on the stimulation of proton uptake by amines. Eur J Biochem 70:569, 1976.

Rittenhouse-Simmons S: Release and metabolism of arachidonate in human platelets. In Dingle JT, Gordon JL (eds): "Platelets in Biology and Pathology," Vol 2. Amsterdam: Elsevier-North Holland, 1981, p 349.

Salama G, Johnson RG, Scarpa A: Spectrophotometric measurements of transmembrane potential and pH gradients in chromaffin granules. J Gen Physiol 75:109, 1980.

Schuldiner S, Avron M: On the mechanism of the energy-dependent quenching of atebrin fluorescence in isolated chloroplasts. FEBS Lett 14:233, 1971.

Schuldiner S, Rottenberg H, Avron M: Determination of pH in chloroplasts. 2. Fluorescent amines as a probe for the determination of pH in chloroplasts. Eur J Biochem 25:64, 1972.

Schwartz DB, Norman NE, Simons ER: Effect of sulfinpyrazone and its thioether metabolite on platelets and their response to thrombin. Circulation 62:III:1051 (abstract), 1980.

Sepersky S, Simons ER: Inhibition of platelet membrane potential changes by metabolic inhibitors. Fed Proc 39:1863, 1980 (abstract).

Shen SS, Steinhardt RA: Direct measurement of ultracellular pH during metabolic depression of the sea urchin egg. Nature 272:253, 1978.

Steinhardt RA, Mazia D: Developmental K^+ conductance and membrane potentials in unfertilized sea urchin eggs after exposure to NH_4OH. Nature 241:400, 1973.

Tam SW, Detwiler TC: Binding of thrombin to human platelet plasma membranes. Biochim Biophys Acta 543:194, 1978.

Tam SW, Fenton JW, Detwiler TC: Platelet thrombin receptors. J Biol Chem 255:6626, 1980.

Thomas JA, Buchsbaum RN, Zimniak A, Racker E: Intracellular pH measurements in Ehrlich ascites tumor cells utilizing spectroscopic probes generated in situ. Biochemistry 18:2210, 1979.

Tollefsen DM, Majerus PW: Evidence for a simple class of thrombin binding sites on platelets. Biochemistry 15:2144, 1976.

Whitin JC, Chapman CE, Simons ER, Chovaniec ME, Cohen HJ: Correlation between membrane potential changes and superoxide production in human granulocytes stimulated by phorbol myristate acetate. J Biol Chem 2555:1874, 1980.

Whitin JC, Clark RA, Simons ER, Cohen HJ: Effects of the myeloperoxidase system on fluorescent probes of granulocyte membrane potential. J Biol Chem 256:8904, 1981.

Winkler MM, Grainger JL: Mechanism of action of NH_4Cl and other weak bases in the activation of sea urchin eggs. Nature 273:536, 1978.

Workman EF Jr, White GC II, Lundblad RL: Structure–function relationships in the interaction of thrombin with blood platelets. J Biol Chem 252:7118, 1977.

Zieler K, Rogus EM: Effects of peptide hormones and adrinergic agents on membrane potentials of target cells. Fed Proc 40:121, 1981.

The Role of Protons in Glucose-Induced Stimulus-Secretion Coupling in Pancreatic Islet B Cells

Caroline S. Pace, John T. Tarvin, and Joel S. Smith

Department of Physiology and Biophysics, University of Alabama in Birmingham, Diabetes Hospital, 1808 Seventh Avenue, South, Birmingham, Alabama 35294

I.	Introduction	483
	A. Influence of Protons on the B Cell Plasma Membrane	483
	B. Influence of Glucose on the Proton Gradient in Secretory Granules of B Cells	485
II.	Materials and Methods	485
	A. Animals, Isolation, and Culture of Islets	485
	B. Electrophysiological Studies	486
	C. $^{86}Rb^+$ Efflux	488
	D. Islet Perifusion System	489
	E. Acridine Orange Experiments	490
	F. Electron Microscopy	490
III.	Results	491
	A. Electrical Activity	491
	B. $^{86}Rb^+$ Efflux	495
	C. Insulin Release	497
	D. Acridine Orange	498
IV.	Discussion	502
	A. Influence of Protons on Electrical Activity and $^{86}Rb^+$ Efflux	503
	B. Influence of pH on Glucose-Induced Insulin Release	507
	C. Glucose-Induced Proton Gradient in Secretory Granules	508
V.	References	510

I. INTRODUCTION

A. Influence of Protons on the B Cell Plasma Membrane

The pancreatic B cell is unique among other endocrine cells in that nutrients elicit not only a secretory response, but also a characteristic pattern of electrical activity [Atwater, 1980]. Glucose occupies a preeminent position as the primary physiological stimulant and metabolic substrate of the B cell and has been used frequently to clarify the primary mechanisms underlying the glucosensor system. Whatever the nature of the glucosensor system, whether it is generation of a signal via glucose metabolism or glucose interaction with a receptor in the plasma membrane [Ashcroft, 1976; Malaisse et al, 1979b; Matschinsky et al, 1975], it must be coupled to a transduction device for the generation of intracellular messengers. It has been generally thought that these secondary signals (ie, Ca^{++} and cyclic AMP) emanate from changes taking place at the level of the plasma membrane. A number of recent studies suggested that the generation of protons (H^+) from glucose metabolism serves as a crucial coupling factor between metabolic and ionic events associated with changes in the permeability of the plasma membrane to K^+ and Ca^{++} [Malaisse et al, 1980]. To what extent H^+ alter the glucose-induced electrical events has not yet been determined. Up to this point alterations in Ca^{++} and K^+ permeabilities (P_{Ca} and P_K) and perhaps the activity of a Na^+ pump have been shown to be the primary mechanisms underlying the electrical events. It is clear that membrane depolarization elicited by glucose is due primarily to a decrease in P_K and that cyclic changes in P_{Ca} and P_K are important in controlling the glucose-induced oscillatory pattern of electrical activity recorded from mouse B cells [Atwater, 1980]. Briefly, this entails a decrease in P_K leading to depolarization to a plateau level (active phase) upon which fast spikes are generated due to activation of a voltage-sensitive increase in P_{Ca}, which in turn leads to an increase in the intracellular level of Ca^{++}. The depolarization and increase in cytosolic [Ca^{++}] activates P_K resulting in repolarization to the silent phase. As the concentration of Ca^{++} is decreased by intracellular buffering systems the P_K is again decreased, leading to a repeat of the cycle. In order to fulfill the proposed role of H^+ as a coupling factor between metabolic and cationic events, it is necessary to document the influence of changes in intracellular pH on the pattern of oscillatory electrical activity evoked by moderate concentrations of glucose.

Metabolic generation of H^+ is the likely cellular control system for triggering or modulating cationic fluxes rather than generation of reduced pyridine nucleotides or adenine nucleotides [Malaisse, 1979a, b; 1980]. Within the noninsulinotropic range of glucose concentrations (ie, 2–5 mM) there are substantial changes in the efflux of K^+ or Ca^{++}, but little or no change in the level of NAD(P)H or adenine nucleotides [Malaisse et al, 1979b; Carpinelli et al, 1980]. However, at low concentrations of glucose there is an increased output of H^+ from the islets [Carpinelli et al, 1980]. Although there is no detectable change in intracellular pH (pH_i) accompanying glucose-induced insulin release [Carpinelli et al, 1980; Hellman et al, 1978], it has been calculated, taking into consideration the buffering capacity of the islets, that there is about a 0.09 unit decrease in pH_i [Hutton et al, 1980]. It is possible that small changes in pH_i may alter the P_K as has been found to occur in nerve fibers [Hille, 1968; Wanke et al, 1979]. The resulting depolarization may then initiate a voltage-dependent increase in P_{Ca}. It has been found recently that, in the absence of glucose, alterations of extracellular pH changed $^{86}Rb^+$ efflux in a manner consistent with P_K changes observed in nerve [Henquin, 1981]. In view of the potential significance of metabolic-induced changes in pH in controlling cationic fluxes in the plasma membrane, we have undertaken a series of studies to examine the influence of changes in pH on the electrical activity of B cells in relation to changes in $^{86}Rb^+$ (substitute for K^+) efflux and insulin release.

B. Influence of Glucose on the Proton Gradient in Secretory Granules of B Cells

The ultimate role of the glucosensor and the subsequent ionic events is to initiate translocation and exocytosis of secretory granules. The secretory granules not only are intimately involved in the storage and subsequent release of insulin, but also may participate in the regulation of intracellular Ca^{++} [Hellman et al, 1980; Abrahamsson et al, 1981]. Accordingly, it has been found that secretory granules contain 30% of the readily exchangeable Ca^{++} in the B cell, and that glucose stimulates $^{45}Ca^{++}$ uptake by this organelle. The presence of Mg-ATP enhances not only $^{45}Ca^{++}$ uptake, but also the loss of $^{45}Ca^{++}$ from the granules without a concomitant change in the insulin content [Abrahamsson et al, 1981]. It is possible that an ATP-dependent H^+ pump exists in the secretory granule membrane and as a consequence exchanges Ca^{++} for H^+. This postulate is supported by the recent finding that the interior of the granule is maintained at a low pH as indicated by accumulation of the fluorescent weak base 9-aminoacridine [Abrahamsson and Gylfe, 1980]. These observations suggest that the pump not only may serve in a chemiosmotic mechanism resulting in lysis of the granule [Somers et al, 1980] as occurs in granules of other cells, notably the chromaffin cell [Pollard et al, 1979], but may also serve as a Ca^{++}-buffering mechanism when the B cell is stimulated by glucose. However, there is no evidence to indicate that the accumulation of H^+ by the secretory granules is enhanced by glucose. We have pursued this issue by utilizing the pH-sensitive metachromatic fluorescent dye acridine orange in conjunction with monolayer cultures of rat islet cells and fluorescent microscopy.

II. MATERIALS AND METHODS

A. Animals, Isolation, and Culture of Islets

Islets were isolated from male Sprague-Dawley rats (250–300 g) by the collagenase technique [Lacy and Kostianovsky, 1967]. For the $^{86}Rb^+$ efflux and insulin secretion studies the islets were hand picked. For islet culture, Ficoll (Pharmacia) gradients were used [Kostianovsky et al, 1974]. The Ficoll was prepared in concentrations of 25%, 23%, 20.5%, and 11% (w/v) in Hanks solution containing 25 mM HEPES (N-2-hydroxyethylpiperazine-N'-2-ethanesulfonic acid). The washed collagenase-treated pancreatic tissue was mixed with 4 ml of the 25% Ficoll, and 2 ml of each of the different Ficoll concentrations were layered in a centrifuge tube that was then centrifuged at 1,800g for 10 minutes. The islets were located in the 20.5% level and at the interfaces between the 23% and 20.5% and the 20.5% and 11% levels. The islets (200–400 per rat) were transferred via a Pasteur pipette to a centrifuge tube containing Hanks. They were then placed in 2.5 ml of prewarmed (37°C) Ca^{++}:Mg^{++}-free buffer containing 0.25% trypsin (GIBCO). The isolated islets were disrupted by pipetting for 1 minute to obtain small clumps of islet cells. After washing twice with tissue culture medium, the islet cell aggregates were placed in 35-mm plastic Petri dishes containing a 25-mm glass coverslip. The coverslips were washed in 1% DMSO (dimethylsulfoxide) and rinsed thoroughly with distilled H_2O before autoclaving. The coverslip had been previously treated with 2 μg/cm fibronectin (Collaborative Research) to facilitate attachment of the islet cells and formation of a monolayer. This usually occurred by 4 to 5 days.

The medium used was RPMI 1640 (GIBCO) with the addition of 10% fetal calf serum, 100 U/ml penicillin, 100 μg/ml streptomycin sulfate, and 0.1 mM IBMX (3-isobutyl-1-methylxanthine). RPMI 1640 was chosen because the levels of glucose (11.1 mM) and myoinositol (35 μg/ml) have been found to maintain the functional parameters of B cells [Pace and Clements, 1981]. The islet cells were maintained in a humidified atmosphere of 95% air and 5% CO_2 at 37°C for 5–7 days. The medium was changed 1 or 2 days before use in an experiment.

For the electrophysiological studies the islets were obtained from fed albino Swiss-Webster mice. The isolated pancreas was placed in a chilled Hanks solution containing 5.6 mM glucose. Using a syringe and a 26-gauge needle, the pancreas was inflated with Hanks solution. This procedure rendered the islets clearly visible as opaque, ovoid bodies under low magnification (7–30 ×). The inflated pancreas was pinned to black wax contained in an ice-chilled Petri dish. Using fine-pointed forceps and iris scissors the acinar tissue was trimmed around two or more large islets leaving partially dissected islets attached to blood vessels and a small amount of acinar tissue. Care must be taken during dissection, since disruption of the connective tissue surrounding the islet results in loss of islet cells. This tedious procedure usually takes from 15 to 45 minutes.

B. Electrophysiological Studies

A block diagram of the perifusion and recording setup is shown in Figure 1. The section of excised pancreas was pinned in a plastic perifusion chamber of 0.5 ml capacity. The preparation was perifused at 37°C with modified Krebs-Ringer solution (mM): 16 HEPES, 136 Na^+, 5.0 K^+, 2.5 Ca^{++}, 1.2 Mg^{++}, 120 Cl^-, 25 HCO_3^-, 1.2 SO_4^{--}, 1.2 $H_2PO_4^-$. The pH of this solution was adjusted by the addition of 6.5 ml of 1 N NaOH/l, such that gassing with 95% O_2 and 5% CO_2 resulted in a final pH of 7.4. The final osmolarity of this solution was

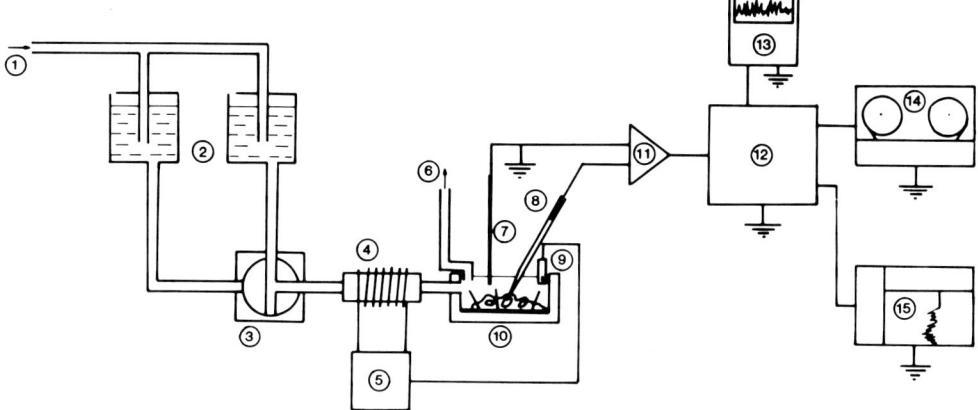

Fig. 1. Diagram of setup for electrophysiological studies. 1) Inlet for gas, 95% O_2/5% CO_2. 2) Beakers for holding solutions. 3) Four-way stopcock for selection of solution. 4) Heating coil. 5) Active temperature regulator. 6) Vacuum for maintaining level of solution in chamber. 7) Ag-AgCl reference electrode. 8) Glass microelectrode with Ag-AgCl internal electrode. 9) Thermistor for monitoring chamber temperature. 10) Lucite chamber, showing tissue pinned to black silicone sealant base. 11) FET-input high-impedance unity-gain isolation amplifier. 12) Control amplifier. 13) Oscilloscope monitor. 14) FM tape recorder. 15) Pen recorder.

290 mOsm as measured with an osmometer. Addition of glucose and other agents was made without adjusting for osmolarity changes. The perifusion of the chamber was achieved with a gravity feed system. Two beakers with attached IV drips were positioned above the perifusion chamber. The maximum flow rate was determined by a small length of 0.015 inch i.d. polyethylene (PE) tubing connecting the IV drip to the perifusion chamber. The height of the beakers was adjusted to obtain a flow rate of 2ml/min. By connecting more than one beaker through a stopcock, rapid changes of solution could be performed. The pH of the various bathing solutions was continuously monitored in the beakers with a Radiometer model 85 Research pH Meter. The perifusion solution was warmed to 37°C by means of a 2 inch long coil of nichrome heating wire. A small thermistor inserted close to the islet preparation was connected to a feedback system regulating current flow in the heating coil. This system was capable of maintaining chamber temperature constant to within 0.5°C. The level of the solution within the chamber was maintained by adjusting the height of a small glass tube attached to a vacuum line. The microelectrodes were prepared with 0.86 mm ID glass capillary tubing with an inner fiber (WP Instruments). The electrodes were pulled with a Narishige model PE2 electrode puller and were filled with 2 M potassium citrate, usually a few hours before needed. Electrode

resistances varied between 100 and 125 megohms. The membrane potentials were measured using Ag-AgCl electrodes, one in contact with the bathing solution and the other inserted in the microelectrode filling solution. The potentials were amplified by a Dagan model 8500 intracellular preamp clamp and were displayed on a Tektronix model 2601 storage oscilloscope. For permanent storage, a Gould series 2400 ink recorder and/or a Tandburg series 155 instrumentation recorder were used. The electrode was advanced using a Leitz micromanipulator for coarse positioning and a David Kopf model 607W stepping hydraulic drive for the final impalement attempt. The preparation was observed through a Zeiss dissection scope (20–100 ×). Initial impalements were made at a glucose level of 11.1 mM, since B cells were then easily recognized by their characteristic pattern of electrical activity [Atwater, 1980]. In general, impalement of B cells proved to be tedious, as the impaled cell often failed to seal properly around the electrode as evidenced by a "fading" of the spike activity with a concomitant depolarization of the membrane baseline potential to zero potential; such "fades" occurred either immediately or within 10 seconds to 2 minutes. It was found that the number of successful impalements was greatly enhanced by the use of positive feedback introduced electronically by advancing the "negative capacitance" control until preamp oscillation occurred. Best results were obtained when this period of artificially introduced oscillation was kept to less than 1 second. Following the initial impalement, the electrical activity was monitored for about 10 minutes on the oscilloscope; this examination period was generally sufficient to ascertain whether or not the impaled cell had successfully sealed around the microelectrode. It was found that a Faraday cage was not required for this setup. Rather, the chamber, micromanipulator, and dissection microscope were set on a large grounded metal plate. By proper grounding of all metal parts close to the preparation, minimal electrical interference was obtained. Illumination was provided by an American Optical fiber optic illuminator and was used only for positioning of the electrode near the islet.

C. ^{86}Rb$^+$ Efflux

For the determination of ^{86}Rb$^+$ efflux, groups of 10 to 15 islets were placed in microcentrifuge tubes, each containing 0.2 ml of Krebs-Ringer solution with the addition of 2.8 mM glucose. The pH was altered by adjusting the concentration of NaHCO$_3$ and equilibrating the solutions with 95% O$_2$ and 5% CO$_2$. The NaCl concentration was adjusted accordingly. After 30 minutes of preincubation at 37°C in an atmosphere of 95% O$_2$ and 5% CO$_2$, the tubes were centrifuged and the medium was withdrawn. A second aliquot of medium containing, in addition, 0.2 mM ^{86}RbCl (5–15 mCi/mmole) and 0.1 mM 6,6′n ^{3}H-sucrose (26 mCi/mmole), was added to each tube (Amersham). The islets were preloaded with the radioisotopes for 90 minutes for the efflux studies. The medium was subsequently removed, the islets were washed briefly, and 0.5 ml of nonradioactive medium containing the desired additions was added to each tube. Each tube was centrifuged at selected intervals in a model 152 Beckman Microfuge to deposit the islets in the tip of the tube. Di-n-butyl phthalate was then layered on top of the medium, and a second centrifugation separated the medium from the islet pellet. The bottom of the tube containing the islet pellet was cut off and placed in a scintillation vial. Hyamine was placed on top of the islets and the vials were incubated for 2 hours at 37°C. After mixing the islets with a neutralizing cocktail, the islets were examined for their $^{86}Rb^+$ and ^3H content. Appropriate corrections were made for 3% spillover of $^{86}Rb^+$ counts to the ^3H channel; the spillover of the ^3H counts to the $^{86}Rb^+$ channel was negligible. The $^{86}Rb^+$ remaining in the islets was corrected for the amount of $^{86}Rb^+$ residing in the extracellular space occupied by ^3H-sucrose. Samples of incubation media were used as external standards. Blanks without islets did not differ from the background of the counter.

D. Islet Perifusion System

The dynamic release of insulin from islets was studied in a perifusion system. The setup consists of a LKB 2132 Microperpex double-headed peristaltic pump with silastic tubing, i.d. 1.3 mm, connected to a 13 mm Swinnex filter chamber via PE tubing, 1.19 mm i.d. A 12 mm Nitex filter cloth with 10 μ pores (previously boiled in 0.7% $NaHCO_3$ and then distilled water) was placed on the surface of the plastic mesh in the chamber. The top of the chamber through which the perifusate was introduced was then screwed tightly in place. The chamber was filled with medium containing 100 islets. The chamber and attached tubing were placed in a water bath, and the perifusate was collected via a fraction collector. The medium used was Krebs-Ringer containing 0.3% bovine serum albumin. Changes in the pH were achieved by equilibrating the medium with different concentrations of $NaHCO_3$ against the same gas mixture of 95% O_2 and 5% CO_2. The NaCl concentration was modified to maintain isosmolarity and total Na^+ concentration. Changes in the medium were accomplished by means of two three-way stopcocks aligned in series. These in turn were connected to 100 ml plastic breakers via PE tubing inserted into and glued to their bases. The medium was drawn by the peristaltic action of the pump through the perifusion chamber at a flow rate of 1 ml/min. Lag time for the system (the time necessary for the medium to reach the fraction collector after switching solutions) was 1 minute as determined by passing trypan blue through the system. The islets were initially exposed to 2.8 mM glucose and samples were collected at 5 minute intervals for 30 minutes to establish a baseline of insulin release. The medium was then switched to one containing 16.7 mM glucose, and samples were collected at 1 minute intervals for 30–40 minutes, 2 minute intervals for 40–50 minutes, and 5 minute intervals for 50–60 minutes to determine if the islets were glucose-sensitive. The medium was then switched to one containing 16.7 mM glucose plus the test substance, and samples were collected in a manner identical to that described for 16.7 mM glucose. Samples were stored at $-20°C$ until assayed for insulin content over the range of 0–25 $\mu U/0.1$ ml using the alcohol precipitation and single antibody method [Wright et al, 1971]. Porcine insulin was used as the standard (courtesy of Eli Lilly & Co.) and pork insulin labeled with ^{125}I was obtained from New England Nuclear.

E. Acridine Orange Experiments

The coverslip with islet cells in monolayer formation was placed in a stainless steel Dvorak-Stotler controlled environment microperifusion chamber (Nicholson Precision Instruments). The medium used was Krebs-Ringer with HEPES equilibrated with 95% O_2 and 5% CO_2. The preparation was maintained at 37°C by means of a hotplate placed on the microscope stage. To the medium were added 25 μM acridine orange (AO) (Eastman Kodak) and the desired agents. The medium was placed in beakers and an IV drip was attached to the bottom of the beaker. The medium was perifused through 1.19 mm i.d. PE tubing into the chamber at a gravity flow rate of 1 ml/min. The preparation was subjected to each test solution for 10–20 minutes. Solutions were changed by means of a stopcock designed to prerinse the dead space without introducing bubbles [Kilb and Stampfli, 1974]. The approximate volume of the chamber was 0.5 ml.

AO not only is distributed into cells or organelles accumulating H^+, but also has a fluorescence spectrum that allows one to identify microscopically the response of cells to agents that modify H^+ gradients [Rabon et al, 1978; DiBona et al, 1979; Berglindh et al, 1980]. AO, a weak base, enters cellular compartments as a function of pH gradients. At high AO concentrations the fluorescence emission at 530 nm (green) quenches, and the fluorescence intensity at 624 nm (red) becomes prominent. Fluorescence was monitored with a Zeiss epiillumination system and an HBO 50 light source. Excitation was achieved by Hg lines 404.7 and 435 nm; emission was observed and photographed via a high-pass reflecting filter with cutoff at 510 nm.

F. Electron Microscopy

After appropriate treatment, the monolayer of islet cells was prefixed by perifusing with 2.5% glutaraldehyde in 0.2 M cacodylate buffer and postfixed in osmium tetroxide solution. After dehydration in ethanol, the islets were embedded in Polybed 812 (Polysciences) on the glass coverslip. Additional embedding medium was placed in a vial, which was then inverted on top of each monolayer. After hardening overnight in a 60°C oven, the glass coverslip was removed by subjecting it to liquid nitrogen immersion and gently removing the cracked glass with forceps and scalpel. The thin sections were stained with uranyl acetate and lead citrate for observation and photography with a Jeol 100CX-TEMSCAM electron microscope.

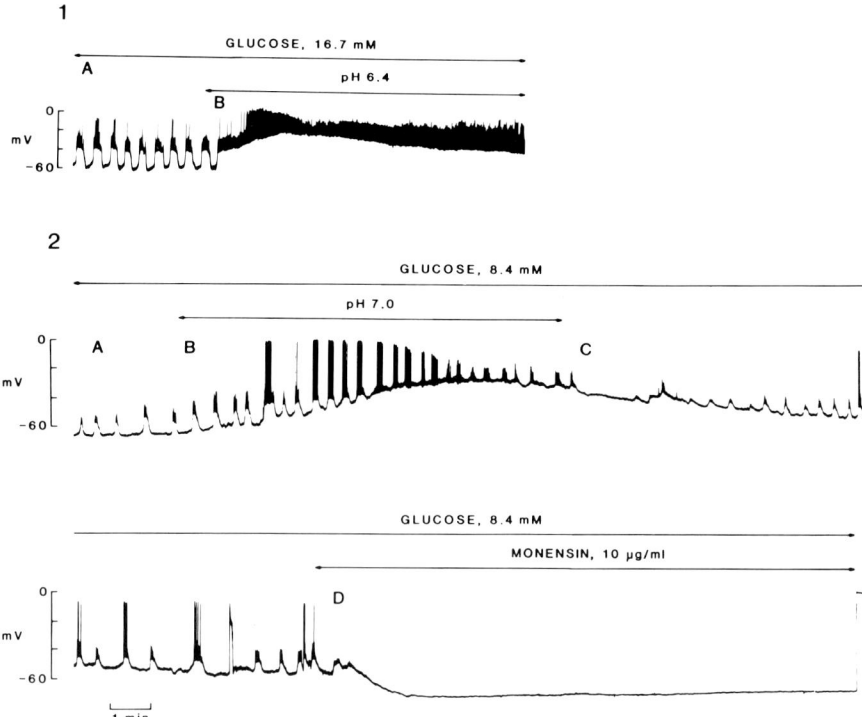

Fig. 2. Influence of extracellular acidification and monensin on glucose-induced electrical activity. Cell 1: At 16.7 mM glucose, effect of changing from pH 7.4 (A) to pH 6.4 (B). Cell 2: At 8.4 mM glucose, effect of changing from pH 7.4 (A) to pH 7.0 (B) followed by return to pH 7.4 (C), and addition of 10 μg/ml monensin (D).

III. RESULTS

A. Electrical Activity

The influence of pH on the electrical activity of B cells was examined by 1) changing the pH of the extracellular medium, and 2) modifying intracellular pH by means of a permeable weak base, imidazole, a permeable weak acid, glycodiazine, and monensin, an electroneutral Na:H antiporter.

In the presence of glucose, acidification of the extracellular medium resulted in enhancement of the electrical activity (Fig. 2). At 16.7 mM glucose a decrease in extracellular pH from 7.4 to 6.4 resulted in a change from a regular oscillatory pattern of electrical activity (Fig. 2–1A) to constant spike activity, including depolarization of the plateau level from which the fast spikes originated (Fig. 2–1B). Upon return to pH 7.4, the electrical activity returned to a pattern of regular oscillations (data not shown). The effect of a smaller decrease in intra-

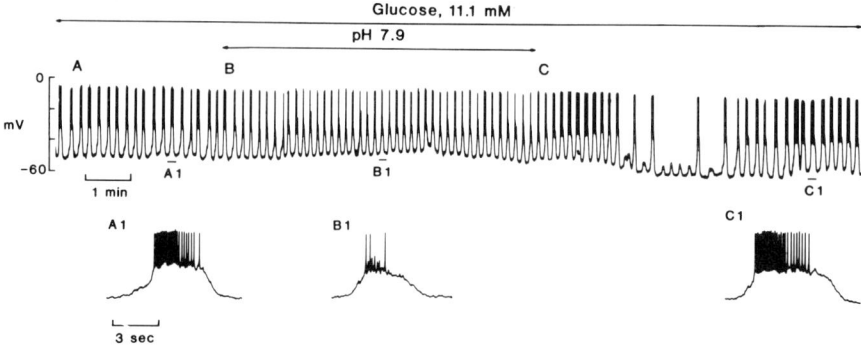

Fig. 3. Influence of extracellular alkalinization on glucose-induced electrical activity. At 11.1 mM glucose, effect of changing from pH 7.4 (A) to pH 7.9 (B) followed by return to pH 7.4 (C). A1, B1, and C1 show expanded records.

cellular pH at 8.4 mM glucose is shown in Fig. 2–2B. When the extracellular pH was lowered to 7.0 (Fig. 2–2B), there was an initial slight increase in the frequency of the oscillatory activity and the fast spike activity. After about 2 minutes the baseline potential started to decrease slowly or to depolarize until the bursts of spike activity were generated without a significant rapid depolarization phase. As this slow baseline depolarization progressed, there was also a diminution of spike amplitude. Following a return to pH 7.4 (Fig. 2–2C), the membrane potential slowly increased or hyperpolarized with a concomitant return to the pattern of electrical activity observed previously at pH 7.4. Subsequent addition of 10 μg/ml monensin resulted in disappearance of the electrical activity accompanied by baseline hyperpolarization (Fig. 2–2D). This hyperpolarization and inhibition of electrical activity was not reversed either by the removal of monensin or by the addition of 0.1 mM ouabain (data not shown).

The effect of extracellular alkalinization is shown in Figure 3. At 11.1 mM glucose an increase in pH from 7.4 (Fig. 3A) to 7.9 (Fig. 3B) resulted in a decrease in the number of fast spikes generated during each slow wave of depolarization. Averaging over 1 minute spans, the number of spikes decreased from 20.5 ± 0.8 (M ± SE) to 6.9 ± 0.4 per active phase (burst), and the frequency of the bursts increased from 5.0 to 6.4 per minute. Following the return to pH 7.4 (Fig. 3C), the number of spikes generated during the active phases increased to 23 ± 1. In other cells, similar results were observed with alkaline pH values of up to 8.2 (data not shown). The brief period of inhibition of burst activity observed a few minutes after return to pH 7.4 was consistent with results obtained during recovery from two other pH changes with this cell, but was not observed in other cells (data not shown), suggesting that this effect may have been peculiar to this cell.

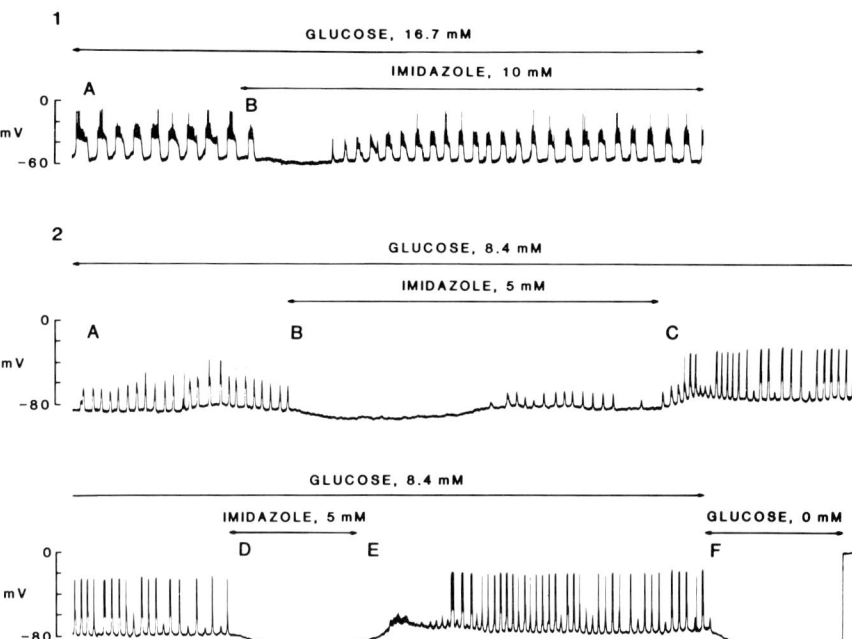

Fig. 4. Influence of the weak base imidazole on glucose-induced electrical activity. Cell 1: At 16.7 mM glucose and pH 7.4 (A), effect of 10 mM imidazole (B). Cell 2: At 8.4 mM glucose and pH 7.4 (A), effect of 5 mM imidazole (B) followed by the removal of imidazole (C). The addition (D) and removal (E) of imidazole is repeated a second time. At F, glucose is removed.

In another series of experiments, the effects of permeable buffers on electrical activity were investigated. When 10 mM imidazole, a weak base (pK_a, 6.9), was added in the presence of 16.7 mM glucose, there was an initial loss of electrical activity, followed by escape and return to an oscillatory pattern of electrical activity (Fig. 4–1B). Similar effects were observed with 8.4 mM glucose and 5 mM imidazole as shown in Fig. 4–2B and 2D. The silent phase elicited by imidazole was usually accompanied by a small increase in the membrane potential (Fig. 4–1B, 4–2B, 4–2D, and Fig. 5D). In the presence of 16.7 mM glucose there was no significant difference in the response elicited by 5 or 10 mM imidazole (data not shown). As shown in Figure 4F, the initial effect of imidazole was qualitatively similar to that elicited by removal of glucose. However, removal of glucose results in a sustained loss of electrical activity (data not shown) [see Atwater, 1980].

The effect of 5mM glycodiazine, a permeable weak acid (pK_a, 5.7), in the presence of 11 mM glucose is shown in Fig. 5B. As observed previously with

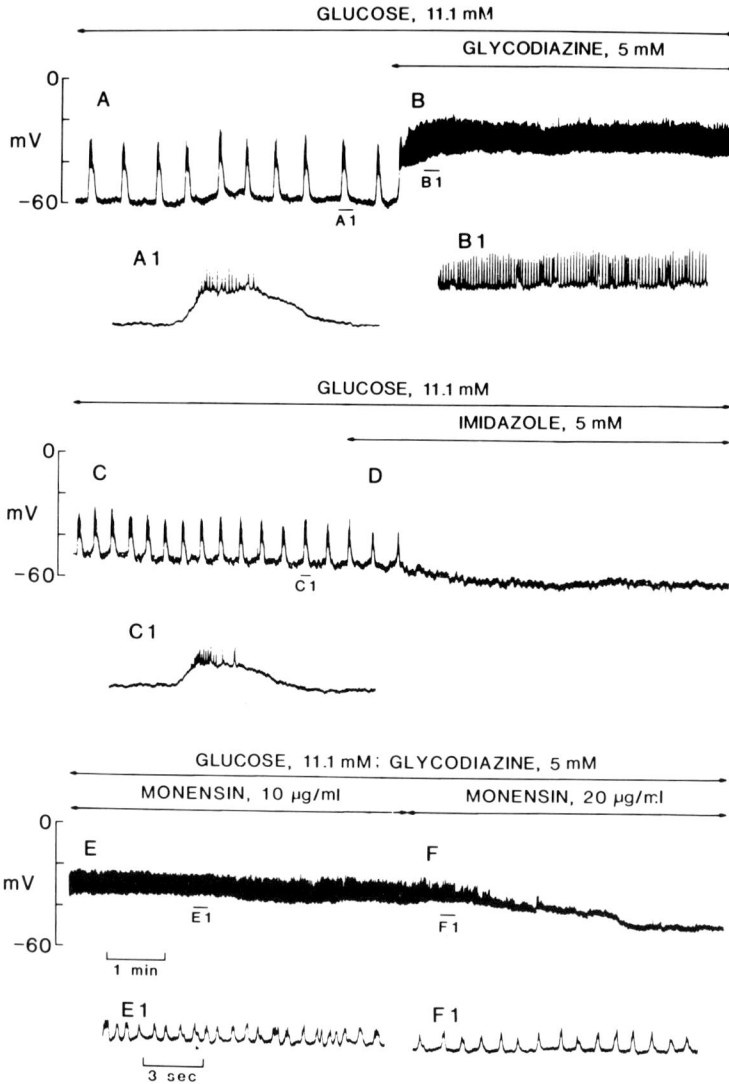

Fig. 5. Influence of the weak acid glycodiazine, the weak base imidazole, and monensin on glucose-induced electrical activity. At 11.1 mM glucose and pH 7.4 (A), effect of 5 mM glycodiazine (B). Following removal of glycodiazine and return to oscillatory electrical activity (C), effect of 5 mM imidazole (D). Following removal of imidazole and elicitation of constant spike activity by glycodiazine (not shown), effect of 10 μg/ml (E) and 20 μg/ml (F) monensin. A1 through F1 show expanded records.

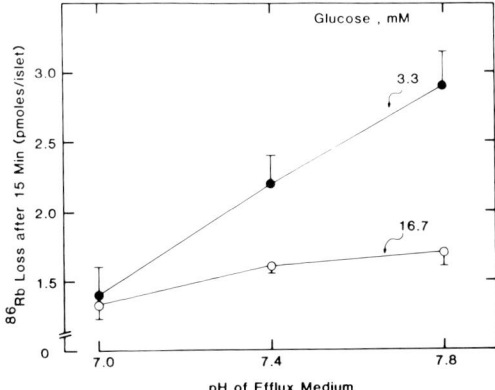

Fig. 6. Influence of pH on $^{86}Rb^+$ efflux from islet cells. Islets preloaded with $^{86}Rb^+$ and ^3H-sucrose were exposed to medium containing 3.3 or 16.7 mM glucose. The medium was adjusted to pH 7.0, 7.4, or 7.8 by altering the concentration of HCO_3^- and equilibrating the medium with 5% CO_2. Samples were taken at 0, 2, 5, 10, and 15 minute intervals. The $^{86}Rb^+$ content of the islets at 0 time was determined by extrapolation of a line drawn according to single exponential curve fits obtained by the method of least squares. The $^{86}Rb^+$ loss was calculated by subtracting the $^{86}Rb^+$ content of the islets at 15 minutes from the 0 time value. The number of experiments for 3.3 mM glucose and pH 7.0 was 4, 8 for pH 7.4, 4 for pH 7.8; and for 16.7 mM glucose and pH 7.0 was 5, 15 for pH 7.4, 5 for pH 7.8. Each value with the vertical bar represents the mean ± SE.

a decrease in extracellular pH (Fig. 2–1B), glycodiazine induced a rapid change from oscillatory electrical activity (Fig. 5A) to plateau depolarization and constant spike activity (Fig. 5B). Following the removal of glycodiazine, the oscillatory pattern of electrical activity returned within 15 minutes (Fig. 4C). Subsequent addition of 5 mM imidazole induced a sustained inhibition of electrical activity and baseline hyperpolarization. Following removal of imidazole, constant spike activity was again elicited with glycodiazine (data not shown). The effect of monensin on this constant spike activity is shown in Figure 5E and F. Monensin, 10 μg/ml, resulted in a decrease of spike frequency (compare Fig. 5–E1 to B1; the spike frequency just preceding Fig. 5E was comparable to that shown in Fig. 5B). Elevation of monensin to 20 μg/ml (Fig. 5F) resulted in a further decrease of spike frequency (Fig. 5F1) and an eventual loss of spike activity accompanied by hyperpolarization. The loss of electrical activity with monensin was not reversible.

B. $^{86}Rb^+$ Efflux

The influence of altering the pH of the extracellular medium on $^{86}Rb^+$ efflux from preloaded islets in the presence of 3.3 or 16.7 mM glucose is shown in Figure 6. At pH 7.0 the amount of $^{86}Rb^+$ lost from the islet cells during a 15

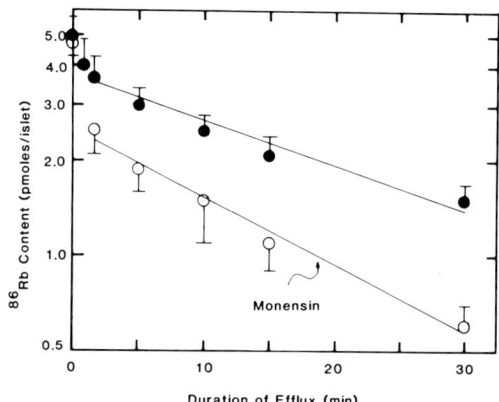

Fig. 7. Influence of monensin on the rate of $^{86}Rb^+$ efflux from islets in the presence of glucose. Islets preloaded with $^{86}Rb^+$ and ^3H-sucrose were incubated in a medium containing 16.7 mM glucose in the absence or presence of 10 µg/ml monensin for 0–15 minutes. Each value represents the mean ± SE of at least five individual experiments. The lines are single exponential curve fits obtained by the method of least squares.

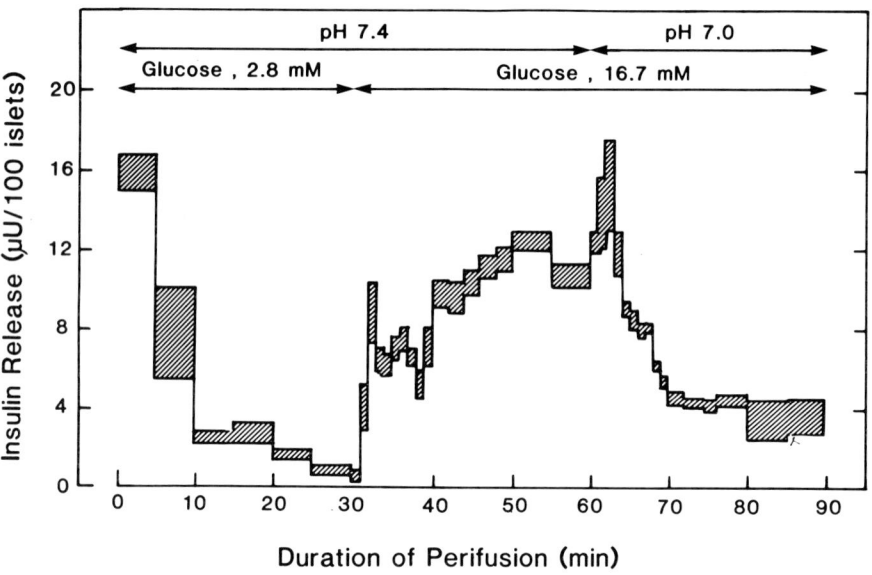

Fig. 8. Influence of acidification of the extracellular medium on glucose-induced insulin release from perifused islets. The values for insulin release are the means of three experiments, and the shaded area under the curve is the SE of the mean.

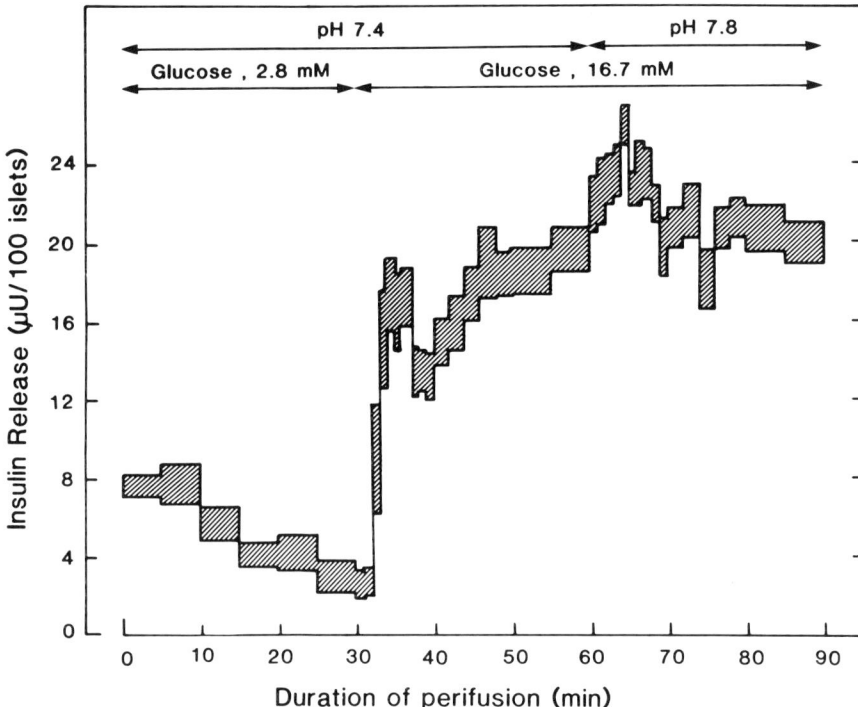

Fig. 9. Influence of alkalinization of the extracellular medium on glucose-induced insulin release from perifused islets. The shaded area under the curve is the SE of the mean of four experiments.

minute period was identical whether the efflux medium contained low or high glucose. With low glucose, increasing the medium pH from 7.0 to 7.4 resulted in a substantial ($P < 0.05$) increase in the exit of $^{86}Rb^+$ from the cells; with high glucose there was a smaller but significant ($P < 0.05$) increment in the loss of $^{86}Rb^+$. A medium pH of 7.8 further increased the exit of $^{86}Rb^+$ from the islet cells exposed to low glucose, but not high glucose.

The addition of 10 μg/ml monensin to a medium containing 16.7 mM glucose (pH 7.4) increased the rate constant for the efflux of $^{86}Rb^+$ from islet cells from 0.03 to 0.05 min^{-1} (Fig. 7).

C. Insulin Release

The influence of altering the pH of the medium from 7.4 to 7.0 or 7.8 on the secretory response of islets exposed to 16.7 mM glucose is shown in Figures 8 and 9, respectively. Alteration of the pH to 7.0 elicited no change in the amount of insulin released during the first 10 minutes (first phase), but substantially

TABLE I. Influence of Changes in Extracellular pH on Glucose-Induced Insulin Release

Experiment	Glucose mM	Medium pH	Period	Relative value for total insulin release (M ± SE)	P value
A N = 3	2.8	7.4	Control	1.0 ± 0.3	
	16.7	7.4	1st Phase	5.2 ± 1.0	
	16.7	7.4	2nd Phase	16.4 ± 1.6	
	16.7	7.0	1st Phase	7.5 ± 1.2	NS
	16.7	7.0	2nd Phase	6.3 ± 1.8	< 0.001
B N = 4	2.8	7.4	Control	1.0 ± 0.4	
	16.7	7.4	1st Phase	3.8 ± 0.6	
	16.7	7.4	2nd Phase	8.6 ± 1.1	
	16.7	7.8	1st Phase	5.4 ± 0.5	NS
	16.7	7.8	2nd Phase	9.6 ± 1.0	NS

Values for insulin release were obtained by integrating the areas under the curves for the secretory responses shown in Figures 8 and 9. All data are expressed as normalized values relative to the mean control value found between 20 and 30 minutes of islet perifusion with 2.8 mM glucose. First phase represents the total insulin released during the first 10 minutes after changing the solution, and second phase represents the total insulin released during the ensuing 10 to 30 minutes. P values indicate the level of significance of the difference between the indicated phase and the corresponding phase before the pH was changed. NS = not significant.

inhibited the total amount of insulin released 20–40 minutes (second phase) after changing the solution (Table I). Changing the medium pH from 7.4 to 7.8 did not change the amount of insulin released during first or second phase (Table I). Monensin (10 μg/ml) was found to inhibit the secretory response to 11.1 and 16.7 mM glucose (Fig. 10). This inhibition was manifested during second-phase insulin release. (The decline in the amount of insulin release to basal values from 0 to 30 minutes represents washout of insulin accumulated in the extracellular space during the isolation procedure.)

D. Acridine Orange

In initial studies the concentration of AO in glucose-free medium was varied, and it was found that with a concentration equal to or greater than 10^{-4} M the cytoplasm of the B cells in monolayer culture showed green to reddish-orange fluorescence, whereas the granules fluoresced red. However, at a concentration of 10–25 μM AO, the granules and cytoplasm were green in the absence of medium glucose (Plate IIA) (see page xxx). A level of 25 μM AO was chosen for further studies. The addition of 16.7 or 27.8 mM glucose to the medium perifusing the islet cell cultures resulted in an increased uptake of the dye into intracellular compartments assumed to be secretory granules (Plate IIB). The red fluorescence was observed to be contained in granular structures upon examination of the cells with a combination of fluorescent and/or interference

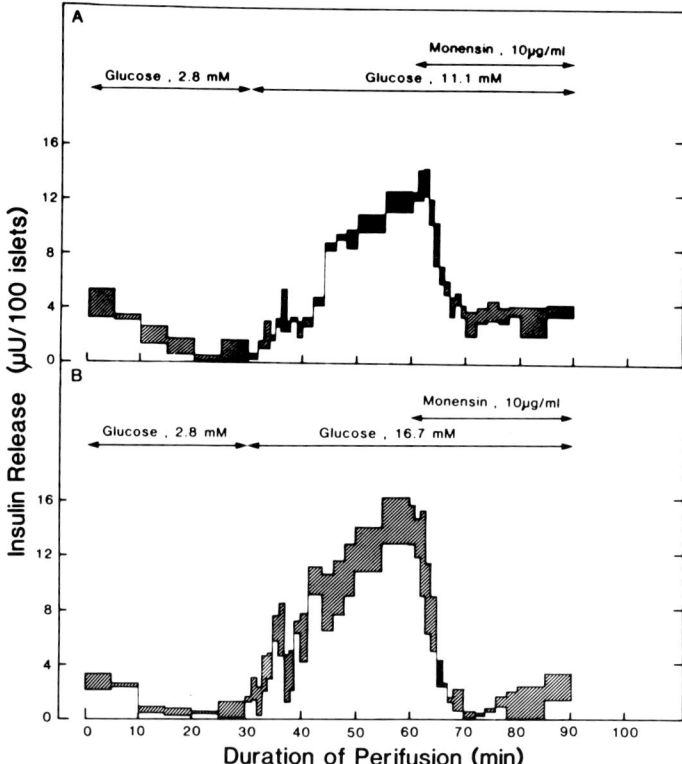

Fig. 10. Influence of monensin on glucose-induced insulin release. The secretory response is the mean of two experiments for each of the curves in A and B. The shaded area represents one-half of the range of the mean.

phase contrast optics (data not shown). These results indicated that glucose induces an acidification and therefore a fluorescence shift of accumulated AO in what appears to be the secretory granules. If a decrease in pH in the granule compartment were a major determinant of the AO gradient, the addition of monensin, an electroneutral Na:H antiporter, would be expected to dissipate the H^+ gradient and accumulated AO. This prediction was supported by the results shown in Plate IIC and IID which show the influence of 10 μg/ml monensin on B cells exposed to 27.8 mM glucose 10 and 30 minutes after application, respectively. Actually, monensin was found to induce a change from red to green fluorescence within 2 minutes. Nigericin, 1 μg/ml, a K:H antiporter, also was found to induce the loss of glucose-induced red fluorescence (data not shown).

The weak base, aminopyrine, has been found to accumulate in the gastric parietal cell as a function of acid secretion. The site of aminopyrine accumulation

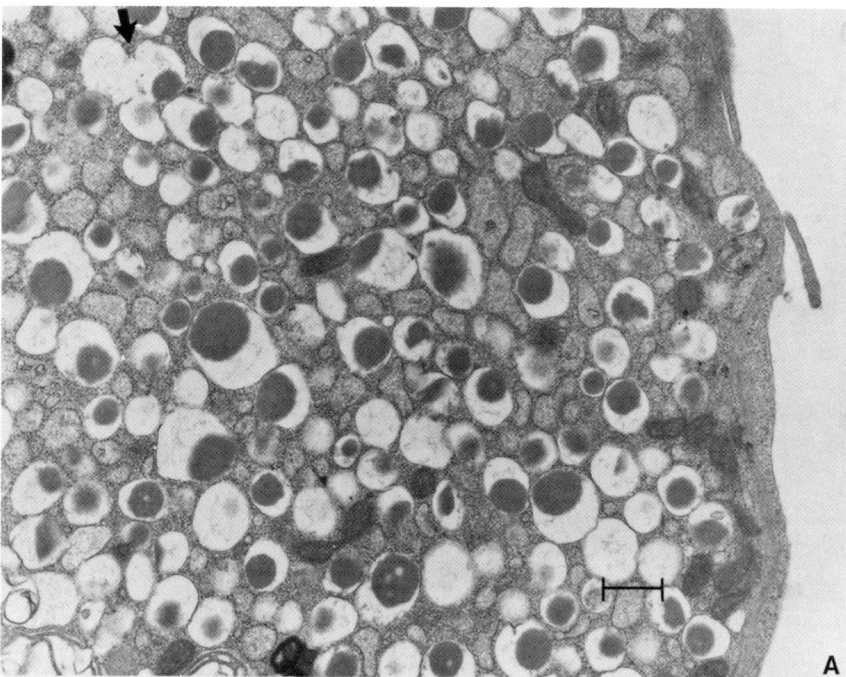

Fig. 11. Electron micrographs of B cells in monolayer culture treated with glucose and benzylamine. The islet cells were exposed to a solution containing 27.8 mM glucose and 25 μM AO for 10 minutes, which resulted in red fluorescence of the secretory granules. Subsequent addition of 1.0 mM benzylamine to the solution induced swelling of the secretory granules (see Plate IIE). After 10 minutes the islet cells were fixed for electron microscopy. On examination of several sections it was clear that the secretory granules had swollen. The mean diameter of the granules was 0.41 μ with a range of 0.21 μ to 0.81 μ. It was also evident that the space occupied by the secretory granules could account for the space containing the red fluorescent granules as shown in Plate II. A) This section illustrates the numerous secretory granules, many of which have swollen to a diameter of 0.7 μ. The arrow points to three granules that have apparently fused. Calibration bar is 0.5 μ. B) The enlarged size and close apposition of the secretory granules are illustrated. Fusion of membranes of contiguous secretory granules is indicated by the arrows. Calibration bar is 0.2 μ.

has been identified with the use of AO to be within an expanded vesicular compartment called the secretory canaliculus [DiBona et al, 1979]. In fact, the addition of 1.0 mM aminopyrine in the presence of high K^+ was found to induce a morphological transformation of the parietal cell to the secretory state, eg, expansion of the secretory canaliculus apparently via osmotic expansion of an intracellular membrane system. This phenomenon was thought to occur as a result of the formation of the impermeant protonated form of aminopyrine.

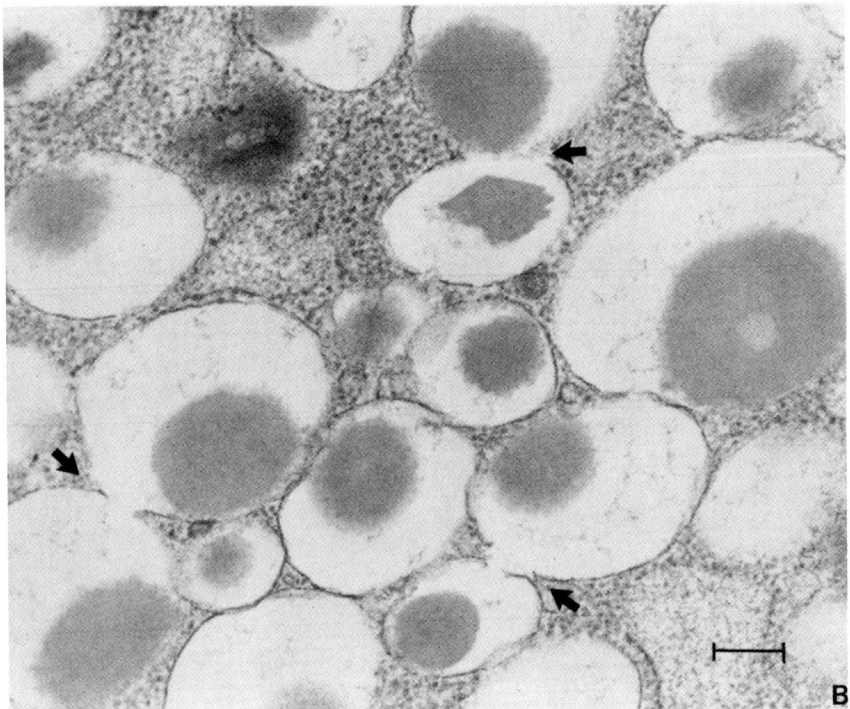

Aminopyrine buffering of the H^+ would be expected to stimulate H^+ pump activity, in order to maintain a fixed H^+ gradient resulting in an increase in osmolarity due to the continued formation of RNH^+ and consequent swelling of the aminopyrine space. With this in mind, we decided to use a permeable weak base, benzylamine, in conjunction with glucose-stimulated islet cell cultures to determine whether the sites of AO accumulation would undergo osmotic expansion. Application of 1.0 mM benzylamine resulted in swelling of the red fluorescent compartments into which AO had accumulated in response to 27.8 mM glucose (Plate IIE). Monolayer cultures subjected to this treatment were fixed for electron microscopic examination in order to determine whether the expanded spaces occupied by the AO could be accounted for by the presence of secretory granules. The structure of the secretory granules is shown in Figure 11. It is clear that the space occupied by the secretory granules can account for the accumulation of AO into the B cells (compare Fig. 11A with Plate IIE). Furthermore, the granules are swollen to a mean diameter of 0.41 μ. Secretory granules of B cells have been reported to have a mean diameter of 0.29 μ [Dean, 1973]. This compares favorably with the 0.33 μ granule diameter of B cells of intact islets preserved and prepared for electron microscopy in our laboratory (electron micrographs were the courtesy of Dr. M. Greider).

IV. DISCUSSION

Recognition of glucose by the pancreatic B cell initiates a sequence of cationic, metabolic, and motile events leading to the release of insulin. In view of the widely held view that factors generated from glucose metabolism are intricately involved in triggering insulin release, it is essential to determine what metabolic coupling factor serves to trigger or modulate ionic events in the B cell membrane associated with depolarization and spike activity. A leading candidate has been proposed to be the increased level of H^+ stemming from metabolism of nutrients [Malaisse et al, 1980]. Although it has not been possible to detect changes in the pH_i due to glucose using ^{14}C-DMO [Carpinelli et al, 1980; Sener et al, 1978], glucose has been found to provoke dose-related increments in H^+ net output from islets [Malaisse et al, 1979a]. In fact, the glucose dose-response curves for H^+ output, the decrease in $^{86}Rb^+$ fractional outflow rate, and phase one of $^{45}Ca^{++}$ efflux display a half maximal change at a glucose concentration of about 3 mM [Malaisse, et al, 1979a, b]. A level of 3 mM glucose is subthreshold for the glucose-induced changes in the generation of NAD(P)H, the second phase of $^{45}Ca^{++}$ efflux, and insulin release. Hence, based on glucose sensitivity it is possible that remodeling of selected ionic events are closely associated with the generation of H^+. Despite the inability to document changes in pH_i induced by glucose it has been calculated, taking into consideration the buffering capacity of islet homogenates, that a fall of pH_i by a unit of 0.09 below the basal value may support coupling between metabolic and ionic events [Hutton et al, 1980]. It is also possible that H^+ generated from glucose metabolism are buffered by passive or active transport into secretory granules, since these intracellular compartments have been shown to have an acidic pH [Abrahamsson and Gylfe, 1980] compared to a cystolic pH of about 7.1 [Malaisse et al, 1979a]. Furthermore, it has been proposed that H^+ may participate in Ca^{++} buffering in secretory granules by an electroneutral exchange mechanism [Hellman et al, 1980]. Based on this hypothesis, the prediction can be made that glucose augments H^+ uptake into the granule compartment to maintain the H^+ concentration for subsequent exchange with an increased challenge to cytosolic Ca^{++} that also occurs as a consequence of glucose stimulation. We have examined this possibility by using AO as an indicator of changes in pH gradients in subcellular compartments.

In essence, H^+ may be involved in many phases of stimulus-secretion coupling, thereby serving a multifactorial role similar to that documented for Ca^{++}.

A. Influence of Protons on Electrical Activity and $^{86}Rb^+$ Efflux

Moderate concentrations of glucose initiate a reduction in P_K resulting in cyclic depolarization of the plasma membrane to a plateau level from which Ca^{++}-dependent spike activity is generated [Atwater, 1980]. High concentrations of glucose evoke sustained depolarization and constant spike activity. There is no evidence to support the postulate that cyclic changes in the level of a factor stemming from glucose metabolism paces the cyclic nature of the electrical activity. However, glycolysis occurs in an oscillatory manner via sensitivity of key enzymes to pH in yeast and ascites cells [Rapp, 1979]. The pH_i in red blood cells [Tomoda et al, 1977] and leukocytes [Halperin et al, 1969] also influences glycolysis such that a rise in pH increases the rate of glycolysis due primarily to activation of phosphofructokinase. It is possible, but not yet documented, that glycolysis in B cells occurs in a pH-sensitive oscillatory manner. Thus far it has been shown that intracellular acidification decreases glucose oxidation after an initial lag [Carpinelli et al, 1980]. Some recent observations suggest that insulin may modulate the glycolytic rate of skeletal muscle via changes in pH_i that may occur by stimulation of Na:H exchange in the sarcolemma [Moore et al, 1979; Moore, 1979, 1981 and this volume]. This hypothesis is supported by the finding that amiloride or a decrease in extracellular Na^+ blocked insulin-induced increases in pH and the rate of glycolysis [Moore et al, 1979]. In fact, the absence of extracellular Na^+ transformed the effect of insulin to a decrease in pH and an inhibition of glycolysis [Moore, 1981]. In the B cell activation of Na:H exchange would serve to increase pH_i thereby increasing P_K and hyperpolarizing the cell as well as increasing the glycolytic rate. The increase in the generation of H^+ would then reinitiate the burst pattern by decreasing P_K (see model in Fig. 12).

Exposure of islets to extracellular acidosis (pH 6.3) elicits a ^{14}C-DMO detectable decrease in pH_i (from 7.2 to 6.8) [Sener et al, 1978]. Our results demonstrate that lowering medium pH augments glucose-induced electrical activity to constant spike activity. Since glucose may produce only a modest change in pH_i, it is possible that a larger change may prevent the oscillatory nature of electrical activity and reduce P_K to the extent that depolarization and constant spike activity are obtained. Alkalinization of the medium produced a decrease in the spike frequency, but did not result in transient or sustained cessation of spike activity as obtained with imidazole and monensin, respectively. The addition of a permeant weak base to the medium at pH 7.4 such as imidazole with a pK_a of 6.9 will result in rapid entry of the neutral form of the base, and rapid alkalinization. Further entry of the neutral form will occur until the ratio of charged to neutral form reflects the pH of the intracellular compartment. The effect of the process should be to alkalinize the cell interior transiently, resulting in a transient hyperpolarization and cessation of spike activity as was found.

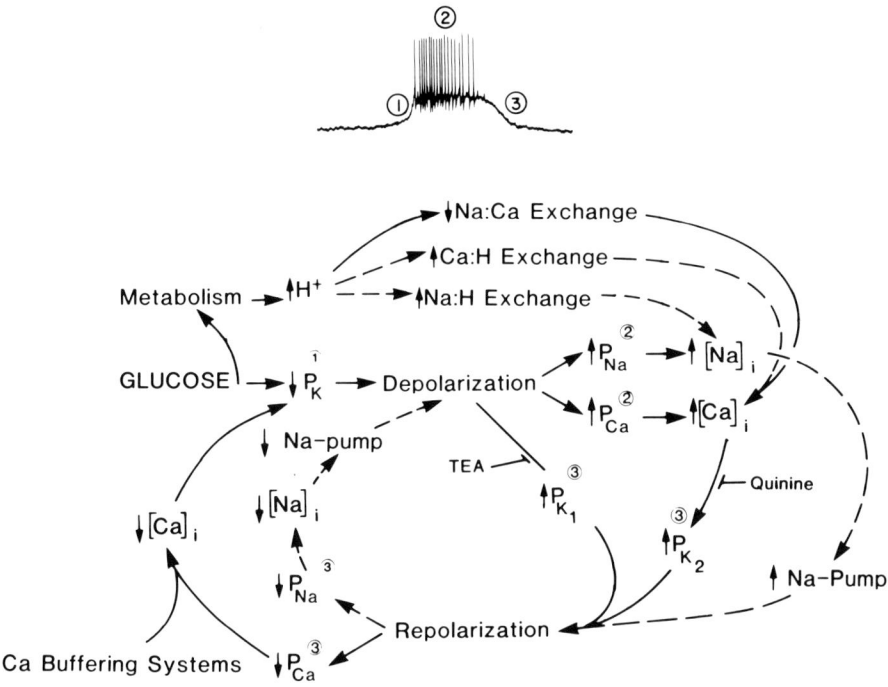

Fig. 12. Model to account for the mechanism underlying the cyclic pattern of electrical activity induced by glucose. The silent phase is terminated by a reduction in P_K leading to depolarization (1). The production of H^+ due to glucose metabolism may further decrease P_K by binding to sites on the K^+ channel. Generation of protons may also inhibit Na:Ca exchange, resulting in a decrease in Ca^{++} efflux [Carpinelli et al, 1980]. This may be accompanied by augmented Ca:H and Na:H exchange. Depolarization activates P_{Ca} and P_{Na}, initiating the active phase of spike activity (2). The voltage-dependent component of P_K is blocked by tetraethylammonium (TEA); the increase in $[Ca^{++}]_i$ activates the Ca-sensitive component of P_K which is blocked by quinine. The Na^+ pump is stimulated by an increase in $[Na^+]_i$ resulting in electrogenic pump current. Repolarization (3) occurs as a consequence of the increase in P_K and the electrogenic pump current. This results in a decrease in P_{Ca} and P_{Na} and a cessation of spike activity. $[Ca^{++}]_i$ is reduced by intracellular buffering systems, such as the mitochondria and secretory granules resulting in a decrease in P_K. A decrease in $[Na^+]_i$ reduces the activity of the Na^+ pump. The relative importance of the Na^+ pump to the depolarization and repolarization phases 1 and 3 is in question. Activity of the Na^+ pump may serve to provide background electrogenic current and maintain ionic gradients, both of which allow optimal operation of a voltage- and Ca-sensitive P_K.

Conversely, the addition of a weak acid such as glycodiazine, with a pK_a of 5.7, will result in equilibration of the neutral form, with release of protons intracellularly until the ratio of the neutral to the charged form of the weak acid reflects the intracellular pH. In this case, the reduction in pH_i superimposed on the hypothetical glucose-induced reduction of pH_i should result in depolarization and constant spike activity, as was observed.

A recent study demonstrated that the rate of $^{86}Rb^+$ efflux from islet cells, in the absence of glucose, was altered by changes in extracellular pH [Henquin, 1981]. This has been confirmed by our observations showing that $^{86}Rb^+$ efflux from islets increased in a linear manner in relation to changes of medium pH from 7.0 to 7.8. Specifically, the rate of efflux of $^{86}Rb^+$ was decreased by acidification of the medium, whereas alkalinization enhanced the rate of $^{86}Rb^+$ efflux. In other studies, manipulations intended to alter pH_i induced only transient changes in $^{86}Rb^+$ efflux, but these changes were in the expected direction [Henquin, 1981]. The transient changes in $^{86}Rb^+$ or P_K are reminiscent of the transient effect of imidazole on the electrical activity.

The glucose dose-response change in $^{86}Rb^+$ efflux begins to plateau above a glucose level of 5.6 mM and reaches a maximum level at about 16.7 mM [Malaisse et al, 1979b]. In view of this, it is not surprising that there is little or no change in the rate of $^{86}Rb^+$ efflux from islets when the pH of the medium is altered from 7.0 to 7.8. It is not known to what extent the pH_i is altered by a change in extracellular pH over this range, but it is evident that 16.7 mM glucose elicits a nearly pH-insensitive decrease in P_K. It is possible that glucose metabolism is altered due to modifications of pH_i and obscures any direct influence of alterations in pH_i on P_K. It has been anticipated that the K^+ channel in the B cell plasma membrane is only indirectly influenced by pH_i, and is primarily responsive to the intracellular concentration of Ca^{++} (see Fig. 12). Accordingly, a low pH was found to decrease Ca^{++} uptake in islet cells [Carpinelli et al, 1980], and this would be expected to lead to a decrease in P_K [Atwater, 1980]. However, the influence of pH on $^{86}Rb^+$ efflux from islet cells persists despite the absence of extracellular Ca^{++}, suggesting a direct action of H^+ on K^+ channels [Henquin, 1981].

In the slow muscle fiber of crayfish a decrease in pH_i to 6.4 elicits all-or-none Ca^{++} action potentials. This occurs owing to a substantial reduction of delayed rectification due to outward K^+ current. This can be predicted if the permeation of the channel to K^+ depends on binding to a negatively charged group with a pK_a of about 6.4. A reduction of pH_i would result in neutralization of this group and a decrease in P_K. The reduction of the delayed outward K^+ current by tetraethylammonium (TEA) also elicits Ca^{++} action potentials in the crayfish muscle fibre [Moody, 1980]. Blockade of voltage-sensitive changes in P_K in the B cell plasma membrane also results in the obliteration of cyclic changes in P_K and the generation of Ca^{++} spikes of enhanced magnitude [Atwater, 1980]. This parallel suggests that the K^+ channel in the B cell is qualitatively similar to that in the crayfish muscle fiber. As such, changes in pH_i may play an important role in the regulation of Ca^{++} spikes and the entry of Ca^{++} into the cytosol via Ca^{++} channels. Monensin was used as another mechanism to

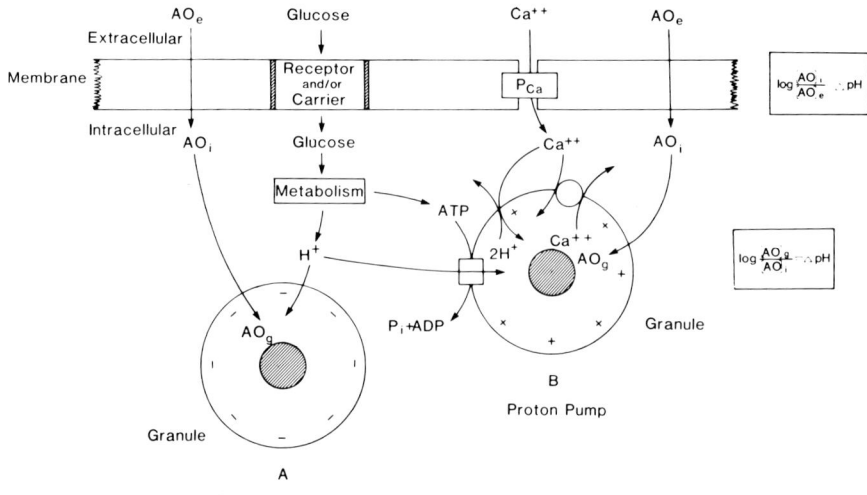

Fig. 13. Model to account for glucose-induced accumulation of protons (H^+) and acridine orange (AO) into the secretory granules of B cells. Glucose binds to a receptor in the plasma membrane. Activation of this receptor results in an increase in Ca^{++} influx into the cytoplasm. Glucose metabolism generates H^+ and ATP. The H^+ may enter the secretory granules as a consequence of Donnan distribution. On the other hand, the increased supply of ATP and H^+ may augment the activity of an ATP-dependent H^+ pump, for the active transport of H^+ into the secretory granule against an electrochemical gradient. The H^+ in the granules may then participate in Ca^{++} buffering by means of Ca:H exchange. Ca^{++} may also enter the granule via Ca:Ca exchange. AO added to the extracellular medium at pH 7.4 traverses the plasma membrane and accumulates in the intracellular space of cytoplasm according to the pH gradient. The AO would then redistribute across the membrane separating the cytoplasm and the secretory granule space as a function of the pH gradient. A greater pH gradient across the secretory granule membrane than that across the plasma membrane would lead to accumulation of AO into the granule space and quenching of the green fluorescence yielding red fluorescence.

alter pH_i: With a prevailing inward Na^+ gradient, Na^+ will enter and H^+ will leave the cell, leading to alkalinization of the cytosol. The addition of monensin resulted in hyperpolarization accompanied by cessation of electrical activity induced by high glucose or glucose plus glycodiazine. Whereas this result is consistent with a direct influence of H^+ on K^+ channels, it is also possible that this hyperpolarization resulted from the increase in intracellular Na^+ leading to a stimulation of an electrogenic Na^+ pump. This alternative effect of monensin has been suggested to occur in a neuroblastoma–glioma hybrid [Litchtshtein et al, 1979] and in 3T3 cells [Smith and Rozengurt, 1978]. However, whereas ouabain inhibited the effect of monensin in the neuroblastoma cells [Litchtshtein et al, 1979], the addition of 0.1 mM ouabain did not influence the effect of monensin in our system (data not shown). This level of ouabain has been shown to depolarize glucose-stimulated cells leading to continuous spike activity [Atwater, 1980]. The extent to which an electrogenic Na pump contributes to the electrical events in islet cells is not clear, but it is likely that the Na pump does not directly control the oscillatory pattern of the electrical activity but, instead, maintains the transmembrane potential at a level against which cyclic variations in ionic conductances can control the oscillatory characteristics of the electrical events [Tarvin and Pace, 1981].

Although it is not yet possible to determine the extent to which alteration of the pH in the extracellular or intracellular space is capable of influencing P_K or electrical activity in B cells, it is evident from our results that the ionic conductances subserving the oscillatory nature of the spike activity are profoundly influenced by the concentration of H^+.

B. Influence of pH on Glucose-Induced Insulin Release

Changes in medium pH have been found to have no effect on basal insulin release and to have a varying influence on insulin release depending on the concentration of glucose [Hutton et al, 1980]. At low or 6.7 mM glucose, an extracellular pH of 7.0 elicted a maximal secretory response. The optimal pH for obtaining a maximum secretory response shifted to alkaline values as the concentration of glucose was increased. These results are difficult to explain in terms of coupling of ionic events to the secretory response. We have shown alteration of medium pH from 7.4 to 7.0 to inhibit insulin release due to 16.7 mM glucose, whereas alkalinization of the medium to pH 7.8 has no influence on the glucose-induced secretory response. It is possible that sustained changes in pH prevent the normal coupling of ionic events to the secretory response, and that the coupling is maintained only in the presence of small or oscillatory changes in pH_i, which are prevented by the application of a sustained change in medium pH. Furthermore, it is possible that the reduction in extracellular pH influences exocytosis of the secretory granules based on the chemiosmotic hypothesis. According to this theory, granules fused to the plasma membrane are subjected to anion entry from the extracellular space resulting in osmotic lysis [Somers et al, 1980]. In the parathyroid cell, release of parathyroid hormone (PTH) is inhibited by only 30% when Cl^- is replaced by the impermeant ion isethionate [Pollard et al, 1979]. Lowering the extracellular pH from 7.5 to 7.0 resulted in a substantial inhibition of PTH release and was not increased by raising the concentration of HCO_3^-. Probenecid, which is a competitive inhibitor for the entry of OH^- as well as Cl^-, also resulted in substantial inhibition of PTH release. In fact, it has been found that probenecid inhibits insulin release to a greater extent than replacement of Cl^- with isethionate [Somers et al, 1980]. It is conceivable that the reduction of OH^- ions by a decrease in extracellular pH may partially inhibit insulin release owing to inhibition of OH^- influx and consequently, chemiosmotic lysis of the secretory granules. It is interesting that modification of extracellular pH to alkaline values did not augment insulin release and did not substantially inhibit glucose-induced electrical activity in terms of the percentage of time the electrical events were in the active, as opposed to the silent, phase.

Alteration of pH_i may more closely maintain coupling of ionic and secretory events. This was found to be the case with monensin-induced alkalinization of the intracellular environment. Thus glucose-induced electrical activity and insulin release were both substantially inhibited by monensin. Preliminary data have also indicated that imidazole partially inhibits insulin release due to 16.7 mM glucose, whereas sulfamerazine (weak acid) slightly augments insulin release due to 16.7 mM glucose (Joel Smith, unpublished results). These results are more consistent with the influence of a weak base and acid on glucose-induced electrical activity.

C. Glucose-Induced Proton Gradient in Secretory Granules

The existence of a proton translocation mechanism has been well documented in the chromaffin granule membrane [Scherman and Henry, 1980]. The function of the reduction of granule pH below that of the cytoplasm has been proposed 1) to provide energy for the concentration of amines within the granules, and 2) to participate in osmotic lysis of the granules, resulting in exocytosis of the granule contents. Secretory granules of mast cells and platelets have also been found to maintain an acidic pH. Neurosecretory vesicles isolated from the posterior pituitary and containing peptide hormones have been found to maintain a H^+ gradient which is increased by Mg-ATP [Russell and Holz, 1981]. Another role for maintenance of a low intragranular pH has been proposed for B cell granules in connection with the role of these structures in sequestering intracellular accumulation of Ca^{++} [Hellman et al, 1980]. The fluorescent weak base 9-aminoacridine was found to accumulate in isolated granules B cells and was released in the presence of NH_4Cl, which presumably alkalinized the intragranular space [Abrahamsson and Gylfe, 1980]. The possibility that generation of the H^+ gradient is dependent on ATP and participates in Ca^{++} uptake into the granules has been suggested [Hellman et al, 1979]. Accordingly, it has been found that 0.1–10 μM Ca^{++} resulted in an enhancement of the granule content of $^{45}Ca^{++}$. In the presence of Mg-ATP and increasing levels of Ca^{++}, there was a reduction in granule $^{45}Ca^{++}$ content without a change in insulin content. This was interpreted to support the presence of a Ca:Ca exchange mechanism in response to an increase in the concentration of Ca^{++} in the cytoplasm [Abrahamsson et al, 1981]. This is summarized in the model shown in Figure 13. It is conceivable that the increase in ATP generated by glucose metabolism enhances the activity of an ATP-dependent H^+ pump. The increase in cytosolic levels of Ca^{++} may then be taken up into the granules by exchange for H^+, a process that would further stimulate the activity of the H^+ pump. The increase in the granule content of Ca^{++} may then accelerate Ca:Ca exchange across the granule membrane. This process would experimentally result in a dilution of the $^{45}Ca^{++}$ content of the granule.

As indicated in Figure 13 the H^+ generated by glucose metabolism may also enter the secretory granule membrane in accordance with the principle of Donnan equilibrium.

Our evidence stemming from the changes in the fluorescence of AO contained in islet cells in monolayer culture indicates that glucose induces an accumulation of H^+ into the secretory granules. This is the first evidence that the granules accumulate H^+ in response to a stimulant, and supports the contention that an ATP-dependent H^+ pump subserves the mechanism for maintaining the Ca^{++} buffering capacity of the secretory granules. The dissipation of the H^+ gradient in the granules, as indicated by a change in AO fluorescence upon addition of monensin or nigericin, further indicates that the granules respond to glucose by accumulating H^+. The possibility exists that glucose increases cytosolic acidity, AO accumulation increases inside the cell, and a maintained H^+ distribution gradient between the cytosol and granule space results in the green-to-red fluorescence shift. The ability of monensin to dissipate the red fluorescence may be due to alkalinization of this compartment driven by the inward Na^+ gradient or loss of intragranular acidity due to Na:H exchange across the granule membrane. Partial discrimination between these two alternatives is achieved by the use of nigericin, a K:H electroneutral exchange ionophore. The cationic exchange is driven by the K^+ gradient. In this case, the cytosol would be subjected to an increase in pH. If there were a H^+ gradient across the granule membrane independent of cytosolic pH, then dissipation of the gradient due to K:H exchange would indicate that the action of both monensin and nigericin may be accounted for by the loss of a pH gradient across the granule membrane rather than alteration of cytosolic pH. Our results indicate that this is indeed the case.

The ability of the weak base benzylamine to accumulate in the granule compartment to the extent that it elicits granule swelling indicates that the base continued to cross the granule membrane in response to maintenance of the glucose-induced H^+ gradient by means of an H^+ pump. As benzylamine is protonated, it is no longer permeable, and as the protonated weak base accumulates in the granule compartment, it creates an osmotic gradient resulting in swelling of the granules.

It is possible that the presence of an ATP-dependent H^+ pump is necessary to catalyze the inward movement of H^+ as permeant counterions to Cl^-, as has been found to occur in chromaffin granules [Pollard et al, 1979]. It has been proposed that ATP evokes granule lysis by raising the osmotic content of the granule due to H^+ and Cl^- entry, resulting in lysis and release of epinephrine. Inhibition of anion permeation by substitution of Cl^- by isethionate, an impermeant anion, was found to inhibit lysis of chromaffin granules [Pollard et al,

1979]. Chloride substitution by isethionate has also been found to inhibit partially the second phase of glucose-induced insulin release, but did not influence the first phase of release [Somers et al, 1980]. This indicates that exocytosis in the B cell may not be dependent on extracellular Cl^- or may not be accounted for solely by osmotic lysis of secretory granules. The existence of an ATP-dependent H^+ pump in the B cell secretory granule membrane as well as the role of this pump in subserving a Ca^{++} buffering mechanism or chemiosmotic lysis remains to be documented or fully substantiated.

ACKNOWLEDGMENTS

This work was supported by National Institutes of Health grant AM21973. Caroline S. Pace is a recipient of a Research Career Development Award AM00499, and John T. Tarvin is the recipient of a postdoctoral fellowship from the Juvenile Diabetes Foundation.

We thank Dr. George Sachs for the use of his fluorescent microscope, and for his enthusiastic support, illuminating discussions, and helpful suggestions throughout the duration of this study. We gratefully acknowledge the cooperation of Dr. Don DiBona and Ms. Sandy Silvers for their expertise and facilities for obtaining the electron micrographs, and the excellent technical expertise of Mr. Douglas Swain and Mrs. Felecia Hester.

V. REFERENCES

Abrahamsson H, Gylfe E: Demonstration of a proton gradient across the insulin granule membrane. Acta Physiol Scand 109:113, 1980.

Abrahamsson H, Gylfe E, Hellman B: Influence of external calcium ions on labelled calcium efflux from pancreatic B-cells and insulin granules in mice. J Physiol 311:541, 1981.

Ashcroft SJH: The control of insulin release by sugars. "Polypeptide Hormones: Molecular and Cellular Aspects." New York: Ciba Foundation Symposium 41, 1976, p 117.

Atwater I: Control mechanisms for glucose induced changes in the membrane potential of mouse pancreatic B-cell. Cienc Biol 5:299, 1980.

Berglindh T, DiBona DR, Pace CS, Sachs G: ATP dependence of H^+ secretion. J Cell Biol 85:392, 1980.

Carpinelli AR, Sener A, Herchuelz A, Malaisse WJ: Stimulus-secretion coupling of glucose-induced insulin release. Effect of intracellular acidification upon calcium efflux from islet cells. Metabolism 29:540, 1980.

Dean PM: Ultrastructural morphometry of the pancreatic B-cell. Diabetologia 9:115, 1973.

DiBona DR, Ito S, Berglindh T, Sachs G: Cellular site of gastric acid secretion. Proc Natl Acad Sci USA 76:6689, 1979.

Halperin ML, Connors HP, Relman AS, Karnovsky ML: Factors that control the effect of pH on glycolysis in leukocytes. J Biol Chem 244:384, 1969.

Hellman B, Andersson T, Berggren PO, Flatt P, Gylfe E, Kohnert KD: The role of calcium in insulin secretion. Horm Cell Regulat 3:69, 1979.

Hellman B, Gylfe E, Berggren PO, Andersson T, Abrahamsson H, Rorsman P, Betsholtz C: Ca^{2+} transport in pancreatic B-cells during glucose stimulation of insulin secretion. Uppsala J Med Sci 85:321, 1980.

Hellman B, Sehlin J, Taljedal IB: The intracellular pH of mammalian pancreatic B-cells. Endocrinology 90:335, 1978.

Henquin JC: The effect of pH on ^{86}rubidium efflux from pancreatic islet cells. Mol Cell Endocrinol 21:119, 1981.

Hille B: Changes and potentials at the nerve surface: Divalent ions and pH. J Gen Physiol 51:221, 1968.

Hutton JC, Sener A, Herchuelz A, Valverde I, Boschero AC, Malaisse WJ: The stimulus-secretion coupling of glucose-induced insulin release. XLII. Effects of extracellular pH on insulin release: Their dependency on nutrient concentration. Horm Metab Res 12:294, 1980.

Kilb KH, Stampfli R: A new stopcock for pharmacological purposes. Naunyn-Schmiedeberg's Arch Pharmacol 285:293, 1974.

Kostianovsky M, McDaniel ML, Still MF, Codilla RC, Lacy PE: Monolayer cell culture of adult rat islets of Langerhans. Diabetologia 10:337, 1974.

Lacy PE, Kostianovsky M: Method for the isolation of intact islets of Langerhans from rat islets. Diabetes 16:35, 1967.

Litchtshtein D, Dunlop K, Kaback HR, Blume AJ: Mechanism of monensin-induced hyperpolarization of neuroblastoma–glioma hybrid NG108-15. Proc Natl Acad Sci USA 76:2580, 1979.

Malaisse WJ, Herchuelz A, Sener A: The possible significance of intracellular pH in insulin release. Life Sci 26:1367, 1980.

Malaisse WJ, Hutton JC, Kawazu S, Herchuelz A, Valverde I, Sener A: The stimulus-secretion coupling of glucose-induced insulin release. XXXV. The links between metabolic and cationic events. Diabetologia 16:331, 1979a.

Malaisse WJ, Sener A, Herchuelz A, Hutton JC: Insulin release: The fuel hypothesis. Metabolism 28:373, 1979b.

Matschinsky FM, Ellerman J, Stillings S, Raybaud F, Pace C, Zawalich W: Hexoses and insulin secretion. In Hasselblatt A, Bruchhausen F (eds): "Insulin, Part Two." Heidelberg: Springer-Verlag, 1975, p 79.

Moody W: Appearance of calcium action potentials in crayfish slow muscle fibres under conditions of low intracellular pH. J Physiol 302:335, 1980.

Moore RD: Elevation of intracellular pH by insulin in frog skeletal muscle. Biochem Biophys Res Commun 91:900, 1979.

Moore RD: Stimulation of Na:H exchange by insulin. Biophys J 33:203, 1981.

Moore RD, Fidelman ML, Seeholzer SH: Correlation between insulin action upon glycolysis and change in intracellular pH. Biochem Biophys Res Commun 91:905, 1979.

Pace CS, Clements RS: Myo-inositol and the maintenance of B-cell function in cultured rat pancreatic islets. Diabetes 30:621, 1981.

Pollard HB, Pazoles CJ, Creutz CE, Zinder O: The chromaffin granule and possible mechanisms of exocytosis. Int Rev Cytol 58:159, 1979.

Rabon E, Chang H, Sachs G: Quantitation of hydrogen ion and potential gradients in gastric plasma membrane vesicles. Biochemistry 17:3345, 1978.

Rapp PE: An atlas of cellular oscillators. J Exp Biol 81:281, 1979.

Russell JT, Holz RW: Measurement of ΔpH and membrane potential in isolated neurosecretory vesicles from bovine neurohypophyses. J Biol Chem 256:5950, 1981.

Scherman D, Henry JP: Role of the proton electrochemical gradient in monoamine transport by bovine chromaffin granules. Biochem Biophys Acta 601:664, 1980.

Sener A, Hutton JC, Kawazu S, Boschero AC, Somers G, Devis G: The stimulus-secretion coupling of glucose-induced insulin release. Metabolic and functional effects of NH_4^+ in rat islets. J Clin Invest 62:868, 1978.

Smith JB, Rozengurt E: Serum stimulates the Na^+-K^+ pump in quiescent fibroblasts by increasing Na^+ entry. Proc Natl Acad Sci USA 75:5560, 1978.

Somers G, Sener A, Devis G, Malaisse WJ: The stimulus-secretion coupling of glucose-induced insulin release. XLV. The anion-osmotic hypothesis for exocytosis. Pflugers Archiv 388:249, 1980.

Tarvin JT, Pace CS: Glucose-induced electrical activity in the pancreatic B-cell: Effect of veratridine. Am J Physiol 240:C127, 1981.

Tomoda A, Tsuda-Hirota S, Minakami S: Glycolysis of red cells suspended in solutions of impermeable solutes. J Biochem 81:697, 1977.

Wanke E, Carbone E, Testa PL: K^+ conductance modified by a titratable group accessible to protons from the intracellular side of the squid axon membrane. Biophys J 26:319, 1979.

Wright PH, Makulu DR, Vichick D, Sussmann KE: Insulin immunoassay by back-titration; some characteristics of the technic and the insulin precipitant action of alcohol. Diabetes 20:33, 1971.

Chemotactic Stimuli-Induced Changes in the pH_i of Rabbit Neutrophils

R.I. Sha'afi, P.H. Naccache, T.F.P. Molski, and M. Volpi
Departments of Physiology (R.I.S., T.F.P.M., M.V.) and Pathology (P.H.N.), University of Connecticut Health Center, Farmington, Connecticut 06032

I.	Introduction	513
II.	Materials and Methods	514
III.	Results	515
	A. Effect of f-Met-Leu-Phe on the pH_i of Rabbit Neutrophils	515
	B. The Role of Calcium Ions in the Chemotactic Factor-Dependent Changes in pH_i	516
	C. The Role of Sodium Ions in the Chemotactic Factor-Dependent Changes in pH_i	518
	D. Effect of the Anion Transport Inhibitor DIDS	520
IV.	Discussion	521
V.	References	524

I. INTRODUCTION

The possible role of intracellular pH change as an initiator and/or regulator of cell activation particularly as an early step in the overall sequence of the excitation–response coupling has been and still is of great interest to cell biologists [for review see Roos and Boron, 1981]. It is not entirely unlikely that changes in intracellular pH may be one of the early biochemical responses following cell stimulation. This hypothesis is based on the fact that there are large numbers of enzymes whose activities are pH-dependent, as evidenced for example by the pH changes that trigger the late events of fertilization in sea urchin eggs [Shen and Steinhardt, 1978; Gillies and Deamer, 1979]. In addition, the contractile activity of purified preparations of actin and myosin and microtubule assembly and disassembly have also been shown to be dramatically influenced by relatively small changes in pH [Condeelis and Taylor, 1977; Regula et al, 1981]. Recently, Lynn and Mohapatra [1980] have demonstrated the presence of an activatable-bound neutral esterase that responds to many known leukocyte cytotoxins and appears to be involved in several membrane functions of these cells. They have suggested that this esterase(s) regulates many neutrophil functions by controlling H^+ efflux.

Using rabbit neutrophils as a model system, we have examined the possible involvement of the intracellular pH in the initiation and/or modulation of the responsiveness of these cells to the chemotactic factor formyl-methionyl-leucyl-phenylalanine (f-Met-Leu-Phe). This stimulus not only induces chemotaxis and chemokinesis in neutrophils, but also causes them to aggregage, activates the respiratory burst, and, in the presences of cytochalasin B, causes neutrophils to secrete enzymes from both specific and azurophil granules [Schiffmann et al, 1975, and for reviews see Becker, 1979; Sha'afi and Naccache, 1981]. This synthetic peptide with potent and well-defined biologic activities was recently found to induce specific, time-dependent biphasic changes in the intracellular pH of the neutrophils [Molski et al, 1980]. As will be discussed later, these changes appear to be secondary to the generally observed chemotactic factor-dependent movements of Ca^{2+} and Na^+.

II. MATERIALS AND METHODS

Rabbit peritoneal neutrophils were drained into a heparinized flask through two layers of cheesecloth 4 or 16 hours after the intraperitoneal injection of 200 or 400 ml respectively of 0.1% glycogen in sterile isotonic saline [Naccache et al, 1977; Showell and Becker, 1976]. The cells were washed twice in Hanks' balanced salt solution (HBSS). The composition of HBSS was as follows (in mM): NaCl, 124; KCl, 4; Na_2HPO_4, 0.64; KH_2PO_4, 0.66; $CaCl_2$, 1.6; $NaHCO_3$, 15.2; HEPES (N-2-hydroxyethyl piperazine-N′,2′-ethane sulfonic acid), 10 (pH = 7.0). Magnesium was omitted from the incubation medium in order to minimize spontaneous and chemotactic factor-induced aggregation. After washing, the cells were resuspended at the desired concentration (10^7 cells/ml) in HBSS to which 1 mg/ml glucose was added and allowed to equilibrate at 37°C for at least 15 minutes. The cell suspensions consisted of at least 85% polymorphonuclear leukocytes as judged from Coulter Counter-determined volume distribution. The cells were used within less than an hour.

The intracellular pH of the neutrophils was monitored by following the distribution of the weak acid 5,5-dimethyloxazolidine-2,4-dione (DMO). This method is based on the assumption that only the uncharged form of DMO is permeable to the plasma membrane [Levin et al, 1976; Addanki et al, 1968; Roos and Boron, 1976; Zieve et al, 1967; Gillies and Deamer, 1979; Waddell and Butler, 1959]. The distribution of DMO is thus pH-dependent, and the value of the latter parameter can be calculated using the following Henderson-Hasselbach relationship:

$$pH_i = pK + \log \left\{ [\frac{C_t}{C_e}(1 + \frac{V_e}{V_i})] \times [10^{(pH_e - pK)} + 1] - 1 \right\}$$

where pH_e = pH of extracellular medium (7.0); pK = the dissociation constant of DMO (6.2); C_t = counts taken up by the cells in cpm/ml cell H_2O; C_e =

counts in the extracellular medium in cpm/ml H_2O; V_e = trapped extracellular space; V_i = the volume of cell water.

Specifically, DMO was dried under N_2, resuspended in HBSS and added (0.1 µCi/ml, 2×10^{-6} M) to thermally equilibrated cell suspensions (10^7 cell/ml). Carbon dioxide (5% CO_2, 95% air) was blown over the cell suspensions during the length of the incubation in order to stabilize the extracellular pH. Earlier experiments have shown that DMO equilibrates across rabbit neutrophil membranes in much less than 1 minute [Molski et al, 1980]. Consequently the various stimuli tested were added 1 minute after DMO, and the distribution of DMO followed with time with the rapid sampling silicone oil method [Naccache et al, 1977; Sha'afi et al, 1981]. In this method, at the desired time intervals, aliquots from the cell suspensions (0.8 ml) were deposited above 0.3 ml of a silicone oil layer of density 1.05 g/ml, in 1.5 ml Eppendorf microcentrifuge tubes. The silicone oil used in this study, Versilube F-50, was obtained from Harwich Chemical Corporation, Cambridge, Massachusetts. The tubes were centrifuged for 30–60 seconds in Eppendorf microcentrifuges. The cells were then found pelleted at the bottom of the tubes, and the suspending media floated on top of the oil layer. Earlier experiments showed that no radioactivity could be found associated with the oil layer, and that the amount of protein recovered at the bottom was linearly related to the cell concentration used. The cell pellets and/or supernatants could then be analyzed for radioactivity. Following the removal by suction of the supernatants, the microcentrifuge test tube tips were cut and transferred to counting vials containing 1 ml of 88% formic acid. These vials were then incubated at 37°C for 30 minutes in order to solubilize cell pellets. At the end of the incubation period, 10 ml of liquid scintillation counting fluid was added to each vial and then they were counted. A 0.1-ml sample from the supernatant was also removed and treated in the same manner.

F-Met-Leu-Phe, Met-Leu-Phe, and boc-Phe-Leu-Phe-Leu-Phe were generous gifts of Dr. R.J. Freer (Medical College of Virginia, Richmond). ^{14}C-DMO was purchased from New England Nuclear, Boston. All other reagents were analytical grade.

The major sources of systematic errors are (1) the value taken for trapped space, (2) the value taken for cell water, and (3) the assumption that only the uncharged species crosses the membrane. Since these and other sources of errors [see review by Roos and Boron, 1981] could significantly influence the actual calculation of the absolute value of intracellular pH, we will consider only changes in intracellular pH that are much less sensitive to these sources of errors.

III. RESULTS

A. Effect of f-Met-Leu-Phe on the pH^i of Rabbit Neutrophils

The validity of the DMO technique for use with neutrophils [Levin et al, 1976] and other cell types [Gillies and Deamer, 1979; Roos and Boron, 1981] has been repeatedly confirmed since its original description by Waddel and Butler

[1959]. In addition, the rapidity with which DMO equilibrates across the neutrophil membranes (much less than 1 minute) makes this probe particularly useful for the study of dynamic events. Recently [Molski et al, 1980] we have examined the dependency of the intracellular pH (pH_i) of the neutrophils on the pH of the extracellular medium (pH_e). A linear relationship between these two variables appears to hold in the middle part of the extracellular pH range (tested (6.7–7.6). The slope of this straight line (as determined from linear regression analysis) is 0.67, indicating intracellular buffering. A curvature of the relationship appears at pH_e less than 6.8. The average pH_i determined at pH_e = 7.0 is 6.98 ± 0.01 (120 determinations), a value in agreement with previously published figures for the neutrophils [Levin et al, 1976]. In this study, we have also examined the effects of f-Met-Leu-Phe (1×10^{-9} M) on the pH_i of the neutrophils, and the results are summarized in Figure 1. F-Met-Leu-Phe causes rapid and biphasic changes in the pH_i of the neutrophils. Initially there is a rapid drop of about 0.05 pH units, which is followed by a slower and larger increase. The increase was easy to detect experimentally (reproducibility = 90%). It is important to point out that the initial drop is extremely small (in an actual experiment, it represents about 50–70 counts from a total of 500–800 counts). In addition, the maximum drop occurs between zero and 0.5 minutes after the addition of f-Met-Leu-Phe. Because of these problems the drop in pH_i is difficult to detect all the time (reproducibility = 50%).

The specificity of this chemotactic factor-dependent change in pH_i was ascertained as follows. The effect of the known f-Met-Leu-Phe binding antagonist boc-Phe-Leu-Phe-Leu-Phe [Aswanikumar et al, 1977] on the chemotactic factor-induced changes in pH_i was first examined. As shown in Figure 1, both the decrease and the subsequent increase, although possibly to a lesser extent, in pH_i induced by f-Met-Leu-Phe are inhibited in the presence of boc-Phe-Leu-Phe-Leu-Phe. The latter has no effect of its own on pH_i. In addition, Met-Leu-Phe, a peptide structurally similar to but of considerably less biologic activity [Showell et al, 1976] than f-Met-Leu-Phe was found to have no effect on pH_i at the same concentration as f-Met-Leu-Phe. As previously discussed, these results demonstrate that the effect of f-Met-Leu-Phe on pH_i is not the result of a nonspecific membrane disturbance, as the above three peptides are of similar size and hydrophobicity. In addition, these results strongly suggest that the effects of f-Met-Leu-Phe are mediated through the occupation of the same membrane receptor population that is involved in neutrophil functional and ionic responses.

B. The Role of Calcium Ions in the Chemotactic Factor-Dependent Changes in pH_i

It is generally agreed that the interaction of the chemotactic factor, f-Met-Leu-Phe, with the neutrophil causes a release of calcium from internal membrane sites and enhances the plasma membrane permeability to Ca^{2+} [Sha'afi and

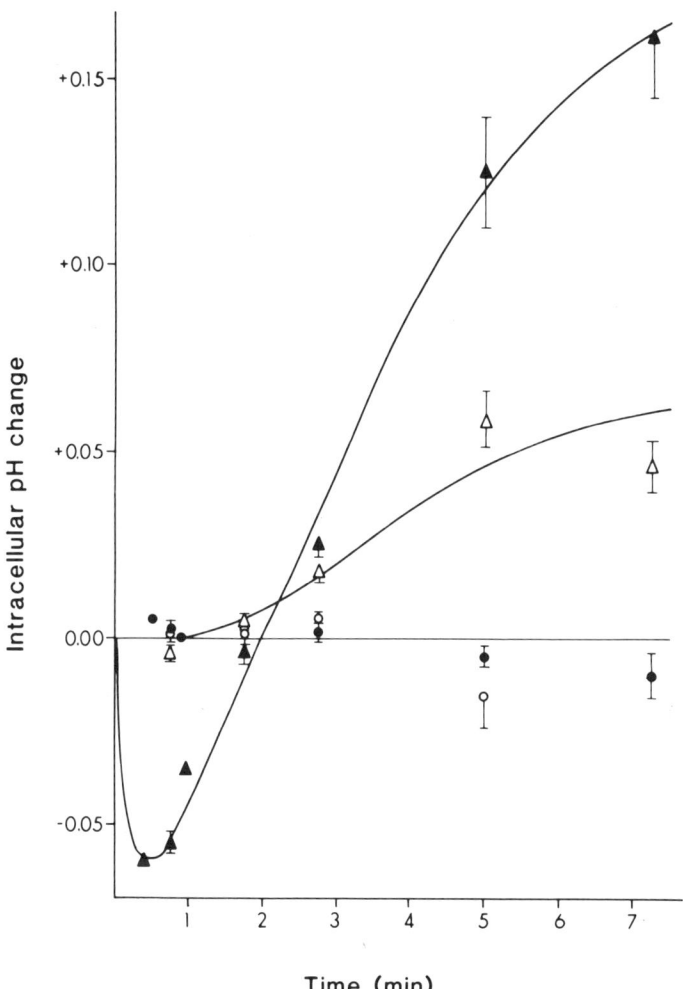

Fig. 1. The time course of the effect of the chemotactic factor f-Met-Leu-Phe on the intracellular pH of the neutrophils and its dependency on the presence of the binding antagonist boc-Phe-Leu-Phe-Leu-Phe. The points represent the means and the standard error of the mean of at least three experiments each performed in duplicate. The concentrations of f-Met-Leu-Phe and boc-Phe-Leu-Phe-Leu-Phe were 10^{-9} M and 2×10^{-5} M, respectively. Filled circles: no additions, open circles: boc-Phe-Leu-Phe-Leu-Phe; filled triangles: f-Met-Leu-Phe; open triangles: boc-Phe-Leu-Phe-Leu-Phe and f-Met-Leu-Phe.

Naccache, 1981]. In the presence of external calcium, the increased permeability results in a net influx of calcium into the cell. The result of these two events, the displacement of previously bound intracellular calcium and the increased permeability to calcium, produces a transient increase in the level of intracellular ionized calcium. As the intracellular concentrations of Ca^{2+} and H^+ often bear a direct relationship to each other [Roos and Boron, 1981] we have sought to determine the role, if any, of the previously described chemotactic factor-induced increase in the level of intracellular calcium on the pH_i of the neutrophils. To this end we have minimized the contribution of the extracellular calcium by removing it from the suspending buffer. In these experiments, cells were incubated for 15 minutes at 37°C in HBSS containing either 1.6 mM $CaCl_2$ or no added calcium (<5 μmolar). At the end of the incubation period DMO was added followed 1 minute later by f-Met-Leu-Phe. The results of these studies summarized in Figure 2 clearly indicate that removal of calcium from the external medium significantly reduces the initial chemotactic factor-dependent drop in pH_i without having any effect on the subsequent increase in pH_i. We have obtained essentially identical results by adding 2 mM EGTA to cell suspensions equilibrated with 1.6 mM $CaCl_2$, that is, a reduction in the F-Met-Leu-Phe-induced drop in pH_i and no effect on the later increase (results not shown).

C. The Role of Sodium Ions in the Chemotactic Factor-Dependent Changes in pH_i

It has been proposed that Na^+ movements are intimately involved in the rise of intracellular pH that occur immediately after fertilization of sea urchin eggs [Johnson and Epel, 1981]. The possible presence of a Na^+-influx–H^+-efflux exchange mechanism driven by the Na^+ concentration gradient across the plasma membranes of various cell types has been suggested by many investigators [for review, see Roos and Boron, 1981]. In addition, it has been found, in the case of the neutrophils, that f-Met-Leu-Phe causes a significant increase in the membrane permeability to Na^+ [Naccache et al, 1977]. With this in mind we have examined the question of whether or not Na^+ movements are involved in the chemotactic factor-dependent cell alkalization ($\delta pH_i > 0$). In the pursuit of this problem, we have not used the standard substitution experiment technique in which the effect of the removal of extracellular Na^+ on changes in pH_i is studied. We avoided this seemingly straightforward and direct approach because such a change in the concentration of Na^+ in the bathing medium results in rapid and significant changes in the concentrations of other intracellular ionic species (Na^+, K^+, Ca^{2+}). This is the case since the permeability of the plasma membrane of the neutrophils to Na^+ and K^+ is relatively high, and the movements of various cations are coupled by many as yet poorly defined mechanisms [Naccache et al, 1977; Sha'afi and Naccache, 1981]. Rarely can the effects of the removal of one ionic species from the extracellular fluid be pinpointed precisely. There-

Stimulus-Induced Changes in pH_i / 519

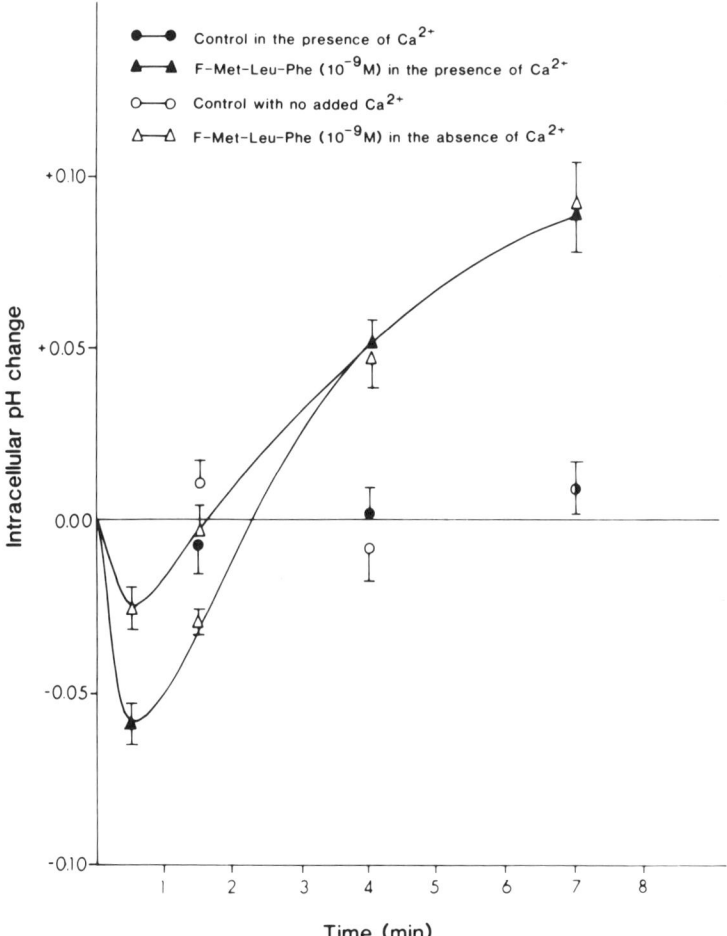

Fig. 2. Effect of the removal of outside calcium on the chemotactic factor-dependent biphasic changes in the internal pH of rabbit neutrophils.

fore, the results from substitution experiments, although informative, are difficult to interpret. Because of the above mentioned difficulties with the substitution experiments, we have instead opted to approach the question of the involvement of Na^+ movement in the regulation of pH_i by examining the effects of two specific inhibitors of Na^+ transport in mammalian cells: amiloride, an inhibitor of hormonally sensitive Na^+ transport across various epithelia, and ouabain, a Na^+/K^+ pump inhibitor.

The effects of the diuretic amiloride on the chemotactic factor-induced rise in cellular pH_i were first examined. Amiloride, a K^+-sparing diuretic of clinical significance, is known to inhibit Na^+ fluxes (as judged by short-circuit measurements) in various epithelia and other cells [Bentley, 1968; Guignard and Peters, 1970; Salako and Smith, 1970; Murer et al, 1976; Johnson et al, 1976; for review see Roos and Boron, 1981]. In addition, amiloride inhibits in a dose-dependent fashion the f-Met-Leu-Phe-stimulated Na^+ influx in rabbit neutrophils but does not affect the f-Met-Leu-Phe-stimulated calcium uptake in these cells [Sha'afi et al, 1981]. In the present experiment, a known amount of amiloride to give a final concentration of 10^{-3} molar was added to the cell suspension 1 minute prior to the addition of DMO (longer preincubation with amiloride did not modify these results). The results of these experiments, which are summarized in Table I, clearly show that amiloride drastically reduces the rise in intracellular pH produced by the chemotactic factor f-Met-Leu-Phe.

The rationale behind the use of ouabain is that the inward movement of Na^+ which is driven by the Na^+ gradient maintained by the Na^+/K^+ pump, may be coupled to the vectorial movement of other ions or compounds. Examples of the latter processes include the Na^+/Ca^{2+} [Blaustein, 1974] and the Na^+/H^+ [Johnson et al, 1976; Murer et al, 1976] exchanges. It is therefore possible that the Na^+ electrochemical gradient may be involved in the regulation of the pH_i in neutrophils. Accordingly ouabain, which, by inhibiting the Na^+/K^+ pump, dissipates the Na^+ gradient and may inhibit the chemotactic factor-dependent increase in pH_i. In preliminary experiments, however, the exposure of neutrophils to 10^{-4} M ouabain was found not to affect significantly the f-Met-Leu-Phe-induced rise in pH_i (results not shown). This suggests that the Na^+/K^+ pump is not involved in stimulated H^+ efflux in rabbit neutrophils.

D. Effect of the Anion Transport Inhibitor DIDS

The amino group reactive agent 4,4'-diisothiocyanostilbene-2,2'-disulfonic acid (DIDS) is a potent inhibitor of anion transport in mammalian red cells

TABLE I. Effect of Amiloride on the Increase in Internal pH in Rabbit Neutrophils Following Stimulation by the Chemotactic Factor f-Met-Leu-Phe*

Experimental condition	$\Delta pH_i = pH_{i,0} - pH_{i,5}$
Control	0.02 ± 0.02 (5)
+ f-Met-Leu-Phe	0.08 ± 0.03 (5)
+ Amiloride	0.02 ± 0.01 (5)
+ f-Met-Leu-Phe + Amiloride	0.01 ± 0.01 (5)

*ΔpH_i changes were calculated as the difference between the values of pH_i at zero time ($pH_{i,0}$) and at 5 minutes ($pH_{i,5}$) after the addition of the compound. The number in parentheses refers to the number of experiments. Each experiment was carried out in duplicate. The errors are standard errors of the means. The concentration of f-Met-Leu-Phe and amiloride were 10^{-9} and 10^{-3} molar, respectively.

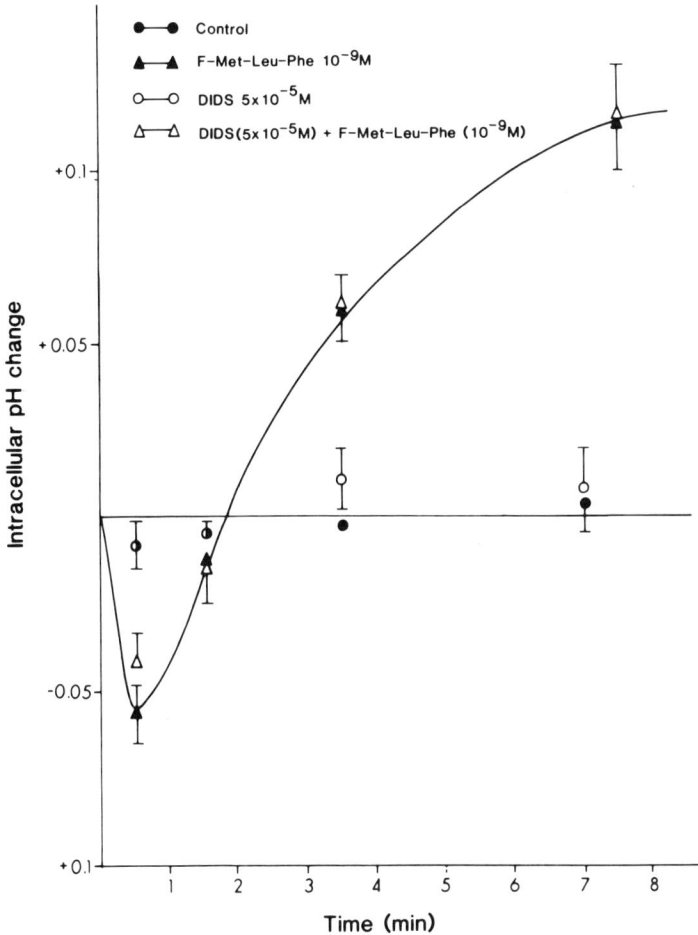

Fig. 3. Effect of DIDS on the time course of the changes in pH_i of rabbit neutrophils following stimulation by the chemotactic factor f-Met-Leu-Phe.

[Cabantchick and Rothstein, 1974]. In addition, this compound has been shown, although at a much higher concentration than those required for the red cells, to affect the responsiveness of the neutrophil to various stimuli including chemotactic factors [Korchak et al, 1980; Tauber and Goetzl, 1981]. As shown in Figure 3, DIDS at a concentration that inhibits anion movement in mammalian red cells has no effect on the pH_i changes produced by f-Met-Leu-Phe.

IV. DISCUSSION

The present studies differ from most other investigations of intracellular pH in that they deal with the question concerning its role in signal transduction in the overall scheme of excitation–response coupling. This is in contrast to most other studies in which the mechanism(s) involved in the regulation of the steady-state level of cellular pH have been probed. Therefore, the aim of the work presented here has been to examine the effect of a specific stimulus on the intracellular pH of its target cell. As such, the mechanisms responsible for the maintenance of the normal or steady-state pH_i have not been investigated. We have attempted rather to establish the ionic basis for the chemotactic factor-induced changes in pH_i that we have previously observed [Molski et al, 1980] and to fit these into the scheme of excitation–response coupling of rabbit neutrophils [Sha'afi and Naccache, 1981]. Two basic sets of observations emerged from the present studies.

The drop in pH_i that rapidly follows the addition of f-Met-Leu-Phe appears to be the result of the simultaneous rise in intracellular calcium. This result is indirectly inferred from the fact that the magnitude of the f-Met-Leu-Phe-induced pH_i drop is reduced in the absence of extracellular calcium. Under these conditions, the stimulus-induced rise in the level of intracellular calcium is diminished in magnitude as it depends partially on a net uptake of calcium from the extracellular medium. The latter event occurs only at concentrations of extracellular calcium greater than 50 μM [Petroski et al, 1979].

The amiloride sensitivity of the f-Met-Leu-Phe-induced rise in pH_i suggests that this event is mediated by a Na^+/H^+ exchange mechanism. This conclusion is supported by the following observations: 1) f-Met-Leu-Phe causes a rapid and large enhancement in the rate of uptake of ^{22}Na into rabbit neutrophils [Naccache et al, 1977], and 2) this increased rate of ^{22}Na uptake is sensitive to amiloride [Sha'afi et al, 1981]. What is still missing is the direct demonstration of an f-Met-Leu-Phe-stimulated H^+ efflux from the neutrophils. In any case, the evidence available at present is strongly supportive of the above suggestion, and is in line with the conclusion reached on similar grounds in renal brush border membrane vesicles [Kinsella and Aronson, 1980] in cells from the proximal tubules [Boron and Boulpaep, 1980] and in sea urchin eggs [Johnson et al, 1976].

The lack of effect of the anion transport inhibitor DIDS suggests that the pH transients observed in neutrophils following stimulation by chemotactic factors are not mediated by a Cl^-–HCO_3^- exchange as demonstrated in squid giant axon [Boron and DeWeer, 1976], snail neuron [Thomas, 1976], and giant barnacle muscle [Boron, 1977].

Our view of the mechanisms underlying the chemotactic factor-induced pH_i changes is summarized in Figure 4. According to this scheme, the increase in the level of intracellular Ca^{2+} concentration, produced by a redistribution of intracellular calcium and by an enhanced membrane permeability, activates various biochemical processes (lipases, proteases, glycolysis, and others) capable of generating hydrogen ions and, in addition, this increase in calcium concentration can act directly to displace or exchange for H^+. The latter could be due

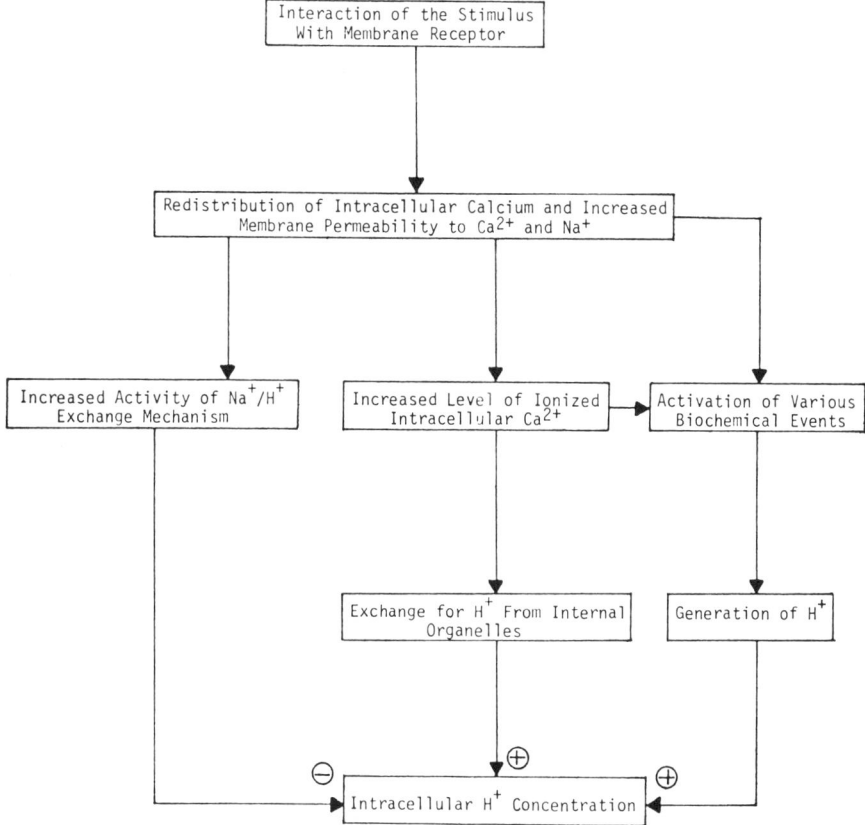

Fig. 4. Schematic representation of the events that are involved in the observed biphasic changes in the pH$_i$ of rabbit neutrophil following the stimulation by the chemotactic factor f-Met-Leu-Phe.

to the uptake of Ca^{2+} by mitochondria or other cellular organelles in exchange for H$^+$ [eg, Vasington et al, 1972]. One or both of these two processes leads to the initial drop in cellular pH. The potential contribution of increased lactic acid production and the exact mechanism of this increase upon stimulation by f-Met-Leu-Phe cannot be evaluated at present.

In addition to mobilizing calcium, f-Met-Leu-Phe also enhances the plasma membrane permeability of the neutrophils to Na$^+$. As the Na$^+$ electrochemical gradient favors an inward movement of Na$^+$, the increased permeability results in a net uptake of Na$^+$, some of which is coupled to the outward-directed movement of H$^+$. The result of this process is to increase pH$_i$.

If the above interpretation is indeed correct, then the observed changes in pH$_i$ are not primary events of the stimulation of the neutrophils but are secondary to the previously described changes in Na^+ and Ca^{2+} movements and distribution. Therefore the pH changes may not be directly involved in signal transduction. This conclusion is consistent with the observation that variations of extracellular pH do not generally abolish the functional responsiveness of neutrophils [DeStefano et al, 1978]. This does not rule out, however, a secondary or modulatory role for the pH changes in neutrophils activation. Changes in pH_i will affect a great many intracellular processes including the behavior of cytoskeletal proteins such as actin and myosin [Condeelis and Taylor, 1977] and tubulin [Regula et al, 1981], two protein systems thought to be intrinsically involved in neutrophil activation.

ACKNOWLEDGMENT

This work was supported by a grant from the National Institutes of Health, No. AI-13734-04.

V. REFERENCES

Addanki S, Cahill ED, Sotos JF: Determination of intramitochondrial pH and intramitochondrial-extramitochondrial pH gradient of isolated heart mitochondria by the use of 5,5-dimethyl-oxazolidine-2,4-dione. J Biol Chem 243:2337, 1968.

Aswankiumar S, Corcoran BA, Schiffmann E, Pert CB, Morell JC, Gross E: Peptides with agonist and antagonist chemotactic activity. In Goodman M, Meishoffer J (eds): "Peptides." New York: Wiley, 1977, p 141.

Becker EL: A multifunctional receptor on the neutrophil for synthetic chemotactic oligopeptides. J Reticuloendothel Soc 26:701, 1979.

Bentley PJ: Amiloride: A potent inhibitor of sodium transport across the toad bladder. J Physiol 195:317, 1968.

Blaustein MP: The interrelationship between sodium and calcium fluxes across cell membranes. Rev Physiol Biochem Pharmacol 70:33, 1974.

Boron WF: Intracellular pH transients in giant barnacle muscle fibers. Am J Physiol 233:C61, 1977.

Boron WF, Boulpaep EL: Intracellular pH regulation in the salamander renal proximal tubule. Kidney Int 18:126A, 1980.

Boron WF, De Weer P: Intracellular pH transient in squid giant axons caused by CO_2, NH_3 and metabolic inhibitors. J Gen Physiol 67:91, 1976.

Cabantchik ZI, Rothstein A: Membrane proteins related to anion permeability of human red blood cells. J Membr Biol 15:207, 1974.

Condeelis JS, Taylor DL: The contractile basis of amoeboid movement. J Cell Biol 74:901, 1977.

DeStefano MJ, Dziezanowski MA, Rabinovitch M: Migration of polymorphonuclear neutrophils under agarose is enhanced by media of low pH or osmolarity. J Cell Biol 79:C1, 1978.

Gillies RJ, Deamer DW: Intracellular pH: Methods and applications. Curr Top Bioenerg 9:63, 1979.

Guignard J-P, Peters G: Effects of triamterene and amiloride on urinary acidification and potassium excretion in the rat. Eur J Pharmacol 10:255, 1970.

Johnson CH, Epel D: Intracellular pH of sea urchin eggs measured by the dimethyloxazolidine (DMO) method, J Cell Biol 89:284, 1981.

Johnson JJ, Epel D, Paul M: Intracellular pH and activation of sea urchin eggs after fertilization. Nature (Lond) 262:661, 1976.

Kinsella JL, Aronson PS: Amiloride: A specific inhibitor of the Na^+–H^+ exchanges in rabbit renal cortical brush border membrane vesicles. Clin Res 28:452, 1980.

Korchak HM, Eisenstat BA, Hoffstein ST, Dunham PB, Weissman G: Anion channel blockers inhibit lysosomal enzyme secretion from human neutrophils without affecting generation of superoxide anion. Proc Natl Acad Sci USA 77:2721, 1980.

Levin GE, Collinson P, Baron DN: The intracellular pH of human leucocytosis response to acid base changes in vitro. Clin Sci Mol Med 50:293, 1976.

Lynn WS, Mohapatra N: Control of leukocyte functions role of internal H^+ concentration and a membrane-bound esterase. Inflammation 4:329, 1980.

Molski TFP, Naccache PH, Volpi M, Wolpert LM, Sha'afi RI: Specific modulation of the intracellular pH of rabbit neutrophils by chemotactic factors. Biochem Biophys Res Commun 94:508, 1980.

Murer H, Hoffer U, Kinne R: Sodium proton antiport in brush-border membrane vesicles isolated from rat small intestine and kidney. Biochem J 154:597, 1976.

Naccache PH, Showell HJ, Becker EL, Sha'afi RF: Transport of sodium, postassium and calcium across rabbit polymorphonuclear leukocyte membranes: Effect of chemotactic factor. J Cell Biol 73:428, 1977.

Petroski RJ, Naccache PH, Becker EL, Sha'afi RI: Effect of chemotactic factors on the calcium levels of rabbit neutrophils. Am J Physiol 237:C43, 1979.

Regula CS, Pfeiffer JR, Berlin RD: Microtubule assembly and disassembly at alkaline pH. J Cell Biol 89:45, 1981.

Roos A, Boron WF: Intracellular pH. Physiol Rev 61:296, 1981.

Salako LA, Smith AJ: Changes in sodium pool and kinetics of sodium transport in frog skin produced by amiloride. Br J Pharmacol 39:99, 1970.

Schiffmann E, Corcoran BA, Wahl SM: N-formylmethionyl peptides as chemoattractants for leukocytes. Proc Natl Acad Sci (USA) 72:1059, 1975.

Sha'afi RI, Molski TFP, Naccache PH: Chemotactic factors activate differentiable permeation pathways for sodium and calcium in rabbit neutrophils. Effect of amiloride. Biochem Biophys Res Commun 99:1271, 1981.

Sha'afi RK, Naccache PH: Ionic events in neutrophil chemotaxis and secretion. In Weissmann G (ed:) "Advances in Inflammation Research," Vol 3. New York: Raven Press, 1981, p 115.

Sha'afi RI, Naccache PH, Alobaidi T, Molski TFP, Volpi M: Effect of arachidonic acid and the chemotactic factor F-Met-Leu-Phe on cation transport in rabbit neutrophils. J Cell Physiol 106:215, 1981.

Shen SS, Steinhardt RW: Direct measurement of intracellular pH during metabolic derepression of the sea urchin egg. Nature (Lond) 272:253, 1978.

Showell HJ, Becker EL: The effects of external K^+ and Na^+ on the chemotaxis of rabbit peritoneal neutrophils. J Immunol 116:99, 1976.

Tauber AI, Goetzl EJ: Inhibition of complement mediated functions of human neutrophils by impermeant stilbene disulfonic acids. J Immunol 126:1786, 1981.

Thomas RC: Ionic mechanism of the H^+ pump in a snail neurone. Nature (Lond) 262:54, 1976.

Vasington FD, Gazzotti P, Tiozzo R, Carafoli E: The effect of ruthenium red on Ca^{2+} transport and respiration in rat liver mitochondria. Biochim Biophys Acta 256:43, 1972.

Waddell WJ, Butler JC: Calculation of intracellular pH from the distribution of 5,5-dimethyl-oxazolidine-2,4-dione (DMO). Application to skeletal muscle of the dog. J Clin Invest 38:720, 1959.

Zieve PD, Hagsheness M, Krevans JR: Intracellular pH of the human lymphocyte. Am J Physiol 212:1099, 1967.

Noninvasive pH$_i$ Measurements of Human Tissue Using ^{31}P-NMR

Peter J. Bore, Lawrence Chan, David G. Gadian, George K. Radda, Brian D. Ross, Peter Styles, and Doris J. Taylor
Department of Biochemistry, University of Oxford, South Parks Road, Oxford OX1 3QU, England (P.J.B., D.G.G., G.K.R., P.S., D.J.T.), and Radcliffe Infirmary, Oxford, England (L.C., B.D.R.)

I.	Introduction	527
II.	Recent Technical Developments	527
	A. Whole Animal Studies	527
	B. Studies of Human Metabolism	528
III.	Studies of Kidney Preservation	529
IV.	Studies of Human Forearm Muscle	529
V.	Conclusions	534
VI.	References	534

I. INTRODUCTION

In an earlier chapter in this volume [Gadian et al, this volume], it was shown how phosphorus nuclear magnetic resonance (^{31}P-NMR) can be used to measure the intracellular pH (pH$_i$) of intact tissue. In addition to these pH$_i$ measurements, information is also available about other metabolic parameters and about the physiological function of the tissue. Recent developments in technology now make it possible to make similar studies of selected tissues and organs within whole animals and humans, and in this article we describe some recent ^{31}P-NMR investigations of healthy and diseased human tissue. We begin, however, with a brief explanation of the technical developments that have made these studies feasible.

II. RECENT TECHNICAL DEVELOPMENTS

A. Whole Animal Studies

There are two main problems associated with whole animal studies. The first is that the animal has to fit within the bore of the magnet. Magnets with a bore of around 10 cm are now routinely available, and this permits the study of small animals such as rats. As we shall see later, magnets with much larger bores are also available for the study of larger animals and human beings.

The second problem relates to signal detection. In conventional NMR experiments, the signals are obtained by placing the sample *within* a radio frequency coil that transmits radiation into the sample and detects the ensuing signal. In this sense, the studies of isolated tissues and organs are reasonably conventional. For studies of whole animals, however, it is neither very practical nor particularly useful to surround the animal by the coil; it is rather pointless to obtain a composite signal originating from all of the various tissues within the animal. Clearly it is necessary to devise a method of localisation that enables signals to be detected from a specific, well-defined region within the animal. Two such methods have recently been developed.

The first method of localisation involves the use of an unusual type of radio frequency coil, which has been termed a surface coil [Ackerman et al, 1980]. In its simplest form, it consists of a flat, circular loop of wire containing one or more turns. If such a coil is placed adjacent to a sample, it will, under normal circumstances, detect signal from an approximately disk-shaped region of the sample immediately in front of the coil, of radius and thickness approximately equal to the radius of the coil. A surface coil therefore provides a very simple method of localising on a selected region that is close to the surface of a sample. For metabolic studies of whole animals, this means that investigations of skeletal muscle and brain are readily feasible [Ackerman et al, 1980]. However, surface coils can be used to localise on internal organs only if coupled with surgery [Bore et al, 1981; Griffiths et al, 1980]. Clearly, there is a need for an alternative method that can localise on internal organs with no requirement for surgery. A method that has been successfully used has been termed topical magnetic resonance (TMR).

In TMR, the magnetic field is profiled in such a way that it is very homogeneous over a central, approximately spherical volume, but elsewhere is very inhomogeneous. As a result, narrow signals are observed only from the central region; the signals from outside this region are very broad and make very little effective contribution to the final spectrum. TMR has been used to localise on the liver of live, anaesthetised rats [Gordon et al, 1980]. In addition, the use of surface coils has been combined with TMR in order to localise on the kidney of live rats [Balaban et al, in press].

B. Studies of Human Metabolism

These studies of live animals can be readily extended to measurements of human metabolism, the main requirement being the availability of suitable magnets large enough to accommodate a human being, or at least a human arm or leg. Magnets with a 20-cm working bore are now in use for this type of work, and we now describe some recent studies that have been performed on human forearm muscle and on isolated human kidneys.

III. STUDIES OF KIDNEY PRESERVATION

Before the development of very wide bore magnets, ^{31}P-NMR had been used to determine adenine nucleotide content and intrarenal pH in kidneys from experimental animals [Sehr et al, 1979; Bore et al, 1981]. These studies have contributed to an understanding of the effects of ischaemia and preservation for transplantation. The emergence of magnets with a 20-cm bore makes it possible to extend this work by studying whole human kidneys in vitro.

Studies have been made of human kidneys that were to be transplanted, and also of kidneys that were removed from patients because they were diseased. For NMR examination, the kidneys were contained within their sterile covering, and were placed in a transparent plastic box in which they were surrounded by ice. The box was positioned above a surface coil of diameter 4.5 cm, in such a way that the coil was adjacent to the region of interest within the kidney [Chan et al, in press].

In contrast to earlier results with the kidneys of small animals (rats and rabbits) in which the ATP signal disappears within minutes of the onset of ischaemia, the human kidney interestingly retains some ATP for up to 24 hours in the cold, as shown in Figure 1. The pH_i falls as anticipated reaching a value of about 6.5 after 24 hours (see Fig. 2). A significant preservative effect has been observed in the human kidney by flushing the donor organ with a hypertonic citrate solution prior to cold storage for 24 hours [Marshall, 1980]. Using ^{31}P-NMR, this flushing solution is seen to exert two effects on the kidney; ATP is conserved for a much longer period of time, and also the pH_i falls considerably more slowly, as shown in Figures 1 and 2. This provides the first indication in the human kidney that pH_i may play a role in renal damage and that preventing its fall could contribute to preservation of function. There is a clinical trial in progress, in which the kidneys are routinely examined by NMR prior to transplantation. One aim of these studies is to correlate the metabolic state prior to transplantation with subsequent graft function, in the hope that the NMR spectra will provide a method of predicting function. On the basis of the preliminary results, the outlook seems promising [Chan et al, in press]. In addition, it should be possible to use NMR to obtain a better understanding of the various factors that influence renal preservation. For example, the results that we have obtained on human and rat kidneys suggest that prevention of the fall in intrarenal pH during ischaemia could prove to be an important factor in successful organ preservation.

IV. STUDIES OF HUMAN FOREARM MUSCLE

Excellent ^{31}P-NMR spectra can be obtained on placing one's forearm adjacent to a surface coil within a wide bore horizontal magnet [Cresshull et al, 1981, a, b; Ross et al, 1981]. Figure 3 shows a spectrum obtained in 1 minute from muscle tissue within the flexor compartment of the forearm of a healthy subject. As anticipated, the spectrum contains signals from ATP, phosphocreatine, and inorganic phosphate within the muscle; in fact, the level of inorganic phosphate is slightly higher than at rest because the subject was performing mild arm exercise during the period of data collection. The pH_i can be deduced from the chemical shift of the inorganic phosphate signal, and the value obtained in our studies to date of 21 healthy subjects at rest is 7.04 ± 0.03 (SD).

During aerobic exercise, changes are observed in the metabolite levels and, for many subjects, in pH_i. However, the magnitude of these changes will depend on a large number of variables, which include the magnitude and duration of the exercise, and the physical fitness, age, etc of the individual. More profound effects are generally observed during anaerobic exercise, when the subject exercises with a sphygmomanometer cuff around his upper arm. Figure 4 shows a series of spectra obtained from a healthy volunteer. Figure 4A is a spectrum

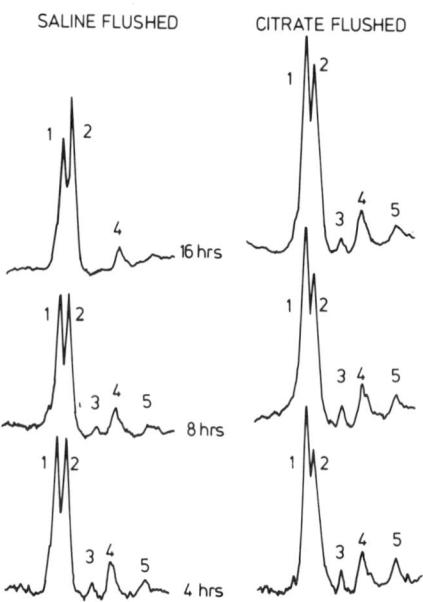

Fig. 1. ^{31}P-NMR spectra of single human kidneys flushed with saline or hypertonic citrate solutions after 4, 8, and 16 hours of cold ischaemia. Peak assignments are as follows: 1) sugar phosphate and AMP; 2) inorganic phosphate; 3) γ-phosphate of ATP and β-phosphate of ADP; 4) α-phosphate of ATP and ADP + NAD/NADH; 5) β-phosphate of ATP.

31P-NMR Studies of Human Tissue / 531

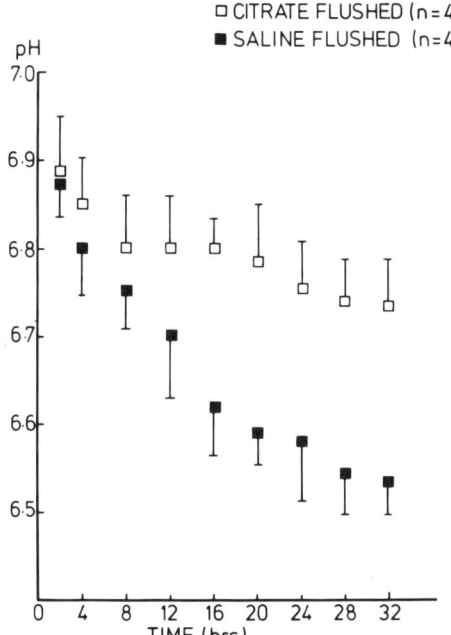

Fig. 2. The change of pH with time of saline or citrate flushed human kidneys during cold ischaemia. pH values are expressed as mean and 1 standard deviation (vertical bars). From 12 hours onwards the differences were statistically significant (P = 0.05).

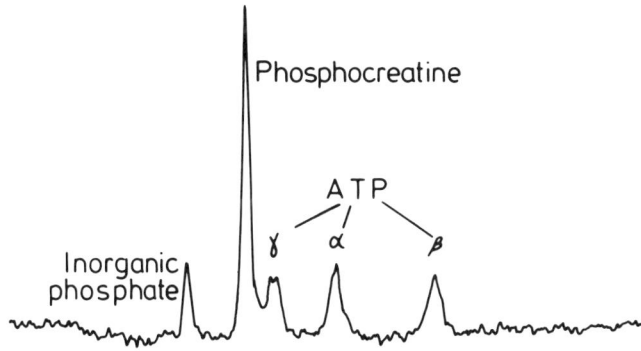

Fig. 3. ^{31}P-NMR spectrum obtained from muscle tissue within the flexor compartment of the forearm of a healthy subject. The spectrum was accumulated using 32 radio frequency pulses applied at 2-second intervals.

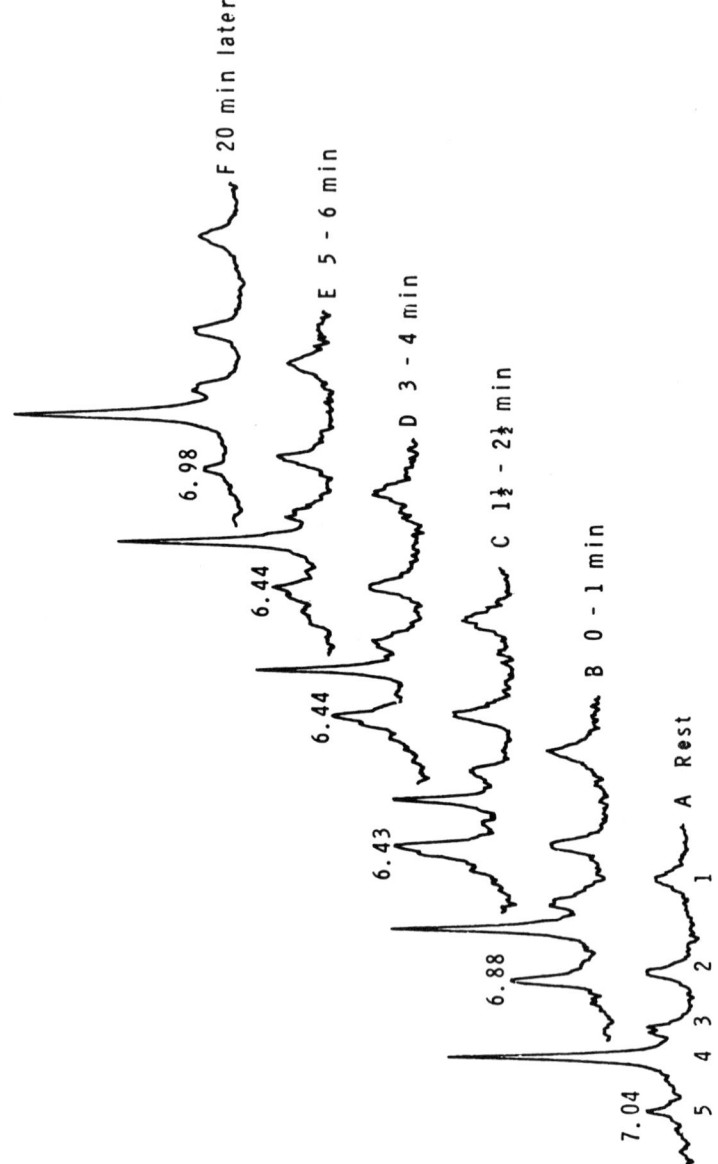

Fig. 4. ^{31}P-NMR spectra from the forearm of a control subject, showing the effects of ischaemic exercise. The signals are assigned as follows: 1, 2, and 3) the β, α, and γ phosphates of ATP; 4) phosphocreatine; and 5) inorganic phosphate. The pH$_i$ value given above each inorganic phosphate signal was determined from the frequency separation of the inorganic phosphate and phosphocreatine signals. The first spectrum (A) was recorded at rest before exercise; subsequent spectra (B–F) were recorded during the periods shown, in which 0 minutes corresponds to the time at which exercise was started. Exercise was maintained during the period from 0 to 1½ minutes, but arterial occlusion was maintained for up to 3 minutes. Arterial flow was restored after this period. [From Ross et al, 1981.]

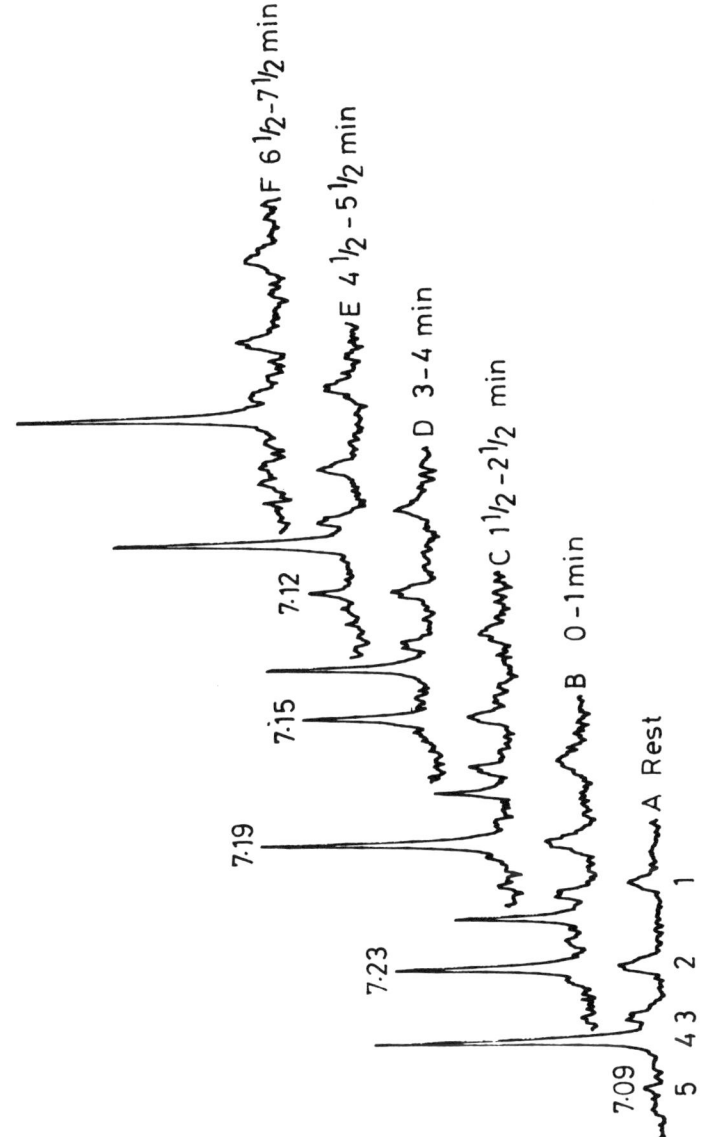

Fig. 5. ^{31}P-NMR spectra of a patient suspected to have McArdle syndrome. Peak assignments are as given in the legend to Figure 4, and pH$_i$ values are given above each inorganic phosphate signal. The first spectrum (A) was recorded at rest before exercise; subsequent spectra (B–F) were recorded during the periods shown, in which 0 minutes corresponds to the time at which exercise was started. Exercise was maintained during the period 0 to ¾ minute, but arterial occlusion was maintained for up to 3 minutes. Arterial flow was restored after this period. [From Ross et al, 1981.]

obtained at rest and contains signals as expected from phosphocreatine, ATP, and inorganic phosphate. The pH_i is estimated to be 7.04. The subsequent spectra, each accumulated in 1 minute, show the metabolic effects of ischaemic exercise and recovery; the phosphocreatine and pH_i decline markedly, and then recover over a period of several minutes. The pH declines because of the accumulation of lactic acid that takes place under anaerobic conditions.

The same protocol was carried out on a patient suspected to have McArdle's syndrome. This is an inborn error of metabolism caused by a lack of glycogen phosphorylase activity in skeletal muscle, and it leads to muscle weakness. Figure 5 shows the series of spectra that were obtained. The most striking abnormality is that the patient's pH_i did not decline despite fatigue during ischaemic exercise; however, his phosphocreatine declined dramatically [Ross et al, 1981]. This observation is entirely consistent with his inability to generate lactic acid from glycogen, and is thus consistent with the absence of glycogen phosphorylase activity. The absence of phosphorylase was confirmed by a direct enzyme assay performed after completion of the ^{31}P-NMR study. The role of ^{31}P-NMR in the diagnosis of this admittedly rare disease suggests that the technique may have wider clinical applications to more common conditions. The dimensions of the current magnet limit examination to tissues of the forearm, and with some difficulty to the calf muscle. Nevertheless, it should be possible to study systemic changes that alter acid–base balance and hence blood pH, for such changes could be reflected in the pH of various tissues, including skeletal muscle.

V. CONCLUSIONS

The pH_i of human tissue can now be measured noninvasively using ^{31}P-NMR. In addition to studying the biochemistry of healthy and diseased muscle, it should be possible to investigate many clinical conditions in which acid–base disturbances are important. With the present instrumentation, measurements are limited to isolated organs and human limbs, but whole body magnets are expected to become available in the near future.

ACKNOWLEDGMENTS

This work is supported by the Science Research Council, the Medical Research Council, and the British Heart Foundation.

VI. REFERENCES

Ackerman JJH, Grove TH, Wong GG, Gadian DG, Radda GK: Mapping of metabolites in whole animals by ^{31}P NMR using surface coils. Nature 283:167, 1980.

Balaban RS, Gadian DG, Radda GK: A phosphorus nuclear magnetic resonance study of the rat kidney in vivo. Kidney Int (in press).

Bore PJ, Sehr PA, Chan L, Thulborn KR, Ross BD, Radda GK: The importance of pH in renal preservation. Transplant Proc 13:707, 1981.

Chan L, French ME, Gadian DG, Morris PJ, Radda GK, Bore PJ, Ross BD, Styles P: Study of human kidneys prior to transplantation by phosphorus nuclear magnetic resonance. In Pegg DE, Jacobson I, Halasz NA (eds): "Organ Transplantation III." MTP Press Ltd (in press).

Cresshull ID, Gordon RE, Hanley P, Shaw D, Gadian DG, Radda GK, Styles P: ^{31}P NMR studies of human forearm muscle. Bull Magn Reson 2:426,, 1981a.

Cresshull I, Dawson MJ, Edwards RHT, Gadian DG, Gordon RE, Radda GK, Shaw D, Wilkie DR: Human muscle analyzed by ^{31}P nuclear magnetic resonance in intact subjects. J Physiol 317:18P, 1981b.

Gadian DG, Radda GK, Dawson MJ, Wilkie DR: pH$_i$ measurements of cardiac and skeletal muscle using ^{31}P NMR. This volume.

Gordon RE, Hanley P, Shaw D, Gadian DG, Radda GK, Styles P, Bore PJ, Chan L: Localization of metabolites in animals using ^{31}P topical magnetic resonance. Nature 287:736, 1980.

Griffiths JR, Stevens AN, Gadian DG, Iles RA, Porteous R: Hepatic fructose metabolism studied by ^{31}P nuclear magnetic resonance in the anaesthetized rat. Biochem Soc Trans 8:641, 1980.

Marshall VC: Transplantation 30:165, 1980.

Ross BD, Radda GK, Gadian DG, Rocker G, Esiri M, Falconer-Smith J: Examination of a case of suspected McArdle's syndrome by ^{31}P nuclear magnetic resonance. New Engl J Med 304:1338, 1981.

Intracellular pH: Its Measurement, Regulation, and
Utilization in Cellular Functions, pages 537-565
© 1982 Alan R. Liss, Inc., 150 Fifth Avenue, New York, NY 10011

The Role of Intracellular pH in the Control of Normal and Ischemic Myocardial Contractility: A ^{31}P Nuclear Magnetic Resonance and Mass Spectrometry Study

William E. Jacobus, Ira H. Pores, Scott K. Lucas, Clayton H. Kallman, Myron L. Weisfeldt, and John T. Flaherty

Peter Belfer Laboratory for Myocardial Research, Department of Medicine (W.E.J., I.H.P., C.H.K., M.L.W., J.T.F.) and Department of Physiological Chemistry (W.E.J.), The Johns Hopkins University School of Medicine, Baltimore, Maryland 21205

I.	Introduction	537
II.	Materials and Methods	539
	A. Nuclear Magnetic Resonance Methods	539
	1. Heart perfusion techniques; 2. Instrumental methods	539
	B. Calibration of NMR pH Measurements	541
	1. Validation of pH$_i$ from Pi chemical shifts; 2. pH$_i$ from the γ-ATP resonance	541
	C. Other Aspects of pH$_i$ Assessment	547
	D. Mass Spectrometry Methods	548
III.	Results	549
	A. Correlation of pH$_i$ to Ventricular Performance in Normal Hearts	549
	B. Contractility and pH$_i$ During Ischemia	551
	C. Mass Spectrometry Studies	554
IV.	Discussion	557
V.	References	562

I. INTRODUCTION

In this chapter we wish to discuss the potential role which intracellular pH (pH$_i$) may play in the regulation of myocardial contractility. This work appropriately complements the dozen or so other cell physiology communications in this volume. The goal of this report is twofold: to illustrate the flexibility of

---

I.H. Pores's current address is Newark Beth Israel Medical Center, 201 Lyons Avenue, Newark, NJ 07112.

S.K. Lucas's current address is University of Oklahoma Health Center, Department of Surgery, P.O. Box 26901, Oklahoma City, OK 73190.

section is somewhat broader in scope than a normal report. Where appropriate, problems, pitfalls, limitations, and methodological details will be extensively outlined.

The central question we have addressed is straightforward: What place do the high energy phosphate metabolites—ie, ATP and phosphocreatine—and pH_i play in the control of both normal and ischemic myocardial contractions? By way of introduction, let us first define the term ischemia. The classical definition of ischemia is an insufficient supply of tissue blood flow as a result of reduced arterial inflow. This definition lacks an indication of the relationships between reduced flow and tissue function. There is a broader, metabolic, and bioenergetic definition of ischemia that incorporates this important concept of performance. Jennings [1976] has suggested that ischemic cell dysfunction results from an oxygen supply/demand imbalance. In other words, cell function is in a delicate balance between metabolic (energetic) supply and its physiological demands.

$$\text{Cell Function} = \text{Supply/Demand} \qquad (1)$$

Normal cell function is observed when supply is equal to or greater than demand. By this definition, ischemic dysfunction of the heart may occur from either a reduction in blood supply at a constant physiological work demand, or from an increase in demand at a constant but limited rate of supply (flow). In cardiology, the latter condition is the cause of the clinical syndrome of angina pectoris, whereas the former initiates acute myocardial infarction, the leading cause of death in the Western world.

Perhaps the most striking result of cardiac ischemia is the almost instantaneous compensatory reduction in ventricular contractility observed at the onset of a decrease in coronary arterial flow. In the intact animal, arterial vasodilation, or autoregulation, protects the heart against moderate changes in perfusion pressure. However, in an isolated, non-blood-perfused, isovolumic working heart, very modest changes in perfusion pressure or flow initiate an immediate change in performance. In most instances these changes are reproducibly noted within 1 or 2 seconds. Although the hypodynamic effect of ischemia has long been noted [Tennant and Wiggers, 1935], the mechanism(s) that so critically and rapidly coordinate ventricular function to coronary flow remain incompletely understood.

At the metabolic or molecular level, two hypotheses related to the Jennings definition have received considerable experimental attention [Kübler and Katz, 1977]. The first theory suggests that ischemic "pump" failure is associated with the lack of supply of some required metabolite. Candidates for this hypothesis include oxygen, ATP, and/or phosphocreatine. On the other hand, reduced flow limits the rate of washout, and toxic compounds may abnormally accumulate. These may then feedback to inhibit metabolism and contraction. Likely candidates for this second hypothesis are the anaerobic generation of lactate, inorganic phosphate (Pi), and H^+ (acidosis). One of the first observations that acidosis depressed heart contractility can be traced to the writings of Sir Isaac Newton (1642–1727) [Roos and Boron, 1981]. Over 100 years ago, Klug [1879] reported on the negative inotropic effect of respiratory acidosis. Since then there has been considerable controversy concerning the importance of acidosis as the factor responsible for early ischemic "pump" failure. Recently, it has been suggested that the ischemic decline in pressure development results from the combined effects of intracellular and extracellular acidosis, with little or no contribution from changes in the cellular high energy phosphates [Steenbergen et al, 1977].

To explore the relative contribution of these two hypotheses, we have used the biophysical technique of ^{31}phosphorus nuclear magnetic resonance [Gadian and Radda, 1981]. As noted in Gadian's introductory presentation in this volume, we are able to obtain data about the tissue quantities of ATP, ADP, phosphocreatine, and Pi from the ^{31}P-NMR spectra of perfused hearts. Moon and Richards [1973] first demonstrated that the chemical shift of the Pi resonance could be used to estimate tissue pH. Since that time the method has been successfully utilized in studies of skeletal muscle [Hoult et al, 1974; Burt et al, 1976; Dawson et al, 1977, 1978] and the perfused heart [Jacobus et al, 1977; Garlick et al, 1977, 1979; Hollis et al, 1977, 1978; Bulkley et al, 1978; Jacobus et al, 1978; Flaherty and Jacobus, 1979; Salhany et al, 1979; Grove et al, 1980; Jacobus et al, 1980, 1982; Fossel et al, 1980; Nunnally and Bottomley, 1981; Flaherty et al, 1982] as well as in other organs in vivo and in vitro [Gadian and Radda, 1981]. Because the method is essentially noninvasive and nondestructive, it offers considerable advantages over biopsy methods for metabolite analysis since spectra may be accumulated sequentially. Since the method is also flow-independent, it can be performed in a no-flow state, conditions when weak acid or weak base distribution methods are not usable [Roos and Boron, 1981].

Our results demonstrate a rather tight coupling between pH_i and contractility in the normal heart during respiratory acidosis. Nevertheless, we show that during total global ischemia, the magnitude of acidification is insufficient to account fully for the observed contractile depression. And finally, using NMR and mass spectrometry, we observe that down to a 50% reduction in flow, pH_i, the high energy phosphates, and tissue oxygen content all remain normal, in spite of a 40% reduction in left ventricular developed pressure. These latter data suggest that an efficient, nonmetabolic, autoregulatory mechanism may protect the heart against ischemic cell damage during moderate reductions in coronary arterial flow.

II. MATERIALS AND METHODS

A. Nuclear Magnetic Resonance Methods

1. Heart perfusion techniques. Hearts (5–6 g) from young female New Zealand white rabbits weighing between 1.2 and 2.0 kg are routinely used in our studies. Although our probe diameter is 25 mm, we are not able to use hearts from larger animals. As a word of caution, the hearts must easily slip into the NMR sample tube. Larger hearts, above 6.0 g, are compressed when inserted into the tube, and a significant Pi peak is noted in control spectra. This results from the partial restriction of the superficial coronary arteries and induces ischemia.

The hearts are perfused retrograde in a modified Langendorff mode previously described in considerable detail [Flaherty et al., 1982]. The perfusion canulla was positioned well above the aortic valve, which is then competent. The Krebs Ringer bicarbonate buffer is phosphate-free and contains 117 mM NaCl, 6.0 mM KCl, 3.0 mM $CaCl_2$, 1.0 mM $MgSO_4$, 0.6 mM EDTA, 16.7 mM glucose, and 24 mM Na bicarbonate, and is vigorously bubbled with 95% O_2/5% CO_2 for a final buffer pH of 7.40. The perfusate temperature is 40°C in the reservoir; perfusate overflow from the heart is 35–37°C. Since perfusate is phosphate-free, all ^{31}P-NMR signals must arrise from endogenous tissue compounds. In a series of control experiments we compared the functional stability of hearts perfused with normal Krebs buffer [Neely and Rovetto, 1975] to those perfused with our phospate-free medium. No significant differences in functional or spectral stability were noted. The hearts were paced at a rate of 150–170 beats per minute. Isovolumic ventricular pressure was measured using a latex balloon positioned in the left ventricular cavity and connected to a Statham P 23 Db transducer, calibrated with a mercury mannometer. The control end diastolic pressure was set close to 10 mm Hg. Care was taken so that the balloon did not herneate into the left atrium. Figure 1 illustrates a heart positioned in the NMR sample tube. Perfusate overflow is removed by vacuum aspiration. The rest of the accessory apparatus required for our studies is shown in Figure 2.

2. Instrumental methods. ^{31}P-NMR spectra were obtained from a wide-bore superconducting magnet (4.23 T) at 72.89 M Hz, using a Bruker WH-180 spectrometer. The stability of the magnetic field was such that field/frequency D_2O lock was not required; the probe diameter was 25 mm. The instrument was operated in the pulsed, Fourier transform mode, and was interfaced to a Bruker 1080 computer. Proton-decoupled spectra were obtained from transients following 25 μsec (45°) pulses delivered at 2-second intervals, conditions resulting in minimal spectral saturation. The data were collected at a 3,000 Hz width with a 2K data table, or at a 5,000 Hz spectral width with a 4K table. A heart spectrum with each peak labeled is shown in Figure 3. Normal spectra required 200–400 transients. Additional details concerning spectral accumulation are detailed in the respective figure legends.

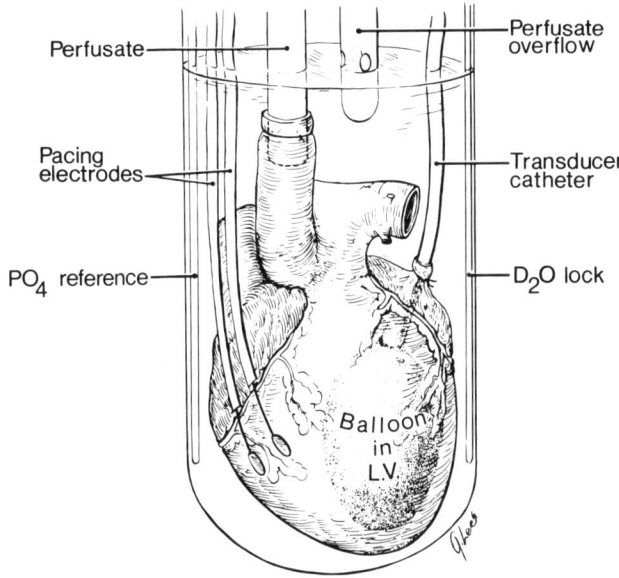

Fig. 1. A view of the perfused heart in the NMR sample tube. Tube diameter is 25 mm. Perfusate flows retrograde into the coronary arteries from the aortic perfusion cannula. The latex balloon in the left ventricle and the transducer catheter used for the measurement of isovolumic pressures are shown. Also illustrated are the pacing electrodes and the overflow line for removal of coronary flow. [Hollis et al, 1978.]

B. Calibration of NMR pH Measurements

Estimates of intracellular pH (pH_i) were determined from the chemical shift (δ_O) of either the Pi peak or the γ-ATP peak according to the following equation:

$$pH_i = pK - \log_{10}(\delta_O - \delta_B/\delta_A - \delta_O) \qquad (2)$$

To minimize the effects of tissue inhomogeity, all chemical shift values were determined relative to the phosphocreatine resonance, which because of its low pK (4.6) is rather insensitive to pH changes above 6.0

1. Validation of pH_i from Pi chemical shifts. The chemical shift of inorganic phosphate is rather insensitive to changes in concentrations of monovalent or divalent cations, altered within physiological limits (Tables I and II). This is an important control since it is known that there are changes in these ions during ischemia, and that the chemical shift of ATP is also sensitive to Mg^{2+} chelation [Cohn and Hughes, 1962].

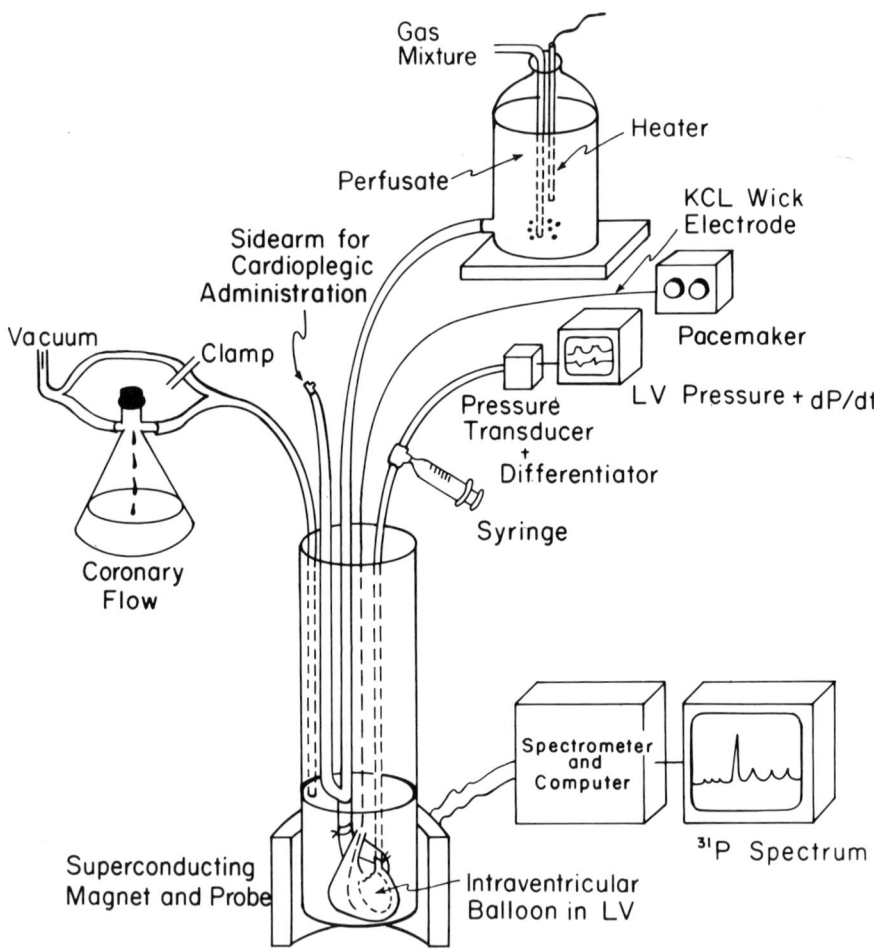

Fig. 2. A drawing of the accessory equipment required for physiological studies using NMR. LV is left ventricle; dP/dt is the first derivative of pressure. [Flaherty et al, 1982.]

In order to calculate pH_i from equation 2, it is necessary to determine the values for the three constants: pK, δ_A, and δ_B. To establish the values for δ_A and δ_B, 25 mM samples of inorganic phosphate were titrated to pH 4.0 and 10.0, respectively. Knowing that the pK of Pi was near 7.0, these samples were 3.0 pH units away from the pK value. Thus, these samples were considered to be 99.9% in the $H_2PO_4^-$ (pH 4.0) and HPO_4^{2-} (pH 10.0) forms. NMR spectra were obtained, and the chemical shift values were measured relative to a sample of phosphocreatine (pH 7.4) contained in a capillary tube vertically aligned in the 25 mm diameter sample tube. From these spectra, δ_A was determined to be 3.290

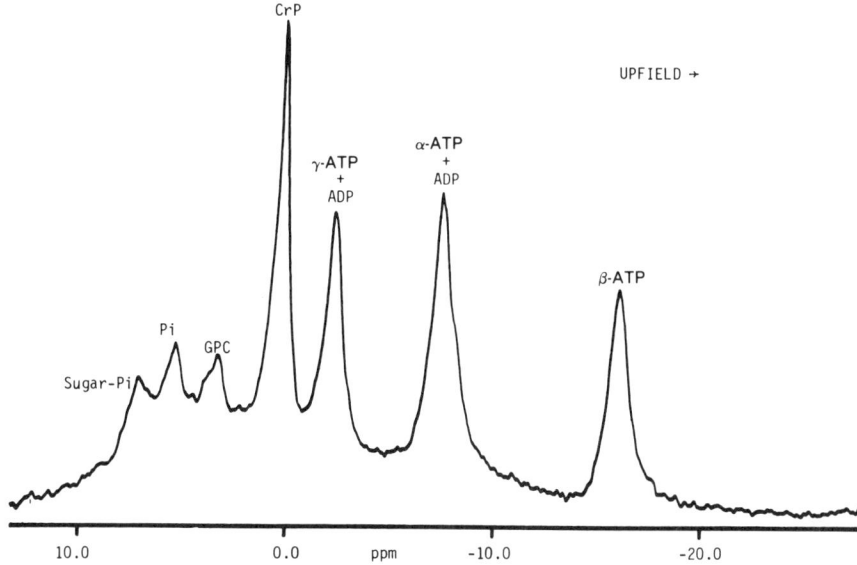

Fig. 3. A ^{31}P-NMR spectrum of a perfused rabbit heart. This is a 9,000-pulse spectrum acquired at a 3,000-Hz spectral width. Peaks are observed for the glycolytic sugar phosphates (Sugar-Pi), inorganic phosphate (Pi), glycerol-3-phosphorylcholine (GPC), phosphocreatine (CrP), and ATP and ADP. The abscissa scale is chemical shift in parts per million (ppm) with 0.0 at the peak of the phosphocreatine resonance.

TABLE I. Effects of Reciprocal Changes of Na$^+$ and K$^+$ on the Chemical Shift of the Inorganic Phosphate Resonance

Concentration (mM)		Chemical shift in ppm
KCl	NaCl	
160	—	+ 1.87
145	15	+ 1.90
130	30	+ 1.92
115	45	+ 1.95
100	60	+ 1.86

The 20-ml solution contained 50 mM K$_2$HPO$_4$ (pH 7.30) plus the indicated concentrations of NaCl and KCl. Spectra were accumulated at 72.89 M Hz as described in Materials and Methods. Chemical shifts were calculated relative to a sample of 0.2 M H$_3$PO$_4$ in 15% HClO$_4$ sealed in a capillary tube vertically mounted in the center of the NMR (25 mm) sample tube.

ppm, and δ_B was 5.805 ppm. Phosphate samples were also titrated in the range of pH 6.5–7.5 to establish the pK value of 6.90. We have previously published the values of these constants [Jacobus, et al, 1978]. With these three constants, it was possible to use them in equation 2 to generate mathematically the theo-

TABLE II. Effects of Mg^{2+} and Ca^{2+} on the Chemical Shift of Inorganic Phosphate

Concentration Mg^{2+} (mM)	Chemical shift in ppm (pH 7.35)
—	+ 1.951
2.0	+ 1.951
5.0	+ 1.971
10.0	+ 1.971
15.0	+ 2.011

Concentration Ca^{2+} (mM)	Chemical shift in ppm (pH 7.04)
5×10^{-8}	+ 1.548
1×10^{-7}	+ 1.548
5×10^{-7}	+ 1.528
1×10^{-6}	+ 1.539
1×10^{-5}	+ 1.548
1×10^{-4}	+ 1.569
1×10^{-3}	+ 1.528

Spectra were collected as described in Table I, except that the samples contained only 10 mM Pi. Spectra were the average of 90 transients.

retical NMR titration curve for Pi, (Fig. 4). In both panels of Figure 4, the solid lines represent the titration curve derived from equation 2. The dots represent NMR chemical shift data, using a Corning 110 digital pH meter with an A. H. Thomas combination pH electrode to determine sample pH values. Figure 4A presents data from samples of 10 mM Pi titrated from pH 5.0 to pH 9.0. Figure 4B presents the data for Pi samples titrated in the supernatant solution of a dog heart homogenate. In these experiments, 100 g of fresh dog heart was minced in 100 ml of 125 mM KCl, homogenized in a Waring blender, and centrifuged at 10,000g for 15 minutes at 2°C. The clear, reddish supernatant solution was used for the Pi titration studies (Fig. 4B). These latter control experiments ruled out the possibility that Pi interacts with cytoplasmic proteins (binding), which could significantly alter its chemical shift characteristics. From the data of Figure 4 and Tables I and II, we conclude that it is valid to estimate pH_i from the Pi chemical shift data. Our current best estimates of the intracellular pH of the perfused rabbit heart under different physiological conditions are given in Table III.

2. pH_i from the γ-ATP resonance. There are several potential problems in estimating cytoplasmic pH values from Pi. These difficulties arise from its mixed cytoplasmic distribution (Table IV). Table IV shows the mitochondrial and cytoplasmic distributions of heart Pi and ATP. From these data we see that 30% of the Pi is localized in the mitochondrial matrix, a unique compartment. However, the bulk of the Pi (70%) is found in the sarcoplasmic fraction. On the

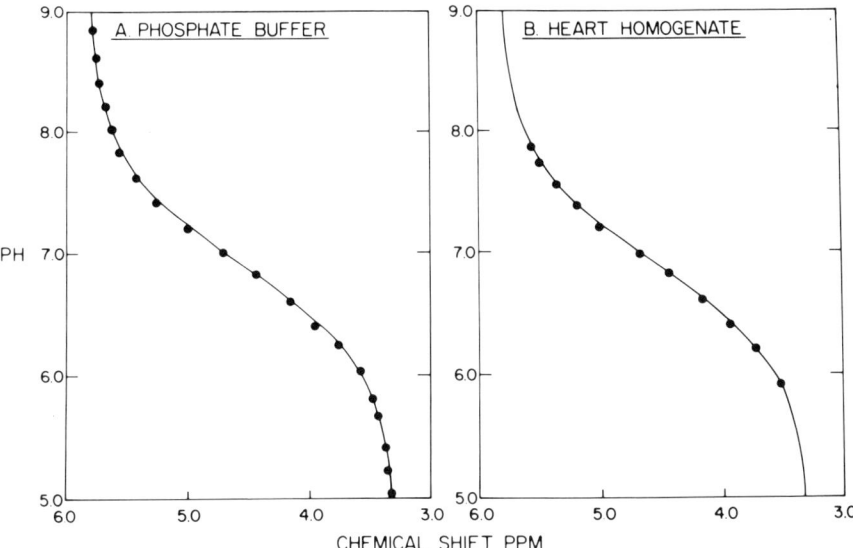

Fig. 4. Titration curve for inorganic phosphate either as a dilute (10 mM) buffer solution (A), or as the endogenous phosphate in a supernatant solution of a heart homogenate (B). In both panels, the line is the theoretical titration curve derived from equation 2, using our values for the constants (Table VI).

TABLE III. Intracellular pH of Perfused Rabbit Hearts

Condition	Intracellular pH
Working (N = 15)	7.18 ± 0.02
KCL arrest (N = 4)	7.22 ± 0.02
Hypoxia (N = 7)	7.19 ± 0.01

Hearts were perfused with phosphate-free Krebs buffer. The working heart was the normal isovolumic model. Hearts were arrested with 37 mM KCl added to the perfusate. Partial hypoxia was induced by bubbling the perfusate with 65% O_2/30% N_2/5% CO_2. In all conditions, buffer pH was 7.45–7.48. Intracellular pH values are given as mean ± SE.

other hand, we see that 97.6% of the total ATP is in the cytoplasm, with only 2.4% in the mitochondrial matrix. It therefore would appear preferable to determine pH_i from the γ-ATP peak. There are, however, some major difficulties with this approach (Fig. 5). Assuming that the predominant cytoplasmic form of ATP is the $MgATP^{2-}$ complex, then the titration data of Figure 5 suggest that the pH_i of the heart is near 6.0. The ventricle line designated O indicates the observed chemical shift for ATP in heart. Although it is true that $MgATP^{2-}$ is the major form, other ionic nucleotide complexes also exist [Kohn et al, 1977].

TABLE IV. Intracellular Distribution of Heart Pi and ATP

	Total	Mitochondrial	Cytoplasmic
Water space (μl/g wet wt)	640	64	576
Phosphate distribution			
Pi content (μmoles)	3.32	1.02	2.30
% Distribution		30%	70%
ATP distribution			
ATP content (μmoles)	4.50	0.11	4.29
% Distribution		2.4%	97.6%

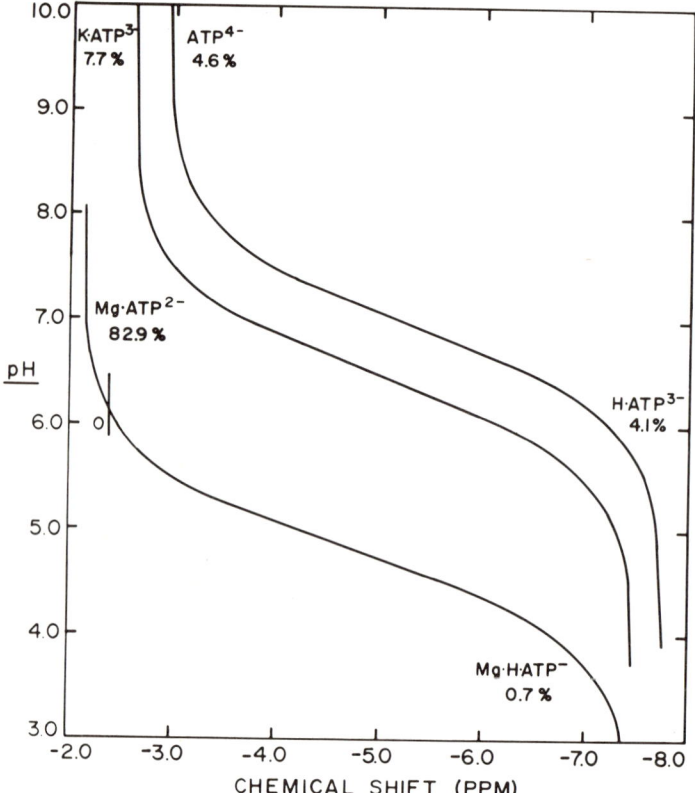

Fig. 5. Titration curve for the gamma-phosphate resonance of ATP. The verticle line designated (○) is the observed chemical shift for this peak in the perfused heart. The percent values indicate the nucleotide distribution for the various ionic species of ATP.

There are substantial amounts of free ATP, as well as potassium ATP (Fig. 5). The titration curves for ATP, K-ATP, and Mg-ATP are shown in Figure 5. Knowing the limit chemical shifts of each form, and the nucleotide distribution, it is possible to calculate the weighted average chemical shift for the γ-ATP resonance (Table V). At a pH value of 7.049 [Kohn et al, 1977], this value is −2.480 ppm. The observed chemical shift value obtained from the average of 15 hearts is −2.410. When the adenine nucleotide distributions were recalculated assuming a pH of 7.18 (Garfinkel, personal communication, 1978), the weighted average chemical shift was now −2.412, or remarkably close to the observed value of −2.410. Therefore, because of the numerous forms of ATP, one can not reliably use it for pH_i estimates. However, the excellent agreement between the calculated and observed values (Table V) lend a moderate degree of support to our estimate that the pH_i of perfused rabbit hearts is in the range of pH 7.2 ± 0.02 (Table III).

C. Other Aspects of pH_i Assessment

The NMR estimate of intracellular pH is a global measurement. In our instrument, the entire heart resides within the sensitive coil region of the probe. Signals arise from both myocyte and nonmyocyte cells. It is possible, however, to obtain NMR signals from specific ischemic regions of the heart [Hollis et al, 1977]. Newer methods using surface coils [Ackerman et al, 1980; Nunnally and Bottomley, 1981] permit direct acquisition of regional data, and "topical magnetic resonance" now allows for highly regional spectral accumulation in whole animals.

The time limitation is variable. In our studies at least 200 transients are required to obtain a spectrum with sufficient signal-to-noise for the accurate determination of the Pi chemical shift. But these 200 transients may be accumulated in a variety of ways. If done as a single continuous spectrum, it requires about 7 minutes for data accumulation. In some of our studies on transient ischemia (Fig. 8), we accumulated 200 separate 10-pulse spectra, for a data

TABLE V. Intracellular Species of ATP and Calculated Weight Average Chemical Shift

Nucleotide form	Chemical shift	Nucleotide distribution	
		pH 7.049	pH 7.180
$MgATP^{2-}$	−2.140	19,464	19,313
$KATP^{3-}$	−2.618	1,987	2,198
ATP^{4-}	−2.950	1,184	1,311
$HATP^{3-}$	−7.743	840	688
$MgHATP^{-}$	−7.446	112	83
Weighted average chemical shift		−2.480	−2.412
Observed heart chemical shift (N = 15)			−2.410

collection period of only 20 seconds. Finally, the NMR spectrometer can be gated to the cardiac cycle [Fossel et al, 1980]. In theory, the ultimate limit is the collection time of the free induction decay, which can be as small as 50 msec. This 50-msec window could be moved to specific phases of the cardiac cycle using standard gating methods.

Resolution and absolute accuracy are also important issues. There is little controvercy over the precision of the method. Measurements can easily be made ± 0.02 pH units. The resolution is also agreed to be about 0.02 pH units. However, there remains some disagreement over the conversion of the NMR data into absolute pH values—ie, its absolute calibration. We estimate pH_i to be about 7.20 (Table III). Grove et al [1980] calculate the pH_i of the in vivo rat heart to be 7.11. Salhany et al [1979] suggest that the value for the perfused guinea pig heart is nearer to 7.00. These differences are not the product of basic NMR methodological or species differences. These discrepencies are generated during the conversion of the NMR data into pH_i data (Table VI). For purposes of illustration, we have collected the published constants used in equation 2 by three laboratory groups. When one inserts the same value for the observed chemical shift (4.982) into these three equations, pH_i values of 7.12, 7.21, and 7.30 are calculated. As yet there is not universal agreement as to the appropriate constants for equation 2. From the standpoint of our studies, however, there is very close agreement on estimates of pH changes. If we reduce the chemical shift of Pi from 4.982 ppm to 4.750 ppm, pH_i is decreased by 0.18 ± 0.01 pH units (Table VI). Therefore, there is solid agreement on the use of NMR to estimate pH_i shifts.

D. Mass Spectrometry Methods

The measurements of tissue gases were done by mass spectrometry methods standard in this laboratory. Briefly, a 22-gauge Teflon-coated needle is inserted tangentially into the free wall of the left ventricle and tightly sutured to prevent leakage. The tissue gas mixture is withdrawn across the Teflon membrane, through stainless steel tubing, and into the mass spectrometer. The time response

TABLE VI. Constants for the Estimation of Tissue pH_i, Equation 2

	Dawson [1977]	Jacobus [1978]	Gadian [personal communication, 1981]
pK for Pi	6.88	6.90	6.75
$_A(HPO_4^{2-})$	3.35	3.290	3.30
$_B(H_2PO_4^-)$	5.60	5.805	5.70
Calculated pH_i			
4.982 ppm	7.302	7.213	7.120
4.750 ppm	7.097	7.041	6.934
Difference	−0.205	−0.172	−0.186

of this method is about 6–8 minutes, and the instrument is calibrated daily using gasses of known composition. Because metal cannot be inserted into the NMR magnetic field, the mass spectrometry data were accumulated from hearts treated in exact parallel to those studied in the NMR. Further details of the mass spectrometry method have been reported [Khuri et al, 1975].

III. RESULTS

A. Correlation of pH_i to Ventricular Performance in Normal Hearts

As studies preliminary to our work with ischemia, it was first necessary to determine the normal relationship between intracellular acidosis and myocardial contractility. Thus, as a calibration study, we used the model of respiratory acidosis. Five rabbit hearts were employed in this protocol. The hearts were initially perfused with normal phosphate-free Krebs Ringer's bicarbonate buffer, gassed with 65% O_2/ 30% N_2/ 5% CO_2, at a buffer pH of 7.48. Control ventricular performance was equivalent to hearts oxygenated with 95% O_2/ 5% CO_2. In all cases control left ventricular developed pressure (LVDP) was greater than 100 mm Hg. A control 400-pulse NMR spectrum was obtained, and the Pi chemical shift suggested that the tissue pH_i was 7.22 ± 0.02 (Fig. 6). A combination glass pH–electrode was then placed into the buffer reservoir, and a second gas used to lower buffer pH. This gas contained 65% O_2/ 35% CO_2, so oxygen remained constant while CO_2 was increased. The flow rate of the second tank

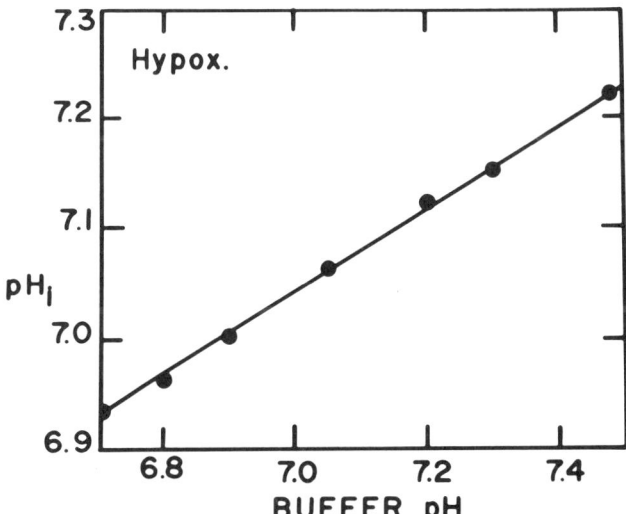

Fig. 6. Relationship between buffer pH and pH_i of the perfused rabbit heart during respiratory acidosis. The chemical shift of Pi was used to estimate pH_i. [Jacobus et al, 1982.]

was set in a stepwise sequence to reduce buffer pH to values of 7.30, 7.20, 7.05, 6.90, 6.80, and 6.70. At each value, buffer pH was stabilized for a period of at least 15–20 minutes. As the buffer pH was reduced we noted a depression in contractility. During these periods of depressed but stable performance, a series of 400-pulse NMR spectra were acquired. The signal-to-noise ratios of these spectra were 100 to 1 for phosphocreatine, 50 to 1 for ATP, and 25 to 1 for Pi. The chemical shift of Pi was used to estimate tissue pH_i (equation 2). The changes in pH_i as a function of buffer pH are illustrated in Figure 6. In Figure 6 we see that over the buffer pH range of 7.48 to 6.70, pH_i decreased from 7.22 to a value of 6.93. At each buffer pH value, we also recorded ventricular performance. It was therefore possible to correlate changes in pH_i to reductions in ventricular contractility.

The correlation between NMR-determined pH_i and ventricular dysfunction is shown in Figure 7. Left ventricular developed pressure (LVDP) was calculated as a percent of control performance. Developed pressure is peak systolic pressure minus end diastolic pressure. The data of Figure 7 shows a quite linear rela-

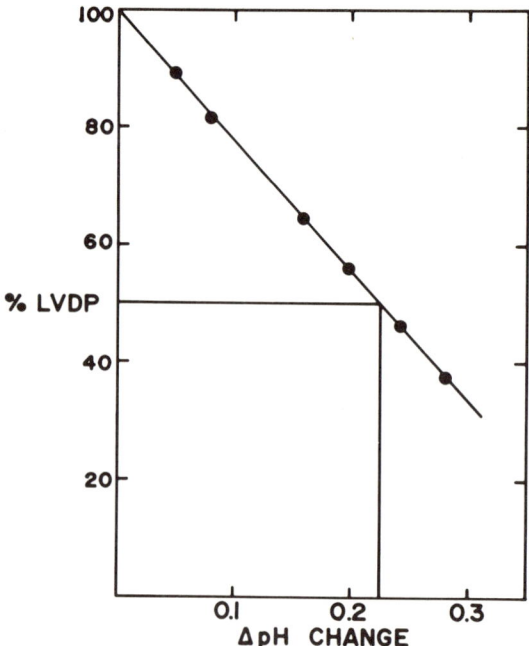

Fig. 7. Relationship between left ventricular developed pressure (LVDP) and intracellular acidification during respiratory acidosis. Data are taken from four hearts whose control pressures exceeded 100 mm Hg. The magnitude of the acidification was calculated from the data of Figure 6. [Jacobus et al, 1982.]

tionship between the depression of LVDP and intracellular acidosis. From Figure 7 it can be estimated that a 50% depression of LVDP correlates with a 0.22 pH unit acidification. In other words, a 50% fall in function is associated with a pH_i shift from 7.22 to 7.00. These data clearly show a very tight coupling between pH_i changes and the contractile state of the heart. However, they do not reveal if this relationship is a primary (direct causative) or secondary (associative) correlation.

In the rat heart, it is well known that extracellular acidosis leads to significant vasoconstriction. To compensate for this effect, Steenbergen et al [1977] used increasing perfusion pressures to maintain constant rates of flow. However, it is possible that there may be considerable arterial–venous shunting under these conditions, leading to endomyocardial ischemia. This effect would not be detected by the surface fluoresence methods they used. In a series of control experiments, the inferior and superior vena cavae were ligated and the pulmonary artery cannulated. Thus, we could directly measure coronary and noncoronary flows. The results showed that over the buffer pH range of 7.4 to 6.7, no significant changes in flow were detected. Although endomyocardial perfusion defects may occur, these data suggest that they are minimal.

B. Contractility and pH_i During Ischemia

Two models were used to examine the relationship between pH_i and contractility in the ischemic myocardium. The first was total global ischemia, achieved by cross-clamping the aortic perfusion input line and inducing a no-flow state. The protocol for these experiments is illustrated in Figure 8. The NMR spectrometer was programmed to accumulate three consecutive spectra, here designated as files 01, 02, and 03. These are only 10-pulse spectra, requiring just 20 seconds per file. Between the first and second spectrum, the aortic line was occluded. After 40 seconds of ischemia reflow was established, indicated as "release." This scheme provided us with data time-averaged during the control period, and at 10 and 30 seconds of ischemia. The performance changes observed during a typical determination are illustrated in the lower panel. One observes the rapid fall of function at the onset of global ischemia, and also notes that shortly after perfusion is restored, function recovered to the control state. Therefore, this brief period of ischemia does not induce persistent contractile defects. In order to obtain sufficient signal, this protocol was repeated 20 times on the same heart. The performance characteristics for a heart subjected to this protocol are given in Table VII. One notes the stability of the control performance, as well as the uniformity of the heart's response to ischemia. Therefore, we felt justified in summating the data from these 20 consecutive experiments to obtain three individual 200-pulse spectra with excellent signal-to-noise. This, then, was a condition of transient ischemia.

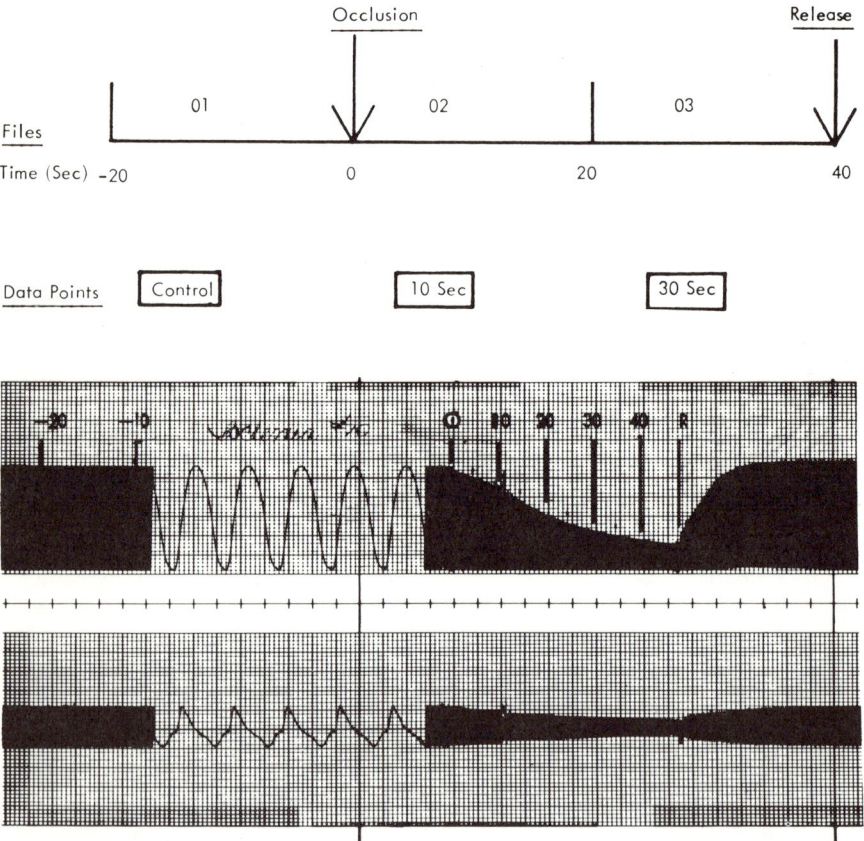

Fig. 8. Protocol for total global ischemia. The upper section illustrates the sequence for the NMR spectrometer. A performance trace for a single determination is shown in the lower portion. The upper trace is ventricular pressure; the lower trace is dP/dt. [Jacobus et al, 1982.]

The second ischemic protocol was that of partial steady-state ischemia (Fig. 9). In these experiments the perfusion pressure was reduced sufficiently to depress contractility by 50%. This was achieved simply by lowering the level of the perfusate reservoir until LVDP was depressed by 50%. If intracellular acidosis fully accounted for the negative inotropic effect of ischemia [Steenbergen et al, 1977], we anticipated from the data of Figure 7 that we would detect a 0.22 pH

TABLE VII. Effects of Repetitive Total Global Ischemia on Left Ventricular Developed Pressure in the Perfused Rabbit Heart

Control	10 sec	20 sec	30 sec	40 sec
106	87	61	45	36
108	89	64	46	36
108	87	60	43	34
108	90	62	45	32
107	86	58	42	32
107	89	61	43	33
106	86	59	42	32
108	88	59	38	33
108	88	59	42	32
107	86	58	42	32
107	90	62	43	33
108	86	58	41	32
107	88	59	42	32
107	86	58	39	30
106	88	60	40	31
106	86	57	39	30
106	87	57	39	30
106	87	56	38	30
106	86	56	38	30
107	85	55	36	30

Data are from a single heart subjected to the protocol detailed in Figure 8. Time in seconds after aortic line occlusion.

unit acidification in this protocol. This, however, was not the case. During 50% steady-state ischemia, intracellular pH declined from 7.15 (spectrum A) to 7.06 (Spectrum B), or only by 0.09 pH units. During reprefusion (spectrum C), intracellular pH and performance returned to control, $pH_i = 7.18$. The fact that these hearts were subjected to ischemia of sufficient magnitude to evoke other contractile abnormalities is shown by the data at the lower right of Figure 9. In the control state, paced hearts normally relaxed to an end diastolic pressure of 4 mm Hg. However, during early postischemic reflow, end diastolic pressures as high as 20 mm Hg were recorded. When the pacer was then turned off, end diastolic pressure fell to 4 mm Hg, indicating that incomplete relaxation was occurring. Incomplete relaxation during early reflow is one of the known effects of ischemia on myocardial performance [Weisfeldt et al, 1974]. Thus, the presence of ischemia in these hearts was evidenced by contractile depression as well as by reflow abnormalities in diastolic properties. Nevertheless, intracellular acidosis was clearly less than expected. To evaluate the role of the high energy phosphates as regulatory metabolites, a difference spectrum was obtained (Fig. 9, spectrum D). The computer subtracted the control spectrum A from the ischemic spectrum B. In control experiments a 10% change in metabolite concentration could be detected in the difference spectrum. However, in spectrum

D we see no changes. This indicates that the reductions in the high energy phosphates were minimal, or less than 10%, an observation in agreement with the conclustions of Steenbergen et al [1977].

The summated pH_i data from both ischemic experiments are illustrated in Figure 10. The solid line is the calibration line determined by respiratory acidosis (Fig. 7). The data of Figure 10 show that for either condition of ischemia, a 50% fall in contractility is associated with only a 0.09 pH unit acidification. In conclusion, these data suggest that intracellular acidosis may account for between 40% and 50% of the immediate negative inotropic effect of ischemia. However, other metabolic and/or physiological mechanisms must also contribute. The minimal changes in the high energy phosphates (Fig. 9, spectrum D) suggest that these metabolites may not be important regulatory signals under these conditions.

C. Mass Spectrometry Studies

From the perspective of heart energy metabolism, oxygen is both the most necessary and also the most limited substrate in the myocardium, having no cytoplasmic reserve. It has been estimated that in less than 5 seconds after coronary artery ligation, the tissue concentration of oxygen falls below the K_m for cytochrome oxidase, and the electron transport chain becomes reduced. Oxidative phosphorylation, the source of 90% of contractile ATP [Neely and Morgan, 1974], stops under these conditions. Physiologically important oxygen chemoreceptors exist in the carotid bodies of the carotid arteries. These receptors monitor blood oxygen content and help ensure adequate blood flow to the brain. In a similar manner, oxygen per se could directly exert a regulatory effect over myocardial contractility. If such were the case, it was possible that the changes in contractility observed at the onset of ischemia might parallel a decline in tissue oxygen tension, or PmO_2. To investigate this hypothesis we used the combined methods of NMR and mass spectrometry (Figs. 11 and 12).

The protocol for these experiments was similar to that used in Figure 9. Hearts were initially perfused at a pressure of 80 mm Hg. In a stepwise manner, perfusion pressure was reduced in 10 mm Hg increments from 80 to 20 mm Hg, or expressed as a percentage reduction to 25% of control pressure. In both Figures 11 and 12, the line of identity is indicated by the dashed line. In Figure 11 we see that three parameters closely parallel the line of identity. These are the rates of coronary blood flow (CBF), left ventricular developed pressure (DP), and the rates of myocardial oxygen utilization (MVO_2). In other words, as perfusion pressure is diminished there is an almost linear decline in coronary blood flow. In conjunction, there is a similar decline in developed pressure, which, because of the reduction in ventricular work, leads to a parallel decline in oxygen utilization. These results were neither new or surprising.

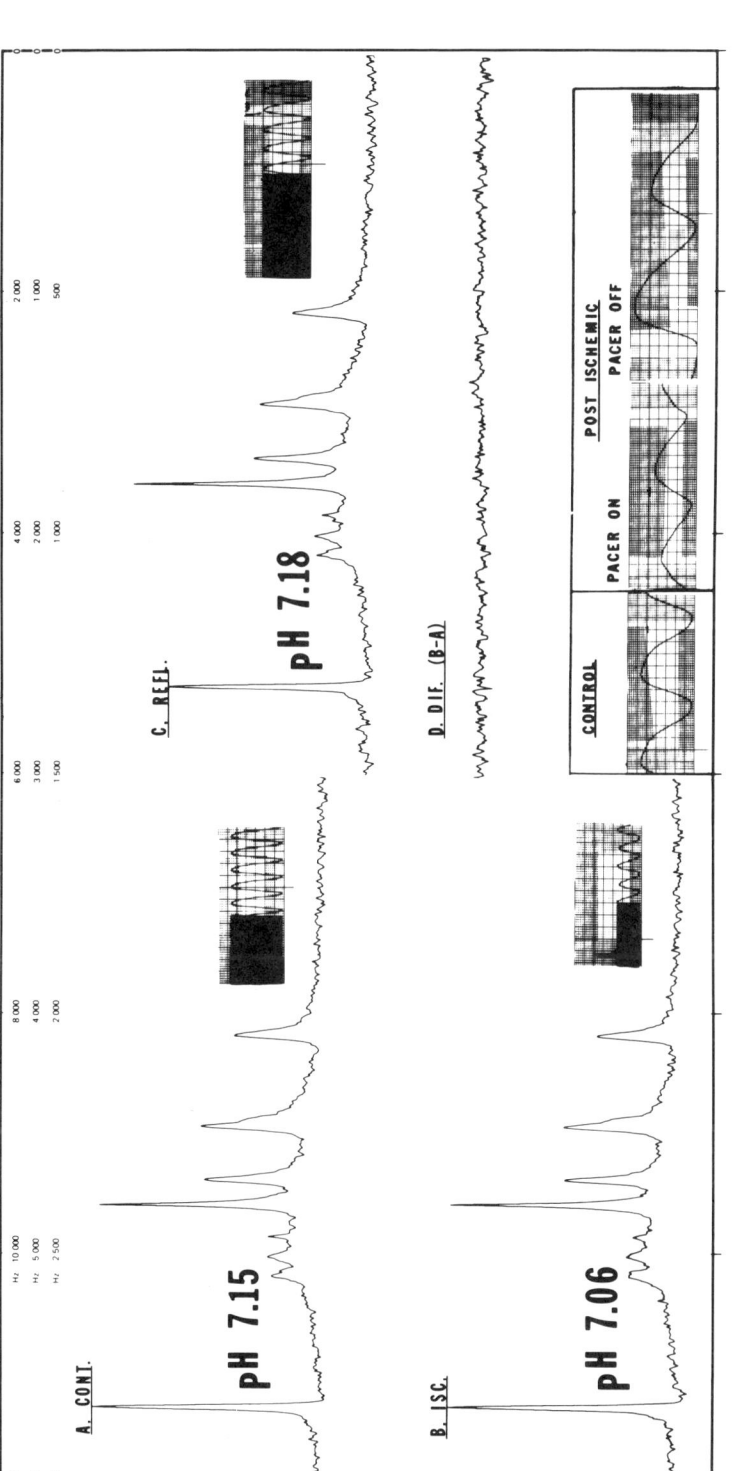

Fig. 9. Protocol for steaty-state partial ischemia. These are 400-pulse spectra at a 5,000-Hz spectral width. A) Control spectrum. B) Spectrum at a flow reduction sufficient to depress function by 50% (see insert pressure trace). C) Spectrum acquired after 15 minutes of reflow. D) Difference spectrum, spectrum B–spectrum A. The pressure traces in the lower right were taken during the preischemic period (control), or during the first minute of reperfusion. [Jacobus et al, 1982.]

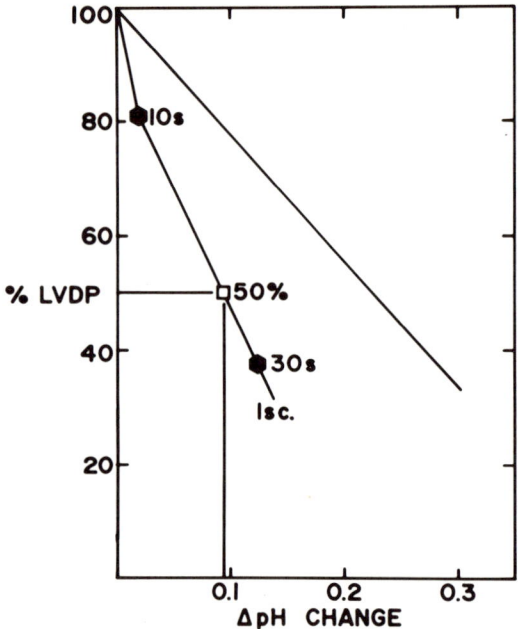

Fig. 10. Relationship between pH_i and LVDP during ischemia. The solid line is the calibration line derived from the respiratory acidosis studies (Fig. 7). Time points of 10 and 30 seconds were derived from the transient ischemia protocol (Fig. 8), and the 50% data point was derived from the protocol outlined in Figure 9. [Jacobus et al, 1982.]

In contrast, a rather striking and unanticipated set of results was seen in the NMR and mass spectrometry data (Fig. 12). If our working hypothesis had been correct, we would have predicted an immediate fall in tissue PmO_2, a sharp rise in tissue $PmCO_2$, with perhaps moderate changes in pH_i. This was not observed (Fig. 12). Initially, PmO_2 increased while tissue $PmCO_2$ fell and pH_i remained constant. As perfusion pressure was further decreased to about 75% of control, PmO_2 then started to decline and $PmCO_2$ began to rise, still at a rather constant pH_i. The gases continued to change until the crossover point, which occurred at about a 44% reduction in perfusion pressure or a 52% reduction in flow (Table VIII). These data suggest that up to 50% reduction in flow, an efficient autoregulatory mechanism so specifically down-regulates contraction and oxygen utilization in response to decreasing flow that oxygen supply apparently exceeds oxygen demand. As a result, tissue PmO_2 increases, presumably allowing aerobic metabolism to continue and intracellular pH to remain in balance (Table VIII). Beyond a 50% reduction in flow, this mechanism fails. Contractile demands then exceed supply, and metabolic indices of ischemia are expressed as high $PmCO_2$, low PmO_2, and intracellular acidosis. Therefore, moderate reductions

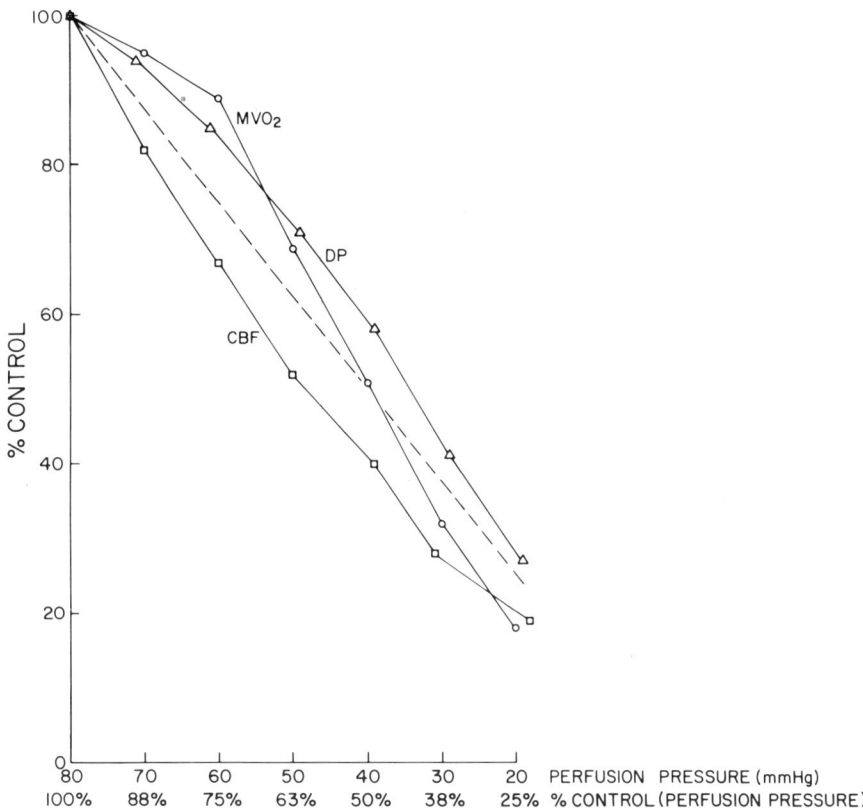

Fig. 11. Changes in coronary blood flow (CBF), developed pressure (DP), and the rate of oxygen utilization (MVO$_2$) as a function of decreased perfusion pressure. Data were collected during a steady-state reduction in perfusion pressure, which was held constant at each point for at least 15 minutes. The dashed line is the line of identity. [Jacobus et al, 1982.]

in coronary flow do not result in a supply/demand imbalance (equation 2), even though left ventricular function is markedly diminished. Since this mechanism also prevents the "ischemic" accumulation of H$^+$ and Pi, agents known to induce cellular damage [Jacobus et al, 1980a; Mukherjee et al, 1979], it appears that this compensatory regulation could protect the heart from "irreversible" damage during moderate reductions in flow.

IV. DISCUSSION

Several new and critical features of cardiac physiology have emerged from the results presented in this volume. Equally important, we have clearly dem-

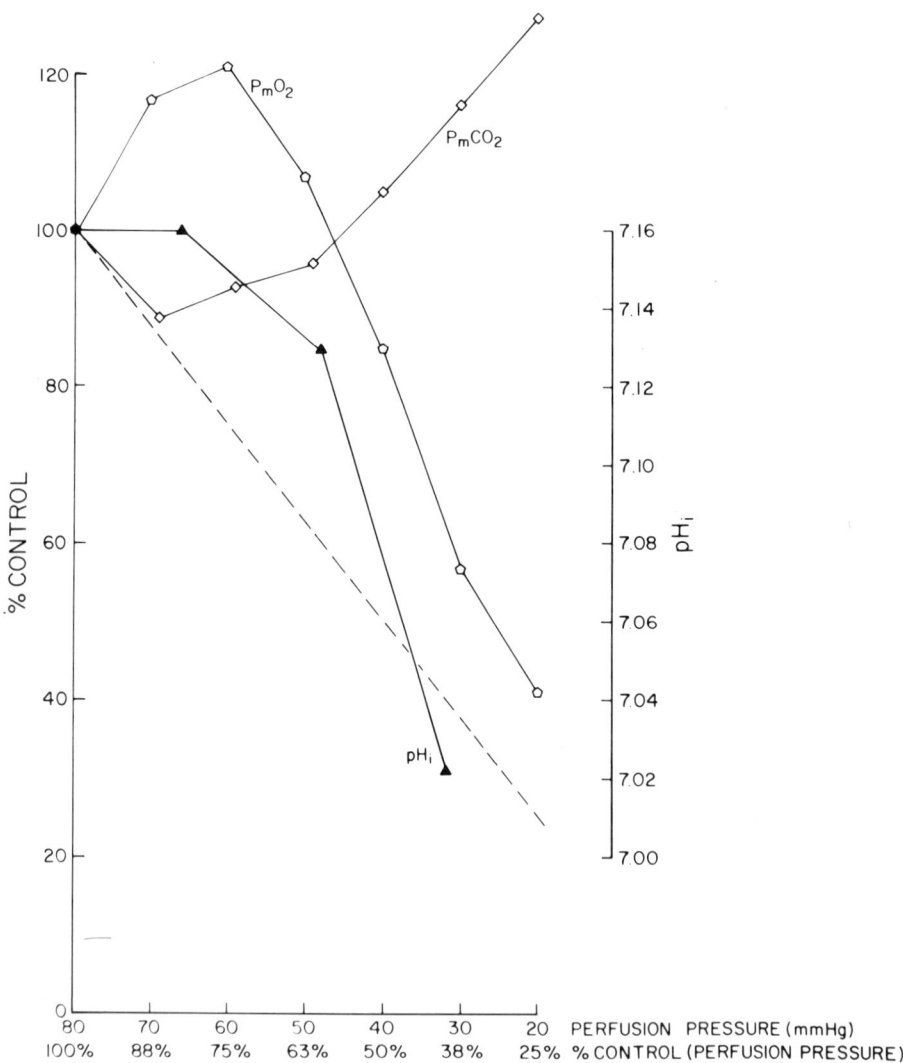

Fig. 12. Changes in tissue oxygen content (PmO$_2$), carbon dioxide content (PmCO$_2$), and intracellular pH (pH$_i$) as a function of decreased perfusion pressure. Conditions were as described in Figure 11. Tissue gases were measured directly by mass spectrometry methods, and pH$_i$ was determined by NMR (see Methods). [Jacobus et al, 1982.]

onstrated the applicability and versitility of the NMR method. Whereas some debate persists concerning the conversion of the NMR data into absolute pH$_i$ values, it is apparent that the method is highly useful for the estimation of intracellular pH shifts (Table VI). The method is noninvasive, nondestructive,

TABLE VIII. Magnitude of the Changes Observed at the Crossover Point for PmO_2 and $PmCO_2$

Perfusion pressure	44% reduction
Coronary blood flow	52% reduction
Left ventricular developed pressure	32% reduction
Rate of O_2 utilization	38% reduction
Intracellular pH	0.04 pH unit acidification
High energy phosphate	No detectable changes
Myocardial PmO_2 and $PmCO_2$	Normal

Data were derived from Figures 11 and 12.

and quite flexible. It thus provides a number of advantages over other widly used methods [Poole-Wilson, 1978]. There has always been some concern that during ischemia, changes in membrane permeability might alter the transmembrane distribution of the weak acids or bases. This would invalidate these methods. The microelectrode methods [Ellis and Thomas, 1976a, b] suffer from the criticism that the method itself may induce damage. Maintenance of electrode position in the beating, moving heart is also a physical limit to this method. On the other hand, NMR measurements can be made under a wide variety of conditions while the heart is functioning physiologically. Sequential steady-state measurements can be made. The collection times may also be manipulated so that transient data can be acquired. And finally, the method may be used when the heart is in a no-flow state, an obviously impossible condition for the distribution techniques.

With respect to cardiac physiology, we have demonstrated the very close coupling between changes in pH_i (acidosis) and myocardial contractility. Steenbergen at al [1977] showed the relationship between perfusate pH and heart performance for four buffer systems. For two of these, bicarbonate and MES, they also graphed pH_i as a function of perfusate (effluent) pH. However, they did not present the correlation between pH_i and contractility. Therefore our data presented in Figure 7 are unique and important. For the first time we appreciate the sensitivity of contractility to changes in pH_i. It is quite striking that a very small change in pH_i (-0.22 pH units) is associated with such a major change in performance, a 50% depression of LVDP.

However, some reservations must be made in the interpretation of these data. We are unable to differentiate whether acidosis is the primary cause for contractile depression, or whether acidosis induces some secondary change which becomes the causative agent. For example, acidosis can depress gap junction conductance [Spray et al, 1981], alter transmembrane ion fluxes [Poole-Wilson and Cameron, 1975; Poole-Wilson and Langer, 1975, 1979; Coraboeuf et al, 1976; Serur et al, 1976; Shine et al, 1977; Lakatta et al, 1979] and thus influence the cardiac action potential [Chesnais et al, 1975; Samulesson and Nagy, 1976; Spitzer and Hogan, 1979]. Within the cell there can be H^+-Ca^{2+} competition at the sarcoplasmic reticulum [Nakamaru and Schwartz, 1970; Fabiato and Fabiato, 1978] or at the Tn-C subunit of troponin [Fuchs et al, 1970]. It is also known that acidosis can regulate the actomyosin ATPase activity [Fabiato and Fabiato, 1978; Donaldson and Hermansen, 1978; Donaldson et al, 1979; Kentish and Nayler, 1979; Donaldson, 1981] as well as depress the rates of cardiac energy metabolism [Wollenberger and Krause, 1968; Neely et al, 1975; Opie, 1976]. Further elucidation of the contribution of nonmetabolic mechanisms is beyond the limits of the ^{31}P-NMR and mass spectrometry methods.

Whereas this work has focused on the role of acidosis, it is also important for readers of this volume to keep in mind the potential regulatory role for intracellular alkalosis. Scheuer and Stezoski [1972] showed that alkalosis protected the myocardium against the depressive effects of hypoxia. More recently Clancy's laboratory has reported that intracellular alkalosis may be a component of the mechanism of action of some of the inotropic drugs [Riegle and Clancy, 1975; Clancy et al, 1976]. Clearly, this is an important area for future investigation.

The contribution of acidosis to the control of contractility in hypoxia and ischemia has been a long debated issue [McElroy et al, 1958; Ng et al, 1967; Cingolani et al, 1970; Effros et al, 1975; Gonzalez and Clancy, 1975; Krug, 1975; Lai and Scheuer, 1975; Weisfeldt et al, 1975; Steenbergen et al, 1977; Cobbe and Poole-Wilson, 1980a, b]. In spite of all this work, there is still some question as to whether the depressive effect is caused by acidosis or by the accumulation of CO_2 under conditions of limited washout [Poole-Wilson. 1975]. It is apparent that changes in intracellular CO_2 could alter rates of mitochondrial ion transport [Elder and Lehninger, 1973]. Nevertheless, our data suggest that neither pH_i or CO_2 can account for the immediate fall in contractility seen at the onset of total global ischemia (Fig. 10), or during moderate steady-state partial ischemia (Fig. 12). These data appear to contradict the conclusions of Steenbergen et al [1977] and Cobbe and Poole-Wilson [1980a]. The role of extracellular acidosis was minimized by the work of Cingolani et al [1970]. And finally, in contrast to the data of Hearse [1979], it appears that contractility can be markedly depressed with little or no changes in the cellular high energy phosphates (Fig. 9, spectrum D). We thereore conclude that neither of the two metabolic hypotheses formulated in the Introduction can explain the instantaneous effects of ischemia on myocardial performance. However, our data are quite consistent with the theory that the physiological effects of reduced ventricular wall tension may be the initial regulatory control mechanism [Apstein et al, 1979]. The phenomenon relating coronary perfusion pressure to the internal stretching of the myocardial muscle fibers has been called the "garden hose" effect [Salisbury et al, 1960, 1962; Arnold et al, 1968].

From the data currently available, it is now possible to speculate on the kinetics of the earliest events occurring at the onset of total global ischemia. Within the first 6 seconds, arterial pressure falls, resulting in decreased ventricular wall tension. The measured decrease in wall tension parallels the decline in contractility [Apstein et al, 1979]. At approximately 6 seconds, pyridine nucleotide fluorescence increases [Brent and Apstein, 1979], indicating changes in the mitochondrial matrix redox state. Presumably, until this time there is sufficient oxygen available to maintain oxidative phosphorylation. Thus, tissue ATP and phosphocreatine concentrations remain normal, and lactate is low. The depression of contractility also reduces work load and oxygen demands, which may act to prolong further the period of aerobic metabolism. During this time (0–6 seconds), pH_i would remain in balance. Oxidative phosphorylation generates net OH^-, which titrates the H^+ produced by the hydrolysis of ATP [Vaghy, 1979]. At approximately 8–9 seconds, anaerobic metabolism begins. However, during the initial period of anaerobiosis, pH_i would remain unchanged. The transphosphorylation of ADP back to ATP via the creatine kinase reaction generates the base creatine, which would act to neutralize ATPase generated protons [Gevers, 1977]. In skeletal muscle, spectrophotometric data have shown a net alkalinization of the cell during a twich, which correlates with phosphocreatine breakdown [MacDonald and Jöbsis, 1976]. Therefore, until the rate of the ATPase reactions exceeds the rate of the creatine kinase reaction, pH_i can be unchanged. Eventually, however, the rate of creatine kinase declines as the phosphocreatine pool is exhausted. At this time ADP will accumulate, anaerobic glycolysis will be initiated, and lactate will be produced. A moderate delay in the activation of glycolysis was reported for ischemia in the brain [Lowry et al, 1964]. It is at this point, a time considerably after the onset of ischemia, that changes in pH_i should be measurable. It thus appears from this discussion that metabolic changes are occurring well after the onset of global ischemia, whereas the physiological alterations in wall tension are immediately observed. In support of this argument it is essential to note that even after 9 seconds of global ischemia, LVDP could be restored to control levels by the injection of viscous dextran, which also returned wall tension to near normal values [Apstein et al, 1979]. The redox state of the pyridine nucleotides remained reduced during these injections. It is also possible that changes in wall tension account for all aspects of the effecient autoregulatory mechanism seen in Figure 12. How wall tension changes are coupled within the cell to depress contractility remains uncertain.

ACKNOWLEDGMENTS

This work was supported by United States Public Health Service grants HL-22080, HL-17655, and HL-19414, and by the resources of the Susan B. Clayton Fund of the Johns Hopkins University School of Medicine. WEJ is an Established Investigator of the American Heart Association.

V. REFERENCES

Ackerman JJH, Grove TH, Wong GG, Gadian DG, Radda GK: Mapping of metabolites in whole animals by ^{31}P NMR using surface coils. Nature 283:167, 1980.

Apstein CA, Ahn J, Briggs L, Shapiro HM: Role of decrease in wall thickness in causing ischemic cardiac failure. Clin Res 27:436a, 1979.

Arnold G, Kosche F, Miessner E, Neitzert A, Lochner W: The importance of the perfusion pressure in the coronary arteries for the contractility and the oxygen consumption of the heart. Pfluegers Archiv 299:339, 1968.

Brent BN, Apstein CS: Kinetics of acute cardiac failure: Comparisons of ischemia, hypoxemia and cyanide. Clin Res 27:156a, 1979.

Bulkley BH, Nunnaly RL, Hollis DP: "Calcium paradox" and the effects of varied temperature on its development. A phosphorus nuclear magnetic resonance and morphologic study. Lab Invest 39:133, 1978.

Burt CT, Glonek T, Barany M: Analysis of phosphate metabolites, the intracellular pH, and the state of adenosine triphosphate in intact muscle by phosphorus nuclear magnetic resonance. J Biol Chem 251:2584, 1976.

Chesnais Jm, Coraboeuf E, Sauviat MP, Vassas JM: Sensitivity to H, Li, and Mg ions of the slow inward sodium current in frog atrial fibers. J Mol Cell Cardiol 7:627, 1975.

Cingolani HE, Mattiazzi AR, Blesa ES, Gonzalez NC: Contractility in isolated mammalian heart muscle after acid–base changes. Circ Res 26:269, 1970.

Clancy RL, Gonzalez NC, Fenton RA: Effect of beta-adrenorecptor blockade on rat cardiac and skeletal muscle pH. Am J Physiol 230:959, 1976.

Cobbe SM, Poole-Wilson PA: The time of onset and severity of acidosis in myocardial ischaemia. J Mol Cell Cardiol 12:745, 1980a.

Cobbe SM, Poole-Wilson PA: Tissue acidosis in myocardial hypoxia. J Mol Cell Cardiol 12:761, 1980b.

Cohn M, Hughs TR Jr: Nuclear Magnetic resonance spectra of adenosine di- and triphosphate. II. Effect of complexing with divalent metal ions. J Biol Chem 237:176, 1962.

Coraboeuf E, Deroubaix E, Hoerter J: Control of ionic permeabilities in normal and ischemic heart. Circ Res 38, Suppl I:I–92, 1976.

Dawson MJ, Gadian DG, Wilkie DR: Contraction and recovery of living muscle studied by ^{31}P nuclear magnetic resonance. J Physiol 267:703, 1977.

Dawson MJ, Gadian DG, Wilkie DR: Muscular fatigue investigated by phosphorus nuclear magnetic resonance. Nature 274:861, 1978.

Donaldson SK: Single skinned skeletal fiber Ca^{2+} and H^+ sensitivities: Comparison of the various histochemical types. Biophys J 33:57a, 1981.

Donaldson S, Bond E, Seeger L, Niles N, Bolles L: Effects of altering [$MgATP^{2-}$] at various pH's on rabbit cardiac disrupted fiber bundle force. Biophys J 25:114a, 1979.

Donaldson SKB, Hermansen L: Differential, direct effects of H^+ on Ca^{2+}-activated force of skinned fibers from the soleus, cardiac and adductor magnus muscles of rabbits. Pfluegers Archiv 376:55, 1978.

Effros RM, Haider B, Ettinger PO, Ahmed SS, Oldewurtel HA, Marold K, Regan TJ: In vivo myocardial cell pH in the dog. Response to ischemia and infusion of alkali. J Clin Invest 55:1100, 1975.

Elder JA, Lehninger AL: Respiration depentent transport of carbon dioxide into rat liver mitochondria. Biochemistry 12:976, 1973.

Ellis D, Thomas RC: Microelectrode measurement of the intracellular pH of mammalian heart cells. Nature 262:224, 1976a.

Ellis D, Thomas RC: Direct measurement of the intracellular pH of mammalian cardiac muscle. J Physiol 262:755, 1976b.

Fabiato A, Fabiato F: Effects of pH on the myofilaments and the sarcoplasmic reticulum of skinned cells from cardiac and skeletal muscle. J Physiol 276:233, 1978.

Flaherty JT, Jacobus WE: Use of ^{31} phosphorus NMR to optimize buffer capacity in cardioplegic solutions. Circulation 60:II-11, 1979.

Flaherty JT, Weisfeldt ML, Bulkley BH, Gardner TJ, Gott VL, Jacobus WE: Mechanisms of ischemic myocardial cell damage assessed by ^{31}phosphorus nuclear magnetic resonance. Circulation 65 (in press), 1982.

Fossel ET, Morgan HE, Ingwall JS: Measurement of changes in high-energy phosphates in the cardiac cycle by using gated ^{31}P nuclear magnetic resonance. Proc Natl Acad Sci USA 77:3654, 1980.

Fuchs F, Reddy Y, Briggs FN: The interaction of cations with the calcium binding site of troponin. Biochim Biophys Acta 221:407 1970.

Gadian DG, Radda GK: NMR studies of tissue metabolism. Ann Rev Biochem 50:69, 1981.

Garlick PB, Radda GK, Seeley PJ, Chance B: Phosphorus NMR studies on perfused heart. Biochem Biophys Res Commun 74:1256, 1977.

Garlick PB, Radda GK, Seeley PJ: Studies of acidosis in the ischaemic heart by phosphorus nuclear magnetic resonance. Biochem J 184:547, 1979.

Gevers W: Generation of protons by metabolic precesses in heart cells. J Mol Cell Cardiol 9:867, 1977.

Gonzalez NC, Clancy RL: Inotropic and intracellular acid–base changes during metabolic acidosis. Am J Physiol 228:1060, 1975.

Grove TH, Ackerman JJH, Radda GK, Bore PJ: Analysis of rat heart in vivo by phosphorus nuclear magnetic resonance. Proc Natl Acad Sci USA 77:299, 1980.

Hearse DJ: Oxygen deprivation and early myocardial contractile failure. A reassessment of the possible role of adenosine triphosphate. Am J Cardiol 44:1115, 1979.

Hollis DP, Nunnally RL, Jacobus WE, Taylor GJ IV: Detection of regional ischemia in perfused beating hearts by phosphorus nuclear magnetic resonance. Biochem Biophys Res Commun 75:1086, 1977.

Hollis DP, Nunnally RL, Taylor, GJ,IV, Weisfeldt ML, Jacobus WE: Phosphorus nuclear magnetic resonance studies of heart physiology. J Mag Res 29:319, 1978.

Hoult DI, Busby SJW, Gadian DG, Radda GK, Richards RE, Seeley PJ: Observations of tissue metabolites using ^{31}phosphorus nuclear magnetic resonance. Nature 252:285, 1974.

Jacobus, WE, Taylor, GJ, Hollis DP, Nunnally RL: Phosphorus nuclear magnetic resonance of perfused working rat hearts. Nature 265:756, 1977.

Jacobus WE, Pores IH, Taylor GJ, Nunnally RL, Hollis DP, Weisfeldt ML: Tight coupling of intracellular pH and ventricular performance. J Mol Cell Cardiol 10, Suppl I:39, 1978.

Jacobus WE, Bittl JA, Weisfeldt ML: Loss of mitochondrial creatine kinase in vitro and in vivo. A sensitive index of ischemic cellular and functional damage. In Jacobus WE, Ingwall JS (eds): "Heart Creatine Kinase: The Integration of Isozymes for Energy Distribution." Baltimore: Williams and Wilkins, 1980a, p 155.

Jacobus W, Lucas S, Flaherty J: Lack of metabolic correlates of ischemic dysfunction. Circulation 62:III-19, 1980b.

Jacobus WE, Pores IH, Lucas SK, Weisfeldt ML, Flaherty JT: Intracellular acidosis and contractility in the normal and ischemic heart as examined by ^{31}P NMR. J Mol Cell Cardiol 14 (in press), 1982.

Jennings RB: Relationship of acute ischemia to functional defects and irreversibility. Circulation 53, Suppl I:I-26, 1976.

Kentish JC, Nayler WE: The influence of pH on the Ca^{2+}-regulated ATPase of cardiac and white skeletal myofibrils. J Mol Cell Cardiol 11:611, 1979.

Khuri SF, Flaherty JT, O'Riordan JB, Pitt B, Brawley RK, Donahoo JS, Gott VL: Changes in intramyocardial ST segment voltage and gas tensions with regional myocardial ischemia in the dog. Circ Res 37:455, 1975.

Klug F: Ueber den Einfluss gasartiger Korper auf die Function des Froschherzens. Archiv Anat Physiol: 435, 1879.

Kohn MC, Achs MJ, Garfinkel D: Distribution of adenine nucleotides in the perfused rat heart. Am J Physiol 232:R158, 1977.

Krug A: Alterations in myocardial hydrogen ion concentration after temporary coronary occlusion. A sign of irreversible cell damage. Am J Cardiol 36:214, 1975.

Kübler W, Katz AM: Mechanism of early "pump" failure of the ischemic heart. Possible role of adenosine triphosphate depletion and inorganic phosphate accumulation. Am J Cardiol 40:467, 1977.

Lai F, Scheuer J: Early changes in myocardial hypoxia: Relations between mechanical function, pH and intracellular compartmental metabolites. J Mol Cell Cardiol 7:289, 1975.

Lakatta EG, Nayler WG, Poole-Wilson PA: Calcium overload and mechanical function in posthypoxic myocardium: Biphasic effect of pH during hypoxia. Eur J Cardiol 10:77, 1979.

Lowry OH, Passonneau JV, Hasselberger FX, Schultz DW: Effect of ischemia on known substrates and cofactors of the glycolytic pathway in brain. J Biol Chem 239:18, 1964.

MacDonald VW, Jöbsis FF: Spectrophotometric studies on the pH of frog skeletal muscle. pH changes during and after contractile activity. J Gen Physiol 68:179, 1976.

McElroy WT Jr, Gerdes AF, Brown EB Jr: Effects of CO_2, bicarbonate and pH on the performance of isolated perfused guinea pig hearts. Am J Physiol 195:412, 1958.

Moon RB, Richards JH: Determination of intracellular pH by ^{31}P magnetic resonance. J Biol Chem 248:7276, 1973.

Mukherjee A, Wong TM, Templeton G, Buja LM, Willerson JT: Influence of volume dilution, lactate, phosphate, and calcium on mitochondrial functions. Am J Physiol 237:H224, 1979.

Nakamaru Y, Schwartz A: Possible control of intracellular calcium metabolism by [H^+]: Sarcoplasmic reticulum of skeletal and cardiac muscle. Biochem Biophys Res Commun 41:830, 1970.

Neely JR, Morgan HE: Relationship between carbohydrate and lipid metabolism and the energy balance of heart muscle. Ann Rev Physiol 36:413, 1974.

Neely JR, Rovetto MJ: Techniques for perfusing isolated rat hearts. Meth Enzymol 39:43, 1975.

Neely JR, Whitmer JT, Rovetto MJ: Effect of coronary blood flow on glycolytic flux and intracellular pH in isolated rat hearts. Circ Res 37:733, 1975.

Ng ML, Levy MN, Zieske HA: Effects of changes of pH and carbon dioxide tension on left ventricular performance. Am J Physiol 213:115, 1967.

Nunnally Rl, Bottomley PA: Assessment of pharmacological treatment of myocardial infarction by phosphorus-31 NMR with surface coils. Science 21:117, 1981.

Opie LH: Effects of regional ischemia on metabolism of glucose and fatty acids. Relative rates of aerobic and anaerobic energy production during myocardiol infarction and comparison with effects of anoxia. Circ Res 38, Suppl I:1–52, 1976.

Poole-Wilson PA: Is the early decline of cardiac function in ischaemia due to carbon-dioxid retention? Lancet 2:1285, 1975.

Poole-Wilson PA: Measurement of myocardial intracellular pH in pathological states. J Mol Cell Cardiol 10:511, 1978.

Poole-Wilson PA, Cameron IR: Intracellular pH and K^+ of cardiac and skeletal muscle in acidosis and alkalosis. Am J Physiol 229:1305, 1975.

Poole-Wilson PA, Langer GA: Effect of pH on ionic exchange and function in rat and rabbit myocardium. Am J Physiol 229:570, 1975.

Poole-Wilson PA, Langer GA: Effects of acidosis on mechanical function and Ca^{2+} exchange in rabbit myocardium. Am J Physiol 236:H533, 1979.

Riegle KM, Clancy RL: Effect of norepinephrine on myocardial intracellular hydrogen ion concentration. Am J Physiol 229:344, 1975.

Roos A, Boron W: Intracellular pH. Physiol Rev 61:296, 1981.

Salhany JM, Pieper GM, Wu S, Todd GL, Clayton FC, Eliot RS: ^{31}P nuclear magnetic resonance measurements of cardiac pH in perfused guinea-pig hearts. J Mol Cell Cardiol 11:601, 1979.

Salisbury PF, Cross CE, Rieben PA: Influence of coronary artery pressure upon myocardial elasticity. Circ Res 8:794, 1960.

Salisbury PF, Cross CE, Rieben PA: Intramyocardial pressure and strength of left ventricular contraction. Circ Res 10:608, 1962.

Samulesson RG, Nagy G: Effects of respiratory alkalosis and acidosis on myocardial excitation. Acta Physiol Scand 97:158, 1976.

Scheuer J, Stezoski SW: The effect of alkalosis upon the mechanical and metabolic response of the rat heart to hypoxia. J Mol Cell Cardiol 4:599, 1972.

Serur JR, Skelton CL, Bodem R, Sonnenblick EH: Respiratory acid–base changes and myocardial contractility: Interaction between calcium and hydrogen ions. J Mol Cell Cardiol 8:823, 1976.

Shine KI, Douglas AM, Ricchiutini N: ^{42}K exchange during myocardial ischemia. Am J Physiol 232:H564, 1977.

Spitzer KW, Hogan PM: The effects of acidosis and bicarbonate on action potential repolarization in canine cardiac purkinje fibers. J Gen Physiol 73:199, 1979.

Spray DC, Harris AL, Bennett MVL: Gap junctional conductance is a simple and sensitive function of intracellular pH. Science 211:712, 1981.

Steenbergen C, Deleeuw G, Rich T, Williamson JR: Effects of acidosis and ischemia on contractility and intracellular pH of rat heart. Circ Res 41:849, 1977.

Tennant R, Wiggers CJ: The effect of coronary occlusion on myocardial contraction. Am J Physiol 112:351, 1935.

Vaghy PL: Role of mitochondrial oxidative phosphorylation in the maintenance of intracellular pH. J Mol Cell Cardiol 11:933, 1979.

Weisfeldt ML, Armstrong P, Scully H, Daggett WM: Incomplete relaxation between beats after myocardial hypoxia and ischemia. J Clin Invest 53:1626, 1974.

Weisfeldt ML, Bishop RL, Green HL: Effects of pH and pCO_2 on performance of ischemic myocardium. In Roy P-E, Rona G (eds): "Recent Advances in Studies on Cardiac Structure and Metabolism," Vol 10. Baltimore: University Park Press, 1975, p 355.

Wollenberger A, Kraus EG: Metabolic control characteristics of the acutely ischemic myocardium. Am J Cardiol 22:349, 1968.

Summary of the Evidence and Discussion Concerning the Involvement of pH_i in the Control of Cellular Functions

Richard Nuccitelli and Jeanne M. Heiple

Zoology Department, University of California, Davis, California 95616 (R.N.), and Department of Biochemistry, Boston University School of Medicine, 80 E. Concord Street, Boston, Massachusetts 02118 (J.M.H.)

I.	Introduction	567
II.	Summary of Presentations and Discussions	568
	A. Activation of Egg Development	568
	B. pH_i Changes During the Cell Cycle	569
	C. Control of Metabolism	578
	D. Control of Plasma Membrane Properties	579
	E. Intracellular pH Changes During Stimulus-Response Coupling	581
	F. Tissue pH_i Measurements: Diagnostic Applications	582
III.	Conclusions	583
IV.	References	585

I. INTRODUCTION

Changes in intracellular pH (pH_i) have been found to accompany a variety of cellular functions, and in some of these the pH_i change is an essential part of the control mechanism. The chapters in this section of the volume describe examples of pH_i changes in a number of cell types, and are grouped by the specific cellular function associated with the pH_i change. Table I lists most of these functions for quick reference. Two striking trends emerge from the data in Table I: 1) In nearly every example listed, the higher pH_i value is associated with increased cellular activity or efficiency. Specifically, intracellular alkalinization accompanies increased rates of glycolysis, protein synthesis, and (in some cases) DNA synthesis, and initiates sperm motility. 2) The magnitudes of the pH_i changes fall into two categories: 13 of the 19 examples listed exhibit a natural pH increase of 0.1–0.5 pH unit, whereas the 4 cases of release from dormancy exhibit a larger pH_i rise of 1.0–1.4 pH units. Such large pH_i increases will dramatically affect the kinetics of many intracellular enzymatic reactions as discussed by Busa and Moore et al (this volume). This observation that dormant cells have relatively low pH_i values also supports the first generalization that increased pH_i favors increased cellular activity. These generalizations stimulated heated discussion and disagreement among the meeting participants, which will be described at the end of this summary.

Some of the papers included in this section are not listed in Table I because they do not report natural pH_i changes, but instead describe the effect of imposed pH_i changes on specific cellular functions. These effects are summarized in Table II. Here we find an exception to the "increased pH_i-increased cellular activity" generalization. Membrane excitability in both crayfish muscle and mouse pancreatic β cells is enhanced at lower pH_i values where the K^+ permeability falls, reducing the short-circuiting outward K^+ current and permitting increased Ca^{2+} spike activity. The reduction in gap junction conductance by reduced pH_i could lead to either an increase or a decrease in cellular activity, although it certainly causes a decrease in intracellular communication.

II. SUMMARY OF PRESENTATIONS AND DISCUSSIONS

A. Activation of Egg Development

Although Loeb's success at NH_3-induced parthenogenesis [1913] and Warburg's discovery that NH_4OH increased pH_i [1910] suggested 70 years ago that a pH_i change was an important step in the activation of development, such pH_i increases have been detected only very recently. The first indication of an alkalinization at fertilization came from measurements on sea urchin egg extracts [Johnson et al, 1976; Lopo and Vacquier, 1977] and then with the more reliable techniques of pH_i microelectrodes, ^{31}P-NMR and DMO distribution [Shen and Steinhardt, 1978; Webb and Nuccitelli, this volume; Johnson and Epel, 1981]. A permanent pH_i rise of 0.3–0.4 unit follows fertilization in both sea urchin and frog eggs, and this alkalinization is important for the increased rate of protein synthesis (Winkler, this volume), increased K^+ permeability [Shen and Steinhardt, 1980], and chromosome replication [Mazia and Ruby, 1974] which follows activation in sea urchin eggs. The protein synthesis rate increase is due to an apparent doubling of the elongation rate and the translation of more mRNA. Although a doubling of the protein synthesis rate parallels the increase in pH_i following fertilization in the frog egg, there is as yet no evidence that these two changes are related.

Another important result of the Webb and Nuccitelli paper was the close agreement in the pH_i values measured with the two separate techniques of recessed-tip pH microelectrode and ^{31}P-NMR. However, Dr. Moody was still concerned with the long period of slow acidification following impalement of the frog egg. This is not a normal characteristic of pH electrode impalement unless injury has occurred. Whereas the close agreement with the noninvasive ^{31}P-NMR results would suggest that the pH microelectrode was measuring an uninjured pH$_i$ value, Dr. Moody raised the possibility that the superfusion rate in the NMR tube may not have been adequate to wash off metabolic CO_2, resulting in a suppressed pH$_i$ value. Dr. Nuccitelli replied that unsuperfused tubes do indeed exhibit an acid drift, but that with the superfusion rate of 4 ml/min, no acid drift was ever observed.

The two-pH electrode study also stimulated discussion since it suggested that the acidification resulting from local electrode penetration injury was fairly localized, and was undetectable in another region of the egg 300 μm away. This surprised some of the microelectrode users who assumed such pH$_i$ gradients could not exist in cells. However, they could find no fault with the technique used.

B. pH$_i$ Changes During the Cell Cycle

Intracellular pH oscillations accompany progression through the cell cycle in a number of different systems (Table III). In general, these oscillations appear to bear some fixed relation to nuclear events (specifically, mitosis and/or DNA synthesis), but there is still insufficient data to attribute to them a causal role (ie, that alkalinization triggers DNA synthesis). The general impression one carries away from the data presented on this topic is that increased proliferation (DNA synthesis, mitosis, increased cell number) is correlated (however loosely) with increased pH$_i$. As shown in Table III, maximal DNA synthesis usually follows maximal alkalinization. However, difficulties with the reproducibility of cell cycle synchronization, as well as the limited sensitivity and temporal resolution of techniques employed to measure DNA synthesis (ie, ^{3}H-thymidine incorporation) make it premature to attach a causal meaning to this correlation. Commitment to initiation of DNA synthesis may occur much earlier in the cycle. It is too early to make generalizations about the role of pH$_i$ in growth control, but the results presented here are intriguing and should encourage further experimentation. For example, in Physarum, mitosis occurs shortly after, or just as a major alkalinization occurs and can be delayed by artificially lowering pH$_i$ (see Steinhardt, this volume). In evaluating the evidence, the reader should not forget to take into account the different pH$_i$ measuring techniques employed. Only the Physarum pH$_i$ measurements represent cytoplasmic pH alone. All of the measurements in the other systems (yeast, Tetrahymena, lymphocytes) inevitably represent averages of more than one intracellular compartment, including the nucleus. The extent to which this problem artifactually obscures or enhances pH$_i$ changes remains unknown, but it suggests the importance of interpretational caution.

TABLE I. Intracellular pH Changes Associated With Cellular Functions

Cellular function	pH$_i$ measurement technique	pH$_i$ change	Net pH$_i$ change	Nature of pH$_i$ change	Reference
Fertilization: activation of development					
Eggs of sea urchin, Lytechinus pictus	Microelectrodes	6.84 to 7.26	+0.42	Permanent pH$_i$ rise follows activation	Shen and Steinhardt, 1978; Winkler, this volume
Eggs of frog, Xenopus laevis	Microelectrodes and ^{31}P-NMR	7.4 to 7.7	+0.3	Permanent pH$_i$ rise follows activation	Webb and Nuccitelli, this volume
Release from dormancy					
Spores of bacterium, Bacillus cereus	^{14}C-methylamine distribution	6.30 to 7.40	+1.1	Permanent pH$_i$ rise accompanies transition from dormant to germinated spore	Setlow and Setlow, 1980
Bacillus megaterium	^{14}C-methylamine distribution	6.38 to 7.40	+1.02	Permanent pH$_i$ rise accompanies transition from dormant to germinated spore	Setlow and Setlow, 1980
Spores of yeast, Pichia pastoris	^{31}P-NMR	6.0 to 7-7.4	+1-1.4	Permanent pH$_i$ rise accompanies transition from dormant to germinated spore	Barton et al, 1980; Shulman, personal communication
Cyst of shrimp, Artemia salina	^{31}P-NMR	6.3–6.7 to ≥ 7.9	≥ +1.2	Permanent pH$_i$ rise accompanies transition from dormancy to developing embryo	Busa, this volume
pH$_i$ influences rate of glycolysis					
Frog, Rana pipiens; muscle	^{14}C-DMO distribution	7 to 7.12	+0.12	Insulin stimulates a pH$_i$ rise, which increases the rate of glycolysis by 43%	Moore et al, this volume
Human, red blood cells	^{14}C-DMO distribution	6.6 to 7.5	+0.9	Sucrose addition stimulates a pH$_i$ rise, which increases rate of glycolysis by 200%	Tomoda et al, 1977

Cell cycle activation					
Lymphocyte	^{14}C-DMO distribution	7.2 to 7.6	± 0.4	A transient pH$_i$ rise occurs with a maximum at roughly the same time that DNA synthesis is maximal	Gerson, this volume
Yeast, Saccharomyces cerevisiae	^{31}P-NMR	7.0 to 7.4	± 0.4	A transient pH$_i$ rise occurs 80 min before DNA synthesis is maximal	Gillies, this volume
Ciliate, Tetrahymena pyriformis	^{14}C-DMO distribution	7.25 to 7.5	+ 0.25	Two transient pH$_i$ rises occur: 1) prior to and 2) at the time of the maximal DNA synthesis	Gillies, this volume
Slime mold, Physarum polycephalum	Microelectrode	7.0 to 7.4	± 0.4	A transient pH$_i$ rise occurs 30 min before mitosis	Steinhardt and Morisawa, this volume
	Antimony microelectrode	6.0 to 6.6	± 0.6	A transient pH$_i$ rise occurs 30 min before mitosis	Gerson and Burton, 1977
Stimulus-response coupling					
Human, platelet, thrombin-stimulated activation	9-Aminoacridine and 6-carboxyfluorescein fluorescence	7.0 to 7.3	+ 0.3	Thrombin stimulates a pH$_i$ rise within 60–90 s	Simons et al, this volume
Rabbit neutrophil, f-Met-Leu-Phe chemotaxis stimulation	^{14}C-DMO distribution	7.0 to 7.1	+ 0.1	F-Met-Leu-Phe stimulates a small, transient, Ca^{2+} dependent pH$_i$ fall followed by a pH$_i$ rise of 0.1 unit by 5 min after addition	Sha'afi et al, this volume

(Table I continued on following page.)

TABLE I. (Continued).

Cellular function	pH$_i$ measurement technique	pH$_i$ change	Net pH$_i$ change	Nature of pH$_i$ change	Reference
Sperm activation					
1. Motility initiation					
sea urchin, Lytechinus pictus	^{14}C-Methylamine distribution	≃ 6.5 to ≃ 7.0	+ 0.5	Na$^+$ addition triggers a permanent pH$_i$ rise initiating sperm motility	Lee and Forte, personal communication, and this volume
Sea urchin, Strongylocentrotus purpuratus	^{14}C-Diethylamine distribution	≃ 7.0 to ≃ 7.4	+ 0.4	Na$^+$ addition triggers a permanent pH$_i$ rise initiating sperm motility	Christen et al, personal communication
2. Acrosome reaction					
Sea urchin, Lytechinus pictus	^{14}C-Methylamine distribution	≃ 7.0 to ≃ 7.1	+ 0.1	A transient pH$_i$ rise accompanies the acrosome reaction	Lee et al, personal communication
Sea urchin, Strongylocentrotus purpuratus	^{14}C-Methylamine -diethylamine -DMO 9-Aminoacridine	≃ 7.4 to ≃ 7.6	+ 0.2	A transient pH$_i$ rise accompanies the acrosome reaction	Schackmann et al, 1981
Myocardial contractility					
Rabbit hearts	^{31}P-NMR	7.2 to 7.0	− 0.2	A 0.2 pH$_i$ unit fall is associated with a 50% depression in left ventricular pressure	Jacobus et al, this volume

TABLE II. Some Cellular Functions Influenced by Imposed pH_i Changes*

Cellular function	pH_i measurement technique	Normal pH_i	Imposed pH_i change	Effect of imposed pH_i change	Reference
Plasma membrane excitability					
Crayfish slow muscle fiber (*Procambarus clarkii*)	pH microelectrode	7.2	−1 to 0	Lowered pH_i lowers K^+ permeability and elicits Ca^{2+} action potentials: At pH_i 6.6, 50% of cells exhibit action potentials	Moody, this volume
Starfish immature oocyte (*Mediaster aequalis*)	pH microelectrode	7.1	−1.3 to +0.5	Lowered pH_i reduces inward rectifying K^+ current with a 50% reduction in current at pH 6.3	Moody, this volume
Mouse pancreatic β cell	—	—	Not measured	Lowered pH_i lowers K^+ permeability and elicits a continuous series of action potentials	Pace et al, this volume
Gap junction conductance					
Teleost (*Fundulus*) and amphibian (*Ambystoma* and *Xenopus*) embryos	pH microelectrode	7.5 to 7.8	−1 to 0	Lowered pH_i reduces gap junctional conductance with a 50% reduction at pH 7.3 and complete closure at 6.9	Spray et al, this volume

*This table lists some of the observations described in this volume and is by no means a comprehensive list of such effects. Imposed pH changes have been shown to influence many other cellular processes such as mitosis [Zetterberg and Engstrom, 1981], cortical actin polymerization [Begg and Rebhun, 1979; Tilney and Jaffe, 1980], microtubule disassembly [Regula et al, 1981], and bacterial chemotaxis [Repaske and Adler, 1981], to name a few.

TABLE III. Summary of pH$_i$ Changes During the Cell Cycle

Cell type	pH$_i$ measuring technique	Synchronized cultures?	Stimulus to initiate cell cycle progression	Initial pH$_i$[a,c]	ΔpH$_i$ maxima[b,c] (approx. time)	Approx. times of onset of/maximal increased DNA synthesis or mitosis[f]	References
Yeast: Saccharomyces cerevisiae	^{31}P-NMR	Yes (starved)	+ Glucose	Not previously oxygenated: 6.75	+ 0.55(10 min)	60 min/90 min (DNA synth.)	Gillies, this volume, Fig. 3
		Yes (starved)	+ Glucose	Previously oxygenated 6.95	+ 0.35(10 min)	NG	
		Yes (starved)	Refeed	7.55	−0.30(100 min) +0.35(180 min) −0.35(280 min) +0.25(350 min) −0.20(400 min)	100 min/350 min (DNA synth.)	Gillies and Deamer, 1979, Fig. 5
Ciliates: Tetrahymena pyriformis	^{14}C-DMO	Yes (heat shock)	Cool to 28°C	NG(7.40)[d]	−0.15(45 min) +0.20(65 min) −0.20(80 min) +0.25(125 min) −0.30(170 min)	90 min/150 min (DNA synth.)	Gillies and Deamer, 1979, Fig. 6

Cell	Method	Notes	Stimulus	Basal pH$_i$	ΔpH$_i$ changes	Event time	Reference
Acellular slime mold: Physarum polycephalum	Recessed-tip pH microelectrode	Plasmodial stage is syncytium—natural mitotic synchrony	Refeed (after starve 9 h)	7.00	+0.40(3½ h) −0.25 (6 h) +0.05(7 h) −0.10(8 h)	4 h (mitosis)	Steinhardt and Morisawa, this volume, Fig. 8
			Refeed (after starve 13 h)	6.70	+0.70(6½ h) −0.20(9 h)	6½ h (mitosis)	
Mammalian cells:	^{14}C-DMO	Yes	con A	7.15[e]	+0.40(52 h) −0.30(78 h)	10 h/48 h (DNA synth.)	Gerson, this volume Fig. 2
Mouse spleen lymphocytes			LPS	7.15[e]	+0.35(72 h) −0.25(78 h)	10 h/54 h (DNA synth.)	

[a] Referred to often as "basal pH$_i$" or "resting pH$_i$," which can be misleading. In each experiment described here, this value is that obtained at t = 0, before addition of nitrogen or release from repression, or the first value given immediately thereafter (as specified).

[b] Changes given are sequential; that is, each subsequent change is with respect to the previous pH$_i$ value (as opposed to the initial value). Eg, if initial pH$_i$ is 7.0, the first ΔpH$_i$ is −0.1 and the second ΔpH$_i$ is 0.2, then the value at that point is 7.1. NG—Not given.

[c] pH$_i$ values are taken from references, as given, approximately to the nearest 0.05 unit.

[d] The earliest time point given is pH$_i$ 7.40 at ~ 10 min after cooling.

[e] Unstimulated cell population.

[f] Via incorporation of ^{3}H-thymidine.

The work presented on pH_i changes during the cell cycle provoked some specific questions and discussions, as summarized below.

Following Dr.Gillies's presentation of pH_i oscillations during the cell cycle of yeast and of Tetrahymena (Gillies, this volume), some concern was expressed about the general health of his yeast suspensions in the NMR tube. He explained that aeration via bubbling maintains O_2-tension above the measured K_m for uptake. That the cells replicate synchronously demonstrates that the conditions are physiological. Dr. Gillies also mentioned that the effect of starvation on pH_i is complex and depends on the conditions used: In controlled starvation buffer, pH_i drops; but in stationary phase, pH_i is high. He went on to say that in the starved condition, both yeast and Tetrahymena are in G_0 and yeast require 30 minutes to "get back in" to G_1 (supported by metabolic data with ^{31}P-NMR and ^{13}C-NMR). In other words, the first ΔpH_i is correlated with the transition from a nonproliferative to a proliferative state (ie, "activation") (in yeast—an alkalinization, in Tetraymena—an acidification). During this transition, there is no phosphorus uptake and very little carbon uptake (Dr. Gillies's unpublished data). It was speculated that the first of the two alkalinizations in Tetrahymena might represent an activation, but Dr. Gillies thinks that the high pH_i at the beginning and end of the heat-shock synchrony may be due to a temperature effect. He has tried, unsuccessfully, to synchronize Tetrahymena with pH shifts. Dr. Gillies also described evidence suggesting that the ΔpH_i in yeast is linked to proton pumping, as opposed to internal metabolism. The pH_i of anaerobic yeast, deprived of any external carbon source and subsequently oxygenated, rises to 7.0 within 5 minutes in starvation medium, but this rise requires $\sim$ 1 hour in distilled water, suggesting dependency on counterions. Adding back K^+, but not Na^+, to the distilled water restores the rate of ΔpH_i. However, in starved cultures of yeast, where a slight (+ 0.1 pH unit) permanent alkalinization can also be detected upon addition of glucose, the ΔpH_i is not inhibited in distilled water. Thus, two components appear to be involved: one ion exchange-dependent (in response to O_2) and one internal (in response to glucose). Dr. Deamer wondered what was happening to all the metabolically-produced protons in the activated yeast. Dr. Gillies reminded us that yeast produce ethanol and CO_2 (which is bubbled out), so that the acid load on the metabolizing yeast is small. Since some protons will be used up in ATP synthesis, it is interesting that anaerobic yeast has undetectable levels of ATP (< 0.1 mM), whereas preoxygenated yeast has the same amount of ATP before and after addition of glucose.

The possible effects of pH gradients or pH changes on differentiation were discussed. Dr. Regelson wondered whether CO_2 could be used to acidify and thus differentially "put to sleep" cells such as tumor cells (cf. the effect of CO_2 on differentiation in hydra [Loomis and Lenhoff, 1956]). However, it is difficult to extrapolate to a single cell from effects on a complex, multicellular organism (such as hydra). Apparently, gamete formation in a malaria parasite is stimulated by host-dependent alkalinization [Carter and Nijhout, 1977], but effects of pH_1 on differentiation at the single cell level (eg, animalization/vegetalization of eggs) are unknown.

A correlation between increased pH_i and DNA synthesis was also reported by Dr. Gerson upon mitogenic stimulation of mouse spleen lymphocytes (Gerson, this volume, and Table III). Participants discussed the problem of interpretation of these correlations in Tetrahymena, in Physarum (see below and Table III) and in lymphocytes. For example, the prevailing view is that alkalinization is required for increased DNA synthesis, and yet in both Physarum and in Tetrahymena, maximum DNA synthesis occurs some time *after* maximum pH_i, during a period of lower pH_i. The suggestion was made that a pH_i transient somehow triggers replication, which, once started, carries on even at lower pH_i. Dr. Gerson mentioned that the cell fraction with highest pH_i often just precedes (temporally) that with the highest rate of DNA synthesis. Examination of this question will require better time resolution, as it takes 1 hour to make the DNA measurements and ½ hour to make the DMO measurements in these experiments (vs 5 minutes for the latter in *Tetrahymena*). The time lag involved makes interpretation of the correlations observed extremely difficult.

It was emphasized that pH_i is the critical factor in all of these experiments, and that although most mammalian cells respond somewhat to changes in extracellular pH (pH_o), many protozoans are extremely resistant to changes in pH_o. Therefore, predictions based on ΔpH_o should not be made unless the relationship between pH_o and pH_i for a particular cell type, under the conditions employed, is well understood.

If ΔpH_i has a causal role, examination of the effect of weak acids or of weak bases on DNA synthesis in lymphocytes will be of interest. Unfortunately, when using 20 mM DMO to reduce the pH gradient across the cell membrane, Dr. Gerson found that this concentration of DMO, in and of itself, inhibited DNA synthesis. A strain of mouse whose lymphocytes lacked the correlation of increased DNA synthesis with increased internal alkalinity would also be helpful, but to date, all strains examined (3 or 4) exhibit the same phenomena. Dr. Deamer pointed out that, though increased pH_i may *permit* increased DNA synthesis, a causal relationship has yet to be established between increased pH_i and increased DNA synthesis. He suggested pulsing cells with pH shifts of varying duration to examine this possibility. In preliminary attempts by Dr. Gerson to stimulate lymphocytes by pH shifts, ΔpH either had no effect or was lethal. Two weeks after the meeting a paper by Zetterberg and Engström [1981] was published reporting that an alkaline pulse stimulates DNA synthesis in serum-starved 3T3 cells.

In the acellular slime mold, Physarum polycephalum, a type of differentiation can be stimulated by starvation. After presentation of data suggesting that a cytoplasmic alkalinization necessarily precedes mitosis in this organism (Steinhardt, this volume), Dr, Steinhardt addressed several questions. Regulation of pH_i by CO_2 perfusion was found to be too difficult (hence the choice of acetate), and application of the acetate to filter paper cultures was accomplished by immersion, followed by decanting. Continuous recordings are impossible because the slime mold moves too much. The buffering capacity and mechanism of pH_i regulation of this organism are unknown, but the former appears to vary throughout the cell cycle, as evidenced by the different concentrations of acetate required to "clamp" pH_i at various points in the cycle (see Steinhardt, this volume). Dr. Gerson added that pH_i regulation in Physarum is very good over a wide range of pH_o.

According to Dr. Steinhardt, DNA synthesis occurs after mitosis, when the pH_i is on its way back down. The possibility of a threshold for activation was discussed (cf. inhibition of mitosis by acetate at pH 6.7). Apparently a pH_i of at least 7.3 is required for subsequent onset of mitosis. Dr. Steinhardt finished on a provocative note, stating that, in his view, the pH_i changes observed upon activation of yeast and of Physarum and sea urchin eggs are phenomena peculiar to the activation of "truly dormant" cells. He feels that they are extreme, special cases, especially in Physarum, where the oscillations continue throughout the cell cycle. "I don't think most cells, even within their cell cycle, use pH to regulate anything," he said.

C. Control of Metabolism

The pH dependence of various metabolic enzymes presents the possibility that pH_i might influence metabolic rates, and two examples of such regulation have been discussed in this volume: 1) Dr. Moore presents strong evidence that insulin increases the rate of glycolysis by stimulating Na^+-H^+ exchange, resulting in increased pH_i. A small, 0.12 unit, pH_i rise increases the rate of glycolysis by 43%. A possible target for this pH_i increase is phosphofructokinase, which is maximally activated by a pH increase of 0.1–0.2 pH unit above the physiological pH_i [Trivedi and Danforth, 1966; Wu and Davis, 1981], and is inhibited at lower pH values. His hypothesis is further supported by the reversal of insulin's effect on glycolysis when the Na^+ gradient is reversed so that stimulation of the Na^+-H^+ exchanger would then decrease pH_i. 2) The principal high-energy phosphate bond store of the Artemia cyst, P^1,P^4-diguanosine-5^1-tetraphosphate (GP_4G), is slowly consumed during anaerobiosis (pH_i 6.3–6.7) when glycolysis is completely shut down. One of the enzymes responsible for this GP_4G consumption is GTP:GTP guanylyltransferase, which has the unusual pH optimum of pH 6 and displays almost no activity at pH 8 (the aerobic pH_i). The Artemia cyst may therefore be utilizing pH_i to switch between glycolysis and GP_4G consumption for its energy source. These are the two strongest cases for pH_i control over metabolism, but others are also discussed by Moore et al in this volume. Unfortunately the Busa paper was not presented at the meeting, but a summary of the discussion following Dr. Moore's presentation follows below.

Dr. Moody pointed out that it is difficult to rule out the possibility that these insulin effects may be mediated by an increase in the internal Na^+ rather than the increased pH_i since even the imposed pH_i change experiments were mainly imposed acidifications that would stimulate Na^+–H^+ exchange for recovery. Dr. Moore's reply was that he could measure no Na^+ increase upon insulin addition. He proposed that since insulin stimulates both the Na^+–K^+ pump and Na^+–H^+ exchange, Na^+ would not accumulate in the cell.

Dr. Gadian suggested that the more interesting question would be to study metabolic regulation during contraction. To this Dr. Moore responded that these frogs hibernate at the bottom of Lake Champlain all winter and they have fantastic glycolytic systems that are quite interesting for the study of cell regulation.

Dr. Nuccitelli asked if this was the first case (except, possibly, the sea urchin egg) where the Na^+–H^+ exchanger, which is well known for its pH_i regulatory role, was being used to change pH_i to influence a cellular function. Dr. Simons responded that thrombin-induced platelet activation exhibited a similar apparent Na^+ involvement in the pH_i increase, as indicated by amiloride sensitivity.

D. Control of Plasma Membrane Properties

The plasma membrane properties of excitability and gap junctional conductance are both pH_i-sensitive as summarized in Table II. The target for the pH_i change in membrane excitability appears to be the K^+ channel. Both the delayed outward K^+ currents in slow muscle fibers and pancreatic β cells and the inward K^+ currents in starfish oocytes are blocked by lowering pH_i. The reduction in outward K^+ currents allows the inward Ca^{2+} current to become regenerative, producing Ca^{2+} action potentials. Half of the cells will exhibit action potentials at pH_i 6.6. While the physiological role for this pH_i-dependent excitability is unclear, the inward K^+ current pH_i sensitivity may be important for the electrical changes accompanying maturation. Specifically, these currents decrease substantially, following 1-methyladenine stimulation of maturation, and this permits an increase in the amplitude of the fertilization potential so that an effective electrical block to polyspermy is possible. This pH_i sensitivity has a pK of 6.3, which means that there is a 50% reduction in the inward K^+ current at that pH_i. When asked about the possibility that this low pH_i was unphysiological, Dr. Moody pointed out that a 20–30% decrease in conductance results from a 0.2–0.3 pH unit decrease from the physiological pH_i of 7.1. Furthermore, changes in other ion concentrations, such as Na^+, could interact with pH_i to give a larger effect on the K^+ channel. Maturation hormone speeds the Na^+ pump 4–5-fold, and the steepest part of the $(Na^+)_i$ vs K^+ conductance curve is exactly at the normal resting internal Na^+ activity. "I think the hypothesis that (these effects) are caused at least in part by a change in some intracellular ion is quite viable," Dr. Moody concluded. This was challenged by Dr. Steinhardt who pointed out that in both the maturing starfish oocyte [Johnson and Epel, 1980] and Xenopus oocyte [Lee and Steinhardt, 1981], the pH_i changes are not important for maturation, but that it may be a cyclic nucleotide event. He later added that he thought the outward K^+ current increased in Pateria during maturation. Dr. Moody replied that this is very difficult to be sure of since both activation and inactivation of that conductance are affected by maturation, and the shunting of inward currents by this outward K^+ current will vary with the kinetics.

Dr. de Hemptinne asked if acidification depressed the crayfish muscle calcium current at all, and Dr. Moody replied that because of the geometry, the cell could not be adequately voltage-clamped so he could not be sure about slight depression. At lower pH_i values the threshold for the Ca^{2+} spike moves closer and closer to the resting potential. "You get what looks for all the world like a molluscan pacemaker neurone popping away, ..." Dr. Moody added.

The second plasma membrane property that is clearly affected by pH_i is gap junctional conductance. Dr. Spray presented some elegant work in which the effects of intracellular Ca^{2+}, pH_i, and transjunctional voltage on junctional conductance were compared. Using recessed-tip pH microelectrodes, aequorin luminescence for Ca^{2+} detection (with Dr. Joel Brown), perfusion, and voltage-clamp methods for transjunctional voltage studies, all three parameters were studied in the same systems, embryos of Fundulus and Ambystoma. While either Ca_i^{2+} or H_i^+ increases will uncouple independently, H^+ is about 10,000 times more effective than Ca^{2+} (apparent pKs 7.3 and 3.3, respectively). Since the normal pH_i for these embryos is 7.5–7.7 and the normal pCa_i is 7.1–7.3 [Rink et al, 1980], small deviations of pH_i from its normal level can sensitively modulate junctional conductance, but very large, unphysiological changes in Ca^{2+} are required. The sensitivity of gap junctional conductance to transjunctional voltage is also quite dramatic in amphibian blastomeres. The conductance falls by about 50% with a 14-mV transjunctional voltage, and this effect is independent of any conductance change resulting from a pH_i change.

Dr. Moore suggested that if one wanted to fill a large volume with high conductivity, a pH_i increase would be the signal of choice. When Dr. Moody pointed out that pH_i increases in this system do not increase junctional conductance, Dr. Spray replied that in at least one other system, cardiac cells, pH_i is normally lower and small increases in pH_i do increase conductance [Weingard and Reber, 1979].

Dr. Pace asked about the conflicting results of Birgit Rose, who finds that small Ca^{2+} changes uncouple cells. Dr. Spray replied that previous, semiquantitative estimates of cytoplasmic Ca^{2+} concentrations necessary for uncoupling were only lower than the values reported here by a factor of 2. Quantitative estimates require buffering both pH and Ca_i^{2+} to known levels which internal perfusion allows.

Dr. Webb asked how one could get a pH_i change with virtually no Ca_i^{2+} change since they are so intimately related. "I think it has to be a quick change. Certainly if you leave CO_2 on it for awhile you get a change in Ca^{2+}," Dr. Spray replied. When then asked about morphological correlations, Dr. Spray's eyes lit up and he eagerly described some recent work with Dr. Robert Hanna and Dr. Thomas Reese using fast-freezing, ultralow temperature, freeze-fracture electron microscopy. Unlike previous freeze-fracture studies (most of which used glutaraldelyde-fixed tissue) indicating regularization of the particle distribution in the junctional lattice, they have found no change in interparticle spacing in uncoupled cells. Studies on fixed tissues are suspect because not only does glutaraldelyde close gap junction channels, but it also abolishes their pH dependence in very low concentration.

E. Intracellular pH Changes During Stimulus-Response Coupling

Three papers in this volume present data linking pH_i changes to stimulus-response coupling. The clearest case for involvement is in human platelet activation where thrombin stimulates a 0.3 pH unit rise within 60–90 seconds of application (Simons et al, this volume). The rabbit neutrophil also exhibits an alkalinization but the time course (5 minutes) may be too slow to be involved in activation. Pace et al made no pH_i measurements but did find that low pH_o (which may well reduce pH_i), as well as monensin application, inhibited glucose-stimulated insulin secretion in rat pancreatic β cells. Dr. Sha'afi did not present the rabbit neutrophil work at the meeting, so only the discussion following the other two presentations will be summarized below.

Dr. Simons presented data documenting a dose-dependent pH_i increase (within 1 minute, pH_i increases from 7.0 to 7.3 when pH_o is 7.35) upon stimulation of serotonin secretion in human platelets by thrombin. Since the earliest known event after addition of thrombin is a membrane depolarization, interest was expressed in the effect of ΔpH_i on membrane potential (4). The human platelet will tolerate only a narrow range of pH_o (7.0–7.4), and the effect of ΔpH_o and/or ΔpH_i on $\Delta\psi$ has not been examined. The focus of Dr. Simons' work has been on the dissection of the effects of various drugs on the platelet activation process. Some of these drugs appear to act as ionophores. The relationship (if any) between the pH change and the voltage change remains unknown, though both appear to involve sodium ions [Horne et al., in press]. For example, both $\Delta\psi$ and ΔpH_i can be blocked 100% by amiloride, though secretion is blocked only 60% under the same conditions. Apparently, at least one other mechanism ("bypass") exists for the stimulation of secretion. Even at doses of thrombin above saturation (no further $\Delta\psi$ or ΔpH_i), some continued increase in secretion is observed. The nature of this secretion bypass is unknown. Neither collagen, glass, nor any of the large surface stimuli for platelet activation has any effect on membrane potential. Activated platelets release other factors that stimulate chemotaxis, secretion, and aggregation of other platelets.

Dr. Pace presented data suggesting that pH_i modifies secretion of insulin in response to glucose stimulation, in pancreatic β cells, by changing the electrical properties of these cells. At high concentrations of glucose, so much insulin is secreted that the effects of ΔpH_i are partially masked. However, changing pH_o from 7.4 to 7.0 appears to inhibit insulin secretion. It was pointed out that the higher the glucose concentration, the greater the deactivation of potassium channels. Acidification of the intracellular environment appears to activate the membrane, but to prevent it from exhibiting normal potential oscillations. Alkalinization does not change electrical events as much. Dr. Pace feels that it is proton output across the membrane, not necessarily reflected in ΔpH_i, which is the important parameter. Apparently pH_i has been measured upon stimulation of glucose metabolism or of insulin release in these cells [Malaisse et al, 1979], and is found to remain constant. Dr. Pace suggests that the secretory granules may act as a proton sink and may be involved in pH homeostasis in the cytosol.

F. Tissue pH_i Measurements: Diagnostic Applications

One of the great advantages of the ^{31}P nuclear magnetic resonance technique for pH_i measurement is that along with pH_i one automatically detects important phosphorous compounds such as phosphocreatine and ATP. The intracellular levels of these important metabolites can be easily determined, and diagnosis of some pathologies is possible. Dr. Gadian described ongoing projects applying this technique to human kidneys prior to transplantation, and to human forearm muscle (Bore et al, this volume). Dr. Jacobus has studied whole rabbit hearts and has found a close correlation between pH_i and contractility. Specifically, a 0.2 pH unit fall is associated with a 50% depression in left ventricular contractility. The discussion following these two presentations will be summarized below.

When asked how CO_2 and O_2 were measured, Dr. Jacobus explained that a 22-gauge needle was inserted into the free wall of the left ventricle to collect samples, which were fed directly into a mass spectrometer through 6 feet of steel tubing for on-line analysis with a 6-minute time resolution. Dr. Deamer then raised the problem of spatial resolution with ^{31}P-NMR: "You might be getting rather dramatic pH changes at a sensitive site which are not really being sensed by the phosphate (NMR)." Dr. Jacobus agreed that he could not separate endomyocardial from epimyocardial signals but felt that reduction in contractility was not likely to be due to localized changes. This measure of average tissue pH may be more pertinent to a tissue-level phenomenon such as pressure development. The NMR technique provides a mass average over the right and left ventricles, but the right ventricle is probably only one-sixth to one-eighth the total mass of the myocardium.

Dr. Gadian's presentation provoked a discussion of preservation methods for organ transplants. He explained that the kidneys to be transplanted are preperfused with a preservation solution and stored in a plastic bag on ice. One important feature of these NMR studies is that they provide a method for directly comparing the metabolite levels present following the various preservation methods. Dr. Jacobus commented that NMR has also been used to improve the heart trasplant techniques. Now, by injecting KCl plus glucose into the coronary artery to arrest the myocardium while monitoring pH_i and ATP with ^{31}P-NMR, much less damage results. Dr. Thomas asked if blood was detected by NMR, and Dr. Gadian replied that they could not detect it since they would expect to see a big signal from diphosphoglycerate. The only discussion of the forearm muscle studies concerned the role of lactic acid in muscle pain. Dr. Gadian felt that muscle pain was more closely associated with nerves, since "if you tie a cuff around your arm and hardly exercise, you may eventually experience pain with relatively little reduction in your high-energy phosphates." With regard to the role of lactic acid, people with McArdle's Syndrome generate very little lactic acid, but can certainly develop pain on exercise.

III. CONCLUSIONS

The last session of the meeting was devoted to discussing the overall implications of the results to date associating pH_i changes with cellular functions. The tone of the discussion was set by Dr. Nuccitelli, who showed a summary slide (see Table I) of pH changes during cellular activity in a variety of systems and who proposed a controversial unifying theme. He suggested that small pH_i alkalinizations are associated with activation of cell cycle progression, with increased glycolysis, with general metabolic activation, with release from dormancy, and with activation of development after fertilization, whereas pH_i acidifications are associated with depression or repression of cell processes. This proposed generalization provoked a heated response from the participants. Dr. Gillies pointed out that it is still unclear whether a ΔpH_i transient or a permanent pH_i change is required, which may depend on the cell type or process in question. Dr. Simons emphasized that, as activation of metabolism is several steps away from primary events at the plasma membrane (including $\Delta\psi$), it is dangerous to conclude that there is any causal relationship. Both she and Dr. R. Thomas agreed that pH changes following activation may be the effect of earlier events, rather than the cause of later ones. As regards "repression" by acidification, Dr. Pace suggested that acidification below the point at which a cell can maintain enzyme activity will, of course, uncouple the cell in a very general way. Dr. Gillies spoke for many by objecting to the implication that "alkalinization is good." Since many cell activities have pH *optima*, which means that pH can be too high, this is an oversimplification. Along these lines, Dr. Jacobus reminded us that lysosomal proteases require acid pH for activation and that, therefore, alkalinization is not universally good for all things. (NB: Aside from Dr. Jacobus' remark, throughout this discussion the importance of compartmentalization in eucaryotic cells was largely ignored. Clearly proton gradients across membranes can do work, as has been amply demonstrated in mitochondria and chloroplasts, but the possible relevance of pH_i changes to these phenomena was not heavily considered by most participants. Although constantly referring to "*intracellular* pH changes", most speakers were apparently really thinking of, and by implication referring to, "*cytoplasmic* pH changes".)

At this point, Dr. Steinhardt livened the discussion by claiming that the proposed unifying theme was a useless generalization. He made the point that activation is not a *thing,* that it means different things to different cell types. Hormonal activation, for example, may proceed by pH-independent pathways. Dr. Moore added that alkalinizations may be necessary but not sufficient for activation or stimulation of many processes (eg, ΔpH_i stimulates glycolysis, but *not* mitosis, in frog muscle; Moore, this volume). So pH_i may have a permissive, but not causal, action, as suggested previously by Dr. Deamer. Dr. Moore went on to say that changes in pH_i may involve changes in stored potential energy in forms other than covalent bonds. A pH increase of 1.0 unit will increase the free energy of ATP hydrolysis by $\geq 10\%$, which could easily have threshold effects on other cell processes. Also, if pH_i is going to exert a regulatory effect, one should see a pattern to the pH profiles of (cytoplasmic) enzymes. The situation could be even more complex, since there may be pH-sensitive effectors interacting with pH-insensitive enzymes (note analogy to Ca^{2+}: eg, calmodulin and kinases [see Means and Dedman, 1980]). According to Dr. Gerson, important enzymatic information may be buried in biochemists' notebooks, since most enzymologists only report pH maxima, whereas what we would like to know is where these enzymes are *most* pH-sensitive. It is important to know the rates of change of activity with respect to pH in various pH ranges. (Biochemists, are you listening?)

As the discussion drew to an enthusiastically disorderly close, Dr. Boron suggested that as pH is so important and has such broad effects, it is not likely to regulate specific functions, but is more likely to exert a "pancellular" effect (eg, egg or spore activation), turning everything on or off. Many participants feel that calcium ions are a more likely candidate for specific, short-term regulatory effects. No better concluding remarks can be made than those of Drs. R. Thomas and R. Moore, who elegantly captured the combination of caution, misgiving, excitment, and sense of possibility generated by this symposium. Dr. Thomas expressed "a nasty feeling that intracellular pH is such a fundamental property that it probably isn't, in fact, used as a regulatory factor except in rare cases. I think we are perhaps falling into a trap of thinking that, since we can measure it, it is bound to be doing something interesting." To which Dr. Moore quickly replied, "Although I think your point is well taken, I am going to make the reverse point that we don't want to fall in the trap that I think has happened in the past, to think that pH cannot be a regulatory agent simply because it is fundamental and is well regulated itself."

IV. REFERENCES

Barton JK, den Hollander JA, Lee TM, MacLaughlin A, Shulman RG: Measurement of internal pH of yeast spores by ^{31}P nuclear magnetic resonance. Proc Natl Acad Sci USA 77:2470, 1980.

Begg DA, Rebhun LI: pH regulates the polymerization of actin in sea urchin egg cortex. J Cell Biol 83:241, 1979.

Carter R, Nijhout M: Control of gamete formation (exflagellation) in malaria parasites. Science 195:407, 1977.

Gerson DF, Burton AC: The relation of cycling of intracellular pH to mitosis in the acellular slime mould Physarum polycephalum. J Cell Physiol 91:297, 1977.

Gillies RJ, Deamer DW: Intracellular pH changes during the cell cycle in Tetrahymena. J Cell Physiol 100:23, 1979

Horne WC, Norman NE, Schwartz DB, Simons ER: Changes in cytoplasmic pH and in membrane potential in thrombin-stimulated human platelets. Eur J Biochem (in press).

Johnson CH, Epel D: Intracellular pH does not regulate protein synthesis in starfish oocytes. J Cell Biol 87:142a, 1980.

Lee S, Steinhardt RA: pH changes associated with meiotic maturation in oocytes of Xenopus laevis. Dev Biol 85:358, 1981.

Loeb J: "Artificial Parthenogenesis and Fertilization." Chicago: University of Chicago Press, 1913.

Loomis WF, Lenhoff HM: Growth and sexual differentiation in hydra mass culture. J Exp Zool 132:555, 1956.

Lopo A, Vacquier VD: The rise and fall of intracelluar pH of sea urchin eggs after fertilization. Nature 269:540, 1977.

Malaisse WJ, Hutton JC, Kawazu S, Herchuelz A, Valverde I, Sener A: The stimulus-secretion coupling of glucose-induced insulin release. XXXV. The links between metabolic and cationic events. Diabetologia 16:331, 1979.

Mazia D, Ruby A: DNA synthesis turned on in unfertilized sea urchin eggs by treatment with NH_4OH. Exp Cell Res 85:167, 1974.

Means AR, Dedman JR: Calmodulin—An intracellular calcium receptor. Nature 285:73, 1980.

Regula CS, Pfeiffer JR, Berlin RD: Microtubule assembly and disassembly at alkaline pH. J Cell Biol 89:45, 1981.

Repaske DR, Adler J: Change in intracellular pH of Escherichia-coli mediates the chemotactic response to certain attractants and repellants. J Bacteriol 145:1196, 1981.

Rink TJ, Tsien RY, Warner AE: Free calcium in Xenopus embryos measured with ion-selective microelectrodes. Nature 283:660, 1980.

Schackmann RW, Christen R, Shapiro BM: Membrane potential depolarization and increased intracellular pH accompany the acrosome reaction of sea urchin sperm. Proc Natl Acad Sci USA 78: October 1981.

Setlow B, Setlow P: Measurements of the pH within dormant and germinated bacterial spores. Proc Natl Acad Sci USA 77:2474, 1980.

Shen SS, Steinhardt RA: Direct measurement of intracellular pH during metabolic derepression of the sea urchin egg. Nature 272:253, 1978.

Shen SS, Steinhardt RA: Intracellular pH controls the development of new potassium conductance after fertilization of the sea urchin egg. Exp Cell Res 125:55, 1980.

Tilney LG, Jaffe LA: Actin, microvilli, and the fertilization cone of sea urchin eggs. J Cell Biol 87:771, 1980.

Tomoda A, Tsuda-Hirota S, Minakami S: Glycolysis of red cells suspended in solutions of impermeable solutes. J Biochem 81:697, 1977.

Trivedi B, Danforth WH: Effect of pH on the kinetics of frog muscle phosphofructokinase. J Biol Chem 241:4110, 1966.

Warburg O: Uber die oxydationen in lebenden Zellen nach Versuchen am Seeigelei. Zeitschr Physiol Chem 66:305, 1910.

Weingart R, Reber WR: Influence of internal pH on r_i of Purkinje fibers from mammalian heart. Experientia 35:929, 1979.

Wu TL, Davis EJ: Regulation of glycolytic flux in an energetically controlled cell-free system: The effects of adenine nucleotide ratios, inorganic phosphate, pH and citrate. Arch Biochem Biophys 209:85, 1981.

Zetterberg A, Engström W: Mitogenic effect of alkaline pH on quiescent serum-starved cells. Proc Natl Acad Sci USA 78:4334, 1981.

Index

Acetate, 367-8, 371-72
Acid extrusion, 249, 251
Acidification, transient, 306, 318-19
Acid loading, 205-18, 228, 231, 233, 253, 260-62, 278, 432
Acidosis, 72-74
Acidosis, respiratory, 538-39, 549-51, 553-54, 556, 559-60
Acridine orange, 138, 144, 149, 154-56, 158, 490, 498-502, 509
Acrosome reaction, 138, 146, 149
Actomyosin, 560
Acute myocardial infarction, 538
Adaptation, 373
Adenine nucleotides, 484, 547
ADP, 62-63, 68, 71-72, 74, 539, 561
Alkalization, cytoplasmic, 270-80
Alkalosis, 560
Ambystoma tigrinum, 253-66
Amiloride, 246, 249, 251, 306, 391, 394, 397-99, 407, 519-20
Aminopyrine, 499-501
Ammonia, 269, 278-81, 349-50
Ammonia prepulse technique, 248-50
Angina pectoris, 538
Anoxia, 437
Artemia cysts, 419-25
Asci spores, 97
Atebrin, 138, 144
ATP, 99-100, 167, 303, 561
 -dependent H pump, 484, 508-10
 in gastric microsomes, 142-43, 154
 in human metabolism, 529-30, 534
 and insulin, 410, 412
 and myocardial contractility, 538-39, 541, 544-47, 561
 and Na pump, 190
 in squid, 233-36
 and tissue metabolism, 62-65, 71-72
ATPase, 142, 144-45, 386, 560-61

Bacillus acidocaldarius, 106, 120
Bacteria, 94-95, 106, 120, 375, 418-19
Balanus nubilus, 205-18
Barnacle muscle, 205-18
Basolateral HCO_3 transport, 260-66
B cells, 483-510; see also lymphocyte
Benzylamine, 501, 509
Bicarbonate, 198-99, 253-66; see also HCO_3^-
Blood
 Coronary, 554-57
 K^+, 386, 387
 red cells, 503
Bromthmol blue, 125
Buffering power, intracellular, 229 245-46

Ca^{++}, 146, 149, 508
 buffering, 502
 :Ca^{++} exchange, 508-9
 and chemotaxis, 514, 516-18, 522-24
 current, 441

-dependent action potential, 431, 434, 435–37, 441
and fertilization, 269–70, 294 302, 316, 318–19, 356–58
in gap junctions, 445–58
and insulin, 386
ionophore, 294, 299, 306–9, 312, 316, 319
permeability, 484
and protein synthesis, 325, 328–32, 339
uptake, 505
Calibration, 108, 113, 120, 122, 143, 166, 168, 194, 541–48
Carbon dioxide, see CO_2
Carboxyfluorescein, 164, 464, 471–77; see also fluorescein
Cardiac
action potential, 559–60
cycle, 548
tissue, 239–51
see also heart, myocardial
Cell
activation sequence, 382
cycle, 309–12
cycle, and insulin, 409–11
cycle, pH regulation of, 341–58, 417–25
cycle, and starvation, 361–73
damage, ischemic, 539, 557
depolarization, 452
dormancy, 417–25
electrical properties of, 427–42
gastric parietal, 499–500
proliferation, 341–58
synchronization, 369–70
tumor, 106–8, 113, 115, 121
see also specific cells
Chemical shift, 68, 79–81, 82, 94, 97
Chemiosmotic hypothesis, 507, 510
Chemotaxis, 513–24
Chloride, see Cl^-
Chlorobutanol, 299, 305, 311
Chloroplasts, 174, 176
Chromaffin granule, 508–9
Chromic acid, 5, 6

Circadian clock, 373
Circumesophageal ganglion, 191
Cl^-, 197–98, 202, 280
flux, 226–227
-HCO_3^- exchange, 239–51, 306–9
-sensitive microelectrodes, 242, 257
Colorometric indicators, 125
Concanavalin A, 128–31, 375, 380
Contractility, 22–24, 29, 72, 537–62
Coronary blood flow, 554–57
CO_2, 349–51, 354–58, 390, 402–3, 407
in crayfish muscle, 435
/HCO_3^-, 398–99, 402
Coupling, 447, 451
Crayfish, 428–37, 441–42
Creatine kinase, 68, 71–72, 561
Crustaceans, 419–25
Cryptobiosis, 418–25
Cytokinesis, 372
Cytoplasmic alkalization, 270–80
Cytoplasmic pH, 39–43, 48, 114–18, 122, 428, 452

ΔG_{Na}, 388–89, 397, 400, 410
$\Delta G_{Na:H}$, 388–89, 395–407
Depolarization, 378, 431–32, 435, 441, 452, 464, 478, 480, 484, 491–92, 503
Derepression, 270, 372–73
Derivatized plastic beads, 92
Dialysis, internal, 222–37
DIDS, 309, 520–22
Digitonin, 470
DMO, 56–58, 128, 131–32, 166–67, 343, 350, 376–78, 387, 390–93, 401–2, 514–15, 520
DNA synthesis
and insulin, 386, 408–9, 411–12
and pH, 128, 325, 344, 346–48, 375–82
Donnan equilibrium, 509
Dormancy, 371, 373, 417–25

Ehrlich ascites tumor cells, 106-8, 113, 115, 121
Electric activity, 491-95, 508
Electrical properties, cell, 427-42
Electrode, double-barrelled pH, 7-19; see also microelectrodes
Electroneutrality, 195, 230
Electron microscopy, 490
Elongation rate, 332
Endocytosis, 24, 33, 166
Energy, 373, 410
Enzymes, intracellular, 411
Epithelial cells, 253-66
Extracellular markers, 56

Fatigue, 68-71
FCCP, 97
Fertilization
    frog egg, 293-322
    sea urchin egg, 155-56, 270-81, 325-39
    starfish egg, 441-42
Flow microfluorometry, 125-32, 164-169; see also microfluorometry
Fluoroscein, 125-26, 164; see also fluorometric pH indicators
Fluoroscein diacetate, 107-8, 109, 129
Fluoroscein-ovalbumin injection, 166
Fluorescence, 490
    -activated cell sorter, 164, 169; see also flow microfluorometry
    microscopy, 105-22
    quenching, 138-40, 148, 157-58
    spectra shift, 149-57
Fluorescent amines, 135-58
Fluorometry, 125-32, 164, 169, 465-77; see also microfluorometry
Formyl-methionyl-leucyl-phenylalanine, 514-16, 521-23
4-methylumbelliferone, 125-32
^{14}C-Inulin, 377
Freeze-fracture, 445

Frogs, 293-322, 390-412
Fructose-1, 84, 97
FTC-ovalbumin, 30, 33-34, 39-44, 46

Gap junction, 445-58, 559
Garden hose effect, 560
Gastric microsomes, 137-45, 153-54, 157
Gastric parietal cell, 499-500
Gating, 447, 457, 548
Germination, 373
Glucose, 87, 89, 386-87, 391, 483-510
Glucosensors, 484-510
Glucose oxidation, 503
Glycodiazine, 495
Glycogen phosphorylase, 534
Glycogen synthesis, 408
Glycogen synthetase, 387, 409
Glycolysis, 97-103, 386-92, 398-404, 406-7, 503, 561
Growth, 343-58
GTP:GTP guanylyltransferase, 425

H^+, 156, 158, 457, 506-7, 513, 622-23
    pump, 485, 508-9, 510
    transport, 142-44, 146-47, 253-66
HCO_3^--Cl^- exchange, 233-37
HCO_3^- flux, 223-37; see also bicarbonate
Heart muscle, 67, 69, 72-74; see also cardiac; myocardial
Heart perfusion techniques, 539-40
Histone H1 phosphorylation, 372
Human
    forearm muscle, 529-34
    peripheral lymphocytes, 378
    platelets, 463-80
    skeletal muscle, 71-72, 74
    tissue, 527-34
Hydrated beads, 92
Hydrogen, see H^+
Hydroxyurea, 380

Imidazole, 493, 495, 503, 505, 508
Injury, 312–13
Inorganic phosphate, 62–68, 71, 74, 80, 82, 89, 303, 530, 541–44
Inotropic drugs, 560
Insulating glass, 2, 4
Insulin, 385–412, 484–85, 497–98, 503, 507–8
Intercellular
   channels, 445–447
   communication, 447
Invertebrates, 428–42
Ion
   flux, 245–46
   -sensitive microelectrodes, 430, 437, 442
   transport, 221–37
Ionic
   atmosphere, 412
   current, 447
   environment, 427–28
   strength, 82
Ischemia, 12–19, 72–74, 529, 534, 538–39, 551–54, 557, 560–61
Isethionate, 507

Junctional conductance, 447–53, 456–58, 559

K^+, 143–44, 147, 149
   blood, 386–87
   channel, 505
   currents, 431, 434–35, 437, 441
   :H^+ antiporter, 499
   :H^+ exchange, 508
   permeability, 484
Kidney, 529; see also renal

Lactate, 390, 392–93, 404, 561
Lactic acid, 68–69, 71, 74, 89, 534
Lactic dehydrogenase, 392
Leukocytes, 378, 503
Li, 408
Lipopolysaccharide, bacterial, 375
Liposomes, 136–42, 149–54, 157, 173–83
Litmus, 125

Loligo pealei, 221–37
Lymphocytes, 125–32, 375–82; see also B cells
Lysis, 507–10
Lytechinus pictus, 269–71, 325–39

Magnesium, See Mg^{++}
Mammalian cells, 89–94
Masked message hypothesis, 326, 372
Mass spectrometry, 548–49, 554–57, 560–61
Mast cell, 508
McArdle syndrome, 534
Membrane, 173–86
   depolarization, 464, 478, 480, 484
   ionic channel, 427–28
   lipid, 173, 176–86
   plasma, 483–84, 507, 523
   potential, 196–97, 378
Metabolic derepression, 270
Metabolic inhibitors, 200–201, 203
Metabolism, 62–65, 68, 75
   and NMR, 86–94
   and pH, 527–34
   pH regulation of, 417–25
Metabolites, 553–54
Methylesculetin, 125
Methyl phosphonate, 82, 94
Mg^{++}, 64, 143, 302, 316, 396–97
Microelectrodes, 193–95, 387–88
   Cl^--sensitive, 242, 257
   exposed tip, 165
   ion-sensitive, 430, 437
   liquid sensor, 442
   Na^+-sensitive, 196, 256–57
   NaCl-sensitive, 197
   problems with, 589
   recessed tip, 1–6, 166, 168, 190, 241, 256, 362–64
   reference, 193
   see also electrode: pH-sensitive microelectrodes
Microfluorometry, 27, 30–49; see also flow microfluorometry; fluorometry
Microforge, 3

Microinjections, 27–31, 39, 46–47
Microphotometry, 26, 47
Micropipette puller, 3–4
Mitochondria, 95–97, 114–18, 121–22
Mitochondrial matrix, 544–45, 561
Mitochondrial membrane, 174, 176, 186
Mitogen, 378–80, 382
Mitogenesis, 408–9
Mitosis, 361–62, 365–71, 408–9
Monensin, 495, 497–99, 503, 505–6, 508–9
Monolayer cultures, 490
Monovalent cations, 386, 409
Muscle
    barnacle, 205–18
    crayfish, 428–37, 441–42
    fatigue of, 68–71
    forearm, 529–34
    heart, 67, 69, 72–74
    and insulin, 390–412
Myocardial
    contractility, 537–62
    infarction, acute, 538
    see also cardiac; heart

Na^+, 146–49, 199–200, 202
    /Ca^{++}, 520, 524
    /Ca^{++} exchange, 321
    and chemotaxis, 514, 518–20, 524
    /CO_3^{2-} cotransport, 397
    in crayfish muscle, 435–37
    and fertilization, 276–77, 279
    flux, 226–27
    free-energy gradient, 388, 400, 406, 409
    :H^+ antiporter, 478–79, 499
    :H^+ counter transport, 246, 250–51
    :H^+ exchange, 261, 263, 266, 320–21, 387–89, 394–401, 404, 406–10, 520, 522
    /HCO_3^- exchange, 306–9
    /HCO_3^-/H/Cl transport system, coupled, 262
    K pump, 386–87, 518–20
    pump, 190, 195, 197, 387, 389, 397, 408–9, 435, 484, 506–7
    -sensitive microelectrode, 196, 256–57
NAD, 62–63, 80
Neurosecretory vesicles, 508
Neutral red, 125
Neutrophil, 513–24
NH_4Cl, 327–32, 432–35, 508
NH_4 prepulse technique, 260
Nigericin, 473, 499, 509
9-aminoacridine, 138, 140, 144, 146, 148, 174–78, 464, 466–71, 477, 508
NMR, 61–75, 164, 167–68, 347, 539–48, 558–59
    ^{1}H-, 67–68
    high density cultures, 86–89
    ^{31}P-, 61–75, 79–103, 164, 167–68, 299–302, 313–16, 387, 396, 408, 421–22, 527–34, 539–62
    signal-to-noise, 86
    time resolution, 97–100
NMU-3 cells, 113
Nonionic mechanisms, 409
Nuclear divisions, 372
Nuclear magnetic resonance. *See* NMR
Nucleoside di- and triphosphate, 80
Nucleotide triphosphate, 101

Oocytes, 428–29, 437–41
Organic acids, 16–18
Orthophosphate, 80, 97, 99–100
Oscillations, 488, 491–95, 503, 507
Ouabain, 226, 387, 390, 393–94, 398, 408–9, 455, 506–7, 519–20
Oxidative phosphorylation, 554, 561
Oxygenation, 87, 89, 554–56
Oxygen supply/demand imbalance, 538, 557

Pancreatic islets, 483–510
Parathyroid hormone, 507
Pasteur effect, 100, 102
PEP carboxykinase, 100
Perchloric acid, 79, 392

Peripheral lymphocytes, human, 378
Permeability, 120
   Ca^{++}, 484
   coefficients, 173, 177, 179, 182, 184–85
   K, 484
   plasma membrane, 523
   proton, 173–86
pH
   -dependent shift in fluorescence spectra, 149–54
   electrode, double-barreled, 7–19
   fluorometric indicators, 125
   in gap junctions, 445–58
   gradient, 146–49
   mitochondrial, 114–118, 121, 122
   phagosomal measurement, 44–45, 48
   -sensitive glass, 1, 2, 4, 5
   -sensitive microelectrode, 1–6, 24–25, 166, 168, 241, 256, 296–99, 319, 429–30, 452–53
   -sensitivity, 47
   spectrophotometric measurement, 105–22
pH_e, 516
pH_i
   and cell cycle, 344–51
   and cell proliferation, 342–43
   and cell starvation, 361–73
   and contractility, 22–24, 537–62
   cytoplasmic, 39–43, 48, 74, 94, 114–18, 122, 165, 428, 452
   and DNA synthesis, 375–82
   and endocytic events, 24
   excursions, large, 418–25
   and growth, 343–44, 347–51
   importance of, in development, 322
   increase at fertilization, 319–20
   and injury, 312–13
   and insulin, 385–412
   in living systems, 79–103
   manipulation, 320–21
   measurement, 24–29, 65–68, 161–69, 135–58
   metabolic regulation by, 417–25
   and myocardial contractility, 537–62
   oscillations, 309–12, 321–22
   in platelets, 463–80
   and protein synthesis, 325–39
   rabbit neutrophil, 513–24
   regulation, 189–203, 283–89
   regulation, in barnacles, 205–18
   regulation, in salamanders, 253–66
   regulation, in sea urchins, 269–81
   regulation to cell electrical properties, 427–42
   renal, 529
   of unfertilized eggs, 316–18
Phagosome, 44–45, 48
Phosphate nuclear magnetic resonance. See NMR, ^{31}P-NMR
Phosphates, high energy, 553–54, 560
Phosphocreatine, 62, 66–68, 71, 74–75, 530, 534, 538–39, 542, 550
Phosphofructokinase, 99–100, 102, 387–90, 404, 406–7, 409, 412, 503
Phosphomonesters of sugars, 80, 83
Physarum, 125
   plasmodium, 361–73
   polycephalum, 376
Piezoelectric devices, 257
Planar lipid membranes, 173, 176–86
Plasma membranes, 483–84, 507, 523
Platelets, 463–80
Polyphosphate, 80, 99
Polysome, 326, 332–33, 334
P^1,P^4-diguanosine-5'-tetraphosphate, 425
Potassium, See K^+
Probenecid, 507
Protein synthesis, 325–39, 342, 409
   and insulin, 386
   and pH, 361–73
Protons, 411–12, 485–510
Proximal tubule, 253–66
Pyranine, 125–26, 180

Pyridine nucleotides, 484, 561

Quinine, 149, 151, 154, 158

Rabbit neutrophil, 513–24
Rana pipiens, 390–412
Rat heart, 67
Rat soleus muscle, 12–16
$^{86}Rb^+$ efflux, 485, 488–89, 495–97, 503–7
Recess configurations, 1–2
Rectification, 437–38, 441
Red blood cells, 503
Redundancy, 409
Renal proximal tubule, 253–66; see also kidney
Repression, 372–73
Respiratory acidosis. See acidosis, respiratory
Ribosome, 332–33
RNA, 326, 332, 336–37, 338, 372
  and insulin, 386, 408–9
  polymerase, 376
  synthesis, 375, 382, 386, 408–9

Salamander, 253–66
Sarcoplasmic reticulum, 560
Sea urchin egg, 137, 145–49, 155–57, 269–81, 325–39, 353–58, 375–76
Secretory canaliculus, 500
Secretory granules, 485, 498, 500–502, 507–10
7-hydroxycoumarins, 125, 127
Sheep Purkinje fiber, 15–18, 239–51
SITS, 208, 215, 226, 235, 243, 245, 262, 265
6-biphosphatase, 100
6-carboxyfluorescein, 464, 471–77; see also carboxyfluorescein; fluorescein
Slow flexor muscle, 429–37
Snails, 189–203
Sodium, See Na^+
Spatial resolution, 164
Sperm motility, 138, 145–49
Squid giant axons, 221–37

Starfish, 428–29, 437–41
Sucrose, 393–94
Sugar phosphates, 80, 83, 87, 99, 101
Surface coil, 528–29, 547

Temporal resolution, 164–65
Tetraethylammonium, 505
Tetrahymena, 343–47
Thermodynamics, 411–12
Thrombin, 463–80
Thymidine, 378, 380
Tiger salamander, 253–66
Tissue
  cardiac, 239–51
  gas, 548–49
  metabolism, 62–65, 68, 75
  oxygen tension, 554–56
  pH measurements, 527–34
Topical magnetic resonance, 528, 547–48
Toxic metabolic byproducts, 89
TPP, 377
Transjunctional voltage, 447
Transmembrane ion flux, 559
Transmembrane potential, 507
Troponin, 560
TTX, 226
2-deoxyglucose-6-phosphate, 83–84
2,3-diphosphoglycerate, 65
Tumor cells, 106–8, 113, 115, 121

Umbelliferone, 125, 127

Vacuolar P_i, 97
Vacuoles, 95–97
Valinomycin, 97, 116, 139, 144, 149, 181, 184
Ventricular dysfunction, 549–51, 554, 559
Ventricular wall tension, 560–61
Voltage clamp, 437–38, 442, 453

Water flux, 184–85

Weak acid-base, 55–59, 165–66, 280, 309, 508
Weak acids, 451–52, 454, 503–4
Wound-healing, 42–43, 47–48

Xenopus laevis, 293–322

Yeast, 100–102, 342, 344, 347, 419